T0177075

Geophysical Monograph Series

Including
IUGG Volumes
Maurice Ewing Volumes
Mineral Physics Volumes

Geophysical Monograph Series

Geophysical Monograph 165

Solar Eruptions and Energetic Particles

Natchimuthukonar Gopalswamy
Richard Mewaldt
Jarmo Torsti
Editors

American Geophysical Union
Washington, DC

Library of Congress Cataloging-in-Publication Data

Solar eruptions and energetic particles / Natchimuthukonar Gopalswamy, Richard Mewaldt, Jarmo Torsti, editors.
     p. cm. -- (Geophysical monograph ; 165)
  Includes bibliographical references and index.
  ISBN-13: 978-0-87590-430-6 (alk. paper)
  ISBN-10: 0-87590-430-0 (alk. paper)
  1. Coronal mass ejections. 2. Solar energetic particles. 3. Solar activity. I. Gopalswamy, N. II. Mewaldt, R. A. III. Torsti, Jarmo.
  QB529.S623 2006
  523.7'5--dc22

                    2006022105

  ISBN-13: 978-0-87590-430-6 (alk. paper)
  ISBN-10: 0-87590-430-0 (alk. paper)

ISSN 0065-8448

CONTENTS

Flares and Energetic Particles

CME-Driven Shocks and SEPs in the Heliosphere

Space Weather

PREFACE

Research over the last three decades identifies coronal mass ejections (CMEs) as the most energetic events in the heliosphere. Although studies of solar energetic particle (SEP) events and nonthermal radio bursts have a longer history, the close connection between CMEs and energetic particles has become much clearer thanks to the large armada of spacecraft observing these phenomena since the mid-1990s. Indeed, understanding the most violent forms of solar eruptions – CMEs, flares, and SEPs – is of fundamental importance to the physics involved and our ability to predict and mitigate disruptive space weather episodes. Questions, of course, remain: We do not fully understand how CMEs and SEPs are accelerated but we do know that they affect space weather in several significant ways. The magnetized plasma of CMEs impacts Earth's magnetosphere, causing large geomagnetic storms. Energetic CMEs also drive shocks that accelerate electrons (observed as type II radio bursts) and ions (detected by spaceborne instruments). SEPs ionize the upper atmosphere driving atmospheric chemistry and present a radiation hazard to humans and hardware in space.

This volume reviews extensive observations of solar eruptions and SEPs by instruments on board a number of spacecraft, including the Solar and Heliospheric Observatory (SOHO), Wind, the Advanced Composition Explorer (ACE), the Ramaty High Energy Solar Spectroscopic Imager (RHESSI), and the Transition Region and Coronal Explorer (TRACE). Highly sensitive coronagraphs on board SOHO image CMEs with unprecedented sensitivity. Several thousand CMEs have been observed, measured and cataloged for the current solar cycle, but only about 1% of these are associated with SEPs. Radio instruments on the Wind spacecraft obtain signatures of solar energetic electrons injected into the heliosphere within minutes of their release near the Sun and also track MHD shocks driven by CMEs. A majority of space instruments detect SEPs in situ and measure their elemental, ionic charge state, and isotopic compositions. Thus, it has become possible to link the evolution of SEP events to CME-driven shocks as they propagate from the Sun to geospace and beyond. In early 2002, RHESSI began complementing these observations with high-resolution imaging of x-rays and gamma rays from flares associated with these events. These multi-spacecraft, multi-instrument and multi-wavelength observations have raised more pointed questions about the origin, acceleration, and interplanetary propagation of SEPs. This volume records advances made in the understanding of solar eruptions with significant consequences in the heliosphere.

The volume is organized into five topical areas, with an introductory review of the early development and current state of CME and energetic particle studies. Topical areas include: CMEs, SEPs, connection to flares, associated phenomena, and space weather. In-depth reviews on solar eruptions and energetic particles also contain observational studies, discussion of theoretical developments, and modeling results. The papers on associated phenomena deal with flares, type II radio bursts, and shock waves. After considering the interplanetary propagation of CMEs and energetic particles, space weather implications are discussed, including the arrival of energetic particles at geospace and their impact on Earth's radiation belts. The review papers cover all important aspects of CMEs and energetic particles making the volume largely self-contained.

Most of the papers in this volume were presented at an AGU Chapman Conference, entitled "Solar Energetic Plasmas and Particles," held at the University of Turku, Finland, August 2-6, 2004. Several additional papers were solicited to make the volume as complete a survey of the subject as possible. Two experts provided peer review for each paper. The editors appreciate the constructive and timely reviews by many members of the international space weather, and Living with a Star, communities that have greatly enhanced the quality of this volume. Finally, the editors are very grateful for the excellent conference arrangements made by the local organizing committee headed by Eino Valtonen from the University of Turku.

Natchimuthukonar Gopalswamy
NASA Goddard Space Flight Center, Greenbelt
Maryland

Richard Mewaldt
California Institute of Technology, Pasadena
California

Jarmo Torsti
University of Turku, Turku
Finland

Solar Eruptions and Energetic Particles
Geophysical Monograph Series 165
This paper is not subject to U.S. copyright. Published in 2006 by the
American Geophysical Union
10.1029/165GM01

Solar Eruptions and Energetic Particles: An Introduction

N. Gopalswamy[1], R. Mewaldt[2], and J. Torsti[3]

This introductory article highlights current issues concerning two related phenomena involving mass emission from the Sun: solar eruptions and solar energetic particles. A brief outline of the chapters is provided indicating how the current issues are addressed in the monograph. The sections in this introduction roughly group the chapters dealing with coronal mass ejections (CMEs), solar energetic particles (SEPs), shocks, and space weather. The concluding remarks include a brief summary of outstanding issues that drive current and future research on CMEs and SEPs.

1. OVERVIEW

The Sun loses mass in three different ways: the Solar Wind, Coronal Mass Ejections (CMEs), and Solar Energetic Particles (SEPs). These phenomena are signatures of solar variability from the matter point of view; electromagnetic radiation in the form of quiescent and flare emissions represents the other major variability. The mass emissions from the Sun also signal fundamental physical processes such as acceleration of plasmas to high energies, magnetic field reconnection and interaction between large-scale plasma structures. Fast solar wind originates from the open field regions on the Sun known as coronal holes while CMEs originate from the closed field regions such as active regions and filament regions. The interplanetary plasma is the solar wind, and the propagation of CMEs through this plasma represents interaction between them. CMEs often attain super-Alfvenic speeds in the magnetized coronal and interplanetary plasmas and hence are bound to drive fast mode shocks, which in turn are known to accelerate the

[1]NASA Goddard Space Flight Center, Greenbelt, Maryland
[2]California Institute of Technology, Pasadena, California
[3]University of Turku, Turku, Finland

Solar Eruptions and Energetic Particles
Geophysical Monograph Series 165
10.1029/165GM02

charged particles in the upstream medium to very high energies. Even sub-Alfvenic CMEs interact with the solar wind resulting in momentum exchange described by the aerodynamic drag. Thus, the three types of mass emissions are not independent. Even the flare process is not independent: Almost all CMEs are associated with flares, which accelerate electrons and ions that flow towards the Sun and away from the Sun producing electromagnetic radiation at various wavelengths and injecting flare-heated plasmas into the CMEs. Flares are also associated with the smaller and more frequent SEP events, known as impulsive SEP events. It is thought that impulsive SEP events can provide seed particles for CME-driven shocks causing an interesting mix of SEP compositions and charge states. Studies of these mass outflows from the Sun are also of significant practical importance: they can cause severe space weather effects such as disruption of power systems on Earth, and can pose radiation hazards to our space travelers and spacecraft systems. Various chapters in this volume provide a review of the current status of our understanding of these energetic phenomena, the underlying physics, and why humans care about them.

The concept of mass leaving the Sun was contemplated well over 100 years ago from when high-speed prominence eruptions from the Sun were observed. *R.A. Howard* describes the historical development culminating in the discovery of CMEs in the early 1970s, and the subsequent milestones in CME research including the pre-eminent position occupied by CMEs as major drivers of geospace disturbances. The discovery of SEPs preceded that of CMEs by three decades, when Forbush reported the first event.

J.R. Jokipii takes us through the early observations and the development of a theoretical understanding of particle acceleration. He points out the universal nature of the energetic particle distribution, whether it is from the Sun or from the rest of the universe as cosmic rays.

2. CORONAL MASS EJECTIONS

S.W. Kahler reviews the current status of CME research drawing heavily from the results of analyzing more than 10,000 CMEs observed by the Solar and Heliospheric Observatory (SOHO) mission. In addition to describing the physical and statistical properties of CMEs, he explains the basic magnetic structures involved in CME initiation and subsequent propagation. He points out that progress in understanding CMEs has been accelerated by the inner coronal imaging in X-rays, EUV and radio, but more observations are needed of spectral lines that can provide information on the physical conditions in CMEs. The outstanding issues concern the magnetic field geometries and topologies of CMEs, and the way in which CME magnetic fields reconnect to become open magnetic fields in the heliosphere. The better sensitivity and extended field of view of the SOHO coronagraphs have resulted in detecting a record number of CMEs. The data size will grow even further when future coronagraphs such as on the STEREO mission observe CMEs with higher cadence. It is therefore necessary to develop automatic techniques for CME detection. *E. Robbrecht and D. Berghmans* discuss the current status of such techniques for various solar features. *R.L. Moore and A.C. Sterling* focus on how CMEs are initiated, especially the fast ones that significantly disturb the heliosphere. They identify three fundamental mechanisms that can explain the CME initiation: (1) runaway internal tether-cutting reconnection, (2) runaway external tether-cutting reconnection, and (3) ideal MHD instability. The three mechanisms may operate singly or in combination. Magnetograms and movies of chromospheric and coronal features are used to tell which one or which combination of these mechanisms is the trigger for eruption. One of the points they emphasize is the production of a flux rope as a result of the eruption. The flux-rope structure has become a basic entity in CME research, used extensively by CME modelers. Flux-ropes are directly observed in the interplanetary medium as magnetic clouds. The fact that CMEs carry twisted magnetic structures must be related to the removal of accumulated helicity in the corona, as discussed by *A. Nindos*. The study of helicity in CMEs has only started recently, but the importance can be felt in the hemispheric dependence of chirality of filaments, helicity of active regions and the handedness of interplanetary magnetic clouds. *R.A. Harrison* discusses the connection between flares and CMEs. While opposing views prevail from the

flare explosion causing CMEs to flares as a byproduct of CMEs, he points out the common cause for the explosive emission of mass and flare radiation.

I. Roussev and I. Sokolov review current CME modeling efforts, focusing on the flux-rope and shear-arcade models. In particular they discuss the salient features and major weaknesses of the two models. Current observations may not be able to discriminate between the two models. They conclude that future prospects for CME modeling are more optimistic, mainly because of the opportunities the STEREO mission will provide for CME theorists and modelers to answer some of the outstanding issues on CME initiation and propagation.

3. SEPs AND THEIR RELATION TO FLARES AND CMEs

The SEPs that reach Earth originate in both flares and CME-driven shocks, and there is on-going debate as to the relative roles of these processes in large SEP events. *T. von Rosenvinge and H.V. Cane* provide an overview of the SEP phenomenon and of our current state of understanding. They focus on the current paradigm of SEP events and suggest that some modification is needed in the two-class (impulsive and gradual SEP events) picture. They conclude that our understanding of the SEP and related phenomena is still not complete despite the extensive body of observations currently available.

R.A. Mewaldt, C.M.S. Cohen, and G.M. Mason provide an authoritative account of the current state of affairs in SEP physics, covering a wide range of topics related to the source material for large SEP events. They review the recent work on the particle sources contributing to the seed population for the accelerated particles. The gradual SEP events differ in composition from the bulk solar wind in several key respects, implying that solar wind does not supply the principal seed population for these events. The suprathermal pool that is accelerated by CME-driven shocks seems to be made up of ions from impulsive solar flares and previous gradual events, CIR events, pickup ions, CME ejecta, and the suprathermal tail of the solar wind.

The ionic charge states of SEPs reflect the thermal history of the material that is accelerated and possible non-thermal processes that may remove electrons. Observations by ACE, SAMPEX and SOHO show that the mean charge state often increases with energy over the energy range from ~50 keV/nucleon to 50 MeV/nucleon. *M. Popecki* reviews these observations and discusses some of the interpretations that have been offered for this unexpected behavior, including mixtures of heated-flare and coronal material, and stripping of electrons during the acceleration and transport of the ions.

L. Kocharov presents the results of a model of the evolution of the ionic charge states of Fe during acceleration and transport and discusses what can be learned from ionic charge state observations. *B. Klecker* reviews recent observations of the ion charge state composition in SEP events and discusses the implications for our understanding of different acceleration scenarios. There is considerable progress in our knowledge of the elemental and isotopic composition, energy spectra and ionic charge states in gradual and impulsive SEP events. These new results show that the classification into two distinct types of events was oversimplified, and they provide important new constraints on theoretical models for SEP acceleration and transport.

During Solar Cycle 23 considerable progress in understanding particle acceleration in SEP events resulted from new remote-sensing observations by missions such as RHESSI, SOHO, TRACE and Yohkoh. Some fraction of the particles accelerated in flares escape into interplanetary (IP) space along open field lines, while others, trapped on closed field lines, interact in the chromosphere and photosphere. *A. MacKinnon* gives an overview of the use of X-rays, γ-rays and radio observations in obtaining direct information on flare site fast electrons and ions. He points out that observations of the SEPs detected in the IP medium as well as the flare-site populations inferred from radiative diagnostics are ultimately needed to fully understand the eruptive events. *G. Share* presents RHESSI observations of the gamma-ray and continuum radiation in some of the largest solar events of this past solar maximum, and reviews how these observations can be used to study the composition and energy spectra of particles accelerated in the flare, as well as the composition of the solar atmosphere. One puzzling result is that the proton/alpha ratio derived from the gamma-ray data is considerably smaller than is observed with *in situ* SEP measurements. *M.J. Aschwanden* uses data from Yohkoh, SOHO, TRACE, and CGRO to illustrate the magnetic topologies in the accelerating region, the asymmetry of upward versus downward acceleration, and particle access to interplanetary space. Of particular interest are the magnetic topologies that lead to reconnection between open and closed field lines. Along similar lines, *R.P. Lin* uses recent observations from Wind and RHESSI to study solar electron events observed both *in situ* at 1 AU and remotely at radio and x-ray wavelengths. While the properties of the interacting and escaping electrons are correlated, the correlation is not explained by simple models.

4. CME-DRIVEN SHOCKS AND
SEPs IN THE HELIOSPHERE

One of the greatest successes of space physics is the theory of shock acceleration, which has been successfully applied to particle acceleration at shocks ranging from planetary bow shocks and to interplanetary shocks, to supernova shocks. CME-driven shocks accelerate both electrons and ions. The most common signature of shock-accelerated electrons is the radio emission known as type II radio bursts, which can be used to track the shocks from the Sun to all the way to Earth and beyond. Although an additional shock source of flare blast waves is often assumed, *N. Gopalswamy* has provided arguments that the type II phenomenon over the entire Sun-Earth distance can be explained in terms of CME-driven shocks. He uses the hierarchical relationship between CME kinetic energy and the wavelength range of type II emission, the weakening of evidence for CME-less type II bursts, the realistic profile of the fast-mode speed in the corona and IP medium, and the close connection between SEP events and type II radio bursts to support his arguments. *G. Mann* explores the relation between coronal shocks (inferred from coronal type II bursts), Moreton waves, and EUV transient waves (known as EIT waves because of their detection by the Extreme-ultraviolet Imaging Telescope (EIT) on board SOHO). He also extends the Alfven speed profile in the corona to two dimensions and attributes the inability of flare-generated shocks to propagate beyond the Alfven-speed maximum to explain the metric type II bursts. *K.-L. Klein* compares the particle acceleration by flares and CME-driven shocks. He calls for a much closer interrelationship between flares and CMEs as sources of SEPs, rather than assuming a clear-cut distinction between flare-related and CME shock-related SEP events.

M.A. Lee and R. Vainio review the present status of the theory of accelerating energetic ions at CME-driven shocks. Although the theory can generally account for the observed features of these events, there remain outstanding issues to be addressed, including ion injection into the acceleration process, an improved description of the wave intensity excited by the accelerated particles, and injection and acceleration at quasi-perpendicular shocks. *A. Tylka and M.A. Lee* investigate the correlations among spectral and compositional characteristics, which serve as powerful constraints for their models of SEP events (both gradual and impulsive). Starting with CME-driven shocks as the accelerator for gradual SEPs, they were able to account for the correlated spectral and compositional variability in SEPs by including the shock normal angle and a compound seed population as variables. On the other hand, they point out the lack of understanding of the spectra and composition of impulsive SEP events in terms of flare parameters, such as size, duration, and magnetic topology. *C.M.S. Cohen* reviews research on energetic storm particles (ESP) events, which are same as the SEP events but occur when the accelerating shock passes the observing spacecraft. She provides an overview of the observations and theories of ESPs from their first detection to our current understanding. ESP events are testimony to the shock acceleration process since parameters of the shock can be measured directly. SEP events with particle energy high enough to penetrate Earth's atmosphere are known as ground level enhancement (GLE) events. *C. Lopate* presents an overview of how ground-based instruments have been used to study GLE events over the past fifty years. Despite of the relative rarity of GLEs, he argues that they are key to our full understanding of the processes involved in high-energy solar phenomena, both on the Sun itself and during the travel through interplanetary space to Earth. The CME-shock vs. flare reconnection debate is also applicable to the acceleration of GLE events. Particles accelerated near the Sun reach Earth by propagating through the dynamical, turbulent solar wind. The paper by *W. Droege* presents the results of a dynamical theory of particle transport in the interplanetary medium that models the evolution of SEP events using transport coefficients derived solely from observed solar-wind plasma observations. The model is successful at reproducing observations from keV to GeV energies.

The interplanetary counterparts of CMEs are known as interplanetary CMEs (ICMEs). In addition to driving shocks that can accelerate energetic particles, there are other interplanetary signatures of ICMEs, including bi-directional ion flows, and the capability of ICMEs to confine (or exclude) energetic particles. *D. Lario* reviews observations of the effects of ICMEs on energetic particles from a variety of spacecraft, using archival data extending from inside 1 AU to the outer heliosphere. *A. Struminski and B. Heber* report on a comparison between two sets of SEP events observed by Ulysses in the heliosphere (a set of eight SEP events in the ecliptic plane around 5 AU and another set of six events at high heliographic latitudes). They compare these events with GOES observations at 1 AU and infer that >30 MeV SEPs are observed over an angular range of 120° in both longitude and latitude. They suggest that CME-driven shocks may not be able to account for these events.

5. SPACE WEATHER

CMEs can be both geoeffective (CME plasma impacts Earth's magnetosphere producing geomagnetic storms) and SEP-effective (accelerated particles from near the Sun extend far into the heliosphere). It is the very fast, Earthward-directed CMEs that produce the highest-intensity SEP events at Earth and these same CMEs are often very effective at triggering geomagnetic storms once they reach Earth. CME-driven shocks reaching Earth are also responsible for the ESP events, which can be a direct radiation hazard for space-based assets. *E. Valtonen* reviews geoeffective and SEPeffective CMEs and describes how energetic particle observations can be used to identify interplanetary CMEs and shocks approaching Earth. As a feasibility demonstration, he

gives two examples illustrating the methods of using energetic particle observations for forecasting geomagnetic storms. He also summarizes the observed geoeffectiveness of halo CMEs and, in particular, magnetic clouds.

During the largest geomagnetic storms SEP ions and electrons may suddenly have access to the inner magnetosphere, where they can become trapped, forming a new radiation belt that stands out against the quiescent trapped particle populations. *J.E. Mazur* and collaborators present observations of new radiation belts during Solar Cycle 23 that contain SEP heavy ions, and it discuss some of the conditions that determine when such new belts are created, and how long they survive. *I. Roth, M.K. Hudson, B.T. Kress, and K.L. Perry* survey geomagnetic modifications due to solar/heliospheric drivers, geomagnetic eigen oscillations and the mechanisms of trapped particle dynamics. They explain how sudden deformation of the field configuration of the magnetosphere can cause trapping of SEPs on a drift time scale to form transient proton and heavy ion belts deep in the magnetosphere. They also point to recent observations which have significantly furthered our understanding of solar wind control of the magnetospheric energetic particle environment.

SEPs have recently taken on new importance in space weather as a result of NASA's decision to once again send astronauts beyond the protective cover of Earth's magnetosphere. *R.E. Turner* reviews the hazards that SEPs could pose to astronauts during transit to the Moon or Mars, and on the lunar surface. He also discusses how research by the space science community can mitigate the effects of SEPs by providing improved forecasts of the onset, evolution, and intensity of large SEP events.

H.E.J. Koskinen and K.E.J. Huttunen provide a critical overview of the worldwide efforts on space weather and explain the complexities in this multidisciplinary research. They emphasize the need for physical understanding of CME/flare production in solar sources and the propagation of CMEs before developing accurate forecasting. They call for the need to understand the structure of ICMEs (shock, sheath and magnetic cloud) for the prediction of geomagnetic storms and to understand their variability at high and low latitudes. They also identify several key areas in which progress is badly needed: prediction of the final intensity of a given storm, the acceleration of electrons to relativistic energies, and the creation of extremely strong induction effects between the ionosphere and the ground.

6. CONCLUDING REMARKS

To conclude, the set of articles presented in this monograph provides an overview of the broad range of issues involved in the generation of solar eruptions and energetic particles and how they propagate to Earth and other destinations in the heliosphere. The rapid progress made in the recent years has been possible thanks to a new generation of instruments and theoretical models, as well as to the ready availability of extended and uniform data sets from various ground and space-based data sources. Yet, there is much more to be done. Some of the key drivers of current research and areas that need more work can be summarized as follows:

The uniform and extended observations by the SOHO mission have revealed that only the fastest of the 10000+ CMEs are important for heliospheric impact. Most of the current numerical simulation efforts can deal with only slow CMEs. What is needed is the development of realistic models of fast CMEs. Recent results on the composition and charge state of material in ICMEs has brought out an undeniable link between the flare process and CME initiation. This connection needs to be pursued to understand the formation and final configuration of CMEs in interplanetary space.

Predicting the existence and magnitude of the southward magnetic field component of ICMEs based on measurements near the solar surface is a difficult but a crucial step in understanding the complete structure of CMEs. Future missions such as STEREO and Solar-B are likely to provide key information towards such an understanding of the magnetic properties of CMEs near the Sun. A related issue is the magnetic structure of ICMEs: one view is that all ICMEs are magnetic clouds and the differences between them arise from the vantage point of the observation. Proving or disproving this view will go a long way toward validating numerical models of CMEs, most of which have a flux-rope structure as the basic ingredient. With STEREO, supplemented by near-Earth observations, there will be multipoint in situ measurements of CME structure in addition to STEREO images.

One characteristic of SEP-associated CMEs is their ability to drive shocks. Shocks ahead of geoeffective CMEs also mark the sudden commencement of geomagnetic storms. Radio dynamic spectra of type II bursts are the only way to remotely sense the shocks when they are still near the Sun. It is highly desirable that white-light signatures of shocks are identified corresponding to the times of type II bursts.

As for SEPs, understanding the deviation from the familiar two-class picture (impulsive and gradual events) is very much needed. While there is no question of particle acceleration by CME-driven shocks (e.g., the ESP events and type II radio bursts), what we lack is a realistic assessment of the relative importance of flare and CME related SEP events. Multi-point measurements by STEREO should make it possible to separate the contributions from these two acceleration processes. Another issue closely related to the two-class scenario is the nature of the source material accelerated by CME-driven shocks. A proper understanding of the relation between SEPs and the progenitor CMEs is crucial for predicting the intensity, duration, and composition of SEPs in the interplanetary medium.

There is a spread of orders of magnitude in the intensity of solar particles that result from CME-driven shocks of any given shock velocity. Are these differences due to shock geometry (quasi-perpendicular vs. quasi-parallel), to differences in wave activity, to pre-conditioning by previous CMEs, or to differences in the seed-particle populations? To understand why some shocks are more efficient accelerators than others can best be answered by future missions that will send spacecraft closer to the Sun, including the ESA Solar Orbiter mission (a single spacecraft to 0.22 AU), the Inner Heliospheric Sentinels (4 spacecraft in orbit between 0.25 and 0.72 AU) and Solar Probe (a single spacecraft to traverse the solar corona at a distance of 4 solar radii from the Sun). These exploratory observations, supplemented by remote sensing data, should form the basis for theoretical models that can forecast the onset and evolution of SEP events with much greater accuracy.

A Historical Perspective on Coronal Mass Ejections

R.A. Howard

E.O. Hulburt Center for Space Research, Naval Research Laboratory,
Washington DC, USA

The concept of mass leaving the Sun was thought possible over 100 years ago from the observations of prominence material that was seen to be moving outward at speeds in excess of the escape velocity. While the direct observation was elusive, the coupling between solar activity and geomagnetic storms became quite apparent. In the 1940's the concept of corpuscular radiation from the sun was proposed and then in the 1950's used to explain the discontinuity in a comet tail. *Parker's* theory in 1957 predicted a continuous outflow from the sun that was then observed by in-situ spacecraft less than 10 years later. The first optical observations of a transient event showing mass moving through the solar corona in 1971 were accompanied by excitement, fascination and speculation. Two questions at that time were: What causes the CME eruption? What is their significance? These questions and others are still with us. In this paper, the coronal mass ejection is viewed in its historical context.

INDIRECT OBSERVATIONS

Coronal mass ejections are large eruptions of mass and magnetic field from the Sun. Although only discovered in the early 1970's [*Tousey*, 1971; *MacQueen*, 1973], the effects of CMEs have been seen indirectly at Earth for many thousands of years. The impact of a CME on the Earth can generate an aurora. Such aurorae were reported in ancient literature in both eastern and western cultures, including the Old Testament, Greek and Chinese literature, and must have been a source of awe, fright and indeed wonderment. They have also been an inspiration for paintings and woodcuts for centuries. While common at northern latitudes such as here in Turku, they would have been quite rare in the most of the regions contributing to this literature. Plate 1 is a composite of a variety of auroral pictures.

The next major milestone was the discovery that the Earth had a magnetic field. The earliest indication of the existence of a geomagnetic field is from the Chinese in the 11th Century. They recognized that certain stones, "lodestones", had a strange property in that they could attract other substances.

Further, an iron needle stroked with such a stone would always point in the north-south direction when freely suspended. The concept of the compass spread to Europe and was used by Christopher Columbus in his voyage across the Atlantic in 1492.

In 1600, *William Gilbert* published a treatise on the magnetic properties of the Earth, "De Magnete". By postulating the Earth's magnetic field, he was able to explain the behavior of the compass to always point along the same direction.

Then in 1722, *George Graham* noticed that the compass needle would suddenly change its direction of pointing by a small angle and would remain that way for up to several days. About 20 years later, *Anders Celsius* and his student *Olaf Peter Hiorter* discovered the diurnal variation of the Earth's magnetic field fluctuations and that the occurrence of the aurora were correlated to magnetic field deflections.

As a result of these discoveries, *Baron Alexander van Humbolt* in 1836 called for a worldwide network of magnetometers to be established to record global magnetic

Solar Eruptions and Energetic Particles
Geophysical Monograph Series 165

10.1029/165GM03

Plate 1. Composite of Auroral Images.

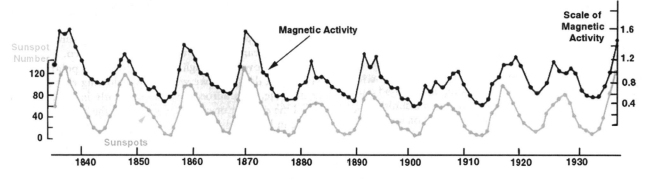

Plate 2. Comparison of Magnetic Activity and Sunspot Number.

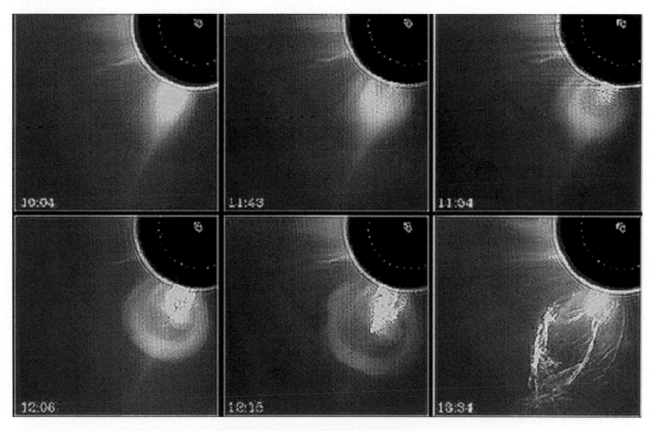

Plate 3. 3-Part CME on 18 Aug 1980.

fluctuations. In the 1840s *van Humbolt* called such disturbances "magnetic storms" and associated such periods with
the occurrence of the aurora. In 1852, *Sir Edward Sabine*
showed that geomagnetic variations are a world-wide
phenomenon.

SOLAR ACTIVITY CYCLE

In 1609 *Galileo Galilei* modified a design for a telescope
to produce one of excellent quality that could see the moons
of Jupiter, the phases of Venus and the surface of the moon,
a capability that was not matched for several years. But he
also turned it to the Sun and charted the location and numbers of sunspots, which had been done with the naked eye by
the ancient Chinese. The use of the telescope for looking at
the heavens marked the beginning of a new era of astronomy.
The interest in tracking sunspots continued for a few more
years, but then they virtually disappeared between 1645 and
1715, during what is now known as the Maunder minimum.

Sunspot tracking resumed in earnest, after *Heinrich
Schwabe*, an amateur astronomer in Dessau Germany,
reported in 1843 on his observations of a cycle in the number
of sunspots. He measured the number of sunspots over a
period of 17 years beginning in 1826. His paper was noticed
by the Swiss astronomer, *Rudolf Wolf* [1851], who in 1847
then began recording sunspots, and computing the Zurich
sunspot number, still in use today.

In 1852 *Sabine* showed that global magnetic fluctuations
synchronized with the sunspot cycle. This relationship is
shown in Plate 2, which demonstrates the excellent correspondence of the periodicity of the Zurich sunspot number
with geomagnetic activity.

In the late 1800's and early 1900s, *E. Walter Maunder* was
studying the relationship of solar phenomena to geomagnetic
storms. In 1892 and more fully in 1904, he stated that the
rare, large geomagnetic storms are associated with large
sunspots near the center of the visible disk, but that for
smaller sunspot groups the association broke down. Later
in 1904, he found that many geomagnetic storms occur at
27-day intervals. *H.W. Newton* [1943] found an association
between large solar flares and geomagnetic storms. He studied 37 large flares and found that a storm followed 27 of the
largest ones within 2 days of the flare as long as the flare was
within 45° of the central meridian.

SOLAR PLASMA EMISSION

The idea of corpuscular radiation being expelled from the
Sun became advanced in the early 1930s. *Chapman and
Ferraro* [1930] proposed that the Sun ejected a neutral
plasma associated with eruptive solar prominences. The
direct evidence for corpuscular emission from the Sun came

from an analysis of the orientation of comet tails [*Biermann*,
1951]. The orientation was consistent with a wind blowing
continuously away from the Sun.

Eclipse observations have been carried out for thousands
of years. During the eclipse of 1860 over Spain the drawing
in Figure 1 was made by *G. Tempel*. *Eddy* [1974] showed that
those observers to the east of Tempel's location did not record
any structures other than streamers, but those in the approximate location of Tempel drew similar structures. While such
drawings were probably greeted with derision, we now know
that it could have been a coronal mass ejection.

Prominence eruptions from the limb of the Sun were commonly observed. It wasn't clear if the material escaped the
solar atmosphere, especially since prominence material was
seen to drain back. A significant observation of outward
motion far from the Sun was observed at 80 MHz by the
Culgoora radioheliograph [*Riddle*, 1970] and is shown in
Figure 2. It was a moving type IV radio burst, which is emission from a dense plasma cloud.

CORONAL MASS EJECTIONS

The advent of the space age saw the development of exter-
nally occulted white light coronagraphs. During the 1960s,
balloon and rocket coronagraphs were observing the outer
corona. Then in Sept. 1971, the first orbiting coronagraph
was launched on OSO-7. Table 1 gives a chronology of the
spaceborne coronagraphs, dates of operation, some characteristics, primary results and weaknesses.

The first CME observed optically is shown in Figure 3. On
13-14 Dec 1971, a bright streamer in the southeast participated
in the "coronal transient" that traveled outward at over 1000
km/s [*Tousey*, 1973].

The OSO-7 and Skylab coronagraphs operated with some
overlap between 1971 and 1974. The major discoveries from
these instruments include:

- Discovery of mass expulsions escaping gravity – no mater-
  ial seen to fall back [*Tousey*, 1973; *MacQueen et al.*, 1974]
- Relation to EPL: If a prominence reached 0.3R above the
  limb it always resulted in an eruption [*Munro et al.*, 1979]
- A good relation to solar activity cycle was established
  [*Hildner*, 1976]
- Kinematic properties identified (speed, size, mass)
  [*MacQueen*, 1980]
- Morphology: planar loops were in vogue
  - Attempt to use polarization to indicate depth [*Crifo
    et al.*, 1983]
  - Associated orientation of loop to the orientation of fil-
    ament [*Trottet and MacQueen*, 1980]
- Flare associated [*Rust et al.*, 1980]
- Modeled by MHD flare pulse [*Wu and Han*, 1974]

Figure 1. 1860 Eclipse Drawing Showing a Possible CME.

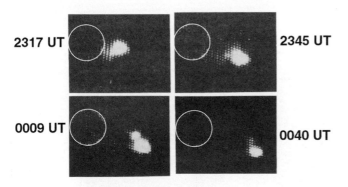

Figure 2. 80 MHz Radio Observation of an Expulsion of a Dense Plasma Cloud from the Sun on 1 Mar 1969.

Table 1. Spaceborne Coronagraphs

NASA Orbiting Solar Observatory 7 (1971-1973)
   3.0 - 10 Rs; SEC Vidicon detector (3 arc min resolution)
   First discovery of coronal transient (CME) 14 Dec 1971
   Weakness - 4 full images per day (~30 CMEs observed)
NASA Skylab (1973-1974)
   2.0 - 6 solar radii; Film detector (5" resolution)
   ~100 CMEs observed, established importance (and beauty);
     statistics; associations
   Weakness: limited film capacity, 3 short duration missions
USAF P78-1 (Solwind) (1979-1985)
   Same characteristics as OSO-7
   CME Statistics, solar cycle dependence, relation to shocks, first
     halo event
   German Helios mission presented in-situ measurements of solar
     wind in quadrature to Sun-Earth line
   Weakness: limited spatial resolution, field of view
NASA Solar Maximum Mission (SMM) (1980, 1984-1989)
   1.6 - 6 solar radii
   5 cm SEC Vidicon detector, (30 arc second resolution)
   CME statistics, 3-part structure to CMEs
   Weakness: quadrant field of view, cadence
ESA Solar and Heliospheric Observatory – SOHO (1995-current)
   EIT/LASCO provide wide field of view & dynamic range
   EIT: UV Disk Imager, (2.5 arc sec pixels)
   C1: 1.1-3 solar radii (5.6 arc sec pixels)
   C2: 2.-7 solar radii (12 arc sec pixels)
   C3: 4-32 solar radii (60 arc sec pixels)
   CCD Imagers (1024 x 1024)
   Initiation of CME, Helical flux rope model, shocks and CMEs,
     geomagnetic effects
   Weakness: Cadence, single viewpoint

During the next decade two coronagraphs again were flying, one on the USAF Space Test Program P78-1 satellite and one on the NASA Solar Maximum Mission satellite. Both of these lasted much longer than the previous missions. The prototypical CME was observed on 18 Aug 1980 (Plate 3). In this event the three-part CME [*Illing and Hundhausen*, 1983] was identified in which the front is followed by a cavity of reduced density and a bright core. The bright core is likely to be prominence material. This became the prototypical CME structure.

Another significant observation was the discovery of the "halo" CME on 27 Nov 1979 [*Howard et al.*, 1982]. Figure 4 shows in difference images a circular band of emission surrounding the occulting disk moves outward. It was followed by a geomagnetic storm a few days later.

These instruments observed many CMEs over 9 years. Other significant results from the 1980's instruments included:

- Streamer blowout [*Howard et al.*, 1985; *Illing and Hundhausen*, 1986]
- Disconnection events [*Illing and Hundhausen*, 1983]
- Associations with IP shocks, LDE X-ray, radio Type IIs, energetic particle emissions [*Sheeley et al.*, 1985, 1983, 1984; *Kahler*, 1985]
- Prominence associated events accelerate in low corona (with Mauna Loa K-coronameter) [*MacQueen and Fisher*, 1983]
- Indirect evidence for acceleration in upper corona [*Woo et al.*, 1985]
- Association with SSN [*Howard et al.*, 1985, 1986]
- Rates as high as 3 per day [*Howard et al.*, 1985]
- Kinematics well established (speed, span, mass, energy) [*Howard et al.*, 1985; *Hundhausen et al.*, 1993]

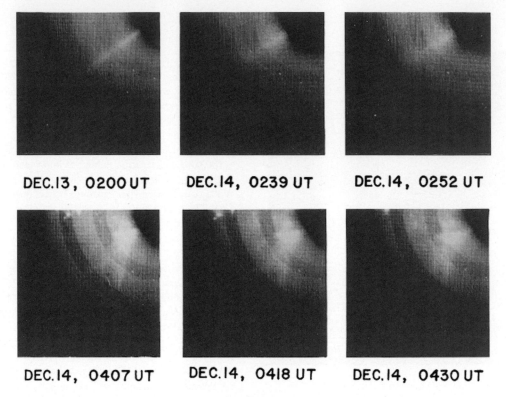

Figure 3. CME Observed on 13-14 Dec 1971.

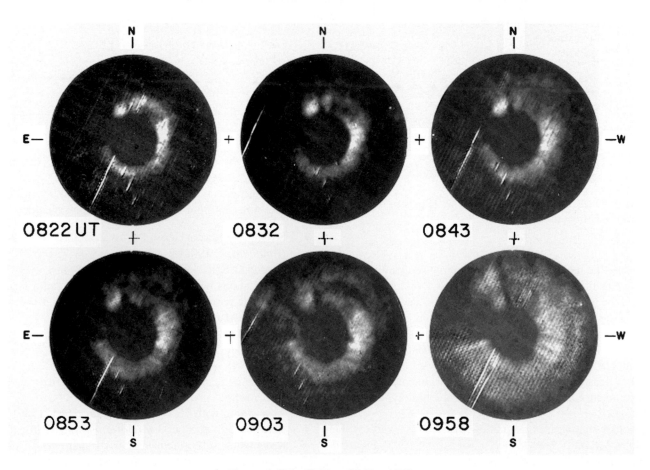

Figure 4. Halo CME on 27 Nov 1979.

The third decade has observations from the SOHO mission, launched in 1995. This has been the best mission for CME studies because of the increased resolution, dynamic range and cadence over previous missions, but also because of a large array of ground-based instruments. The results are still coming. I only list a few here.

- More CME observations than all previous missions [*Yashiro et al.*, 2004]
- Acceleration profile [Howard et al., 1997], flux rope [*Chen et al.*, 1997], interacting CMEs [*Gopalswamy et al.*, 2001]
- Established Halo CMEs source of geomagnetic storms [*Gosling*, 1993; *Brueckner et al.*, 1998]
- Higher Occurrence Rates [*St. Cyr et al.*, 2000]

FUTURE

The next mission with a coronagraph will the NASA STEREO mission to be launched in 2006. The mission objective is to understand the 3D nature of CMEs, their initiation and propagation. A drawback of previous missions is that the observations of the optically thin Thomson scattering give little information on the distribution along the line of sight. Thus the internal structure of CMEs is uncertain.

STEREO will send two identically instrumented spacecraft into a heliocentric orbit, one leading Earth and one trailing. Optical and radio remote sensing instruments as well as in-situ particle energy and solar wind composition measurements, will use modeling to determine the 3D structure and to couple the remote sensing to in-situ observations. The spacecraft will drift away from Earth at an average rate of about 22°/year resulting in varying science objectives. At the end of the two-year nominal mission the spacecraft will be 90° from each other. This orientation the coronagraphs from one spacecraft will observe the solar corona above the EUV disk observed by the other spacecraft.

This work was supported by NASA and the Office of Naval Research. The SOHO/LASCO data used here are produced by a consortium of the Naval Research Laboratory (USA), Max-Planck-Institut for Sonnensystemforschung (Germany), Laboratoire d'Astronomie Spatiale (France), and the University of Birmingham (UK). SOHO is a project of international cooperation between ESA and NASA.

REFERENCES

Biermann, L., Kometenschweife und solare Korpuskularstrahlung, *Z. Astrophys.*, 29, 274-286, 1951.

Brueckner, G.E., J.-P. Delaboudiniere, and 8 co-authors, Geomagnetic storms caused by coronal mass ejections (CMEs): March 1996 through June 1997, *Geophys. Res. Let.*, 25, 3019-3022, 1998.

Chapman, S, V. Ferraro, A new theory of magnetic storms, *Nature*, 126, 129-, 1930.

Chen, J., R.A. Howard, G.E. Brueckner, R. Santoro, and 7 co-authors, Evidence of an Erupting Magnetic Flux Rope: LASCO Coronal Mass Ejection of 1997 April 13, *Astrophys. J.*, 490, L191, 1997.

Crifo, F., J.P. Picat, M. Cailloux, Coronal transients - Loop or bubble, *Solar Phys.*, 83, 143-152, 1983.

Eddy, J.A., A Nineteenth-century Coronal Transient, *Astron. Astrophys.*, 34, 235-40, 1974.

Gopalswamy, N., S. Yashiro, M.L. Kaiser, R.A. Howard, J.-L. Bougeret, Radio Signatures of Coronal Mass Ejection Interaction: Coronal Mass Ejection Cannibalism?, *Astrophys. J.*, 548, L91-L94, 2001.

Gosling, J.T., The solar flare myth, *J. Geophys. Res.*, 98, 18937-18950, 1993.

Hildner, E., J.T. Gosling, and 4 co-authors, Frequency of coronal transients and solar activity, *Solar Phys.*, 48, 127-135, 1976.

Howard, R.A., D.J. Michels, N.R. Sheeley, Jr., M.J. Koomen, The observation of a coronal transient directed at Earth, *Astrophys. J.*, 263, L101-L104, 1982.

Howard, R.A., N.R. Sheeley, Jr., D.J. Michels, M.J. Koomen, Coronal Mass Ejections, 1979-1981, *J. Geophys. Res.*, 90, 8173-8191, 1985.

Howard, R.A., N.R. Sheeley, Jr., D.J. Michels, M.J. Koomen, The solar cycle dependence of coronal mass ejections, *in ASSL Vol. 123: The Sun and the Heliosphere in Three Dimensions*, pp. 107-111, 1986.

Howard, R.A., G.E. Brueckner, O.C. St Cyr, and 16 co-authors, CMEs observed from LASCO, in *Coronal Mass Ejections*, edited by N. Crooker, J.A. Joselyn, and J. Feynman, American Geophysical Union, Washington, D.C., pp. 17-26, 1997.

Hundhausen, A.J., Sizes and locations of coronal mass ejections - SMM observations from 1980 and 1984-1989, *J. Geophys. Res.*, 98, 13177, 1993.

Illing, R.M.E., A.J. Hundhausen, Possible observation of a disconnected magnetic structure in a coronal transient, *J. Geophys. Res.*, 88, 10210-10214, 1983.

Illing, R.M.E., A.J. Hundhausen, Observation of a coronal transient from 1.2 to 6 solar radii, *J. Geophys. Res.*, 90, 275-282, 1985.

Illing, R.M.E., A.J. Hundhausen, Disruption of a streamer by an eruptive prominence and coronal mass ejection, *J. Geophys. Res.*, 91, 10951-10960, 1986.

Kahler, S. and 5 co-authors, A comparison of solar helium-3-rich events with type II bursts and coronal mass ejections, *Astrophysical Journal*, 290, 742-747, 1985.

MacQueen, R.M. Coronal transients - A summary, Royal Society of London Philosophical Transactions Series A, 297, 605-620, 1980.

MacQueen, R.M. and 7 co-authors, The Outer Solar Corona as Observed from Skylab: Preliminary Results, *Astrophys J.*, 187, L85-L88, 1974.

MacQueen, R.M., R.R. Fisher, The kinematics of solar inner coronal transients, *Solar Phys.*, 89, 89-102, 1983.

Maunder, E.W., Connection between solar activity and magnetic disturbances, etc. on the Earth, *PASP*, 6, 125-125, 1892.

Maunder, E.W., The "great" magnetic storms, 1875 to 1903, and their association with sun-spots, *MNRAS*, 64, 205-222, 1904.

Maunder, E.W., Distribution of magnetic disturbances, *MNRAS*, 65, 18-25, 1904.

Munro, R.H., J.T. Gosling, E. Hildner, R.M. MacQueen, A.I. Poland, C.L. Ross, The association of coronal mass ejection transients with other forms of solar activity, *Solar Phys.*, 61, 201-215, 1979.

Newton, H.W., Solar Flares and Magnetic Storms, *MNRAS*, 103, 244-257, 1943.

Riddle, A.C., 80 MHz observations of a moving type IV solar burst, March 1, 1969, Solar Phys., 13, 448-457, 1970.

Rust, D.M., E. Hildner, and 8 co-authors, Mass ejections, in *Skylab Solar Workshop II*, pp. 273-339, 1980.

Schwabe, H., Sonnenbeobachtungen im Jahre 1843, *Astronomische Nachrichten*, 20, 495, 1843.

Sheeley, N.R., Jr., R.A. Howard, M.J. Koomen, D.J. Michels, Associations between coronal mass ejections and soft X-ray events, *Astrophysical Journal*, 272, 349-354, 1983.

Sheeley, N.R., Jr. and 5 co-authors., Associations between coronal mass ejections and metric type II bursts, *Astrophysical Journal*, 279, 839-847, 1984.

Sheeley, N.R., Jr., R.A. Howard, and 5 coauthors, Coronal mass ejections and interplanetary shocks, *J. Geophys. Res.*, 90, 163-175, 1985.

St.Cyr, O.C., R.A. Howard, N.R. Sheeley, Jr., S.P. Plunkett, and 10 more co-authors, Properties of coronal mass ejections: SOHO LASCO observations from January 1996 to June 1998, *J. Geophys. Res.*, 105, 18169-18185, 2000.

Tousey, R., The Solar Corona, in *Space Research XIII*, edited by M.J. Rycroft and S.K. Runcorn, Akademie-Verlag, Berlin, p713, 1973.

Trottet, G., R.M. MacQueen, The orientation of pre-transient coronal magnetic fields, *Solar Phys.*, 68, 177-186, 1980.

Wolf, M.R. Universal sunspot numbers, *Naturf. Gesell. Bern. Mitt.*, 1, 89-95, 1851.

Woo, R., J.W. Armstrong, N.R. Sheeley, Jr., R.A. Howard, D.J. Michels, M.J. Koomen, R. Schwenn, Doppler scintillation observations of interplanetary shocks within 0.3 AU, *J. Geophys. Res.*, 90, 154-162, 1985.

Wu, S.T., S.M. Han, in *Solar Wind Three; Proceedings of the Third Conference, Pacific Grove, Calif., March 25-29, 1974*, University of California, Los Angeles, pp. 144-146, 1974.

Yashiro, S., and 6 co-authors, A catalog of white light coronal mass ejections observed by the SOHO spacecraft, *Journal of Geophys. Res.*, 109, 07105, 2004.

Solar Energetic Particles and Coronal Mass Ejections: A Perspective

J.R. Jokipii

Depts of Planetary Sciences and Astronomy, University of Arizona, Tucson, Arizona

Energetic charged particles are ubiquitous in space, and are apparently present wherever the ambient gas density is low enough. Solar energetic particles were first observed in the 1940's and have been the subject of intense study ever since. The essential similarity of their spectral slopes from event to event and with those observed in many different astrophysical sites places significant general constraints on the mechanisms for their acceleration and transport. Diffusive shock acceleration is at present the most successful acceleration mechanism proposed, and, together with transport in broadband turbulence, can account naturally for the very similar specta. It is suggested that acceleration of solar energetic particles, *both* the CME-associated gradual events *and* the impulsive events are accelerated mostly by shocks. ^{3}He-rich events may be the exception, although selection during the injection process may cause the enhancement of ^{3}He.

1. INTRODUCTION

Cosmic rays, or more generally, energetic charged particles, are found in many places in nature, including the Sun, heliosphere, our galaxy, external galaxies and galaxy clusters. Although the energetic particles are found in these quite disparate places, where the parameters are expected to be quite different, the observed energy spectra are for the most part very nearly power laws, with quite similar slopes. Departures from the power laws generally occurs in the form of high-energy turnovers or cutoffs. This similarity of the spectra suggests a common mechanism for the bulk of the acceleration, with the high-energy cutoff being due to time-dependent or other site-specific effects. *Syrovatsky* [1961] originally pointed out this observational fact, and pointed out that it strongly implied a common or universal acceleration mechanism.

The Sun accelerates energetic particles (SEPs), in events associated with energy releases such as coronal mass ejections or flares. As noted above, the resulting SEP events are often power laws in energy with a similar power-law index. The first solar energetic particle events were observed in the mid 1940s by Forbush [*Forbush*, 1954]. One early event is shown in Figure 1. The energy spectra of various energetic particle species during a typical solar particle event is shown in Figure 2.

Figure 3 shows the observed spectrum of galactic cosmic rays at Earth over the energy interval 10^5–10^{20} eV, compiled from a number of sources. Apparent at low energies, below $\simeq 1$ GeV are a turnover and various other features caused by the interaction of cosmic rays with the sun and the solar wind. The energetic particle spectrum, at the very lowest energies, which can be observed only *in situ*, is observed to merge smoothly into the background thermal plasma distribution. The energy spectrum of galactic cosmic rays is a remarkably-smooth power law above some 10^9 eV, with only a minor change in slope occurring between 10^{15} and 10^{16} eV (the "knee"), and possibly a small flattening at 10^{19} eV (the "ankle"). The anomalous cosmic-ray oxygen, seen in Figure 3, is a consequence of the interaction of the heliosphere with the interstellar medium.

Photons emitted from distant galaxies, between galaxies and other regions of our galaxy as a consequence of the interaction of cosmic rays with ambient matter or electromagnetic fields, show that cosmic rays exist in many places (*Kronberg* [1996]). Moreover, the inferred energy spectra

Solar Eruptions and Energetic Particles
Geophysical Monograph Series 165
Copyright 2006 by the American Geophysical Union
10.1029/165GM04

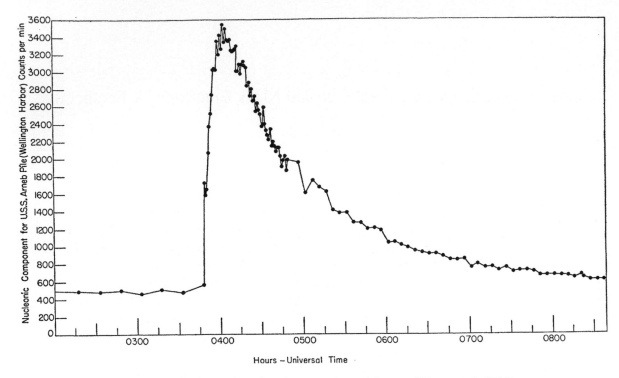

Figure 1. Early observation of a solar energetic particle event [*Meyer et al.*, 1956].

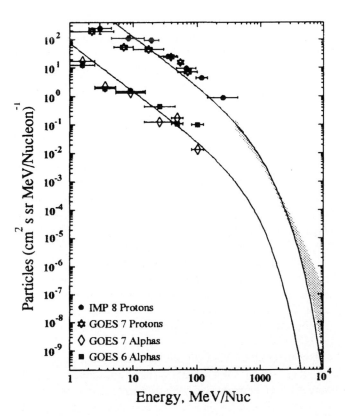

Figure 2. Energy spectra during a typical SEP event [*Reames*, 1999].

of the cosmic rays in these remote sources are also power laws with slopes not much different from that observed at Earth. This fact, together with the smoothness of the observed spectrum at Earth over a large range in energy, suggests a common acceleration mechanism for the bulk of the cosmic rays.

2. ACCELERATION OF ENERGETIC CHARGED PARTICLES

The change in energy ΔT of a particle having energy T and velocity $\mathbf{w}$, moving an electric field $\mathbf{E}(\mathbf{r}, t)$ in the time interval Δt may be written in general as:

$$\Delta T = q\int_{t}^{t+\Delta t} \mathbf{w} \cdot \mathbf{E}(\mathbf{r}, t)dt, \qquad (1)$$

where the integrand must be evaluated along the actual particle trajectory.

From this we see that in order to evaluate the energy change, we must know the particle trajectory in the electromagnetic field. This leads to the general requirements that acceleration and spatial transport be intimately coupled and both $\mathbf{E}(\mathbf{r}, t)$ and $\mathbf{B}(\mathbf{r}, t)$ must be considered together.

This paper concerns energetic particles, for which the relevant spatial plasma scales are the energetic-particle gyro-radii. At these scales the ambient plasmas are generally

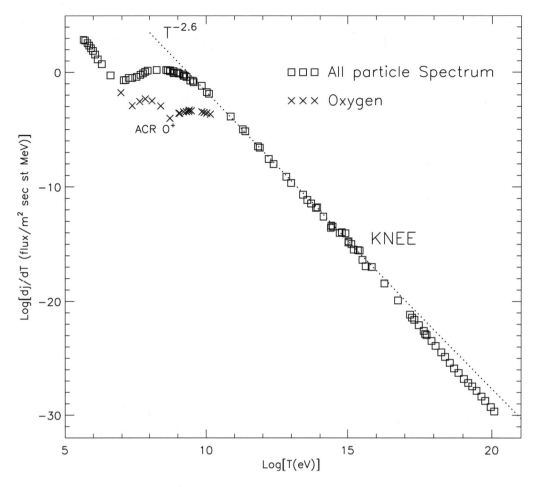

Figure 3. The quiet-time cosmic-ray energy spectrum, compiled from a variety of sources.

hydromagnetic, and the electric field is determined from the fluid velocity **U** and magnetic field by the relation

$$\mathbf{E} = -\frac{\mathbf{U} \times \mathbf{B}}{c} \qquad (2)$$

where c is the speed of light. The physical problem of acceleration, then, consists of solving equations 1 and 2 for a given situation. Note that since the electric field **E** is determined by **U** and **B**, it need not appear in the transport equation.

A robust and widely applicable transport theory, applicable to particles whose speed w is significantly larger than U has been developed and applied extensively over the past four decades. It may be considered as an expansion in powers of the small ratio U/w. Unfortunately, a corresponding general transport theory for lower-energy particles for which U/w is *not* small is not available – for such particles one must use numerical simulations. The available theory utilizes the fact that the plasmas in space are turbulent, with broadband fluctuations over scales including those comparable with the particle

gyro-radii. The particle motions are described statistically – the turbulent magnetic fluctuations "scatter" the particles in angle, driving them to near-isotropy in the fluid frame (or, more generally, the frame of the scatterer) with a time scale τ_{scat}. This scattering occurs much more rapidly than energy change.

To first order in U/w, the transport equation derived by *Parker* [1965] has proved to be robust and applicable to many situations. It makes use of the fact that the ambient turbulence scatters the particles in pitch angle to make the distribution nearly isotropic. The equation may be written:

$$\frac{\partial f}{\partial t} = \frac{\partial}{\partial x_i}\left[\kappa_{ij}\frac{\partial f}{\partial x_j}\right] - U_i\frac{\partial f}{\partial x_i} + \frac{1}{3}\frac{\partial U_i}{\partial x_i}\frac{\partial f}{\partial \ell n(p)} + Q, \qquad (3)$$

where $f(\mathbf{r}, p, t)$ is the distribution function as a function of momentum magnitude p, position **x** and time t, and K_{ij} is the particle diffusion tensor determined by the magnetic scattering.

In this equation we have, in addition to the diffusion, convection and acceleration/deceleration caused by the electric field. Note that the electric field does not appear explicitly – it is nonetheless contained in the terms containing the flow velocity **U** by making use of the relationship in equation (2).

If we carry out the expansion to second order in U/w, we obtain new effects involving the viscosity of the cosmic rays and the acceleration of the flow (*Earl et al.* [1988]). In many cases these can be neglected because U/w is small.

2.1. Diffusive Shock Acceleration

Consider a steady, plane shock propagating in a uniform medium. Define the x-direction as the direction of propagation and let particles be introduced uniformly and steadily at the shock, at an injection momentum p_0. Work in the shock-normal coordinate system, with the shock at the fixed position $x = x_{sh}$. The shock ratio r is defined as the ratio of upstream to downstream flow speed U_1/U_2. It is readily found that the steady solution to the Parker equation in this case is given by

$$f(p) = Ap^{-3r/(r-1)}H(p-p_0)F(x,p) \qquad (4)$$

where $H(p)$ is the Heaviside step function and the spatial dependence of $F(x, p)$ is independent of x and p at the shock, is independent of x behind the shock, and decreases exponentially upstream as $\exp(-U_1(x_{sh}-x)/k_{xx}(p))$. Note that in the limit of a strong shock, where $r \to 4$ the momentum dependence becomes $f(p) \propto p^{-4}$, which corresponds to an energy spectrum $dj/dT = p^2 f \propto p^{-2}$ which is not far from the observed spectrum at relativistic energies (e.g., Figure 2). This energy dependence is independent of shock speed, diffusion coefficients and other parameters. Since shocks in astrophysics tend to be strong, this may be the desired "universal" spectrum. This precise spectrum is not seen in nature, because of the idealizations involved, but these cause relatively small corrections to the basic power law, and also tend to be similar in many places. In particular, there will be a high-energy cutoff. In addition, in some cases, the accelerated particles will modify the shock (for a discussion, see *Jones and Ellison* [1991]).

2.1.1. Rate of diffusive shock acceleration. The acceleration by the shock is not instantaneous, of course. Solving the time-dependent version of equation (3), with the injection at momentum p_0 turned on at a time t_0, reveals that the spectrum above p_0 is still the universal power law given in equation (4), but with a high-momentum cutoff, p_c which increases at a rate

$$dp_c/dt \approx 4U_1^2/\kappa_{xx}. \qquad (5)$$

Hence, since for quasi-perpendicular shocks $\kappa_{xx} = \kappa_\perp$ which is generally significantly smaller than $\kappa_\parallel$, quasi-perpendicular shocks will in general accelerate particles significantly faster than will quasi-parallel shocks. In fact, the physics of acceleration at quasi-perpendicular shocks is quite different than that at quasi-parallel shocks [*Jokipii*, 1982]. Nonetheless, the mathematics is the same, and the conclusions concerning the energy spectrum remain unchanged.

The acceleration rate in the lower solar atmosphere, where the magnetic field can be 100 G or larger, is high enough to accelerate 1 GeV protons in a time of the order of a few seconds or less.

2.2. 2nd-Order Fermi Acceleration

The above approximations neglect the possible random motions of the scattering magnetic irregularities relative to the flow velocity. If these are included (e.g., forward and backward moving Alfvén waves), further acceleration is introduced. This is the well-known 2nd-order Fermi acceleration, introduced by *Fermi* [1949], long before the present transport equation was developed. *Fermi* spoke in terms of randomly moving magnetic clouds, but it is straightforward to generalize it to the more modern random spectrum of Alfvén waves.

It is quite often the mechanism of choice for possible diffuse *re*-acceleration of cosmic rays in the interstellar medium and solar flares. Nonetheless, it has at least one considerable disadvantage. It does not produce a power law spectrum with the desired slope in a robust way. The shape of the spectrum depends sensitively on the transport parameters in the acceleration region. Nonetheless, it remains popular. The effect of 2nd-order Fermi acceleration on the distribution function f may be written

$$\left(\frac{\partial f}{\partial t}\right)_{2nd\text{-}order\ Fermi} = \frac{1}{p^2}\frac{\partial}{\partial p}\left\{p^2 D_{pp}\frac{\partial f}{\partial p}\right\} \qquad (6)$$

and represents diffusion in momentum, with momentum diffusion coefficient, D_{pp}. This term may simply be added to the right side of equation (3) and is often invoked where shocks are believed not to be present. Since the Alfvén speed and fluid speed are often comparable, the viscous acceleration and 2nd-order Fermi acceleration are formally of the same general order of magnitude, although the Fermi term is the one most-often discussed.

One difficulty for this mechanism was seen and very clearly discussed many decades ago [*Syrovatsky*, 1961]. This is the fact that the energy spectrum is not in general a power law, and if the parameters are chosen to produce a power law, the spectral slope is quite sensitive to parameters which are not well known, and which are expected to vary considerably from event to event and from location to location, contrary to observations. Hence the similar power laws in many places are not easy to understand. We will see that this problem does not occur for diffusive shock acceleration. A common expression for the momentum diffusion may be written in terms of the momentum p, the Alfvén speed V_a and the time for scattering by magnetic irregularities, τ_{scatt} as:

$$D_{pp} = \frac{1}{2} < \frac{(\Delta p)^2}{\Delta t} > = p^2 / \tau_{acc} = (V_a^2 / c^2)(p^2 / \tau_{scatt}), \qquad (7)$$

where the acceleration time scale $\tau_{acc} = \tau_{scatt} c^2 / V_a^2$. In the standard leaky box model with loss time τ_{loss}, and if τ_{coll} and τ_{loss} are constant, we obtain a power law.

$$f(p) = Ap^{-\alpha} \qquad (8)$$

where A is a constant and

$$\alpha = \frac{3}{2} + \sqrt{\frac{9}{4} + \frac{\tau_{acc}}{\tau_{loss}}} \qquad (9)$$

Unfortunately, the highly uncertain and variable parameters, V_a, τ_{acc} and τ_{loss}, must vary considerably from location to locatin and event to event, producing quite variable spectral slopes. As emphasized by *Syrovatsky ibid*, who presented a quite similar analysis, this argues against this form of statistical acceleration.

However, it is also important to remember that the only plausible mechanism for producing the extremely large enhancements of ^{3}He relative to ^{4}He which has been worked out is statistical acceleration. For this reason, if for no other, we must consider statistical acceleration as a possibility. Another possibility is that a statistical, resonant, process accelerates the particles to a superthermal energy, enhancing ^{3}He, and these are then further accelerated to energetic-particle energies by a shock, thus producing the desired energy spectrum.

3. SUMMARY

Diffusive shock acceleration has so many attractive aspects – it is quite fast (especially at quasi-perpendicular shocks), it naturally produces a power-law energy spectrum which is quite close to that observed in many places, and the shocks which can do the acceleration are quite common – that it is regarded by many as possibly the only important acceleration mechanism. Certainly, it is highly likely it is the most important mechanism outside of solar flares, and it may play an important role in flares, as well.

Acknowledgments. This work was supported, in part, by NASA under grants NAG5-6620, NAG5-7793, NAG5-12919, and by the NSF under grant ATM9616547 and ATM0327773. I acknowledge helpful discussions regarding many of these matters with my colleagues J. Kóta and J. Giacalone.

REFERENCES

Earl, J.A., J.R. Jokipii, G. Morfill, Cosmic-Ray Viscosity, *Ap. J. Lett.*, 331, L91-L94, 1988.

Fermi, E., On the Origin of Cosmic Rays, *Phys. Rev.*, 75, 1169-1174, 1949.

Forbush, Scott E., Three Unusual Cosmic-Ray Increases Possibly Due to Charged Particles from the Sun, *Phys. Rev.*, 70, 771-772, 1954.

Jokipii, J.R., Particle drift, diffusion, and acceleration at shocks, *Astrophys. J.*, 255, 716-720, 1982.

Jones, Frank C., and Ellison, Donald C., The plasma physics of shock acceleration, *Space Sci. Rev.*, 58, 259-346, 1991.

Kronberg, Phillip P., Intergalactic magnetic fields and some connection with cosmic rays, *Space Sci. Rev.*, 75, 387-399, 1996.

Meyer, P., E.N. Parker, J.A. Simpson, Solar Cosmic Rays of February, 1956 and Their Propagation through Interplanetary Space, *Phys. Rev.*, 104, 768-783, 1956.

Parker, E.N., On the Passage of Cosmic Rays Through Interplanetary Space, *Planet. Sp. Sci.*, 13, 9, 1965.

Reames, Donald V., Particle Acceleration at the Sun and in the Heliosphere *Space Sci. Rev.*, 90, 413-491, 1999.

Syrovatsky, S.A., Spectrum of Galactic and Solar Cosmic Rays, *J. Exptl. Theoret. Phys. (U.S.S.R.)*, 40, 1788-1793, 1961.

Observational Properties of Coronal Mass Ejections

S.W. Kahler

Air Force Research Laboratory, Space Vehicles Directorate, Hanscom AFB, Massachusetts

Coronal mass ejections (CMEs) have been known and observed for over 30 years. The total number of observed CMEs is now approaching 10,000, most of them detected with the LASCO coronagraph on the SOHO spacecraft. We review statistical work on CME widths, latitudes, accelerations, speeds, masses, and rates of occurrence. Solar-cycle variations of these parameters are presented. Recent work has focused on CME internal properties and compositions and on CME dynamics, particularly at low (< 3 R$_\odot$) altitudes. The challenges to understand the magnetic topology of narrow ($< 20°$ width) CMEs, to determine the relationship of coronal holes to CMEs, and to observe magnetic reconnection that effects magnetic disconnections of CMEs from the Sun are discussed.

1. INTRODUCTION

Coronal mass ejections (CMEs) have now been observed for over three decades. The earlier observations from the OSO-7, Skylab, Solar Maximum Mission (SMM), and Solwind coronagraphs have been reviewed and compared with the current coronagraph observations from the LASCO coronagraphs on SOHO by *Gopalswamy et al.* [2003] and *Yashiro et al.* [2004], who emphasized the LASCO CME annual variations, and by *Gopalswamy* [2004], who emphasized the mission-cumulative statistics. Here we briefly review the statistical properties of CMEs presented in those works using the large database of ~7000 LASCO CMEs, whose parameters are measured from sequences of running difference images and are given in the web-based catalog of CMEs provided by the Catholic University of America (CUA) [*Yashiro et al.*, 2004]. *Robbrecht and Berghmans* [2004 and this volume] discuss an automated CME recognition program to provide an objective and more comprehensive method of selecting and measuring CMEs.

The CME images are projections on the plane of the sky, so some measured properties such as CME speeds and masses can be only lower limits and others such as CME widths and latitudes only upper limits, rather than true values [*Burkepile et al.*, 2004]. For each CME the speed of only the fastest part of the leading edge is measured with both first and second order fits. Angular widths around the coronagraph occulting disk and position angles (PA) measured counterclockwise from north are also given. Halo CMEs (Figure 1) are defined as CMEs with angular widths $\geq 120°$ and are treated separately. Masses of LASCO CMEs have only recently been compiled statistically and are reviewed. The CME statistics from solar minimum in 1996 through the recent maximum in 2000 have allowed us to compare the CME properties of the two extreme periods of solar activity.

We also review recent work on more detailed observations to describe the early dynamics and structures of CMEs and then discuss three research topics that appear ripe for new advances. These are the magnetic topologies of narrow ($< 20°$) CMEs; coronal magnetic reconnection in CMEs; and the relationship of coronal holes (CHs) to CMEs.

2. CME STATISTICS

2.1. Occurrence Rates of CMEs

The CME occurrence rate is an index of solar activity. It has long been known to track the sunspot number and other indices of solar activity. Figure 2 [*Gopalswamy*, 2004] compares the smoothed LASCO CME rate with the sunspot number, showing good agreement, although with some lag

Solar Eruptions and Energetic Particles
Geophysical Monograph Series 165
10.1029/165GM05

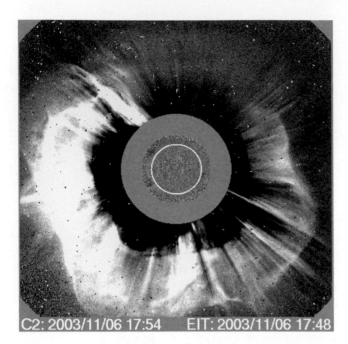

Figure 1. A LASCO 360° halo CME of 6 November 2003 shown as a subtracted image. The CUA website gives a speed of 1523 km s^{-1} measured at a PA of 100° and an acceleration of −59.5 m s^{-2}. The white circle is the solar disk, and the C2 coronagraph inner limit of the field of view lies at 2.2 R$_\odot$ The dark circular band is caused by subtracting the previous LASCO image taken 24 minutes earlier.

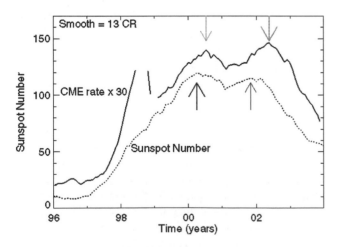

Figure 2. The LASCO CME rate smoothed over 13 Carrington rotations and compared with the solar sunspot number. Arrows indicate the two maxima in CME rate and sunspot number. Large data gaps occurred during June 1998 to February 1999. From *Gopalswamy* [2004].

near the two sunspot number peaks. The lag is probably related to the fact that high latitude (> 60°) CMEs, arising in the polar crown filaments, are also important near solar maximum but are not directly related to the sunspot number

[*Gopalswamy et al.*, 2003]. The low-latitude CME rate increased in a step-like fashion in 1998 and remained relatively constant while the high latitude CME rate was much more variable. The CME occurrence rate is less than 1 CME/day at solar minimum and about 6 CME/day at solar maximum.

2.2. Latitudes of CMEs

CMEs tend to be confined to low-latitude streamer belts at solar minimum and to range over a broad latitude range at solar maximum. *Yashiro et al.* [2004] have compiled the annual solar latitude distributions of the LASCO CME PAs from 1996 at solar minimum through 2002. The solar latitude ranges that contain 80% of the CME PAs for those years range from (~22°) in 1996 to (~63°) at maximum in 2000. Halo CMEs were not included in the statistics.

2.3. Widths of CMEs

Angular widths of all LASCO CMEs observed from 1996 through August 2003 have been compiled by *Gopalswamy* [2004]; the average width of all non-halo (width ≤ 120°) CMEs is 47° (Figure 3). It is convenient to divide CMEs into three width groups: narrow (< 20°), normal (20° to 120°) and wide, or halos, (≥ 120°). *Yashiro et al.* [2004] found that over the years 1996-2002 the three width groups were distributed as: narrow, 18%; normal, 70%; and halo, 12%. There was an increase in the median widths of normal CMEs from 43° at the 1996 solar minimum to 58° in 1999 and then a decrease to 49° in 2002. During the period 1998-2000 a bimodal distribution appeared, with peaks at ~15° and ~50°. The average widths do not appear to vary with latitude [*Michalek and Mazur*, 2002].

CME angular widths may vary with height. *St. Cyr et al.* [1999] compared the widths of 132 CMEs observed in both the 1.15 to 2.4 R$_\odot$ and the 1.7 to 6 R$_\odot$ fields of view of the Mark III coronameter and the SMM coronagraph, respectively. They found an average 12° increase in the SMM widths, which they interpreted as a 20% to 30% increase in angular span as the CMEs traveled from the inner to the middle corona. However, *Stockton-Chalk* [2002] found only a modest average nonradial expansion of 1.84°, corresponding to a width increase of 3.7°, in 50 CMEs measured at various points in the ~5 to 25 R$_\odot$ range of the LASCO C2 field of view.

Inclusion of adjacent pre-existing streamer deflections or wave-like coronal disturbances as parts of the CMEs can result in over-estimations of intrinsic CME widths [*Cremades and Bothmer*, 2004], particularly for halo CMEs [*Michalek et al.*, 2003; *St. Cyr et al.*, 2005; Sec. 2.6]. In those cases the smaller widths measured when the CME leading edges are still in the LASCO C2 field of view may provide more realistic CME widths. In their study of structured CMEs [Sec. 3.2] *Cremades and Bothmer* [2004] found an average width of 85°, much smaller than the 155° value derived from the same CME widths of the CUA catalog.

2.4. Speeds and Accelerations of CMEs

The CME linear-speed distribution through 2003 is shown in Figure 3, where the average value of 489 km s^{-1} is indicated [*Gopalswamy*, 2004]. Only 25 of the 7567 CME speeds exceeded 2000 km s^{-1}; the fastest CME speed measured thus far was 2657 km s^{-1} on 4 November 2000. When compiled on an annual basis, there is a very clear increase in average annual speeds from 281 km s^{-1} in 1996 to 560 km s^{-1} in 2003. CME speeds as a function of width declined slightly over the range 0° to ~65° and then increased with considerable scatter over larger widths [*Yashiro et al.*, 2004].

CME accelerations are important for the insights they can provide into the balance among the Lorentz, gravitational, and drag forces [*Vršnak et al.*, 2004]. When 5 or more data points were available in the LASCO height-time plots, second-order fits could be done to obtain accelerations. The results showed an anticorrelation between acceleration and speed in that slower CMEs generally accelerated and faster CMEs decelerated [*Yashiro et al.*, 2004; *Gopalswamy*, 2004], a result attributed to aerodynamic drag [*Vršnak et al.*, 2004]. The average deceleration of the fastest (> 900 km s^{-1}) CME group is −16 m s^{-2}, comparable to that of solar gravity

at 5 R$_\odot$ (−11 m s^{-2}). The accelerations decline for CMEs with larger widths and at greater solar distances such that CMEs are nearly at solar wind speeds within the 30 R$_\odot$ LASCO field of view. However, a small (≤ 10% fraction of CMEs, generally broader and faster than the rest of the CME sample, continue to be propelled beyond 25 R$_\odot$ [*Vršnak et al.*, 2004].

2.5. Masses and Energies of CMEs

The most recent distribution of LASCO CME masses includes 4297 CMEs observed through the end of 2002 and with widths between 10° and 150°. The average and median values for those CMEs are 1.57 × 10^{15} gms and 6.67 × 10^{14} gms, respectively [*Vourlidas*, 2004)]. We compare the LASCO mass distribution with those from the Solwind [*Jackson and Webb*, 1994] and SMM [*Burkepile et al.*, 2004] coronagraphs in Figure 4. The earlier measurements were more characteristic of solar maximum, and the characteristic values are higher. However, the ~15% of CMEs with masses below 10^{14} gms appears to be the result of the higher LASCO sensitivity [*Vourlidas et al.*, 2002]. *Vourlidas et al.* [2002] found a positive correlation between CME masses and accelerations, i.e., small mass CMEs tended to decelerate while the large-mass CMEs accelerated.

The CME kinetic energies can also be calculated from the LASCO measured masses and speeds. The average (median)

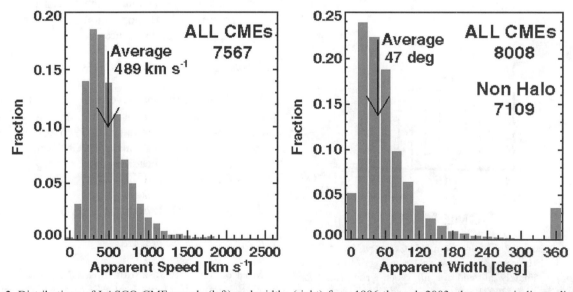

Figure 3. Distributions of LASCO CME speeds (left) and widths (right) from 1996 through 2003; the arrows indicate distribution averages. Apparent speeds are measured in the plane of the sky at the PA of the fastest moving part of the CME leading edge. Speeds could be measured for only 7567 of the total 8008 detected CMEs. The average width of 47° corresponds only to the 7109 nonhalo (≤ 120°) CMEs. From *Gopalswamy* [2004].

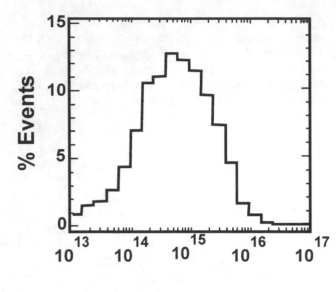

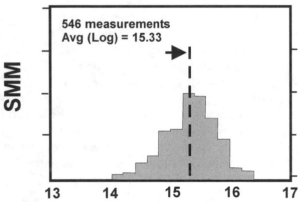

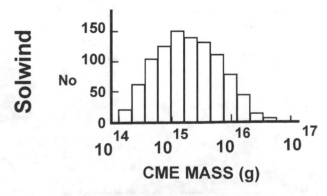

Figure 4. CME mass distributions from LASCO (top), SMM (middle), and Solwind (bottom). The mass scales in logs of masses in grams are approximately aligned for all plots. From *Vourlidas* [2004], *Burkepile et al.* [2004], and *Jackson and Webb* [1994].

kinetic energy is 2.4×10^{30} ergs (5.0×10^{29} ergs) [*Vourlidas*, 2004].

2.6. Halo CMEs

Halo CMEs (Figure 1) can be full (360°) or partial (> 120°). Based on a study of the widths of SMM limb CMEs *Burkepile et al.* [2004] concluded that halo CMEs are typical CMEs originating close to disk center and directed preferentially along the Sun-Earth line, either toward or away from the Earth. However, the average speed of LASCO halo CMEs during 1996-2003 was 1004 km s^{-1}, well above the average speed of 489 km s^{-1} for the general CME population [*Gopalswamy*, 2004]. Those directed toward the Earth are often associated with shocks, solar energetic particle (SEP) events, and geomagnetic storms and therefore merit consideration as a more energetic group of CMEs [*Michalek et al.*, 2003; *Gopalswamy*, 2004], contrary to the conclusion of *Burkepile et al.* [2004]. Three different types of halo CME are now recognized [*St. Cyr et al.*, 2005]: 1. eruptions lying near the Sun-Earth line; 2. eruptions causing deflections of pre-existing coronal structures; and 3. the coalescence of multiple CMEs. Those of the second type can lie near the limb. All types are illustrated at the CUA LASCO website.

Full halos are ~3.5% [*Gopalswamy*, 2004] and partial halos (> 120°) are ~12% [*Yashiro et al.*, 2004] of all CMEs. Most full halo CMEs are faster than 900 km s^{-1} and therefore a subset of a group of fast (> 900 km s^{-1}) and wide (> 60°) CMEs, which are particularly efficient in driving shocks [*Gopalswamy*, 2004] and producing SEPs [*Kahler and Reames*, 2003]. The fast-and-wide CMEs constitute about 4.7% of the total CME population and generally track the total CME population in occurrence rate.

2.7. Solar Cycle Variations of CMEs

We summarize the solar-cycle variations of CME characteristics in Table 1. Those CME parameters generally increase from solar minimum in 1996 to solar maximum around 2001 as the contribution of CMEs to the solar wind and to space weather becomes more important around solar maximum.

3. CME EARLY DYNAMICS AND STRUCTURES

3.1. Early Dynamics of CMEs

Most of the acceleration of most CMEs occurs below the ~2.2 R$_\odot$ inner limit of the LASCO C2 coronagraph. The LASCO C1 coronagraph and other space or ground-based instruments have been used to understand the initial speed and acceleration profiles of CMEs. *St. Cyr et al.* [1999]

Table 1. Solar-Cycle Variations of CME Parameters

Parameter	Minimum	Maximum	Notes
Occurrence	< 1/day	~6/day	
Solar Latitude Range	22°	63°	Includes 80% of CMEs.
Median Widths[a]	43°	58°	1999 Maximum.
Average Widths[a]	47°	61°	1999 Maximum.
Median Speeds (km s^{-1})	250	495	2003 Maximum.
Average Speeds (km s^{-1})	281	560	2003 Maximum.
Accelerations	lower	higher	Not measured separately.
Masses	smaller	larger	Not measured separately.
Kinetic Energies	smaller	larger	Not measured separately.

[a] > 20° to ≤ 120° CMEs only.

combined ground-based MK3 and SMM CME observations to track 76 features in 55 CMEs above 1.15 R$_\odot$ Thirty features were consistent with a constant speed profile and 46 with a constant acceleration, the median value of which was 44 m s^{-2}. Those features associated with active regions were more likely to have constant speeds or, if accelerating, to have larger accelerations and to have higher final speeds than the features associated with prominence eruptions. On the basis of 4 flare-associated CMEs observed close to the limb in the LASCO C1 field of view (1.1-3 R$_\odot$) *Zhang et al.* [2001] found a three-phase kinematic profile. A slow rise (< 80 km s^{-1}) over tens of minutes constitutes the first phase; in the second phase a rapid acceleration of 100-500 m s^{-2} occurs in the height range ~1.4 to ~4.5 R$_\odot$ during the flare rise phase; the final phase is a propagation at a constant or declining speed. Subsequent detailed CME studies combining the LASCO C2 with other coronal observations [*Maričič et al.*, 2003; *Shanmugaraju et al.*, 2003; *Gallagher et al.*, 2003] have narrowed the strong (> 200 km s^{-1}) acceleration region of impulsive CMEs to ~1.5 to 3 R$_\odot$ However, *Zhang et al.* [2004] have shown that CMEs may not easily fall into the gradual/impulsive categories. Their three CMEs span a range from high speed with short and strong accelerations to low speed with long and weak accelerations.

A controversy about CME speeds began when *Sheeley et al.* [1999] distinguished two dynamical classes of CMEs: gradual CMEs, which are slower, accelerate in the coronagraph fields of view, and are preferentially associated with prominence eruptions; and impulsive (or fast) CMEs, which are faster, decelerate in the coronagraph fields of view, and are preferentially associated with solar flares. The two CME

dynamical classes had been suggested earlier by *Gosling et al.* [1976] and *MacQueen and Fisher* [1983]. The basic question is whether there are two physically different processes that launch CMEs or whether all CMEs belong to a dynamical continuum with a single physical initiation process. Recent observational studies of CME and coronal flare/prominence associations and timings have claimed to support the two-class view [*Moon et al.*, 2002; *Zhang et al.*, 2002; *Zhang and Golub*, 2003]. Several cases of flare-associated CMEs with large accelerations in the range 5 to 15 R$_\odot$ were explained by *Moon et al.* [2004] in terms of effects of destabilization of helmet streamers and of subsequent flare/CME events rather than as evidence against two classes of CMEs. In addition, *Low and Zhang* [2002] have proposed a model of two kinds of CMEs originating from normal and inverse magnetic geometries in prominences. They found that CMEs arising in normal polarity eruptions have more energy and higher speeds, a result confirmed by numerical modeling [*Liu et al.*, 2003].

Evidence for a single dynamical CME class was presented by *Feynman and Ruzmaikin* [2004], who discussed a CME with a high (> 1500 km s^{-1}) speed, acceleration, and an erupting prominence association, thereby combining attributes of both dynamical types. In a recent comprehensive statistical analysis of 545 flare-associated and 104 non-flare CMEs *Vršnak et al.* [2005] found considerable overlap of accelerations and speeds between the two CME groups. While flare-associated CMEs are generally faster than those without flares, there is also a correlation between CME speeds and flare X-ray peak fluxes, in which CMEs associated with the small B and C class flares are similar to the CMEs associated with filament eruptions. Thus, they argue for a CME continuum and against the two-class concept.

Lin [2004] has discussed the two classes of CMEs based on a single catastrophe model for eruptions that treats the CME, flare, and prominence as constituents of a process that depends on the magnetic field intensity and structure and plasma density. Earlier, *Chen and Krall* [2003] found that their 3-D flux rope model could explain the observed distribution of CME accelerations in terms of one mechanism with two distinct phases of acceleration. A more compelling argument is based on the observational result that the speeds of both accelerating and decelerating LASCO CMEs are distributed lognormally [*Aoki et al.*, 2003; *Yurchyshyn et al.*, 2005], implying that the speeds of both groups result from many simultaneous processes or from sequential series of processes, as discussed by e.g., *Campbell* [2003] and *Bogdan et al.* [1988]. *Yurchyshyn et al.* [2005] interpret these processes as a multiple magnetic reconnection process; hence, there is no physical distinction between accelerating and decelerating CMEs. However, it is not obvious that a scheme of two CME acceleration classes, each of which undergoes further coronal processes that modify their CME speeds, is precluded. The controversy is not yet settled, but the one-class continuum model should be regarded as preferred. CME initiation and numerical modeling are explored in detail in the reviews in this volume by *Moore and Sterling* and *Roussev*, respectively.

3.2. CME Structures

A large fraction of CMEs show a three-part structure consisting of a bright leading edge, a dark void, and a bright core [*Hundhausen*, 1999; *Gopalswamy*, 2004], as shown in the CME of Figure 5. The leading edge is compressed overlying coronal material; the void is assumed to originate in the prominence cavity and to be a magnetic flux rope; and the bright core corresponds to the erupting prominence. These features are readily distinguished in white light, but any analysis of the physical conditions within those structures, the prominence in particular, must be done with spectral observations. The SMM coronagraph included an Hα-band filter, which was used for studies of a few CMEs with large prominences. Comparisons of Hα and white light images from eight prominence/CMEs established that some CME prominence masses exceed 10^{15} gm, thus constituting a large fraction of total CME masses [*Illing and Athay*, 1986]. However, not all white-light prominences were enhanced in Hα. The evolution of the Hα brightness gradients in

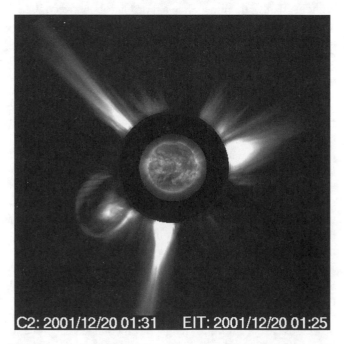

Figure 5. A LASCO CME over the southeast limb showing a three-part structure: a bright leading edge, a dark void, and a bright core. The inner image is the 195 Å solar disk image from the SOHO EIT.

prominence features gives clues to the expansion and heating rates in those features [*Athay et al.*, 1987]. Comparison of ground-based Hα prominence observations with LASCO white light observations confirms that prominences can be followed into the C2 field of view, where they form the trailing edges of the coronal cavities of the CMEs [*Plunkett et al.*, 2000], and that their speeds are always less than the speeds of the white light leading edges [*Simnett*, 2000]. However, the leading edge and prominence accelerate simultaneously [*Maricic et al.*, 2004].

Most coronagraph studies of CME structures are based on unpolarized, broad-band white light observations. However, CMEs have been observed with coronagraph polaroid [*Sheeley et al.*, 1980; *MacQueen et al.*, 1980; *Brueckner et al.*, 1995] filters. Coronagraph polaroid images can help to determine the average distances between CME masses and the plane of the sky. Because the polarization of CME Thomson-scattered light ranges from linear and tangential to the solar limb for CMEs that lie in the plane of the sky to unpolarized for CMEs lying near the Sun-Earth line, the ratios of polarized to unpolarized brightness can be used to construct topographic maps of the structures and positions of CMEs. The first detailed treatment of LASCO C2 polaroid CME observations [*Moran and Davila*, 2004] showed loop arcades and filamentary structure in two halo CMEs and one backside CME. A recent LASCO polaroid analysis [*Dere et al.*, 2005] of three CMEs in 2002 August also showed filamented loop arcade structures (Figure 6), one of which appeared to be a flux rope. Those results appear challenging for recent work to determine CME structure based on a simple cone model [*Zhao et al.*, 2002; *Xie et al.*, 2004] or CME evolution using a flux rope model [*Chané et al.*, 2005]. The review by *Nindos* in this volume explores the role of magnetic helicity in CMEs.

Cremades and Bothmer [2004] recently presented a simple scheme to relate CME structure to the heliographic position and orientation of the underlying magnetic neutral line. Working with a set of 124 LASCO CMEs showing intricate fine structure and for which relevant information about the CME source region could be determined, they found the following rule. When the neutral line is approximately parallel to the solar limb, the CME appears as a linear feature parallel to the limb and having a broad, diffuse inner core. When the neutral line is approximately perpendicular to the solar limb, the CME is observed along its symmetry axis, and the core material lies along the line of sight. Joy's law implies that the frontside neutral line lies predominately perpendicular to the east limb and parallel to the west limb, as indicated schematically in Figure 7. The neutral line and CME orientations are reversed for the solar backside, so the backside CMEs are viewed predominately orthogonally to frontside CMEs at each limb. These CME orientations are generally

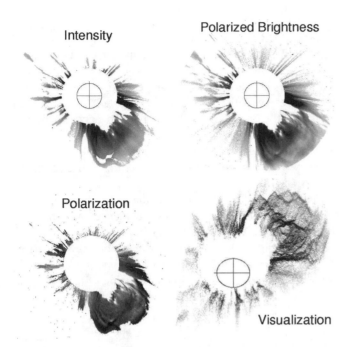

Figure 6. The intensity, polarized brightness, and polarization LASCO images of the CME of 7 August 2002 at 11:48 UT. The derived structure was rotated 90° counterclockwise in the sky plane and then tilted backwards by 30° along the new *x* axis for the visualization. The filamentary structure is common to other reconstructed CMEs. From *Dere et al.*, 2005.

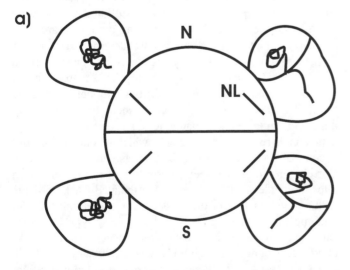

Figure 7. Schematic relating frontside neutral lines, oriented according to Joy's Law, to the envelopes of associated CMEs. CMEs on the east limb are observed along the symmetry axis; CMEs on the west limb perpendicular to the symmetry axis. The orientations reverse for backside neutral lines. From *Cremades and Bothmer* [2004].

valid only for CMEs with source regions in the active regions, below latitudes of ~50°. The CME orientations will be different for polar crown filaments [*McAllister et al.*, 2002; *Gopalswamy et al.*, 2003] or for CME source regions outside the active regions, where the neutral lines do not obey Joy's law.

The definition of a CME involves material in a magnetic field that is expelled from the corona [*Hundhausen*, 1999], so we assume that all the material observed in coronagraphs escapes the corona. However, in a few CMEs with relatively slow speeds material in bright cores collapses back to the Sun with speeds of ~50 to 200 km s^{-1} [*Wang and Sheeley*, 2002a]. These collapses have been interpreted in terms of gravitational and magnetic tension forces as well as the drag forces of the ambient solar wind. It is not clear whether these collapses are only a minor part of some CMEs or more generally important for the CME dynamics.

EUV spectral observations from the UVCS, CDS, and SUMER instruments on SOHO have defined CME densities, temperatures, ionization states, and Doppler velocities [*Raymond*, 2002]. Most CME material observed in UVCS is cool (< 10^{55} K) and concentrated into small regions [*Akmal et al.*, 2001], although this is not the case for fast CMEs associated with X-class flares [*Raymond et al.*, 2003]. In one well observed case the prominence core reached coronal temperatures at the top and was cool at the base [*Ciaravella et al.*, 2003], in agreement with earlier SMM results [*Illing and Athay*, 1986]. In addition, heating rates inferred from models using UVCS observations show that heating of the material continues out to 3.5 R$_\odot$ and is comparable to the kinetic and gravitational potential energies gained by the CMEs [*Akmal et al.*, 2001]. The Doppler information from UVCS combined with the EIT and LASCO images has shown in one case the unwinding of a helical structure [*Ciaravella et al.*, 2000]. See the review by *Ciaravella and Raymond* in this volume for further discussion of spectroscopic investigations of CMEs and coronal shocks.

4. SELECTED QUESTIONS ABOUT CMES

4.1. Narrow CMEs

The possibility that narrow (5°-40°) CMEs are physically distinct from the general population of all CMEs was addressed by *Kahler et al.* [1989], who found that 22% of all impulsive > M1 X-ray flares were associated with Solwind CMEs. A common view [*Svestka*, 1986] was that all flares either are confined and not associated with CMEs or are CME-associated eruptive events. Since confined flares were presumed to be impulsive, the CME associations of some impulsive flares violated the basic confined-eruptive flare paradigm. A correlation between associated CME width and X-ray flare duration implied [*Kahler et al.*, 1989] a single continuous class of CMEs rather than two CME classes, although the enhanced energetics and lack of postflare loop associations for impulsive flares with CMEs suggested a different kind of CME for the impulsive events.

The question of whether narrow CMEs are a physically distinct class of CMEs arose again when *Kahler et al.* [2001] reported an association of a narrow LASCO CME with an impulsive flare and SEP event on 1 May 2000 and found narrow CMEs associated with other impulsive SEP events. They drew on recent modeling work of *Shimojo and Shibata* [2000] to suggest that especially energetic coronal reconnection events can both accelerate particles and propel mass outward along open field lines as CMEs. The open magnetic-field topology of those CMEs would differ fundamentally from that of the larger, closed-field CMEs [*Reames*, 2002], and no magnetic flux would be expelled by the open-field CMEs. Figure 8 shows a cartoon of magnetic reconnection leading to the narrow CME along open field lines.

Gilbert et al. [2001] compared properties of 15 narrow (< 15°) CMEs with those of more typical CMEs to understand how the narrow CMEs were generated. The facts that their narrow CMEs were well associated with active prominences but not with surges led them to conclude that narrow CMEs originate in closed-field regions as do the larger CMEs. Models of properties of 5 of those narrow CMEs based on UVCS spectral observations were consistent with both reconnection jets on open fields and with closed-field CMEs [*Dobrzycka et al.*, 2003]. The speeds, widths, and flare and filament associations of the narrow CMEs of the *Gilbert et al.* [2001] study are very similar to those of the white-light

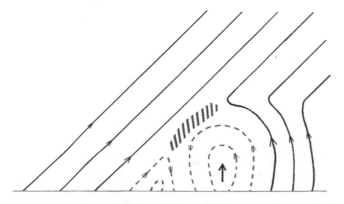

Figure 8. Schematic of magnetic reconnection between a closed-field region (dashed lines) and an overlying open-field region (solid lines). Reconnection occurs in the shaded region; mass on previously closed fields is ejected outward along open field lines. Adapted from *Reames* [2002] and *Shimojo and Shibata* [2000].

coronal jets studied by *Wang and Sheeley* [2002b] during a comparable 1999 time period. The jets were associated with bipolar magnetic regions located near CH boundaries, similar to the good source associations of narrow CMEs with sharp bends in magnetic polarity-reversal lines [*Gilbert et al.*, 2001]. However, the jets were interpreted by *Wang and Sheeley* [2002b] in terms of reconnection between the CH open fields and bipolar magnetic-region closed fields with subsequent ejection of material along open field lines, contrary to the conclusion of *Gilbert et al.* [2001] regarding narrow CMEs.

Yashiro et al. [2003] surveyed the properties of narrow (< 20°) LASCO CMEs and found evidence for a bimodal distribution of CME widths during 1998-2000 as well as differences in speed distributions between normal and narrow CMEs. *Yashiro et al.* [2003] and *Wang and Sheeley* [2002b] argue that the jets are a separate population of narrow open-field CMEs; they deserve further study.

4.2. Reconnection in CMEs

The early realization that CMEs were continually injecting new magnetic flux into the interplanetary medium, although the magnetic flux there varies by less than a factor of 2 over the solar cycle [*Gosling*, 1975; *Wang and Sheeley*, 2002c], implied that magnetic reconnection must accompany or follow CMEs to detach the CME field lines from the Sun. Yohkoh Soft X-ray Telescope (SXT) observations have provided substantial X-ray evidence of post-CME reconnection, such as cusp-shaped loops [*Shibata*, 1999] and supra-arcade downflows [*McKenzie*, 2000] over long-duration flares. An extensive survey of post-eruptive arcades in SOHO 195 Å images has shown that every arcade is associated with a LASCO CME [*Tripathi et al.*, 2004]. These observations have been interpreted in terms of a basic model (named the CSHKP model to reflect its provenance) of reconnecting magnetic fields behind a magnetic flux rope and over a magnetic arcade (e.g., *Lin et al.*, 2004), which results in a disconnection of CME fields from the Sun, as shown in Figure 9. Correlations found between inferred magnetic reconnection rates in arcades and the speeds of associated CMEs provide further confirmation of the model [*Jing et al.*, 2005]. Radio imaging of the moving and quasi-stationary type IV bursts can provide upper limits to the current sheet length by bracketing the reconnection region [*Pick et al.*, 2005].

An observational challenge is to detect coronal white-light signatures of reconnection in the wakes of CMEs. *Webb and Cliver* [1995] looked in pre-LASCO coronagraphs for *Y*-shaped or concave-outward CME structures in which the vertical line of the *Y* is the reconnecting current sheet that

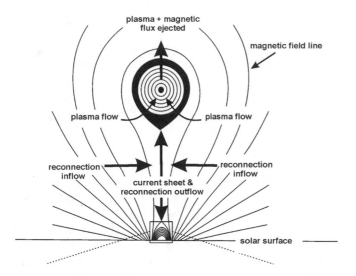

Figure 9. Basic CSHKP model of CMEs looking parallel to the magnetic neutral line. Solid lines are magnetic field lines. An inflow behind the ejected mass and magnetic flux produces the current sheet and post-eruptive arcade at bottom center, which forms a Y-shaped structure. Details of the flare and arcade are omitted here. Adapted from *Lin et al.* [2004].

appears as a bright ray (Figure 9). They concluded that > 10% of CMEs showed such structures. With the advent of LASCO, cases of Y-shaped CMEs were reported by *Simnett et al.* [1997], and *St. Cyr et al.* [2000] found such features in one third to one half of all LASCO CMEs. *Webb et al.* [2003] studied 26 SMM CMEs followed by narrow (~2.5°) rays extending to beyond 5 $R_\odot$ and concluded that they were consistent with two CME models predicting an extended long-lived current sheet. Bright narrow features with enhanced temperatures (3-6 × 10^6 K), densities (~5 × 10^7 cm^{-3} at 1.5 $R_\odot$) and abundances of elements with low first ionization potentials (FIPs) were observed with the UVCS following slow (~180 km s^{-1}; *Ciaravella et al.* [2002]) and very fast (≥ 1800 km s^{-1}; *Ko et al.* [2003]; *Lin et al.* [2005]) CMEs. Those features were also interpreted in terms of reconnecting current sheets.

Several considerations complicate the above picture of field-line disconnection in a large-scale current sheet. In the simple 2-dimensional view disconnection leads to interplanetary magnetic fields completely unattached to the corona, for which solar wind signatures are rarely, if ever, seen. *Crooker et al.* [2002] have proposed that an interchange reconnection between open coronal fields and closed CME fields, shown in Figure 10, will prevent the addition of new magnetic flux to the interplanetary medium from CMEs.

The product will be small closed coronal loops and large open CME fields. Note that disconnection in this 3-dimensional version may still act to form the CME flux rope itself and underlying arcade [*Dere et al.*, 1999], but eventually the flux rope itself must reconnect to prevent an interplanetary magnetic-flux buildup. One question is whether we have any observational evidence of this interchange reconnection, in coronagraph, EUV, soft X-ray, or radio images. We should expect to find interchange reconnection signatures at the peripheries, rather than the centers, of the CME source regions where reconnection with ambient open coronal magnetic fields can occur. Supposing that such reconnection does occur, we are left with two other problems. The first is that the interchange reconnection must occur at only one leg of the CME to produce a resulting flux tube or rope open at only one end [*Crooker et al.*, 2002], which seems very unlikely. The second is that we have seen interplanetary

a. Disconnection

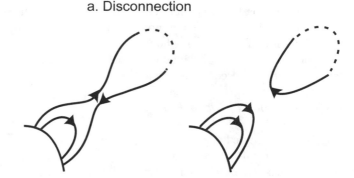

b. Interchange reconnection

Figure 10. Schematic of the interchange reconnection proposed for the primary method of disconnecting CMEs from the Sun. This reconnection would occur near the periphery of CMEs, away from the central post-eruptive arcade shown in Figure 9. From *Crooker et al.* [2002].

CMEs in which the interior parts of the flux rope are magnetically open and the outer parts are closed, as interpreted from heat-flux electron measurements [*Larson et al.*, 1997]. Those flux-rope open fields would appear to result from a very unlikely reconnection with ambient fields of the same polarity as the flux rope itself if we accept the common interpretation of flux rope geometry [*Burlaga*, 1995]. It remains to find the solar signatures of magnetic reconnection that convert the closed CME fields to open fields and then to understand how that reconnection leads to the complex combinations of open and closed fields observed in interplanetary CMEs.

Coronal reconnection may cause the coronal inflows observed in the LASCO C2 coronagraph [*Sheeley et al.*, 2001; *Sheeley and Wang*, 2001, 2002]. The faint blobs may be described as falling curtains, sinking columns, or inflowing/outflowing pairs [*Simnett*, 2004] with speeds generally < 100 km s^{-1}. The blobs are confined to heights below 5.5 R$_{\odot}$ and track sector boundaries where oppositely directed fields can reconnect [*Sheeley and Wang*, 2002]. Perhaps surprising is that inflows are not well associated with CMEs [*Sheeley et al.*, 2001], even though the observed inflow rates can reach 30 per day. Are these inflows showing us the characteristic temporal and spatial scales of coronal magnetic reconnection? Why are they not more commonly observed in the aftermath of CMEs? If they represent reconnections between open fields, then why do we not see the interplanetary signatures of disconnected fields?

4.3. Coronal Holes (CHs) and CMEs

CHs are long-lived open-field regions clearly discerned in soft X-ray, He I 10830 $\mathring{A}$, and some EUV images. A long-standing question is whether the open fields of CHs may somehow initiate or enhance CMEs from surrounding regions of closed fields. A survey of interplanetary disturbances detected by radio scintillations by *Hewish and Bravo* [1986] showed that all corotating and transient interplanetary shocks originated in CHs. In particular, their transient events, the erupting streams manifested as CMEs, originated in mid-latitude CHs. However, *Harrison* [1990] surveyed 95 CMEs observed on SMM and concluded that CMEs were associated with active regions and not with CHs. Based on another survey showing that CMEs producing interplanetary shocks were associated with CHs, *Bravo* [1995] offered the following scenario. Newly emerging magnetic flux reconnects with opposite polarity fields of an adjacent coronal helmet streamer, producing a CME from the overlying streamer. The area of a CH adjacent to the active region expands, causing the magnetic expansion factor of the CH to decrease with a transient increase in the solar wind flow speed. Other studies by *Bravo and Rivera* [1994] and by *Gonzalez et al.* [1996] showed that the solar sources of the most intense interplanetary disturbances

and geomagnetic storms were active regions near low-latitude CHs. *Gonzalez et al.* [1996] called these sources of active regions (with flares and/or filament eruptions) occurring close to the streamer belt and to growing low-latitude CHs CHARCS (for CH-Active Region-Current Sheet). *Bravo et al.* [1998] found a solar-cycle correlation between active regions near CHs and intense geomagnetic storms, further strengthening the connection of shock-producing CMEs and CHARCS.

A more direct connection between CMEs and CHs was found in Yohkoh SXT images by *Bhatnagar* [1996], who studied 15 large post-eruptive X-ray loop arcades outside active regions. These events, called X-ray blowouts, were not associated with observed chromospheric activity, but all were located at or near the boundaries of CHs. *Bhatnagar* [1996] proposed for the blowouts an interaction between the open CH fields and opposite polarity fields of adjacent closed field regions. *Webb et al.* [1978] had earlier found a significant, but not strong, tendency for Skylab X-ray arcades outside active regions to occur over neutral lines forming the borders of CHs. *Lewis and Simnett* [2000] performed a statistical study of CME source locations near solar minimum (1996-97) and found the centroid of CME sources to be located about 45° west of an active region complex, in the vicinity of a polar CH extension to low latitudes. A more extreme example is that of an erupting filament and CME on 28 December 1999 observed with Yohkoh/SXT, EIT, and LASCO by *Chertok et al.* [2002], who interpreted the source to lie inside a large transequatorial CH.

These observations suggest that low-latitude CHs may be important for at least some CMEs. The adjacent open fields of the CHs may interact with closed fields of the CMEs either by magnetic reconnection or by deflecting the courses of the CMEs away from the CHs. A significant equatorward deflection of CME trajectories, attributed to polar coronal holes is observed around solar minimum [*Cremades and Bothmer*, 2004]. However, we do not yet have a good observational understanding of CME source regions and CHs. Among the questions to answer are the following. How often are CME source regions adjacent to CHs? If so, what are the CH characteristics in terms of their sizes and growth rates? Are more energetic CMEs more likely to lie adjacent to CHs? What is the effect of CHs on CMEs - reconnection with closed field regions or deflections or modifications of CME trajectories?

5. SUMMARY

We now have white-light CME observations over several solar cycles from coronagraphs of increasing capabilities in terms of dynamic range, cadence, and field of view. The > 10,000 CMEs thus far observed have been statistically analyzed for their average and median properties and the variations of the properties over the solar cycle. The coronagraph observations have been supplemented with ground-based coronameter and Hα observations, as well as EUV, X-ray, and radio observations that have allowed us to study details of the structures, early dynamics and source regions of CMEs. However, with this abundance of white light observations there is still a real dearth of spectral line observations needed to assess temperature, velocity, and density distributions that can tell us about the spatial and temporal dynamics of CMEs.

As is the case with solar flares, our increasing wealth of observations has been accompanied by a slow progress in understanding the fundamental questions posed by CMEs. We still have difficulty determining the exact source regions of CMEs even when the sources are near central meridian and well observed in EUV and X-rays. The CME magnetic field geometries and topologies are unclear, and the way in which the expelled magnetic fields of CMEs reconnect to convert themselves to open magnetic fields has yet to be defined.

Acknowledgments.   I thank the organizers for my travel support and for arranging a superb Chapman Conference.

REFERENCES

Akmal, A., *et al.*, SOHO observations of a coronal mass ejection, *Astrophys. J.*, 553, 922, 2001.

Aoki, S.I., S. Yashiro, and K. Shibata, The log-normal distributions of coronal mass ejection-related solar flares and the flare/CME model of gamma-ray bursts, *Proc. 28th Int. Cosmic Ray Conf.*, 5, 2729, 2003.

Athay, R.G., B.C. Low, and B. Rompolt, Characteristics of the expansion associated with eruptive prominences, *Sol. Phys.*, 110, 359, 1987.

Bhatnagar, A., Solar mass ejections and coronal holes, *Astrophys. Space Sci.*, 243, 105, 1996.

Bogdan, T., P.A. Gilman, I. Lerche, and R. Howard, Distribution of sunspot umbral areas: 1917-1982, *Astrophys. J.*, 327, 451, 1988.

Bravo, S., A solar scenario for the associated occurrence of flares, eruptive prominences, coronal mass ejections, coronal holes, and interplanetary shocks, *Sol. Phys.*, 161, 57, 1995.

Bravo, S., and A.L. Rivera, The solar causes of major geomagnetic storms, *Ann. Geophys.*, 12, 113, 1994.

Bravo, S., J.A.L. Cruz-Abayro, and D. Rojas, The spatial relationship between active regions and coronal holes and the occurrence of intense geomagnetic storms throughout the solar activity cycle, *Ann. Geophys.*, 16, 49, 1998.

Brueckner, G.E., *et al.*, The Large Angle Spectroscopic Coronagraph (LASCO), *Sol. Phys.*, 162, 357, 1995.

Burkepile, J.T., A.J. Hundhausen, A.L. Stanger, O.C. St. Cyr, and J.A. Seiden, Role of projection effects on solar coronal mass ejection properties: 1. A study of CMEs associated with limb activity, *J. Geophys. Res.*, 109, A03103, doi:10.1029/2003JA010149, 2004.

Burlaga, L.F., *Interplanetary Dynamics,* p.89, Oxford University Press, New York, 1995.

Campbell, W.H., *Introduction to Geomagnetic Fields,* p.293, Cambridge U. Press, Cambridge, 2003.

Chané, E., C. Jacobs, B. Van der Holst, S. Poedts, and D. Kimpe, On the effect of the initial magnetic polarity and of the background wind on the evolution of CME shocks, *Astron. Astrophys.*, 432, 331, 2005.

Chen, J., and J. Krall, Acceleration of coronal mass ejections, *J. Geophys. Res.*, 108(11), 1410, doi:10.1029/2003JA009849, 2003.

Chertok, I.M., E.I. Mogilevsky, V.N. Obridko, N.S. Shilova, and H.S. Hudson, Solar disappearing filament inside a coronal hole, *Astrophys. J.*, 567, 1225, 2002.

Ciaravella, A., *et al.*, Solar and Heliospheric Observatory observations of a helical coronal mass ejection, *Astrophys. J.*, 529, 575, 2000.

Ciaravella, A., J.C. Raymond, J. Li, P. Reiser, L.D. Gardner, Y.-K. Ko, and S. Fineschi, Elemental abundances and post-coronal mass ejection current sheet in a very hot active region, *Astrophys. J.*, 575, 1116, 2002.

Ciaravella, A., J.C. Raymond, A. van Ballegooijen, L. Strachan, A. Vourlidas, J. Li, J. Chen, and A. Panasyuk, Physical parameters of the 2000 February 11 coronal mass ejection: ultraviolet spectra versus white-light images, *Astrophys. J.*, 597, 1118, 2003.

Cremades, H., and V. Bothmer, On the three-dimensional configuration of coronal mass ejections, *Astron. Astrophys.*, 422, 307, 2004.

Crooker, N.U., J.T. Gosling, and S.W. Kahler, Reducing heliospheric magnetic flux from coronal mass ejections without disconnection, *J. Geophys. Res.*, 107(A2), doi:10.1029/2001JA000236, 2002.

Dere, K.P., G.E. Brueckner, R.A. Howard, D.J. Michels, and J.P. Delaboudiniere, LASCO and EIT observations of helical structure in coronal mass ejections, *Astrophys. J.*, 516, 465, 1999.

Dere, K.P., D. Wang, and R. Howard, Three-dimensional structure of coronal mass ejections from LASCO polarization measurements, *Astrophys. J.*, 620, L119, 2005.

Dobrzycka, D., J.C. Raymond, D.A. Biesecker, J. Li, and A. Ciaravella, Ultraviolet spectroscopy of narrow coronal mass ejections, *Astrophys. J.*, 588, 586, 2003.

Feynman, J., and A. Ruzmaikin, A high-speed erupting-prominence CME: a bridge between types, *Sol. Phys.*, 219, 301, 2004.

Gallagher, P.T., G.R. Lawrence, and B.R. Dennis, Rapid acceleration of a coronal mass ejection in the low corona and implications for propagation, *Astrophys. J.*, 588, L53, 2003.

Gilbert, H.R., E.C. Serex, T.E. Holzer, R.M. MacQueen, and P.S. McIntosh, Narrow coronal mass ejections, *Astrophys. J.*, 550, 1093, 2001.

Gonzalez, W.D., B.T. Tsurutani, P.S. McIntosh, and A.L. Clua de Gonzalez, Coronal hole-active region-current sheet (CHARCS) association with intense interplanetary and geomagnetic activity, *Geophys. Res. Lett.*, 23, 2577, 1996.

Gopalswamy, N., A global picture of CMEs in the inner heliosphere, in *The Sun and the Heliosphere as an Integrated system,* edited by G. Poletto and S. Suess, ASSL Kluwer, Dordrecht, 2004.

Gopalswamy, N., A. Lara, S. Yashiro, S. Nunes, and R.A. Howard, Coronal mass ejection activity during solar cycle 23, in *Proc. Solar variability as an input to the Earth's environment. International Solar Cycle Studies (ISCS) Symposium,* edited by A. Wilson, p.403, ESA SP-535, 2003.

Gosling, J.T., Large-scale inhomogeneities in the solar wind of solar origin, *Rev. Geophys.*, 13, 1053, 1975.

Gosling, J.T., E. Hildner, R.M. MacQueen, R.H. Munro, A.I. Poland, and C.L. Ross, The speeds of coronal mass ejection events, *Sol. Phys.*, 48, 389, 1976.

Harrison, R.A., The source regions of solar coronal mass ejections, *Sol. Phys.*, 126, 185, 1990.

Hewish, A., and S. Bravo, The sources of large-scale heliospheric disturbances, *Sol. Phys.*, 106, 185, 1986.

Hundhausen, A., Coronal mass ejections, in *The Many Faces of the Sun: a summary of the results from NASA's Solar Maximum Mission,* edited by K.T. Strong, et al., p.143, Springer, New York, 1999.

Illing, R.M.E., and R.G. Athay, Physical conditions in eruptive prominences at several solar radii, *Sol. Phys.*, 105, 173, 1986.

Jackson, B.V., and D.F. Webb, The masses of CMEs measured in the inner heliosphere, in *Proc. Third SOHO Wkshp,* p. 233, ESA SP-373, 1994.

Jing, J., J. Qiu, J. Lin, M. Qu, Y. Xu, and H. Wang, Magnetic reconnection rate and flux-rope acceleration of two-ribbon flares, *Astrophys. J.*, 620, 1085, 2005.

Kahler, S.W., N.R. Sheeley, Jr., and M. Liggett, Coronal mass ejections and associated X-ray flare durations, *Astrophys. J.*, 344, 1026, 1989.

Kahler, S.W., D.V. Reames, and N.R. Sheeley, Jr., Coronal mass ejections associated with impulsive solar energetic particle events, *Astrophys. J.*, 562, 558, 2001.

Kahler, S.W., and D.V. Reames, Solar energetic particle production by coronal mass ejection-driven shocks in solar fast-wind regions, *Astrophys. J.*, 584, 1063, 2003.

Ko, Y.-K., J.C. Raymond, J. Lin, G. Lawrence, J. Li, and A. Fludra, Dynamical and physical properties of a post-coronal mass ejection current sheet, *Astrophys. J.*, 594, 1068, 2003.

Larson, D.E., et al., Tracing the topology of the October 18-20, 1995, magnetic cloud with ~0.1-10^2 keV electrons, *Geophys. Res. Lett.*, 24, 1911, 1997.

Lewis, D.J., and G.M. Simnett, The occurrence of coronal mass ejection at solar minimum and their association with surface activity, *Sol. Phys.*, 191, 185, 2000.

Lin, J., CME-flare association deduced from catastrophic model of CMEs, *Sol. Phys.*, 219, 169, 2004.

Lin, J., J.C. Raymond, and A.A. van Ballegooijen, The role of magnetic reconnection in the observable features of solar eruptions, *Astrophys. J.*, 602, 422, 2004.

Lin, J., Y.-K. Ko, L. Sui, J.C. Raymond, G.A. Stenborg, Y. Jiang, S. Zhao, and S. Mancuso, Direct observations of the magnetic reconnection site of an eruption on 2003 November 18, *Astrophys. J.*, it 622, 1251, 2005.

Liu, W., X.P. Zhao, S.T. Wu, and P. Scherrer, Effects of magnetic topology on CME kinematic properties, in *Proc. Solar variability as an input to the Earth's environment. International Solar Cycle Studies (ISCS) Symposium,* edited by A. Wilson, p.459, ESA SP-535, 2003.

Low, B.C., and M. Zhang, The hydromagnetic origin of the two dynamical types of solar coronal mass ejections, *Astrophys. J.*, 564, L53, 2002.

MacQueen, R.M., A. Csoeke-Poeckh, E. Hildner, L. House, R. Reynolds, A. Stanger, H. Tepoel, and W. Wagner, The high altitude observatory coronagraph/polarimeter on the Solar Maximum Mission, *Sol. Phys.*, 65, 91, 1980.

MacQueen, R.M., and R.R. Fisher, The kinematics of solar inner coronal transients, *Sol. Phys.*, 89, 89, 1983.

Maricic, D., B. Vršnak, A.L. Stanger, D. Rosa, and D. Hrzina, Initiation and development of two coronal mass ejections, in *Proc. Solar variability as an input to the Earth's environment. International Solar Cycle Studies (ISCS) Symposium*, edited by A. Wilson, p.441, ESA SP-535, 2003.

Maričić, D., B. Vršnak, A.L. Stanger, and A. Veronig, Coronal mass ejection of 15 May 2001: I. evolution of morphological features of the eruption, *Sol. Phys.*, 225, 337, 2004.

McAllister, A.H., D.H. McKay, and S.F. Martin, The skew of high-latitude X-ray arcades in the declining phase of cycle 22, *Sol. Phys.*, 211, 155, 2002.

McKenzie, D.E., Supra-arcade downflows in long-duration solar flare events, *Sol. Phys.*, 195, 381, 2000.

Michalek, G., and J. Mazur, Properties of coronal mass ejections, in *Proc. 10th Eur. Solar Phys. Mtg, Prague, 1*, edited by A. Wilson, p.181, ESA SP-506, 2002.

Michalek, G., N. Gopalswamy, and S. Yashiro, A new method for estimating widths, velocities, and source location of halo coronal mass ejections, *Astrophys. J.*, 584, 472, 2003.

Moon, Y.-J., G.S. Choe, H. Wang, Y.D. Park, N. Gopalswamy, G. Yang, and S. Yashiro, A statistical study of two classes of coronal mass ejections, *Astrophys. J.*, 581, 694, 2002.

Moon, Y.-J., K.S. Cho, Z. Smith, C.D. Fry, M. Dryer, and Y.D. Park, Flare-associated coronal mass ejections with large accelerations, *Astrophys. J.*, 615, 1011, 2004.

Moran, T.G., and J.M. Davila, Three-dimensional polarimetric imaging of coronal mass ejections, *Sci.*, 305, 66, 2004.

Pick, M., P. Démoulin, S. Krucker, O. Malandraki, and D. Maia, Radio and X-ray signatures of magnetic reconnection behind an ejected flux rope, *Astrophys. J.*, 625, 1019, 2005.

Plunkett, S.P., *et al.*, Simultaneous SOHO and ground-based observations of a large eruptive prominence and coronal mass ejection, *Sol. Phys.*, 194, 371, 2000.

Raymond, J.C., Spectroscopic diagnostics of CME material, in *Proceedings of the SOHO 11 Symposium on From Solar Min to Max: Half a Solar Cycle with SOHO*, edited by A. Wilson, p.421, ESA SP-508, 2002.

Raymond, J.C., A. Ciaravella, D. Dobrzycka, L. Strachan, Y.-K. Ko, M. Uzzo, and N.-E. Raouafi, Far-ultraviolet spectra of fast coronal mass ejections associated with X-class flares, *Astrophys. J.*, 597, 1106, 2003.

Reames, D.V., Magnetic topology of impulsive and gradual solar energetic particle events, *Astrophys. J.*, 571, L63, 2002.

Robbrecht, E., and D. Berghmans, Automated recognition of coronal mass ejections (CMEs) in near-real-time data, *Astron. Astrophys.*, 425, 1097, 2004.

Shanumgaraju, A., Y.-J. Moon, M. Dryer, and S. Umapathy, On the kinematic evolution of flare-associated CMEs, *Sol. Phys.*, 215, 185, 2003.

Sheeley, N.R., Jr., D.J. Michels, R.A. Howard, and M.J. Koomen, Initial observations with the Solwind coronagraph, *Astrophys. J.*, 237, L99, 1980.

Sheeley, N.R., Jr., J.H. Walters, Y.-M. Wang, and R.A. Howard, Continuous tracking of coronal outflows: Two kinds of coronal mass ejections, *J. Geophys. Res.*, 104, 24739, 1999.

Sheeley, N.R., Jr., T.N. Knudson, and Y.-M. Wang, Coronal inflows and the Sun's nonaxisymmetric open flux, *Astrophys. J.*, 546, L131, 2001.

Sheeley, N.R., Jr., and Y.-M. Wang, Coronal inflows and sector magnetism, *Astrophys. J.*, 562, L107, 2001.

Sheeley, N.R., Jr., and Y.-M. Wang, Characteristics of coronal inflows, *Astrophys. J.*, 579, 874, 2002.

Shibata, K., Evidence of magnetic reconnection in solar flares and a unified model of flares, *Astrophys. Space Sci.*, 264, 129, 1999.

Shimojo, M., and K. Shibata, Physical parameters of solar X-ray jets, *Astrophys. J.*, 542, 1100, 2000.

Simnett, G.M., The relationship between prominence eruptions and coronal mass ejections, *J. Atmos. Terr. Phys.*, 62, 1479, 2000.

Simnett, G.M., Evidence for magnetic reconnection in the high corona, *Astron. Astrophys.*, 416, 759, 2004.

Simnett, G.M., *et al.*, LASCO observations of disconnected magnetic structures out to beyond 28 solar radii during coronal mass ejections, *Sol. Phys.*, 175, 685, 1997.

St. Cyr, O.C., J.T. Burkepile, A.J. Hundhausen, and A.R. Lecinski, A comparison of ground-based and spacecraft observations of coronal mass ejections from 1980-1989, *J. Geophys. Res.*, 104, 12493, 1999.

St. Cyr, O.C., Properties of coronal mass ejections: SOHO LASCO observations from January 1996 to June 1998, *J. Geophys. Res.*, 105(A8), 18169, 2000.

St. Cyr, O.C., *et al.*, The last word: the definition of halo coronal mass ejections, *EOS*, 86(30), 281, 2005.

Stockton-Chalk, A., The limit of non-radial expansion of coronal mass ejections, in *Proc. SOLSPA: The second solar cycle and space weather euroconference, ESA SP-477*, 277, 2002.

Svestka, Z., On the varieties of solar flares, in *The lower atmosphere of solar flares; Proc. Solar Maximum Mission Symp.*, edited by D.F. Neidig, p.332, National Solar Observatory, 1986.

Tripathi, D., V. Bothmer, and H. Cremades, The basic characteristics of EUV post-eruptive arcades and their role as tracers of coronal mass ejection source regions, *Astron. Astrophys.*, 422, 337, 2004.

Vourlidas, A., D. Buzasi, R.A. Howard, and E. Esfandiari, Mass and energy properties of LASCO CMEs, in *Proc. 10th Eur. Solar Phys. Mtg, Prague, 1*, edited by A. Wilson, p.91, ESA SP-506, 2002.

Vourlidas, A., private comm., 2004.

Vršnak, B., D. Ruždjak, D. Sudar, and N. Gopalswamy, Kinematics of coronal mass ejections between 2 and 30 solar radii: What can be learned about forces governing the eruption?, *Astron. Astrophys.*, 423, 717, 2004.

Vršnak, B., D. Sudar, and D. Ruždjak, The CME-flare relationship: Are there really two types of CMEs?, *Astron. Astrophys.*, 435, 1149, 2005.

Wang, Y.-M., and N.R. Sheeley, Jr., Observations of core fallback during coronal mass ejections, *Astrophys. J.*, 567, 1211, 2002a.

Wang, Y.-M., and N.R. Sheeley, Jr., Coronal white-light jets near sunspot maximum, *Astrophys. J.*, 575, 542, 2002b.

Wang, Y.-M., and N.R. Sheeley, Jr., Sunspot activity and the long-term variation of the Sun's open magnetic flux, *J. Geophys. Res.*, 107, 1302, doi:10.1029/2001JA000500, 2002c.

Webb, D.F., P.S. McIntosh, J.T. Nolte, and C.V. Solodyna, Evidence linking coronal transients to the evolution of coronal holes, *Sol. Phys.*, 58, 389, 1978.

Webb, D.F., and E.W. Cliver, Evidence for magnetic disconnection of mass ejections in the corona, *J. Geophys. Res.*, (A4), 5853, 1995.

Webb, D.F., J. Burkepile, T.G. Forbes, and P. Riley, Observational evidence of new current sheets trailing coronal mass ejections, *J. Geophys. Res.*, 108, 1440, doi:10.1029/2003JA009923, 2003.

Xie, H., L. Ofman, and G. Lawrence, Cone model for halo CMEs: application to space weather forecasting, *J. Geophys. Res.*, 109, A03109, doi:10.1029/2003JA010226, 2004.

Yashiro, S., N. Gopalswamy, G. Michalek, and R.A. Howard, Properties of narrow coronal mass ejections observed with LASCO, *Adv. Space Res.*, 32, 2631, 2003.

Yashiro, S., N. Gopalswamy, G. Michalek, O.C. St.Cyr, S.P. Plunkett, N.B. Rich, and R.A. Howard, A catalog of white light coronal mass ejections observed by the SOHO spacecraft, *J. Geophys. Res.*, 109, A07105, doi:10.1029/2003JA010282, 2004.

Yurchyshyn, V., S. Yashiro, V. Abramenko, H. Wang, and N. Gopalswamy, Statistical distributions of speeds of coronal mass ejections, *Astrophys. J.*, 619, 599, 2005.

Zhang, J., K.P. Dere, R.A. Howard, M.R. Kundu, and S.M. White, On the temporal relationship between coronal mass ejections and flares, *Astrophys. J.*, 559, 452, 2001.

Zhang, J., K.P. Dere, R.A. Howard, and A. Vourlidas, A study of the kinematic evolution of coronal mass ejections, *Astrophys. J.*, 604, 420, 2004.

Zhang, M., and L. Golub, The dynamical morphologies of flares associated with the two types of solar coronal mass ejections, *Astrophys. J.*, 595, 1251, 2003.

Zhang, M., L. Golub, E. DeLuca, and J. Burkepile, The timing of flares associated with the two dynamical types of solar coronal mass ejections, *Astrophys. J.*, 574, L97, 2002.

Zhao, X.P., S.P. Plunkett, and W. Liu, Determination of geometrical and kinematical properties of halo coronal mass ejections using the cone model, *J. Geophys. Res.*, 107(A8), 1223, doi:10.1029/2001JA009143, 2002.

S. Kahler, AFRL/VSBXS, 29 Randolph Rd., Hanscom AFB, MA 01731. (Stephen.kahler@hanscom.af.mil)

A Broad Perspective on Automated CME Tracking: Towards Higher Level Space Weather Forecasting

Eva Robbrecht and David Berghmans

We discuss our current capabilities to deliver the CME parameters required for the space weather forecasting process. The ever growing importance of space weather has lead to new requirements on the timeliness and objectiveness of CME detection. It has become indispensable to report the occurrence of Earth-directed CMEs and to predict their possible impact on the geospace environment. Early 2005, we are on the eve of a new era in space weather forecasting. We point out the restricted accuracy on the current forecasts and discuss a chance for amelioration. This invokes data-driven models (empirical and numerical), triggered by a real-time CME disturbance, simulating the propagation and interaction of the ejection with the ambient solar wind. We discuss the link between the direct observable parameters (like the CME *projected* speed and angle around the occulter) and the required input parameters (like radial speed, direction, …). The only way to guarantee the real-time value of the simulations is by employing software which autonomously detect CME parameters in a variety of data. This paper focusses on the automated CME detection algorithms that are currently available. Automated CME tracking is yet in its infancy, therefore this 'review' will be an outlook on the potential of this field rather than looking back on already achieved milestones.

1. INTRODUCTION

Since their discovery in the seventies in coronagraphic observations on the OSO-7 sattelite [*Tousey et al.*, 1974] and the Skylab mission [*Gosling et al.*, 1974] coronal mass ejections have been subject to numerous studies. Recent reviews can be found in *Forbes* [2000]; *Low* [2001]; *Low and Zhang* [2002]. With the increase in knowledge on the physics of CMEs, there has also been an increase in the number of ways a CME can be observed. CMEs are now known to be complex events linked to other phenomena such as flares and filament/prominence eruptions; however also more subtle events like dimmings, EIT waves and radio type II bursts are valuable CME indicators. For reviews on the different CME manifestations and related events we refer to *Hudson*

and Cliver [2001]; *Cliver and Hudson* [2002]; *Munro et al.* [1979]. *In situ*, the passage of an interplanetary CME (ICME) appears as a shock in the solar wind parameters. They are generated by compressive interaction regions formed as the CME overtakes slower plasma [*Gosling et al.*, 1994]. CMEs are the most energetic eruptions on the Sun, are the primary cause of major geomagnetic storms and are believed to be responsible for the largest solar energetic particle events [*Gosling et al.*, 1990; *Tsurutani et al.*, 1997]. These events can highly affect radio HF communication throughout the polar region (no other communication is possible above 82 degrees latitude, e.g. for polar flights), the reliability of power systems (voltage control problems, blackout or collapse of grid, damage to transformers), induce currents on pipelines and cause spacecraft surface charging. Hence, to guarantee the reliability of many technologically dependant systems and activities, timely and accurate forecasting of the most energetic space weather events - CMEs - have become indispensable. A good forecast includes information on the onset time, the duration

Solar Eruptions and Energetic Particles
Geophysical Monograph Series 165
10.1029/165GM06

and the strength of the storm. In order to allow companies to take protective action in preparation of the storm, this forecast should be done as timely as possible. Unlike terrestrial weather conditions, which are monitored routinely at thousands of locations around the world, the conditions in space are monitored by only a handful of space-based and ground-based facilities. Space weather forecasters are required to predict conditions in space and on Earth using a very limited guidance from actual measurements.

From the moment an indication for CME occurrence is observed, space weather forecasters try to gather as much details as possible on the nature, origin and evolution of the CME. This involves usually combined observations in the EUV and X-rays as well as coronagraphic movies, magnetograms and irradiance data. Given the occurrence of a CME, the main task of a space weather forecaster is to predict if, when, for how long and how strong it will impact the earth. Translated in terms of *in situ* data, this means we need to estimate the solar wind speed and density and the north-south orientation of the IMF (B_z), since these are the most important parameters deciding the geo-efficiency [recent studies are *Gopalswamy*, 2003; *Yurchyshyn et al.*, 2003, 2004; *Zhang et al.*, 2004]. Two stages can be discerned in the forecasting process. Using remote sensing data one can obtain a 2-4 day warning time (depending on the speed of the CME) with rough estimations on the CME time of arrival and the expected impact. This is done by assigning probabilities on the future geomagnetic activity levels. When the perturbation induced by the CME actually passes by at L1 (215 $R_\odot$) new *in situ* data becomes available (ACE, CELIAS onboard SOHO, Wind when available) allowing amelioration of the forecast ~1 hour ahead of the magnetospheric disturbances. It is obvious that this data gap between 32 (LASCO field-of-view) and 215 $R_\odot$ makes accurate forecasting only possible on the short term. The challenge for space weather forecasting is to enhance the accuracy of the forecast 2-4 days before the onset of the storm despite the limited amount of data available. New types of remote sensing instruments make it in principle possible to observe CMEs continuously from the Sun till 1 AU: SMEI, [Solar Mass Ejection Imager *Eyles et al.*, 2003], the Heliospheric Imagers [HI *Defise et al.*, 2001] onboard STEREO (Solar Terrestrial Relations Observatory) or interplanetary scintillation techniques [IPS *Watanabe and Schwenn*, 1989]. These instruments are however far from being usable in an operational context.

The vision for future space weather forecasting (illustrated in Figure 1) is to utilize a sequence of real-time data-driven models that simulate the CME evolution from the Sun via L1 to the Earth and eventually calculating its effect on the geo-space. Ideally, heliospheric models should deliver *in situ*-like parameters quantifying the perturbation at L1 like the solar wind speed, density, temperature and magnetic field parameters,

which can on their turn serve as input for magnetospheric and ionospheric models. This will allow the formulation of a more accurate 2-4 day forecast. Hence, CME surveillance implies identifying CME parameters, which can serve as input for CME propagation models. To describe a CME and its interaction with background solar wind as completely and correctly as possible there is the need to assimilate all relevant data (magnetograms, EUV movies, coronagraphic sequences, radio observations). The only way to extract the required CME parameters from these data in near-real-time is by integrating automated software in the forecasting process. This appears to be the best approach to significantly improve space weather forecasting in the near future.

In this paper we focus on the current status of the *automated nowcast capabilities of the various CME manifestations*. In the next section we illustrate the model-for forecast principle enlisting existing efforts. We argue that the next step includes CME driven models predicting its manifestation at 1 AU. We compare the directly measured parameters with the needed input parameters. In section 3 we discus the available algorithms for automated CME detection which are needed to ensure the timeliness of the required data.

2. THE MODEL-FOR-FORECAST PRINCIPLE

The concept of real-time data-driven modelling is young but not new. At the NOAA-SEC Rapid Prototyping Center (RPC, http://www.sec.noaa.gov/rpc/), various models are tested on their operational use. However, up till now it has been common to use measurements form the Lagrangian point (L1) to drive magnetospheric and ionospheric forecast models. These allow accurately forecasting the upcoming geomagnetic activity during the next few hours, but not the next few days. In order to enhance the quality of the 2-4 day forecasts we should focus on integrating heliospheric models in the forecasting process. Modeling a CME event from the sun to the Earth is a complex task. Even starting from above the barely understood CME initiation, modeling the evolution, propagation and interaction of a CME with the background solar wind is a big challenge.

A new approach to this problem is the coupled-model approach, adopted and assessed by the Center for Integrated Space weather Modeling [CISM *Spence et al.*, 2004]. The CISM simulate the Sun-Earth system by coupling state-of-the-art codes modeling the solar corona, solar wind, magnetosphere and upper atmosphere/ionosphere using interfaces that exchange parameters specifying each component of the solar-terrestrial system. In addition, new efforts involve techniques continuously assimilating new data to keep the forecast 'on track'. As to proof the coupled-model-for-forecasting principle *Luhmann et al.* [2004] have simulated a Sun-to-Earth space weather event from the Solar corona to the Earth's

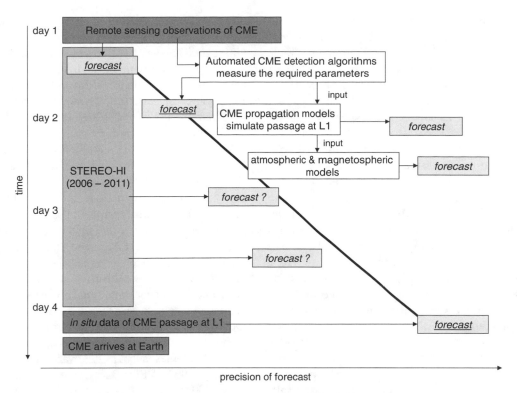

Figure 1. Concept map illustrating the value of real-time data-driven models regarding space weather forecasting. The vertical bar indicates the timeline. Note the current gap of new data between day 1 and day 4 (for example). This is expected to be filled with STEREO data in 2006. The horizontal bar indicates the precision of the forecast. The more to the right, the more reliable is the forecast. With the current data limitation to 30 $R_{\odot}$, only better understanding and models simulating the CME evolution from the Sun to L1 can push the forecast to the right. The underlined forecasts represent the actual forecasting process.

ionosphere. Many recent examples of modeling space weather events from the Sun to 1 AU (L1 or Earth) exit [e.g., *Wu and Guo*, 1999; *Odstrčil et al.*, 2004; *Manchester et al.*, 2004; *Intriligator et al.*, 2005].

Given the challenge to produce effective and efficient physics based models, forecasters now mostly rely on empirical models based on large statistics. The information extracted from these models is mostly restricted to kinematic properties (speed, acceleration, travel time), given the assumption of an earth directed CME. Despite their limited accuracy and information, they are of great value for forecasters and as reference models against which the newly developed models are compared. Recent examples of such an empirical approach are found in *dal Lago et al.* [2003]; *Xie et al.* [2004]; *Gopalswamy et al.* [2005]. The strength of the empirical model is its operational simplicity and therefore its ability to make quickly a good *estimate* of the arrival of a CME. The more advanced empirical models do even incorporate ambient parameters, like the average solar wind speed (~ 400 km s^{-1}). Numerical models are expected to be slightly better in predicting arrival times of CMEs, because they can use more realistic ambient solar wind. Further, numerical

simulations provide global context, predictions whether shock and/or ejecta will arrive to Earth and also are expected to predict the connectivity of the IMF line between Earth and eventual shock [*Odstrčil et al.*, 2004].

Directly connected to the call for inputs, there is the question of the operational capacity to deliver these parameters and with what accuracy and timeliness. At present our capacity to measure the time-dependent boundary-conditions near the sun are very limited. Therefore various approximations are utilized. CMEs are classically viewed with coronagraphs in space above the scattered light form the Earth's atmosphere. The observed white light signal is roughly proportional to the integrated mass along the line of sight. Ironically, the most geo-effective CMEs (halo and partial halo CMEs) are the most difficult to measure: they are very faint (due to Thomson scattering and a short integration length in the line of sight direction) and their structure is often very complicated. Measurements of the CME's *projected* speed, launch-angle and mass can be deduced from this data.

We need thus an 'interface' translating the measured parameters into input-parameters. *Schwenn et al.* [2001] have introduced the 'lateral expansion speed' of a CME and derived its relation with the radial speed [*dal Lago et al.*, 2003]. Under the assumption that halo CMEs propagate radially and with constant angular widths, *Zhao et al.* [2002] can reproduce some useful geometric and kinematic properties of halo CMEs using a simple geometrical model of a CME as a cone. Assuming also a constant traveling speed, a simpler cone-model is described by *Michalek et al.* [2003]. Applying their model on 72 asymmetric halo CMEs (observed with LASCO) they showed that the average corrected speeds only differ ~20% from the projected speeds. This is not so surprising, since for large cone-angles (they find an average cone angle of 120°), the edges of the CME cone have still a large component in the plane of the sky. Assuming a symmetric CME with cone-angle 120° erupting from the central meridian gives $v\_proj = \cos(90 - (120/2))°$ $v\_rad = \cos 30°$ $v\_rad \approx 0.7 \, v\_rad$. Using the fact that most halo CMEs are not symmetric this implies $v\_proj \geq 0.7 \, v\_rad$ for wide halo CMEs.

CME propagation models very often superpose a density blob on a given background. Nevertheless, in spite of this, the essential physics of the CME phenomenon probably resides in the magnetic field or its associated current systems [*Hudson and Cliver*, 2001; *Chané et al.*, 2005], which we cannot observe so directly as we observe the mass itself. Recent progress has been made in determining the magnetic field orientation of CMEs deduced from their on-disk counter-parts. For example the chirality of the pre-eruption filament is strongly correlated with the orientation of the magnetic field at L1 associated with the CME and hence can be used to predict the probability of a geomagnetic storm [*Yurchyshyn et al.*, 2001]. In addition to the importance of Hα and EUV images showing pre-eruption magnetic field structures, they are also useful for determining whether a halo CME is Earth-directed, even though we can still make mistakes especially when the first detection of the CME comes long after (or before) the solar signatures. It is thus essential to observe different manifestations of CMEs in different types of data. For an elaborate review on the current monitoring capabilities we refer to *Hochedez et al.* [2005].

It is empirically [e.g., *Crooker and Cliver*, 1994] and numerically [e.g., *Odstrčil and Pizzo*, 1999] shown that the motion and appearance of a CME in interplanetary space is strongly affected by its interaction with the ambient heliospheric pattern. Besides, the combination of sequential CMEs can produce complex structures at 1 AU. It is thus important to include the influence of the background solar wind in the model. Some use the actual solar wind speed measured *in situ* as an approximation for the ambient wind speed. But the solar wind speed is a highly variable parameter and on the spatial and temporal timescale of an IP ejection it can vary as much as 100 km s^{-1} [e.g. *Burlaga et al.*, 1987]. Using such a value as model input can introduce in some cases larger errors than using an average wind speed [*Vršnak and Gopalswamy*, 2002]. The Wang-Sheeley-Arge (WSA) model [*Wang and Sheeley*, 1992; *Arge and Pizzo*, 2000] provides the background solar wind up to 1 AU extrapolating the photospheric magnetic field into the heliosphere using the potential field source surface (PFSS) model [*Schatten et al.*, 1969; [*Altschuler and Newkirk*, 1969]. It works quite well regarding the simplifying assumptions.

The open data policy common to solar physics makes a large quantity of data available, but some, e.g., LASCO [Large Angle and Spectrometric Coronagraph; *Brueckner et al.*, 1995], are truly unique and therefore critical. On the other hand, some observations of great monitoring value are not easily or regularly accessible due to organizational problems. At the moment CMEs and their by-products are monitored by a number of different people independently, each one limited to its own instruments and specialization. As a consequence, their results remain usually disconnected. The challenge now is to link the sparse observations from instruments distributed worldwide and in space missions, in order to develop a global approach to CME monitoring. New internet technologies such as the various grid efforts: EGSO [*Bentley*, 2002]; VSO [*Davey et al.*, 2003]; coSEC [*Hurlburt et al.*, 2004] might greatly contribute to this attempt.

3. EXISTING SOFTWARE FOR AUTOMATED CME DETECTION IN REMOTE SENSING DATA

Early 2005, various algorithms for the detection of CME manifestations are either up and running or under development. Depending on the targeted CME manifestation, different types of data are 'scanned' by automated procedures. Automated software has the ability of analyzing data continuously (i.e. 24hr/day) and objectively. Its implications for forecasting are twofold: (1) The software tools can send out near-real-time alerts to the solar and space weather community reporting on the detection and providing important CME characteristics. These details together with observations enable the forecaster to send out timely space weather warnings. However, the limited range ($0-32 R_{\odot}$) of early CME observations limits the accuracy of these 2-4 day predictions. (2) Automatically generated alerts could also launch data-driven models simulating the response of the solar wind or the magnetosphere to the reported disturbance. For this purpose the automated software packages should calculate the most important parameters at 1 AU such as the solar wind bulk speed and density, ion temperature and magnetic field strength and orientation. to feed the model with a real-time disturbance. Hence the benefit of automated monitoring exceeds the level of immediate response, but can act as the 'missing link' between observation and modelling. In what follows we describe the existing detection algorithms on events linked to CME occurrence. Several of these techniques have been described in *Robbrecht and Berghmans, 2005]* focussing on image processing.

3.1. Flare Detection in X-ray and EUV Data

A CME is often (but not always!) accompanied by a flare, a 'large' and 'sudden' increase in radiation, mostly in X-rays and ultraviolet wavelengths observed in image and irradiance data [see e.g., *Hochedez et al.*, 2005]. SXI (Solar X-ray Imager) observations from GOES-12 (the 12th Geostationary Operational Environmental Satellite) provide valuable flare location and other information, especially when no optical (i.e. white light) observations are available. SEC (Space Environment Center) developed the SXI flare algorithm, triggered by GOES X-ray events, which finds the brightest area in the latest SXI image and assigns the region number of the closest active solar region. Since Jan. 2004 this information (XFL) is added to the Edited Solar Events Lists available online (http://www.sec.noaa.gov/ftpdir/indi-ces/events/ events.txt). It groups several reports into a single event, as determined by the SEC forecaster. It includes information on radio bursts, filament disappearances and prominence eruptions, flares in Hα and even Forbush decreases. However, due to human interference the SEC reports are sometimes lagging behind. Therefore, fully automated software developed for flare-monitoring form a valuable addition to this such as the SolarSoft Latest Events page (http://www.lmsal.com/solarsoft/ latest_events/). X-ray flux enhancements above the B-class flare level are recorded automatically from GOES-10 irradiance data. For every flare a detailed page is setup providing the start- peak- and end-time of the flare and also its peak-intensity. The software also determines its location on the disk using EIT high cadence wavelength or SXI (GOES12) movies.

In the frame of their automatic alerting system the 'Solar Influences Data analysis Center' [*Berghmans et al.*, 2005] has developed software to automatically detect flares, using GOES-12 irradiance data as input. The detector has been tuned as to produce the same flare parameters as NOAA/SEC. First tests (few tens of examples) show that the output is indeed identical to the NOAA/SEC flare report. The output is an ASCII table listing in a line per flare, its main parameters. It is also available online (http://sidc.oma.be/ GOESdata).

3.2. Disappearing Filaments in Hα

Over the last five years a number of groups successfully developed codes and algorithms for the automated detection of filaments, best observed in Hα images, where they appear as dark elongated threads. There exist different levels in output.

The first detection schemes [*Gao et al.*, 2002; *Shih and Kowalski*, 2003; *Fuller and Aboudarham*, 2004; *Zharkova et al.*, 2003] generate black-and-white images with the black areas indicating the filaments. Besides a graphical display this allows determination of parameters like total filament area (pixel area) and location on the disk (e.g., coordinates of center of mass). This parameterized 'filament description' is important in view of automated interpretation of these outputs. However the above mentioned references do not report on the generation of output-tables containing relevant filament parameters. Automated filament detection schemes have to deal with a number of sub-problems: finding the right intensity threshold, developing appropriate preprocessing techniques optimized for filament tracking, merging filament parts belonging to the same body, etc. These are extensively discussed in *Robbrecht and Berghmans* [2005]. For space weather purposes, we are interested in filament disappearances. This implies that the technique is able to 'recognize' a filament in subsequent images. Very recently *Qu et al.* [2005] and *Bernasconi et al.* [2005] introduced spines to characterize the filaments. A Spine marks the 'skeleton' of a filament consisting of several footpoints mutually connected. *Qu et al.* [2005] use adaptive edge linking [introduced by *Shih and Cheng*, 2004] to trace the filament's spine. Based on evolution in size and intensity a filament is reported to have disappeared or significantly shrunk. *Bernasconi et al.* [2005] apply a principal curve algorithm to find the spine; this is a multi-step iterative technique developed from Kégl's algorithm [*Kégl et al.*, 2000]. For every filament and in every image a central latitude and longitude is determined. If there is no close match for these coordinates in subsequent images, the filament is regarded as (partially) disappeared. Important to note here is the difference between thermal disappearance (DBt) and dynamical disappearance (DBd) [*Mouradian et al.*, 1995]. Only the latter one involves plasma ejections and leads to the permanent disappearance of filaments. *Bernasconi et al.* [2005] tersely take this into account by only allowing the label 'disappeared' for filaments that are not detected during 3 subsequent days. This limits of course its value for space weather warnings.

Analyses of Hα filaments and photospheric magnetograms have revealed two chiralities of filaments, dextral and sinistral [*Martin et al.*, 1994]. It can be determined using the orientation of barbs relative to the filament's main axis [*Pevtsov et al.*, 2003]. Using this technique *Bernasconi et al.* [2005] automatically determine a filaments' chirality and list it with the other filament parameters (see Figure 2). Chirality is an important parameter, since together with solar magnetic field data (e.g., MDI) the filaments' magnetic helicity can be determined. Under assumption of magnetic helicity conservation this means the polarity of the associated ICME can be estimated. Many observational studies report on this correlation

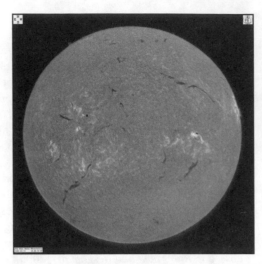 

Figure 2. *a:* Example of an processed Hα image from the Big Bear Solar Observatory (BBSO). Standard BBSO processing includes dark current and flat field correction, contrast enhancement and limb darkening removal. *b:* Example of a filament detection by Bernasconi & Rust, indicating the spine and barbs. Next to each barb is a letter indicating whether it is left-bearing (L), right-bearing (R) or undetermined (?). Since this filament has more left-bearing barbs than right-bearing ones, the code determines it to have sinistral chirality.

[e.g., *Bothmer and Schwenn*, 1994; *McAllister et al.*, 2001; *Yurchyshyn et al.*, 2001]. These findings mean a great step forward in the predictability of the geo-effectiveness of the related CME.

The above described methods are applied on Hα observations from Big Bear Solar Observatory (BBSO, California) and the Meudon Observatory (in France). They are part of the global high resolution Hα network (http://www.bbso.njit.edu/Research/Halpha/).

3.3. Prominence Eruptions in Radio Images

Unique observations of solar eruptions are acquired with the Nobeyama Radio-heliograph-NoRH [*Nakajima et al.*, 1995] at the National Astronomical Observatory of Japan since April, 1992. During the daily 8-hours of observations (from 23 to 07 UT), full disk solar images are produced at the 17 GHz and 34 Ghz radio frequencies. In what follows we only consider the 17 Ghz images which have a spatial resolution of 15 arc sec and a time resolution of 10 min (but they are reconstructible down to a 50 ms resolution).

An operational tool was developed to detect so-called limb-events, including mostly prominence eruptions but also prominence activity and limb flares. The technique tracks the center of mass of bright pixel groups (off-limb) in subsequent images. If the group is detected in three subsequent images, the structure is defined as an event and listed on the website (http://solar.nro.nao.ac.jp/norh/html/prominence/). The detection program runs once per day after the daily observation period (at 9 UT), after which the output is checked manually.

Bad detections are removed from the list. The output per event contains the start-, peak- and end-time, the location of the event in cartesian coordinates and a rough estimate of its pixel-size. No velocity is measured, although *Mouradian et al.* [1995] report that speeds from radio data show good correspondence with CME speeds in initial phase.

3.4. EIT Waves and Dimmings

After the discovery of the so-called 'EIT wave' phenomenon *Thompson et al.* [1998] it was soon realised that these waves are strongly associated with earth-directed CMEs [e.g., *Plunkett et al.*, 1998; *Biesecker et al.*, 2002]. A dimming is usually observed behind an EIT wave and is most likely due to the evacuation of mass during the CME [*Thompson et al.*, 1998; *Harra and Sterling*, 2001]. Automatization of EIT wave detection is a hard problem given their large variety in physical appearance and their weak intensity variation. Recently *Podladchikova and Berghmans* [2005] developed software to automatically detect these EIT waves, illustrated in Figure 3. Bearing in mind the above mentioned problems, the detection is based on the characteristics of the histogram-distribution of running difference images. The presence of large-scale coherent structures, such as dimmings and EIT waves, strongly influences the higher order moments (skewness, kurtosis) of the distribution. Deviations therein can be used as triggers for the occurrence of an EIT wave and associated dimming. Its output is location, timing, structure and dynamics of the EIT wave.

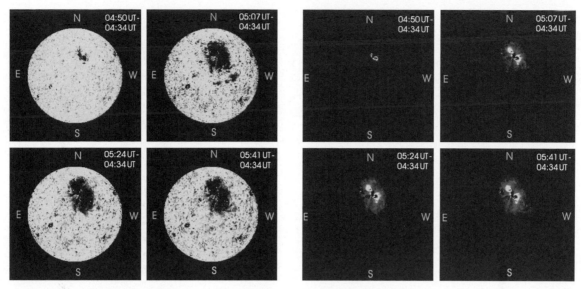

Figure 3. Illustration of a dimming extraction on EIT 195Å running difference images on 12 May 1997, by *Podladchikova and Berghmans* [2005] (see this paper for a color version). After a dimming is detected its location and size is found by applying a region growing method.

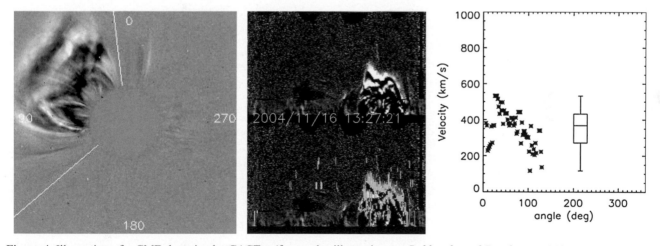

Figure 4. Illustration of a CME detection by CACTus (for a color illustration see *Robbrecht and Berghmans*, 2004). *a:* C2 running difference image, the white lines indicate the measured angular width of the CME. *b:* The CME in polar view. The top panel shows the original running difference image. The bottom panel shows the CACTus CME detection. *c:* CME velocity profile as a function of the angle, which runs counterclockwise from the north. A boxplot is drawn: the box itself contains the middle 50% of the measured speeds. The whiskers at both ends indicate respectively the minimal and maximal detected speeds. Note that the whiskers are no error-bars, but indicate the range of the measured speeds.

3.5. CMEs in Coronagraphic White Light Images

The classical white-light picture of a CME is that of a bright leading edge followed by a dark cavity and a central core representing the erupted prominence [*Hundhausen et al.*, 1987]. However, CMEs are very variable in appearance and often do *not* show clearly this three part structure. This is particularly true for halo CMEs, when the erupted matter is seen around the entire occulting disk. Automated detection and measurements of their physical properties (speed, density, mass, magnetic field) is therefore difficult. *Berghmans et al.* [2002] developed software which automatically detects CME propagations in coronagraphic image sequences. Instead of making morphology assumptions, this technique basically detects bright ridges in [time, height]-maps [see [*Sheeley et al.*, 1999] indicating features moving radially outward from the Sun. Recently [*Robbrecht and Berghmans,* 2004] have enhanced the performance significantly; 96% of all CMEs during a 6 day test-period was recovered. The program now runs near-real-time ant the results are put immediately on-line. The output includes the first time of occurrence in the c2 field-of-view, a linear speed estimation, principal angle and angular width. An illustration can be found in Figure 4. The software sends out alerting messages whenever a partial halo CME candidate is detected.

4. CONCLUSIONS

Early 2005, we are on the eve of a new era in space weather forecasting, in which intelligent software may autonomously detect significant space weather events, launch model simulations and sends out warnings. This invokes incorporating real-time data-driven models in the forecasting process. Existing efforts on magnetospheric data-driven models serve as valuable examples. The next step towards automated space weather forecasting is to fill the data gap between early observations of a CME and its *in situ* manifestation using CME propagation models. We have argued than this can greatly improve the accuracy when forecasting the space weather a few days ahead. Hence, CME surveillance implies defining CME and ambient solar wind parameters, which can serve as input for data-driven models. We believe that only automated CME detection software is able to deliver these parameters in a timely manner.

We have given an overview of the existing software for automated CME detection. Their output parameters are valuable for both, the space weather forecaster and the models simulating the propagation of the detected CME. Hence the benefit of automated monitoring exceeds the level of immediate response, but can act as the 'missing link' between observation and modelling.

To ensure the success of the model-for-forecasting principle a collaborative effort is needed from both, the automated detection community and the CME modelling community. The detection software have to be tuned in order to deliver the required inputs in a computer-readable format.

Acknowledgments.   The authors hereby acknowledge the organizers of the conference for the invitation to present this review. We thank our colleague Jean-François Hochedez for inspiring discussions which have broadened the perspective of this paper.

REFERENCES

Altschuler, M.D., and G. Newkirk, Magnetic Fields and the Structure of the Solar Corona. I: Methods of Calculating Coronal Fields, *Sol. Phys.*, 9, 131, 1969.

Arge, C.N., and V.J. Pizzo, Improvement in the prediction of solar wind conditions using near-real time solar magnetic field updates, *J. Geophys. Res.*, 105, 10,465-10,480, doi:10.1029/1999JA900262, 2000.

Bentley, R.D., EGSO - the next step in data analysis, in *ESA SP-477: Solspa 2001, Proceedings of the Second Solar Cycle and Space Weather Euroconference*, pp. 603-606, 2002.

Berghmans, D., B.H. Foing, and B. Fleck, Automated detection of CMEs in LASCO data, in *ESA SP-508: From Solar Min to Max: Half a Solar Cycle with SOHO*, pp. 437-440, 2002.

Berghmans, D., *et al.*, Solar Activity: nowcasting and forecasting at the SIDC, *Annales Geophysicae*, 23, pp. 3115-3128.

Bernasconi, P.N., D.M. Rust, and D. Hakim, Advanced Automated Solar Filament Detection And Characterization Code: Description, Performance, And Results, *Sol. Phys.*, 228, 97-117, doi:10.1007/s11207-005-2766-y, 2005.

Biesecker, D.A., D.C. Myers, B.J. Thompson, D.M. Hammer, and A. Vourlidas, Solar Phenomena Associated with "EIT Waves", *Astrophys. J.*, 569, 1009-1015, 2002.

Bothmer, V., and R. Schwenn, Eruptive prominences as sources of magnetic clouds in the solar wind, *Space Science Reviews*, 70, 215, 1994.

Brueckner, G.E., *et al.*, The Large Angle Spectroscopic Coronagraph (LASCO), 162, 357-402, 1995.

Burlaga, L.F., K.W. Behannon, and L.W. Klein, Compound streams, magnetic clouds, and major geomagnetic storms, *J. Geophys. Res.*, 92, 5725-5734, 1987.

Chané, E., C. Jacobs, B. van der Holst, S. Poedts, and D. Kimpe, On the effect of the initial magnetic polarity and of the background wind on the evolution of CME shocks, *Astron. Astrophys.*, 432, 331-339, doi:10.1051/0004-6361:20042005, 2005.

Cliver, E.W., and H.S. Hudson, CMEs: How do the puzzle pieces fit together?, *Journal of Atmospheric and Terrestrial Physics*, 64, 231-252, 2002.

Crooker, N.U., and E.W. Cliver, Postmodern view of M-regions, *J. Geophys. Res.*, 99, 23,383, 1994.

dal Lago, A., R. Schwenn, and W.D. Gonzalez, Relation between the radial speed and the expansion speed of coronal mass ejections, *Advances in Space Research*, 32, 2637-2640, doi:10.1016/j.asr.2003.03.012, 2003.

Davey, A.R., R.S. Bogart, G. Dimitoglou, J.B. Gurman, F. Hill, P.C. Martens, K.Q. Tian, and S. Wampler, First Steps Towards a VSO, *AAS/Solar Physics Division Meeting*, 34, 2003.

Defise, J., J. Halain, E. Mazy, P.P. Rochus, R.A. Howard, J.D. Moses, D.G. Socker, G.M. Simnett, and D.F. Webb, Design of the Heliospheric Imager for the STEREO mission, in *Proc. SPIE Vol. 4498, p. 63-72, UV/EUV and Visible Space Instrumentation for Astronomy and Solar Physics, Oswald H. Siegmund; Silvano Fineschi; Mark A. Gummin; Eds.*, pp. 63-72, 2001.

Eyles, C.J., *et al.*, The Solar Mass Ejection Imager (Smei), *Sol. Phys.*, 217, 319-347, 2003.

Forbes, T.G., A review on the genesis of coronal mass ejections, *J. Geophys. Res.*, pp. 23,153-23,166, 2000.

Fuller, N., and J. Aboudarham, Automatic detection of solar filaments versus manual digitization, *Lecture notes in computer science*, 2004.

Gao, J., H. Wang, and M. Zhou, Development of an Automatic Filament Disappearance Detection System, 205, 93-103, 2002.

Gopalswamy, N., Coronal mass ejections as a source of space Weather, *EGS - AGU - EUG Joint Assembly, Abstracts from the meeting held in Nice, France, 6 - 11 April 2003, abstract #4456*, p. 4456, 2003.

Gopalswamy, N., S. Yashiro, Y. Liu, G. Michalek, A. Vourlidas, M. L. Kaiser, and R.A. Howard, Coronal mass ejections and other extreme characteristics of the 2003 October-November solar eruptions, *Journal of Geophysical Research (Space Physics)*, 110, 9, doi:10.1029/ 2004JA010958, 2005.

Gosling, J.T., E. Hildner, R.M. MacQueen, R.H. Munro, A.I. Poland, and C.L. Ross, Mass ejections from the sun - A view from SKYLAB, *J. Geophys. Res.*, 79, 4581-4587, 1974.

Gosling, J.T., M.F. Thomsen, S.J. Bame, R.C. Elphic, and C.T. Russell, Cold ion beams in the low latitude boundary layer during accelerated flow events, *Geophys. Res. Lett.*, 17, 2245-2248, 1990.

Gosling, J.T., D.J. McComas, J.L. Phillips, L.A. Weiss, V.J. Pizzo, B.E. Goldstein, and R.J. Forsyth, A new class of forward-reverse shock pairs in the solar wind, *Geophys. Res. Lett.*, 21, 2271-2274, 1994.

Harra, L.K., and A.C. Sterling, Material Outflows from Coronal Intensity "Dimming Regions" during Coronal Mass Ejection Onset, *Astrophys. J.*, 561, L215-L218, 2001.

Hochedez, J.-F., A. Zhukov, E. Robbrecht, R. Van der Linden, D. Berghmans, P. Vanlommel, A. Theissen, and F. Clette, Monitoring capabilities for solar weather nowcast and forecast, *Annales Geophysicae*, 23, pp. 3149-3161.

Hudson, H.S., and E.W. Cliver, Observing coronal mass ejections without coronagraphs, *J. Geophys. Res.*, pp. 25,199-25,214, 2001.

Hundhausen, A.J., T.E. Holzer, and B.C. Low, Do slow shocks precede some coronal mass ejections?, *J. Geophys. Res.*, 92, 11,173-11,178, 1987.

Hurlburt, N., P. Bose, S. Freeland, M. Woodward, and G. Slater, Collaborative Virtual Observatories using CoSEC, *American Astronomical Society Meeting Abstracts*, 204, 2004.

Intriligator, D.S., W. Sun, M. Dryer, C.D. Fry, C. Deehr, and J. Intriligator, From the Sun to the outer heliosphere: Modeling and analyses of the interplanetary propagation of the October/November (Halloween) 2003 solar events, *Journal of Geophysical Research (Space Physics)*, 110, 9—+, doi:10.1029/2004JA010939, 2005.

Kégl, B., A. Krzyżak, T. Linder, and K. Zeger, Learning and Design of Principal Curves, *IEEE Transactions on Pattern Analysis and Machine Intelligence*, 2000.

Low, B.C., Coronal mass ejections, magnetic flux ropes, and solar magnetism, *J. Geophys. Res.*, pp. 25,141-25,164, 2001.

Low, B.C., and M. Zhang, The Hydromagnetic Origin of the Two Dynamical Types of Solar Coronal Mass Ejections, *Astrophys. J.*, 564, L53-L56, 2002.

Luhmann, J.G., S.C. Solomon, J.A. Linker, J.G. Lyon, Z. Mikic, D. Odstrčil, W. Wang, and M. Wiltberger, Coupled model simulation of a Sun-to-Earth space weather event, *Journal of Atmospheric and Terrestrial Physics*, 66, 1243-1256, doi:10.1016/j.jastp.2004.04.005, 2004.

Manchester, W.B., T.I. Gombosi, I. Roussev, A. Ridley, D.L. De Zeeuw, I.V. Sokolov, K.G. Powell, and G. Tóth, Modeling a space weather event from the Sun to the Earth: CME generation and interplanetary propagation, *Journal of Geophysical Research (Space Physics)*, p. 2107, 2004.

Martin, S., R. Billimoria, and P. Tracadas, in *Solar Surface Magnetism*, p. 303, 1994.

McAllister, A.H., S.F. Martin, N.U. Crooker, R.P. Lepping, and R.J. Fitzenreiter, A test of real-time prediction of magnetic cloud topology and geomagnetic storm occurrence from solar signatures, *J. Geophys. Res.*, pp. 29,185-29,194, 2001.

Michalek, G., N. Gopalswamy, and S. Yashiro, A New Method for Estimating Widths, Velocities, and Source Location of Halo Coronal Mass Ejections, *Astrophys. J.*, 584, 472-478, doi:10.1086/345526, 2003.

Mouradian, Z., I. Soru-Escaut, and S. Pojoga, On the two classes of filament-prominence disappearance and their relation to coronal mass ejections, *Sol. Phys.*, 158, 269-281, 1995.

Munro, R.H., J.T. Gosling, E. Hildner, R.M. MacQueen, A.I. Poland, and C.L. Ross, The association of coronal mass ejection transients with other forms of solar activity, *Sol. Phys.*, 61, 201-215, 1979.

Nakajima, H., *et al.*, NEW NOBEYAMA RADIO HELIOGRAPH, *Journal of Astrophysics and Astronomy Supplement*, 16, 437, 1995.

Odstrčil, D., P. Riley, and X.P. Zhao, Numerical simulation of the 12 May 1997 interplanetary CME event, *Journal of Geophysical Research (Space Physics)*, 109, 2116, doi:10.1029/2003JA010135, 2004.

Odstrčil, D., and V.J. Pizzo, Three-dimensional propagation of coronal mass ejections in a structured solar wind flow 2. CME launched adjacent to the streamer belt, *J. Geophys. Res.*, 104, 493-504, doi:10.1029/1998JA900038, 1999.

Pevtsov, A.A., K.S. Balasubramaniam, and J.W. Rogers, Chirality of Chromospheric Filaments, *Astrophys. J.*, 595, 500-505, 2003.

Plunkett, S.P., B.J. Thompson, R.A. Howard, D.J. Michels, O.C. St. Cyr, S.J. Tappin, R. Schwenn, and P.L. Lamy, LASCO observations of an Earth-directed coronal mass ejection on May 12, 1997, *Geophys. Res. Lett.*, 25, 2477-2480, 1998.

Podladchikova, O., and D. Berghmans, Automated Detection Of Eit Waves And Dimmings, *Sol. Phys.*, 228, 265-284, doi:10.1007/s11207-005-5373-z, 2005.

Qu, M., F.Y. Shih, J. Jing, and H. Wang, Automatic Solar Filament Detection Using Image Processing Techniques, *Sol. Phys.*, 228, 119-135, doi:10.1007/s11207-005-5780-1, 2005.

Robbrecht, E., and D. Berghmans, Automated recognition of coronal mass ejections (CMEs) in near-real-time data, *Astron. Astrophys.*, 425, 1097-1106, 2004.

Robbrecht, E., and D. Berghmans, Entering The Era Of Automated CME Recognition: A Review Of Existing Tools, *Sol. Phys.*, 228, 239-251, doi:10.1007/s11207-005-5004-8, 2005.

Schatten, K.H., J.M. Wilcox, and N.F. Ness, A model of interplanetary and coronal magnetic fields, *Sol. Phys.*, 6, 442-455, 1969.

Schwenn, R., A. Dal Lago, W.D. Gonzalez, E. Huttunen, C.O. St.Cyr, and S.P. Plunkett, A Tool For Improved Space Weather Predictions: The CME Expansion Speed, *AGU Fall Meeting Abstracts*, pp. A739+, 2001.

Sheeley, N.R., J.H. Walters, Y.-M. Wang, and R.A. Howard, Continuous tracking of coronal outflows: Two kinds of coronal mass ejections, *J. Geophys. Res.*, 104, 24,739-24,768, 1999.

Shih, F.Y., and S. Cheng, Adaptive mathematical morphology for edge linking, *Information sciences*, 167, 9-21, 2004.

Shih, F.Y., and A.J. Kowalski, Automatic Extraction of Filaments in Hα Solar Images, *Sol. Phys.*, 218, 99-122, 2003.

Spence, H., D. Baker, A. Burns, T. Guild, C.-L. Huang, G. Siscoe, and R. Weigel, Center for integrated space weather modeling metrics plan and initial model validation results, *Journal of Atmospheric and Terrestrial Physics*, 66, 1499-1507, doi:10.1016/j.jastp.2004.03.029, 2004.

Thompson, B.J., S.P. Plunkett, J.B. Gurman, J.S. Newmark, O.C. St. Cyr, and D.J. Michels, SOHO/EIT observations of an Earth-directed coronal mass ejection on May 12, 1997, *Geophys. Res. Lett.*, 25, 2465-2468, 1998.

Tousey, R., R.A. Howard, and M.J. Koomen, The Frequency and Nature of Coronal Transient Events Observed by OSO-7*, BAAS, 6, 295, 1974.

Tsurutani, B.T., W.D. Gonzalez, Y. Kamide, and J.K. Arballo, *Preface*, pp. D9+, Magnetic Storms, Geophysical Monograph Series, Vol. 98, 1997.

Vršnak, B., and N. Gopalswamy, Influence of the aerodynamic drag on the motion of interplanetary ejecta, *Journal of Geophysical Research (Space Physics)*, 107, 2-1, doi:10.1029/2001JA000120, 2002.

Wang, Y.-M., and N.R. Sheeley, On potential field models of the solar corona, *Astrophys. J.*, 392, 310-319, 1992.

Watanabe, T., and R. Schwenn, Large-scale propagation properties of interplanetary disturbances revealed from IPS and spacecraft observations, *Space Science Reviews*, 51, 147-173, 1989.

Wu, S.T., and W.P. Guo, Generation and propagation of solar disturbances: a magneto-hydrodynamic simulation, *Journal of Atmospheric and Terrestrial Physics*, 61, 109-117, 1999.

Xie, H., L. Ofman, and G. Lawrence, Cone model for halo CMEs: Application to space weather forecasting, *Journal of Geophysical Research (Space Physics)*, 109, 3109, doi:10.1029/2003JA010226, 2004.

Yurchyshyn, V., H. Wang, and V. Abramenko, How directions and helicity of erupted solar magnetic fields define geoeffectiveness of coronal mass ejections, *Advances in Space Research*, 32, 1965-1970, 2003.

Yurchyshyn, V., H. Wang, and V. Abramenko, Correlation between speeds of coronal mass ejections and the intensity of geomagnetic storms, *Space Weather*, 2, 2001, 2004.

Yurchyshyn, V.B., H. Wang, P.R. Goode, and Y. Deng, Orientation of the Magnetic Fields in Interplanetary Flux Ropes and Solar Filaments, *Astrophys. J.*, 563, 381-388, 2001.

Zhang, J., M.W. Liemohn, J.U. Kozyra, B.J. Lynch, and T.H. Zurbuchen, A statistical study of the geoeffectiveness of magnetic clouds during high solar activity years, *Journal of Geophysical Research (Space Physics)*, p. 9101, 2004.

Zhao, X.P., S.P. Plunkett, and W. Liu, Determination of geometrical and kinematical properties of halo coronal mass ejections using the cone model, *Journal of Geophysical Research (Space Physics)*, 107, 13-1, doi:10.1029/2001JA009143, 2002.

Zharkova, V.V., S.S. Ipson, S.I. Zharkov, A. Benkhalil, J. Aboudarham, and R.D. Bentley, A full-disk image standardisation of the synoptic solar observations at the Meudon Observatory, *Sol. Phys.*, 214, 89-105, 2003.

Initiation of Coronal Mass Ejections

Ronald L. Moore and Alphonse C. Sterling

NASA Marshall Space Flight Center, XD12/Space Science Branch, Huntsville, Alabama

This paper is a synopsis of the initiation of the strong-field magnetic explosions that produce large, fast coronal mass ejections. The presentation outlines our current view of the eruption onset, based on results from our own observational work and from the observational and modeling work of others. From these results and from physical reasoning, we and others have inferred the basic processes that trigger and drive the explosion. We describe and illustrate these processes using cartoons. The magnetic field that explodes is a sheared-core bipole that may or may not be embedded in surrounding strong magnetic field, and may or may not contain a flux rope before it starts to explode. We describe three different mechanisms that singly or in combination can trigger the explosion: (1) runaway internal tether-cutting reconnection, (2) runaway external tether-cutting reconnection, and (3) ideal MHD instability or loss or equilibrium. For most eruptions, high-resolution, high-cadence magnetograms and chromospheric and coronal movies (such as from TRACE or Solar-B) of the pre-eruption region and of the onset of the eruption and flare are needed to tell which one or which combination of these mechanisms is the trigger. Whatever the trigger, it leads to the production of an erupting flux rope. Using a simple model flux rope, we demonstrate that the explosion can be driven by the magnetic pressure of the expanding flux rope, provided the shape of the expansion is "fat" enough.

1. INTRODUCTION

An intense, days-long solar energetic particle (SEP) storm in interplanetary space, one that could kill astronauts and spacecraft, is generated by the bow shock of a large, fast coronal mass ejection (CME) (*Kahler*, 1992; *Reames*, 1999, 2001). The driving CME is a magnetic explosion that also produces a flare in the source region on the Sun (*MacQueen & Fisher*, 1983; *Moore et al.*, 2001; *Falconer et al.*, 2002). The magnetic fields that explode to drive the fastest CMEs and the most dangerous particle storms are in active regions with large sunspots, the regions of the strongest magnetic fields found on the Sun (e.g., *Gopalswamy et al.*, 2004).

Solar Eruptions and Energetic Particles
Geophysical Monograph Series 165
10.1029/165GM07

Plate 1 shows a CME that is representative of those that drive SEP storms. A SOHO/LASCO difference image captures this CME as it blasts through the outer corona. Here, and in the original (non-differenced) LASCO images, this CME shows the three-part bubble structure (bright envelope around a dark void with a bright core) typical of large, fast CMEs. Plate 1 also shows the onset of the magnetic explosion as observed by SOHO/EIT in Fe XII images. Although the magnetic field was too weak to have sunspots, and hence was weaker than in many still stronger CME explosions, the explosion produced a large CME that was fast enough [900 km/s (SOHO/LASCO CME Catalog)] to drive a bow shock and SEP storm. No SEP storm from this CME was observed at SOHO because SOHO was shielded from the particles by the interplanetary magnetic field: the CME occurred in a sector of the interplanetary magnetic spiral far behind that of SOHO and Earth.

The magnetogram and Fe XII coronal images in Plate 1 show that the exploding field initially was a closed arcade

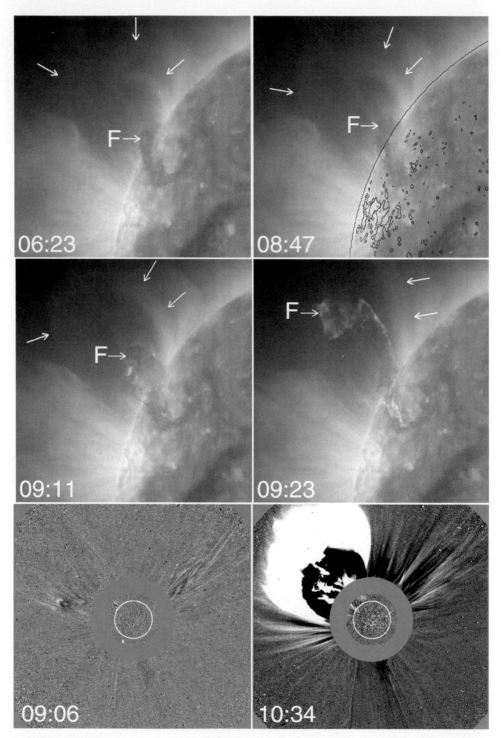

Plate 1. Onset of a typical filament-eruption magnetic explosion and the resulting large, fast CME. This eruption was observed by SOHO on 2002 January 4, and has been studied by *Sterling and Moore* (2004b). The universal time of each image is in the lower left corner. The upper four panels are EIT 195Å Fe XII coronal images. The arrow labeled F points to the filament before it started to erupt (first panel), during the slow-rise onset of its eruption (second panel), and during the explosive fast-rise phase (third and fourth panels). The other arrows point to the bright arcade that envelops the filament cavity and filament, and that is blown open by their eruption. The superposed MDI magnetogram in the second panel shows the quadrupolar arrangement of positive polarity (red) and negative polarity (blue) magnetic flux in which the filament and cavity are seated. The bottom two panels are LASCO/C2 coronagraph running-difference images of the outer corona at the start of the fast-rise phase of the filament eruption (left) and well after the CME had emerged from behind the occulting disk (right). The white circle outlines the solar disk centered behind the occulting disk.

with a dark filament of chromospheric-temperature plasma suspended in its core. The filament stands above the neutral line (polarity dividing line) straddled by the arcade and shows that the magnetic field in the core of the arcade is strongly sheared: the direction of the field threading the filament is nearly parallel to the neutral line rather than nearly orthogonal as it would be if the field were in its minimum-energy potential configuration. That is, the magnetic location and form of the filament show that the core field is greatly deformed from its potential configuration and hence has a large store of nonpotential (free) magnetic energy that in principle is available for release in an explosion.

Over the past three decades, there has been a synthesis of various complementary observations of many ejective solar eruptions similar to that in Plate 1. These observations include chromospheric images, line-of-sight magnetograms, vector magnetograms, and coronal X-ray and EUV images of the eruption region, and chromospheric, coronal, and hard X-ray movies of the erupting filament-carrying sheared core field and the ensuing heated foot points and loops of the flare. From these observations a strong case has been made that (1) the pre-eruption configuration of the field that explodes to become a fast CME is typically a sheared-core arcade like that in Plate 1, (2) there is enough free magnetic energy in the sheared core field to produce the explosion, and (3) the energy going into the CME and flare comes from the expanding sheared core field as it erupts in the explosion (*Hirayama*, 1974; *Kopp & Pneuman*, 1976; *Heyvaerts et al.*, 1977; *Moore & LaBonte*, 1980; *Moore et al.*, 1980, 1987, 1995, 1997, 2001; *Moore*, 1988, 2001; *Shibata et al.*, 1995; *Shibata*, 1998; *Canfield et al.*, 1999; *Sterling et al.*, 2000, 2001a,b; *Sterling & Moore*, 2001a,b, 2003, 2004a,b).

The three-dimensional configuration of a sheared-core arcade prior to eruption is sketched in the first panel of Figure 1. In some cases, the sigmoidal shape of the sheared core field is as symmetric and as obvious in coronal X-ray images as it is in this drawing, but in many cases it is distorted, asymmetric, and much less obvious [e.g., see examples in *Sterling et al.* (2000)]. This drawing shows only the closed bipolar field that has the sheared field and filament in its core. Any sheared-core bipole on the Sun is embedded in other magnetic fields rooted around it more or less as in Plate 1. If the surrounding fields are of strength and span comparable to those of the sheared-core bipole, they can strongly influence whether and how the sheared-core bipole explodes (*Antiochos*, 1998; *Antiochos et al.*, 1999). If the surrounding fields are weak enough compared to the bipole, their effect on the initiation and growth of the explosion can reasonably be ignored. This is the case depicted in Figure 1.

At chromospheric and low coronal heights in sunspot active regions, the magnetic field strongly dominates the weight and pressure of the plasma (*Gary*, 2001). As a result,

the equilibrium configuration of the field is very nearly the force-free configuration that the field would have if the plasma had no weight and no pressure, the configuration determined by the balance between the outward push of magnetic pressure and the inward pull of magnetic tension (e.g., *Cowling*, 1957). So, in an active-region sheared-core bipole the cool filament material and the hot coronal plasma are constrained to trace the field lines, and do not appreciably deform the field from its force-free configuration. The elongation and striation of the filament show that the core field is strongly sheared, and the coronal plasma often illuminates the overall form of the sheared-core bipole, showing that it typically has the sigmoidal character of the initial sketch in Figure 1 (*Canfield et al.*, 1999; *Sterling et al.*, 2000).

In this paper, we consider the onset of CME explosions in which, until it begins to explode, the driving magnetic field is a nearly force-free sheared-core bipole of the sigmoidal form sketched in Figure 1. The feet of all the field lines are locked to the massive body of the Sun. The magnetic pressure pushes against the photosphere and outward in all directions, trying to explode the field, but is held in check by the magnetic tension pulling back toward the feet. So, for the case of force-free equilibrium in the part of the pre-eruption field that is the source of the energy released in the explosion, the problem of the initiation of the CME explosion is that of how this equilibrium is destabilized or broken, so that the magnetic pressure is unleashed to drive the explosion.

We present three basic alternatives for triggering the explosion: internal tether cutting, external tether cutting (breakout), and ideal MHD instability or loss of equilibrium. The presentation outlines our current view of the eruption onset, based on our own observations and on the observations and theoretical work of others. In our view, for the strong-field (initially force-free) case we are considering, the three presented alternatives cover the range of possibilities for triggering the explosion. Each alternative is illustrated by cartoons representing the driving field, a surrounding field, and their interaction. No observations other than in Plate 1 and no numerical simulations of CME onsets are shown, but published observations and modeling studies are cited and briefly discussed in relation to each alternative. Our exposition of CME initiation is similar in approach to both that of *Moore and Roumeliotis* (1992) and that of *Klimchuk* (2001). It is an extension of *Moore and Roumeliotis* in that they did not consider the role of surrounding magnetic fields. Also in contrast to *Moore and Roumeliotis*, we do not explicitly portray the role of emerging magnetic fields in destabilizing the sheared-core bipole; in our scheme, reconnection between crossed strands of the sheared core field has basically the same role. Our considerations of CME initiation are narrower in scope than those of *Klimchuk* (2001), which included the possibilities of either plasma weight or plasma pressure being

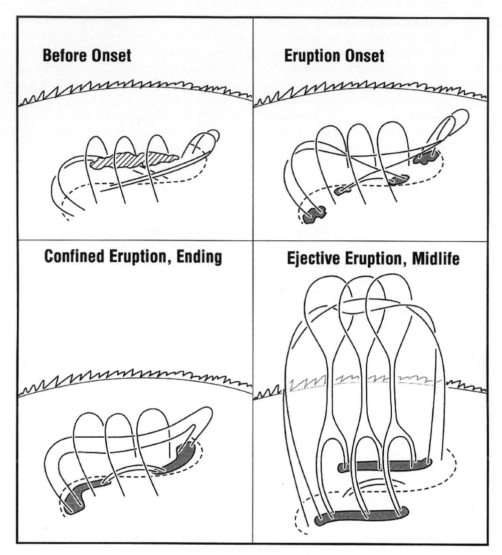

Figure 1. Progression of the three-dimensional configuration of a sheared-core bipolar magnetic field and its internal reconnection in a CME explosion (from *Moore et al.*, 2001). The ragged arc in the background is the chromospheric limb of the Sun. The solid curves are magnetic field lines. The dashed curve is the magnetic polarity dividing line (neutral line). The elongated cross-hatched feature in the first panel is a filament of cool material suspended in the sheared core field. This material is carried in the erupting flux rope (as in Plate 1), but is not shown in the other panels for clarity of the field configuration. The shaded areas are flare ribbons at the feet of reconnected field lines. The third panel shows the conclusion of a confined eruption, a "failed-CME" explosion.

important in causing the explosion, and the possibility that the energy for the explosion comes from below the photosphere during the explosion. We limit our considerations to the force-free situation appropriate for the initiation of fast CMEs that come from strong-field regions and that are driven by the release of magnetic energy stored above the photosphere. Within this scope, our approach is a refinement of the approach of *Klimchuk* (2001). In *Klimchuk* (2001), the basic alternatives for CME initiation are illustrated by cartoons of analogous simple mechanical systems (springs, weights, strings, and pulleys) and thermal explosions

(bombs). We illustrate our alternatives by cartoons that more directly represent the magnetic field in observed CME explosions in strong-field regions.

For each of the three trigger mechanisms, we describe the initiation of the CME explosion with the aid of a sequence of three cartoons of the magnetic field (before eruption onset, just after eruption onset, and after the eruption is well underway) for the case in which the erupting sheared core field is in the central lobe of a quadrupolar field. It is observed that when a sheared core field (traced by a filament) erupts, the eruption often begins with a relatively long-lasting slow-rise phase (having little acceleration and during which the filament noticeably ascends) and then rather abruptly transitions to an explosive fast-rise phase of strong acceleration (e.g., *Kahler et al.*, 1988; *Sterling and Moore*, 2004a,b, 2005). In the quadrupolar case that we depict in our cartoons, there are two places that reconnection can occur: (1) at the magnetic null (X-point) above the erupting central lobe, and (2) between the sheared legs of the erupting core field (below the erupting filament). Reconnection at the null is expected to produce coronal and chromospheric brightening in the two side lobes of the quadrupole, and reconnection below the filament is expected to produce coronal and chromospheric brightening in the sheared core field. The expected brightenings have been observed in many eruptions, and are taken as evidence of the reconnection (e.g., *Aulanier et al.*, 2000; *Gary and Moore*, 2005; *Sterling and Moore*, 2001a,b, 2004a,b, 2005). In the cartoons, we use the label "slow runaway reconnection" for reconnection that occurs during the slow-rise phase of the eruption, and we use the label "explosive reconnection" for reconnection that occurs during the explosive phase of the eruption. These terms are not intended to mean that there is necessarily different physics in the reconnection in the two phases, but only that, as observed brightenings strikingly indicate, the reconnection is much faster and stronger in the explosive phase than in the slow-rise phase of the eruption.

In each of our initiation alternatives, sooner or later, reconnection within the sheared core field creates a flux rope that erupts upward, carrying the filament (if present) within it. Observations indicate that the explosion is driven by the release of magnetic energy from the erupting flux rope via its expansion (*Moore*, 1988). We point out that in order for this to work, that is, in order for there to be a net decrease of the magnetic energy in the exploding flux rope, the shape of the expansion must be sufficiently "fat."

2. CANNONICAL PRE-ERUPTION FIELD CONFIGURATION

In Plate 1, the pre-eruption filament and sheared core field run along the middle neutral line of a triplet of roughly parallel neutral lines. These three neutral lines divide the polarity domains of a quadrupolar arrangement of magnetic flux that spans about a solar radius laterally from southeast to northwest. This magnetic setting of a sheared-core bipole is shown schematically in Plate 2. The style of the depiction is that used by *Antiochos et al.* (1999) in displaying their simulation of CME initiation by breakout reconnection. Whereas their simulation was for a 2-D global quadrupolar field that straddled the equator and circled the Sun, our sketch is for an analogous 3-D quadupolar field of the scale of that in Plate 1 or smaller. The 3-D form of the central sheared-core bipole is

the same as that in Figure 1 (except that the sense of the magnetic shear and twist in the sigmoid is right-hand rather than left-hand).

The situation depicted in Plate 2 is the ideal symmetric one, much more symmetric than in the example observed case in Plate 1, but having the same topology. Nearly symmetric pre-eruption quadrupolar configurations of this topology do occur (e.g., *Sterling & Moore*, 2004a). In other observed quadrupolar situations, the sheared-core bipole is one of the side lobes rather than the central lobe (e.g., *Sterling & Moore*, 2004b). However, the situation in Plate 2 encompasses our three basic alternatives for CME initiation in basically the same way as for any multi-polar configuration in which there is an embedded sheared-core bipole with an external magnetic null somewhere above it. In this canonical configuration, there are two places where reconnection can begin: within the sheared core field as depicted in Figure 1, or at the magnetic null between the envelope of the sheared-core central lobe and the oppositely-directed overarching envelope of the quadrupole. In the following section, for each of our CME-initiation alternatives, we depict the 3-D field configuration of Plate 2 with 2-D cartoons representing this configuration viewed horizontally along the direction of the neutral lines.

3. ERUPTION ONSET

In this section we describe how the three alternatives for CME initiation would occur in our canonical single-bipole and quadrupolar field configurations, and consider their observable signatures.

3.1. Eruption Triggered by Internal Tether-Cutting Reconnection

Our first alternative for CME initiation is depicted by the 2-D cartoons in Plate 3. Here, before eruption onset, the 3-D sigmoidal sheared core field sketched in 3-D in Figure 1 and Plate 2 is represented by the innermost loop of the central lobe of the quadrupole. We have drawn this loop with a symbolic spiral dip in it to indicate that the core field is twisted and sheared along the neutral line and that it is thereby able to suspend cool filament material within it. Before eruption onset, the quadrupole is in overall force-free equilibrium. In particular, the magnetic pressure of the sheared core field is balanced by the combination of its own magnetic tension and the magnetic pressure and tension in the rest of the quadrupole. Also before eruption onset, the field around the magnetic null above the central lobe is sufficiently relaxed that there is no current sheet at the null. But we do have a current sheet low in the sheared core field, between the inner legs of the two elbows of the sigmoid

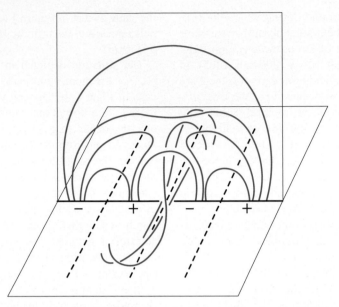

Plate 2. The three-dimensional topology of the quadrupolar magnetic field configuration for CME explosions such as in Plate 1. This is the canonical configuration, which has the exploding sheared core field in the central lobe of the quadrupole. As in Figure 1, the solid curves are field lines and the dashed lines are polarity dividing lines. The polarity of the field on each side of each of the three neutral lines is specified by a plus sign or a minus sign, and matches the polarity arrangement in Plate 1.

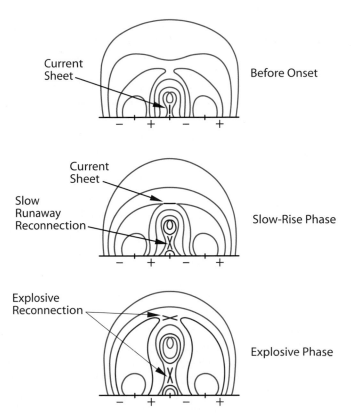

Plate 3. Onset of a quadrupolar CME explosion triggered by runaway internal tether-cutting reconnection. Here and in Plates 4 and 5, the drawings are 2-D renditions of the 3-D configuration in Plate 2 viewed from the front end.

where they shear past each other under the filament (as sketched in Figure 1). We suppose that this current sheet has been formed by the legs being slowly pushed together by photospheric flows, perhaps involving flux cancellation at the neutral line as in the pre-eruption slow tether cutting of *Moore and Roumeliotis* (1992). The current sheet is the contact interface between the two legs. Across this interface the vertical component of the sheared core field reverses direction. As it forms, the current sheet itself may or may not become non-force-free before it is thin enough for reconnection to begin across it.

When the current sheet becomes sufficiently thin, reconnection begins as depicted in the second cartoons in Figure 1 and Plate 3. This is a "runaway tether-cutting" process in the following sense. The reconnected field lines above the reconnection site have had their number of footpoints cut from four to two (Figure 1), so that they are no longer tied to the photosphere under the filament and now run the length of the sigmoid. This releases them to erupt upward, which allows the inner legs of the sigmoid to further collapse together and drive more reconnection. The reconnection and eruption begin slowly (the "slow-rise phase" in Plate 3), but due to this positive feedback the flux rope that is built and released by the reconnection becomes progressively farther out of force balance, and the eruption speed and reconnection rate progressively increase. In this manner, the CME explosion is initiated by the onset and runaway growth of tether-cutting reconnection internal to the sheared core field (*Moore and LaBonte*, 1980; *Sturrock et al.*, 1984; *Sturrock*, 1989; *Moore and Roumeliotis*, 1992).

As Figure 1 and Plate 3 indicate, this alternative for CME initiation is a possibility for either an isolated sheared-core bipole or a quadrupolar field having a sheared-core bipole within it. For a single-bipole CME explosion, the eruption of the unleashed flux rope is strong enough to overcome the restraint of the envelope of the bipole, the envelope field is blown out with the flux rope inside it as in Figure 1, and the legs of the "opened" envelope field re-close by reconnection in the wake of the CME expulsion. Early in the eruption, when the bipole is still closed, the reconnection produces a "four-ribbon" flare. As the reconnection seamlessly progresses from being between the closed sheared magnetic loops of the sigmoid early in the eruption to being between the stretched legs of the extruded envelope field late in the eruption, the flare loops that are formed and heated by the reconnection and issue downward from the rising reconnection site progress from being low and strongly sheared across the neutral line to becoming an increasingly less-sheared, growing arcade rooted in two separating flare ribbons (Figure 1). Such progression from four to two flare ribbons has been observed in ejective filament-eruption flares (e.g., *Moore et al.*, 1995).

No matter how the eruption of the sheared core field is triggered, and whether or not the exploding bipole is embedded in strong surrounding field, flare brightening indicative of internal tether-cutting reconnection is usually observed to begin early in the eruption (e.g., *Moore and LaBonte*, 1980; *Kahler et al.*, 1988; *Moore et al.*, 1984, 1995, 2001; *Sterling and Moore*, 2004a,b, 2005). Because this reconnection produces closed flare loops and begins while the exploding bipole is still closed, only part of the bipole's field is "opened" in producing the CME (as in Figure 1). This is compatible with the Aly-Sturrock theorem: no stressed closed magnetostatic field has enough free energy to explode itself entirely open (*Aly*, 1991; *Sturrock*, 1991).

Is internal tether-cutting reconnection observed to be the trigger of single-bipole CME explosions? It appears to be in cases in which flare brightening (presumably from reconnection) is observed to begin low in the sheared core field in near synchrony with the onset of the slow rise of the filament/flux rope (e.g., *Moore and LaBonte*, 1980; *Moore et al.*, 2001; *Sterling and Moore*, 2005). However, because of finite time resolution and brightness noise thresholds, it remains uncertain whether (1) the internal reconnection begins at the start of the rising motion (which would be evidence for internal tether-cutting as the trigger) or (2) the rising motion begins first. That is, observations of any single-bipole eruption have not yet ruled out the possibility that the sheared core field first begins to erupt via an ideal MHD instability or loss of equilibrium, and that this soon drives the production of a current sheet and reconnection low in the erupting field (e.g., *Rust and Kumar*, 1996; *Rust and LaBonte*, 2005). This possibility is our third alternative for CME initiation. Many magnetostatic and magnetohydrodyamic modeling studies of single-bipole CME onsets based on prescribed evolution of the magnetic flux and/or magnetic shear have found results favoring an ideal MHD trigger (e.g., *Isenberg et al.*, 1993; *Titov & Demoulin*, 1999; *Amari et al.*, 2000; *Linker and Mikic*, 1995; *Chen and Shibata*, 2000; *Roussev et al.*, 2003; *Gibson et al.*, 2004). Triggering of eruption in a single-bipole configuration by spontaneous onset of tether-cutting reconnection in the sheared core field has not yet been demonstrated by MHD modeling (*Antiochos*, 2005). Observations of filament-eruption flares with early brightening at sites of emerging magnetic flux near the neutral line suggest that such flux emergence may be required for internal tether-cutting reconnection to be the trigger in practice (e.g., *Heyvaerts et al.*, 1976; *Moore et al.*, 1984; *Sterling and Moore*, 2005). That is, if the configuration is initially MHD stable, local flux emergence may be required to produce the internal current sheet and start the runaway reconnection.

When the sheared-core bipole is embedded in a quadrupolar field as in Plate 3, once an explosion is triggered in this bipole, it erupts upward, compresses the null, and soon drives reconnection there. Before the explosion is triggered, the null may or may not be so relaxed as to have no current sheet as we have drawn it in Plate 3. Either way, for our present alternative for CME initiation (triggering by internal tether cutting), there is no reconnection at the null until it is further compressed by the eruption from below. The eruption from below is triggered and starts to grow in the same way as in the single-bipole case. Once reconnection at the null begins, it amounts to external tether cutting and is a runaway (explosive) process in the same way as the internal tether-cutting reconnection. Before reconnecting, the field lines of the envelope of the sheared-core bipole and the field lines of the overall envelope of the quadrupole act to tie down the sheared-core bipole. The null-point reconnection cuts these tethers and produces reconnected field lines that sling themselves out the way. Hence, the external reconnection further unleashes the exploding bipole, which makes the explosion stronger and drives the reconnection faster.

In the present scenario (triggering by internal tether cutting), the external reconnection at the null, once it gets started, is the same as the breakout reconnection in the model of *Antiochos et al.* (1999) and in our second alternative for CME initiation, and could equal or exceed the internal reconnection in further unleashing and growing the explosion. The essential difference between the present scenario and the breakout scenario is in the cause and effect relation between the onset of breakout reconnection and the onset of internal tether-cutting reconnection. In the present scenario, internal tether cutting begins first and leads to breakout reconnection, whereas the opposite occurs in the breakout scenario.

In each of our three alternatives for the canonical quadrupolar situation, before and during its eruption, the central lobe has the form of the erupting bipole in Figure 1. The external reconnection strips away some of the outer envelope of the erupting bipole. While the erupting bipole is thereby breaking through the envelope of the quadrupole, internal reconnection is growing the flux rope above it and the flare arcade below it. Once the erupting bipole has broken out of the quadrupole, it explodes on out to become a CME and its legs continue to re-close, further growing the arcade as in Figure 1.

As indicated in Plate 3, reconnection driven at the null produces hot plasma on the reconnected field lines, resulting in bright coronal loops in the side lobes and remote flare brightening at the feet of these loops. If runaway internal tether-cutting reconnection is the trigger of the explosion, and not breakout reconnection, then flare heating in the sheared core field should begin together with the onset of the filament eruption and before the onset of the heating effects of the breakout reconnection. We know of no published example of an observed embedded-bipole CME explosion onset in which this is clearly the case. However the number of published well-observed events that have been studied in this respect is

still small (<~10). While most of these show evidence for breakout reconnection starting early in the eruption, some also show brightening in the core field early in the eruption, and so leave open the possibility that either internal tether cutting or MHD instability is the trigger instead of breakout (e.g., *Sterling and Moore*, 20004a,b). It is also possible that any two or all three of our CME-initiation alternatives play together so closely from the beginning in some events that the trigger should not be assigned to only one alternative. Since internal tether-cutting reconnection begins early in the onsets of what appear to be single-bipole CME explosions (*Moore et al.*, 2001), it will be surprising if no embedded-bipole CME explosions are triggered in the same way, with no initial help from breakout reconnection.

3.2. ERUPTION TRIGGERED BY EXTERNAL TETHER-CUTTING RECONNECTION (BREAKOUT)

Our second alternative for CME initiation is depicted in Plate 4. This is similar to the first alternative in that the explosion of the sheared core field is again triggered by runaway tether-cutting reconnection, but this time the reconnection is at the external null rather than between the crossed legs of the core field. Unlike the other two alternatives, this alternative is not an option for the initiation of single-bipole CME explosions because single bipoles, by definition, have no appreciably strong surrounding fields and hence can have no significant external tether cutting. When the exploding bipole is embedded in an arrangement of strong field giving an external null, as in our canonical quadrupolar configuration, the explosion can be triggered by reconnection at the null as follows.

It is supposed that photospherically-driven evolution of the sheared core field, via shearing flows or further emergence of sheared core field, gradually inflates the middle lobe of the quadrupole without producing a current sheet within the sheared core field. This gradually compresses the field at the null and produces a current sheet there that becomes progressively thinner. Eventually reconnection begins at the current sheet. This reconnection may begin slowly, but because it renders the sheared-core bipole progressively farther out of force balance, it is a runaway process. As the filament/flux rope rises up, the stretched legs of the central-lobe core and envelope field begin to collapse together under it and form a current sheet at their interface (Plate 4, second cartoon). As the external reconnection and the core-field eruption continue to grow, the internal current sheet grows and thins and runaway tether-cutting reconnection is soon driven there as well, which further unleashes the explosion.

The sequence of events described above is that demonstrated by *Antiochos et al.*, (1999) in their numerical MHD 2-D simulation of the breakout scenario for CME initiation

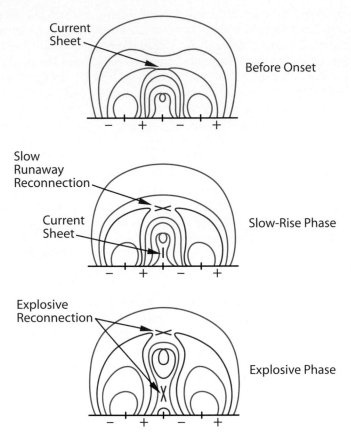

Plate 4. Onset of a quadrupolar CME explosion triggered by breakout reconnection (runaway external tether-cutting reconnection).

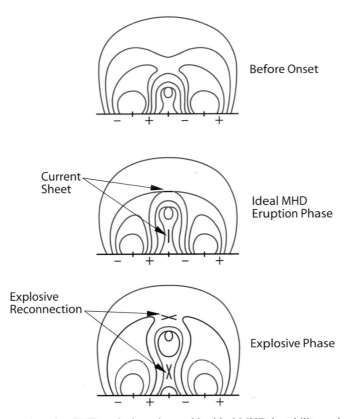

Plate 5. Onset of a quadrupolar CME explosion triggered by ideal MHD instability or loss of equilibrium.

conceived by *Antiochos* (1998). In a CME explosion from a sheared-core bipole embedded in strong surrounding field with an opposing polarity arrangement, there will be a null more or less above the exploding bipole, and, regardless of how the explosion is triggered, the explosion will drive breakout reconnection at the null. Clear evidence for this reconnection has been observed in several such explosions (*Aulanier et al.*, 2000; *Sterling et al.*, 2001a,b; *Sterling and Moore.*, 2001a,b, 2004a,b; *Gary and Moore.*, 2004; *Li et al.*, 2005). Among these, the strongest cases for the explosion being triggered by breakout reconnection, that is, the cases having the most compelling evidence that remote flare brightening occurred before the onset of the filament/flux rope eruption, are the event studied by *Aulanier et al.*, (2000) and the event studied by *Gary and Moore* (2004) and by *Li et al.* (2005). Even in these two cases, the observations do not rule out that the breakout reconnection started in response to an unnoticed ideal MHD convulsion in the sheared-core bipole. In the other cases, either no clear tracer of the onset of the core-field eruption (such as a slowly rising filament) is present or to within the time resolution of the observations the remote brightening begins together with the slow-rise phase of the filament eruption. So, in these other cases the observations are consistent with either that the breakout reconnection triggered the slow-rise phase of the core-field explosion or that the slow-rise onset was triggered by one or both of the other alternatives and initially drove the breakout reconnection.

3.3. Eruption Triggered by Ideal MHD Instability or Loss of Equilibrium

Our third alternative for CME initiation is shown in Plate 5. This alternative differs from the other two in that the first step in the eruption, the trigger, does not involve reconnection. Like the internal reconnection alternative, it is a possibility for the trigger whether or not the sheared-core bipole is embedded in strong surrounding field. In the quadrupolar case in Plate 5, before eruption onset there is no current sheet either at the null or in the sheared legs of the core field, or if there is, neither current sheet is yet thin enough for reconnection to start. It is supposed that photospherically-driven evolution of the flux distribution, perhaps including flux emergence and/or cancellation, gradually evolves the field configuration in the central lobe until the field in the filament/flux rope is so twisted and/or untethered that its force-free magnetostatic equilibrium becomes unstable (e.g., to kinking) or untenable (the core field looses its equilibrium and seeks a new equilibrium by erupting upward). Magnetostatic and magnetohydrodymamic 2-D and 3-D modeling studies of this scenario have shown (at least in single-bipole situations) that such dynamic behavior can be

initiated without resistive dissipation (reconnection) at current sheets (e.g., *Isenberg et al.*, 1993; *Linker and Mikic*, 1995; *Titov and Demoulin*, 1999; *Amari et al.*, 2000; *Chen and Shibata.*, 2000; *Roussev et al.*, 2003; *Gibson et al.*, 2004). In these bipolar models, the instability or loss of equilibrium leads to the formation of a current sheet below or low in the erupting flux rope, between the legs of the envelope field or in the fold of a kink in the flux rope. In the quadrupolar situation in Plate 5, the upward eruption in the central lobe also produces a current sheet at the null. As soon as either current sheet is thin enough, runaway reconnection occurs there. As in the other two alternatives, this tether cutting further unleashes the explosion of the central lobe and helps the eruption open the quadrupole and become a CME (Plate 5, third cartoon). (Again, we emphasize the difference between an ideal MHD trigger and either of the two tether-cutting options for the trigger: only in the ideal MHD option does the eruption begin without any pre-existing current sheets, or if any current sheets are present, without any reconnection at these current sheets. Ideal MHD instability or loss of equilibrium starts the eruptive motion, even though, as the above modeling studies indicate, this soon produces one or more reconnection current sheets and the subsequent reconnection is essential for the production of a full-blown CME.)

The amount of twist displayed by sigmoidal core fields before or during eruption onset suggests that the eruption may be triggered by kink instability, and many erupting filament/flux ropes do appear to kink as they erupt (*Rust and Kumar*, 1996; *Rust and LaBonte*, 2005). While these observations are consistent with triggering by ideal MHD instability, in observed eruption onsets this alternative is difficult to distinguish from the other two, unless the formation of the tether-cutting current sheets is sufficiently delayed from the start of the ideal MHD eruption. As noted in the Introduction, observed filament/flux rope eruptions often begin with a slow-rise phase that persists with little acceleration until the filament has ascended markedly and then rapidly transitions to an explosive fast-rise phase of strong acceleration. During the slow-rise phase, if there is internal tether-cutting reconnection or external breakout reconnection in progress, it is apparently slow and its heating effects (flare brightening) should be relatively weak [as is seen in *Moore et al.* (2001) and in *Sterling and Moore* (2004a)] or perhaps below detection threshold (*Sterling and Moore*, 2003). In the few cases we have studied so far, the brightening becomes obvious before or soon after the start of the fast phase of the eruption. As a rule, the stronger the erupting magnetic field, the faster the eruption develops, and the stronger the flare heating. If reconnection heating is occurring in the slow-rise phase, it should be easier to detect in active-region eruptions than in quiet-region eruptions, given adequate time resolution.

So, if an active-region filament eruption were observed to enter its fast phase with no sign of flare heating anywhere in the active region, that would be strong evidence for initiation by an ideal MHD process. To our knowledge, this has not yet been observed in either a bipolar or multi-polar setting. If MHD-triggered eruptions do occur, they will be easier to detect in bipolar situations than in multi-polar situations if, as seems likely, the MHD eruption takes longer to produce a reconnecting current sheet in the legs of the erupting bipole than at an external null.

4. SHAPE OF EXPANSION OF THE ERUPTING FLUX ROPE

In CME explosions of sigmoidal sheared core fields, whether the pre-eruption sigmoid is in an isolated bipole as in Figure 1 or in an embedded bipole as in Plate 2, and regardless of which of our three alternatives or any combination of these triggers the eruption, observations show that by the time the filament-carrying field has erupted to a height of order the original length of the sigmoid, internal tether-cutting reconnection is in progress between the legs of the erupting bipole (e.g., *Sterling and Moore*, 2004b, 2005). For example, the last Fe XII image in Plate 1 captures the erupting filament at about this height and shows flare ribbons brightening along the neutral line. Before the eruption, it is reasonable to consider the field lines holding the filament to comprise a flux rope, but these field lines may or may not run the length of the sigmoid without rooting in the photosphere. By the time internal reconnection is well underway and producing flare ribbons as in Plate 1, any ties of the filament field to the photosphere under it have apparently been cut, and the filament can be considered to be carried in an erupting flux rope then and after, if not before. It appears to be the expansion of the field in this flux rope that drives the explosion, depleting the magnetic energy in the flux rope (*Moore*, 1988).

For the magnetic energy content of the flux rope to decrease as it erupts, it must expand in cross-sectional area enough faster than it expands in length. That is, the shape of the expansion must be sufficiently "fat." We will demonstrate this requirement by using the simple cylindrical model flux rope shown in Figure 2. For simplicity, and because sigmoids and filaments in active regions are only mildly twisted, having no more than a full turn from end to end, we ignore the twist. We also ignore the curvature of the erupting rope and the plasma pressure in the rope, and take the magnetic field to be uniform inside the model rope and the cross section to be constant along the length. At any instant in the eruption, the magnetic energy content E_{mag} of the model flux rope is the magnetic energy density times the volume:

$$E_{mag} = (B^2/8\pi)Al = (\Phi^2/8\pi)l/A, \qquad (1)$$

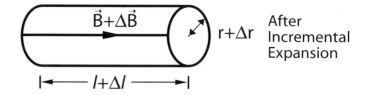

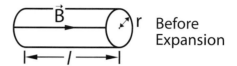

Figure 2. Simple (untwisted) model for the erupting flux rope in a CME explosion.

where B is the field strength, A and l are the cross-sectional area and length of the rope, and Φ is the magnetic flux in the rope ($\Phi = BA$). Due to the high electrical conductivity, the magnetic field in the rope obeys the frozen-in condition and defines the flux rope, and Φ remains constant as the rope expands. So, differentiation of Equation (1) gives

$$(\Delta E_{mag})/E_{mag} = (\Delta l)/l - (\Delta A)/A \qquad (2)$$

for the incremental change in the magnetic energy content from incremental changes in the length and area of the rope as it expands. For the expansion of the flux rope to be driving the explosion, the magnetic energy content of the rope must be decreasing ($\Delta E_{mag} < 0$), which requires

$$(\Delta A)/A > (\Delta l)/l. \qquad (3)$$

Thus, the shape of the expansion, $[(\Delta A)/A]/[(\Delta l)/l]$, must be "fatter" than unity for the explosion to proceed. If the area expands at the same fractional rate as the length or slower, the magnetic energy content of the rope remains constant or increases, and the explosion has to be driven by something other than the expansion of the flux rope. Twist in the flux rope field would lower the above limit on the expansion fatness, because the energy in the component of the field perpendicular to the length of the rope would be decreased by expansion of the rope along its length.

In Plate 1, the expansion of the erupting flux rope (gauged by the erupting filament and filament cavity) appears to be roughly isotropic; that is, the diameter of the filament cavity and the length of the filament appear to increase at roughly the same rate. For isotropic expansion, $r \propto l$, where r is the radius of the flux rope, $A \propto l^2$, and $E_{mag} \propto 1/l$. So, in the CME explosion in Plate 1, the magnetic energy of the flux rope rapidly decreases as it erupts, consistent with the explosion being driven by the magnetic pressure of the flux rope.

5. SUMMARY AND DISCUSSION

From observations of CME explosions from strong sigmoidal sheared core fields traced by filaments, it is nearly certain that the explosion is driven by the magnetic pressure in the erupting filament-carrying flux rope that is unleashed as the explosion is triggered and grows. Before it explodes, the sheared core field is in force-free equilibrium; its magnetic pressure is balanced by its own magnetic tension together with the tension and pressure of surrounding fields. Gradual evolution of the arrangement and amount of magnetic flux in and around the sheared core field can eventually upset this equilibrium and trigger the explosion. When the sheared field is the core of an isolated bipole, there are two different possibilities for the triggering process: (1) runaway tether-cutting reconnection could begin inside the core field, or (2) the equilibrium could become MHD unstable or impossible without an abrupt change in the field configuration (loss of equilibrium). When the sheared-core bipole is embedded in surrounding strong fields arranged in polarity so that there is a magnetic null above the embedded bipole, the explosion could be triggered by either of the above two alternatives for single bipoles or by a third alternaive: breakout reconnection, which is runaway tether-cutting reconnection at the external null.

So far as we know, good evidence for breakout reconnection being the trigger has been found in only a couple of observed quadrupolar eruptions. In many observed eruptions in multi-polar configurations, while it is clear that external tether-cutting reconnection occurs, the observations permit the explosion to be triggered by one or both of the other two alternatives. Likewise, in observed single-bipole eruptions, there is often clear evidence for internal tether cutting reconnection starting early in the eruption, but in no case yet have the observations ruled out that the eruption was initiated by ideal MHD instability or loss of equilibrium. There is no a priori reason that pairs or all three of our alternative mechanisms could not act in concert as the trigger in some situations. For most eruptions, sorting out from observations which of these various possibilities is the trigger apparently requires (at least) high-cadence, high-resolution movies in chromospheric, transition-region, and coronal emission, such as are provided by TRACE and are expected from Solar-B, along with high-cadence, high-resolution magnetograms.

However the eruption is triggered, the erupting sheared core field is or soon becomes a flux rope that carries the erupting filament within it and continues to be built and further unleashed by tether-cutting reconnection below it.

The expanding flux rope can drive the explosion only if its expansion results in a decrease in its magnetic energy content. This occurs if the shape of the expansion is sufficiently "fat." Specifically, the rate of logarithmic increase in the cross-sectional area of the flux rope $[(1/A)dA/dt]$ must be enough faster than the rate of logarithmic increase in the length of the flux rope $[(1/l)dl/dt]$. In observed active-region filament eruptions, the expansion often appears to be roughly isotropic [e.g., see the filament eruptions shown in *Moore* (1987), *Kahler et al.* (1988), or *Sterling and Moore* (2004b, 2005)], which indicates that the magnetic energy of the erupting flux rope rapidly goes into the explosion. If an erupting flux rope were observed to have an expansion shape of only about unity or less $([(1/A)dA/dt]/[(1/l)dl/dt] < {\sim}1)$, this would indicate that the eruption of the flux rope was not driven by the flux rope acting on its surroundings but by the surroundings acting on the flux rope.

Acknowledgments. This work was supported by NASA's Science Mission Directorate through the Solar and Heliospheric Physics Program and the Sun-Solar System Connection Guest Investigator Program of its Earth-Sun System Division. We acknowledge the use of the CME catalogue generated and maintained by the Center for Solar Physics and Space Weather of the Catholic University of America in cooperation with the Naval Research Laboratory and NASA.

REFERENCES

Aly, J.J., How much energy can be stored in a three-dimensional force-free magnetic field?, *Astrophys. J.*, 375, L61-L64, 1991.

Amari, T., J.F. Luciani, Z. Mikic, and J. Linker, A twisted flux rope model for coronal mass ejections and two-ribbon flares, *Astrophys. J.*, 529, L49-L52, 2000.

Antiochos, S.K., The magnetic topology of solar eruptions, *Astrophys. J.*, 502, L181-L184, 1998.

Antiochos, S.K., private communication, 2005.

Antiochos, S.K., C.R. DeVore, and J.A. Klimchuk, A model for solar coronal mass ejections, *Astrophys. J.*, 510, 485-493, 1999.

Aulanier, G., E.E. DeLuca, S.K. Antiochos, R.A. McMullen, and L. Golub, The topology and evolution of the Bastille Day flare, *Astrophys. J.*, 540, 1126-1142, 2000.

Canfield, R.C., H.S. Hudson, and D.E. McKenzie, Sigmoidal morphology and eruptive solar activity, *Geophys. Res. Lett.*, 26, 627-630, 1999.

Chen, P.F., and K. Shibata, An emerging flux trigger mechanism for coronal mass ejections, *Astrophys. J.*, 545, 524-531, 2000.

Cowling, T.G., *Magnetohydrodynamics*, Interscience, New York, 1957.

Falconer, D.A., R.L. Moore, and G.A. Gary, Correlation of the coronal mass ejection productivity of solar active regions with measures of their global nonpotentiality from vector magnetograms: Baseline results, *Astrophys. J.*, 569, 1016-1025, 2002.

Gary, G.A., Plasma beta above a solar active region: rethinking the paradigm, *Sol. Phys.*, 203, 71-86, 2001.

Gary, G.A., and R.L. Moore, Eruption of a multiple-turn helical flux tube in a large flare: Evidence for external and internal reconnection that fits the breakout model of solar magnetic eruptions, *Astrophys. J.*, 611, 545-556, 2004.

Gibson, S.E., Y. Fan, C. Mandrini, G. Fisher, and P. Demoulin, Observational consequences of a magnetic flux rope emerging into the corona, *Astrophys. J.*, 617, 600-613, 2004.

Gopalswamy, N., S. Yashiro, A. Vourlidas, A. Lara, G. Stenborg, M.L. Kaiser, and R.A. Howard, Coronal mass ejections when the Sun went wild, *Bull. Am Astron. Soc.*, 36, 738, 2004.

Heyvaerts, J., E.R. Priest, and D.M. Rust, An emerging flux model for the solar flare phenomenon, *Astrophys. J.*, 216, 123-137, 1977.

Hirayama, T., Theoretical model of flares and prominences. I: Evaporating flare model, *Sol. Phys.*, 34, 323-338, 1974.

Isenberg, P.A., T.G. Forbes, and P. Demoulin, Catastrophic evolution of a force-free flux rope: A model for eruptive flares, *Astrophys. J.*, 417, 368-386, 1993.

Kahler, S.L., R.L. Moore, S.R. Kane, and H. Zirin, Filament eruptions and the impulsive phase of solar flares, *Astrophys. J.*, 328, 824-829, 1988.

Klimchuk, J.A., Theory of coronal mass ejections, in *Space Weather*, edited by P. Song, H.J. Singer, and G.L. Siscoe, pp. 143-157, 2001.

Kopp. R.A., and G.W. Pneuman, Magnetic reconnection in the corona and the loop prominence phenomena, *Sol. Phys.*, 50, 85-98, 1976.

Li, J., D.L. Mickey, and B.J. LaBonte, The X3 flare of 2002 July 15, *Astrophys. J.*, 620, 1092-1100, 2005.

Linker, J.A., and Z. Mikic, Disruption of a helmet streamer by photospheric shear, *Astrophys. J.*, 438, L45-L48, 1995.

MacQueen, R.M., and R.R. Fisher, The kinematics of solar inner coronal transients, *Sol. Phys.*, 89, 89-102, 1983.

Moore, R. *et al.*, The thermal X-ray plasma, in *Solar Flares*, edited by P.A. Sturrock, pp. 341-409, Colorado Associated University Press, Boulder, 1980.

Moore, R., Solar prominence eruption, in *Encyclopedia of Astronomy and Astrophysics*, edited by P. Murdin, pp. 2691-2695, Institute of Physics Publishing, Bristol, 2001.

Moore, R.L., Observed form and action of the magnetic field in flares, *Sol. Phys.*, 113, 121-124, 1987.

Moore, R.L., Evidence that magnetic energy shedding in solar filament eruptions is the drive in accompanying flares and coronal mass ejections, *Astrophys. J.*, 324, 1132-1137, 1988.

Moore, R.L., M.J. Hagyard, and J.M. Davis, Flare research with the NASA/MSFC vector magnetograph: Observed characteristics of sheared magnetic fields that produce flares, *Sol. Phys.*, 113, 347-352, 1987.

Moore, R.L., G.J. Hurford, H.P. Jones, and S.R. Kane, Magnetic changes observed in a solar flare, *Astrophys. J.*, 276, 379-390, 1984.

Moore, R.L., and B.J. LaBonte, The filament eruption in the 3B flare of July 29, 1973: Onset and magnetic field configuration, in *Solar and Interplanetary Dynamics*, edited by M. Dryer and E. Tandberg-Hanssen, pp. 207-210, Reidel, Dordrecht, 1980.

Moore, R.L., T.N. LaRosa, and L.E. Orwig, The Wall of Reconnection-Driven Magnetohydrodynamic Turbulence in a Large Solar Flare, *Astrophys. J.*, 438, 985-996, 1995.

Moore, R.L., and G. Roumeliotis, Triggering of eruptive flares: Destabilization of the preflare magnetic field configuration, in *Eruptive Solar Flares*, edited by Z. Svestka, B.V. Jackson, and M.E. Machado, pp. 69-78, Springer-Verlag, Berlin, 1992.

Moore, R.L., B. Schmieder, D.H. Hathaway, and T.D. Tarbell, 3-D magnetic field configuration late in a large two-ribbon flare, *Sol. Phys.*, 176, 153-169, 1997.

Moore, R.L., A.C. Sterling, H.S. Hudson, and J.R. Lemen, Onset of the magnetic explosion in solar flares and coronal mass ejections, *Astrophys. J.*, 552, 848-883, 2001.

Reames, D.V., Particle acceleration on the Sun and in the heliosphere, *Space Sci. Rev.*, 90, 413-491, 1999.

Reames, D.V., SEPs: Space weather hazard in interplanetary space, in *Space Weather*, edited by P. Song, H.J. Singer, and G.L. Siscoe, pp. 101-107, AGU, Washington, D.C., 2001.

Roussev, I.I., T.G. Forbes, T.I. Gombosi, I.V. Sokolov, D.L. De Zeeuw, and J. Birn, A three-dimensional flux rope model for coronal mass ejections based on a loss of equilibrium, *Astrophys. J.*, 588, L45-L48, 2003.

Rust, D.M., and A. Kumar, Evidence for helically kinked magnetic flux ropes in solar eruptions, *Astrophys. J.*, 464, L199-L202, 1996.

Rust, D.M., and B.J. LaBonte, Observational evidence of the kink instability in solar filament eruptions and sigmoids, *Astrophys. J.*, 622, L69-L72, 2005.

Shibata, K., A unified model of solar flares, in *Observational Plasma Astrophysics: Five years of Yohkoh and Beyond*, edited by T. Watanabe, T. Kosugi, and A.C. Sterling, pp. 187-196, Kluwer, Dordrecht, 1998.

Shibata, K., S. Masuda, M. Shimojo, H. Hara, T. Yokoyama, S. Tsuneta, T. Kosugi, and Y. Ogawara, Hot-plasma ejections associated with compact-loop solar flares, *Astrophys. J.*, 451, L83-L85, 1995.

Sterling, A.C., H.S. Hudson, B.J. Thompson, and D.M. Zarro, Yohkoh SXT and SOHO EIT observations of "Sigmoid-to-arcade" evolution of structures associated with halo CMEs, *Astrophys. J.*, 532, 628-647, 2000.

Sterling, A.C., and R.L. Moore, Internal and external reconnection in a series of homologous solar flares, *J. Geophys. Res.*, 106, 25,227-25,238, 2001a.

Sterling, A.C., and R.L. Moore, EIT crinkles as evidence for the breakout model of solar eruptions, *Astrophys. J.*, 560, 1045-1057, 2001b.

Sterling, A.C., and R.L. Moore, Tether-cutting energetics of a solar quiet-region prominence eruption, *Astrophys. J.*, 599, 1418-1425, 2003.

Sterling, A.C., and R.L. Moore, Evidence for gradual external reconnection before explosive eruption of a solar filament, *Astrophys. J.*, 602, 1024-1036, 2004a.

Sterling, A.C., and R.L. Moore, External and internal reconnection in two filament-carrying magnetic cavity solar eruptions, *Astrophys. J.*, 613, 1221-1232, 2004b.

Sterling, A.C., and R.L. Moore, Slow-rise and fast-rise phases of an erupting solar filament, and flare emission onset, *Astrophys. J.*, 630, 1148-1159, 2005.

Sterling, A.C., R.L. Moore, J. Qiu, and H. Wang, Hα proxies for EIT crinkles: Further evidence for preflare "breakout"-type activity in an ejective solar eruption, *Astrophys. J.*, 561, 1116-1126, 2001a.

Sterling, A.C., R.L. Moore, and B.J. Thompson, EIT and SXT observations of a quiet-region filament eruption: First eruption, then reconnection, *Astrophys. J.*, 566, L219-L222, 2001b.

Sturrock, P.A., The role of eruption in solar flares, *Sol. Phys.*, 121, 387-397, 1989.

Sturrock, P.A., Maximum energy of semi-infinite magnetic field configurations, *Astrophys. J.*, 380, 655-659, 1991.

Sturrock, P.A., P. Kaufman, R.L. Moore, and D.F. Smith, Energy release in solar flares, *Sol. Phys.*, 94, 341-357, 1984.

Titov, V.S., and P. Demoulin, Basic topology of twisted magnetic configurations in solar flares, *Astron. & Astrophys.*, 351, 707-720, 1999.

Ronald L. Moore, Marshall Space Flight Center, XD12, Space Science Branch, Huntsville, AL 35812

Alphonse C. Sterling, Marshall Space Flight Center, XD12, Space Science Branch, Huntsville, AL 35812

Magnetic Helicity and Coronal Mass Ejections

A. Nindos

Section of Astrogeophysics, Physics Department, University of Ioannina, Ioannina GR-45110, Greece

Magnetic helicity is a quantity that descibes the chiral properties of magnetic structures. It has the unique feature that it is probably the only physical quantity which is approximately conserved even in resistive MHD. This makes magnetic helicity an ideal tool for the exploration of the physics of coronal mass ejections (CMEs). CMEs carry away from the Sun twisted magnetic fields and the concept of helicity can be used to monitor the whole history of a CME event from the emergence of twisted magnetic flux from the convective zone to the eruption and propagation of the CME into interplanetary space. I discuss the sources of the helicity shed by CMEs and the role of magnetic helicity in the initiation of CMEs.

1. INTRODUCTION

Coronal mass ejections (CMEs) are large-scale expulsions of magnetized plasma from the Sun. They are observed with a coronagraph above the occulting disk as projections on the plane of the sky. CMEs have attracted significant attention lately because they are the primary cause of the largest and most damaging space weather disturbances (e.g. *Gosling*, 1993). In an average event, 10^{14}-10^{16} gr of plasma is ejected into the heliosphere with speeds ranging from 100 to 2000 km/s (e.g., *Howard et al.*, 1985; *Hundhausen, Burkepile and StCyr*, 1994). The occurence of CMEs shows a strong solar cycle dependence: during solar minimum there is one CME every 2-3 days on average, while during solar maximum they are more frequently observed with several to more than 10 CMEs every day. CMEs sometimes, but not always, are associated with flares. Overall the occurence rate of flares is larger than the occurence rate of CMEs (averaged over a solar cycle, there are 2-3 CMEs per day while the corresponding occurence rate of GOES X-ray flares is 5-6 events per day).

Observations with a coronagraph do not reveal the correlation of CMEs with the configuration and evolution of the underlying solar structures. However, soft X-ray observations obtained with the soft X-ray telescope (SXT) on board the Yohkoh satellite and EUV observations with the EUV Imaging Telescope (EIT) on board the SOHO satellite have eased the tasks of identifying CME counterparts near the solar surface and of following the early development of the eruptions. The low corona CME counterparts (see the article by *Hudson and Cliver*, 2001 for a detailed review) may include coronal waves, EUV and/or soft X-ray "dimmings", and long-duration soft X-ray events that occur after the CME eruption. Also SXT data reveal a relationship between coronal X-ray sigmoids and eruptivity. A flare may also destabilize an adjacent transequatorial loop structure, thus launching a CME. Note, however, that not every CME is accompanied by all the noncoronagraphic signatures listed here.

Coronagraphs observe CMEs via sunlight Thompson-scattered from free electrons and the observed signal is roughly proportional to the integrated mass along the line of sight. Nevertheless, in spite of this, the essential physics of CMEs probably resides in the magnetic field which cannot be observed directly. This argument is based on the fact that the magnetic field dominates the plasma throughout the corona because the coronal plasma $\beta << 1$ (β is the ratio of plasma to magnetic pressure). Therefore the most important forces which determine the eruption initiation and dynamics are magnetic. For the same reason, the energy required to lift the mass of a CME against gravity and accelerate it to the observed velocities is believed to be magnetic (e.g., *Vourlidas et al.*, 2000). The fraction of magnetic energy

Solar Eruptions and Energetic Particles
Geophysical Monograph Series 165
10.1029/165GM08

which is converted to other forms yielding the CME initiation and motion comes from non-potential magnetic field because energy cannot be extracted from a current-free potential field. The magnetic field carried away by CMEs can be obviously helical at times. A typical example is given in Figure 1: the CME's stressed helical structure shows clearly in the bottom row of the figure. Fig. 1 also demonstrates another interesting property of the magnetic field evolution during CMEs: a part of the pre-CME magnetic configuration may contain closed field lines that open up as a result of the eruption (compare for example, the top left and the bottom right frames of fig. 1).

In this article, I focus on the physics of CMEs as a magnetic phenomenon. For such treatment the concept of magnetic helicity is of great importance. Magnetic helicity quantifies the chiral properties of magnetic structures and has some unique features which can help us understand several aspects of the CME physics. Note that I do not attempt to provide an exhaustive review of either CMEs or helicity; the interested reader is refered to the articles in the "Coronal Mass Ejecions" monograph (*Crooker, Joselyn and Feynman*, 1997) and to the articles in the "Magnetic Helicity in Space and Laboratory Plasmas" monograph (*Brown, Canfield, Pevtsov*, 1999), respectively. All I will try to do is to review the aspects of the CME phenomenon that are relevant to the concept of magnetic helicity. In section 2, the definition of magnetic helicity is given. In section 3, I discuss the sources of helicity carried away by CMEs and in section 4, I discuss what magnetic helicity can tell us about CME initiation. I present conclusions and suggestions for future work in section 5.

2. THE CONCEPT OF MAGNETIC HELICITY

2.1. Definition

Magnetic helicity is a quantity which describes nonpotential magnetic fields. For a magnetic field **B** within a volume V, it is defined as:

$$H_m = \int_V \mathbf{A} \cdot \mathbf{B} dV \qquad (1)$$

where **A** is the magnetic vector potential ($\mathbf{B} = \nabla \times \mathbf{A}$). Eq. (1) is physically meaninful only when the magnetic field is fully contained inside V (i.e. at any point of the surface S surrounding V, the field's normal component B_n vanishes); this is so because the vector potential is defined through a gauge transformation ($\mathbf{A}' = \mathbf{A} + \nabla\Phi$), then H_m is gauge-invariant only when $B_n = 0$.

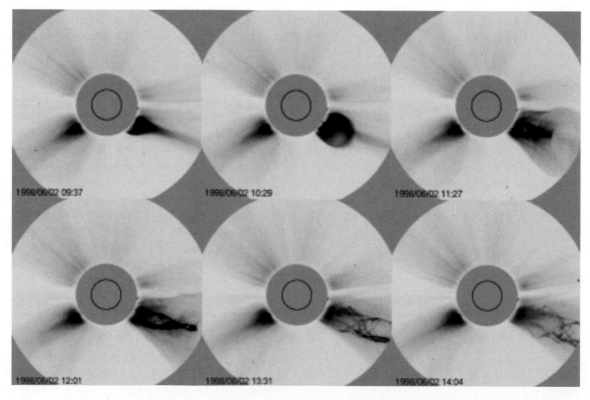

Figure 1. The evolution of a CME observed on June 2, 1998 with LASCO's C2 coronagraph.

In the case of the solar atmosphere, magnetic flux passes through S (especially in the photosphere) and therefore the above condition is not satisfied. However, *Berger and Field* (1984) and *Finn and Antonsen* (1985) have shown that when $B_n \neq 0$ on S, we can define a gauge-invariant relative magnetic helicity H (hereafter refered to as helicity) of **B** with respect to the helicity of a reference field $\mathbf{B_p}$ having the same distribution of normal magnetic flux on the surface S surrounding V:

$$H = \int_V \mathbf{A} \cdot \mathbf{B}dV - \int_V \mathbf{A_p} \cdot \mathbf{B_p}dV \qquad (2)$$

where $\mathbf{A_p}$ is the vector potential of $\mathbf{B_p}$. The quantity H does not depend on the common extension of **B** and $\mathbf{B_p}$ outside V. Being a potential field it is a convinient choice for $\mathbf{B_p}$. If in addition $\nabla \cdot \mathbf{A_p} = 0$ and $(A_p)_n = 0$ on S then the term $\int_V \mathbf{A_p} \cdot \mathbf{B_p}dV$ of eq. (2) vanishes (*Berger*, 1988).

It is worth noting that the "natural" unit of helicity is the square of magnetic flux (Mx^2) and therefore the helicity of a twisted flux tube with N turns and magnetic flux equal to unity is simply N. For more complex magnetic topologies, helicity can be regarded as a measure of the topological complexity of the magnetic field (e.g. linkage and twistedness in the field).

2.2. Flux of Magnetic Helicity

Generally, the amount of helicity within V can change either due to helicity flux crossing S or/and due to dissipation within V. *Berger* (1984) has demonstrated that the helicity dissipation rate is negligible in all processes taking place in the corona, including reconnection and all non-ideal processes. Helicity's dissipation time scale is the global diffusion time scale. As an example, I note that *Berger* (1984) found that the minimum helicity dissipation time in a typical coronal loop is about 10^5 years. Consequently, helicity can be regarded as an (almost) conserved quantity even in resistive MHD. The fact that it is one of the few (probably the only) quantities that it is preserved in the absence or near absence of resistivity makes it a powerful tool that can help us trace the transport of magnetic field from the sub-photospheric layers to the corona, and then its ejection into the interplanetary medium.

For convenience, the temporal evolution of helicity across the photospheric boundary S_p can be separated into a tangential term $dH/dt|_t$ and a normal term $dH/dt|_n$. Then according to *Berger* (1999) we get

$$\left.\frac{dH}{dt}\right|_t = -2\oint (\mathbf{v_t} \cdot \mathbf{A_p})B_n dS_p \qquad (3)$$

$$\left.\frac{dH}{dt}\right|_n = 2\oint (\mathbf{A_p} \cdot \mathbf{B_t})v_n dS_p \qquad (4)$$

where B_t and B_n are the tangential and normal components of the magnetic field on the photosphere and $\mathbf{v_t}$ and $\mathbf{v_n}$ the tangential and normal components of the photospheric plasma velocity. Eq. (3) gives the change of helicity due to horizontal motions on the photospheric surface. Such motions may come either from differential rotation and/or from transient photospheric shearing flows. Eq. (4) gives the change of helicity due to the emergence of twisted field lines that cross the photosphere.

3. THE SOURCES OF HELICITY CARRIED AWAY BY CMES

3.1. CMEs as a Way of Removing Helicity from the Corona

The solar cycle dependence of the CME occurence rate may imply that CMEs are somehow connected to the solar dynamo. Over the years, it has been realized that in several occasions magnetic fields emerging from the solar interior to the photosphere are twisted (e.g., *Leka et al.*, 1996, *Zhang*, 2001). This indicates that a significant fraction of active region's helicity is created by the dynamo and then transported into the corona through the photosphere with the emerging magnetic flux. Since helicity is not destroyed under reconnection (see section 2.2) this process accumulates helicity in the corona. Furthermore, on the global scale, helicity emerges predominantly negative in the northern hemisphere and predominantly positive in the southern hemisphere (*Seehafer*, 1990, *Pevtsov, Canfield and Metcalf*, 1995). And also this hemispheric helicity sign pattern does not change from solar cycle to solar cycle (*Pevtsov, Canfield and Latushko*, 2001). Consequently, on the global scale, mutual cancellation of helicity of opposite signs cannot relieve the Sun from excess accumulated helicity. It has been suggested (e.g., *Low*, 1996) that CMEs, as expulsions of twisted magnetic fields, consist an important process through which accumulated helicity is removed from the corona.

The above statement is supported from solar wind observations. In situ magnetic field measurements show that there are sporadic intervals when magnetic fields are twisted in either a right-handed or left-handed sense, corresponding to positive and negative helicity, respectively. Such disturbances are called magnetic clouds (MCs) when they show enhanced magnetic field whose vector direction rotates gradually through 180°. MCs have been often associated with disappearing filaments on the Sun (e.g., *Marubashi*, 1986; *Rust*, 1999) and have been interpreted as magnetic flux ropes ejected from the corona as CMEs. Consequently, MCs can be regarded as the interplanetary manifestations of some CMEs (note, however, that not all CMEs give MCs at 1 AU). The observations show (*Rust*, 1994, *Bothmer and Schwenn*, 1994) that the number of MCs with positive chirality (i.e. handedness) is roughly equal to the number of MCs with negative chirality. But CMEs that originate from the northern hemisphere tend strongly to have negative chirality in the corresponding MCs, and those from the southern hemisphere have positive chirality in their MCs.

3.2. The Helicity Budget of CME-Productive Active Regions

Once it was realized that CMEs remove helicity from the corona, the obvious question was where that helicity comes from. In section 2.2, I mentioned that on the photospheric surface, helicity may change either due to shearing horizontal motions or/and due to the emergence of twisted field lines that cross the photosphere. For the computation of the helicity budget of an active region, one needs to compute the helicity injection rate from eq. (3) and (4) and also the helicity stored in the corona and the helicity ejected into interplanetary space.

The first process which was studied was differential rotation (*DeVore*, 2000). *Démoulin et al.* (2002a) and *Green et al.* (2002) studied the long-term evolution of the helicity injected by differential rotation into the coronal part of two active regions which were followed from their birth until they decayed. The helicity injection rate from differential rotation was calculated as the sum of the rotation rate of all pairs of elementary fluxes weighted with their magnetic flux. The coronal magnetic field was modeled under the force-free field assumption ($\nabla \times \mathbf{B} = \alpha \mathbf{B}$, e.g. see *Alissandrakis*, 1981). The best value of α, α_{best}, is determined by comparing the computed field lines with the observed soft X-ray or EUV coronal structures. Then the computation of the coronal helicity is relatively straightforward (*Berger*, 1985). The helicity carried away by CMEs cannot be directly computed. In order to overcome this problem one assumes that each CME ejected from the active region under study produces a MC, and that the helicity carried away by each CME is equal to the helicity in the corresponding MC. The helicity per unit of length in a MC can be calculated if we know the axial magnetic field B_0 and the radius R of the cloud's flux rope (*DeVore*, 2000). These parameters can be calculated using a magnetic field model (*Lepping, Burlaga and Jones*, 1990) that assumes that the magnetic field's twist along the interplanetary flux tube is uniform and the field within the MC is described by the first harmonic of a linear force-free field. For the calculation of the total MC helicity, the cloud's flux tube axis ℓ is needed which cannot be obtained observationally; usually a lower limit of $\ell = 0.5$ AU and an upper limit of $\ell = 2$ AU are used. *Démoulin et al.* (2002a) and *Green et al.* (2002) found that the helicity injected by differential rotation is at least about an order of magnitude smaller than the helicity stored in the corona and the helicity carried away by CMEs.

When high-cadence photospheric longitudinal magnetograms are available, the horizontal velocity appearing in eq. (3) can be computed using the local correlation tracking (LCT) technique (*November and Simon*, 1988). Several authors have computed the corresponding helicity injection rate (*Chae*, 2001, *Nindos and Zhang*, 2002; *Moon et al.*, 2002a,b; *Nindos, Zhang and Zhang*, 2003, *Chae et al.*, 2004). However, as *Démoulin and Berger* (2003) have pointed out, with the magnetograms one follows the photospheric intersection of the magnetic flux tubes but not the evolution of the plasma (generally the two velocities are different). Consequently, from the observed magnetic evolution we obtain the flux tube motion and not the plasma motion parallel to the photosphere. If v_t is the tangential component of the photospheric plasma velocity and v_n the velocity perpendicular to the photosphere, the LCT method detects the velocity of the footpoints of the flux tube which is

$$\mathbf{u} = \mathbf{v_t} - \frac{v_n}{B_n}\mathbf{B_t} \qquad (5)$$

If we insert this expression for the velocity in eq. (3) we get the whole helicity flux density (see eqs [3] and [4]). Consequently, one may use the quantity $G = -2u \cdot \mathbf{A_p}B_n$ as a proxy to the helicity flux density. Recently, *Pariat et al.* (2005) have proposed a different proxy to the helicity flux density. Its concept is based on the fact that the helicity injection rate can be understood as the summation of the rotation rate of all pairs of elementary fluxes weighted with their magnetic flux (*Berger*, 1984). The resulting quantity can be derived from observations but such calculations have not been done yet.

A typical sequence of computed velocity vectors and the corresponding maps of G is given in Figure 2. The velocity vectors reflect mainly two large-scale systematic flow patterns: (1) a radial outflow of moving magnetic features from the sunspot's moat and (2) the initially westward and then northwestward motion of the whole sunspot. A purely westward motion of a symmetric sunspot gives no net helicity injection because G has opposite sign contribution in its northern and southern part (see the bipolar structure of G above the sunspot). The bulk of the net helicity injection in AR 8375 comes from deviations from such situation (shearing and twisting components of the motions of the sunspot and moving magnetic features with respect to opposite flux concentrations, polarity deformations and asymmetries in B_n repartition). Despite the spatial incoherence of the computed G maps, *Nindos et al.* (2003) were able to partially reconcile the amount of helicity injected into the corona with the helicity carried away by the CMEs in the 6 six active regions they studied. When they assumed that the length of the MC's flux tube ℓ is determined by the condition for the initiation of the

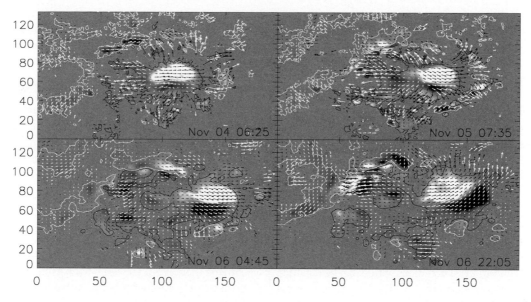

Figure 2. One–hour averages of the computed velocity vectors and the corresponding $G = -2u \cdot \mathbf{A}_p B_n$ (gray–scale images) for active region 8375. The middle of each time interval is indicated in the panels. The maximum arrow length measures velocity of 0.7 km/sec. The white and black contours represent longitudinal magnetic field strengths of -200 and 200 G, respectively. The axis labels denote arcseconds on the solar photosphere (from *Nindos et al.*, 2003).

kink instability in the coronal flux rope or $\ell = 0.5$ AU then the total CME helicities and the total helicities injected into the corona were broadly consistent. However, for $\ell = 2$ AU, the total helicities injected into the corona were a factor of 2.9-4 lower than the total CME helicities.

It is important to note that the LCT method has serious limitations that lead to underestimation of the computed helicities (*Démoulin and Berger*, 2003). Furthermore, the method described above, cannot separate the contribution of the shearing term (eq. 3) from the contribution of the advection term (eq. 4) to the helicity injected into the corona. The theoretical work by *Démoulin et al.* (2002b) who separated the helicity injection into two terms, twist and writhe, indicates that shearing motion is an inefficient way of providing helicity on the active region scale which will be subsequently removed by CMEs. Recently, alternative approaches have been developed which attempt to compute separately both the shearing and advection term using photospheric vector magnetograms. *Kusano et al.* (2002) proposed a method which uses the vertical component of the induction equation. In fact the velocity of flux tubes cannot be deduced fully from the induction equation and part of the velocity is still computed from the LCT method (*Welsch et al.*, 2004). When the transverse component of the magnetic field is available, *Kusano et al.* (2004) developed a method which minimizes the input from the LCT technique. *Longcope* (2004) introduced a technique which demands that the photospheric flow agree with the observed photospheric field evolution

according to the magnetic induction equation. It selects, from all consistent flows, that with the smallest overall flow speed by demanding that it minimize an energy functional. *Georgoulis and LaBonte* (2005) introduce a minimum structure reconstruction technique to infer the velocity field vector. Their analysis simultaneously determines the field-aligned flows and enforces a unique cross-field velocity solution of the induction equation. All these methods have not been tested extensively with data. Also the comparison of their results when applied to the same set of simulated data shows differences (Georgoulis, private communication).

4. MAGNETIC HELICITY AND CME INITIATION

CMEs carry away coronal plasma and open up magnetic field lines which were closed before the eruption. Therefore, the pre-eruption state must contain enough magnetic energy to account for the observed gravitational potential and kinetic energies of the CMEs and also the energy required to open up the magnetic field. The latter imposes a difficulty to all CMEs models that comes from the Aly-Sturrock theorem (*Aly*, 1991, *Sturrock*, 1991). This theorem establishes that in the case of an open field in which all of the field lines extend to infinity, the magnetic energy increases during the eruption. Consequently, one needs a process capable of opening the field and at the same time decrease its energy by the amount required to power the mass motion. However, as *Forbes* (2000) has summarized, there are several ways to bypass this constraint: (1) the CME involves flux from several flux systems so most of the field involved is not opened (*Antiochos, DeVore and Klimchuk*, 1999), (2) the CME involves a detached flux rope (*Low*, 1996), (3) an ideal MHD process may extend field lines such that they do not open all the way to infinity, (4) the pre-CME corona is not force-free and cross-field currents are present (*Wolfson and Dlamini*, 1997), (5) non-ideal processes, specifically reconnection, might be important.

The processes leading to CMEs require helicity accumulation to the magnetic structure that will subsequently erupt. This can be done by shearing the magnetic field or by assuming a pre-existing flux rope. Several mechanisms are capable of shearing the pre-eruption field: emergence of twisted flux, hearing, twisting, and convering motions (the fact that the large amount of shear applied to several models is not observed in the photosphere poses a problem in our understanding of CME initiation). Then the eruption can be regarded as a failure of field confinement. Overall, the current status of our understanding of CME initiation gives a very important role to magnetic helicity and reconnection. In order to appreciate fully the role of helicity in the initiation of CMEs we discuss in the next subsection two important theoretical arguments.

4.1. Woltjer Theorem and Taylor Relaxation

Woltjer (1958) showed that for a perfectly conducting plasma, the total magnetic helicity remains invariant during the evolution of any closed flux system and the minimum energy state of this system corresponds to a linear force-free field configuration. This statement has been proved mathematically and is known as Woltjer theorem. *Taylor* (1974) applied that theorem to the subject of plasma relaxation by proposing that turbulent magnetic reconnection will occur in a plasma of small but finite resistivity to change field topology and transport helicity from one part of the plasma to another, until the field reaches its minimum energy state which is the linear force-free configuration, according to Woltjer theorem. Therefore, while in the limiting case of ideal MHD the helicity of each field line will be an invariant of motion, Taylor's conjecture suggests that only the total helicity of the flux system will be approximately invariant during its evolution to the minimum energy state. A way to understand Taylor's theory is that the magnetic energy should be selectively dissipated faster than the magnetic helicity and that the final state of the linear force-free field might be self-organized.

The relevance of Taylor's theory to solar plasmas is still under debate. While some 3D numerical MHD simulations of relaxation processes in the corona indeed show such a relaxation toward a constant α-field (*Kusano et al.*, 1994), other simulations (*Amari and Luciani*, 2000, *Antiochos and DeVore*, 1999) show a more complicated behavior at variance with Taylor's theory. *Antiochos and DeVore* (1999) have argued that Taylor's theory requirement that complete reconnection occurs, i.e. reconnection continues until magnetic energy reaches its lowest possible state, is wrong because complete reconnection requires the formation of numerous current sheets that do not form easily in the corona due to photospheric line-tying. However, *Nandy et al.* (2003) report the detection of a Taylor-like plasma relaxation process in the corona: their statistical study of highly flare-productive active regions implies that the magnetic field relaxes toward a constant-α configuration.

4.2. The Hydromagnetic Origin of CMEs

Low and Zhang in a series of papers (*Low*, 1996, 1999, *Low and Zhang*, 2002, *Zhang and Low*, 2001, 2003) provided a unified theoretical view of CMEs as the last physical mechanism in the chain of processes that transfer magnetic flux and helicity from the base of the convective zone into the interplanetary medium. In this section I summarize their results with emphasis on the role of helicity in the initiation of CMEs. New active regions emerge with twisted magnetic field and magnetic polarities opposite to that of the pre-existing field. When the new field enters the corona, repeated reconnections between the new and pre-existing field take place. This process simplifies the magnetic topology and the dissipated magnetic energy produces flares. Finally the field achieves its minimum-energy state which is the constant-α force-free configuration required by the Woltjer-Taylor theories. An example of such process is given in Figure 3 taken by *Zhang and Low* (2003). The left panel of fig. 3 shows the initial state of a current-sheet field whose interface (i.e. the current sheet denoted by the thick solid line) separates two axisymmetric force-free fields. The final minimum-energy state of the field resulting from Taylor relaxation is given in the right panel of fig. 3. Note that the relaxation resulted in the formation of a flux rope: the redistribution of helicity, whose total amount is conserved, yields a minimum-energy state where a significant part of the total helicity of the system is contained within the flux rope.

The fate of the flux rope is determined by the efficiency of its confinement by its surrounding anchored field. Flux rope ejection occurs when the magnetic energy it contains is sufficient to drive an outward expansion against the confining field. MHD simulations by *Zhang, Stone and Low*, (2003) explored the conditions required for the failure of confinement of a flux rope. Their work shows that the ratio of the emerging flux to the preexisting flux determines the dynamics of the flux rope: if this ratio is below a critical value, the flux rope remains in a quiescent state in the corona. But if the ratio of the emerging flux to the pre-existing flux is high then the flux rope escapes from the computational box.

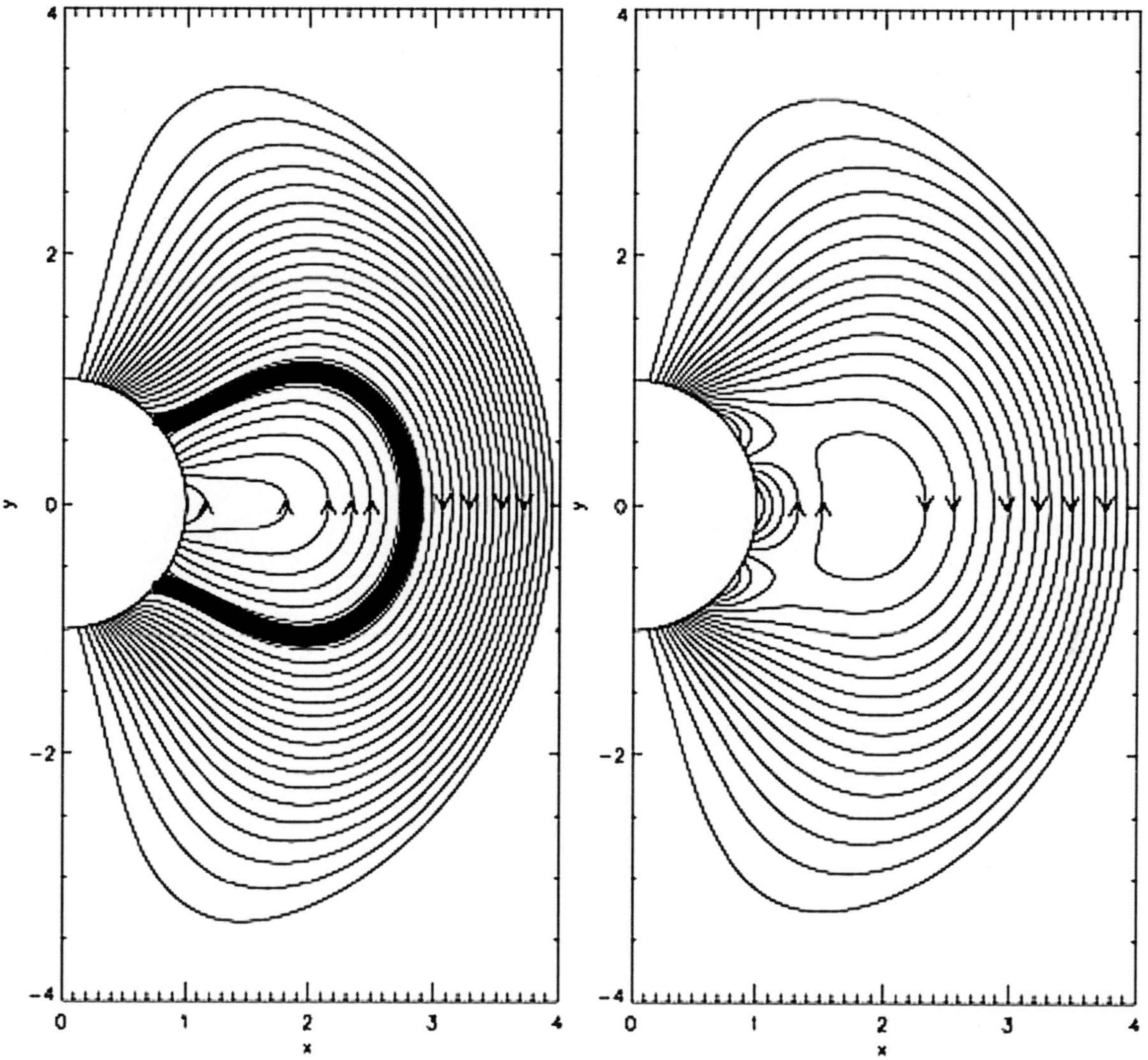

Figure 3. Contours of the stream function of an initial current-sheet magnetic field (left panel) and the final minimum-energy state of the field (right panel). The thick solid line in the current-sheet field shows the location of the current sheet (from *Zhang and Low*, 2003).

We may generalize the above findings as follows. Reconnection liberates magnetic energy by changing the field topology, but not all the magnetic free energy can be so released due to the conservation of the total helicity. Therefore, so long as a magnetic structure is confined to the corona during its relaxation through reconnection and its total helicity is significant, its relaxed state is a twisted field that conserves that total helicity. Trapped with that total helicity is a fraction of the magnetic free energy that cannot be released by reconnection. The removal of the trapped helicity and magnetic energy is achieved by the bodily expulsion of the twisted structure, i.e. the CME (*Low*, 1994, 1996).

The physical view by Low and Zhang presented in this subsection is supported by the work by *Nindos and Andrews*

(2004). The starting point of their work was the study by *Andrews* (2003) who found that approximately 40% of M-class flares between 1996 and 1999 are not associated with CMEs. *Nindos and Andrews*, (2004) modeled under the constant α force-free field approximation, the pre-flare coronal magnetic field of 78 active regions from *Andrews's* (2003) data set. Then from the derived values of α_{best} (see ß3.2 for its definition), they computed the corresponding coronal helicites H_c. Their results appear in Figure 4 and Figure 5. In both figures, both the scatter plots and histograms show clearly that several ARs that give big flares without CMEs tend to have smaller absolute values of α_{best} and H_c than those producing CME-associated flares. They found that this result is statistically significant: their analysis demonstrated

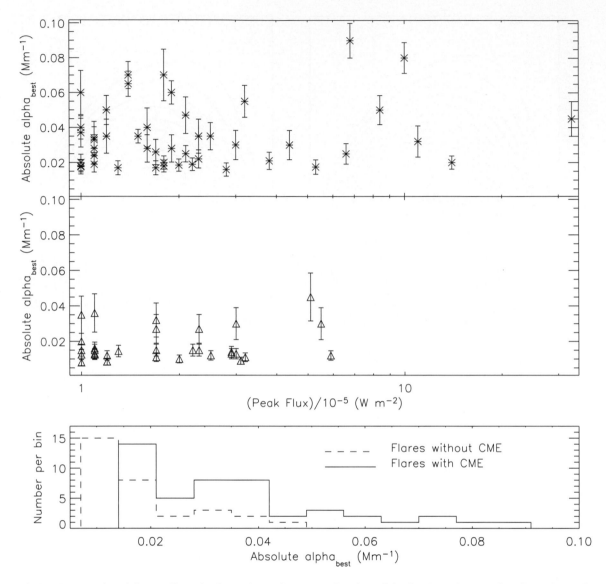

Figure 4. *Top*: Scatter plot of the pre-flare absolute values of α_{best} as a function of the flare's peak X-ray flux for active regions producing CME-associated flares. *Middle*: Same as top panel, but for the active regions producing flares that do not have associated CMEs. *Bottom*: Histograms of the values of α_{best} appearing in the top and middle panels. The solid line is the histogram of α_{best} of the active regions which give CME-associated flares and the dashed line is the histogram of α_{best} of the active regions which produce flares that do not have CMEs (from *Nindos* and *Andrews*, 2004).

that in a statistical sense, both the preflare absolute value of α and the corresponding coronal helicity of the active regions producing CME-associated big flares are larger than the absolute α and coronal helicity of those that do not have associated CMEs. The study by *Nindos and Andrews*, (2004) indicates that the amount of the preflare stored coronal helicity may determine whether a big flare will be associated with a CME or not.

In this section the Low and Zhang unified picture of CMEs as an hydromagnetic phenomenon was summarized. However, the consequences of the helicity conservation and

Taylor relaxation have been applied to oversimplified magnetic configurations. More realistic magnetic topologies may reveal new aspects and/or modify the picture. Furthermore, the analysis of observations may reveal new results and supply the modeling efforts with valuable input.

4.3. Other Approaches

Several other approaches to the problem of CME initiation have appeared in the literature. The purpose of this article is not to present an exhaustive review of all CME models;

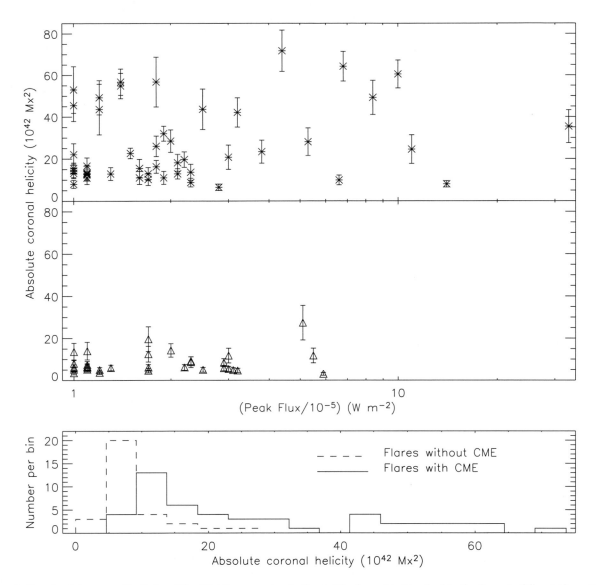

Figure 5. The absolute coronal helicity of the 78 ARs appearing in fig. 4. The format is identical to the format of fig. 4 (from *Nindos and Andrews*, 2004).

the interested reader may refer to the excellent articles by *Forbes* (2000) and *Klimchuk* (2001). Here I focus on the role of helicity in selected representative CME models. Overall, in most models the initiation of a CME comes from the interplay between the accumulation of helicity into the corona and reconnection. In most models the pre-eruption topology is either a sheared arcade (bipolar or quadrupolar configuration) or a flux rope. It has been suggested that in CMEs showing the "three-part structure" (i.e. a bright frontal loop, a dark cavity underneath, and an embedded bright core), cavities correspond to flux ropes seen edge-on. However, not all CMEs have a clear three-part structure. Additional evidence of the flux rope topology in some events is the in

situ observation of a rotating magnetic field pattern in magnetic clouds. In §4.2, we discussed that in the framework of Taylor relaxation, flux ropes may be regarded as the inevitable outcome in a structure with sufficient accumulation of helicity. But the observation of a flux rope in a CME does not necessarily suggest that the flux rope was part of the initial configuration because reconnection may cause flux ropes to form from erupting sheared arcades (*Gosling*, 1990).

Antiochos, DeVore and Klimchuk (1999) developed the "breakout model" which follows the evolution of a quadrupolar magnetic field in which the inner part of the central arcade is sheared by antiparallel footpoint motions near the neutral line. As a result, reconnection between the sheared field and its neighboring field triggers an eruption. Reconnection removes the overlying unsheared field to allow the low-lying sheared field to open up. *MacNeice et al.* (2004) have studied the evolution of helicity under the break-out model (see Plate 1). In their simulation, the model has been driven by a shear flow that injects both free energy and net helicity into the corona. Their results show that the helicity shed by the plasmoid ejection is at least 80% of the total originally injected into the system. They interpreted this result as an indication that although CMEs remove the bulk of the coronal helicity, some fraction remains behind (in the bottom row of Plate 1, this is indicated by the red color appearing in the closed field corona after the plasmoid's ejection). *MacNeice et al.* (2004) suggest that some other mechanism (possibly small-scale diffusion) might be responsible for dissipating the rest of the helicity. Furthermore, *Phillips et al.* (2005) presented simulations of the breakout model where eruption occurs even when no net helicity is injected into the corona. In their simulations the eruption occurs at a fixed magnitude of free energy in the corona, independent of the value of helicity. It would be desirable to check these results against computations of the helicity evolution in observed eruptions that appear to be due to breakout.

Amari et al. (2003a, 2003b) have constructed a set of force-free fields having different magnetic flux and helicity contents and used them as initial conditions by applying converging motions or a diffusion-driven evolution. These processes can trigger eruptive events that may be either confined or unconfined, depending on the value of the initial helicity. *Amari et al.* (2003b) concluded that the helicity cannot be the only parameter controlling the triggering of an ejection: having a large enough helicity seems a necessary condition for an ejection to occur, but not a sufficient one.

Other CME models include the catastrophe model developed by Forbes and colleagues (e.g. *Forbes*, 1990, *Lin, Forbes and Isenberg*, 2001) where the CME comes from a catastrophic loss of equilibrium. In the model by *Wu et al.* (1999) the CME is produced from the interaction between a pre-existing flux rope and the overlying helmet streamer. In the model by *Linker and Mikic* (1995), the CME initiation requires reconnection at the base of a sheared arcade. The model by *Chen* (1996) involves an equilibrium flux rope which contains a low-density hot component and a denser cold component. The eruption of the entire flux rope is triggered by an increase in the azimuthal magnetic flux of the structure.

5. CONCLUSIONS AND FUTURE WORK

Magnetic helicity provides an important tool for the study of CME physics. Its use is justified by two key properties of CMEs: (1) the pre-eruption magnetic topology is non-potential (either a sheared arcade or flux rope) and (2) CMEs carry away twisted magnetic fields. Of course, there are also other physical quantities that describe non-potential fields, for example the α parameter of the force-free field approximation. But helicity is superior because of its unique feature of being conserved even in resistive MHD on time-scales less than the global diffusion time-scale. This makes helicity probably the only physical quantity which can monitor the entire history of the event: from the transfer of magnetic field from the convective zone all the way to the eruption and the escape of the CME into the interplanetary medium. On the other hand, the calculations of helicity are difficult and only recently attempts have been made to measure helicity using solar observations.

Once the importance of helicity was realized, a lot of effort was put on the helicity budget of CME-productive active regions and on the mechanisms which provide the helicity shed by CMEs. Theoretical considerations have demonstrated rigorously that shearing motions (either differential rotation and/or transient flows) on the photospheric surface is an inefficient way of providing helicity on the active region scale. However, computations using high-cadence longitudinal magnetograms give the total helicity flux and cannot separate the shearing term from the advection term. Furthermore, the computation of velocities using the LCT method has serious limitations. Attempts for the computation of the shearing and advection term separately have been made using vector magnetograms. But the algorithms that have been developed have not been applied extensively to observations. Even more serious uncertainties are associated with the computation of the helicity carried away by the CMEs. The use of MCs as proxies for the calculation of CME helicity is only a zero-order approximation. All the above problems contribute to the discrepancies concerning the helicity budget of active regions. At this point, these uncertainties have been cleared up only partially and much work needs to be done on this issue.

The new generation of ground-based and space-born vector magnetograms (e.g., SOLIS and the instruments on board the coming "Solar B" and SDO missions) will provide important new data that should trigger the development of improved algorithms for the computation of the helicity injected into the solar atmosphere. Concerning the helicity of magnetic clouds, we need to constrain more accurately the length of magnetic cloud flux tubes and investigate whether the twist along the interplanetary flux tube is uniform. Hopefully, the coming STEREO mission will provide important observational input to such tasks.

The role of helicity in the initiation of CMEs is a theoretical subject of intense debate. There is a general consensus that for the CME initiation, helicity must be accumulated to the pre-eruption topology. However, diverse opinions exist about

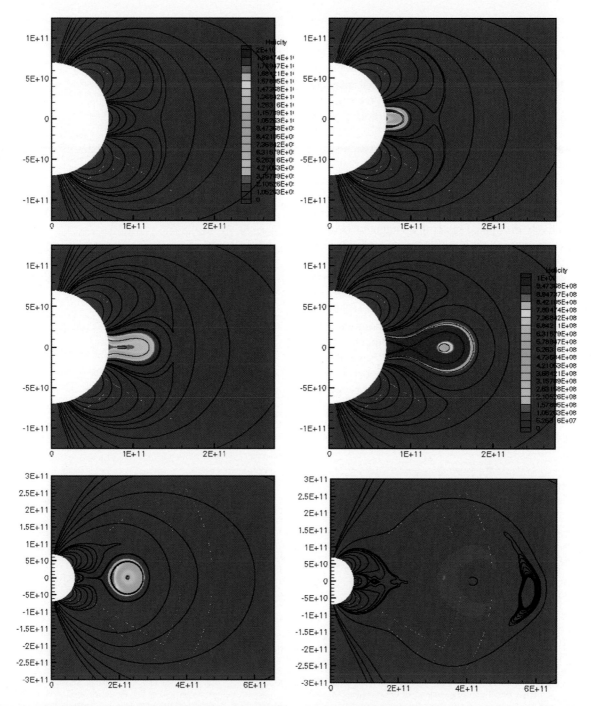

Plate 1. Helicity density evolution under the breakout model. The color map is altered for the last three frames to highlight the lower helicity densities in the expanding flux rope: in the first three frames the color scale represents helicity values from 0 to 2×10^{10} with a step of 1.052×10^9 while in the last three frames the color scale represents helicity values from 0 to 10^9 with a step of 5.263×10^7 (from *MacNeice et al.*, 2004).

the importance of helicity in the overall coronal evolution which leads to a CME. It has been argued that conservation of global helicity plays a minor role in determining the nature and consequences of reconnection in the Sun. The basic argument that supports this point of view is that the coronal magnetic field violates Taylor-theory's requirement that reconnection continues until the magnetic energy reaches its lowest possible state.

Helicity is a crucial element in the other approach which views CMEs as the last chain of a unified process that starts with emergence of twisted magnetic flux from the sub-photospheric layers. This point of view exploits the conservation of total helicity and accepts that magnetic energy should be selectively dissipated faster than magnetic helicity; it takes these arguments to their full logical conclusions and implications. Flux emergence charges the corona with helicity and the excess magnetic energy is removed by reconnection events (i.e. flares). But not all the magnetic energy can be so released due to the conservation of total helicity. The relaxed state is a twisted structure that conserves the total helicity. The CME is the removal of the trapped helicity, enabling the field to reach its minimum-energy state. Recent results from observations give indirect support to that picture. On the theoretical front, the picture should be refined using more realistic magnetic topologies. Observationally, it will be desirable to test whether there is a helicity threshold above which the accumulated helicity results to CME initiation.

Acknowledgments. I thank the organizers of the AGU Chapman Meeting for inviting me to give this review. I also thank M. Georgoulis for his useful comments on the manuscript. Thanks are also due to C.E. Alissandrakis, P. Démoulin, H. Zhang, J. Zhang, S.K. Antiochos, D. Rust and N. Gopalswamy for useful discussions and/or help.

REFERENCES

Alissandrakis, C.E., On the computation of constant alpha force-free magnetic field, *A&A*, 100, 197-200, 1981.

Aly, J.J., How much energy can be stored in a three-dimensional force-free magnetic field?, *ApJ (Lett)*, 375, L61-64, 1991.

Amari, T., and J.F. Luciani, Helicity redistribution during relaxation of astrophysical plasmas, *Phys. Rev. Lett.*, 84, 1196-1199, 2000.

Amari, T., J.F. Luciani, J.J. Aly, Z. Mikic, and J. Linker, Coronal mass ejection: initiation, magnetic helicity, and flux ropes: II. boundary motion-driven evolution, *ApJ*, 585, 1073-1086, 2003a.

Amari, T., J.F. Luciani, J.J. Aly, Z. Mikic, and J. Linker, Coronal mass ejection: initiation, magnetic helicity, and flux ropes: II. Turbulent diffusion-driven evolution, *ApJ*, 595, 1231-1250, 2003b.

Andrews, M.D., A search for CMEs associated with big flares, *Sol. Phys.*, 218, 261-279, 2003.

Antiochos, S.K., and C.R. DeVore, The role of helicity in magnetic reconnection: 3D numerical simulations, in *Physics of Magnetic Flux Ropes, Geophys. Monogr. Ser.*, vol. 111, edited by M.R. Brown, R.C. Canfield, and A.A. Pevtsov, pp. 187-196, AGU, Washington DC, 1999.

Antiochos, S.K., C.R. DeVore, and J.A. Klimchuk, A model for solar coronal mass ejections, *ApJ*, 510, 485-493, 1999.

Berger, M.A., Magnetic helicity in the solar corona, PhD thesis, Harvard University, Cambridge MA., 1984.

Berger, M.A., Structure and stability of constant-alpha force-free fields, *ApJ (Suppl.)*, 59, 433-444, 1985.

Berger, M.A., An energy formula for nonlinear force-free magnetic fields, *A&A*, 201, 355-361, 1988.

Berger, M.A., Magnetic helicity in space physics, in *Physics of Magnetic Flux Ropes, Geophys. Monogr. Ser.*, vol. 111, edited by M.R. Brown, R.C. Canfield, and A.A. Pevtsov, pp. 1-9, AGU, Washington DC, 1999.

Berger, M.A., and G.B. Field, The topological properties of magnetic helicity, *Journal of Fluid Mech.*, 147, 133-148, 1984.

Bothmer, V., and R. Schwenn, Eruptive prominences as sources of magnetic clouds in the solar wind, *Space Sci. Rev.*, 70, 215, 1994.

Brown, M.R., R.C. Canfield, A.A. Pevtsov (Eds.), *Helicity in Space and Laboratory Plasmas,*, 304 pp., *Geophys. Monogr. Ser.*, vol. 111, AGU, Washington DC, 1999.

Chae, J., Observational determination of the rate of magnetic helicity transport through the solar surface via the horizontal motion of field line footpoints, *ApJ*, 560, L95-98, 2001.

Chae, J., Y.-J. Moon, and Y.-D. Park, Determination of magnetic helicity content of soalr active regions from SOHO/MDI magnetograms, *Sol. Phys.*, 223, 39-55, 2004.

Chen, J., Theory of prominence eruption and propagation: interplanetary consequences, *JGR*, 101, 27499-27520, 1996.

Crooker, N., J.A. Joselyn, and J. Feynman (Eds.), *Coronal Mass Ejections,* 299 pp., *Geophys. Monogr. Ser.* vol. 99, AGU, Washington DC, 1997.

Démoulin, P., and M.A. Berger, Magnetic energy and helicity fluxes at the photospheric level, *Sol. Phys.*, 215, 203-215, 2003.

Démoulin, P., C.H. Mandrini, L. van Driel-Gesztelyi, B.J. Thompson, S. Plunkett, Zs. Kovari, G. Aulanier, and A. Young, What is the source of the magnetic helicity shed by CMEs? The long-term helicity budget of AR 7978, *A&A*, 382, 650-665, 2002a.

Démoulin, P., C.H. Mandrini, L. van Driel-Gesztelyi, M.C. Lopez Fuentes, and G. Aulanier, The magnetic helicity injected by shearing motions, *Sol. Phys.*, 207, 87-110, 2002b.

DeVore, C.R., Magnetic helicity generation by solar differential rotation, *ApJ*, 539, 944-953, 2000.

Georgoulis, M.K., and B.J. LaBonte, Reconstruction of inductive velocity field vector from doppler motions and a pair of solar vector magnetograms, *ApJ*, submitted, 2004.

Gosling, J.T., The solar flare myth, *JGR*, 98, 18, 18937-18949, 1993a.

Finn, J.H., and T.M. Antonsen, Magnetic helicity: what is it, and what is it good for?, *Comments on Plasma Phys. and Contr. Fusion*, 9, 111, 1985.

Forbes, T.G., Numerical simulation of a catastrophe model for coronal mass ejections, *JGR*, 95, 11919-11931, 1990.

Forbes, T.G., A review on the genesis of coronal mass ejections, *JGR*, 105, 23153-23165, 2000.

Gosling, J.T., Coronal mass ejections and magnetic flux ropes in interplanetary space, in *Physics of Magnetic Flux Ropes, Geophys. Monogr. Ser.*, vol. 58, edited by C.T. Russell, E.R. Priest, and L.C. Lee, pp. 343-364, AGU, Washington DC, 1990.

Gosling, J.T., The solar flare myth, *JGR*, 98, 18, 18937-18949, 1993.

Green, L.M., M.C. Lopez-Fuentes, C.H. Mandrini, P. Démoulin, L. van Driel-Gesztelyi, and J.L. Culhane, The magnetic helicity budget of a CME-prolific active region, *Sol. Phys.*, 208, 43-68, 2002.

Howard, R.A., N.R. Sheeley, Jr, M.J. Koomen, and D.J. Michels, Coronal mass ejections 1979-1981, *JGR*, 90, 8173-8191, 1985.

Hudson, H.S., and E.W. Cliver, Observing coronal mass ejections without coronagraphs, *JGR*, 106, 25199-25213, 2001.

Hundhausen, A.J., J.T. Burkepile, and O.C. St. Cyr, Speeds of coronal mass ejections: SMM observations from 1980 and 1984-1989, *JGR*, 99,6543-6552.

Klimchuk, J.A., Theory of coronal mass ejections, in *Space Weather, Geophys. Monogr. Ser.*, vol. 125, edited by P. Song, H.J. Singer, and G.L. Siscoe, pp. 143-158, AGU, Washington DC, 2001.

Kusano, K., Y. Suzuki, H. Kubo, T. Miyoshi, and K. Nishikawa,Three-dimensional simulation study of the magnetohydrodynamic relaxation process in the solar corona. 1: Spontaneous generation of Taylor-Heyvaerts-Priest state, *ApJ*, 433, 361-378 1994.

Kusano, K., T. Maeshiro, T. Yokoyama, and T. Sakurai, Measurement of magnetic helicity injection and free energy loading into the colar corona, *ApJ*, 577, 501-512, 2002.

Kusano, K., T. Maeshiro, T. Yokoyama, and T. Sakurai, in ASP Conference Series, in press.

Leka, K.D., R.C. Canfield, A.N. McClymont, and L. van Driel-Gesztelyi, Evidence for current-carrying emerging flux, *ApJ*, 462, 547-560, 1996.

Lepping, R.P., L.F. Burlaga, J.A. Jones, Magnetic field structure of interplanetary magnetic clouds at 1 AU, *JGR*, 95, 11957-11965, 1990.

Lin, J., T.G. Forbes, P.A. Isenberg, Prominence eruptions and coronal mass ejections triggered by newly emerging flux, *JGR*, 106, 25053-25074, 2001.

Linker, J.A., and Z. Mikic, Disruption of a helmet streamer by photospheric shear, *ApJ*, 438, L45-L48, 1995.

Longcope, D.W., Inferring a photospheric velocity field from a sequence of vector magnetograms: the minimum energy fit, *ApJ*, 612, 1181-1192, 2004

Low, B.C., Magnetohydrodynamic processes in the solar corona: flares, coronal mass ejections, and magnetic helicity, *Phys. Plasmas*, 1, 1684-1690, 1994.

Low, B.C., Solar activity and the corona, *Sol. Phys.*, 167, 217-265, 1996.

Low, B.C., and M. Zhang, The hydromagnetic origin of the two dynamical types of solar coronal mass ejections, ApJ, 564, L53-56, 2002.

MacNeice, P., S.K. Antiochos, A. Phililips, D.S. Spicer, C.R. DeVore, and K. Olson, A numerical study of the breakout model for coronal mass ejection initiation, ApJ, 614, 1028-1041.

Marubashi, K., Structure of the interplanetary magnetic clouds and their solar origins, *Adv. Space Res.*, 6, 335-338, 1986.

Moon, Y.-J., J. Chae, H. Wang, G.S. Choe, Y.D. Park, Impulsive variations of the magnetic helicity change rate associated with eruptive flares, ApJ, 580, 528-537, 2002a.

Moon, Y.-J., *et al.*, Flare activity and magnetic helicity injection by photospheric horizontal motions, ApJ, 574, 1066-1073, 2002b.

Nandy, D., M. Hahn, R.C. Canfield, and D.W. Longcope, Detection of a Taylor-like plasma relaxation process in the Sun, ApJ, 597, L73-L76, 2003.

Nindos, A., and H. Zhang, Photospheric motions and coronal mass ejection productivity, ApJ, 573, L133-136, 2002.

Nindos, A., and M.D. Andrews, The association of big flares and coronal mass ejections: what is the role of magnetic helicity?, ApJ, 616, L175-178, 2004.

Nindos, A., J. Zhang, and H. Zhang, The magnetic helicity budget of solar active regions and coronal mass ejections, ApJ, 594, 1033-1048, 2003.

November, L.J., and G.W. Simon, Precise proper-motion measurement of solar granulation, ApJ, 333, 427-442, 1988.

Pariat, E., P. Démoulin, and M.A. Berger, Photospheric flux density of magnetic helicity, *A&A*, in press.

Pevtsov, A.A., R.C. Canfield, and T.R. Metcalf, Latitudinal variation of helicity of photospheric magnetic fields, ApJ, 440, L109-111, 1995.

Pevtsov, A.A., R.C. Canfield, and S.M. Latushko, Hemispheric helicity trend for solar cycle 23, ApJ, 549, L261-263, 2001.

Phillips, A.D., P.J. MacNeice, and S.K. Antiochos, The role of magnetic helicity in coronal mass ejections, ApJ, 624, L129-132, 2005.

Rust, D.M., Spawning and shedding helical magnetic fields in the solar atmosphere, *GRL*, 21, 241-244, 1994.

Rust, D.M., Magnetic helicity in solar filaments and coronal mass ejections, in *Physics of Magnetic Flux Ropes, Geophys. Monogr. Ser.*, vol. 111, edited by M.R. Brown, R.C. Canfield, and A.A. Pevtsov, pp. 221-227, AGU, Washington DC, 1999.

Seehafer, N., Electric current helicity in the solar atmosphere, *Sol. Phys.*, 125, 219-232, 1990.

Sturrock, P.A., Maximum energy of semi-infinite magnetic field configurations, ApJ, 380, 655-659, 1991.

Taylor, J.B., Relaxation of toroidal plasma and generation of reverse magnetic fields, *Phys. Rev. Lett.*, 33, 1139-1141, 1974.

Vourlidas, A., P. Subramanian, K.P. Dere, and R.A. Howard, Large-angle spectrometric coronagraph measurements of the energetics of coronal mass ejections, ApJ, 534, 456-467, 2000.

Welsch, B.T., G.H. Fisher, W.P. Abbett, and S. Regnier, ILCT: recovering photospheric velocities from magnetograms by combining the induction equation with local correlation tracking, ApJ, 610, 1148-1156, 2004.

Wolfson, R., and B. Dlamini, Cross-field currents: An energy source for coronal mass ejections?, ApJ, 483, 961-971, 1997.

Woltjer, L.A., A theorem on force-free magnetic fields, *Proc. Natl. Acad. Sci. USA,*, 44, 489, 1958.

Wu, S.T., W.P. Guo, D.J. Michels, L.F. Burlaga, MHD description of the dynamical relationships between a flux rope, streamer, coronal mass ejection, and magnetic cloud: An analysis of the January 1997 Sun-Earth connection event, JGR JGR JGR, 104, 14789-14802, 1999.

Zhang, H., Formation of current helicity and emerging magnetic flux in solar active regions, *MNRAS*, 326, 57-66, 2001.

Zhang, M., and B.C. Low, Magnetic flux emergence into the solar corona: III. the role of magnetic helicity conservation, ApJ, 584, 479-496, 2003.

Zhang, M., and B.C. Low Magnetic flux emergence into the solar corona. I. its role for the reversal of global coronal magnetic fields, ApJ, 561, 406-419, 2001.

Zhang, M., J.M. Stone, and B.C. Low, Formation of electric current sheets and magnetic flux ropes during magnetic flux emergence, AGU Fall meeting 2003, abstract SH42B-0517, 2003.

A. Nindos, Section of Astrogeophysics, Physics Department, University of Ioannina, Ioannina GR-45110, Greece (anindos@cc.uoi.gr)

Bursting the Solar Bubble: The Flare-Coronal Mass Ejection Relationship

Richard Harrison

*Space Science and Technology Department, Rutherford Appleton Laboratory,
Chilton, Didcot, Oxfordshire OX11 0QX, UK*

One of the fundamental problems in solar physics research today is our understanding of the flare-coronal mass ejection (CME) relationship. On the face of it, a basic investigation of the association between the two most energetic transient phenomena in the solar system would appear to be a relatively simple prospect. However, observational limitations in particular have served to limit our understanding of the flare-CME phenomenon, and thus our understanding of the CME onset mechanisms. Here, we review briefly flare-CME research prior to the Solar and Heliospheric Observatory (SOHO) era, and, through a series of illustrations of recent observations, and reference to several new directions of investigation, summarise our current thinking regarding the flare-CME scenario and the CME onset.

THE PRE-SOHO VIEW

This is something of a personal review, which explores a very basic question: what is the relationship between the two most powerful transient phenomena in the solar system? The first of these is the solar flare, an explosive release of energy in the solar atmosphere, with larger events releasing in excess of 10^{25} J in some tens of minutes. These events are confined to regions where magnetic fields have become stressed to the extent that breakdown occurs and energy is transferred to the local plasmas, through magnetic reconnection. The second event-type is the coronal mass ejection, or CME, where material is ejected into interplanetary space. CME events frequently expel up to 10^{12}–10^{13} kg of matter typically at speeds of several hundreds of km/s. The average CME spans about 45 heliographic degrees but events can span anything from just a few degrees to a full 360 degrees.

In the most basic picture of the flare-CME relationship one may consider the CME to be a coronal response to the flare. If such a scenario were true, we might expect the following:

Solar Eruptions and Energetic Particles
Geophysical Monograph Series 165
10.1029/165GM09

- The flare would sit under the core of the CME;
- The onset of the flare and the CME launch would coincide in time and space;
- The scale of the flare and CME would be consistent with one another;
- We would expect a near one-to-one flare-CME occurrence.

Early CME observations suggested that these criteria are not met. Skylab observations hinted that there was not a one-to-one flare-CME association, even when one took beyond-the-limb flares into account (e.g., *Munro et al.*, 1979). However, from the mid-1980's, with the Solar Maximum Mission (SMM) era, the finding that the four criteria listed above were not consistent with the observations was underlined (see e.g., *Harrison*, 1986). Numerous studies were made using SMM and these were reviewed by *Harrison* (1995), who also presented a statistical study of the flare-CME relationship. This was not the only such study, but it was a thorough review combined with a statistical analysis and, thus, represented the view just prior to the launch of the Solar and Heliospheric Observatory (SOHO). The principal conclusions from that study were:

- There is a strong statistical association between flares and CMEs, but there is NOT a one to one association between flares and CMEs.

- The onset of a CME associated with a flare appears to occur at any time within several tens of minutes of the flare onset; i.e. either can appear to lead the other.
- The scale sizes of CMEs and flares are very different; the average CME spans some 45 degrees whereas active regions are typically much smaller than 10 degrees in size.
- The flare tends to lie anywhere within the span of an associated CME, and may often lie to one side.

Whether or not these conclusions have stood the test of time is the subject of the later sections of this review. However, it was these observations that led *Harrison* (1995, 1996) to conclude:

"The flare and CME are both consequences of the same magnetic 'disease'. They do not cause one another but are closely related. Their characteristics are the results of local conditions, and thus, we may witness a spectrum of flare and CME properties which are apparently unrelated, even resulting in events without the flare or CME component."

Putting this another way, let us consider the source region as a complex magnetic hierarchy. A single driver, such as magnetic shear, may generate a situation where the response in different parts of the hierarchy results in the CME and the flare. The individual characteristics of the flare and CME, and, indeed, the very existence of each, is dependent on the original magnetic configuration. Given a picture like this, it makes no sense to talk of cause and effect; the flare and CME are closely associated but are different manifestations of the same driver.

The Solar and Heliospheric Observatory (SOHO; *Domingo et al.*, 1995) has, in recent years, been providing a wonderful opportunity to investigate the launch and propagation of CMEs with a combination of coronagraph and coronal multi-wavelength imaging and spectroscopy, with no eclipses. The interest in using SOHO in particular is due to (a) the fact that it carried the first spacecraft-borne coronagraph since SMM, and (b) the fact that it carried two spectrometers capable of plasma diagnostics in the corona across a broad range of temperatures.

Many multi-instrument campaigns were designed for the investigation of CMEs well before the launch of SOHO, and many such campaigns have been run using SOHO since 1996. The results from these have been more rewarding as solar activity increased and as experience in using the SOHO instrumentation was gained. Increasingly, such campaigns also involved other spacecraft, such as the Transition Region and Coronal Explorer (TRACE; *Handy et al.*, 1998) and Yohkoh (*Ogawara et al.*, 1991). So, given such resurgence in CME investigations, since the mid to late 1990s, where are we now? What have we learnt in recent years? This will now be addressed through the illustration of a series of observations and a summary of some aspects of recent work.

SAMPLE CME OBSERVATIONS IN THE SOHO ERA

First, we will describe three events which were observed in some detail using SOHO instrumentation, namely the events of 25 July 1999, 23 September 2001, and 24 January 2001. These events demonstrate different aspects of the flare-CME relationship. They are used simply as illustrations of the CME onset work and of the nature of the results we are obtaining.

A large, loop-like CME was detected over the solar north-west limb using the LASCO coronagraph (*Brueckner et al.*, 1995) aboard SOHO on 25 July 1999. An image taken from the LASCO observation sequence is shown in Figure 1. The loop foot-points appeared to extend from near the western equator to within ten degrees of the pole. A back-projection in time of the CME loop, to zero altitude at the limb, suggests an onset time of 13:10 UT. This assumes no acceleration or deceleration under the occulting disc and a source region exactly on the limb. Indeed, an associated flare occurred within an underlying active region located precisely on the north-western limb. The flare was observed using the CDS instrument (*Harrison et al.*, 1995) aboard SOHO, thus allowing extreme ultraviolet (EUV) spectroscopic observations revealing plasmas at specific temperatures in the range

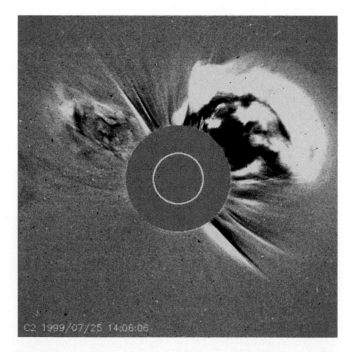

Figure 1. The CME of 25 July 1999 as detected by the LASCO coronagraph aboard the SOHO spacecraft. The image is produced by subtracting a pre-event image. This procedure highlights changes with darker regions and lighter regions denoting areas of decreased and increased density respectively. The occulting disc is 3 solar radii across, centred on Sun-centre.

20,000 K to 2,000,000 K. CDS has a field of view of 4 arcmin × 4 arcmin, which was directed to the active region from 11:03 UT. This field of view included the active region and the flare, and revealed a sequence of events apparently relevant to the CME onset. The CDS data were taken in six selected emission lines, from ions of He, O, Si, Mg and Fe, providing simultaneous views at five different temperatures of ion formation. The images were taken with a raster procedure, using a slit and interlacing exposures and scanning motions to build up the images every 16 minutes. Selected images from the sequence are shown in Plate 1.

From 11:03 UT the CDS images reveal diffuse 1-2 million K emission filling the off-limb corona, made up of weak loops, which are ascending above the active region, at a speed of about 20 km/s. At 12:07 and in the subsequent frames at 12:23 and 12:39 UT, a bright, single loop is detected in the O V 630 Å emission line, characteristic of 250,000 K. This is clearly visible in Plate 1. This loop, and the hotter diffuse loops appear to span the active region. The O V loop is fading but just visible in the 12:55 UT frame. By 13:11 UT the coronal loops have gone; the corona off-limb in the CDS frames has the appearance of being depleted in intensity relative to earlier images, and this is particularly striking in the hottest emission line, the Fe XVI data, characteristic of 2 million K, though it is evident in the million K Mg and Si data. A glance down the right-hand column of the plate shows a clear darkening in the bottom two frames. This decline in off-limb emission took place between the frames at 12:55 UT and 13:11 UT and this process is known commonly as coronal dimming, a phenomenon that is often associated with CMEs. From the frame at 13:28 UT a flare is seen in all CDS wavelengths (temperatures), which develops into a clear arcade which is still evident at the end of the image sequence at 17:13 UT.

We note that although the CME is centred on the flaring active region, its legs do not appear to separate whilst being viewed by the coronagraph, suggesting that the source region is much larger than the flare or active region unless there was considerable lateral expansion restricted to a period very early in the event whilst it is under the area of the occulting disc. This apparent inconsistency between the flare-scale and CME-scale is commonly the case for CME events.

The size of the CDS images means that we are most likely only viewing a small portion of the dimming region which we take to be the CME source region. We note also that the CME event appears to have caused significant disturbances to the adjacent streamers and that there is a smaller CME to the north-east of the Sun's disc. This is demonstrated by the difference image technique shown in Figure 1. This global activity, in fact, led to the event originally being classified as a halo CME. However, although the main loop is clearly associated with the north-west limb, the coincident activity in the north-east may be a sympathetic eruption.

The important features of this event in the context of flare-CME modelling are the following: There is evidence of pre-flare, million K ascending loops above an active region, which span the active region; The corona shows significant dimming at the time of the projected CME onset – suggesting that we have witnessed the ejection of at least part of the original CME material from the corona; The projected CME onset and the onset of the dimming, are pre-flare, and any additional early acceleration of the CME will put its onset even earlier; Based on a simple back-projection, the CME source region appears to be much larger than the flare and the active region, and extends over almost 90 heliographic degrees; There is a suggestion of sympathetic CME activity.

Some would point out that since this event is apparently above the limb we do not see the full picture. What about beyond the limb activity? The active region is just on the north-west limb and we witness a flare. The pre-flare activity and the flare show apparent associations with the CME and if there is additional activity beyond the limb, this would imply that the source region of the CME is even larger. Any flaring from beyond the limb associated with the same active region would have been evident above the limb. Thus, there is no evidence for over the limb activity; there is no evidence to suggest that the CME is not centred on the limb.

The events of 23 September 2001 are, again, revealed by the LASCO and CDS instruments aboard SOHO. This event is illustrated in Plate 2. LASCO data reveal a narrow CME event just above the eastern limb (left hand side, feature labelled 'C'). With a projected velocity of 370 km/s and a projected onset time of 13:13 UT, the CME occupies position angles 55-90 degrees. Other features are identified in the LASCO image (private communication, G. Simnett, 2002) with possible loop-like ejecta labelled 'A' and 'D' and an inflection in the streamer structure, signalling some transient activity, labelled 'B'. As with the July event, the significance of this coincident activity is unknown, but some have argued that such coincident activity signals sympathetic activity or a large-scale response to CME activity. Nevertheless, we concentrate on the eastern limb event.

For this event, the CDS instrument was in a CME onset study mode, making 4 arcminute × 12 arcminute rasters of the western limb, from a mosaic of 4 arcminute × 4 arcminute rasters, with a range of emission lines, at a cadence of 45 minutes. In this case, the field of view is appropriate for CME onset research, and it covers the region under the CME in question, from 12:13 UT to 15:33 UT – which spans the projected CME onset time. Images taken in the 2 million K Fe XVI 360 Å emission line are shown in Plate 2 (left panel) for the times, 12:13 UT, 13:03 UT, 13:53 UT, 14:43 UT and 15:33 UT. The raw images are shown along the top of the left panel of Plate 2. The same images, with the first image subtracted are shown underneath. These reveal significant coronal dimming over the limb in 2 million K plasma. Similarly,

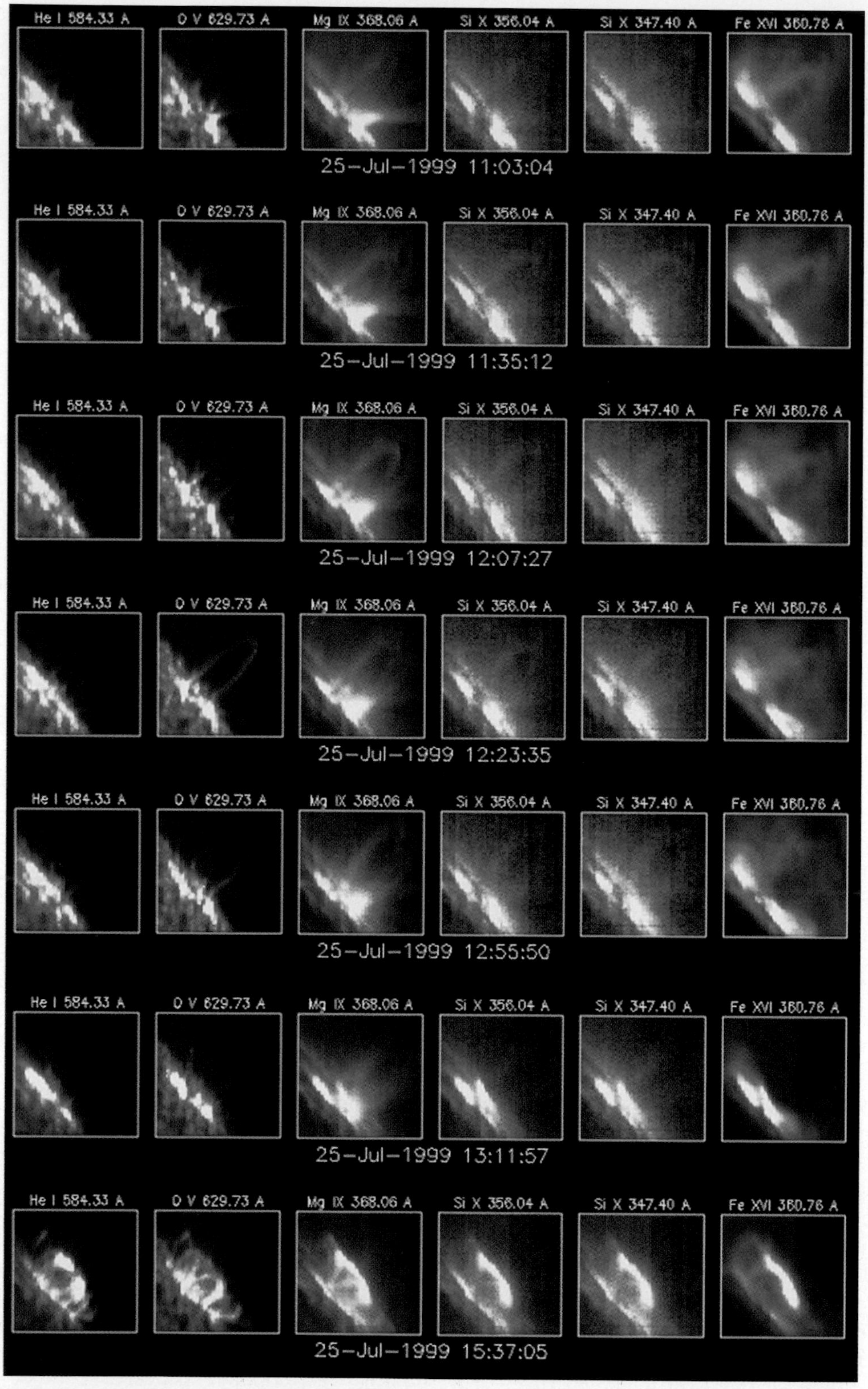

Plate 1. Extreme-UV images of the flare of 25 July 1999, taken using the CDS instrument aboard SOHO. Images in six different emission lines are produced by rastering a series of exposures using a narrow slit. The six images represent different temperature regimes (see text). Such images, taken simultaneously in the six emission lines were taken at a cadence of 16 minutes from 11:03 UT. A selection of the rasters is shown here.

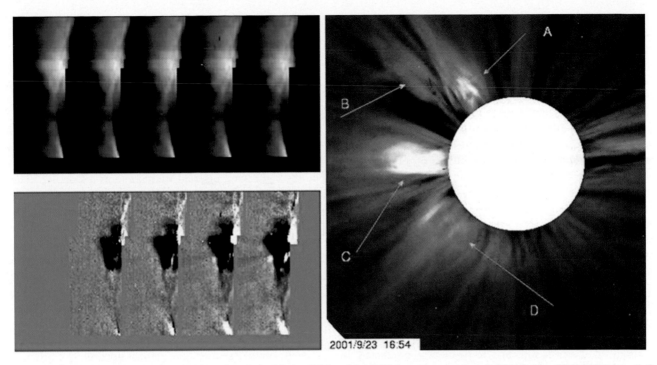

Plate 2. The event of 23 September 2001 as detected with the CDS and LASCO instruments aboard SOHO. The CDS EUV data (left hand panel) show evidence of coronal dimming in a limb region under the CME in 2 million K plasmas. The LASCO data (right hand panel) show the associated CME event above the eastern equator. For details, see text.

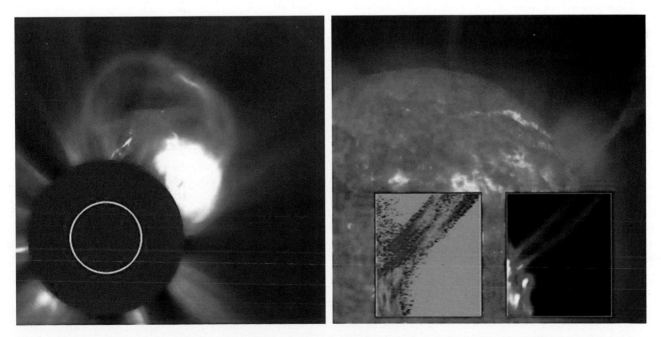

Plate 3. The events of 14 January 2001, as detected using the CDS, LASCO and EIT instruments aboard SOHO. The CME is shown using a LASCO image in the left hand panel. The right hand panel shows an EUV image from the EIT imager on SOHO and the two images inserted into the right hand panel are observations of the southern leg of the CME using CDS. For details, see text.

the Mg IX 368 Å emission line images show clear evidence for dimming on the limb for million K plasma. The dimming region is directly under the ascending CME. The position angle spread of the dimming and the CME are identical and the images show the dimming onset as between 12:13 UT and 13:03 UT. Thus, the dimming onset and the projected CME onset are consistent in time to within the errors on the projections and the uncertainty due to the image cadence. Observations in the He I 584 Å emission line, characteristic of 20,000 K plasma, reveals an activated prominence in the same region, but it does not appear to erupt, and none of the emission line observations show any flare activity. Indeed, there are no reports of flares at any location, in H-alpha or X-rays, in the period 12:00-13:00 UT. The last recorded flare, prior to the projected CME onset, was an H-alpha SN event at S18 E40 from 11:26 to 11:36 UT. This was associated with an X-ray M1.1 event. The intensity of the X-ray event relative to the H-alpha flare suggests that there may have been activity on the limb, with the H-alpha component occulted. However, we do not know the location of such an event, if it occurred, and given the projected CME onset time of 13:13 UT, it seems that this event is too early to be clearly associated. In addition, the following flare was at 13:49 to 13:59 UT, recorded as an SF H-alpha event and C3.4 X-ray event, at S16 E36. Given this, and the CDS data, we conclude that the CME was not flare associated. With the active region on the limb, and clearly close enough to the limb to reveal flare activity in X-rays, there is no evidence for a flare association.

For this event, we conclude the following salient points for CME onset investigation: The CME appears to not be associated with a flare; EUV dimming, consistent with the position angle location and size of the overlying CME, is observed. This suggests that the source region of the CME is detected in the low corona. The dimming onset and CME onset are apparently coincident in time.

The events of 14 January 2001 were observed by several instruments aboard SOHO. A CME was detected by the LASCO instrument from 06:30 UT. An image of the CME is shown in the left panel of Plate 3. It was a large event, with footpoints spanning the north-western limb from the western equator to the north polar region, and this spread was consistent with prominence material seen in EUV images from the SOHO/EIT instrument (*Delaboudiniere et al.*, 1995), shown in the right hand panel of Plate 3. The EIT data are taken in emission from the He II 304 Å line, which is characteristic of 60,000 K. The data show two legs of what appears to be a prominence after eruption, spanning the extent of the CME shown by LASCO data. However, the LASCO images also show very bright, complex prominence material concentrated near the southern leg of the CME in the coronagraph data. The only flare reported within hours of the CME onset is an

H-alpha SF event at N08 W74, from 04:35 UT, lasting about 10 minutes, which was not even listed as an X-ray event in GOES listings.

The CDS instrument was directed at the southern leg of the CME event, as shown in the inserts of the right hand panel of Plate 3. The CDS data show strands of emission from the O V emission line at 630 Å, characteristic of 250,000 K, and showing what appears to be the southern leg of the event seen by EIT. This is centred on the active region which flared before the CME was observed. The CDS data are shown raw, in the right hand inserted frame, and as a Doppler image in the left hand inserted frame, revealing significant red-shifted material at the base of the event. The data from LASCO, EIT and CDS show many interesting features, but the most important ones to be noted here are the following:

This event sequence clearly shows asymmetry and inconsistency of scales; The CME is associated with an active region, and possibly a flare, adjacent to the southern leg of the CME; they do not lie under the centre of the CME; The CME legs appear to remain fixed indicating that the original CME source region is much larger than the size of the associated active region or flare. This is apparently confirmed by consistency with the EIT images as shown in Plate 3.

Having provided some illustrations we now review a few key areas of CME onset research before drawing any general conclusions.

CORONAL DIMMING

Dimming of the corona in X-rays and EUV in likely CME source regions of the low corona has been reported in studies using SOHO and Yohkoh (e.g., *Sterling and Hudson*, 1997; *Gopalswamy and Hanaoka*, 1998; *Zarro et al.*, 1999; *Harrison and Lyons*, 2000, and refs. therein), although the first observation of such an event was reported by *Rust and Hildner* (1976) using Skylab data.

The dimmings have been identified using X-ray and EUV imagers and EUV spectrometers. Analyses of Yohkoh soft X-ray dimming observations have been interpreted as evidence for mass-loss (e.g., *Hudson and Webb*, 1997) but many of the associations with CMEs were implied due to the lack of a flying coronagraph at the time of the observations. It is the author's opinion that in such analyses the detection of a CME using a coronagraph is an essential element.

The use of spectroscopic observations have been particularly useful for confirming that the dimming is due to mass loss rather than temperature changes; as pointed out by *Hudson and Webb* (1997), the broad-band X-ray calculations of mass-loss were limited by lack of knowledge of the differential emission measure. However, the conclusions supported by the imagers and spectroscopic interpretations are consistent, which is encouraging.

Harrison and Lyons (2000) examined the spectroscopic details of a dimming region, using the SOHO CDS instrument under a modest CME event detected by LASCO. The analysis of a number of emission lines confirmed that the mass-loss due to dimming was approximately equal to the mass of the overlying CME, that the position angle and width of the CME were consistent with that of the dimming region, and that the dimming onset and CME onset were near in time. This analysis was for one event over the western limb. A careful consideration of the conditions in the vicinity of the western limb led *Harrison and Lyons* to conclude that the event was not flare-related.

Recognising the value of the spectroscopic data in interpreting the dimming phenomenon, this work was extended considerably by *Harrison et al.* (2003). They studied five dimming events identified under CMEs using EUV spectroscopy. Spectral analyses confirmed that the dimmings were due to mass loss in each case, rather than heating or cooling of plasma, consistent with mass being ejected from the low corona. The mass-loss in each case was calculated using spectral line diagnostic methods and compared to the estimated mass of the overlying CME. The dimming mass and corresponding CME mass for four of the events were within an order of magnitude of each other. Given the errors of the calculated masses from coronagraph data and the EUV observations, the most logical interpretation is that the dimming mass represents mass ejected in the overlying CME. The fifth event showed more mass lost in the dimming event than the overlying CME, i.e. the dimming mass could easily provide the mass detected in the outer corona.

In addition to their basic observations of the five dimming events, *Harrison et al.* (2003), showed that the dimming occurred over many hours in each case, that the dimming onsets and corresponding projected CME onsets, and the relative locations and spreads, were consistent with one another.

The five dimming events reported in the papers referred to, above, have been supplemented by studies of more events, such as those of *Harra and Sterling* (2001), *Howard and Harrison* (2004) and, indeed, the event of 23 September 2001, described above, which has not been published previously.

Why is the dimming so important, and, what relevance has it to the flare-CME relationship? The importance of the dimming question was reviewed in section 1 of *Harrison et al.* (2003). It effectively allows an identification of the source region of a CME and, thus, a thorough analysis of such a source region prior to, during and after the dimming/eruption, could provide major insights to the CME onset processes.

That being the case, can these events help us to understand the relationship of the CME to the flare, where the two events are associated? In the *Harrison et al.* (2003) study, only one of the dimming events was apparently associated with a flare, that of 25 July 1999, as described above, and this serves to stress the fact that many CME events do not appear to be flare related at all. As shown above, the 25 July event occurred on the solar north-western limb where a bright flare arcade was observed. The flare can be seen clearly late in the He I sequence, from the SOHO EUV data, as a bright arcade. As mentioned above, prior to the flare, weak, large EUV arches could be seen, in the million K Mg IX data, gradually ascending above the flare-site at approximately 20 kms^{-1}. The coronal dimming is characterised by the weak loops suddenly disappearing. The dimming is at the projected onset time of the CME and appears to lead the flare onset in time. So, this event was seen as a sudden dimming of the corona, off-limb, above an active region and flare, using spectroscopic EUV observations, but with a limited field of view. The CME itself is clearly associated with the site of the flare but is almost certainly involving a much larger source region. We will discuss the significance of the sequence of events later.

Dimming observations reported by *Sterling and Hudson* (1997) and *Zarro et al.* (1999) also include a flare, that of 7 April 1997. This was a so-called halo CME event, i.e. it was directed towards the Earth and seen as a bright cloud emerging from behind the occulting disc over a wide angular range. X-ray dimming was identified in pockets on either extreme of a sigmoidal shaped active region and flare on the solar disc. These observations were made using the Yohkoh Soft X-ray Telescope. A mass was calculated from the missing intensity, and this was much smaller than the anticipated mass for the overlying CME. However, from the current author's point of view, mass calculations using wide-band imaging can never be as accurate as spectroscopic studies, to identify missing mass detected in X-rays and EUV wavelengths. In addition, dimming of the sort detected off-limb for the 25 July 1999 event, shown above, would probably not have been identified well against the disc, so the full extent of any dimming may not be known accurately at all. Furthermore, the *Sterling and Hudson*, and *Zarro et al.*, dimming 'patches' are not only smaller than the associated active region but are tiny compared to the scales of the associated CME. It is difficult to see how the dimming in such small patches can possibly relate to the eruption of large-scale phenomena. It is the author's belief that any coronal dimming associated with CME source regions ought to be large-scale as illustrated by the 23 September 2001 and 25 July 1999 events, above. In short, we have established that coronal dimming is a CME-related phenomenon, but we must take care in interpreting such observations in low coronal data; observations of small dimming patches within active regions may be irrelevant to the CME process given the scale differences and may simply reflect mass displacement within active region structures.

To understand the CME source region better, the improved observation of the coronal dimming under CMEs would seem to be a very high priority. Current spectroscopic techniques provide small fields of view, e.g., for the 25 July 1999 event, or, if rastered to produce larger fields, are limited in temporal resolution. On the other hand, imagers can provide larger field images, but with very limited temperature and density information. Thus, it is clear that we need a combination of imaging and spectroscopy to understand the dimming phenomenon significantly better. As a natural consequence of such observations, any associated flare activity will be viewed. In determining the flare's association with the dimming phenomenon, we may understand better the flare-CME relationship.

SIGMOIDAL MORPHOLOGY

Using soft X-ray observations from Yohkoh, *Canfield et al.* (1999) announced that active regions with 'S'- shaped or reverse 'S'-shaped morphologies possessed higher probabilities for eruption than other regions. This resulted in a great deal of excitement regarding the potential for CME onset prediction and the modelling of CME onsets, and generated a period of intense study into the nature of sigmoidal shaped active regions. *Canfield et al.*, selected active regions from 2 years of data, which could be well viewed over long periods, 117 in all, and classified each as either sigmoidal or non-sigmoidal. Some 52% of the active regions were classified as sigmoidal. All of the active regions were also classified as eruptive or non-eruptive, as defined by the appearance of X-ray arcades or cusps. From this, it could be seen that 65% of the eruptive active regions were sigmoidal. Stating the figures in another way, 84% of the sigmoidal regions were classed as eruptive and 50% of the non-sigmoidal regions were classed as eruptive. Thus, there is an association between sigmoidal structure and eruptive activity.

Given this, is the existence of the sigmoidal shape so intimately associated with eruptions that the study of such magnetic topologies would provide the key to understanding CMEs? Also, can the observation of sigmoidal active regions be used to provide predictions of CME onsets?

One important point must be understood. The *Canfield et al.*, study did not use coronagraph data. The eruptive nature was defined by the morphology and evolution of soft X-ray structures, and it is not known how many of the eruptive events were related to CMEs.

Hudson et al. (1998) attempted to address this by including coronagraph data in a sigmoidal structure study. They selected SOHO/LASCO halo events in the period December 1996 to May 1997 in order to identify a set of clear on-disc source regions. Of the 11 halo events, they believed that 7 had identifiable X-ray features at consistent locations on the disc; the remaining events were assumed to be directed away from Earth, i.e. the source region is on the far side of the Sun. *Hudson et al.*, claimed that studies of the 7 events showed evidence for a characteristic sigmoidal pattern, evolving into an arcade, presumably in response to the CME eruption. In all cases, the activity was associated with a flare, with X-ray maximum intensities ranging from A1 to M1. *Hudson et al.*, also identified patches of dimming in the X-ray images (see discussion above).

A demonstration of the imaged structure of four of *Hudson et al.*'s events was provided by *Sterling et al.* (2000). The events were shown using X-ray and EUV data from Yohkoh and SOHO. In each case there was a flare which had a sigmoidal configuration. It has to be said that the classification of an active region or flare as sigmoidal or non-sigmoidal is rather subjective and may depend on the line-of-sight and any foreshortening effects. However, for one event of the *Sterling et al.*, and *Hudson et al.*, studies, that of 19 December 1996, the sigmoidal shape of the X-ray features is beyond doubt, and its evolution to an arcade between approximately 15:45 UT and 16:20 UT is quite apparent. For that event, the halo CME was first detected using the LASCO instrument at 16:30 UT and a straight projection to the flare site suggests a CME onset at about 15:35 UT. The GOES flare onset appears to be about 15:10 UT, with a flare duration of about 3 hours. These events appear to be well related and the sigmoid-to-arcade evolution seems to be closely associated with the eruption of a CME in an active region which is flaring.

However, there are two concerns. First, the size of the sigmoidal active region is shown to be about 200 arcseconds. We have no measure of the true angular spread of a CME which is seen as a halo, but for an average CME of 45 heliographic degrees, the source region may be expected to be of size about 750 arcsec, unless there is considerable lateral expansion early in the eruption. We return to this point later, but, if one demands a source size larger than an active region, then most sigmoidal structures which are being associated with CMEs in the research literature must be only part of the source region and this is not noted or considered by the authors. We note that this is the case for the other three events of the *Sterling et al.*, study as well. Second, the 19 December event is the best example of a sigmoid-to-arcade development and this author for one finds it rather difficult to accept a sigmoidal classification for the other three events shown by *Sterling et al.*, The classification is rather subjective. Having said that, the events shown do demonstrate some X-ray restructuring in association with a CME onset and a flare. It should also be pointed out that the study was deliberately restricted to events showing the sigmoid-to-arcade evolution in order to investigate the physics of such events. This is a fair procedure as long as the community is aware that it is pre-selecting the sigmoidal events in this way. The *Sterling et al.*, study concluded with a cartoon model showing the basic idea for a sigmoidal configuration evolving to an arcade during an eruption.

Sterling (2000) reviewed the sigmoid studies and stressed the following three conclusions:

- Pre-eruption sigmoids are more prominent in soft X-rays than in EUV (suggesting that the hotter plasmas are confined to the basic sigmoidal structures);
- Sigmoidal configurations can be found as precursors to over 50% of CMEs;
- Some CMEs have no associated sigmoidal structure and no prominent soft X-ray signature.

One issue which has been hotly debated is the question of line of sight. Given that the most basic coronal structure is the loop and given the complexity of the solar atmosphere, is it not true that a sigmoidal classification could be made in error, quite frequently, simply because of the particular orientation at the time of observation? This has been a common criticism, which was taken up by *Glover et al.* (2000). Using LASCO, EIT and H-alpha data, *Glover et al.*, attempted to reclassify active regions as 'sigmoidal', 'non-sigmoidal' and 'appearing to look sigmoidal due to projection'. They still came to the conclusion that the sigmoidal regions were well associated with CMEs but stressed a need for a quantitative observational definition of the term sigmoidal.

Despite the concerns of line of sight interpretation, the subjectivity of feature definitions etc... the basic association between the sigmoidal active regions and an increased chance of CME occurrence seems to be sound. However, we must note that even the *Canfield et al.*, study noted that 50% of the non-sigmoidal active regions were classed as eruptive. So, one major question remains. Is it the sigmoidal structure itself of the active region that is related to the chance of an eruption and flare onset, or is the sigmoidal shape simply an indicator of a magnetically complex active region which, due to that complexity, has a greater chance of an association with an eruption and a flare? If the latter is true, then modelling the sigmoidal configuration will not necessarily provide answers to the CME and flare onsets or their relationship, i.e. it is not the sigmoidal shape that is important, just the complexity of the magnetic structure.

THE LONG DURATION EVENT

Many researchers talk about the 'LDE' or long duration event, as though it is a distinct class of flare, and the association between LDEs and CMEs is often taken as read. Many authors appear to hold the opinion that CME onsets are usually associated with an LDE. However, the LDE is an event-type with no agreed definition, and it is this author's belief that there is some confusion about LDEs and the association with the CME process. To explore this fully, we must consider the history of LDE studies.

Kreplin et al. (1962) reported an integrated Sun X-ray event lasting 6 hours, associated in time with an eruptive prominence. Of course, with no positional information, the association could only be inferred on the grounds of timing. Some years later, *Sheeley et al.* (1975) published a report on soft X-ray flare profiles for events associated with CMEs. They examined 19 events from integrated-Sun data, choosing events of durations longer than 4.5 hours in order to study events similar to that of *Kreplin et al.*, Some 16 of the events occurred during the operation of the Skylab coronagraph and, of these, 7 were found to be associated with CMEs in time.

Kahler (1977) chose to examine X-ray integrated-Sun flares of duration with e-folding times of longer than 0.87 hours. He concluded that these long-lived events were associated with prominence eruptions and CMEs. *Sheeley et al.* (1983) studied X-ray flares of duration longer than 30 minutes and reported an increased probability of a CME association with flare duration.

These are just a few of the studies of LDEs from the literature but they provide a snapshot of the approach to the subject. The bottom line seems to be that there is a perceived association between long-duration X-ray flare events and CMEs. However, the way the subject has unfolded has created something of a misconception that (a) the LDE is a subclass of flare which is somehow physically different from the majority of flares, and (b) that the LDE and CME are so closely associated that some researchers almost automatically assume that a long duration event must be associated with a CME. These views were most likely not intended or even suggested by the papers listed above. However, to ensure that the view of LDEs in the CME process is considered correctly, we must raise the following points about the studies above:

- Most of the LDE studies used widely differing definitions of an LDE – there was no consistency about duration; indeed we note immediately that statistical studies of flare duration do not show that there is a distinct class of flare defined by duration – the LDE is simply the long-duration tail of a continuum distribution (see e.g., *Phillips*, 1972, *Drake*, 1971, *Pearce and Harrison*, 1988);
- Most LDE studies have been influenced by the previous studies and even filter out the shorter duration flares to perform the analysis of flare-CME associations; for example, the 30 minute duration filter used by *Sheeley et al.* (1983) would have excluded 28% of the flares used in the statistical flare-CME study of *Harrison* (1995) and, as a consequence, would have rejected 19% of the flare events associated with a CME onset;
- Most early LDE studies used integrated X-ray signatures, i.e. no spatial information was available and the only association was in time;

- The statistical study of *Harrison* (1995) showed that *longer duration flares* have a greater chance of association with a CME, but also show that a CME can be associated with a flare of any duration, or can be associated with no flare at all. This is supported by many analyses made over the years, e.g., *Harrison et al.* (1985), and *Nitta and Hudson* (2001).

One intriguing point to note is that the *Sheeley et al.*, results actually showed that there was an equal chance of an LDE being associated with a CME or an H-alpha flare. The logic used by some would suggest that all H-alpha flares should be accompanied by an LDE!

It should be stressed that any misinterpretations are not necessarily driven by the papers mentioned above, but more by the way people have used their results. In a sense, all of the papers are consistent with one point – the last bullet of the list above.

In conclusion, if we are to use the term LDE, we must (a) accept that it refers to one part of the flare-duration spectrum, and not suggest that it is a different event-type without clear evidence to that effect, (b) we must identify a definition for the LDE which can be commonly agreed, and (c) we must not forget that CMEs can be associated with flares of any duration or even no flares. We must avoid an observational bias.

OTHER LANDMARK STUDIES

To complete this review, before drawing any general conclusions, we now consider a number of papers which ought to be mentioned either because they bring out specific points which are not related to the issues already discussed, or because they encompass many of the major issues in a single study.

First we consider the work of *Zhang et al.* (2001, 2004) which attempts to bring together a unique set of multi-wavelength observations for a set of CMEs. They use data from LASCO and EIT on SOHO, as well as the X-ray data from the GOES spacecraft. The observations are made of CMEs originating near the solar limb, but the key feature of the study is the use of the LASCO C1 coronagraph. C1 is the innermost portion of the LASCO coronagraph, which was lost in mid-1998. It allowed observations down to 1.1 solar radii from Sun-centre. By far the majority of LASCO data-sets used for onset studies are restricted to an inner limit of 2 solar radii (the C2 inner edge) because the loss of C1 occurred prior to the build up of CME activity for the last maximum. There are few events where we have clear CME observations down to 1.1 solar radii combined with 'surface' observations in X-rays or/and the EUV, so the study of *Zhang et al.*, is particularly important both to investigate the lowest altitude activity and to test our understanding of CME onsets and the CME-flare question.

The time-altitude curves for three of the four events reported by *Zhang et al.* (2001) show a three-phase ascent (see Figure 2). First, the 'initiation phase', displays a gradual expansion in the CME loops at speeds of under 80 kms^{-1}. The events under study show initiation phases lasting from half an hour to 2 hours. This phase is followed by an 'impulsive acceleration' phase during which there is a sudden increase in the CME speed of ascent, at altitudes in the range 1.3 to 4.6 solar radii above Sun-centre. For the events in question, this coincides in time with the onset and rapid rise in intensity of an associated flare. This is followed by the 'propagation phase', where the events ascend at constant or near-constant velocities. It should be noted that the fourth event appears to display the impulsive acceleration phase and propagation phase only.

Zhang et al.'s time-altitude profiles assume that the bright C1 loops are identical to those seen in the outer coronagraphs. C1 is an emission line coronagraph, utilising the 2 million K Fe XIV 5303 Å emission line, whereas the outer coronagraphs are white-light coronagraphs sensitive to Thomson scattered photospheric light. Thus, C1 is sensitive to the square of the density of 2 million K plasma, whereas the other coronagraphs are sensitive to electron density alone. Care must be taken in the projection of structures from C1 to the outer coronagraphs.

Zhang et al. (2001) show the flare-sites relative to the C1 CME images (e.g., Figure 2), which demonstrate a flare-CME asymmetry, which is consistent with aspects of the pre-SOHO scenario discussed above. However, they also appear to show evidence for considerable lateral expansion of the CMEs studied at the lowest altitudes. One very important question is, does this expansion mean that the apparent inconsistency in the scale-sizes of the CME and flare sites is simply due to very early lateral expansion of CMEs? This would certainly provide an argument against one of the stated pre-SOHO conclusions. Indeed, *Zhang et al.* (2004) show detailed sequences of some events with coronagraph data from all three of the LASCO coronagraphs. Their interpretation of the 11 June 1998 event sequence (the event shown in Figure 2) is that the CME expanded from large loops enveloping a single active region. The C1 images show a CME structure with footpoints separated by 20 degrees. This in itself is large compared to a flare but is not a typical position angle width for a CME detected higher in the corona. The initial C1 images show a loop which is consistent with EUV dimming detected using the EIT instrument on SOHO. The C1 image at 10:11 appears to show some expansion perhaps by 50% since the first frame of Figure 2 (09:33 UT). However, whereas the C1 CME is centred north of the eastern equator, the CME event reports from the LASCO team describe a subsequent C2 CME of width 156 degrees, centred on position angle 127 degrees, some 50 degrees to the south.

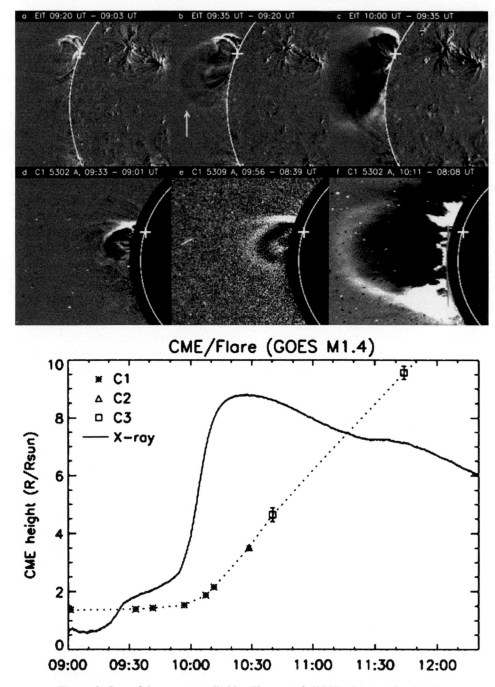

Figure 2. One of the events studied by *Zhang et al.* (2001), See text for details.

So, whereas there is clear evidence for some early expansion in the CME, the required non-radial motion is extreme. Similarly, the sequence from 19 October 1997 also shown by *Zhang et al.* (2004) shows an active region sized event apparently expanding considerably in the early stages, with the lateral expansion extending the event to the south in much the same fashion as the first event.

These apparent lateral expansion observations merit serious consideration. It suggests that we cannot rely completely on the apparent CME source regions provided by back-projection of the higher altitude coronagraph data. Again, observations above a solar radius show little sign of lateral expansion. However, we must take care not to ignore past observations. Evidence from events such as that of Plate 3 clearly involve an eruption from a large structure, much larger than an active region, yet the concentration of mass is from one side, over the active region. This kind of asymmetry is not a new observation by any means. That event would most likely look similar to the 11 June event if C1 had been working yet that would have been misleading. Also, many observations of dimming listed above showed consistency between the projected CME width and the dimming region, without C1 data, and that suggests little expansion in those cases. There are also a number of examples of CMEs which have been attributed to the eruption of active region interconnecting loops (e.g., *Harrison*, 1985; *Khan and Hudson*, 2000).

It is the author's view that the *Zhang et al.* (2004) report may be interpreted in two ways. They may well be evidence for considerable lateral, asymmetrical expansion, limited to the lowest altitudes. However, again, we note that C1 is an emission line coronagraph, sensitive to density and temperature, whereas the outer coronagraphs are white light coronagraphs sensitive to only density. *Zhang et al.*'s (2004) image sequence of the 19 October event shows a dark loop in C1 data north of the equator. That loop can be seen expanding outwards. However, one can track the expansion of that loop and still actually identify additional features to the south, especially from the frame at 04:26 UT. Indeed, the EIT image they show identifies some EUV activity south of the equator on the limb and this is the southern extent of the CME when it has fully developed. So, are these images simply telling us that the greatest density event at the C1 temperatures started in one active region but that the full event is much larger? This would fit with the event of Plate 3. Very similar arguments can be made for the 11 June event. So, despite these results, this author would stress that the jury is out. Given this, this author believes that this aspect of CME research is a major issue for future investigation.

Zhang et al.'s conclusions are the following: CMEs and flares are two different manifestations of the same magnetic process; they have a strongly coupled relationship but not a cause and effect one. This, they point out is consistent with the interpretation of *Harrison* (1995). The results reject the scenario where CMEs are driven by flare-induced coronal responses, because the initiation phase of the CME is clearly pre-flare. They suggest that the initiation phase may be caused by the destabilization and quasi-static evolution of a large-scale coronal magnetic structure. They argue that if a critical point is reached, violent magnetic activity may be triggered that induces the magnetic force to drive the CME, whilst simultaneously triggering the flare. They also believe that the lateral expansion they observe could account for the inconsistency between the flare and CME scales.

The event of 25 July 1999, discussed above shows very similar characteristics to those of *Zhang et al.*, However, in

the absence of available C1 data, the lowest altitude loops, in the initiation phase – as labelled by *Zhang et al.*, – are detected to be ascending slowly using observations in the 1 million K Mg IX 368 Å emission line detected by the SOHO CDS instrument. These slowly ascending loops were detected prior to the flare, ascending at a speed of approximately 20 kms^{-1}. The CME was later measured to be achieving speeds of up to 1118 kms^{-1}. As discussed above, at the time of CME onset the corona above the active region showed significant EUV dimming, which was effectively due to the sudden disappearance from the CDS field of the ascending loops, and this was followed by a flare. This is consistent with *Zhang et al.*'s impulsive acceleration and propagation phases.

The work of *Khan and Hudson* (2000) is also discussed here because it is one of the few studies that considers a large-scale nature of CME source regions. They used Yohkoh X-ray images in conjunction with LASCO data and showed three events where active region interconnecting loops disappeared in association with a flare event in one active region and with an overlying CME. Their interpretation of the morphology and timing was that flare-generated shocks may destabilize an associated active region interconnecting loop, which becomes the source of the CME. The asymmetry is clearly akin to that suggested by the pre-SOHO studies, and often seen in more recent studies. The timing, however, certainly puts the flare onset first; and it is described as the driver of the event sequence - at least for the events in question. Given the fact that many CMEs are not flare related and that some flares certainly start well after the onset of the initial CME ascent, this cannot be a general picture. However, the fact that *Khan and Hudson* demonstrate a link between a coronal feature much larger than an active region, and a CME is important. Again, it is interesting to note that SMM studies had suggested that active region interconnecting loops were a most likely the source of CMEs (*Harrison*, 1986) and that there was evidence for ascending X-ray fronts (*Harrison et al.*, 1985) – these are both conclusions now made by *Khan and Hudson*. The one major difference between the work of *Khan and Hudson* and many other reports is that they believe that the flare initiates the CME onset.

Plunkett et al. (2002) discuss the EUV signatures of CMEs, specifically referring to the EIT instrument on SOHO (*Delaboudiniere et al.*, 1995). They identify the common EUV signatures such as dimming, the formation of bright post-eruption arcades or loops, erupting filaments, the expansion of pre-existing loops or arcades, and large-scale wave disturbances. The nature of dimming associated with CME source regions is discussed above. The EIT observations of filament eruptions, using the He II 304 Å filtered images with typical temperatures in the 60,000 K range, can be identified clearly. However, the overlying CME structure is not so easy to identify and many CMEs do not include filament or prominence eruptions. With regard to loop expansions, post-eruption loops etc… identified in the EUV, as *Plunkett et al.*, note, the precise relationship between the various EUV signatures to the structures observed in white light by the coronagraphs is very unclear at present. Great care must be taken in comparing the two wavelength regions. The EUV observations are restricted to the detection of plasma at specific temperatures whereas the white-light observations are sensitive to density alone. A careful comparison of LASCO CME and EIT eruptive activity has been made by *Delannée et al.* (2000). They did not conduct a study of the CME-flare relationship specifically, though 9 of their events were flare-related. Their principal aim was to identify a set of EUV eruptive events and look for LASCO counterparts. In this, they were somewhat successful, but the precise CME-flare relationship study required that the projected onsets and relative locations be investigated further. However, it is important to note that of 17 EUV eruptions, 13 were found to be closely associated with CMEs.

As *Plunkett et al.*, conclude, 'some CMEs have no EUV signature, even when there is good reason to believe the source region is on the visible side of the solar disc, and other EUV eruptions do not correspond well with white-light CMEs'. This stresses that a CME must never be identified by anything other than a white-light coronagraph, at least until we are clear about the relationship between CMEs and other activity. This is stressed because it is becoming more common to see any outward expanding feature labelled as a CME even without supporting coronagraph data. However, studies such as that by *Delanée et al.*, do suggest that we are making good progress.

The identification of large-scale wave disturbances, or EIT waves, is an important one to consider. These are rapidly travelling coronal disturbances, identified primarily using the SOHO/EIT instrument, which appear to propagate from flare/CME sites and often wrap around the solar globe (see *Thompson et al.*, 1998; *Delanée et al.*, 2000). *Plunkett et al.* (2002) suggest that they are probably fast-mode MHD waves that propagate outward from the CME initiation site. Perhaps the most significant study on this topic, relevant to the discussion here is that of *Biesecker et al.* (2002). They studied 173 EIT wave events and looked for associated transient phenomena. They reported that in every case, having accounted for observational biases, a CME occurred in association with the wave. Flares occurred in association with the waves less frequently. This suggests that the EIT waves are not directly related to Moreton waves, which are considered to be flare-related. Thus, the picture suggested by *Plunkett et al.*, would be consistent and, in fact, for the particular topic under discussion here, it sheds little light on the precise relationship between the flare and CME, and more on the associations

with other transient phenomena. One thing is absolutely clear, the wave does not represent the footprint of the overlying, outwardly propagating CME; the EIT waves can wrap around the solar disk, but CME footpoints do not expand to that extent.

DISCUSSION AND CONCLUSIONS

The idea that the flare and CME do not cause one another but are different responses to the same driver has become a common conclusion as more sophisticated post-SMM instrumentation has been applied to the study of CME onsets, with a few exceptions (e.g., *Khan and Hudson*, 2000). The close association between flares and CMEs was never in question but there are always 'flare-less' CME events, and the recent studies appear to confirm these views. There are also many more flares than CMEs and this has to be taken into account in any CME-flare model. The view that longer duration flares have a greater chance of CME association appears to be upheld by recent work; we must always keep in mind that a CME onset can occur in association with a flare of any duration or no flare at all. The models have to cope with this. We note also, that the pre-SOHO conclusions about relative flare-CME locations and asymmetry are consistent with many recent studies.

The identification of sigmoids as potential CME sources may have generated some excitement, but to establish that this is not simply another way of saying that the more complex active regions have a greater chance of CME generation, there needs to be a more quantitative way of defining a sigmoidal region and thorough statistical analyses must be performed on their association with CME and flare activity to establish whether it is the sigmoidal shape or just the complexity that is important.

The detection of EUV dimming is extremely important because it suggests that we can identify the CME source region. If this is the case, we can study the pre-event activities of the region and obtain information on the nature of the flare and CME activity from that region. This is an area which deserves much study, especially using spectroscopic plasma diagnostic interpretation.

One recent aspect of the CME-flare picture of great interest is the three phase CME ascent shown by *Zhang et al.* (2001), which made use of the low-altitude C1 observations. Their suggestion of a CME initiation phase, followed by an impulsive acceleration phase and a propagation phase fits well with many aspects of the pre-SOHO picture. They concluded that the flare and CME events were driven by the same magnetic driver. They also demonstrated that the CME activity, for three of their events, was certainly initiated before the associated flare. This is consistent with other studies such as the July 1999 observation reported above.

However, they also stress that the impulsive acceleration phase is intimately related to the flare intensity increase. This is an important result. Even if one assumes that the flare and CME do not drive one another, that one can occur without the other, that there can be asymmetry and even a very different scale-size, the intimacy between the CME acceleration and the flare intensity rise in the events studied by *Zhang et al.*, do highlight a close relationship which must not be ignored. It suggests that at least for the events described the flare onset and CME onset association is perhaps more intimate than some pre-SOHO conclusions would have suggested. Although some pre-SOHO conclusions would allow for coincident events, they did suggest that the onsets could be loosely associated in time. This question merits much further investigation.

The *Zhang et al.* (2001, 2004) studies also suggested that there is very significant lateral expansion of CMEs in the earliest phases and that this may explain an apparent difference in scale-size between flares and CME source regions. Again, this is a very important observation, though, as explained above, the current author believes that the jury is still out on this one, which stresses that there is a great need to explore this particular issue in greater detail as a matter of some urgency.

There is one major point which need to be made about current trends in CME onset research. Many authors claim to detect signatures of CME onsets without detailed correlations with coronagraph data or without evidence to provide convincing links between the surface activity and the CME events. This is especially worrying. Good examples of thorough attempts to perform detailed correlations between CMEs and underlying activity are *Delanée et al.* (2000) and *Zhang et al.* (2001). Claims of associations must be supported by clear demonstrations of links in time or space; for example, it is not sufficient to say that there was a CME and we see an EUV outflow so it must be the CME onset.

Having explored the recent CME onset studies, as is often the case, we find ourselves basically supporting past results and conclusions and raising more questions. Given the tools we have available, with SOHO in particular, there is no reason why we cannot expand the onset studies, in particular to investigate the dimming, to confirm the three-phase scenario of *Zhang et al.*, to extend the EUV and white-light associations started by *Delanée et al.*, and others, to investigate the nature of active region interconnecting loops, etc…. We have the ability to do this now.

New aspects of CME-onset research will emerge with the advent of the upcoming STEREO and Solar-B missions, planned for launch in 2006, especially if we still have SOHO in operation at the same time. Most of the studies mentioned here involve near-limb observations with the problems of foreshortening. In most CME onset studies, for example, there is no chance of using magnetograph data. However, with spacecraft at different viewing angles, on and off the Sun-Earth line, which the future mission programme will provide, we may be able to detect CMEs near the limb whilst observing the source regions on the disc from other platforms. This will open up totally new aspects of CME onset research, but will also introduce complex problems of event and feature recognition from different angles. Nevertheless, this will open up the next chapter in CME onset research, to which we can look forward.

A FLARE-CME MODEL?

So, where does this all leave us in the quest to develop a valid flare-CME model?

The first point to make is that we are well aware that when studying the solar atmosphere, we are studying a complex magnetic environment and not a set of discrete, isolated magnetic loops. Images and movies, in particular from the SOHO and TRACE missions, show complex hair-like hierarchies of magnetic structures in the corona writhing in response to photospheric motions. It is clear to us all that magnetic complexities can build up in such a system, and we can witness a breakdown of local conditions, perhaps leading to magnetic reconnection and a restructuring of the magnetic configuration. Such responses will depend on the local conditions and the magnetic configuration.

Figure 3 shows a magnetic hierarchy in cartoon form, produced for a review of CME onsets in 1991 (*Harrison*, 1991). The cartoon shows a range of magnetic structures – the smallest scales (the arcade on the right hand side of Figure 3a) may represent an active region and the larger scales indicate features including active region interconnecting loops and streamers. In such a starting configuration, there is complexity and asymmetry. The sequence shown in Figures 3b to 3e show one suggested sequence of events. The smaller scale features are subject to shear, driven by photospheric motion, and this results in ascending magnetic structures which must interact with the overlying, larger-scale features. One could imagine scenarios where magnetic reconnection could occur in such a configuration, and indeed, the location of a current sheet is indicated in Figure 3c for this scenario. However, whether there is reconnection or simply a loss of equilibrium in the larger-scale structures, the result would be a CME, as those structures ascend. A more complete description is given by *Harrison* (1991).

The basic ideas of this cartoon model still hold true. The point of the scenario outlined by the cartoon is that the CME and flare are consequences of the same magnetic driver – neither drives the other. Their characteristics are the result of local conditions. Thus, we may witness a spectrum of flare and CME parameters; we can have event sequences with no flare or no CME, and we can have small flares with large

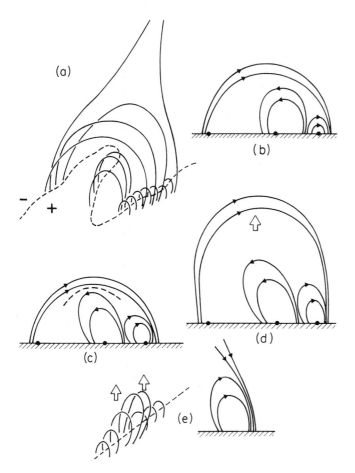

However, it has to be noted that this is in direct response to the observational results.

Having presented the multi-scale hierarchy cartoon of Figure 3 as a 'standard model', we must note that researchers frequently refer to the so-called CSHKP model as a standard for flare-CME onsets. The model has developed from the work of several authors, whose names form the CSHKP acronym. This includes *Carmichael* (1964), *Sturrock* (1966), *Hirayama* (1974) and *Kopp and Pneuman* (1976). The basic approach is one of an arcade, containing prominence material, which erupts. The footpoints are associated with flare H-alpha ribbons and the hard X-ray footpoints of the flare impulsive phase, and the ascending structure, including the embedded prominence, represent the overlying CME. Detailed discussions of the CSHKP model can be found in a number of papers, e.g., *Magara et al.* (1996). However, it is this author's opinion that several of the key features of the CME-flare relationship are not covered by the CSHKP approach. For example, the CSHKP model demands that the flare and CME onset timings are identical and would suggest a flare-CME symmetry, both of which appear to be at odds with many observations. We have discussed the issue of relative flare and CME scale sizes, above, and that could also be a problem for the CSHKP model. In addition, the multi-scale approach of Figure 3 allows a spectrum of flare and CME responses which seems well suited to the observations.

Figure 3. A cartoon model of the CME onset presented by *Harrison* (1991). The basic configuration includes a hierarchy of magnetic scales including a streamer and active region interconnecting loops, a filament channel and at least one active region. This is shown schematically in image (a).

CMEs and vice versa. In other words, we can envisage situations where the large-scale eruption component does not occur, or where the smaller-scale structures do not flare in a traditional sense.

This scenario would be the 'standard model' supported by this author. The lesson given by the observations, which is supported by this scenario is one of not making things too simple and not making biased assumptions. We have to cater for the fact that the flare-CME relationship is not one-to-one and has numerous complexities as described above. The consequence of this is that the modeller is being asked to forget simple isolated loop systems and consider a complex, 3D magnetic model with asymmetry and a hierarchy of loop sizes. It could be argued that this is a 'cop out', i.e. it is a woolly scenario which caters for a wildly varying range of possibilities and including asymmetry, 3D and hierarchical structures that are almost impossible to model at present.

REFERENCES

Biesecker, D.A., D.C. Myers, B.J. Thompson, D.M. Hammer, and A. Vourlidas, (2002), Solar phenomena associated with "EIT waves", *Astrophys. J.*, 569, 1009-1015.

Brueckner, G., R.A. Howard, and 13 co-authors (1995), The Large Angle Spectroscopic Coronagraph (LASCO), *Solar Phys.*, 162, 357-402.

Canfield, R.C., H.S. Hudson, and D.E. McKenzie (1999), Sigmoidal morphology and eruptive solar activity, *Geophys. Res. Lett.*, 26, 627-630.

Carmichael, H. (1964), in AAS-NASA Symp. On Solar Flares, ed. W.N. Hess, NASA SP-50, 451.

Delaboudiniere, J.-P., G.E. Artzner, and 26 co-authors (1995), EIT: Extreme Ultraviolet Imaging Telescope for the SOHO mission, *Solar Phys.*, 162, 291-312.

Delannée, C., J.-P. Delaboudiniere, and P. Lamy (2000), Observation of the origin of CMEs in the low corona, *Astron. Astrophys.*, 355, 725-742.

Domingo, V., B. Fleck, and A.I. Poland (1995), The SOHO Mission: An Overview, *Solar Phys.*, 162, 1-37.

Drake, J.F. (1971), Characteristics of soft X-ray bursts, *Solar Phys.*, 16, 152.

Glover, A., N.D.R. Ranns, L.K. Harra, and J.L. Culhane (2000), The onset and association of CMEs with sigmoidal active regions, *Geophys. Res. Lett.*, 27, 2161-2164.

Gopalswamy, N., and Y. Hanaoka (1998), Coronal dimming associated with a giant prominence eruption, *Astrophys. J.*, 498, L179-L182.

Handy, B.N., M.E. Bruner, T.D. Tarbell, A.M. Title, C.J. Wolfson, M.J. LaForge, and J.J. Oliver (1998), UV Observations with the Transition Region and Coronal Explorer, *Solar Phys.*, 183, 29-43.

Harra, L.K., and Sterling, A.C. (2001), Material outflows from coronal intensity dimming regions during coronal mass ejection onset, *Astrophys. J.*, 561, L215.

Harrison, R.A. (1986), Solar coronal mass ejections and flares, *Astron. Astrophys.*, 162, 283-291.

Harrison, R.A. (1991), Coronal mass ejection, *Phil. Trans. R. Soc. Lond.* A., 336, 401-412.

Harrison, R.A. (1995), The nature of solar flares associated with CMEs, *Astron. Astrophys.*, 304, 585-594.

Harrison, R.A. (1996), Coronal magnetic storms: A new perspective on flares and the 'solar flare myth' debate, *Solar Phys.*, 166, 441-444.

Harrison, R.A. (2003), SOHO observations relating to the association between flares and coronal mass ejections, *Adv. Space Res.*, 32, 2425-2437.

Harrison, R.A., and M. Lyons (2000), A spectroscopic study of coronal dimming associated with a coronal mass ejection, *Astron. Astrophys.*, 358, 1097-1108.

Harrison, R.A., P.W. Waggett, R.D. Bentley, K.J.H. Phillips, M. Bruner, M. Dryer, and G.M. Simnett (1985), The X-ray signature of solar coronal mass ejections, *Solar phys.*, 97, 387-400.

Harrison, R.A., P. Bryans, G.M. Simnett, and M. Lyons (2003), Coronal dimming and the coronal mass ejection onset, *Astron. Astrophys.*, 400, 1071-1083.

Harrison, R.A., E.C. Sawyer, M.K. Carter, and co-authors (1995), The Coronal Diagnostic Spectrometer for the Solar and Heliospheric Observatory, *Solar Phys.*, 162, 233-290.

Hirayama, T. (1974), Theoretical model of flares and prominences, *Solar Phys.*, 34, 323

Howard, T.A., and Harrison, R.A. (2004), On the coronal mass ejection onset and coronal dimming, *Solar Phys.*, 219, 315-342.

Hudson, H.S., and Webb, D.F. (1997), Soft X-ray signatures of coronal ejections, in 'Coronal Mass Ejections', (eds) Nancy Crooker, Jo Ann Jocelyn and Joan Feynman, AGU Monograph, pp. 27-38.

Hudson, H.S., J.R. Lemen, O.C. St Cyr, A.C. Sterling, and D.F. Webb (1998), X-ray coronal changes during halo CMEs, *Geophys. Res. Lett.*, 25, 2481-2484.

Kahler, S.W. (1977), The morphological and statistical properties of solar X-ray events with long decay times, *Astrophys. J.*, 214, 891-897.

Khan, J.I., and H.S. Hudson (2000), Homologous sudden disappearances of transequatorial interconnecting loops in the solar corona, *Geophys. Res. Lett.*, 27, 1083-1086.

Kopp, R.A. and Pneuman, G.W. (1976), *Solar Phys.*, 50, 85.

Kreplin, R.W., T.A. Chubb, and H. Friedman (1962), X-ray and Lyman-alpha emission from the Sun as measured from the NRL SR-1 satellite, *J. Geophys. Res.*, 6, 2231.

Magara, T., S. Mineshige, T. Yokoyama, and K. Shibata, (1996), Numerical simulation of magnetic reconnection in eruptive flares, *Astrophys. J.*, 466, 1054-1066.

Munro, R.H., J.T. Gosling, E. Hildner, R.M. MacQueen, A.I. Poland, and C.L. Ross (1979), The association of coronal mass ejection transients with other forms of solar activity, *Solar Phys.*, 61, 201-215.

Nitta, N.V., and H.S. Hudson (2001), Recurrent flare/CME events from an emerging flux region, *Geophys. Res. Lett.*, 28, 3801-3804.

Ogawara, Y., T. Takano, T. Kato, T. Kosugi, S. Tsuneta, T. Watanabe, I. Kondo, and Y. Uchida (1991), The Solar-A Mission: An Overview, *Solar Physics*, 136, 1-16.

Pearce, G., and R.A. Harrison (1988), A statistical analysis of the soft X-ray profiles of solar flares, *Astron. Astrophys.*, 206, 121-128.

Phillips, K.J.H. (1972), PhD thesis, University College London.

Plunkett, S.P., D.J. Michels, R.A. Howard, G.E. Brueckner, O.C. St Cyr, B.J. Thompson, G.M. Simnett, R. Schwenn, and P. Lamy (2002), New insights on the onsets of CMEs from SOHO, *Adv. Space Res.*, 29, 1473-1488.

Rust, D.M., and E. Hildner (1976), *Solar Phys.*, 48, 381-387.

Sheeley, N.R., J.D. Bohlin, G.E. Brueckner, and co-authors (1975), Coronal changes associated with a disappearing filament, *Solar Phys.*, 45, 377-392.

Sheeley, N.R., R.A. Howard, M.J. Koomen, and D.J. Michels (1983), Associations between coronal mass ejections and soft X-ray events, *Astrophysical J.*, 272, 349-354.

Sterling, A.C. (2000), CME source regions at the Sun: Some recent results, *J. Atmosph. and Solar Terrestr. Phys.*, 62, 1427-1435.

Sterling, A.C., and H.S. Hudson (1997), Yohkoh SXT observations of X-ray dimming associated with a halo coronal mass ejection, *Astrophys. J.*, 491, L55-L58.

Sterling, A.C., H.S. Hudson, B.J. Thompson, and D.M. Zarro (2000), Yohkoh SXT and SOHO EIT observations of sigmoid-to-arcade evolution of structures associated with halo CMEs, *Astrophys. J.*, 532, 628-647.

Sturrock, P.A. (1966), Model of high-energy phase of solar flares, *Nature*, 211, 695.

Thompson, B.J., S.P. Plunkett, J.B. Gurman, J.S. Newmark, O.C. St Cyr, and D.J. Michels (1998), SOHO/EIT observations of an Earth-directed CME on May 12 1997, *Geophys. Res. Lett.*, 25, 2465-2468.

Zarro, D.M., A.C. Sterling, B.J. Thompson, H.S. Hudson, and N. Nitta (1999), *Astrophys. J.*, 520, L139-L142.

Zhang, J., K.P. Dere, R.A. Howard, M.R. Kundu, and S.M. White (2001), On the relationship between coronal mass ejections and flares, *Astrophys. J.*, 559, 452-462.

Zhang, J., K.P. Dere, R.A. Howard, and A. Vourlidas (2004), A study of the kinematic evolution of coronal mass ejections, *Astrophys. J.*, 604, 420-432.

Models of Solar Eruptions:
Recent Advances from Theory and Simulations

Ilia I. Roussev and Igor V. Sokolov

Center for Space Environment Modeling, University of Michigan, Ann Arbor, Michigan, USA

Coronal mass ejections and solar flares are the most stunning manifestations of solar activity. They are of great scientific and practical interest because solar eruptions and related energetic particle events endanger human life in outer space and pose major hazards for spacecraft in the inner solar system. Coronal mass ejections and flares are presently recognized as two different consequences of a single physical process, which involve a catastrophic loss of mechanical equilibrium of coronal magnetic fields. The actual trigger mechanisms for solar eruptions are, however, not established yet. These mechanisms are the subject of active research and debate by both solar theorists and computational modelers. To improve present knowledge on solar eruptions, any related physical model has to account for two basic properties of the eruption process. These are the fundamental cause of the eruption itself and the nature of morphological features associated with it. This paper summarizes the current understanding of the origin and dynamics of coronal mass ejections through theory and modeling. A strong emphasis is put on the research studies being conducted at the Center for Space Environment Modeling at the University of Michigan. The future prospects for realistic, data-driven modeling of solar eruptions and related solar energetic particle events are also discussed. Once perfected, the models should lead to significant improvements in our ability to forecast those violent solar disturbances that have great impact on the near-Earth environment.

1. INTRODUCTION

Coronal mass ejections (CMEs) are one of the most astonishing manifestations of solar activity in which vast amounts of magnetic flux ($\sim10^{21-23}$ Mx) and solar plasma ($\sim10^{15-16}$ g) are ejected from the low corona into interplanetary space [*Gosling*, 1990]. Because of their large scale and high energetics, CMEs are thought to be important for reconfiguring the large-scale fields of the solar corona over the Solar Cycle [*Low*, 2001]. CMEs also play a major role in the Sun-Earth connection. For a long time, it was thought that solar flares

are the dominant sources of geomagnetic activities [e.g., *Dryer*, 1982]. A different paradigm was put forward by *Gosling* [1993] in which CMEs, not flares, were regarded as the main drivers of non-recurrent geomagnetic storms. Today's view is that CMEs and solar flares are interrelated. The two phenomena are recognized to be two different manifestations of a single physical process that involves a major restructuring of coronal magnetic fields. Statistical studies of CMEs based on SoHO and TRACE observations indicate that a large fraction of solar flares ($\geq20\%$) are associated with CMEs, and that more than 50% of the CMEs are represented by erupting prominences [see, e.g., *Gosling*, 1990]. At present, the main requirement for CME models put forth by observations is two-fold. In the first place, any physical model developed to explain solar eruptions has to account for the fundamental trigger of the eruption. Secondly, these

Solar Eruptions and Energetic Particles
Geophysical Monograph Series 165
Copyright 2006 by the American Geophysical Union
10.1029/165GM10

models must account for the wide variety of features that form and develop in the eruptive process, such as bright H_α ribbons on the solar disk, rising soft X-ray and H_α loop systems in the corona, and EIT dimmings, among other phenomena [see *Priest and Forbes*, 2002]. These features, as revealed by space- and ground-based solar observations, manifest the complex nature of CMEs. The related joint observations also indicate that a solar eruption is not an isolated transient feature seen by a particular instrument, but a full fledged solar phenomenon, spawning a rich variety of other processes on the Sun.

The physical mechanisms of CME initiation are the subject of active research and debate. To date, there is no CME model sufficiently well developed to explain the real events of solar eruptions and related phenomena. Nevertheless, significant progress has been made in understanding the basic physical processes that are involved in those events. The various CME models that currently exist in the literature differ by the details in which the eruption is achieved. These models, however, agree that a CME is the result of a catastrophic loss of mechanical equilibrium of solar plasma confined by the coronal magnetic field. The CME models can be organized into two large groups depending on the state of the assumed coronal magnetic field prior to the eruption. The first class of models [e.g., *Chen*, 1996; *Forbes and Isenberg*, 1991; *Forbes and Priest*, 1995; *Gibson and Low*, 1998; *Titov and Démoulin*, 1999; *Roussev et al.*, 2003a; *Wu et al.*, 1999] assumes that a magnetic flux rope exists prior to the eruption. The rope is suspended in the solar corona by a balance between magnetic compression, hoop, and tension forces associated with the magnetic field of the rope and the background field. The gravity and pressure gradient forces are important for the mechanical equilibrium in flux-rope models comprising the prominence material. Both theoretical and numerical studies of magnetic flux ropes suggest that they may suddenly lose mechanical equilibrium and erupt due to footpoint motions, injection of magnetic helicity, or draining of heavy prominence material. The second group of models [e.g., *Antiochos et al.*, 1999; *Amari et al.*, 1999, 2000; *Linker and Mikić*, 1995; *Manchester*, 2003; *Roussev et al.*, 2004] relies on the existence of sheared magnetic arcades, which become unstable and erupt once some critical state is reached in the solar corona. Unlike the previous class of models, here a flux rope does not exist *prior* to the disruption of the coronal magnetic field, but is formed *in the course* of the eruption by magnetic reconnection. As of today, there is no convincing observational evidence that proves or disproves either class of CME models, and no model is advanced enough to explain real observations.

In the pioneering numerical studies, heliospheric disturbances, such as interplanetary CMEs, were modeled by the injection of dense plasma through an inner boundary placed upstream of the critical point of the solar wind (at $R > 20 R_\odot$) [e.g., *Odstrčil and Pizzo*, 1999]. These models provide physical insight into how a large solar disturbance travels and interacts with the large-scale solar wind. However, these models rely on *ad hoc* inputs to the solar wind without considering a self-consistent propagation of such a structure through the corona. Only recently has the propagation of a CME from the inner corona to 1 AU (and beyond) been numerically modeled in two-dimensional [e.g., *Odstrčil et al.*, 2002; *Riley et al.*, 2002; *Vandas et al.*, 1995; *Wu et al.*, 1999] and three-dimensional [e.g., *Groth et al.*, 2000; *Vandas et al.*, 2002] geometries. The time-dependent magnetic structure of over-expanding CMEs and the dynamic forces acting on them as they propagate through the bimodal solar wind were also studied in length [e.g., *Cargill et al.*, 2000; *Cargill*, 2004; *Riley and Gosling*, 1998]. Although idealized, these numerical studies reproduced many generic features of CMEs seen in observations.

1.1. CME Modeling at CSEM

The principal aim of solar and heliospheric research in the Center for Space Environment Modeling (CSEM) at the University of Michigan is the development of numerical models that account for the initiation [*Manchester et al.*, 2004a; *Roussev et al.*, 2003a, 2004] and evolution of solar eruptions and associated shock waves [*Manchester et al.*, 2004b,c], and the diffusive acceleration of solar energetic particles (SEPs) in the inner corona and interplanetary space by CME-driven shock waves [*Sokolov et al.*, 2004]. The models developed so far have succeeded in capturing many of the general characteristics of CMEs, such as properties of pre-event structures, interaction of the ejecta with the ambient solar wind, and shock formation and development in the inner corona. Future advanced models of CMEs, based on real observations, should result in an improved understanding of: (1) how solar eruptions are initiated; (2) how they evolve and interact with the structured solar wind; and (3) how solar energetic particles are produced in these eruptions.

Recently, we have developed a fully three-dimensional numerical model of a solar eruption that incorporates solar magnetogram data and a loss-of-equilibrium mechanism [*Roussev et al.*, 2004]. The study was inspired by the CME event that took place on May 2, 1998, in NOAA AR 8210 and is one of the SHINE Campaign Events. The CME model has demonstrated that a CME-driven shock wave can develop close to the Sun ($\sim 3 R_\odot$), and is sufficiently strong enough to account for the prompt appearance of high-energy solar protons (~ 1 GeV) at the Earth. Furthermore, we have developed a new Field-Line-Advection Model for Particle Acceleration (FLAMPA) [*Sokolov et al.*, 2004], which was successfully coupled with the global MHD model of *Roussev et al.* [2004]. The follow-up study was intended to quantify the diffusive acceleration and transport of solar protons at the shock wave from the MHD calculations. The coupled CME-SEP simulation has demonstrated that the theory of diffusive shock acceleration alone can account for the production of GeV protons during solar eruptions.

This paper summarizes the current state of understanding the origin and dynamics of CMEs, through theory and modeling, at the CSEM and elsewhere. In the next section, we discuss the physical requirements, which a CME model should meet. The following two sections describe a few representative models of each CME category to illustrate the basic principles involved: the flux-rope models are presented in Section 3 and the sheared-arcade models in Section 4. The present status of modeling gradual SEP events in conjunction with CME simulations is discussed in Section 5. The concluding remarks and future prospects for CME and SEP modeling are outlined in Section 6.

2. PHYSICAL REQUIREMENTS FOR CME MODELS

A physical model of solar eruptions has to meet the following requirements for it to agree with relevant observations. In the first place, a CME model has to produce: (1) explosive mass acceleration and result in ejecta speeds ranging from 100 km/s (slow events) to about 2,500 km/s (fast events); (2) CME mass of about 10^{15-16} g; (3) bulk kinetic energy of $\sim 10^{31-32}$ ergs; and (4) advected away magnetic flux of the order of 10^{21-23} Mx [see, e.g., *Gosling*, 1990]. Secondly, any CME model has to account for the opening, at least partially, of the coronal magnetic field all the way to infinity, while the magnetic energy of the system should decrease [see *Sturrock et al.*, 1984]. This requirement, however, faces the limitations of the Aly-Sturrock theorem [*Aly*, 1991; *Sturrock*, 1991], which states that, as the field opens up in space the magnetic energy of the system must increase. In order to overcome this constraint of the theorem, which applies only to simply connected field lines, it is necessary for the erupting field configuration to possess knotted field lines and/or disconnected field lines (i.e., not attached to the Sun), and/or to have a non-ideal MHD process at work, such as magnetic reconnection. The latter would reconfigure the coronal magnetic field in the course of the eruption process so that the CME becomes energetically possible.

Further, in the requirements for CME models, a boundary condition should fix the normal component of the magnetic field at the Sun. This condition results from the observational fact that the photospheric magnetic field sources remain virtually unchanged in the eruption process. This also hints that the photospheric sources are unlikely to play a role in energizing CMEs. Vice versa, any coronal disturbance produced by a CME has practically no effect on the massive photosphere because of the enormous difference in mass density

($\sim 10^9$ times) between the photosphere and the solar corona. Particularly, the photospheric footpoints of coronal magnetic fields are not affected by the eruption itself on its characteristic time-scale (some tens of minutes). This physical effect is called "inertial line-tying" of the field.

Last, but not least, on the list of necessary requirements is the observational fact that the photospheric magnetic field is driven quasi-statically during the eruption process. The typical value of photospheric motions is of the order of a few km/s, which is only 0.1-1% of the Alfvén speed in the corona. Therefore, if magnetic stresses are continuously built in the corona at this characteristic speed, it would take a few days to store sufficient energy in the corona to power a large CME (bulk energy of $\sim 10^{32}$ ergs) [see *Forbes*, 2000].

Because the solar corona is magnetically dominated ($\beta \ll 1$), the electric currents producing the excess magnetic energy from the potential limit must be either force-free, i.e., field-aligned (except in prominences, to be explained below), or confined to thin current sheets (where $\beta \sim 1$). The latter are thought to be important in providing the energy release during small compact flares on the Sun, as argued by *Forbes* [2000], and particle acceleration during impulsive SEP events [see *Reames*, 2002], but they cannot explain the ejections of mass and magnetic flux observed during CMEs. This implies that the energy for CMEs is most likely stored in the non-potential magnetic fields generated by volumetric, field-aligned coronal currents. It should be noted here that in solar prominences not all volumetric currents are force-free. This is because some finite Lorentz forces (due to cross-field currents) are required to offset the pressure gradient and gravity forces associated with the presence of cold and dense prominence material. The presence of cross-field electric currents, however, is not essential for powering a CME, but could be important for triggering the eruption if, for example, most prominence material has been drained by some buoyant instability mechanism [see *Gibson and Low*, 1998]. In any case, the cross-field currents increase the magnetic energy available for a CME by not more than 10% above the corresponding force-free portion [see *Forbes*, 2000].

It is evident from this discussion that the energy needed to power an eruption is most likely stored in the non-potential magnetic fields of volumetric coronal currents. In the study of *Canfield et al.* [1980], it was demonstrated quantitatively that only the coronal magnetic field has sufficient energy density (~ 100 ergs/cm^3), among the other forms of energy in the corona, to explain the observed CME energetics. For example, if one imagines a cube of coronal volume with a size of 2×10^5 km (typical length-scale of a CME source region) and assumes that the average pre-eruption value of the magnetic field is 30 G, then a decrease in the field strength by 20% (i.e., change by 6 G) is sufficient to energize a fast CME with a bulk kinetic energy of 10^{32} ergs.

The discussion so far favors the so-called "storage models" of solar eruptions [see *Forbes*, 2000]. Other types of models, for instance the "flux injection" model of *Chen* [1989], will not be addressed here. The basic philosophy in the "storage models" is that magnetic stresses (i.e., volumetric electric currents) are continuously built in the coronal volume by the omnipresence of flux emergence and shuffling of footpoints of coronal loops, among other processes. The energy buildup proceeds up to a point when the mechanical equilibrium of the coronal magnetic field is no longer attainable and, as a result, a CME occurs. The eruption process releases some (not all) magnetic energy stored in the non-potential field and associated with volumetric coronal currents.

3. FLUX-ROPE MODELS

The basic magnetic topology in the flux-rope models is a twisted flux rope suspended in the solar corona by a balance between magnetic compression, hoop, and tension forces associated with the magnetic field of the rope and the background field. However, some models argue that the rope has emerged from below the convection zone as a coherent magnetic feature [see review by *Forbes*, 2000]. The majority of these models do not discuss the means by which the flux rope is formed in the corona prior to the eruption. Rather, they concentrate on exploring its stability properties in the context of CME production. Recent numerical simulations by *Gibson et al.* [2004] and *Manchester et al.* [2004a] have demonstrated the emergence of a twisted magnetic flux tube from below the photosphere into a pre-existing coronal magnetic field. These studies have predicted solar features and dynamics that are consistent with observations.

Assuming the background coronal field to be a superposition of a dipole field plus the field of a line-current, *Titov and Démoulin* [1999] have shown that the equilibrium of an axially symmetric flux rope with total electric current I has two possible equilibrium positions, provided that the current is not too large. This configuration is shown in the left panel of Plate 1. The equilibrium loop location closest to the Sun is stable, but the one farthest from the Sun is unstable. As the current in the flux rope is increased, the two equilibrium states approach one another, and they meet when the current reaches a critical value. There are no equilibria for ropes with current above this value [*Titov and Démoulin*, 1999]. The difference in the stability properties of the two equilibria comes from the fact that one sits in a magnetic energy valley, while the other one sits on a hill. In other words, for the unstable equilibria, a small outward displacement of the flux rope leads to an outward force, which acts to further increase the displacement. Alternatively, the mechanical balance of a flux rope with the surrounding field can be disrupted if the photospheric sources of the background field are either reduced

below a critical point [e.g., *Lin and Forbes*, 2000], or are moved closer to one another without reduction of the photospheric field [e.g., *Forbes and Priest*, 1995]. The loss of equilibrium could also occur due to the emergence of new magnetic flux on the Sun.

When the system comprising the flux rope is brought up to a critical point where a stable equilibrium no longer exists, the growth of MHD perturbations leads to an ideal (kink-like) instability [see *Kliem et al.*, 2004; *Török et al.*, 2004], or a lack of equilibrium and a vertical current sheet forms [e.g., *Lin et al.*, 1998b; *Roussev et al.*, 2003a]. The electric current in the vertical sheet is of the same sign as the flux rope current, so that the sheet attracts the current loop. Therefore, the erupting flux rope cannot escape unless a non-ideal process, such as magnetic reconnection, dissipates the current sheet fast enough; the reconnection process manifests a solar flare. If the current does not dissipate fast enough [see *Roussev et al.*, 2003a], then the flux rope is doomed to fail and no CME occurs, only a flare. There are numerous examples of failed eruptions on the Sun to support this scenario, for instance the event observed by TRACE on May 27, 2002, in association with a M2-class flare [*Ji et al.*, 2003].

Below we discuss in detail the analytical flux-rope model of *Titov and Démoulin* [1999] (T&D), which has been proposed to explain flares and CMEs. The T&D model uses a fully three-dimensional magnetic field geometry, which looks remarkably realistic when compared with observations. Inspired by this model, *Roussev et al.* [2003a], *Kliem et al.* [2004], and *Török et al.* [2004] have carried out numerical studies to test the T&D model in the context of CME initiation. Previous two-dimensional flux-rope models [e.g., *Forbes and Priest*, 1995] suggest than a CME will occur provided that the anchoring of the ends of the rope to the Sun does not prevent it, as argued by *Antiochos et al.* [1999].

The T&D model consists of a circular flux rope with total current I and major radius R, which is embedded in a line-tying surface, as shown in Plate 1 (left panel). The hoop force of the rope ($F_{hoop} \sim I^2/R$) [*Shafranov*, 1966] is counterbalanced by the Ampére force acting on the rope current by the outer dipole field from two point magnetic charges $\pm q$ buried at a depth d below the surface and located at $x = \pm L$ ($F_{Amp} \sim IqL/(L^2 + R^2)^{3/2}$). In addition to the external field generated by the point charges, the model allows a contribution from an infinitely long line-current, I_0, which coincides with the x-axis and also lies below the photosphere at a depth d. The purpose of the toroidal field produced by the imaginary line-current below the surface is to reduce the number of turns of the field lines within and outside the flux rope. Without the line-current, the field lines at the surface of the flux rope are purely poloidal, and they have an infinite number of turns in a finite length (see right panel of Plate 1). However, some observations [e.g., *Leroy et al.*, 1983] imply

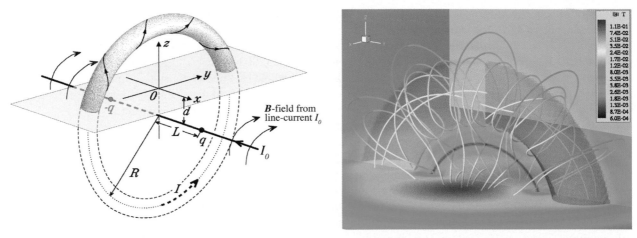

Plate 1. Basic magnetic field topology in the model of *Titov and Démoulin* [1999] (left panel) and in the numerical study by *Roussev et al.* [2003a] (right panel). *Left Panel:* Sources of the coronal magnetic field are: a force-free circular flux tube (shown as the shaded torus) with total current I; a pair of magnetic charges $\pm q$; and an infinite line-current I_0. The central axis of the torus, the line-current, and the two point charges are buried at a depth d below the solar surface. This configuration has no physical meaning below the photospheric plane. *Right panel:* Three-dimensional view of the flux rope configuration with the line-current removed. The solid lines are magnetic field lines and the false color code visualizes the strength of the magnetic field in Tesla. Left panel: Figure 2 from Titov, V.S., and P. Démoulin, Basic Topology of Twisted Magnetic Configurations in Solar Flares, *Astronomy and Astrophysics*, 351, 707-720, 1999. Used with permission. Right panel: Figure 1 from Roussev, I.I., T.G. Forbes, T.I. Gombosi, I.V. Sokolov, D.L. DeZeeuw, and J. Birn, A Three-Dimensional Flux-Rope Model for Coronal Mass Ejections Based on a Loss of Equilibrium, *Astrophys. J.*, 588, L45-L48, 2003. Used with permission.

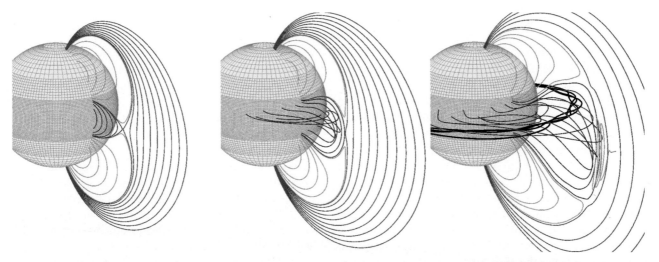

Plate 2. Basic elements of the magnetic field topology and dynamics in the model by *Antiochos et al.* [1999]. The initial state is shown in the left panel. As the purple field lines are sheared along the solar equator (middle panel), field-aligned volumetric currents are created and the flux system begins to expand. As a result of the outward expansion of the sheared region, a curved, horizontal current sheet is formed between the red and the purple field lines. The sheared field lines begin to open up (erupt) once the unsheared field lines from above begin to reconnect in the sheet (right panel). The flux rope formed during the "breakout" is visualized as the thick black lines (plate templates courtesy of B. Lynch).

that the maximum number of turns is less than two, but these observations do not include the cavity region which may comprise most of the flux rope. Incorporating the line-current eliminates this problem by creating a toroidal field, which ensures that no field lines are highly twisted. By adjusting the strength of the line-current, one can achieve a reasonable amount of twist everywhere.

Titov and Démoulin [1999] have also considered the stability of their configuration and found that it may be unstable if the large radius, R, of the flux rope exceeds $\sqrt{2}L$. This is, however, only a necessary condition for instability because their analysis does not include the effects of the line-tying of the poloidal field circulating around the flux rope. Indeed, the numerical study by *Roussev et al.* [2003a] has demonstrated that this condition is not a sufficient one, and that larger values of R in excess of $5L$ are required to achieve an eruption. They have also found that even when the initial equilibrium is unstable, the flux rope cannot escape, unless the static arcade field associated with I_0 is removed. Thus, although the T&D model can produce a CME-like eruption, it cannot do so without requiring a highly twisted field at the surface of the flux rope, as shown in the right panel of Plate 1. It may be possible in the future to mitigate this by using a configuration in which the infinitely long line-current is replaced by a current source whose magnetic field falls off more rapidly (i.e., quadratically) with height.

If the line-current I_0 is present in the magnetic configuration, as in the numerical studies by *Kliem et al.* [2004] and *Török et al.* [2004], the T&D flux rope is found to be unstable with respect to the ideal kink mode. (This suggests the kink instability as a mechanism for the initiation of flares.) The two studies have found that the threshold for instability increases as the major radius of the rope (also aspect ratio of the rope) becomes larger. The growth of kink perturbations is eventually slowed down by the surrounding potential field produced by the line-current. In agreement with the results of *Roussev et al.* [2003a], a global CME-like eruption is not plausible unless I_0 is removed. Once the eruption starts and the flux rope begins to lift off, a vertical current sheet forms beneath the rope, which acts to decelerate it. In other words, the symbiosis between the flux rope current and the sheet current may inhibit a successful CME. Then, magnetic reconnection plays a crucial role in dissipating the current sheet, thus helping the flux rope to escape. The dissipation of the current sheet manifests a two-ribbon flare. It is also important to mention that in the three-dimensional case, the torsional Alfvén waves play an important role in the eruption process since they transport helicity from the footpoints to the top of the flux rope. This process acts to sustain the eruption. The effects of the line-tying at the ends of the rope, however, act to decelerate it. In future investigations of the T&D model, or similar models, a more realistic treatment of the

reconnection process in the current sheet and the incorporation of a spherical geometry may greatly reduce the deceleration that is observed in the simulations of *Roussev et al.* [2003a]. At least, this study has demonstrated that, in principle, a three-dimensional, line-tied flux rope can lose equilibrium, and that the anchoring of the flux rope does not automatically prevent an eruption.

3.1. Flux-Rope Models Including Prominence Mass

In the flux-rope models that include the presence of prominence material, i.e., the non-force-free-field models, an eruption can be triggered via some buoyant instability mechanism in which most of the prominence mass has been drained through the footpoints of the flux rope. An example is the analytical model by *Gibson and Low* [1998] (G&L), which has a complex magnetic topology of a spheromak-type flux rope distorted into a three-dimensional teardrop shape. In the model, the Lorentz forces associated with the cross-field currents of the flux rope support the weight of a prominence. The rope itself possess sufficient free energy needed for an eruption. In a series of numerical studies, *Manchester et al.* [2004b,c] have investigated the dynamics of the eruption predicted from the analytical model. In these studies, the G&L flux rope was superimposed onto a global MHD solution of the solar corona, with bi-modal solar wind, and it was embedded in a helmet steamer at the equatorial region of the Sun. The MHD simulations have shown that by removing 20% of the prominence mass, the resulting eruption yields the observed values for ejected mass ($\sim 1.0 \times 10^{15}$ g) and kinetic energy ($\sim 4.0 \times 10^{31}$ ergs) during typical CME events. The fast ejecta (with speed in excess of $\sim 1,000$ km/s near the Sun) drives a shock wave and interacts with a structured, bi-modal solar wind. The modeled CME is found to propagate faster in the fast background wind than in the region occupied by the slow wind. As a result, a "dimple" is formed at the leading edge of the ejecta. *Manchester et al.* [2004c] have reported on the presence of post-shock compression due to the deflection of magnetic field lines. There is also a magnetic field enhancement on the back side of the CME, creating a region of reverse shock compression. The authors have argued that these two effects have important implications for the diffusive acceleration of solar particles by the interplanetary ejecta. In these simulations, the CME was propagated out to 1 AU, and it was shown that the interplanetary transient arrives at the Earth with geoeffective properties. Thus, the authors have demonstrated how an analytical flux-rope model can be incorporated within the framework of a global MHD model of the solar wind to investigate the dynamics of CMEs in interplanetary space, the consequences for the production of SEPs, and the impact of the ejecta on geospace.

4. SHEARED-ARCADE MODELS

The basic topology in this class of models is a sheared magnetic arcade containing free energy in the form of volumetric electric currents. These models invoke a non-ideal MHD process, i.e., magnetic reconnection, to achieve an abrupt loss of equilibrium of the coronal magnetic field and a subsequent eruption. Unlike the previous group of models, here magnetic reconnection is *not* a consequence of the eruption process, but the fundamental trigger of the eruption (i.e., responsible for its onset and growth in time). Another main difference is that in the sheared-arcade models a flux rope is formed *in the course* of the eruption, not before. Depending on the height at which the reconnection process occurs in these models, there are three sub-classes namely, flux-cancellation, tether-cutting, and breakout models.

In the flux-cancellation models [e.g., *Amari et al.*, 1999, 2000; *Linker et al.*, 2001; *Roussev et al.*, 2004], magnetic reconnection takes place at the photosphere or near the base of the solar corona. Here a flux rope is formed by reconnecting the opposite polarity feet of a shared magnetic arcade. The dynamics of the eruption proceed in much the same manner as in the flux-rope models discussed above. To achieve an eruption here, however, the "photospheric" boundary condition is usually changed at a rate that is too high to be considered quasi-static.

In the tether-cutting models [e.g., *Sturrock*, 1989; *Moore et al.*, 2001], a CME is triggered by reconnection inside a coronal filament in the low corona – a process referred to as "runaway tether-cutting". The filament here is comprised of a number of magnetic strands and the reconnection process occurs between the threads of opposite polarity that contact each other. The internal reconnection starts at the very beginning of the filament eruption and grows in time as the eruption advances. In this process, all connections (tethers) of the filament with the photosphere are broken (cut), except for those at the ends of the erupting flux rope.

In the breakout model of *Antiochos et al.* [1999], the eruption is again triggered by magnetic reconnection, but here this process occurs in a curved, horizontal current sheet situated above the magnetic arcade being sheared. A sketch of the magnetic topology involved in the model, which has a spherical symmetry with respect to the Sun's rotation axis, is shown in Plate 2. The basic concept in this two-dimensional model is as follows. As the central arcade above the equator (shown in purple) is sheared, it begins to expand outward in the coronal volume. During the expansion, the sheared field lines push against the X-line from below. As a result, a curved, horizontal current sheet forms between the purple and the red field lines. This sheet acts to confine the sheared arcade underneath so that it cannot open in the volume above the sheet. The only way for the sheared arcade to open up is

by dissipating the current sheet. Once the reconnection process starts, the field lines above the sheet are moved out of the way of the expanding arcade and the newly formed field lines join the two flux systems on the side (shown in green). However, rapid reconnection in the sheet does not occur at first. Rather, as the shear increases in the central arcade, the diffuse current transforms into a thin sheet, which then experiences fast reconnection. The rate of reconnection accelerates as the dissipating current sheet is pushed further away from the Sun. The nature of the transition from slow to fast reconnection in this model remains to be investigated. This requires the development of a more comprehensive theory of magnetic reconnection in three-dimensions, which also addresses the nature of micro-instabilities in current sheets leading to enhanced, or anomalous, resistivity. Future investigations are required to determine whether the breakout model of *Antiochos et al.* [1999] can produce an eruption in a realistic, fully three-dimensional geometry.

Recalling the flux-cancellation models, there is a series of fully three-dimensional numerical studies by *Amari et al.* [1999, 2000, 2003], and most recently by *Roussev et al.* [2004], which demonstrate how a flux rope can be formed by reconnecting the opposite polarity footpoints of a sheared magnetic arcade. This mechanism was originally proposed by *van Ballegooijen and Martens* [1989], and was first simulated in two-dimensional geometry by *Inhester et al.* [1992]. (Note that the three-dimensional flux-cancellation models are closely related to the T&D model discussed above since both cases involve line-tied flux ropes.) In these models, continued reconnection of the feet of the arcade weakens the ability of the overlying, unreconnected field lines to hold the line-tied flux rope next to the surface and, eventually, the rope erupts. In a numerical simulation with continually evolving boundary conditions, however, some care needs to be taken to determine whether the subsequent evolution is actually due to a loss of equilibrium or simply a consequence of driving by the boundary conditions.

The CME studies by *Amari et al.* [1999, 2000, 2003] outlined a practical numerical recipe of how to create a line-tied flux rope in an arbitrary, three-dimensional geometry. This recipe is illustrated in Plate 3. As a first step, let us apply shear-type horizontal motions along the polarity inversion line (PIL) separating two magnetic spots (left panel). These motions can resemble, for example, the apparent rotation (in the same direction) of the two spots of a magnetic dipole. There are many examples from observations to support the sunspot-rotation scenario [e.g., *Zhao and Kosovichev*, 2003]. As the spots rotate and introduce shear along the PIL, the initially potential field evolves through a sequence of states close to equilibrium to a non-potential, force-free field. In this "storage" phase, a free energy is built in the coronal magnetic field in the form of field-aligned currents. This is

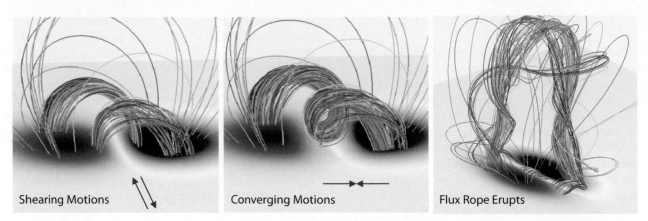

Plate 3. Formation of a three-dimensional flux rope by flux cancellation and subsequent eruption [after *Amari et al.*, 1999, 2000]. *Left Panel:* The rotation of two magnetic spots shears the field lines along the PIL and builds field-aligned currents in the corona; the initially potential magnetic field evolves to a non-potential, force-free field. This is the "storage" phase in which free energy is monotonically built in the solar corona needed to power a CME. *Middle Panel:* The application of converging motions towards the PIL triggers magnetic flux cancellation near the solar surface. The sheared field lines on either side of the PIL begin to reconnect and a flux rope forms. *Right Panel:* The weakening of the overlying field due to flux cancellation, while the flux rope increases in size and strength (twist), leads to the point of loss of confinement. As a result of this run-away process the rope erupts, manifesting a CME. Figure 2 from Amari, T., J.F. Luciani, Z. Mikić, and J. Linker, Three-Dimensional Solutions of Magnetohydrodynamic Equations for Prominence Magnetic Support: Twisted Magnetic Flux Rope, *Astrophys. J.*, 518, L57-L60, 1999, and Figure 3(b) from: Amari, T., J.F. Luciani, Z. Mikić, and J. Linker, A Twisted Flux Rope Model for Coronal Mass Ejections and Two-Ribbon Flares, *Astrophys. J.*, 529, L49-L52, 2000. Both used with permission.

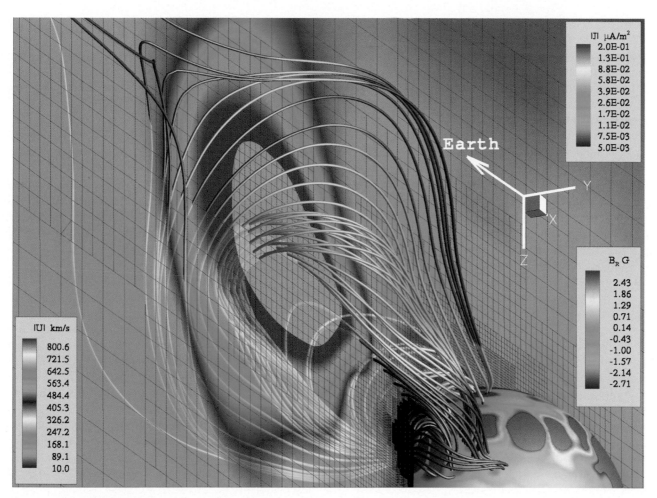

Plate 4. Three-dimensional view of the modeled CME from May 2, 1998, at 1.1 hrs after the initiation [from *Roussev et al.*, 2004]. The solid lines are magnetic field lines and the false color shows the magnitude of the current density in units of $\mu A/m^2$ (see color legend at top right). The magnitude of flow velocity, in units of km/s, is shown on a translucent plane (see color legend to the left). Values in excess of 1,000 km/s are blanked and shown in light grey. The grid-structure on this plane is also shown as the black frame. The inner sphere corresponds to $R = R_\odot$. The color shows the distribution of radial magnetic field in units of Gauss (see color legend at bottom right). Regions with field strength greater than 3 G are blanked and appear in grey. Figure 3 from Roussev, I.I., I.V. Sokolov, T.G. Forbes, T.I. Gombosi, M.A. Lee, and J.I. Sakai, A Numerical Model of a Coronal Mass Ejection: Shock Development with Implications for the Acceleration of GeV Protons, *Astrophys. J.*, 605, L73-L76, 2004. Used with permission.

the energy that will eventually power a solar eruption. At some later stage, the process of flux cancellation may occur if, for instance, new magnetic flux of opposite polarity begins to emerge in the vicinity of the PIL. This can be modeled by applying converging-type horizontal motions towards the PIL (see middle panel of Plate 3). As illustrated, a flux rope forms as a result of continuing reconnection between the sheared field lines on either side of the PIL. While during the "storage" phase the distribution of the normal component of the magnetic field is preserved, in this second phase the normal component changes, i.e., it decreases by absolute value. In essence, this leads to the weakening of the field surrounding the newly formed flux rope. This is a run-away process in which the rope at one point can no longer be held in equilibrium with the surrounding field. As a result, the rope erupts, manifesting a CME (see right panel). It should be noted that not all the energy built during the "storage" phase is released in the eruption process, but only some 8% to 17% depending on the global magnetic geometry. The released energy is converted into heat and kinetic energy of plasma bulk motions. Some energy is also carried out by the accelerated solar particles during the eruption.

Roussev et al. [2004] have utilized the above numerical recipe to model a real event, namely the CME from May 2, 1998, which occurred in active region (AR) 8210. This is a fully three-dimensional, compressible MHD calculation, which uses the global, data-driven model of the solar corona developed by *Roussev et al.* [2003b]. The CME model incorporates real magnetogram data from the Wilcox Solar Observatory (WSO), and it adopts the loss-of-equilibrium scenario described above to initiate the eruption. The CME is achieved by slowly evolving the boundary condition for the horizontal magnetic field at the Sun to account for the rotation of the main sunspot of AR 8210 – the energy "storage" phase – followed by an episode of flux cancellation. Although *Roussev et al.* [2004] did not model the actual photospheric motions inferred from the high-resolution vector magnetograms [see *Welsch et al.*, 2004], their model is generally consistent with the observed field dynamics, while allowing the CME initiation in a physically plausible manner. That is, one that extracts magnetic energy stored in the corona by evolving the low-resolution field from the WSO magnetic data up to a point when a loss of confinement occurs and the eruption starts.

In the model, the excess magnetic energy that is built in the sheared coronal field prior to the eruption is 4.1×10^{31} ergs, which is in close agreement with the EIT data for the dimming. The eruption itself takes place in a multi-polar type magnetic field configuration, which explains why a significant fraction (nearly 16%) of the energy stored prior to the CME is converted at the time of the eruption into thermal and kinetic energy of the ejecta. A view of the CME at 1.1 hrs after the initiation is shown in Plate 4. The modeled ejecta

achieves a maximum speed in excess of 1,000 km/s, which is in close agreement with the LASCO observations for the event. The CME drives a shock wave and interacts with a highly structured solar wind. The shock wave reaches a fast-mode Mach number in excess of four and a compression ratio of nearly three at a distance of $4R_\odot$ from the Sun. The fact that a strong shock wave can develop so close to the Sun is very important from the perspective of solar particle acceleration by a Fermi process, as discussed in the next Section. To summarize, the numerical model of *Roussev et al.* [2004] has succeeded in capturing many of the observed properties of the May 2, 1998 event and also general characteristics of CMEs (i.e., properties of pre-event structures, interaction of the ejecta with the ambient solar wind, and shock formation and development in the inner corona).

5. SOLAR ERUPTIONS, ENERGETIC PARTICLES AND SPACE WEATHER

From the perspective of space weather, CMEs and the solar energetic particle (SEP) events associated with them are of particular importance since they endanger human life in outer space and pose major hazards for spacecraft in the inner solar system. High-energy solar protons (>100 MeV) can be accelerated within a short period of time (≤1 hr) after the initiation of CMEs, which makes them difficult to predict and poses a serious concern for the design and operation of both manned and unmanned space missions. Recent theories [e.g., *Lee*, 1997; *Ng et al.*, 1999, 2003; *Zank et al.*, 2000] and related observations [e.g., *Cliver et al.*, 2004; *Kahler*, 1994; *Tylka et al.*, 1999, 2005], suggest that these high-energy particles are the result of Fermi acceleration processes at a shock wave driven by a solar eruption, so called diffusive shock acceleration (DSA), in the Sun's proximity (distances of 2-15 $R_\odot$). These theories are being debated within the community [e.g., *Reames*, 1999, 2002; *Tylka*, 2001], since very little is known from observations about the dynamical properties of CME-driven shock waves in the inner corona soon after the onset of the eruption. The main argument against the shock origin is that near the Sun the ambient Alfvén speed is so large, due to the strong magnetic fields there, that a strong shock wave is difficult to anticipate [*Gopalswamy et al.*, 2001]. How soon after the onset of a CME the shock wave forms, and how it evolves in time depends largely on how this shock wave is driven by the erupting coronal magnetic field. As recently proposed by *Tylka et al.* [2005], the shock geometry plays a significant role in the spectral and compositional variability of SEPs above ~30 MeV/nuc. Therefore, in order to explain the observed signatures of gradual SEP events, the global models of solar eruptions are required to explain the time-dependent changes in the strength and geometry (quasi-parallel *versus* quasi-perpendicular) of shocks during these events.

To address the issue of shock origin during CMEs requires that real magnetic data are incorporated into global MHD models of the solar corona, as in *Roussev et al.* [2004]. Recently, *Sokolov et al.* [2004] have carried out a numerical investigation in which they have studied the diffusive acceleration of solar particles at a realistically driven shock wave by a solar eruption. For this purpose, they have used the data-driven CME model of *Roussev et al.* [2004]. Note here that for the CME being studied (i.e., May 2, 1998), there was a prompt SEP event associated with it and observed by the NOAA GOES-8 satellite. A ground-level event was also detected by the CLIMAX neutron monitor.

The SEP study of *Sokolov et al.* [2004] has demonstrated that the theory of DSA alone can account for the production of GeV protons during solar eruptions, provided that a sufficiently strong shock wave forms near the Sun, as in the CME study. This was the first numerical calculation in which a realistic coupling of a global CME model with a SEP model was accomplished. Previous studies of particle acceleration at CME-driven shocks employed either an idealized, spherically symmetric shock wave [e.g., *Li et al.*, 2003; *Zank et al.*, 2000], or one taken from a snapshot of the compressible MHD simulation of a CME event [e.g., *Heras et al.*, 1995], and extended in time assuming self-similarity of the flow. *Sokolov et al.* [2004] have taken a different approach, following the time-dependent, three-dimensional evolution of a shock wave as the distribution of solar protons (subject to DSA) is advanced in time. For this purpose, they have performed frequent dynamical coupling between a MHD code, BATS-R-US, and a kinetic code, FLAMPA, to capture time-scales and spatial gradients of dynamical importance for the DSA of solar protons.

Plate 5 demonstrates key features predicted from the theory of DSA, namely: existence of an exponential tail of particles upstream of the shock wave (length of the tail $\approx D/U_{sh}$, where D is the scalar diffusion coefficient and U_{sh} is the shock speed); elevated proton intensities at high energies behind the shock wave; and escape of high-energy (>10 MeV) particles upstream of the shock wave. The latter is due to the fact that $D \propto E$ (where E is the kinetic energy of particles), meaning that once the particles are accelerated to high energies, their upstream tail becomes very large. These particles are of particular importance for space weather and are the ones considered as precursors of disruptive events in the near-Earth environment.

The results of the coupled CME-SEP simulation by *Sokolov et al.* [2004] clearly demonstrate the capability to provide a high-performance simulation of the acceleration and transport of solar protons observed early during gradual SEP events. There are, however, some discrepancies between the observed proton fluxes by GOES-8 on May 2, 1998, and those obtained from the model. Also, the onset of the observed SEP event happens ~30-60 mins. after the onset of the flare, while in the simulation this occurs ~30 mins. after the shock wave forms (or ~60 mins. after the onset of the eruption). This implies that it is important to derive from a CME model when, or if, a shock wave forms on the magnetic field line connecting the Sun with the Earth. This also shows that a coupled model is very informative and allows one to use real SEP data to verify it.

The kinetic code used by *Sokolov et al.* [2004] is a relatively new tool that incorporates many essential features of the theory of shock acceleration and transport of solar particles. Further extensions of this code, however, are required to address the wide range of SEP variability, which is a key issue for developing a predictive capability for SEPs. One example of such extensions is the role of self-generated Alfvén waves, which have been demonstrated to have important consequences for SEP elemental abundance variations [see *Ng et al.*, 1999; *Tylka et al.*, 1999] and the evolution of SEP anisotropies [*Reames et al.*, 2001]. Another example is the possible role of the shock geometry in controlling the spectral shapes of the highest energy SEPs [*Tylka et al.*, 2005] and the injection threshold [*Jokipii*, 1987; *Webb et al.*, 1995]. Extensions such as these are very important for understanding the observed variability during gradual SEP events and should be present in the relevant particle codes.

6. CONCLUSIONS AND FUTURE PROSPECTS FOR CME MODELS

To date, the majority of proposed CME models agree in describing the general properties of solar eruptions. Still, very few of them are sufficiently developed to explain the real events. The existing models imply that at least some eruptions, perhaps all, occur because coronal magnetic fields suffer a sudden loss of mechanical balance or stability. The source of energy for CMEs most likely comes from the free energy stored in volumetric electric currents in the corona.

At present, the two classes of competing CME models are the flux-rope and the sheared-arcade models. These groups differ on two major points. First, in the flux-rope models, it is assumed that a flux tube comprising a region in the corona of strong concentration of volumetric electric currents exists *prior* to the eruption. In the sheared-arcade models, it is assumed that a flux rope forms *during* the eruption. Second, while both classes of models invoke a non-ideal MHD process to explain CMEs, in the flux-rope models magnetic reconnection is a *consequence* of the eruption, while in the other group reconnection is the fundamental *trigger* of the ejection. To date, there is no convincing observational evidence that proves or disproves either class of CME models. The existence of flux ropes prior to CMEs is still unclear. To determine this, one needs precise measurements of the coronal magnetic field, but this data collection has just started

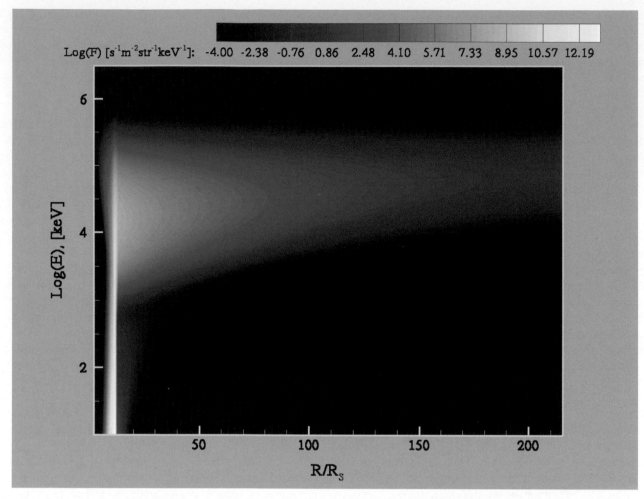

Plate 5. Snapshot of the differential proton intensity as a function of radial distance from the Sun and particle energy at 1.6 hrs from the shock formation time. The shock wave is located at $12.2R_\odot$ at this time [from *Sokolov et al.*, 2004]. Figure 2 (left panel) from Sokolov, I.V., I.I. Roussev, T.I. Gombosi, M.A. Lee, J. Kóta, T.G. Forbes, W.B. Manchester, and J.I. Sakai, A New Field-Line-Advection Model for Solar Particle Acceleration, *Astrophys. J.*, 616, L171-L174, 2004. Used with permission.

[e.g., *Lin et al.*, 1998a]. Regarding the shear-arcade models, the origin of shear is still a mystery, even though the rotation of magnetic sunspots evident from observations is one way to introduce shear along PILs required in the models.

In the sheared-arcade models, the fundamental trigger of the CME, magnetic reconnection, somehow disrupts the force balance of coronal magnetic field, thus starting the eruption. However, a delay in the turn-on of reconnection is required to store energy in the corona needed for powering the CME. This naturally raises the question of the nature of micro-instabilities in current sheets leading to enhanced, or anomalous, resistivity. The nature of this resistivity is poorly understood, and to improve the present understanding of this process requires the development of a comprehensive theory of magnetic reconnection in three-dimensions.

Improved understanding of CMEs also requires developing more realistic coronal magnetic field models in which photospheric field measurements (e.g., high quality MDI data) are used as a boundary condition for the magnetic field at the Sun, and the field is reconstructed in the coronal using one of the standard extrapolation techniques (e.g., potential field source surface model). Apart from this, the plasma properties at the photosphere and in the solar corona are also needed to drive the models. However, presently, this is far from being realistic. Future improved solar observations (e.g., those from STEREO and SDO) are expected to provide a more complete and realistic set of initial and boundary conditions to drive these computational models of CMEs, so that reliable tools for space weather predictions become available to the community.

Nonetheless, the future prospects for CME modeling are more optimistic, primarily because the forthcoming STEREO mission will provide an ideal opportunity for CME theorists and modelers to answers some of the outstanding issues raised here. The multi-dimensional solar observations by STEREO will help determine the magnetic field geometries that exist prior to solar eruptions, because once CMEs take off STEREO will observe the erupting filaments and coronal structures from multiple viewpoints. These observations will follow the propagation of solar and interplanetary transients all the way from the Sun to the Earth. This is important for testing and validating the numerical models of CME propagation. This mission will also enable modelers to couple photospheric with coronal magnetic field measurements and it will provide direct tests for the SEP models. Overall, the STEREO mission will provide a new, multi-dimensional level of coupling between numerical models and observations of solar eruptions. The realistic data-driven modeling of CMEs and related SEP events is just about to become a reality – a reality that will ultimately lead solar scientists to an improved understanding of this magnificent and complex solar phenomenon.

Acknowledgments. The authors thank T.G. *Forbes* and A. *Tylka* for their comments. The research work of I. R. and I. S. in the CSEM at the University of Michigan is supported by the following grants: DoD MURI F49620-01-1-0359, NSF ATM-0454469 (SHINE), and NASA-ESS CT cooperative agreement NCC5-614.

REFERENCES

Aly, J.J., How Much Energy Can be Stored in a Three-Dimensional Force-Free Field?, *Astrophys. J.*, 375, L61-L64, 1991.

Amari, T., J.F. Luciani, Z. Mikić, and J. Linker, Three-Dimensional Solutions of Magnetohydrodynamic Equations for Prominence Magnetic Support: Twisted Magnetic Flux Rope, *Astrophys. J.*, 518, L57-L60, 1999.

Amari, T., J.F. Luciani, Z. Mikić, and J. Linker, A Twisted Flux Rope Model for Coronal Mass Ejections and Two-Ribbon Flares, *Astrophys. J.*, 529, L49-L52, 2000.

Amari, T., J.F. Luciani, J.J. Aly, Z. Mikić, and J. Linker, Coronal Mass Ejection: Initiation, Magnetic Helicity, and Flux Ropes. II. Turbulent Diffusion-Driven Evolution, *Astrophys. J.*, 595, 1,231-1,250, 2003.

Antiochos, S.K., C.R. Devore, and J.A. Klimchuk, A Model for Solar Coronal Mass Ejections, *Astrophys. J.*, 510, 485-493, 1999.

Canfield, R.C., C.-C. Cheng, K.P. Dere, G.A. Dulk, D.J. McLean, R.D. Robinson Jr., E.J. Schahl, and S.A. Schoolman, Radiative Energy Output of the 5 September 1973 Flare, in Solar Flares: A Monogram From the Skylab Solar Workshop, edited by P.A. Sturrock, pp. 451-469, Colo. Assoc. Univ. Press, Boulder, Colorado, 1980.

Cargill, P.J., J. Schmidt, D.S. Spicer, and S.T. Zalesak, Magnetic Structure of Over-Expanding Coronal Mass Ejections: Numerical Models, *J. Geophys. Res.*, 105, 7,509-7,520, 2000.

Cargill, P.J., On the Aerodynamic Drag Force Acting on Interplanetary Coronal Mass Ejections, *Sol. Phys.*, 221, 135-149, 2004.

Chen, J., Effects of Toroidal Forces in Current Loops Embedded in a Background Plasma, *Astrophys. J.*, 338, 453-470, 1989.

Chen, J., Theory of Prominence Eruption and Propagation: Interplanetary Consequences, *J. Geophys. Res.*, 101, 27,499-27,520, 1996.

Cliver, E.W., S.W. Kahler, and D.V. Reames, Coronal Shocks and Solar Energetic Proton Events, *Astrophys. J.*, 605, 902-910, 2004.

Dryer, M., Coronal Transient Phenomena, *Space Sci. Rev.*, 33, 233-275, 1982.

Forbes, T.G., and P.A. Isenberg, A Catastrophe Mechanism for Coronal Mass Ejections, *Astrophys. J.*, 373, 294-307, 1991.

Forbes, T.G., and E.R. Priest, Photospheric Magnetic Field Evolution and Eruptive Flares, *Astrophys. J.*, 446, 377-389, 1995.

Forbes, T.G., A Review on the Genesis of Coronal Mass Ejections, *J. Geophys. Res.*, 105, 23,153-23,166, 2000.

Gibson, S.E., and B.C. Low, A Time-Dependent Three-Dimensional Magneto-hydrodynamic Model of the Coronal Mass Ejection, *Astrophys. J.*, 493, 460-473, 1998.

Gibson, S.E., Y. Fan, C. Mandrini, G. Fisher, and P. Démoulin, Observational Consequences of a Magnetic Flux Rope Emerging into the Corona, *Astrophys. J.*, 617, 600-613, 2004.

Gopalswamy, N., A. Lara, M.L. Kaiser, and J.-L. Bougeret, Near-Sun and near-Earth Manifestations of Solar Eruptions, *J. Geophys. Res.*, 106, 25,261-25,278, 2001.

Gosling, J.T., Coronal Mass Ejections and Magnetic Flux Ropes in Interplanetary Space, *AGU Monograph Series*, 58, 343-364, 1990.

Gosling, J.T., The Solar Flare Myth, *J. Geophys. Res.*, 98, 18,937-18,949, 1993.

Groth, C.P.T., D.L. DeZeeuw, T.I. Gombosi, and K.G. Powell, Global Three-Dimensional MHD Simulation of a Space Weather Event: CME Formation, Interplanetary Propagation, and Interaction With the Magnetosphere, *J. Geophys. Res.*, 105, 25,053-25,078, 2000.

Heras, A.M., B. Sanahuja, D. Lario, Z.K. Smith, T. Detman, and M. Dryer, Three Low-Energy Particle Events: Modeling the Influence of the Parent Interplanetary Shock, *Astrophys. J.*, 445, 497-508, 1995.

Inhester, B., J. Birn, and M. Hesse, The Evolution of Line-Tied Coronal Arcades Including a Converging Footpoint Motion, *Sol. Phys.*, 138, 257-281, 1992.

Ji, H., H. Wang, E.J. Schmahl, Y.-J. Moon, and Y. Jiang, Observations of the Failed Eruption of a Filament, *Astrophys. J.*, 595, L135-L138, 2003.

Jokipii, J.R., Rate of Energy Gain and Maximum Energy in Diffusive Shock Acceleration, *Astrophys. J.*, 313, 842-846, 1987.

Kahler, S., Injection Profiles of Solar Energetic Particles as Functions of Coronal Mass Ejection Heights, *Astrophys. J.*, 428, 837-842, 1994.

Kliem, B., V.S. Titov, and T. Török, Formation of Current Sheets and Sigmoidal Structure by the Kink Instability of a Magnetic Loop, *Astron. & Astrophys.*, 413, L23-L26, 2004.

Lee, M.A., Particle Acceleration and Transport at CME-driven Shocks, In Crooker, N., J.A. Joselyn, and J. Feynman, editors, Coronal Mass Ejections, pages 227-234, AGU: Washington DC, 1997.

Leroy, J.L., V. Bommier, and S. Sahal-Brechot, The Magnetic Field in the Prominences of the Polar Crown, *Sol. Phys.*, 83, 135-142, 1983.

Li, G., G.P. Zank, and W.K.M. Rice, Energetic Particle Acceleration and Transport at Coronal Mass Ejection-Driven Shocks, *J. Geophys. Res.*, 108, 10-21, 2003.

Lin, H., M.J. Penn, and J.R. Kuhn, HeI 10830 Angstrom Line Polarimetry: A New Tool to Probe the Filament Magnetic Fields, *Astrophys. J.*, 493, 978-995, 1998a.

Lin, J., T.G. Forbes, P.A. Isenberg, and P. Démoulin, The Effect of Curvature on Flux-Rope Models of Coronal Mass Ejections, *Astrophys. J.*, 504, 1,006-1,019, 1998b.

Lin, J., and T.G. Forbes, Effects of Reconnection on the Coronal Mass Ejection Process, *J. Geophys. Res.*, 105, 2,375-2,392, 2000.

Linker, J.A., and Z. Mikić, Disruption of a Helmet Streamer by Photospheric Shear, *Astrophys. J.*, 438, L45-L48, 1995.

Linker, J.A., R. Lionello, Z. Mikić, and T. Amari, Magnetohydrodynamic Modeling of Prominence Formation Within a Helmet Streamer, *J. Geophys. Res.*, 106, 25,165-25,176, 2001.

Low, B.C., Coronal Mass Ejections, Magnetic Flux Ropes, and Solar Magnetism, *J. Geophys. Res.*, 106, 25,141-25,163, 2001.

Manchester IV, W.B., Buoyant Disruption of Magnetic Arcades with Self-Induced Shearing, *J. Geophys. Res.*, 108, 1,162, 2003.

Manchester, W.B., T.I. Gombosi, D.L. De Zeeuw, and Y. Fan, Eruption of a Buoyantly Emerging Magnetic Flux Rope, *Astrophys. J.*, 610, 588-596, 2004a.

Manchester, W.B., T.I. Gombosi, I. Roussev, D.L. De Zeeuw, I.V. Sokolov, K.G. Powell, G. Tóth, and M. Opher, Three-Dimensional MHD Simulation of a Flux Rope Driven CME, *J. Geophys. Res.*, 109, 1,102-1,119, 2004b.

Manchester, W.B., T.I. Gombosi, I. Roussev, A. Ridley, D.L. De Zeeuw, I.V. Sokolov, K.G. Powell, and G. Tóth, Modeling a Space Weather Event from the Sun to the Earth: CME Generation and Interplanetary Propagation, *J. Geophys. Res.*, 109, 2,107-2,122, 2004c.

Moore, R.L., A.C. Sterling, H.S. Hudson, and J.R. Lemen, Onset of the Magnetic Explosion in Solar Flares and Coronal Mass Ejections, *Astrophys. J.*, 552, 833-848, 2001.

Ng, C.K., D.V. Reames, and A.J. Tylka, Effect of Proton-Amplified Waves on the Evolution of Solar Energetic Particle Composition in Gradual Events, *Geophys. Res. Lett.*, 26, 2,145-2,148, 1999.

Ng, C.K., D.V. Reames, and A.J. Tylka, Modeling Shock-Accelerated Solar Energetic Particles Coupled to Interplanetary Alfvén Waves, *Astrophys. J.*, 591, 461-485, 2003.

Odstrčil, D., and V.J. Pizzo, Three-Dimensional Propagation of Coronal Mass Ejections in a Structured Solar Wind Flow 2. CME Launched Adjacent to the Streamer Belt, *J. Geophys. Res.*, 104, 493-504, 1999.

Odstrčil, D., J.A. Linker, R. Lionello, Z. Mikić, P. Riley, V.J. Pizzo, and J.G. Luhmann, Merging of Coronal and Heliospheric Numerical Two-Dimensional MHD Models, *J. Geophys. Res.*, 107, doi:10.1029/ 1998JA900038, 2002.

Priest, E.R., and T.G. Forbes, The Magnetic Nature of Solar Flares, *Astron. & Astrophys. Rev.*, 10, 313-377, 2002.

Reames, D.V., Particle Acceleration at the Sun and in the Heliosphere, *Space Sci. Rev.*, 90, 413-491, 1999.

Reames, D.V., C.K. Ng, and D. Berdichevsky, Angular Distributions of Solar Energetic Particles, *Astrophys. J.*, 550, 1,064-1,074, 2001.

Reames, D.V., Magnetic Topology of Impulsive and Gradual Solar Energetic Particle Events, *Astrophys. J.*, 571, L63-L66, 2002.

Riley, P., and J.T. Gosling, Do Coronal Mass Ejections Implode in the Solar Wind?, *Geophys. Res. Lett.*, 25, 1,529-1,532, 1998.

Riley, P., J.A. Linker, Z. Mikić, D. Odstrčil, V.J. Pizzo, and D.F. Webb, Evidence of Post-Eruption Reconnection Associated with Coronal Mass Ejections in the Solar Wind, *Astrophys. J.*, 578, 972-978, 2002.

Roussev, I.I., T.G. Forbes, T.I. Gombosi, I.V. Sokolov, D.L. DeZeeuw, and J. Birn, A Three-Dimensional Flux-Rope Model for Coronal Mass Ejections Based on a Loss of Equilibrium, *Astrophys. J.*, 588, L45-L48, 2003a.

Roussev, I.I., T.I. Gombosi, I.V. Sokolov, M. Velli, W. Manchester, D.L. DeZeeuw, P. Liewer, G. Tóth, and J. Luhmann, A Three-Dimensional Model of the Solar Wind Incorporating Solar Magnetogram Observations, *Astrophys. J.*, 595, L57-L61, 2003b.

Roussev, I.I., I.V. Sokolov, T.G. Forbes, T.I. Gombosi, M.A. Lee, and J.I. Sakai, A Numerical Model of a Coronal Mass Ejection: Shock Development with Implications for the Acceleration of GeV Protons, *Astrophys. J.*, 605, L73-L76, 2004.

Shafranov, V.D., Plasma Equilibrium in a Magnetic Field, *Rev. Plasma Phys.*, 2, 103-151, 1966.

Sokolov, I.V., I.I. Roussev, T.I. Gombosi, M.A. Lee, J. Kóta, T.G. Forbes, W.B. Manchester, and J.I. Sakai, A New Field-Line-Advection Model for Solar Particle Acceleration, *Astrophys. J.*, 616, L171-L174, 2004.

Sturrock, P.A., P. Kaufmann, R.L. Moore, and D.F. Smith, Energy Release in Solar Flares, *Sol. Phys.*, 94, 341-357, 1984.

Sturrock, P.A., The Role of Eruption in Solar Flares, *Sol. Phys.*, 121, 387-397, 1989.

Sturrock, P.A., Maximum Energy of Semi-Infinite Magnetic Field Configurations, *Astrophys. J.*, 380, 655-659, 1991.

Titov, V.S., and P. Démoulin, Basic Topology of Twisted Magnetic Configurations in Solar Flares, *Astron. & Astrophys.*, 351, 707-720, 1999.

Török, T., B. Kliem, and V.S. Titov, Ideal Kink Instability of a Magnetic Loop Equilibrium, *Astron. & Astrophys.*, 413, L27-L30, 2004.

Tylka, A.J., D.V. Reames, and C.K. Ng, Observations of Systematic Temporal Evolution in Elemental Composition During Gradual Solar Energetic Particle Events, *Geophys. Res. Lett.*, 26, 2,141-2,144, 1999.

Tylka, A.J., New Insights on Solar Energetic Particles from Wind and ACE, *J. Geophys. Res.*, 106, 25,333-25,352, 2001.

Tylka, A.J., C.M.S. Cohen, W.F. Dietrich, M.A. Lee, C.G. Maclennan, R.A. Mewaldt, C.K. Ng, and D.V. Reames, Shock Geometry, Seed Populations, and the Origin of Variable Elemental Composition at High Energies in Large Gradual Solar Particle Events, *Astrophys. J.*, 625, 474-495, 2005.

van Ballegooijen, A.A., and P.C.H. Martens, Formation and Evolution of Solar Prominences, *Astrophys. J.*, 343, 971-984, 1989.

Vandas, M., S. Fischer, M. Dryer, Z. Smith, and T. Detman, Simulation of Magnetic Cloud Propagation in the Inner Heliosphere in Two-Dimensions. 1: A Loop Perpendicular to the Ecliptic Plane, *J. Geophys. Res.*, 100, 12,285-12,292, 1995.

Vandas, M., D. Odstrčil, and S. Watari, Three-Dimensional MHD Simulation of a Loop-Like Magnetic Cloud in the Solar Wind, *J. Geophys. Res.*, 107, doi:10.1029/ 2001JA005068, 2002.

Webb, G.M., G.P. Zank, C.M. Ko, and D.J. Donohue, Multidimensional Green's Functions and the Statistics of Diffusive Shock Acceleration, *Astrophys. J.*, 453, 178, 1995.

Welsch, B.T., G.H. Fisher, W.P. Abbett, and S. Regnier, ILCT: Recovering Photospheric Velocities from Magnetograms by Combining the Induction Equation with Local Correlation Tracking, *Astrophys. J.*, 610, 1,148-1,156, 2004.

Wu, S.T., W.P. Guo, D.J. Michels, and L.F. Burlaga, MHD Description of the Dynamical Relationships Between a Flux Rope, Streamer, Coronal Mass Ejection, and Magnetic Cloud: An Analysis of the January 1997 Sun-Earth Connection Event, *J. Geophys. Res.*, 104, 14,789-14,802, 1999.

Zank, G.P., W.K.M. Rice, and C.C. Wu, Particle Acceleration and Coronal Mass Ejection Driven Shocks: A Theoretical Model, *J. Geophys. Res.*, 105, 25,079-25,096, 2000.

Zhao, J., A.G. and Kosovichev, Helioseismic Observation of the Structure and Dynamics of a Rotating Sunspot Beneath the Solar Surface, *Astrophys. J.*, 591, 446-453, 2003.

Ilia I. Roussev and Igor V. Sokolov, Center for Space Environment Modeling, Space Physics Research Laboratory, University of Michigan, 2455 Hayward St, Ann Arbor, MI 48109, USA. (iroussev@umich.edu; igorsok@umich.edu)

Solar Energetic Particles: An Overview

Tycho von Rosenvinge and Hilary V. Cane

Laboratory for High Energy Astrophysics, NASA/Goddard Space Flight Center, Greenbelt, Maryland, USA

This paper presents an overview of what we know about energetic particles from the sun, discusses where progress still needs to be made, and briefly enumerates what steps need to be taken next in order to better understand how and where solar electrons and ions are accelerated to high energies.

1. INTRODUCTION

Instruments on-board spacecraft beyond the earth's magnetosphere observe energetic particles - electrons and ions - with intensities which vary substantially in time, space, energy, elemental composition, and origin. Some of these particles, called Galactic Cosmic Rays (GCRs), are known to originate from beyond the solar system. Others, called Anomalous Cosmic Rays (ACRs), are known to be accelerated at the solar wind termination shock. Yet other energetic particles in space are associated with planetary magnetospheres and with quasi-stationary interplanetary shocks (Corotating Interaction Regions (CIRs)). The remaining particles are the subject of this paper: Solar Energetic Particles (SEPs). SEPs are energetic electrons (~1 keV to tens of MeV) and ions (~50 keV/nucleon to ~10 GeV) accelerated in association with events initiated at the sun. This definition includes particles which are accelerated well away from the sun by interplanetary shock waves driven by solar Coronal Mass Ejections (CMEs). When such a shock passes an observer at 1 AU, a particle increase lasting for some hours and loosely associated with the shock passage may be observed. The particles in this increase are usually being accelerated locally at the shock. These particles are referred to as Energetic Storm Particles (see *Cohen*, this volume). SEP events appear as intensity increases which decay away on time scales of the order of hours to days. At one time such events were referred to as solar flare events. The term 'SEP event' came into use with the recognition that the particles in many such events are accelerated by shock waves driven by

CMEs and not by flares. The once popular term 'solar cosmic rays' has all but disappeared.

The sun has a magnetic activity cycle with a period of approximately 11 years. GCR and ACR intensities have corresponding long-term variations, being depressed at times of maximum solar magnetic activity. By contrast, solar flares, CMEs, and SEP events are more numerous when the sun is magnetically active.

Some characteristics of SEP events are illustrated in Plate 1 for a period in late August, 2002. The upper panel shows the intensities of 0.175 – 0.315 MeV electrons as 5-minute averages. These data were obtained by the EPAM instrument on the Advanced Composition Explorer (ACE) spacecraft [*Gold et al.*, 1998]; all the energetic particle data in Plate 1 are available on-line via the ACE Science Center (see Table 1). Also shown in this panel are vertical lines at the times CMEs were observed to leave the sun. These lines are labeled with the CME sky-plane speeds. Other vertical lines, labeled S, indicate times of shock passage at 1 AU. The lower panel shows 1-hour averages for He, O, and Fe at 14 MeV/nucleon (measured by the ACE/SIS instrument [*Stone et al.*, 1998]). The corresponding Fe/O ratios are shown in the middle panel. This period is dominated by a series of six SEP events all from a single active region (NOAA Region 10069). Note that all events but the second are associated with a CME. The electron event time profiles are typically shorter with faster rise times than the ion event time profiles. Thus it is easier to discern individual events in the electron data than in the ion data, where individual events may overlap. The first four events were short-lived (all ≤15 hours) and these and the fifth event had Fe intensities which exceeded the O intensities. By contrast, the coronal abundance of Fe relative to that of O is thought to be 0.134 [*Reames*, 1999, Table 9.1]. We will refer to events with Fe/O greater than two times this value at a specific energy as being Fe-rich at that energy.

Solar Eruptions and Energetic Particles
Geophysical Monograph Series 165
Copyright 2006 by the American Geophysical Union
10.1029/165GM11

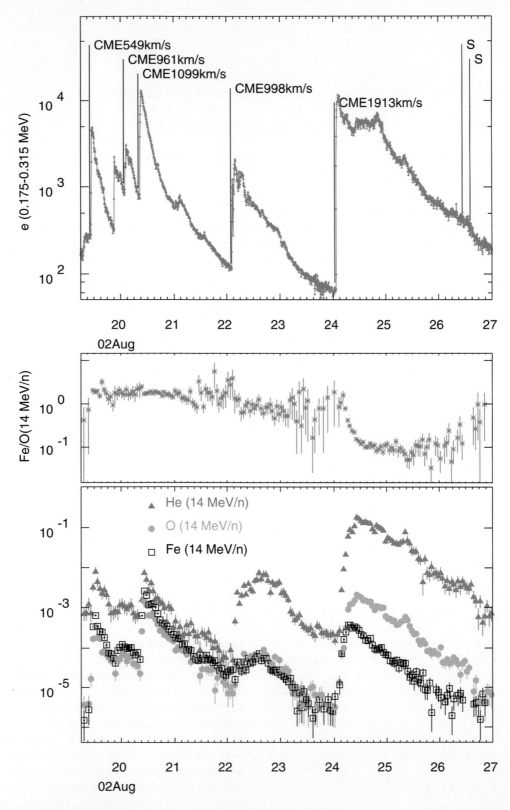

Plate 1. Electron data from ACE/EPAM and ion data from ACE/SIS for six SEP events in August, 1992. The middle panel shows the Fe/O ratio observed at 14 MeV/nucleon.

Table 1. Major Solar Missions of the Current Solar Cycle.

Spacecraft	Dates	Brief Description	Web Site
Ulysses	6 Oct 1990 – present	Particles and fields instruments; only space-craft to have achieved high solar latitudes	http://helio.esa.int/ulysses/archive/
Compton Gamma-Ray Observatory (CGRO)	5 April 1991 – 4 June 2000	Observed nuclear and e^+-e^- annihilation lines in solar flare events	http://cossc.gsfc.nasa.gov/archive/
Yohkoh	30 August 1991 – December, 2001	Soft X-ray imager + harder x-ray measurements	http://www.lmsal.com/SXT/Yohkoh
WIND	1 Nov 1994 – present	Particles and fields instruments; plasma wave instrument.	http://spds.gsfc.nasa.gov/spds.html http://lep694.gsfc.nasa.gov/waves/waves.html
Solar Heliospheric Observatory (SoHO)	2 Dec 1995 – present (out of commission from 6/1998 – 2/1999)	2 Coronagraphs, UV telescope, particles and fields instruments; in halo orbit about L1.	http://sohowww.nascom.nasa.gov/ http://cdaw.gsfc.nasa.gov/ CME_list/
Advanced Composition Explorer (ACE)	25 August 1997 – present	Observes elemental, isotopic and charge-state abundances over wide energy range; in halo orbit about L1.	http://www.srl.caltech.edu/ACE/
Transition Region and Coronal Explorer (TRACE)	2 April 1998 – present	Full sun image mosaics; beautiful images of flaring loops and arcades; in orbit around earth.	http://hea-www.harvard.edu/ SSXG/ kathy/flares/flares.html
Ramaty High Energy Solar Spectrometer Imager (RHESSI)	5 February 2002 – present	High resolution Ge detectors for observing gamma-ray lines; coded aperture mask for precise imaging	http://hessi.ssl.berkeley.edu/ http://hesperia.gsfc.nasa.gov/hessi http://sprg.ssl.berkeley.edu/~krucker/hessi_plots/

Note that abundances can be energy dependent (e.g., the event of 20 April 1998 was Fe-rich at low energies [*Tylka, Reames and Ng*, 1999] and Fe-poor at higher energies [*von Rosenvinge et al.*, 1999]). The fourth event has a much lower He/Fe ratio than any of the other events in Plate 1 and Fe/O (at 14 MeV/nucleon) is also >1 for this event. The fifth event was also Fe-rich at 14 MeV/n (Fe/O ~0.7) and lasted roughly 40 hours. The sixth event was a Ground Level Event (GLE) and lasted for about 3 days. A GLE is an event which is detected by neutron monitors at sea level. The event-averaged Fe/O ratio for this event increased with increasing kinetic energy, becoming Fe-rich above 14 MeV/nucleon.

Our challenge is to make sense out of the variety of features shown in SEP events, of which only a small (but rich) sample is shown in Plate 1. What differences in seed-particle populations, acceleration mechanisms, and locations give rise to such varied observations at 1 AU? Is the behavior shown by the final event of Plate 1 at all affected by the occurrence of prior events? We shall see that answering such questions is not easy.

We are presently near the end of the most recent period of maximum solar magnetic activity (Cycle 23 nominally started in June, 1996). This period has been actively studied by modern instruments on multiple spacecraft which are summarized briefly in Table 1: Yohkoh, Ulysses, CGRO, WIND, SoHO, TRACE, ACE, and RHESSI. As a result, this is a good time to review recent progress made in understanding SEPs. At the same time, we have to acknowledge that progress has been rather slow. We will begin by considering some of the reasons for this.

1.1. Why Has Progress Been Slow?

First of all, the sun is far away from an observer at 1 AU, spanning only half a degree in the sky. Secondly, events at the sun can be complex. Electromagnetic emissions associated with solar events include Type II, III, and IV radio bursts, visible wavelengths (e.g. the H_α hydrogen line), ultraviolet, X-rays, and γ-rays. While these emissions can in principle be interpreted to tell us what is happening at the sun, this is difficult because different emissions come from different altitudes in the solar atmosphere and thermal equilibrium generally doesn't apply. Figure 1, taken from *Yokoyama and Shibata* [2001], illustrates a magnetic reconnection model for

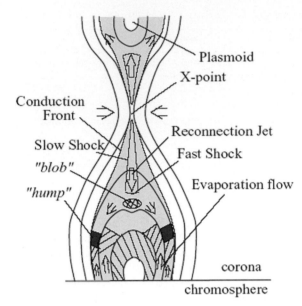

Figure 1. Flare model of *Yokoyama and Shibata* [2001].

an eruptive solar flare. Magnetic reconnection near the X-point drives a fast downward shock which can accelerate particles at the top of the magnetic loop structure, and there is an upward shock as well. Alternatively, turbulence from this event can produce stochastic acceleration. Particles can possibly be accelerated by electric fields created by reconnection [*Holman*, 1996]. What's more, the plasmoid at the top of the figure will be ejected as a CME, which in turn drives a magnetic shock wave which can accelerate particles. The purpose of Figure 1 at this juncture is only to illustrate the level of complexity of what may be occurring at the sun for many events. Whether any of this complexity is relevant to SEPs observed in space will be addressed later.

Associating particles at 1 AU with events at the sun is an art. For example, particles are sometimes seen from events which are largely invisible because they occur behind the visible disk. Propagation of particles from the sun to the observer can vary significantly from event to event. The average interplanetary magnetic field is an Archimedean spiral which favors particle propagation from the sun to earth for particles which leave the sun at ~W50 ±30 degrees (e.g. [*Van Hollebeke et al.*, 1975]). Successive CMEs from active regions can however lead to interplanetary magnetic fields which are significantly different from average, connecting the observer to unexpected longitudes. Finally, there is significant velocity dispersion associated with particle energy, pitch angle, and field geometry, leading to variable propagation times from the sun to 1 AU. Sometimes different observers associate the same particle event with different events at the sun. Such discrepancies, when they occur, make it difficult to draw correct conclusions. The Plasma Waves

instrument on the WIND spacecraft has proven particularly useful for identifying flares associated with specific particle events. By following the associated Type III burst from the lowest frequencies, that are seen near particle onset, to the highest frequencies provides a time marker for the event at the sun.

Another difficulty is that our views of CMEs, so far, are 2-dimensional projections onto the sky. Reported CME speeds are sky-plane speeds. Similarly, *in situ* observations are frequently limited because they are made at only a single point in space, most often at 1 AU. The Helios 1 and 2 spacecraft, launched in 1974 and 1976, went as close to the sun as 0.28 AU, but even then they spent most of their time near 1 AU. The STEREO Mission, to be launched in early 2006, will provide the first stereoscopic views of CMEs by carrying coronagraphs on two separate spacecraft.

Yet another difficulty is that there are various kinds of gaps in data coverage. There are gaps between missions, gaps due to saturation of instrument response in large events, and gaps due to spacecraft being on the backside of the earth or in the South Atlantic Anomaly (a region of intense magnetospheric radiation). There can also be long intervals without specific types of instruments being available in space. Finally, solar active periods come and go approximately every 11 years. Spacecraft missions are only loosely timed to this cycle. Some researchers are unable to bridge funding gaps between active periods.

The next section will consider different models for SEP acceleration. It will be seen that many fundamental ideas were first introduced a long time ago but validating them has been slow.

2. MODELS FOR SEP ACCELERATION

In 1963, *Wild, Smerd and Weiss* proposed a two phase model for SEP acceleration. This model was based upon their studies of solar radio bursts. The two phases were as follows:

1. The first phase consists of "a succession of bursts of electrons (~100 keV), the acceleration of each burst being accomplished in a very short time (~1 sec). …. The acceleration of protons to high energies need not be involved in this phase."
2. "The second phase, occurring only in large flares, is initiated directly by the first: the sudden release of energy sets up a magnetohydrodynamic shock wave which travels out through the coronal plasma and creates conditions suitable for Fermi acceleration of protons and electrons to very high energies (<~BeV)."

The second phase was inferred from metric Type II bursts. The framework of this model is similar to what many believe today but with the two phases giving rise to two classes of particle events and now knowing that ions, including protons, are indeed accelerated in the first phase. The definitive indication of the latter was found by *Forrest and Chupp* in 1983. They showed that the time profile of 4.2-6.4 MeV gamma-rays (due to energetic ions >10 MeV) was delayed from the time profile of 35-114 keV hard X-rays (due to energetic electrons) by only 1-2 seconds. *Forrest and Chupp* [1983] pointed out that energetic particle intensities in space are poorly correlated with gamma-ray flare intensities and suggested that SEPs may be due to second phase acceleration. Another modification is that second phase acceleration is often attributed to a CME-driven shock rather than to a flare wave as proposed by *Wild, Smerd and Weiss*; CMEs were unknown in 1963. Opinion is divided, however, as to whether the metric Type II bursts are created by CME-driven shocks (e.g. compare *Cliver* [1999] with *Gopalswamy et al.* [1997], *Cane et al.* [2002], and *Cane and Erickson* [2005]). Furthermore a second phase of gamma rays has been seen in some events (the 11 June 1991 event produced gamma-rays above 50 MeV for 11 hours [*Kanbach et al.*, 1993]). As pointed out by *Hudson and Ryan* [1995], the prolonged acceleration of the hard proton spectrum responsible for photons >50 MeV cannot be produced by a coronal shock because such a shock would soon leave the sun. In a further complication, *Cane et al.* [2002] have pointed out that long lasting type III emissions, and not type II bursts, are the dominant feature in the dynamic spectra of SEP-associated flare events. These type III emissions, which they called Type III-*l*, start at relatively low frequencies, last a long time, and occur after the flare impulsive phase. They suggested that Type III-*l* emission is indicative of ion acceleration, is associated with reconnection of magnetic field lines below the CME, and is unlikely to be due to shock acceleration. It is therefore likely that second phase acceleration is basically the same as first phase acceleration but in a different environment: a lower density, higher loop system (e.g., see *Qiu et al.* [2004]). This would then mean that acceleration at the CME-driven shock would constitute a third phase of acceleration (cf. [*Klecker et al.*, 1990]). Note that impulsive phase Type III bursts are often absent in large SEP events [*Švestká and Fritzova-Švestková*, 1974; *Cane and Reames*, 1988], implying that impulsive phase particles do not escape to the interplanetary medium.

Pallavacini et al. [1977], based upon their study of soft X-ray limb flares observed from Skylab, proposed that there are two classes of solar flares:

- "Events of class I are compact flares with smaller volume ($\sim 10^{26} - 10^{27}$ cm^3), and lower height ($\leq 10^4$ km); faster rise and decay times, and shorter duration ($\sim$tens of minutes). They have higher energy density ($\sim 10^2 - 10^3$ ergs cm^{-3}), do not appear to be associated with white-light coronal transients, and are located very low at the base of active regions; ..."

- "Events of class II are long-enduring events ($\sim$hours) with longer rise and decay times and greater height ($\sim 5 \times 10^4$ km). They have larger volume ($\sim 10^{28} - 10^{29}$ cm^3) and lower energy density ($10 - 10^2$ ergs cm^{-3}) and are associated with prominence eruption or activation and with white-light coronal transients."

These two classes of soft X-ray events were subsequently described as impulsive and gradual events, respectively. *Cane et al.* [1986] divided >3 MeV electron events into two classes based on the soft X-ray flare time scales i.e. with impulsive and gradual soft X-ray events. They found that the particle events had different properties consistent with different acceleration processes: 'flare' acceleration for impulsive events and shock acceleration for gradual events. This division of events was similar to that discussed by *Lin* [1970, 1974] who described electron and proton events and found that the electron (proton-poor) events were associated with small flares. The relevance of flare time scale was first found by *Sarris and Shawhan* [1973], later noted by *Kocharov et al.* [1983], and also discussed by *Bai* [1986]. *Reames* [1988] further associated ^{3}He/^{4}He and Fe/O enhancements with impulsive events, and found that at low energies gradual events had coronal abundances. Up until this point it had been unclear whether the two classes of SEP events were opposite ends of a continuum or were distinct. However *Reames* [1988] reported a bimodal distribution of Fe/O, which suggested that the two classes were distinct.

2.1. Properties of Impulsive and Gradual Events

Based on the foregoing, the generally accepted model at the beginning of the most recent solar maximum was the one championed by *Reames* [1997]. In this model there are two different kinds of events, impulsive and gradual, with the characteristics given in Table 2. Impulsive events are generally small, whereas gradual events are larger (i.e. last longer and have higher peak intensities particularly for protons). The particles in impulsive events are considered to be solar flare particles produced by resonant stochastic acceleration. *Mandzhavidze et al.* [1999] determined that in gradual flares the prompt gamma rays originate by the same mechanism as in impulsive events. In gradual SEP events ions are accelerated by a CME-driven shock and in the Reames paradigm there are no flare particles because they cannot escape (see *Reames* [2002]). Earlier *Cane et al.* [1991] (see also *Reames et al.* [1990]) had suggested that they could escape: "The abundances as a function of longitude can be understood in terms of two particle sources, namely (1) particles originating out of flare-heated material and (2) particles accelerated at coronal and interplanetary shocks." We shall return to this point later.

Table 2. Properties of Impulsive and Gradual Solar Particle Events [*Reames*, 1997] in italics. Modifications discussed in this paper are shown in red in parentheses.

	Impulsive	Gradual
Particles	*Electron-rich*	*Proton-rich*
$^3He/^4He$	*~1*	**(>)** *~0.0005*
Fe/O	*~1*	*~0.1* **(0.01-1)**
(Enhancements of (Z > 33)/O)	**(~x100-10,000[a])**	**(x 0.2 – 20[a])**
H/He	*~10*	*~100*
Q_{Fe}	*~20*	**(>)** *~14*
Duration	*Hours*	*Days*
Longitude Cone	*< 30°*	*~180°*
Radio Type	*III, V* **(II)**	*II, IV* **(III-*l*)**
X-Rays	*Impulsive*	*Gradual*
Coronagraph	*–*	*CME [96%]*
	(CMEs OFTEN)	**(ALWAYS)**
Solar Wind	*–*	*IP Shock*

[a] relative to solar corona; from *Reames and Ng* [2004].

Returning to Plate 1, the first 5 events appear to be impulsive events based on the Fe/O ratio. The original version of Table 2 implied that there were no CMEs associated with impulsive events. Of the five impulsive events in Plate 1, however, four are accompanied by CMEs. *Kahler et al.* [2001] conceded that impulsive events might occasionally be associated with CMEs but dismissed them as "narrow CMEs". Event 4 is listed in the SoHO CME catalog as >122° in width; the CME of the 5th event is listed as a halo event (360°). Event 4 was the largest impulsive event above 10 MeV/nucleon during the recent solar maximum period [*Leske et al.*, 2003].

The entries in Table 2 which are presented in parentheses were not in the original table. They represent modifications which were made to accommodate new observations. These will be discussed shortly. In actual fact it had been recognized much earlier that large (gradual) events sometimes had significant enhancements or depletions of heavy elements relative to coronal abundances (e.g. *McGuire et al.* [1986]; *Meyer* [1985]; *Breneman and Stone* [1985]), a fact which was ignored in constructing the original table. This oversight no doubt contributed to the confusion which ensued when gradual events first observed by ACE frequently had heavy-element enhancements.

Note that some researchers (e.g. *Cliver* [1996] and *Kocharov and Torsti* [2002]) have suggested that there are hybrid or mixed events in which there are particles from both acceleration mechanisms. Although the concept is reasonable, and under debate (see Section 3), the more limited proposal that first phase particles escape and make a significant

contribution in gradual events seems less likely. The primary reason for this is the absence of Type III emission coincident with the impulsive phase, suggesting that there are no open field lines (also see *Reames* [2000]).

2.2. Computational Models for SEP Events

A complete mathematical model for SEP acceleration and propagation is not currently available. Various individuals are working to develop such models, but the models are incomplete and are not readily available. Given the lack of good predictive models, early analyses of ACE observations utilized two different approaches.

The first approach was based upon two empirical observations: (a) SEP event-averaged abundances are related to coronal abundances via a power law in Q/M as shown in Figure 2 [*Breneman and Stone*, 1985], and (b) coronal abundances are related to photospheric abundances via a function of First Ionization Potential (FIP) [*Hovestadt*, 1974; *Webber*, 1975]. The latter function is frequently taken to be a step function, with elements with FIP below ~10 eV being ~4 times more abundant than elements above ~10 eV. One problem with applying this picture was that the Q/M ratios were not well known for any given event. *Cohen et al.* [1999] turned the problem around and used the event-averaged abundances from the SEP event of November 6, 1997, to

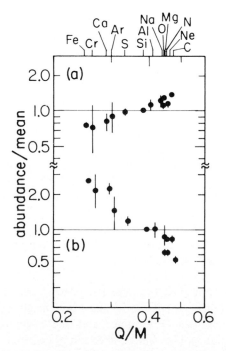

Figure 2. Shows the dependence of elemental abundances on Q/M for two different SEP events. From *Breneman and Stone* [1985].

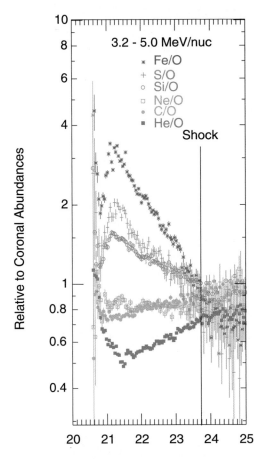

Plate 2. LEMT data for the 20 April 1998 event from *Tylka et al.* [1999].

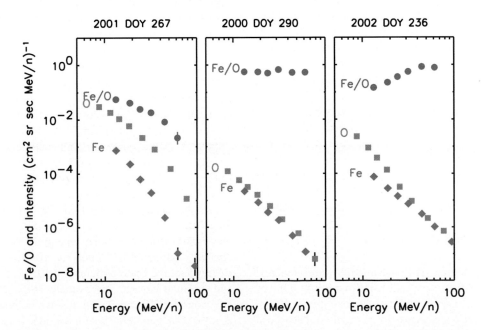

Plate 3. Shows Fe and O spectra and the corresponding Fe/O ratio versus kinetic energy for three different SEP events. The three events start on September 24, 2001, October 16, 2000, and August 24, 2002, respectively. (Courtesy, *C.M.S. Cohen*).

infer the Q/M values and thence a source region temperature of $3 - 6 \times 10^6$ K, considerably higher than that inferred at lower energies. Notice that Figure 2 shows two extreme events: the upper event has a considerably lower Fe/O ratio than the lower event.

Breneman and Stone [1985] concluded that Q/M is the principal organizing parameter for the fractionation of SEPs by acceleration and propagation processes and for event-to-event variability. Such a dependence, they said, is "not unexpected since acceleration and propagation processes depend on the rigidity of the ions which is inversely proportional to Q/M". They did not address the fact that protons, the one component for which the value of Q/M seems certain (= 1), did not fit on the curves shown in Figure 2. Helium also does not fit well. Today we would ascribe this to the fact that protons generate waves in a highly non-linear process and the relative abundances depend not only on Q/M but also on the wave spectrum.

The second approach was based upon an *ansatz* proposed by *Ellison and Ramaty* [1985], in which shock acceleration was pictured as a power law in kinetic energy/nucleon with an exponential roll-off at higher energies. This approach was successfully followed by, for example, *Mazur et al.* [1992] and by *Tylka et al.* [2000]. The roll-off was intended to take into account that above some rigidity particles would escape from the shock and no longer be accelerated. Hence, for element X, one expected

$$dj/dE_X \sim C_X \cdot E^{-\gamma} \cdot \exp(-E/E_{0X}),$$

where $E_{0X} = E_{0H} \cdot Q_X / M_X$ and E = kinetic energy/nucleon. The subscript H refers to hydrogen and Q_X / M_X is the charge to mass ratio for element X. Following *Mazur et al.* [1992], it was presumed that the seed particle population was the solar wind, so C_X was taken to be a measure of the coronal abundance.

The expected Fe/O ratios for two different values of Q_{Fe}, 15 and 20, are shown in Figure 3. It may be seen that in this picture the Fe/O ratio never goes above the coronal abundance (~0.134). Figure 3 also shows that Fe/O monotonically declines with increasing energy, even given a combination of different charge states. Thus event 6 in Plate 1 is inconsistent with the Ellison and Ramaty *ansatz* (see also Plate 3, panel 3).

3. OBSERVATIONS

A number of notable new SEP instruments were launched prior to the onset of the most recent solar maximum. One of these was the Low Energy Matrix Telescope (LEMT) launched on the WIND spacecraft in November, 1994 [*von Rosenvinge et al.*, 1995]. The combination of a large geometry factor and high-speed on-board particle identification

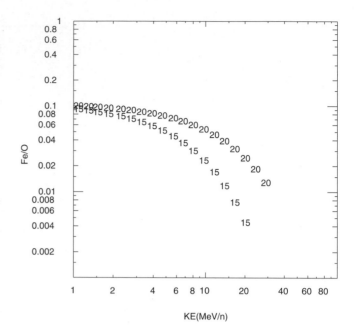

Figure 3. Shows the expected Fe/O ratios as a function of kinetic energy for two values of Q_{Fe}, 15 and 20, following *Ellison and Ramaty* [1985].

made possible the detection of high Z elements with good statistics as a function of time during solar events. This is illustrated in Plate 2 for the event of 20 April 1998. The ordering of these data by Q/M is readily apparent. In addition, LEMT was the first instrument in almost 30 years to confirm the acceleration of nuclei with Z>60 [*Reames*, 2000] after the first reported observation by *Shirk and Price* [1973], who used the Apollo16 module window as a detector. In particular, *Reames* [2000] was able to clearly associate these increases with impulsive events and to show monotonically larger enhancements relative to coronal abundances with increasing Z. We have accordingly added a whole new row to Table 2. For example, the elements observed above Z = 50 in impulsive events show an enhancement relative to O of ~1000 (see also *Reames and Ng* [2004]). *Shirk and Price* [1973] lacked time resolution and associated the ultraheavies they observed with a gradual event.

The Advanced Composition Explorer (ACE) spacecraft, launched in August, 1997, also carried a number of notable new instruments: the Solar Energetic Particle Ionic Charge Analyzer (SEPICA) [*Möbius et al.*, 1998], the Ultra Low Energy Isotope Spectrometer (ULEIS) [*Mason et al.*, 1998], and the Solar Isotope Spectrometer (SIS) [*Stone et al.*, 1998]. For example, SIS can cleanly resolve individual isotopes up to Z = 30. The geometry factor of SIS is about the same as for LEMT but its energy range is higher, starting at about 10 MeV/nucleon as compared to ~2 MeV/nucleon for LEMT.

Mason et al. [2004], using ULEIS, have confirmed the findings of *Reames* [2000]. They conclude that the enhancements of ultraheavies pose a problem for understanding flare acceleration. In particular, they conclude that cascading turbulence, a possible cause of Fe enhancements in impulsive events, cannot account for the enhancements of ultraheavies.

Prior to Solar Cycle 23, it was difficult to measure ^{3}He intensities less than 10% of the ^{4}He intensity due to background due to ^{4}He and limited isotope resolution. Since the coronal abundance of ^{3}He is approximately 1 part in 2500 of coronal ^{4}He, there is a lot of room for events to have ^{3}He less than 10% of ^{4}He and still have large enhancements of ^{3}He. The entry in Table 2 indicating that gradual events have ^{3}He/^{4}He ~0.0005 was not based on measurements at SEP energies but on measurements at solar wind energies together with the idea that gradual events would reflect the composition of a solar wind seed population. As a result of actual measurements it has become clear that many gradual events have ^{3}He/^{4}He considerably greater than 0.0005 [*Mason et al.*, 1999; *Wiedenbeck et al.*, 2003; *Torsti et al.*, 2003].

Plate 3 shows the event-averaged differential intensity spectra from SIS for Fe and O together with the Fe/O ratio as a function of kinetic energy/nucleon for each of three different SEP events. The first event, that of 24 September 2001, is consistent with the Ellison and Ramaty *ansatz* shown in Figure 3. The second event (16 October 2000) shows an Fe/O ratio which is flat as a function of energy. One could imagine that the spectral shapes are consistent with the Ellison and Ramaty *ansatz* if, for example, the exponential breaks occur at higher energies than are shown in the figure. On the other hand the ratio is about 5 times the coronal value, which is unexpected. Finally, the third event (24 August 2002, the last event in Plate 1) has Fe/O starting out at about the coronal value at about 12 MeV/nucleon but then increasing to a much higher value at higher energies, distinctly unexpected in the Ellison and Ramaty formulation.

von Rosenvinge et al. [2001] compared the event-averaged abundances measured by SIS for 5 different events compared with gradual event average composition and with average impulsive event composition. The gradual and impulsive event average compositions were taken from *Reames* [1999, Table 9.1]. These 5 events all show impulsive-like composition but otherwise have the characteristics of gradual events, i.e. association with large CMEs and evidence of shocks. They are all associated with Type III-*l* bursts. In fact, gradual events are almost always associated with Type III-*l* bursts [*Cane et al.*, 2002]. We have therefore added Type III-*l* in parentheses to Table 2 for gradual events.

Figure 4 shows the ratio Fe/O versus time for 72 different SIS events, all relative to the Reames reference value. Each of these events is related to a relatively large, fast CME and would be considered to be a gradual event. Each data point

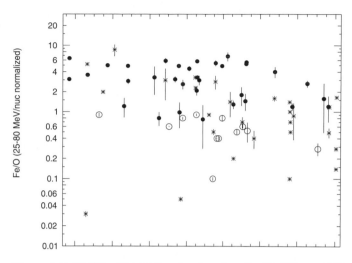

Figure 4. Fe/O (25 – 80 MeV/nucleon) vs time for 72 different SEP events; each point is normalized to 0.134, the nominal coronal Fe/O value; see text for a description of the different symbols.

represents Fe and O integrated over the event from 25 MeV/nucleon to 80 MeV/nucleon. The filled circles are events in which the particles peak promptly after the associated flare and the unfilled circles are events in which the particles peak at the 1 AU passage of the associated shock. The remaining events have a variety of different profiles. Note that the percentage of shock associated events increased with time. The lowest Fe/O ratio corresponds to the 20 April 1998 event. Note also that the normalized Fe/O ratio for this event, as shown in Plate 2 at a lower energy, is well above 1. These results, particularly prior to 2000, led to considerable confusion as to how they fit into the prevailing picture of gradual and impulsive events. Whereas Table 2 originally said that gradual events have Fe/O ~0.1 (approximately the Reames coronal value), the observations show that Fe/O at high energies varies from <0.1 to ~8 times coronal.

It's clear from Figure 4 that events with strong shock effects at 1 AU are the events for which the Fe/O ratio is depressed, as expected by *Ellison and Ramaty* [1983]. *McGuire et al.* [1986] had already noticed that heavy element abundances appeared to be depressed in some shock-related events. In fact, although *McGuire et al.* [1986] reported on only 15 events, they also saw several events with Fe/O increasing with respect to energy and events with Fe/O abundances substantially enhanced relative to coronal abundances, similar to the more extensive SIS results reported here.

So how do we interpret these results? If we stick to the picture where the particles in large SEP events are caused only by acceleration at a CME-driven shock this obviates the need to consider the complications associated with the features shown in Figure 1. It has been suggested [*Reames*, 2002] that transient shock effects could explain Fe/O enhancements as measured in the 20 April 1998 event discussed by *Tylka et al.* [1999]. However the event of 6 November 1997 had a rather constant value of Fe/O over a period of two and a half days. It is hard to explain such a sustained enhancement of Fe/O on a transient basis.

At about the same time, *Mason et al.* [1999] proposed the idea that ^{3}He/^{4}He enhancements in large SEP events could be due to shock acceleration of suprathermal remnants from preceding impulsive events. *Desai et al.* [2001] observed such acceleration at 1 AU. This led to the suggestion that heavy element enhancements could have the same source. This shall be referred to as the 'suprathermal remnants model'. Papers proposing this view include *Tylka et al.* [2000], and *Tylka et al.* [2001]. An alternative model, which we shall call the 'concomitant flare model', proposes that heavy element enhancements are due to particles from the associated flares. Papers espousing this view include *Van Hollebeke et al.* [1990], *Cane et al.* [1991], *von Rosenvinge et al.* [2001], *Torsti et al.* [2003], and *Cane et al.* [2003]. This could include acceleration due to reconnection behind the CME in a flare process separate from, but similar to, the impulsive phase as previously discussed in section 2.1. This distinguishes the concomitant flare model from the direct flare model critiqued by *Tylka et al.* [2005]. *Tylka et al.* [2005] have proposed that perpendicular shocks preferentially accelerate particles with 'impulsive' composition and charge states. These could be suprathermal remnants, but *Tylka et al.* [2005] have also acknowledged the possibility that they might come from the associated flare if open field lines connect the flare site to the shock front. Thus the *Tylka et al.* [2005] concept of preferential acceleration by perpendicular shocks can be incorporated into either the suprathermal remnant model or the concomitant flare model.

Yet another idea has been that it might be possible, given the right turbulence spectrum, for shock acceleration to produce harder spectra for heavier elements than for lighter elements, thereby leading to heavy element enhancements at higher energies [*Cohen et al.*, 2003]. One observation which favors the suprathermal remnant and concomitant flare models appears in Figure 5 (from Table 1, *Leske et al.*, [2001]). This shows a correlation between the Fe charge state and the ratio Fe/O. Note that the largest values of Fe/O are well above the coronal value and the corresponding charge state is ~20; these correspond to the expected values for impulsive material (see Table 2).

Attempts to select between the remnant suprathermal model and the concomitant flare model have both raised important questions. *Mewaldt et al.* [2003b] questioned whether there were enough remnant suprathermals to account for the observed intensities of heavy elements. They estimated that, for all but one out of 26 different large gradual

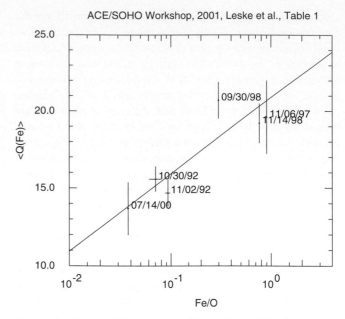

Figure 5. $\langle Q_{Fe} \rangle$ vs Fe/O as reported by *Leske et al.* [2001].

events, the number of heavy ions required exceeds the upper bound to the number which would have been available at random 75% of the time. Unless the remnants available are not randomly produced prior to large gradual events (a definite possibility), then it would appear that there are not enough suprathermals to account for the enhanced heavies in Figures 7 and 8. This would favor the concomitant flare model.

Mewaldt et al. [2003a] provide an alternative attempt to differentiate between the remnant suprathermal model and the concomitant flare model by examining the velocity dispersion of the first arriving particles in a number of different events. They plotted the arrival time of the very first particles of a given energy to arrive at ACE versus c/v for the given energy. The good fit to a straight line indicates that particles of all energies were first released at the same time. The event of 1 May 2000 was an impulsive event with a CME. The earliest particles left the sun at about the same time as the corresponding flare, as one might expect for an impulsive event. This time approximately coincided with the CME departure time. The 6 May 1998 event was similar, however it would be an unusual impulsive event since it is a GLE and had a fast CME (see *von Rosenvinge et al.* [2000] for further discussion). On the other hand, other particle events showed delays on the order of 40-50 minutes after the flare. One way to interpret these delays would be to say that the CME-driven shock is delayed in forming due to the high Alfvén speed close to the sun. This would seem to support the suprathermal remnant model. Alternatively, particles accelerated in the reconnection region of the concomitant flare model could traverse the CME-driven shock on their way to the observer, receiving further acceleration and delay. Near-relativistic electrons are also delayed relative to the flare start times/CME departure times. *Haggerty et al.* [2002] have interpreted these delays as implying that the electrons are accelerated and released by the CME-driven shock. Also see *Krucker and Lin* [2000] for a similar interpretation. The situation is not clear. Certainly it is unlikely that the high energy particles would be accelerated immediately at the time flare emissions are first seen, so the relevant fiducial should be some time afterwards; some flares last for hours. Another complication is the extent of interplanetary scattering.

In summary, the situation is not entirely favorable for either the suprathermal remnants model or for the concomitant flare model.

4. CONCLUSION

The goal of this paper was to provide an overview of Solar Energetic Particle phenomena and of our current state of understanding. While there is an extensive body of observations, our understanding of these phenomena is still not complete. RHESSI (see Table 1) is a relative newcomer on the scene and collaborative data analysis of data from WIND, ACE, and RHESSI can be expected to shed new light on the open questions discussed in this paper. Steps which need to be taken in the future to improve our understanding include:

Take observations much closer to the sun in order to greatly reduce the confusion due to slow arrival of low energy ions, interplanetary scattering, and the unknown topology of the interplanetary field.

Develop much better computational models which include phenomena taking place behind CME-driven shocks.

Work towards understanding acceleration better. Specifically, we need to understand how particles can be accelerated to 5 GeV in a matter of ~10 minutes, and how gamma-rays above 50 MeV can be continuously emitted from the sun for periods over 10 hours.

Maintain ACE, RHESSI, SoHO, TRACE, and WIND into the STEREO era so that the complementary data needed for STEREO remain available.

Acknowledgments. TvR would like to acknowledge various contributions from his ACE/SIS team-mates: A.C. Cummings, C.M.S. Cohen, R.A. Leske, R.A. Mewaldt, E.C. Stone, and M.E. Wiedenbeck. This work was supported by NASA at the Goddard Space Flight Center and at the California Institute of Technology and JPL under NASA Grant NAG5-12929.

REFERENCES

Bai, T., Two classes of gamma-ray/proton flares: Impulsive and gradual, *Astrophys. J.* 308, 912, 1986.

Breneman, H.H. and E.C. Stone, Solar coronal and photospheric abundances from solar energetic particle measurements, *Astrophys. J.* 299, L57-L61, 1985.
Cane, H.V. and W.E. Erickson, Solar type II radio bursts and IP type II events, *Astrophys. J.* 623, 1180, 2005.
Cane, H.V., W.E. Erickson, and N.P. Prestage, Solar flares, type III radio bursts, coronal mass ejections, and energetic particles, *J. Geophys. R.* 107, 1315, doi:10.1029/2001JA000320, 2002.
Cane, H.V., R.E. McGuire, T.T. and von Rosenvinge, Two classes of solar energetic particle events associated with impulsive and long-duration soft X-ray flares, *Astrophys. J.* 301, 448, 1986.
Cane, H.V., and D.V. Reames, Soft X-ray emissions, meter wavelength radio bursts and particle acceleration in solar flares, *Astrophys. J.* 325, 895, 1988.
Cane, H.V., D.V. Reames, and T.T. von Rosenvinge, Solar particle abundances at energies greater than 1 MeV per nucleon and the role of interplanetary shocks, *Astrophys. J.* 373, 675, 1991.
Cane, H.V., T.T. von Rosenvinge, C.M.S. Cohen, and R.A. Mewaldt, Two components in major solar particle events, *Geophys. Res. Lett.* 30, 8017, doi:10,1029/2002GL016580, 2003.
Cliver, E.W., Solar flare gamma-ray emission and energetic particles in space, AIP Conf. Proc. 374, *High Energy Solar Physics*, Ramaty, R., Mandzhavidze, N., and Hua, X.-M., Eds., 45, 1996.
Cliver, E.W., Comment on "Origin of coronal and interplanetary shocks: A new look with Wind spacecraft data" by N. Gopalswamy et al., *J. Geophys. Res.* 104, 4743, 1999.
Cohen, C.M.S., A.C. Cummings, R.A. Leske, R.A. Mewaldt, E.C. Stone, B.L. Dougherty, M.E. Wiedenbeck, E.R. Christian, and T.T. von Rosenvinge, Inferred charge states of high energy solar particles from the Solar Isotope Spectrometer on ACE, *Geophys. Res. Lett.* 26, 149, 1999.
Cohen, C.M.S., R.A. Mewaldt, A.C. Cummings, R.A. Leske, E.C. Stone, T.T. von Rosenvinge, and M.E. Wiedenbeck, Variability of spectra in large solar energetic particle events, *AdSpR* 32, 2649-2654, 2003.
Desai, M.I., G.M. Mason, J.R. Dwyer, J.E. Mazur, C.W. Smith, and R.M. Skoug, Acceleration of ³He nuclei at interplanetary shocks, *Astrophys. J.* 553, L89, 2001.
Ellison, D., and R. Ramaty, Shock acceleration of electrons and ions in solar flares, *Astrophys. J.* 298, 400, 1985.
Forrest, D.J., and E.L. Chupp, Simultaneous acceleration of electrons and ions in solar flares, *Nature* 305, 291, 1983.
Gold, R.E., S.M. Krimigis, S.E. Hawkins, D.K. Haggerty, D.A. Lohr, E. Fiore, T.P. Armstrong, G. Holland, and L.J. Lanzerotti, Electron, Proton, and Alpha Monitor on the Advanced Composition Explorer Spacecraft, *Space Sci. Rev.* 86, 541, 1998.
Gopalswamy, N., M.R. Kundu, P.K. Manoharan, A. Raoult, N. Nitta, and P. Zarkas, X-ray and radio studies of a coronal eruption: shock wave, plasmoid, and coronal mass ejection, *Astrophys. J.* 486, 1036, 1997.
Haggerty, D.K., and E.C. Roelof, Impulsive near-relativistic solar electron events: Delayed injection with respect to solar electromagnetic emission, *Astrophys. J.* 579, 841, 2002.
Holman, G.D., Particle acceleration by DC electric fields in the impulsive phase of solar flares, AIP Conf. Proc. 374, *High Energy Solar Physics*, Ramaty, R., Mandzhavidze, N., and Hua, X.-M., Eds., 479, 1996.
Hovestadt, D., in Solar Wind III, C.T. Russell, Ed., (Los Angeles, Institute of Geophysics and Planetary Physics, Univ. of California), 2, 1974.
Hudson, H., and J. Ryan, High-energy particles in solar flares, *Ann. Rev. Astron. Astroph.* 33, 239, 1995.
Kahler, S.W., D.V. Reames, and N.R. Sheeley, Jr., Coronal mass ejections associated with impulsive solar energetic particle events, *Astrophys. J.* 562, 558, 2001.
Klecker, B., E.W. Cliver, S.W. Kahler, and H.V. Cane, Particle acceleration in solar flares, *EOS* 71(39), 1102, 1990.
Kocharov, G.E., G.A. Kovaltsov, and L.G. Kocharov, Generation of accelerated particles and hard radiation during solar flares, Proc. 18th Int. Cosmic Ray Conf. (Bangalore) 4, 105, 1983.
Kocharov, L. and J. Torsti, Hybrid solar energetic particle events observed on board SOHO, *Solar Phys.* 207, 149, 2002.
Krucker, S., and R.P. Lin, On the solar release of energetic particles detected at 1 AU, *AIP Conf. Proc.* 528, 87, 2000.
Lee, M.A., Coupled hydromagnetic wave excitation and ion acceleration at interplanetary traveling shocks, *J. Geophys. Res.* 88, 6109, 1983.
Lee, M.A., Particle acceleration and transport at CME-driven shocks, in *Coronal Mass Ejections*, Crooker, N., Jocelyn, J.A., and Feynman, J., Eds., Geophys. Monograph 99 (AGU Press) 227, 1997.
Leske, R.A., R.A. Mewaldt, A.C. Cummings, E.C. Stone, and T.T. von Rosenvinge, The ionic charge state composition at high energies in large solar energetic particle events in solar cycle 23, *AIP Conf. Proc.* 598, 171, 2001.
Leske, R.A., M.E. Wiedenbeck, C.M.S. Cohen, R.A. Mewaldt, A.C. Cummings, E.C. Stone, and T.T. von Rosenvinge, The unusual solar particle events of August 2002, Proc. 28th Int. Cosmic Ray Conf. (Tsukuba), 3253, 2003.
Lin, R.P., The emission and propagation of ~40 keV solar flare electrons I, *Solar Phys.* 12, 266, 1970.

Lin, R.P., Non-relativistic solar electrons, *Space Sci. Rev.* 16, 189, 1974.

Mandzhavidze, N., R. Ramaty, and B. Kozlovsky, Determination of the abundances of subcoronal ^{4}He and of solar flare accelerated ^{3}He and ^{4}He from gamma-ray spectroscopy, *Astrophys. J.* 518, 918, 1999.

Mason, G.M., R.E. Gold, S.M. Krimigis, J.E. Mazur, G.B. Andrews, K.A. Daley, J.R. Dwyer, K.F. Heuerman, T.L. James, M.J. Kennedy, T. LeFevere, H. Malcolm, B. Tossman, and P.H. Walpole, The Ultra-Low Energy Isotope Spectrometer (ULEIS) for the ACE spacecraft, *Space Sci. Rev.* 86, 409-448, 1998.

Mason, G.M., J.E. Mazur, and J.R. Dwyer, ^{3}He enhancements in large solar energetic particle events, *Astrophys. J.* 525, L133-L136, 1999.

Mason, G.M., J.E. Mazur, J.R. Dwyer, J.R. Jokipii, R.E. Gold, and S.M. Krimigis, Abundances of heavy and ultraheavy ions in ^{3}He-rich solar flares, *Astrophys. J.* 606, 555, 2004.

Mazur, J.E., G.M. Mason, B. Klecker, and R.E. McGuire, The energy spectra of solar flare hydrogen, helium, oxygen, and iron: Evidence for stochastic acceleration, *Astrophys. J.* 401, 398, 1992.

McGuire, R.E., T.T. von Rosenvinge, and F.B. McDonald, The composition of solar energetic particles, *Astrophys. J.* 301, 938-961, 1986.

Mewaldt, R.A., C.M.S. Cohen, D.K. Haggerty, R.E. Gold, S.M. Krimigis, R.A. Leske, R.C. Ogliore, E.C. Roelof, E.C. Stone, T.T. von Rosenvinge, and M.E. Wiedenbeck, Heavy ion and electron release times in solar particle events, 28th Int. Cosmic Ray Conf. (Tsukuba), 3313, 2003a.

Mewaldt, R.A., C.M.S. Cohen, G.M. Mason, M.I. Desai, R.A. Leske, J.E. Mazur, E.C. Stone, T.T. von Rosenvinge, and M.E. Wiedenbeck, Impulsive flare material: A seed population for large solar particle events?, Proc. 28th Int. Cosmic Ray Conf. (Tsukuba), 3229, 2003b.

Meyer, J.-P., The baseline composition of solar energetic particles, *Astrophys. J. Supp.* 57, 151-171, 1985.

Möbius, E., *et al.*, The Solar Energetic Particle Ionic Charge Analyzer (SEPICA) and the Data Processing Unit (S3DPU) for SWICS, SWIMS, and SEPICA, *Sp. Sci. Rev.* 86, 449-495, 1998.

Pallavacini, R., S. Serio, and G.S. Vaiana, A survey of soft X-ray limb flare images: The relation between their structure in the corona and other physical parameters, *Astrophys. J.* 216, 108-122, 1977.

Qiu, J., J. Lee, and D.E. Gary, Impulsive and gradual nonthermal emissions in an X-class flare, *Astrophys. J.* 603, 335, 2004.

Reames, D.V., Bimodal abundances in the energetic particles of solar and interplanetary origin, *Astrophys. J.* 330, L71-L75, 1988.

Reames, D.V., Energetic particles and the structure of coronal mass ejections, AGU *Geophysical Monograph,* 99, 217, 1997.

Reames, D.V., Solar energetic particles: Sampling coronal abundances, *Sp. Sci. Rev.,* 85, 327-340, 1998.

Reames, D.V., Particle acceleration at the sun and in the heliosphere, *Sp. Sci. Rev.* 90, 413-491, 1999.

Reames, D.V., Abundances of trans-iron elements in solar energetic particle events, *Astrophys. J.* 540, L111, 2000.

Reames, D.V., Magnetic topology of impulsive and gradual solar energetic particle events, *Astrophys. J.* 571, L63-L66, 2002.

Reames, D.V., H.V. Cane, and T.T. von Rosenvinge, Energetic particle abundances in solar electron events, *Astrophys. J.* 357, 259, 1990.

Reames, D.V. and C.K. Ng, Heavy-element abundances in solar energetic particle events, *Astrophys. J.* 610, 510, 2004.

Sarris, E.T., and S.D. Shawhan, Characteristics of electron and high-energy proton flares, *Solar Phys.* 28, 519, 1973.

Shirk, E.K., and B.P. Price, Observation of trans-iron solar flare nuclei in an Apollo 16 command module window, Proc. 13th Int. Cosmic-Ray Conf. (Denver), 2, 1474, 1973.

Stone, E.C., C.M.S. Cohen, W.R. Cook, A.C. Cummings, B. Gauld, B. Kecman, R.A. Leske, R.A. Mewaldt, M.R. Thayer, B.L. Dougherty, R.L. Grumm, B.D. Milliken, R.G. Radocinski, M.E. Wiedenbeck, E.R. Christian, S. Shuman, and T.T. von Rosenvinge, The Solar Isotope Spectrometer for the Advanced Composition Explorer, *Space Sci. Rev.* 86, 357-408, 1998.

Švestá, Z., and L. Fritzova-Švestková, Type II radio bursts and particle acceleration, *Solar Phys.* 36, 417, 1974.

Torsti, J., J. Laivola, and L. Kocharov, Common overabundance of ^{3}He in high-energy solar particles, A&A 408, L1–L4 (2003) doi: 10.1051/0004-6361:20031090

Tylka, A.J., P.R. Boberg, R.E. McGuire, C.K. Ng, and D.V. Reames, Temporal evolution in the spectra of gradual solar energetic particle events, AIP Conf. Proc. 528, *Acceleration and Transport of Energetic Particles Observed in the Heliosphere,* R.A. Mewaldt, J.R. Jokipii, M.A. Lee, E. Möbius, and T. Zurbuchen, Eds.,147-152, 2000.

Tylka, A.J., C.M.S. Cohen, W.F. Dietrich, M.A. Lee, C.G. Maclennan, R.A. Mewaldt, C.K. Ng, and D.V. Reames, Shock geometry, seed populations, and the origin of variable element composition at high energies in large gradual solar particle events, *Astrophys. J.* 625, 474, 2005.

Tylka, A.J., C.M.S. Cohen, W.F. Dietrich, C.G. Maclennan, R.E. McGuire, C.K. Ng, and D.V. Reames, Evidence for remnant flare suprathermals in the source population of solar energetic particles in the 2000 Bastille Day event, *Astrophys. J. Lett.* 558, L59, 2001.

Tylka, A.J., D.V. Reames, and C.K. Ng, Observations of systematic temporal evolution in elemental composition during gradual solar energetic particle events, *Geophys. Res. Lett.* 26, 145, 1999.

Van Hollebeke, M.A.I., L.S. Ma Sung, and F.B. McDonald, The variation of solar proton energy spectra and size distribution with heliolongitude, *Solar Phys.* 41, 189-223, 1975.

Van Hollebeke, M.A.I., J.-P. Meyer, and F.B. McDonald, Solar energetic particle observations of the 1982 June 3 and 1980 June 21 gamma-ray/neutron events, *Astrophys. J. Supp.* 73, 285-296, 1990.

von Rosenvinge, T.T., L.M. Barbier, J. Karsch, R. Liberman, M.P. Madden, T. Nolan, D.V. Reames, L. Ryan, S. Singh, H. Trexel, G. Winkert, G.M. Mason, D.C. Hamilton, and P. Walpole, The Energetic Particles: Acceleration, Composition, and Transport (EPACT) investigation on the WIND spacecraft, *Sp. Sci. Rev.* 71, 155-206, 1995.

von Rosenvinge, T.T., C.M.S. Cohen, E.R. Christian, A.C. Cummings, R.A. Leske, R.A. Mewaldt, P.L. Slocum, E.C. Stone, and M.E. Wiedenbeck, Time variations of solar energetic particle abundances observed by the ACE spacecraft, 26th Int. Cosmic Ray Conf. (Salt Lake City) 6, 131, 1999.

von Rosenvinge, T.T., C.M.S. Cohen, E.R. Christian, A.C. Cummings, R.A. Leske, R.A. Mewaldt, P.L. Slocum, O.C. St. Cyr, E.C. Stone, and M.E. Wiedenbeck, Time variations in elemental abundances in solar energetic particle events", AIP Conf. Proc. 598, *Solar and Galactic Composition,* Wimmer-Schweingruber, R., Ed., 343-348, 2001.

von Rosenvinge, T.T., C.M.S. Cohen, E.R. Christian, A.C. Cummings, R.A. Leske, R.A. Mewaldt, P.L. Slocum, E.C. Stone, and M.E. Wiedenbeck, The solar energetic particle event of 6 May 1998, *AIP Conf. Proc.* 528, 111, 2000.

Webber, W.R., Solar and galactic cosmic ray abundances- A comparison and some comments, Proc. 14th Int. Cosmic Ray Conf. (Munich) 5, 1597, 1975.

Wiedenbeck, M.E., R.A. Leske, C.M.S. Cohen, A.C. Cummings, R.A. Mewaldt, E.C. Stone, and T.T. von Rosenvinge, The ^{3}He-rich SEP events of August 2002, 28th Int. Cosmic Ray Conf. (Tsukuba), 3245, 2003.

Wild, J.P., S.F. Smerd, and A.A. Weiss, Solar bursts, *Ann. Rev. Astr. and Astrophys.* 1, 291, 1963.

Yokoyama, T. and K. Shibata, Magnetohydromagnetic simulation of a solar flare with chromospheric evaporation effect based on the magnetic reconnection model, *Astrophys. J.* 549, 1160, 2001.

Tycho von Rosenvinge, Code 661, NASA/Goddard Space Flight Center, Greenbelt, MD 20771, USA. (tycho@milkyway.gsfc.nasa.gov)

Hilary Cane, Department of Physics, University of Tasmania, GPO Box 252-21, Hobart 7001, Australia. (hilary.cane@utas.edu.au)

The Source Material for Large Solar Energetic Particle Events

R.A. Mewaldt and C.M.S. Cohen

Department of Physics, California Institute of Technology, Pasadena, California

G.M. Mason

Applied Physics Laboratory/Johns Hopkins University, Laurel MD

We review evidence regarding the origin of material accelerated in large, gradual solar energetic particle (SEP) events. According to the two-class paradigm in place at the start of solar cycle 23, impulsive SEP events accelerate heated flare material, while gradual SEP events are accelerated out of the solar wind by shocks driven by fast coronal mass ejections (CMEs). However, new data from solar cycle 23 has shown that the energetic ions in gradual events often include composition signatures associated with impulsive events, including enrichments in ^{3}He, heavy elements such as Fe, and ionic charge states indicative of ~10 MK temperatures. In addition, gradual SEP events differ in composition from bulk solar wind in several key respects, implying that solar wind is not the principal seed population for these events. Several lines of evidence show that CME-driven shocks accelerate principally suprathermal ions with velocities several times that of the solar wind. The suprathermal pool incorporates ions from impulsive solar flares and previous gradual events, CIR events, pickup ions, CME ejecta, and the suprathermal tail of the solar wind. This paper reviews evidence for the sources of ions accelerated in gradual SEP events, considers the composition and available densities of suprathermal ions, and describes models that attempt to account for the surprisingly variable composition of gradual SEP events. We find that below ~1 MeV/nucleon almost all events are Fe-rich compared to the average 5 to 12 MeV/nucleon SEP composition. Iron-rich SEP events above 10 MeV/nucleon occur mainly during periods when the intensity of suprathermal iron in the inner heliosphere is high, due mainly to previous gradual events.

1. INTRODUCTION

According to the paradigm that was established during the late 1980's and early 1990's there are two classes of solar energetic particle (SEP) events (*Reames*, 1995*a*). "Impulsive" events are generally small events associated with impulsive x-ray flares that are enriched in ^{3}He and

Solar Eruptions and Energetic Particles
Geophysical Monograph Series 165
Copyright 2006 by the American Geophysical Union
10.1029/165GM12

heavy elements such as Fe. Measurements at ~1 MeV/nucleon showed that the ions in impulsive events had average charge states characteristic of temperatures of ~5 to 10 MK. "Gradual" events on the other hand, are accelerated by CME-driven shocks, with abundances similar to those in the corona, and charge states characteristic of ~2-3 MK. It was generally assumed that impulsive events accelerate flare-heated material, while gradual events accelerate coronal material and the ambient solar wind.

During the past solar maximum it became possible to measure the composition of SEP events accurately over a very broad energy interval that extended down to suprathermal energies (~40 keV/nucleon). These data were supplemented by isotopic and ionic charge-state data of unprecedented accuracy. These new observations showed that gradual events are accelerated from additional seed populations than just solar wind.

In order to understand the energy scales involved, it is useful to refer to Figure 1, which shows the measured fluence of oxygen from solar wind to galactic cosmic-ray energies from September 1997 to June 2000. Also shown are the contributions that various examples of solar and interplanetary events made to this 33-month spectrum. Much of the progress in understanding the source material of SEP events has come from the comparison of composition studies over a broad energy interval.

1.1. New Observations of Impulsive and Gradual Events

Data from this last solar maximum have shown that solar particle events are more complex than indicated in the picture described above [see also *von Rosenvinge and Cane* (2005)].

In particular, with new instruments that could resolve ^3He in much lower concentrations ($^3He/^4He$ from ~0.001 to 0.1), it was found that in the energy range from ~0.1 to ~30 MeV/nucleon most gradual events have $^3He/^4He$ ratios well above that of the solar-wind ratio of 4×10^{-4} (*Cohen et al.*, 1999, *Mason et al.*, 1999a, *Wiedenbeck et al.*, 2000, *Torsti, Laivola, and Kocharov*, 2003).

In addition, *Cohen et al.* (1999) found that a large fraction of gradual events are enriched in Fe and other heavy elements at energies >10 MeV/nucleon, as well as in neutron-rich isotopes of elements such as Ne and Mg (*Leske et al.*, 1999). Both characteristics had been observed in impulsive events (*Mason et al.*, 1994), but were not expected in gradual events. Figure 2 shows the fluence of Si (a measure of the event size) plotted versus the Fe/O ratio for 76 large SEP events from 1997 to 2003. Note the wide spread in Fe/O. About half of the events have Fe/O ratios on either side of the average 5 to 12 MeV/nucleon Fe/O ratio of *Reames* (1995b), but the distribution is not symmetric - the very largest events tend to be Fe-poor, while the Fe-rich events are on average smaller in intensity.

Although there are exceptions, many Fe-poor events have spectral breaks near ~10 MeV/nucleon, with Fe having a smaller break energy than lighter species such as O (e.g., *Tylka et al.*, 2000, *Cohen et al.*, 2005). In contrast,

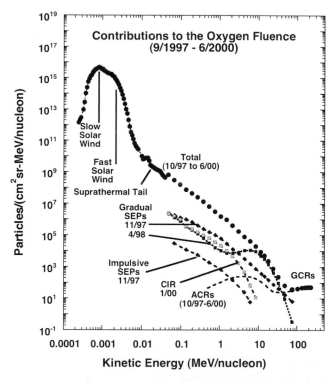

Figure 1. Fluence of energetic oxygen nuclei from solar wind to galactic cosmic ray energies. The fluence spectra >40 keV/nucleon from the ULEIS, SIS, and CRIS instruments on ACE were accumulated from September 1997 through June, 2000. Fluences <0.04 MeV/nucleon from the SWICS instrument were accumulated from January through December 1999 and multiplied by a factor of 11/4. Also shown are the contributions to the overall fluence by various solar and interplanetary events (from *Mewaldt and Mason*, 2005)

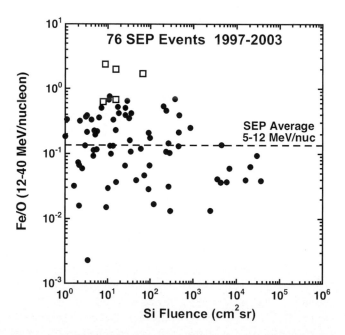

Figure 2. Plot of the fluence of 12 to 40 MeV/nucleon Si versus the Fe/O ratio for 76 large SEP events from late 1997 thru 2003. Data are from the SIS instrument on ACE. The average SEP Fe/O ratio of 0.134 determined from 5 to 12 MeV/nucleon measurements is shown for comparison (*Reames*, 1995b). The diamonds are events identified in the literature as "impulsive".

many Fe-rich events are power-laws (e.g., *Mason et al.*, *1999a, Cohen et al.*, 1999). So, while the relative location of the Fe and O spectral breaks can explain events below the average SEP line, the large number of Fe-rich events with Fe/O ≈0.3 to 1 was not expected, based on studies at lower energy (*Reames*, 1995b). Many Fe-rich events are also enriched in ³He (*Cohen et al.*, 1999, *Cane et al.*, 2003).

Finally, it was also shown that ionic charge-state patterns in SEP events are more complex than previously believed. Prior to solar cycle 23 *Luhn et al.* (1984) found that the mean charge states at ~1 MeV/nucleon in twelve gradual events were characteristic of temperatures of ~2 to 4 MK, with some event-to-event variation. Using geomagnetic techniques, *Leske et al.* (1995) and *Tylka et al.* (1995) measured similar mean charge states ranging from ~15 up to several hundred MeV/nucleon. In contrast, the mean Fe charge-state in 22 impulsive events suggested much higher (~10 MK) temperatures *Klecker et al.* (1984).

Once SEP charge-state measurements in the same SEP events became available over a broad energy interval, it was discovered that the mean charge states in most gradual events are energy-dependent, with higher-energy ions more ionized than lower-energy ions (*Oetliker et al.*, 1997, *Moebius et al.*,

1999, *Mazur et al.*, 1999; *Popecki et al.*, 2003). The mean charge state of Fe at ≥20 MeV/nucleon is often ≈20 (see Figure 3), comparable or greater than that in impulsive events (*Mazur et al.*, 1999, *Leske et al.*, 2001, *Labrador et al.*, 2003; *Dietrich and Tylka*, 2003).

One suggestion for the increase in the mean charge-state with energy in gradual events is electron stripping. Perhaps high-energy particles traverse more material during acceleration and are stripped of additional electrons (*Ruffolo*, 1997, *Reames, Ng, and Tylka*, 1999, *Barghouty and Mewaldt*, 2000; *Kocharov, Kovaltsov, and Torsti*, 2001). The amount of electron stripping depends on the product of the density in the acceleration region (n) and the acceleration time (τ). To strip MeV/nucleon Fe requires $n\tau \approx 10^{10}$ s/cm³ (*Kovaltsov et al.*, 2001). However, there is a correlation between the Fe/O ratio and the mean charge state of Fe (*Moebius et al.*, 2000; *Labrador et al.*, 2003), as shown in Figure 4. It is not obvious why the conditions that lead to stripping in gradual events would favor Fe-rich events.

Following the discovery of energy-dependent charge states in gradual events, it was found that the mean charge states in impulsive events also increase with energy (*Moebius et al.*, 2003, *Klecker et al.*, 2005). Figure 5, includes charge-state

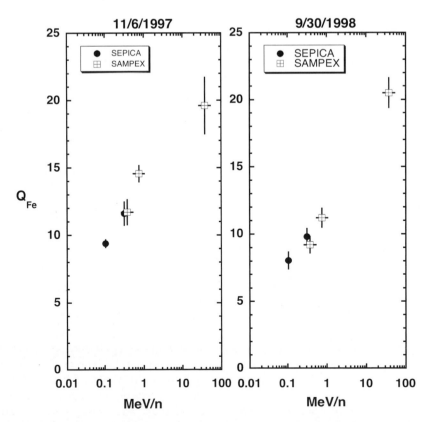

Figure 3. Energy dependence of the Fe charge state measured in two gradual events with data from ACE/SEPICA and from SAMPEX (adapted from *Popecki et al.*, 2003).

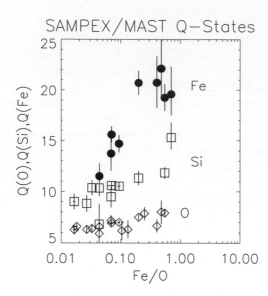

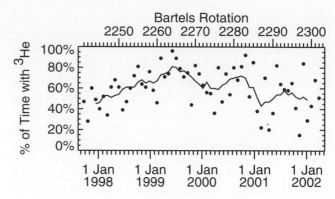

Figure 6. Fraction of time that ^{3}He was present in the interplanetary medium during the years 1998–2001. The solid line is the running mean (based on *Wiedenbeck et al.*, 2002).

Figure 4. Plot of the mean charge state of O, Si, and Fe as measured by SAMPEX versus the Fe/O ratio measured by the SIS/ACE instrument (from *Labrador et al.*, 2003). The mean charge state was determined using the geomagnetic method in the energy ranges from 15 to 60, 20 to 60, and 25 to 90 MeV/nucleon for O, Si, and Fe, respectively.

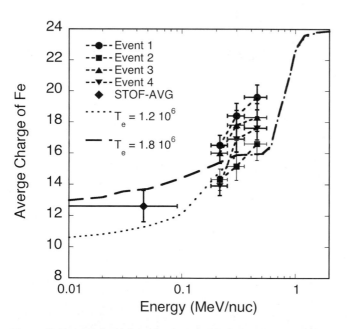

Figure 5. Energy-dependent charge states in impulsive events from *Klecker et al.* (2005). The curves assume thermal equilibrium for the source material, but allow for electron stripping during acceleration.

measurements from four impulsive events and a theoretical calculation of the equilibrium charge-state distribution. *Klecker et al.* (2005) concluded from this comparison that in impulsive events the increase in charge state with energy is

due to stripping in the low corona rather than to acceleration of high-temperature material. For further discussion see papers by *Popecki* (2005) and *Kocharov* (2005) and *Klecker* (2005) in this volume.

The combined effect of these new observations has been to blur many of the distinctions between impulsive and gradual events that are based on composition signatures (*Mewaldt*, 1999; *Cohen*, 2003, *von Rosenvinge and Cane*, 2005). The ^{3}He observations, in particular, led to the conclusion that many (and possibly most) gradual events somehow include impulsive-flare material. Because these observations are reminiscent of Cliver's concept of "Hybrid Events" (*Cliver*, 1996, 2000) that include both flare-accelerated and shock-accelerated particles, this term will be used to denote events that are associated with CME-driven shocks, but have a composition that includes signatures of flare-accelerated particles. The reader is reminded that there are also other properties of SEP events that help to distinguish impulsive and gradual events (*Reames*, 1995a; *von Rosenvinge and Cane*, 2005).

1.2. Suggested Explanations for Hybrid Events

There have been several suggestions as to how mixtures of flare and gradual material may originate. *Mason et al.* (1999) suggested that ^{3}He and Fe enrichments result from remnant interplanetary material from previous impulsive flares that is preferentially accelerated by CME-driven shocks. They showed that ^{3}He is often present in the interplanetary medium (see also *Wiedenbeck et al.*, 2002 and Figure 6).

Tylka et al. (2005) proposed a model that builds on the idea of *Mason et al.* (1999), but also relies on proposed differences between acceleration at quasi-parallel and quasi-perpendicular shocks (*Jokipii*, 1982; *Giacalone*, 2005). *Tylka et al.*, argue that the injection threshold is somewhat greater at quasi-perpendicular shocks than at quasi-parallel shocks, and suggest that only suprathermal particles (e.g., >10 keV/nucleon) are accelerated at quasi-perpendicular shocks. This results in a hybrid event if the suprathermal ions are dominated by remnant flare material. In this model the larger, events with normal or Fe-poor composition are due mainly to acceleration at quasi-parallel shocks, in which the shock velocity is great enough to inject solar-wind material with approximately coronal composition.

Cane et al. (2003, 2005) suggested that hybrid events include a mixture of flare and shock-accelerated material, with flare particles contributing at high energies to events with direct connection to the flare site. As support for this suggestion, they cite the fact that the observed Fe, ^{3}He, and high-charge-state enrichments occur mainly for events originating at longitudes from W20 to W90, where there is relatively good magnetic connection with the flare site.

Finally, one additional possibility is that the CME-driven shock accelerates flare particles from the associated SEP event in addition to accelerating *in situ* coronal and interplanetary seed particles (e.g., *Mewaldt et al.*, 2003; *Tylka et al.*, 2005). *Li and Zank* (2005) have calculated time-intensity profiles and composition patterns that might be expected in such events.

This paper does not attempt to decide between these possibilities, but points out examples of the evidence that should be explained by these models. Any complete model also needs to address other issues including the timing of the associated flare, CME, and particle arrival times, longitudinal and temporal profiles, and observed energy spectra (see *Klecker*, 2005; *Cane et al.*, 2005; *Tylka et al.*, 2005; *Tylka and Lee*, 2005; *von Rosenvinge and Cane*, 2005).

2. DO GRADUAL SEP EVENTS ACCELERATE SOLAR WIND?

2.1. SEP Fractionation According to First Ionization Potential

Early studies of SEP events discovered that the composition of SEPs is fractionated with respect to photospheric abundances by a selection process that depends on first ionization potential (FIP) or some related atomic parameter (*Hovestadt*, 1973, *Cook et al.*, 1979, *Breneman and Stone*, 1985; *Meyer et al.*, 1985; *McGuire et al.*, 1986). Indeed, it was SEP composition studies such as these that first indicated that the coronal composition is fractionated with respect to the photosphere by some ion-neutral selection process (*Mewaldt*, 1980, *Cook, Stone, and Vogt*, 1984). A comparison of average SEP and photospheric abundances is shown in the top panel of Figure 7 using average SEP abundances from *Reames* (1995*b*) and photospheric abundances from *Anders and Grevesse* (1989) – note that species with FIP > ~10 eV are depleted with respect to those with FIP < 10 eV by a factor of ~3.3. Using average SEP abundances from *Breneman and Stone* (1985) the FIP step is similar, about a factor of 3.6.

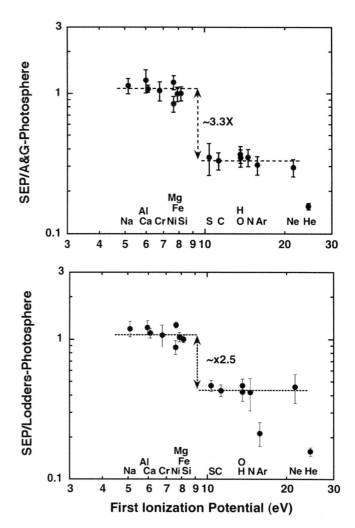

Figure 7. (Top) The ratio of the average SEP abundances with ~5 to ~12 MeV/nucleon from *Reames* (1995*b*) to the photospheric abundances tabulated by *Anders and Grevesse* (1989) is plotted versus first ionization potential (FIP). Note that elements with FIP >10 eV are depleted by a factor of ~3.3 in SEPs. (Bottom) Same as (top), but using photospheric abundances from *Lodders* (2003). Here the FIP-fractionation step is reduced to ~2.5. In addition, Argon no longer fits the pattern of other high-FIP species.

During recent years there have been significant revisions in the photospheric abundances of O and other volatile elements including C and N. These changes are reflected in a new table of solar system and photospheric abundances by *Lodders* (2003). The bottom panel of Figure 7 shows that these revisions reduce the magnitude of the FIP-fractionation step in SEPs from ~3.3 to ~2.5 using the average SEP abundances of *Reames* (1995*b*).

It has also been suggested that the differences between photospheric abundances and *in situ* SEP and solar wind abundances are better organized by first ionization time

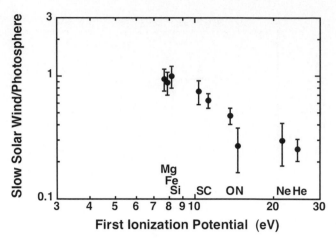

Figure 8. (Top) Ratio of the average slow-speed solar-wind abundances of *von Steiger et al.* (2000) to the photospheric abundances tabulated by *Lodders et al.* (2003), plotted vs FIP. A similar dependence is seen for fast solar-wind abundances (see *Mewaldt et al.*, 2001*b*). Note that CME-driven shocks travel mainly through slow solar wind since there are no active regions in coronal holes.

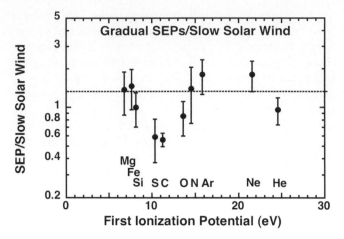

Figure 9. Ratio of the average 5-12 MeV/nucleon SEP abundances from *Reames* (1995*b*) to the slow solar-wind abundances of *von Steiger et al.* (2000), plotted vs FIP. The uncertainties in the average SEP abundances have been added in quadrature with the maximum systematic uncertainties in the SW abundances (statistical uncertainties are negligible for the solar wind measurements).

(FIT) rather than FIP (e.g., *Geiss*, 1998). However, this possibility does not affect the discussion in the next section.

2.2. Comparison of Solar-Wind and Solar-Particle Abundances

Solar-wind abundances also show a FIP fractionation pattern (see Figure 8), but the dependence on FIP appears more like a smooth function than a step function. The differences are seen more clearly in the ratio of average slow solar-wind abundances to SEP abundances (see Figure 9). The ratio of fast solar wind to SEP abundances looks very similar (*Mewaldt et al.*, 2001).

The ratios in Figure 9 are normalized at Si. Note that only 5 of the other 9 ratios are consistent with 1. *Mewaldt et al.* (2002) pointed out that several abundance ratios are consistently different in SEPs and solar wind. For example, the C/O ratio in SEPs is almost always ~0.4, while it is ~0.7 in the solar wind. Similarly, the S/Si and the Ne/O ratios also differ, as summarized in Table 1.

One possibility for explaining this situation might be that SEPs have been fractionated as a result of some Q/M or mass-dependent acceleration process, as observed, for example by *Desai et al.* (2003) at interplanetary shocks. However, plotting the ratios in Figure 9 vs. Z, M, or Q/M does not result any simple dependence (see example in Figure 10 and *Mewaldt et al.*, 2001). As a result of these differences, it was concluded that the majority of SEPs are not simply an accelerated sample of bulk solar wind – SEPs must include substantial contributions from other seed populations (*Mewaldt et al.*, 2001, 2002).

Table 1. Solar Wind and SEP Abundance Ratios.

Ratio	Slow Solar Wind (*von Steiger et al.*, 2000)	Solar Particles (*Reames*, 1995*b*)
C/O	0.67 ± .08	0.46 ± .01
Ne/O	0.10 ± .03	0.15 ± .01
S/Si	0.33 ± .16	0.21 ± .01

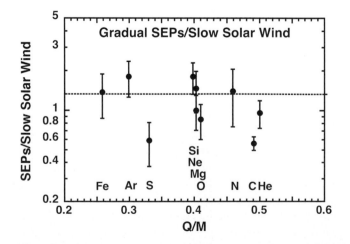

Figure 10. Ratio of average SEP abundances (*Reames*, 1995*b*) to slow solar-wind abundances (*von Steiger et al.*, 2000), plotted versus Q/M values from *Leske et al.* (1995). No simple, monotonic fractionation pattern is seen.

3. ARE THERE ENOUGH REMNANT SUPRATHERMALS?

The interplanetary density of suprathermal seed particles has been measured continuously since late 1997 by the ACE/ULEIS instrument. Figure 11 shows daily-average values of the Fe intensity versus the O intensity in the 0.04 to 0.64 MeV/nucleon interval. The composition in this low energy range is surprisingly Fe-rich relative to the average SEP value of Fe/O = 0.134 at 5 to 12 MeV/nucleon (*Reames*, 1995*b*). It is clear that in addition to a supply of Fe-rich suprathermals from small impulsive events, the material from gradual events is also Fe-rich. Indeed, at all intensity levels, suprathermal SEPs are, on average, Fe-rich. It is therefore not surprising that CME-driven shocks passing through this material would usually accelerate a Fe-rich composition.

The Fe-rich character of low-energy SEP events is demonstrated further in Figure 12, in which the 12 to 40 MeV/nucleon Fe/O ratio is plotted against the 0.04 to 0.64 MeV nucleon ratio for 60 large SEP events. Note that for all but two of the 60 events, Fe/O is greater than 0.13 at low energy, while at high energy about half the events are Fe-rich and half are Fe-poor. Also identified in Figure 12 are nine hybrid events, identified by the presence of ^{3}He and/or Fe with Q ≥ 16 (as well as being Fe-rich at >12 MeV/nucleon). It appears

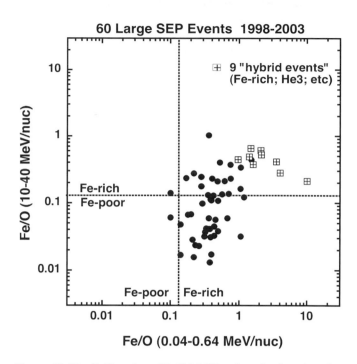

Figure 12. The Fe/O ratio at 12-40 MeV/nucleon is plotted against that at 0.04 to 0.64 MeV/nucleon for 60 large gradual SEP events from 1997–2003. The 10 to 40 MeV/nucleon data are from ACE/SIS and the 0.04 to 0.64 MeV/nucleon data are from ACE/ULEIS. Nine events indicated by square symbols are examples of hybrid events with Fe/O >0.27, and either ^{3}He/^{4}He >0.001 or Q(Fe) ≥ 16. The average SEP abundance ratio of 0.134 (*Reames*, 1995*b*) is indicated on each axis.

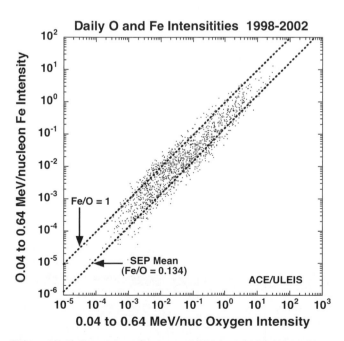

Figure 11. Daily average fluences of 0.04 to 0.64 MeV/nuc Fe are plotted the fluence of 0.04 to 0.64 MeV/nuc oxygen. There are diagonal lines corresponding the average SEP abundance ratio of 0.134 (*Reames*, 1998) and to Fe/O = 1. The data are from the ACE/ULEIS instrument.

that hybrid events are ~4 times more Fe-rich at low energy than are typical gradual SEP events.

If the enhanced abundance of ^{3}He and Fe in hybrid events results from accelerating remnant flare particles from earlier impulsive events (or earlier gradual events), there must be a sufficient density of remnant material available close to the Sun to account for all of the accelerated ions in these events. *Mewaldt et al.* (2003) surveyed the number density and fluence of suprathermal Fe during 1998-2002 (see Figure 13). The Fe fluence in typical hybrid events (including 4 of the 9 in Figure 12) was ~50 times greater than the typical quiet-day fluence, and was also much greater than during days when ^{3}He-rich events occurred. This implies that typical remnant populations during quiet or moderately active periods are not sufficient to account for the Fe fluence in hybrid events—there must be an additional source of Fe. A similar conclusion is reached if we consider the highly-ionized (Q_{Fe} ≥ 15) fraction of impulsive SEPs (Figure 5), and try to account for the highly-ionized fraction of hybrid events (Figure 3).

Mewaldt et al. (2003; see also *Tylka et al.*, 2005) suggested that additional Fe might come from ions accelerated by

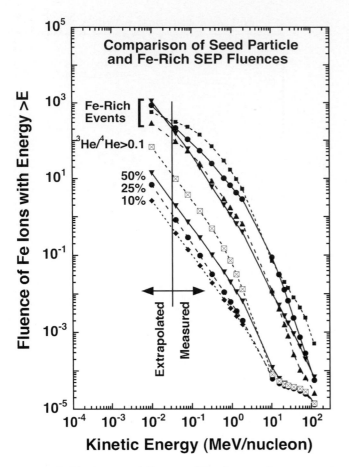

Figure 13. The integrated fluence of Fe above a given energy is shown for selected time periods and SEP events (based on the study of *Mewaldt et al.* (2003). The average fluence during quietest 10%, 20%, and 50% of the days during the period from February 1998 through December 2001 are indicated, as well as the average for days with ^{3}He/^{4}He >0.01. Also shown are Fe fluence spectra from four hybrid events.

the associated flare that are then re-accelerated by the shock. Other possibilities include the injection of even lower-energy ions into the acceleration process, or, the possibility of injection and/or acceleration effects that favor Fe because of its lower Q/M ratio. [However, in interplanetary shocks at 1 AU, Fe is *less* efficiently accelerated than O or other light ions (*Desai et al.*, 2003)].

Although one would like to also compare the fluence of ^{3}He in the remnant-flare source and in impulsive and gradual events, ^{3}He is more difficult to measure over a broad energy range under all conditions. The remnant source is more compelling for ^{3}He than for Fe because impulsive SEP material is the only common source of large ^{3}He/^{4}He enrichments. The quiet-time Fe/^{3}He ratio is ~0.05 at ~0.1 MeV/nucleon (*Mason et al.*, 1999), while in typical ^{3}He-rich events Fe/^{3}He ~0.1 (*Mason et al.*, 2002), so quiet-time material has a

similar composition to impulsive material. Hybrid events and other gradual events with enriched ^{3}He have Fe/^{3}He ~ 1 (*Mewaldt et al.*, 2003), so they have relatively more Fe than impulsive events. In addition, there is apparently a ceiling on the ^{3}He fluence in SEP events (*Ho et al.*, 2005). Thus, if the remnant source is responsible for the ^{3}He in gradual events, the remnant contribution to Fe would again appear to fall short by a factor of ~10 to ~20.

Although it would be revealing if we could measure suprathermal ions close to the Sun to compare with the characteristics of SEP events, this is presently difficult because suprathermals take ~1 day to get to 1 AU. However, we can measure the suprathermal population the day before an event (reflecting conditions near the Sun ~2 days before the event). This comparison relies on the fact that suprathermal ion intensities change rather slowly because these ions are not very mobile. Surprisingly, the >0.04 MeV/nucleon Fe/O ratio on days before hybrid events was not any more Fe-rich than for more normal or Fe-poor events. However, during hybrid events the >0.04 MeV/nucleon Fe/O ratio was much more Fe-rich than normal (see Figure 12).

The other interesting finding is that most hybrid events do not occur during quiet periods. For 6 of 8 hybrid events the >0.04 MeV/nucleon Fe intensity on the previous day was among the top 25% most-active days; for only one event was the previous day among the 50% least-active days (see Figure 13). For other gradual SEP events only 13 of 33 occurred after the 25% most-active days. So, it appears that another solution to the source problem is to accelerate remnant material from earlier, large, gradual events, which happens quite naturally when several SEP events occur within a few days. While the approach used here does not sample conditions close to the Sun at the onset of the event (the previous day is only a rather crude proxy), it does indicate that hybrid events occur mainly during periods of high activity rather than during quiet periods. In this case the most plentiful source of suprathermal ions may be material from previous gradual events, which are inherently Fe-rich on average (see Figures 11 and 12).

4. OTHER SOURCES OF SEED PARTICLES

Other possible sources of seed particles should also be considered. It is known that suprathermal particles are injected more easily into the shock acceleration process than thermal particles (*Gloeckler et al.*, 1994, *Desai et al.*, 2003). Interstellar (IS) singly-charged pickup ions are accelerated at the termination shock, but there does not appear to be efficient acceleration of the bulk solar wind. (*Cummings, Stone and Steenberg*, 2002). At interplanetary shocks near 1 AU IS pickup He is efficiently accelerated, but IS He is not available close to the Sun (*Kucharek et al.*, 2003). It also appears that inner-source pickup ions do not make a significant contribution to gradual SEP events.

ICME material is another possible source of seed particles. There is evidence that larger SEP events often result when a fast CME follows a slower CME (*Gopalswamy et al.*, 2002). Whether this is a result of "CME interactions" (*Gopalswamy et al.*, 2002) or "preconditioning" of the medium by the first ICME (*Kahler et al.*, 2003) is unclear, but in this situation the shock from the 2nd CME may pass through freshly ejected CME material, which often has highly-ionized Fe and other anomalous abundances (e.g., *Lepri et al.*, 2004). *Lepri et al.*, suggest that the highly-ionized material is produced by electron beams from the associated flare or injected directly from the flare site.

Mason (2001) has described the pool of suprathermal ions in the inner heliosphere. In addition to remnant flare particles, there are remnant particles from previous gradual and CIR events, and also CME material. In addition, there are ubiquitous suprathermal tails on the solar wind (*Geiss et al.*, 1995; *Gloeckler et al.*, 2000) with roughly E^{-2} energy spectra starting at several times the solar-wind velocity (see Figure 1). The origin of these tails is not known, but they apparently result from interplanetary acceleration process(es) because there are tails on interstellar pickup He as well as on all common solar-wind species. It is not yet clear if these tails are sufficiently intense to be a viable source of seed particles for the largest SEP events, and it is also not known if their composition can explain the composition differences described in Section 2.2 and Figure 9. Further investigation of the properties of suprathermal solar-wind tails could address several important questions about the source of SEP seed material.

5. SUMMARY

The wealth of new SEP data from solar maximum 23, supplemented by the first comprehensive solar-wind composition measurements, and improved solar observations from SOHO, TRACE, and RHESSI has led to significant evolution of our picture of SEP events. The new data covering multiple species over broad energy ranges has shown that CME-driven shocks accelerate more than the ambient solar wind, which had previously been the presumed source material for shock-associated SEP events. By realizing that the compositional signatures of SEPs are not due solely to acceleration and transport mechanisms, SEP researchers have changed the paradigm for probing the physical processes operating in these events.

To account for the observed enhancement of ^{3}He in gradual events, material from impulsive SEP events must somehow be incorporated. This has generated fresh ideas for models that include mechanisms such as direct contributions of flare-accelerated particles, or the shock acceleration of

either remnant flare material, or flare material from the associated flare. New SEP ionization-state measurements give evidence for stripping of SEPs, which implies acceleration much closer to the Sun. CME-driven shocks sometimes accelerate material more closely resembling that of the corona, but the FIP-fractionation patterns observed in SEP events differ in several key respects from those in the solar wind. Although theoretical studies indicate that fast CME-driven shocks can, in principle, accelerate the bulk solar wind, studies at 1 AU suggest that it is predominantly suprathermal ions that are accelerated. This revolutionizes our interpretation of SEP abundances, and makes clear the need to thoroughly understand the suprathermal populations, and the mechanisms that lead to their preferential acceleration in shock-associated events.

The composition of SEP events varies considerably with energy. In the energy range below 1 MeV/nucleon, it is surprising that essentially all SEP events are Fe rich, with typical Fe/O ratios of ~0.5, compared to typical solar wind values of ~0.2 and average SEP values of ~0.13 at 5 to 12 MeV/nucleon. Because of the higher intensities at lower energy, the bulk SEP composition must also be Fe-rich. This suggests that either (1) the seed material is Fe-rich (possibly derived from flare material), or (2) there are rigidity-dependent injection/acceleration effects operating near the Sun that favor ions with larger M/Q ratios such as Fe.

At higher energies, >10 MeV/nucleon, the composition is more variable, with Fe/O ratios varying by ~2 orders of magnitude from event to event. Much of this variation is due to spectral effects - SEP events commonly exhibit Q/M-dependent spectral breaks, in which the break energy is located at higher energy/nucleon for lighter species with higher Q/M ratios. However, in many cases, the Fe-rich character observed at lower energies continues to higher energies. There is a marked tendency for hybrid events to be more Fe-rich than average below 1 MeV/nucleon. In addition, hybrid events tend to have power-law spectra that preserve the low-energy composition rather than distorting it.

Whatever source of seed material is considered, it is important to establish how it is accelerated to produce the observed fluences in the largest SEP events. Simple considerations show that this is not easy, thus illustrating challenges ahead in the modeling these events. As an example of the issues, the fluence >10 keV/nucleon Fe ions on typical quiet days is ~10 times less than the >10 keV/nucleon fluence in typical hybrid events, and ~100 times less than the fluence during the largest SEP events. Therefore, to supply the needed material in large events the seed particles must either be (1) derived from energies <10 keV/nucleon (e.g., heated solar wind or CME material), or (2) derived from remnant material from previous gradual and impulsive events, or (3) derived directly or indirectly from flare particles originating in the associated flare. It is important that these processes be modeled to determine whether the candidate seed populations can be injected and accelerated to yield the observed spectra and composition.

New missions, both planned or proposed, can give critical insights into these issues. STEREO, supplemented by *in situ* and imaging data from near Earth, will investigate longitudinal variations in the composition and energy spectra and identify the relative contributions of shock and flare-accelerated particles. Missions that venture closer to the Sun (Inner Heliospheric Sentinels, Solar Orbiter, and Solar Probe) will provide measurements of the available seed populations, plasma, and interplanetary conditions closer to the Sun, where most of the acceleration takes place. In addition, global models of particle acceleration by shocks are now exploring more realistic computations of shock acceleration scenarios. Taken together, these advances in observation and theory/modeling should make possible quantitative tests of the entire process of particle acceleration and propagation in the inner heliosphere.

Acknowledgments. This work was supported by NASA under grants NNG04GB55G, NNG04088G, and NAG5-12929. We appreciate discussions with Berndt Klecker, Eberhard Moebius, and Edward Stone, and figures provided by Berndt Klecker, Allan Labrador, Mark Popecki, and Mark Wiedenbeck. We thank the local organizing committee for a very informative and enjoyable conference.

REFERENCES

Anders, E., and N. Grevesse, Abundances of the elements: meteoritic and solar, *Geochim. Cosmochim. Acta*, 53, 197-214, 1989.

Barghouty, A.F., and R.A. Mewaldt, Simulation of charge-equilibrium and acceleration of solar energetic ions, in *Acceleration and Transport of Energetic Particles Observed in the Heliosphere*, AIP Conf, Proc. 528, 71-78, 2000.

Breneman, H.H., and E.C. Stone, Solar coronal and photospheric abundances from solar energetic particle measurements, *Astrophys. J. Lett.* 299, L57-L61, 1985.

Cane, H.V., T.T. von Rosenvinge, C.M.S. Cohen, and R.A. Mewaldt, Two components in major solar particle events, *Geophys. Res. Lett.*, 30, SEP 5-1, doi: 10.1029/2002GL016580, 2003.

Cane, H.V., R.A. Mewaldt, C.M.S. Cohen, and T.T. von Rosenvinge, The role of CMEs and shocks in determining SEP abundances, submitted to *J. Geophys. Res.*, 2005.

Cliver, E.W., Solar flare gamma ray emission and energetic particles in space, in *High Energy Solar Physics*, R. Ramaty *et al.*, Eds, *AIP Conf. Proc.* 374, 45-60, 1996.

Cliver, E.W., Solar flare photons and energetic particles in space, in *Acceleration and transport of energetic particles in the heliosphere*, A.I.P. Conf. Proc. 528, 21-31, 2000.

Cohen, C.M.S., Solar energetic particle acceleration and interplanetary transport, in *Frontiers of Cosmic Ray Science*, T. Kajita *et al.*, Editors, Universal Academy press, Tokyo, pp. 113-134, 2004.

Cohen, C.M.S. *et al.*, New observations of heavy-ion-rich solar particle events from ACE, *Geophys. Res. Lett.*, 26, 2697-2700, 1999.

Cohen, C.M.S., *et al.*, Heavy ion abundances and spectra from the large SEP events of October-November 2003, *J. Geophys. Res.*, 110, doi:101029/2005JA011004, 2005.

Cook, W.R., E.C. Stone, and R.E. Vogt, Elemental composition of solar energetic particles, *Astrophys. J.*, 279, 827-838, 1984.

Cummings, A.C., E.C. Stone, and C.D. Steenberg, Composition of anomalous cosmic rays and other heliospheric ions, *Astrophys. J.*, 578, 194-210, 2002.

Desai, M.I. *et al.*, Evidence for a suprathermal seed population of heavy ions accelerated by interplanetary shocks near 1 AU, *Astrophys. J.*, 588, 1149-1162, 2003.

Dietrich, W.F., and A.J. Tylka, Time to maximum studies and inferred ionic charge sates in the solar energetic particle events of 14 and 15 April, 2001, *28ᵗʰ Internat. Cosmic Ray Conf.*, 6, 3291-3294, 2003.

Ellison, D.C., and R. Ramaty, Shock acceleration of electrons and ions in solar flares, *Astrophys. J.*, 298, 400-408, 1985.

Geiss, J., G. Gloeckler, L.A. Fisk, and R. von Steiger, Pickup ions in the heliosphere and their origin, *J. Geophys. Res.*, 1000, 23373-23377, 1995.

Geiss, J., Constraints on the FIP mechanisms from solar wind abundance data, *Space Sci. Rev.*, 85, 241-252, 1998.

Giacalone, J., *Astrophys. J.*, 628, L37-L40, 2005.

Gloeckler, G., L.A. Fisk, T.H. Zurbuchen, and N.A. Schwadron, in *Acceleration and Transport of Energetic Particles in the Heliosphere*, AIP Conf. Proc. 528, 229-233, 2000.

Gloeckler, G., *et al.*, Acceleration of interstellar pickup ions in the disturbed solar wind observed on Ulysses, *J. Geophys. Res.*, 99(A9), 17637-17643, 1994.

Gopalswamy, N., *et al.*, Interacting coronal mass ejections and solar energetic particles, *Astrophys. J. Lett.*, 572, L103-L106, 2002.

Ho, G., E.C. Roelof, and G.M. Mason, The upper limit on ³He fluence in solar energetic particle events, *Astrophys. J.*, 621, L141-L144, 2005.

Hovestadt, D. (1973), "Nuclear composition of solar cosmic rays", in *Solar Wind Three*, C.T. Russell, Ed., IGPP/UCLA, Los Angeles, pp. 2-25, 1973.

Jokipii, J.R., Particle drift, diffusion, and acceleration at shocks, *Astrophys. J.*, 255, 716-720, 1982.

Klecker, B., *et al.*, Direct determination of the ionic charge distribution of helium and iron in He-3-rich solar energetic particle events, *Astrophys. J.*, 281, 458-462, 1984.

Klecker, B.E., *et al.*, Observation of energy-dependent ionic charge states in impulsive solar energetic particle events, *Adv. Space Res.*, in press, 2005.

Kocharov, L.G., Modeling the energy-dependent charge states of solar energetic particles, this volume, 2005.

Kocharov, L., and J. Torsti, Hybrid solar energetic particle events observed on SOHO, *Solar Phys.*, 207, 149-157, 2002.

Kovaltsov, G.A., A.F. Barghouty, L. Kocharov, V.M. Ostryakov, and J. Torsti, Charge-equilibrium of Fe ions accelerated in a hot plasma, *Astron. Astrophys.*, 375, 1075-1081, 2001.

Kucharek, H., *et al.*, On the source and acceleration of energetic He⁺: A long-term observation with ACE/SEPICA, *J. Geophys. Res.*, 108, A10, doi:10.1029/2003JA009938, 2003.

Lepri, S. T., and T. H. Zurbuchen, Iron charge state distributions as an indicator of hot ICMEs: Possible sources and temporal and spatial variations during solar maximum, *J. Geophys. Res.*, 109(A1), doi:10.1029/2003JA009954, 2004.

Leske, R.A., *et al.*, Unusual isotopic composition of solar energetic particles observed in the November 6, 1997 event, *Geophys. Res. Lett.*, 26, 153-156, 1999.

Leske, R.A., *et al.*, The ionic charge state composition at high energies in large solar energetic particle events in solar cycle 23, in Solar and Galactic Composition, AIP Conf. Proc. 598, 171-176, 2001.

Leske, R.A., *et al.*, Measurements of the ionic charge states of solar energetic particles using the geomagnetic field, *Astrophys. J.*, 452, L149-L152, 1995.

Li, G., and G.P. Zank, Mixed particle acceleration at CME-driven shocks and flares, *Geophys. Res. Lett.*, 32, L02101, doi:10.1029/2004GL021250, 2005.

Li, G., G.P. Zank, and W.K.M. Rice, Acceleration and transport of heavy ions at coronal mass ejection-driven shocks, submitted to *J. Geophys. Res.*, 2005.

Lodders, K., Solar system abundances and condensation temperatures of the elements, *Astrophys. J.*, 591, 1220-1247, 2003.

Luhn, A., *et al.*, Ionic Charge States of N, Ne, Mg, Si, and S in Solar Energetic Particle Events, *Adv. Space Res.*, 4(2), 161-166, 1984.

Mason, G.M., Heliospheric lessons for galactic cosmic-ray acceleration, *Space Sci. Rev.*, 99, 119-133, 2001.

Mason, G.M., J.E. Mazur, and D.C. Hamilton, Heavy-ion isotopic anomalies in ³He-rich solar particle events, *Astrophys. J.*, 425, 843-848, 1994.

Mason, G.M., *et al.*, Particle acceleration and sources in the November 1997 solar energetic particle events, *Geophys. Res. Lett.*, 26, 141-144, 1999a.

Mason, G.M., J.E. Mazur, and J.R. Dwyer, ³He enhancements in large solar energetic particle events, *Astrophys. J. Lett.*, 525, L133-L136, 1999b.

Mason, G.M., *et al.*, Spectral properties of He and heavy ions in ³He-rich solar flares, *Astrophys. J.*, 574, 1039-1058, 2002.

Mazur, J.E., G.M. Mason, B. Klecker, and R.E. McGuire, The energy spectra of solar flare hydrogen, helium, oxygen, and iron: evidence for stochastic acceleration, *Astrophys. J.* 401, 398-410, 1992.

Mazur, J.E. *et al.*, Charge states of solar energetic particles using the geomagnetic cut-off technique: SAMPEX measurements in the 6 November 1997 solar particle event, *Geophys. Res. Lett.*, 226, 173-176, 1999.

McGuire, R.E., T.T. von Rosenvinge, and F.B. McDonald, The composition of solar energetic particles, *Astrophys. J.*, 302, 938-961, 1986.

Mewaldt, R.A., Spacecraft measurements of the elemental and isotopic composition of solar energetic particles, in *Proceedings of the Conference on the Ancient Sun*, eds, R.O. Pepin, *et al.*, Pergamon, New York, pp. 81-101, 1980.

Mewaldt, R.A., New views of solar energetic particles from the Advanced Composition Explorer, *Proceedings of the 26th International Cosmic Ray Conference, Invited, Rapporteur, and Highlight Papers*, AIP Conf. Proc. 516, 265-270, 2000.

Mewaldt, R.A., *et al.*, Long-term fluences of energetic particles in the heliosphere, in *Solar and Galactic Composition*, AIP Conf. Proc. 598, 165-170, 2001.

Mewaldt, R.A., *et al.*, Impulsive flare material: A seed population for large solar particle events? *28th Internat. Cosmic Ray Conf.*, 6, 3329-3332, 2003.

Mewaldt, R.A., *et al.*, Solar energetic particle spectral breaks, in *Physics of Collisionless Shocks*, AIP Conf. Proc., 227-232, 2005.

Mewaldt *et al.*, Fractionation of solar energetic particles and solar wind according to first ionization potential, *Adv. Space Res.*, 30(1), 79-84, 2002.

Mewaldt, R.A., *et al.*, Are solar energetic particles an accelerated sample of solar wind? R.A. Mewaldt, *et al.*, *Proc. 27th Internat. Cosmic Ray Conf.*, 8, 3132-3135, 2001.

Meyer, J.P., The baseline composition of solar energetic particles, *Astrophys. J. Supp.*, 57, 151-171, 1985.

Moebius, E., *et al.*, Strong energy dependence of ionic charge states in impulsive solar events, *28th Internat. Cosmic Ray Conf.*, 6, 3273-3276, 2003.

Moebius, E., *et al.*, Energy dependence of the ionic charge state charge state distribution during the November 7-9 solar energetic particle event observed by ACE SEPICA, *Geophys. Res. Lett.*, 226, 145-148, 1999.

Moebius, E., *et al.*, Survey of ionic charge states of solar energetic particle events during the first year of ACE, in *Acceleration and Transport of Energetic Particles Observed in the Heliosphere*, AIP Conf, Proc. 528, 131-134, 2000.

Oetliker, M., *et al.*, The ionic charge of solar energetic particles with energies of 0.3 – 70 MeV per nucleon, *Astrophys. J.*, 477, 495, 1997.

Popecki, M.A., Observations of energy-dependent charge states in solar energetic particles, this volume, 2005.

Popecki, M.A. *et al.*, *Proc. 28th Internat. Cosmic Ray Conference*, 6, 3283-3286, 2003.

Reames, D.V., Solar energetic particles: A paradigm shift, *Rev. of Geophys.*, 33, 585-589, 1995a.

Reames, D.V., Coronal abundances determined from energetic particles, *Adv. Space Res.*, 15(7), 41-51, 1995b.

Reames, D.V., C.K. Ng, and A.J. Tylka, Energy-dependent ionization states of shock-accelerated particles in the solar corona, *Geophys. Res. Lett.*, 26, 3585-3588, 1999.

Ruffolo, D., Charge states of solar cosmic rays and constraints on acceleration times and coronal transport, *Astrophys. J. Lett.*, 481, L119-L122, 1997.

Torsti, J., J. Laivola, and L. Kocharov, Common overabundance of ^{3}He in high-energy solar particles, *Astron. Astrophys.*, 408, L1-L4, 2003.

Tylka. A.J., *et al.*, The mean ionic charge state of solar energetic Fe ions above 200 MeV per nucleon, *Astrophys. J. Lett.*, 444, L109-L113, 1995.

Tylka, A.J., D.V. Reames, and C.K. Ng, Observations of systematic temporal evolution in elemental composition during gradual solar energetic particle events, *Geophys. Res. Lett.*, 26, 2141-2144, 1999.

Tylka, A.J., P.R. Boberg, R.E. McGuire, C.K. Ng, and D.V. Reames, Temporal evolution in the spectra of gradual solar energetic particle events, in *Acceleration and Transport of Energetic Particles Observed in the Heliosphere*, A.I.P. Conf. Proc. 528, 147-152, 2000.

Tylka, A.J., *et al.*, Shock geometry, seed populations, and the origin of variable elemental composition at high energies in large gradual solar particle events, *Astrophys. J.*, 625, 474-495, 2005.

Tylka, A.J., and M.A. Lee, this volume, 2005.

von Rosenvinge, T.T., and H.V. Cane, Overview of solar energetic particles, this volume, 2005.

von Steiger, R., *et al.*, Composition of quasi-stationary solar wind flows from the Ulysses Solar-Wind Ion Composition Spectrometer, *J. Geophys. Res.*, 105(A12), 27217-27238, 2000.

Wiedenbeck, M.E., *et al.*, Enhanced abundances of ^{3}He in large solar energetic particle events, in *Acceleration and transport of energetic particles observed in the heliosphere*, AIP. Conf. Proc. 528, 107-110, 2000.

Wiedenbeck, M.E., *et al.*, How common is energetic ^{3}He in the inner heliosphere? in *Solar Wind 10*, AIP Conf. Proc. 679, 652-655, 2003.

C.M.S. Cohen, 220-47 Downs Laboratory, Caltech, Pasadena, CA 91125

G.M. Mason, Johns Hopkins University, Applied Physics Laboratory, Johns Hopkins Rd, Laurel, MD 20723-6099

R.A. Mewaldt, 220-47 Downs Laboratory, Caltech, Pasadena, CA 91125

Observations of Energy-Dependent
Charge States in Solar Energetic Particles

M.A. Popecki

The University of New Hampshire, Durham, NH

Ionic charge states of solar energetic particles (SEPs) provide information about their source environment and acceleration process. Recently, some large SEP events have been observed with energy dependent charge states. In these cases, higher energy ions have higher mean charge states. We review observations of events with energy dependent charge states, using data from the ACE, SAMPEX and SOHO spacecraft. Instruments on these spacecraft use various techniques in several energy ranges to measure the ionic charge states of energetic ions. Combining observations from these instruments, the mean iron charge states have been measured for four events over a wide energy range. In these events, the mean Fe charge states increased by 4 to 12 units in the energy range 65 keV/nuc to 47 MeV/nuc. Other events were observed only below 1 MeV/nuc. Some of those were related to flare activity in which there was little known contribution from interplanetary shocks. Others included large interplanetary shocks. All of them displayed energy dependent charge states, even though they were not chosen for that reason. These results will be discussed with respect to models for energy dependent charge states.

1. INTRODUCTION

Ionic charge states of solar energetic particles (SEPs) can provide tracers for their acceleration processes. The observed mean charge state at Earth is the combined result of the source region temperature and the influence of collisional processes. It is also affected by rigidity-dependent escape from the acceleration region. SEP charge states were first measured using the ISEE3/ULEZEQ sensor [*Hovestadt et al.*, 1978; *Hovestadt et al.*, 1981; *Klecker et al.*, 1984; *Luhn et al.*, 1984; *Luhn et al.*, 1987]. They separated their events into two categories: long duration events and 3He-rich events. Long duration events are now generally known as gradual events. These events are typically associated with

coronal mass ejections. The 3He-rich events are associated with flares [*Reames*, 1999a]. *Luhn et al.* [1987] found that the charge states of ions from Si to Fe were higher in the 3He-rich events than in the gradual events. They reported that the average Fe charge state in gradual events was 14.1+ and in impulsive, 3He-rich events it was 20.5+. This result was interpreted as an indication of the source region temperature. They estimated the source temperature at 2-4 MK for the large gradual events, and approximately 10 MK for the 3He-rich events. *Luhn et al.* [1984] were unable to fit their Mg charge states to the same source temperature model as their other ion species. They suggested that the Mg inconsistency might occur if the accelerated ions could not achieve equilibrium with the local plasma, because of insufficient residence time. This suggestion hinted that the observed charge state depended partly on non-equilibrium processes. Nevertheless, these results became part of the definition of these two kinds of SEP events.

The strict categorization of SEP events into two classes, as distinguished by charge state, was altered somewhat

Solar Eruptions and Energetic Particles
Geophysical Monograph Series 165
Copyright 2006 by the American Geophysical Union
10.1029/165GM13

by the observations of *Oetliker et al.* [1997]. They detected an energy dependence in the Fe charge state during an SEP event series of October and November, 1992. In this case, the Fe charge state increased with energy. These observations were made with instruments aboard the SAMPEX spacecraft [*Baker et al.*, 1993], using the geomagnetic cutoff method [*Mason et al.*, 1995; *Leske et al.*, 1995; *Klecker et al.*, 1995]. Later, in the November 6, 1997 event, SEPs with energy dependent charge states were observed once more. This time, both Si and Fe charge states displayed this pattern [*Möbius et al.*, 1999, *Mazur et al.*, 1999]. The energy dependence was observed simultaneously by the instruments on both the SAMPEX and ACE spacecraft.

This work presents charge state measurements for several events over a wide energy range. It combines data from up to four instruments on three spacecraft. The energy range is of unprecedented width, spanning 65 keV/nuc to 47 MeV/nuc.

2. METHODS FOR CHARGE STATE MEASUREMENT

Two basic methods have been used to measure the ionic charge states of solar energetic particles. These methods either use direct detection or make use of the Earth's magnetic field as a geomagnetic filter. Direct ion detection in the interplanetary medium was used by ULEZEQ on ISEE-3 [*Hovestadt et al.*, 1978], STOF on SOHO [*Hovestadt et al.*, 1995], and SEPICA on ACE [*Möbius et al.*, 1998]. In these cases, a high voltage creates an electrostatic deflection, which is proportional to the charge to energy ratio (Q/E) of the ion. Total energy E is then measured. The charge state is derived from this information for each ion.

Ion elemental identification is performed by either of two methods. The SOHO/STOF instrument combines time of flight with residual energy measurements in a solid state detector (SSD) to determine the ion mass. The ULEZEQ and SEPICA designs employ a dE/dx vs. residual energy telescope to determine the nuclear charge Z.

The dE/dx vs. residual E technique is used at higher energies than the TOF/E technique. SOHO/STOF provides Fe charge states in the 10-100 kev/nuc energy range. ACE/SEPICA uses an energy loss/residual energy telescope and electrostatic deflection. It covers the energy range from 65-600 keV/nuc for Fe. The ULEZEQ energy range for Fe was 0.5-2.0 MeV/nuc [*Luhn et al.*, 1984].

The geomagnetic filter approach was used with the LICA and MAST instruments on SAMPEX [*Leske et al.*, 1995; *Oetliker et al.*, 1997; *Mason et al.*, 1995; *Mazur et al.*, 1999]. SAMPEX/LICA [*Mason et al.*, 1993] and MAST [*Cook et al.*, 1993] both measure elemental composition. The SAMPEX spacecraft is in a polar orbit around the Earth. In solar events, the flux of SEPs into the magnetosphere is modulated by their rigidity (momentum per unit charge) and the Earth's magnetic field. The intensity of penetrating ions is highest at the magnetic poles, and then decreases with decreasing latitude. This phenomenon depends on ion rigidity and is independent of species. A cutoff latitude is defined, at which the ion intensity drops to a selected fraction of the polar intensity. The rigidity at this latitude is calculated using protons and helium, which have known charge states. Elemental composition measurements are used to identify ion species such as Fe. The charge state of the species is then derived from the mass, energy and rigidity information. Together, LICA and MAST cover the energy range 0.25-60 MeV/nuc for Fe.

The direct and geomagnetic filter methods are complementary. Using direct detection, small events with just a few ions can be observed. The charge state distribution may be obtained, instead of the mean. The distribution may be compared to the results of models of acceleration and ionization mechanisms. Distributions also permit identification of multiple ion populations with distinctive ionization signatures. The principal disadvantage of the direct detection method is that the energy range is somewhat limited. The geomagnetic filter method has a wider energy range. The disadvantages are that only the mean is determined, and large events are needed.

Indirect methods for estimating the ionic charge state have also been used by several authors. All of them infer charge states by using models to interpret observations. They employ: elemental energy spectra [*Tylka et al.*, 2000], elemental and isotopic composition patterns [*Cohen et al.*, 1999], or temporal profiles of particle intensities in the rise phase of an event [*Dietrich and Lopate*, 1999; *Dietrich and Tylka*, 2003] or the extended decay phase [*Sollitt et al.*, 2003]. This review, however, will concentrate on direct observations of ionic charge states.

3. LARGE EVENTS WITH ENERGY-DEPENDENT CHARGE STATES

Energy dependent charge states were first observed in a 1992 event by SAMPEX instruments [*Oetliker et al.*, 1997]. In this event, the mean Fe charge state was somewhat flat at 11+ in the energy range 0.3-3.0 MeV/nuc. It then increased from 11+ at 10 MeV/nuc to 17+ at 60 MeV/nuc.

A second event also featured an energy dependence. It occurred in November, 1997 and was observed by instruments on both the ACE and SAMPEX spacecraft. Two X-class flares occurred during this event. The second occurred at a solar longitude of W63, which would be nominally magnetically well-connected to the Earth. After the initial X-class flare eruption, a shock passed the Earth, followed by a magnetic cloud CME, and finally a second shock [*Popecki et al.*, 2002].

The mean Fe charge state again increased with energy in this event. In this case, the SAMPEX observations showed an increase that began at a lower energy than in the 1992 event [*Mazur et al.*, 1999; *Leske et al.*, 2001]. Indeed, the iron charge state was already rising at the lowest measured energy of 0.25 MeV/nuc. Altogether, the mean iron charge state rose by 11 units from 9+ at 0.25 MeV/nuc, up to 19+ to 20+ at 40 MeV/nuc.

ACE/SEPICA observations of the November, 1997 event were reported by *Möbius et al.* [1999]. The mean iron charge state increased from 11+ to 14.5+ in the energy range 0.2-0.5 MeV/nuc. The existence of SEPs with energy-dependent charge states was thus confirmed by two spacecraft using two different methods, in two different locations.

The mean charge states of C, O, Ne, Mg and Si also increased with energy at both spacecraft in the November, 1997 event [*Möbius et al.*, 1999; *Mazur et al.*, 1999]. Previously, only Fe displayed this trend in the October, 1992 event [*Oetliker et al.*, 1997].

Möbius et al. [1999] showed iron charge state distributions at different energies. They were formed by integrating ion measurements over the entire event. The distributions changed with increasing energy. At the lowest energy, the peak charge state was near 10+, with a tail toward higher values. Toward the higher energies, the distributions became more symmetrical.

Temporal variations of the iron charge state distributions were also investigated with the SEPICA instrument [*Popecki et al.*, 2002]. Using the four energy bands employed by *Möbius et al.* [1999], charge state distributions were formed for three time periods: before the magnetic cloud CME passage, during the cloud passage, and afterward. In the period before cloud passage, some of the ions were locally accelerated at the first shock passage. In all three periods, the form of the charge state distributions in the four energy bands remained essentially unchanged. In each period, the distributions were similar to those presented by *Möbius et al.* [1999].

Charge states have been obtained from multiple spacecraft for four events altogether. Each produced large quantities of energetic ions. This is a selection effect required by the geomagnetic filter method. Each event displayed an increase in charge state with energy. The first was the November, 1997 event discussed above. The others took place on 9/30/98, 11/6/98 and 7/14/00 (Bastille Day). In these events, the Fe charge states increased from 4 to 12 units in the energy range 65 keV/nuc to 47 MeV/nuc. Charge state variation with energy is plotted in Plate 1.

In the 9/30/98 event, the mean Fe charge state is approximately 8+ at 100 keV/nuc. It increases to 20.5+ at 30 MeV/nuc. This event shows no apparent rollover in charge state up to the maximum energy. In the 0.1-1.0 MeV/nuc energy range, the increase was approximately 3.2 units.

The 11/6/98 event includes data from STOF, SEPICA and LICA. The mean Fe charge state is between 8+ and 9+ at 80 keV/nuc, and increases to 17.5+ at 0.8 MeV/nuc. MAST data are not available for this event. The gradient of charge state with energy is the steepest in this event, at 9 units in the 0.065-1.0 MeV/nuc range.

The fourth event occurred on 7/14/00, which is known as Bastille Day. This period featured several flares and interplanetary shocks. The interplanetary medium was highly disturbed [*Smith et al.*, 2001]. The high voltage on the SEPICA instrument, which is necessary for charge state measurements, went down briefly during this period. Consequently, the integration period for the charge state data is shorter for the SEPICA instrument than for SAMPEX. SAMPEX/MAST integrated from 00/196.5 to 197.5 (7/14-15/00), while SEPICA integrated over the latter part of the period, from 00/197.5 to 197.9 (7/15/00). Interestingly, the increase in average Fe charge state in this event is the smallest of all four events. It increased by less than four units in the energy range 0.3 to 50 MeV/nuc, from 10.4 to 13.8 [*Leske et al.*, 2001; *Smith et al.*, 2001].

4. OBSERVATIONS BELOW ONE MeV/NUC

4.1. SOHO/ACE Observations

Charge state observations are available for many events that were too small to employ the geomagnetic filter method. The widest energy range may be obtained by combining data from the SOHO/STOF and ACE/SEPICA instruments. *Bogdanov et al.* [2000] and [2002] presented four events with O and Fe charge states in the energy range 20 - 600 keV/nuc. The events took place on May 1, 1998 (98/121), November 6, 1998 (98/310), April 6, 2000 (00/097), and June 26, 1999 (99/177). The first three events were associated with an interplanetary shock passage. For example, in the May 1, 1998 event, particle intensities peaked at the shock passage. In the June 26 event, intensities peaked a day before shock passage. Combining data from both instruments, they found increases in the mean Fe charge state of about one to three units in the first three events, and about 2-3 units in the June 26 event over the 0.020-0.400 MeV/nuc energy range. There was no significant change in the mean O charge state. The lowest mean Fe charge state was 8+ at 0.020 MeV/nuc in events 99/177 and 98/310.

Iron charge states from the June 26, 1999 (99/177) STOF-SEPICA event are shown in Plate 2. Two low energy points from SEPICA are included for comparison to the STOF results. The low energy Fe ions are selected by extending the traditional method for identifying elements in the energy

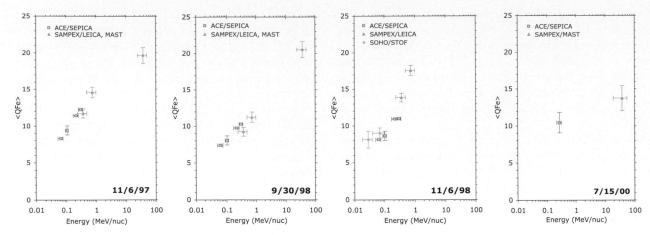

Plate 1. Iron charge state data are available from multiple spacecraft during four SEP events. The plate presents the Fe charge state vs. energy from SOHO/STOF, ACE/SEPICA, SAMPEX/LICA and MAST, as available. In each event, the mean Fe charge state increased with energy. The increases were 4 to 12 charge units in the energy range 65 keV/nuc to 47 MeV/nuc.

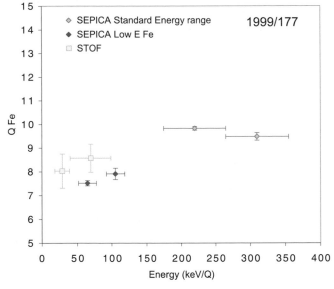

Plate 2. Iron charge states from SOHO/STOF and ACE/SEPICA are compared for an SEP event on June 26, 1999 (99/177). The SEPICA points at 64 and 100 keV/nuc are obtained by extending the traditional method of element identification. This technique separates low energy Fe from C and N. Charge states at these two energies are shown in Plate 3 for events in which STOF data are unavailable.

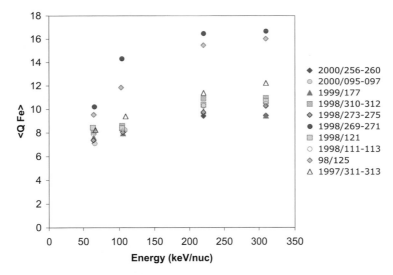

Plate 3. Iron charge states have been observed by SEPICA in several events for which STOF and SAMPEX data are unavailable. All the events display a tendency for the mean Fe charge state to increase in the energy range 64-220 keV/nuc. Considering the uncertainty of the charge state measurements at 64 and 100 keV/nuc, the change is typically at least 1 charge unit over this energy range.

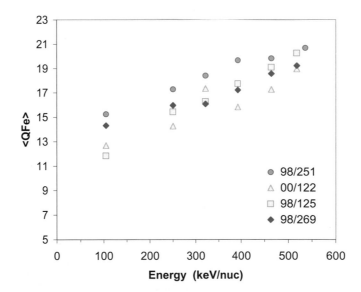

Plate 4. Iron charge states are shown for three flare-related SEP events, for energies below 600 keV/nuc. Data are from SEPICA only. An additional event from May 5, 1998 (98/125) with a high average charge state is also shown. In all events, the Fe charge states tend to increase by 5-8 units in the energy range 100 to 550 keV/nuc.

loss/residual energy telescope. Elements are normally identified by plotting their energy loss against their residual energy and observing the resulting characteristic trails. The Fe trail crosses C, N and O trails at low energies. Where the Fe and N are mixed, the N contribution may be extracted by using a three step process. The pure N charge state distribution is first obtained from a nearby energy where there is no mixing. Then the quantity of mixed N is obtained by projecting the N energy spectrum into the mixture region. Finally, the pure N charge state distribution is normalized and subtracted from the mixed distribution. Thus the Fe charge state may be determined where Fe crosses the N trail at approximately 100 keV/nuc, and where Fe extends below the C trail at about 64 keV/nuc. At 64 and 100 keV/nuc, the mean SEPICA Fe charge states are about one charge unit below the STOF measurement. At these energies, the SEPICA measurement uncertainty is approximately 0.7 units for a single ion. The uncertainty shown on the SEPICA points, however, is the statistical uncertainty in the calculated average. With this in mind, a two unit increase in the average Fe charge state with energy is observed over the energy range 20-355 keV/nuc, from approximately 7.7+ to 9.5+.

4.2. Observations With ACE/SEPICA

Several additional events have been observed by SEPICA alone, for which no STOF data are available due to their relative geometric factors. The variation of Fe charge state with energy for these events is shown in Plate 3. The uncertainties in the mean charge states are similar to those shown for SEPICA data in Plate 2. In all cases, there is a tendency for the Fe charge state to rise with energy, even though the events were not selected with that requirement. The lowest mean charge states approach 7+ to 8+ at 64 keV/nuc. Considering the uncertainties in the two low energy points, the increase is approximately one to two units in the energy range 64 to 310 keV/nuc.

Most of the events are associated with CMEs and interplanetary shocks, however, no effort has been made to sort them according to the type and location of particle acceleration. They may include ions accelerated both close to the Sun and in the interplanetary medium. Two events have a higher average charge state than the others. One event, September 26-28, 1998 (98/269-271), is temporally associated with a series of flares in the western solar hemisphere. In this case, the average Fe charge state rose from approximately 14.3+ at 104 keV/nuc to 16.7+ at 310 keV/nuc.

Iron charge states for two additional flare-related events are shown in Plate 4. The 98/269 event is repeated for comparison. The two flare-related events occurred on Sep. 8, 1998 (98/251) and May 1, 2000 (00/122). They were selected

by *Ho et al.* [2001a] and *Ho et al.* [2001b] using energetic particle and X-ray data. Here, the term "flare-related" means that the observed energetic particles are temporally associated with a flare, with relatively little contribution from interplanetary shock acceleration. An additional event from May 5, 1998 (98/125) with a high average charge state is also shown. All the flare-related events display a tendency for the Fe charge state to increase by 5-8 units in the energy range 100 to 550 keV/nuc.

Klecker et al. [2005] compiled SEPICA charge state measurements for four flare events in the energy range 180-550 keV/nuc. In three cases, STOF data were also available. They compared the SEPICA results to the average STOF Fe charge state for all three events in the energy range 10-100 keV/nuc. The average STOF Fe charge state was 12.5 units. In the higher energy range observed by SEPICA, the average Fe charge state increased from 14+ to 16+ at the lowest energy up to 2-3 units higher at the highest energy. *Klecker et al.* compared these results to the model of *Kocharov et al.* [2000] that describes charge-changing effects by collisions in the low solar corona. They concluded that the increase of charge state with energy was the result of stripping rather than by production in a hot 10 MK source region. *Miller* [1998] proposed a flare acceleration process that would produce the energy spectra and heavy ion abundances observed in flare-related events. His model used a plasma temperature of 3 MK, and thus an average Fe charge state of 13+. Ions at the lowest energies in the STOF/SEPICA observations may have experienced the least acceleration and the least stripping [*Möbius et al.*, 2003], and may therefore be consistent with the *Miller* [1998] source population. An acceleration model that can simultaneously produce the abundance enhancements and energy-dependent charge states could help constrain the location of acceleration region.

An avenue of future work is to compare these low energy measurements to what is seen in the solar wind. Solar wind Fe charge states are available for at least one of the SEPICA events (*Lepri et al.*, 2001). The solar wind Fe charge state distribution for the 98/310-312 event peaks between approximately 7.5+ and 9.5+. This is likely to be the corresponding solar wind time period because the SEPs were detected during a local shock passage at this time. The lowest energy mean Fe charge state from SEPICA for that event is 8.1+ at 65 keV/nuc (Plate 3). The SEPICA measurement is therefore near the lower bound of the solar wind Fe range. It is important to note that the SOHO/STOF Fe charge states at 65 keV/nuc are typically one unit higher than what is observed by SEPICA. Until a careful comparison between SEP and solar wind iron charge states is made, it is not yet possible to conclude whether there is a difference between them.

5. MECHANISMS FOR PRODUCING ENERGY-DEPENDENT CHARGE STATES

The November 1997 event focused discussion on mechanisms that may create energy dependent charge states. These mechanisms generally involve either collisional stripping during acceleration in a dense solar atmosphere, or rigidity-dependent acceleration with an interplanetary shock.

In collisional stripping models, ionization occurs by thermal electron and proton impact. Acceleration may be driven by shock or stochastic mechanisms. Observations have been compared to models involving shock acceleration close to the Sun (*Barghouty and Mewaldt*, 1999, 2000; *Reames et al.*, 1999b; *Stovpyuk and Ostryakov*, 2001) and stochastic mechanisms (*Ostryakov et al.*, 2000; *Kartavykh et al.*, 2001), or both [*Stovpyuk and Ostryakov*, 2003]. *Stovpyuk and Ostryakov* [2003] successfully fit Fe charge state distributions below 0.5 MeV/nuc that were observed by SEPICA in the November, 1997 event. Their model included both shock and stochastic acceleration in the dense solar atmosphere.

The final charge state depends on a density-traversal time product (nt), and the relative timescales for acceleration, trapping, ionization and recombination (see for ex. *Stovpyuk and Ostryakov* [2003]). As the ions are accelerated, the equilibrium charge increases with energy. The ionic charge state will approach the equilibrium value, but may never reach it. *Klecker et al.* [2001] presented a general description of mechanisms that create energy-dependent charge states. A thorough discussion of models that accelerate ions in a collisional environment is presented in *Kocharov* [2005] (this issue).

Energy-dependent charge states may also be produced in the interplanetary medium near a shock [*Ellison and Ramaty*, 1985, *Tylka*, 2001; *Tylka et al.*, 2005; *Klecker et al.*, 2000, 2001, 2003]. In this environment, rigidity-dependent effects are responsible for the increase in charge state with energy, instead of collisions. During shock acceleration, particle losses and/or finite shock size cause a rollover of the energy spectrum. *Ellison and Ramaty* [1985] modeled the energy spectrum as a powerlaw modified by a rigidity-dependent exponential. The rollover energy depends on M/Q, with intensities of higher M/Q particles rolling off at lower energies.

Klecker et al. [2003] considered an SEP event with an interplanetary shock passage on June 22-23 2000. They inferred a rollover energy of 0.4 MeV/nuc from the Fe energy spectra. Using a solar wind Fe charge state distribution as a source, they showed that a modest 1-2 unit increase could occur for Fe between the solar wind and the energy range 0.18-0.43 MeV/n. Their modeled result was generally consistent with SEPICA Fe charge state observations for the event.

6. DISCUSSION

Energy-dependent Fe charge states have been observed in four large SEP events with instruments on up to three spacecraft. Mean iron charge states in those events rose 4-12 units, from 8+ to 10+ up to 20+, in the energy range 65 keV/nuc to 47 MeV/nuc. In other events observed at energies less than 1 MeV/nuc, an energy dependence was often found below 400 keV/nuc. Flare-related events also displayed an energy-dependent charge state behavior, with minimum average Fe charge states as low as 12+ to 13+.

None of the events discussed above were chosen because they exhibited a tendency for charge states to increase with energy. They were instead chosen because they were observed by multiple spacecraft, or were known to be flare-related. The events observed below 1 MeV/nuc only by SEPICA were drawn randomly from the SEPICA mission dataset. All events show some tendency for Fe charge states to increase with energy. None have been found so far in which charge states decrease with energy.

The energy-dependent charge state profiles may be produced by acceleration of ions in the dense solar atmosphere, with charge-changing collisions. They may also be produced by rigidity-dependent acceleration at shocks. In the four large events presented here, both of these mechanisms may have contributed to the observed energy dependence. Early in the event, a shock or a flare may have accelerated ions in a dense region close to the Sun. Moreover, a CME-driven shock could create M/Q-dependent energy spectra. These processes would produce an energy-dependent charge state profile in the promptly accelerated ions. After a shock leaves the corona, it may weaken but still be capable of accelerating ions in the interplanetary medium. During the transit from the corona to 1 A.U., rigidity-dependent acceleration could enhance the energy dependence at energies close to the rollover in the energy spectrum. All of these ion populations would be integrated into the energy-dependent charge state profiles of Plate 1.

In some events, it may be difficult to determine whether the collisional or interplanetary mechanism is the primary cause of energy dependent charge states. *Klecker et al.* [2001] and *Barghouty and Mewaldt* [2000] fit the charge state observations of the October, 1992 event from SAMPEX. This was the first event in which energy dependent charge states were reported. *Klecker et al.* [2001] used a rigidity-dependent acceleration model, while *Barghouty and Mewaldt* [2000] used a collisional acceleration model. Both were able to produce an energy-dependent charge state profile that approximately fit the observations. Acceleration may have occurred both close to the Sun with collisions, and in the interplanetary medium with very few collisions, however.

The resolution of this uncertainty would be aided by charge state measurements over a wider range of energies and longitudes than is presently available. Direct measurements, in which the charge state of each ion could be determined, would deliver the best temporal resolution. They would provide charge state distributions instead of just the mean; and they could be used in both large and small events. Charge states of promptly arriving, high energy ions would supply information about acceleration early in the event, close to the Sun.

In the interplanetary medium, shock characteristics and magnetic connection to solar active regions can vary as a function of heliospheric longitude. Observations of charge state distributions and energy spectra could be obtained at several longitudes. They could help separate the contributions from acceleration sites close to the Sun from those in the interplanetary medium. For example, *Popecki et al.* [2003], compared iron charge state distributions from an event on the central solar disk to an event in the western hemisphere, at 0.3 MeV/nuc. The distribution from the central disk event was similar to what was expected from a source at 1 MK. On the other hand, the distribution from the western event had an extension to higher charge states. *Popecki et al.* [2003] suggested that ions from a flare-associated mechanism may account for the extension in the western event, while ions from a more spatially distributed source, such as an interplanetary shock, contributed to both events.

A complete effort to interpret observations with models must integrate the strong charge-changing effects of acceleration close to the Sun with the effect of acceleration in the interplanetary medium. The observed gradients in charge state with energy may then place limits on the relative contribution of each mechanism, particularly in cases where acceleration in one regime dominates the other.

Acknowledgments. The SEPICA instrument was supported by NASA under contract NAS5-32626. SEPICA data analysis was supported by Grant NAG 5-6912.

REFERENCES

Baker, D.N., G.M. Mason, O. Figueroa, G. Colon, J.G. Watzin, R.M. Aleman, An overview of the Solar, Anomalous, and Magnetospheric Particle Explorer (SAMPEX) mission, *IEEE Trans. Geosci. and Remote Sensing*, 31, 531-541, 1993.

Barghouty, A.F. and R.A. Mewaldt, Charge states of solar energetic iron: non-equilibrium calculation with shock-induced acceleration, *ApJ Letters*, 520, L127, 1999.

Barghouty, A.F., and R.A. Mewaldt, Simulation of charge-equilibration and acceleration of solar energetic ions, in *Acceleration and Transport of Energetic Particles Observed in the Heliosphere*, edited by R.A. Mewaldt, J.R. Jokipii, M.A. Lee, E. Möbius, and T.H. Zurbuchen, Melville, NY, AIP, 71-78, 2000.

Bogdanov, A.T., B. Klecker, E. Möbius, M. Hilchenbach, L.M. Kistler, M.A. Popecki, D. Hovestadt, J. Weygand, Energy Dependence of Ion Charge States in CME Related Solar Energetic Particle Events Observed with ACE/SEPICA and SOHO/STOF, in *Acceleration and Transport of Energetic Particles Observed in the Heliosphere*, edited by R.A. Mewaldt, J.R. Jokipii, M.A. Lee, E. Möbius, and T.H. Zurbuchen, Melville, NY, AIP, 143-146, 2000.

Bogdanov, A.T., B. Klecker, E. Möbius, M. Hilchenbach, M.A. Popecki, L.M. Kistler, D. Hovestadt, Observations of heavy ion charge spectra in CME driven gradual solar energetic particle events, *Adv. Space Res.*, 30, 111-117, 2002.

Cohen, C.M.S., A.C. Cummings, R.A. Leske, R.A. Mewaldt, E.C. Stone, B.L. Dougherty, M.E. Wiedenbeck, E.R. Christian, T.T. von Rosenvinge, Inferred charge states of high energy solar particles from the Solar Isotope Spectrometer on ACE, *GRL*, 26, 149-152, 1999.

Cook, W.R., A.C. Cummings, J.R. Cummings, T.L. Garrard, B. Kecman, R.A. Mewaldt, R.S. Selesnick, E.C. Stone, T.T. von Rosenvinge, MAST - A mass spectrometer telescope for studies of the isotopic composition of solar, anomalous, and galactic cosmic ray nuclei, *IEEE Trans. Geosci. and Remote Sensing*, 31, 557-564, 1993.

Dietrich W.F. and A.J. Tylka, Time-to-Maximum studies and inferred ionic charge states in the solar energetic particle events of 14 and 15 April 2001, *Proceedings of the 28th International Cosmic Ray Conf.*, Tsukuba, Japan, eds. T. Kajita, Y. Asaoka, A. Kawachi, Y. Matsubara, M. Sasaki, 6, 3291-3294, 2003.

Ellison, D.C. and R. Ramaty, Shock acceleration of electrons and ions in solar flares, *ApJ*, 298, 400-408, 1985.

Ho, G.C.; G.M. Mason, R.E. Gold, J.R. Dwyer, J.E. Mazur, Propagation of impulsive solar energetic particle events, in *Solar and Galactic Composition*, edited by R.F. Wimmer-Schweingruber, pp. 353-354, American Institute of Physics, Melville, N.Y., 2001a.

Ho, G.C., E.C. Roelof, S.E. Hawkins III, R.E. Gold, G.M. Mason, J.R. Dwyer, J.E. Mazur, Energetic electrons in 3He-enhanced solar energetic particle events, *ApJ*, 552, 863-870, 2001b.

Hovestadt, D., B. Klecker, M. Scholer, H. Arbinger, G. Gloeckler, F.M. Ipavich, J. Cain, C.Y. Fan, L.A. Fisk, J.J. Ogallagher, The nuclear and ionic charge distribution particle experiments on the ISEE-1 and ISEE-C spacecraft, *IEEE Trans. Geos. Electr.*, GE-16, 166-175, 1978.

Hovestadt, D., H. Hoefner, B. Klecker, M. Scholer, G. Gloeckler, F.M. Ipavich, C.Y. Fan, L.A. Fisk, J.J. Ogallagher, Direct observation of charge state abundances of energetic He, C, O, and Fe emitted in solar flares, *Adv. Sp. Res.*, 1, 61-64, 1981.

Hovestadt, D., M. Hilchenbach, A. Burgi, B. Klecker, P. Laeverenz, M. Scholer, H. Grunwaldt, W.I. Axford, S. Livi, E. Marsch, B. Wilken, H.P. Winterhoff, F.M. Ipavich, P. Bedini, M.A. Coplan, A.B. Galvin, G. Gloeckler, P. Bochsler, H. Balsiger, J. Fischer, J. Geiss, R. Kallenbach, P. Wurz, K.-U. Reiche, F. Gliem, D.L. Judge, H.S. Ogawa, K.C. Hsieh, E. Möbius, M.A. Lee, G.G. Managadze, M.I. Verigin, M. Neugebauer, CELIAS: The charge, element and isotope analysis system for SOHO, *Solar Physics*, 162, 441-481, 1995.

Kartavykh, Yu.Yu., V.M. Ostryakov, D. Ruffolo, E. Möbius, M.A. Popecki, Heavy ions from impulsive SEP events and constraints on the plasma temperature in the acceleration site, *Proceedings of the 27th International Cosmic Ray Conf.*, Hamburg, Germany, eds. W. Droge, H. Kunow, M. Scholer, 8, 3091-3094, 2001.

Klecker, B., D. Hovestadt, G. Gloeckler, F.M. Ipavich, M. Scholer, L.A. Fisk, Direct determination of the ionic charge distribution of helium and iron in He-3-rich solar energetic particle events, *ApJ*, 281, 458-462, 1984.

Klecker, B., M.C. McNab, J.B. Blake, D.C. Hamilton, D. Hovestadt, H. Kaestle, M.D. Looper, G.M. Mason, J.E. Mazur, M. Scholer, Charge States of Anomalous Cosmic-Ray Nitrogen, Oxygen and Neon: SAMPEX Observations, *ApJ*, 442, L69-L72, 1995.

Klecker, B., A.T. Bogdanov, M. Popecki, R.F. Wimmer-Schweingruber, E. Möbius, R. Schaerer, L.M. Kistler, A.B. Galvin, D. Heirtzler, D. Morris, D. Hovestadt, G. Gloeckler, Comparison of ionic charge states of energetic particles with solar wind charge states in CME related events, in *Acceleration and Transport of Energetic Particles Observed in the Heliosphere*, edited by R.A. Mewaldt, J.R. Jokipii, M.A. Lee, E. Möbius, and T.H. Zurbuchen, Melville, NY, AIP, 135-138, 2000.

Klecker, B., E. Möbius, M.A. Popecki, M.A. Lee, A.T. Bogdanov, On the energy dependence of ionic charge states in solar energetic particle events, in *Solar and Galactic Composition*, edited by R.F. Wimmer-Schweingruber, American Institute of Physics, Melville, N.Y., 317-321, 2001.

Klecker, B., M.A. Popecki, E. Möbius, M.I. Desai, G.M. Mason, and R.F. Wimmer-Schweingruber, On the energy dependence of ionic charge states, *Proceedings of the 28th International Cosmic Ray Conference*, Tsukuba, Japan, edited by T. Kajita, Y. Asaoka, A. Kawachi, Y. Matsubara, M. Sasaki, 3277-3280, 2003.

Klecker, B., E. Möbius, M.A. Popecki, L.M. Kistler, H. Kucharek, M. Hilchenbach, Observation of Energy-dependent ionic charge states in impulsive solar energetic particle events, *Adv. Space Res.*, submitted, 2005.

Kocharov, L., G.A. Kovaltsov, G.A. Torsti, V.M. Ostryakov, Evaluation of solar energetic Fe charge states: effect of proton-impact ionization, *Astron. Astrophys.*, 357, 716-724, 2000.

Kocharov, L., submitted, 2005.

Lepri, S.T., T.H. Zurbuchen, L.A. Fisk, I.G. Richardson, H.V. Cane; G. Gloeckler, Iron charge distribution as an identifier of interplanetary coronal mass ejections, *JGR*, 106, 29231-29238, 2001.

Leske, R.A., J.R. Cummings, R.A. Mewaldt, E.C. Stone, T.T. von Rosenvinge, Measurements of the ionic charge states of solar energetic particles using the geomagnetic field, *ApJ (Letters)*, 452, L149-L152, 1995.

Leske, R.A., R.A. Mewaldt, A.C. Cummings, E.C. Stone, T.T. von Rosenvinge, The ionic charge state composition at high energies in large solar energetic particle events in solar cycle 23, in *Solar and Galactic Composition*, edited by R.F. Wimmer-Schweingruber, 171-176, American Institute of Physics, Melville, N.Y., 2001.

Luhn, A., B. Klecker, D. Hovestadt, G. Gloeckler, F.M. Ipavich, M. Scholer, C.Y. Fan, and L.A. Fisk, Ionic charge states of N, Ne, Mg, Si and S in solar energetic particle events, *Adv. Space Res.*, 4, 161-164, 1984.

Luhn, A., B. Klecker, D. Hovestadt, and E. Möbius, The mean charge of Si in 3He-rich solar flares, *ApJ*, 317, 951-955, 1987.

Mason, G.M., D.C. Hamilton, P.H. Walpole, K.F. Heuerman, T.L. James, M.H. Lennard, J.E. Mazur, LEICA - A low energy ion composition analyzer for the study of solar and magnetospheric heavy ions, *IEEE Trans. Geosci. and Remote Sensing*, 31, 549-556, 1993.

Mason, G.M., J.E. Mazur, M.D. Looper, and R.A. Mewaldt, Charge state measurements of solar energetic particles observed with SAMPEX, *ApJ*, 452, 901-911, 1995.

Mazur, J.E., G.M. Mason, M.D. Looper, R.A. Leske, and R.A. Mewaldt, Charge states of solar energetic particles using the geomagnetic cutoff technique: SAMPEX measurements in the 6 November 1997 solar particle event, *GRL*, 26, 173-176, 1999.

Miller, J.A., Particle acceleration in impulsive solar flares, *Space Science Reviews*, 86, 79-105, 1998.

Möbius, E., L.M. Kistler, M.A. Popecki, K.N. Crocker, M. Granoff, S. Turco, A. Anderson, P. Demain, J. Distelbrink, I. Dors, P. Dunphy, S. Ellis, J. Gaidos, J. Googins, R. Hayes, G. Humphrey, H. Křstle, J. Lavasseur, E.J. Lund, R. Miller, E. Sartori, M. Shappirio, S. Taylor, P. Vachon, M. Vosbury, V. Ye, D. Hovestadt, B. Klecker, H. Arbinger, E. Knneth, E. Pfeffermann, E. Seidenschwang, F. Gliem, K.-U. Reiche, K. Stöckner, W. Wiewesiek, A. Harasim, J. Schimpfle, S. Battell, J. Cravens, G. Murphy, The Solar Energetic Particle Ionic Charge Analyzer (SEPICA) and the Data Processing Unit (S3DPU) for SWICS, SWIMS AND SEPICA, *Sp. Sci. Rev.*, 86, 449-495, 1998.

Möbius, E., M. Popecki, B. Klecker, L.M. Kistler, A. Bogdanov, A.B. Galvin, D. Heirtzler, D. Hovestadt, E.J. Lund, D. Morris, W.K.H. Schmidt, Energy dependence of the ionic charge state distribution during the November 1997 solar energetic particle event, *GRL*, 26, 145-148, 1999.

Möbius, E., Y. Cao, M.A. Popecki, L.M. Kistler, H. Kucharek, D. Morris, B. Klecker, Strong energy dependence of ionic charge states in impulsive solar events, *Proceedings of the 28th International Cosmic Ray Conf.*, Tsukuba, Japan, edited by T. Kajita, Y. Asaoka, A. Kawachi, Y. Matsubara, M. Sasaki, 3273-3276, 2003.

Oetliker, M.B., B. Klecker, D. Hovestadt, G.M. Mason, J.E. Mazur, R.A. Leske, R.A. Mewaldt, J.B. Blake, and M.D. Looper, The ionic charge state of solar energetic particles with energies of 0.3-70 MeV/nuc, *ApJ*, 477, 495-501, 1997.

Ostryakov, V.M., Y.Y. Kartavykh, D. Ruffolo, G.A. Kovaltsov, L. Kocharov, Charge state distributions of iron in impulsive solar flares: Importance of stripping effects, *JGR*, 105, 27315-27322, 2000.

Popecki, M.A., E. Möbius, B. Klecker, A.B. Galvin, L.M. Kistler, A.T. Bogdanov, Ionic charge state measurements in solar energetic particle events, *Adv. Space Res*, 30, 33-43, 2002.

Popecki, M.A., E. Möbius, D. Morris, B. Klecker, L.M. Kistler, Iron charge state distributions in large, gradual solar energetic particle events, *Proceedings of the 28th International Cosmic Ray Conf.*, Tsukuba, Japan, eds. T. Kajita, Y. Asaoka, A. Kawachi, Y. Matsubara, M. Sasaki, 6, 3287-3290, 2003.

Reames, D.V., *Sp. Sci. Rev.*, Particle acceleration at the Sun and in the Heliosphere, 90 (3/4), 413-491, 1999a.

Reames, D.V., C.K. Ng, A.J. Tylka, Energy dependent ionization states of shock-accelerated particles in the solar corona, *GRL*, 26, 3585-3588, 1999b.

Smith, C.W., N.F. Ness, L.F. Burlaga, R.M. Skoug, D.J. McComas, T.H. Zurbuchen, G. Gloeckler, D.K. Haggerty, R.E. Gold, M.I. Desai, G.M. Mason, J.E. Mazur, J.R. Dwyer, M.A. Popecki, E. Möbius, C.M.S. Cohen, R.A. Leske, ACE observations of the Bastille Day 2000 interplanetary disturbances, *Solar Physics*, 204, 229-254, 2001.

Sollitt, L.S., E.C. Stone, R.A. Mewaldt, C.M.S. Cohen, R.A. Leske, M.E. Wiedenbeck, T.T. von Rosenvinge, Ionic charge states of high energy solar energetic particles in large events, *Proceedings of the 28th International Cosmic Ray Conf.*, Tsukuba, Japan, eds. T. Kajita, Y. Asaoka, A. Kawachi, Y. Matsubara, M. Sasaki, 6, 3295-3298, 2003.

Stovpyuk, M.F. and V.M. Ostryakov, Heavy ions from the 6 November 1997 sep event: a charge-consistent acceleration model, *Solar Physics*, 198, 163-167, 2001.

Stovpyuk, M.F. and V.M. Ostryakov, A charge-consistent model for the acceleration of iron in the solar corona: nonisothermal injection, *Astronomy Reports*, 47, 343-353, 2003.

Tylka, A.J., P.R. Boberg, R.E. McGuire, C.K. Ng, and D.V. Reames, Temporal evolution in the spectra of gradual solar energetic particle events, in *Acceleration and Transport of Energetic Particles Observed in the Heliosphere*, edited by R.A. Mewaldt, J.R. Jokipii, M.A. Lee, E. Möbius, and T.H. Zurbuchen, Melville, NY, AIP, 147-152, 2000.

Tylka, A.J., New insights on solar energetic particles from Wind and ACE, *JGR*, 106, 25333-25352, 2001.

Tylka, A.J.; C.M.S. Cohen, W.F. Dietrich, M.A. Lee, C.G. Maclennan, R.A. Mewaldt, C.K. Ng, D.V. Reames, Shock Geometry, Seed Populations, and the Origin of Variable Elemental Composition at High Energies in Large Gradual Solar Particle Events, *ApJ*, 625, 474-495, 2005.

M. Popecki, Institute of Earth, Oceans and Space, University of New Hampshire, Durham, NH, 03824. (mark.popecki@unh.edu)

Modeling the Energy-Dependent
Charge States of Solar Energetic Particles

Leon Kocharov

Department of Physics and the Väisälä Institute for Space Physics and Astronomy,
University of Turku, Finland

Energy-dependent charge states of solar energetic ions were observed in a number of particle events. They carry information on the local conditions at the origin of solar energetic particles as well as on the acceleration and transport processes in the solar corona and interplanetary medium. However, this information could not be extracted without numerical modeling. This review is focused on recent achievements in the modeling of the charge-consistent acceleration and transport of solar ions in the energy range from ~0.1 to ~100 MeV/nucleon, as applied to the ionic charge states of iron. A history of acceleration and transport leaves a clear imprint in the energy-charge profiles of iron. The charge-state measurements can play a unique role in an integrated analysis of solar particles events if the time resolution is improved and different observables are consistently included.

1. INTRODUCTION

The energy-dependent charge states of accelerated Fe ions were observed in a number of solar energetic particle (SEP) events [*Oetliker et al.*, 1997; *Möbius et al.*, 1999; *Mazur et al.*, 1999; *Popecki et al.*, 2000; *Möbius et al.*, 2003, *Labrador et al.*, 2003; *Dietrich and Tylka*, 2003], with different charge measurement techniques employed to cover the energy range from a fraction to ~100 MeV/nucleon [*Luhn et al.*, 1984; *Leske et al.*, 1995; *Mason et al.*, 1995; *Tylka et al.*, 1995; *Cohen et al.*, 1999; *Möbius et al.*, 2002]. Ionization-state measurements are expected to play a key role in deciphering the local conditions at the origin of SEPs as well as the processes involved in their selection, acceleration and transport. The observational issues have been reviewed at the Chapman Conference by *M. Popecki* [this volume], whereas the present paper is focused on the development of theoretical calculations of SEP charge states.

The importance of charge-changing processes for Coulomb energy losses and by this expedient for the injection of

different ion species into the acceleration regime was recognized in early days of cosmic ray research [*Ginzburg and Syrovatskii*, 1964]. Based on experimental data for neutral matter, it was accepted that energetic ions rapidly approach an equilibrium value of their charge state after traversing a small amount of matter, so that a certain mean charge could be ascribed to a given ion velocity irrespective of a history of ion acceleration/deceleration processes [*Barkas*, 1963]. The equilibrium charge was also used for calculations of ion stopping in target matter (effective charge for stopping). In application to solar energetic particles such an approach was adopted by *Korchak* [1980]. However already at that time *Korchak and Filippov* [1979] raised a question as to whether an equilibrium charge can be attained during heavy ion acceleration in coronal plasma.

A key contribution to the kinetic modeling of SEP charge states was done by *Luhn and Hovestadt* [1987], who evaluated energy-dependent rates of the energetic ion recombination by averaging the corresponding cross sections over thermal electron distributions in the ion rest frame. Those rates were employed to calculate mean equilibrium charge states of energetic ions from C through Si. However, proton-impact ionization was missing, despite being previously mentioned in the monograph by *Ginzburg and Syrovatskii* [1964]. A pioneering kinetic model of ion acceleration with

Solar Eruptions and Energetic Particles
Geophysical Monograph Series 165
Copyright 2006 by the American Geophysical Union
10.1029/165GM14

allowance for charge-changing processes was proposed by *Kurganov and Ostryakov* [1991]. In that oversimplified model all energy and charge dependencies in the rates of ionization and recombination were neglected and by this expedient an elegant analytic solution was found. More realistic kinetic considerations were done with different numerical techniques [*Kartavykh*, 1999; *Stovpyuk*, 2000; *Barghouty and Mewaldt*, 2000; *Ostryakov et al.*, 2000; *Kocharov et al.*, 2001]. At that time a major distinction existed between the kinetic modeling approach and the equilibrium charge-state approach (the latter was adopted for the last time by *Reames, Ng, and Tylka* [1999]). The present review is focused on the most recent results of kinetic modeling.

2. PROCESSES OF CHARGE EQUILIBRATION IN PLASMA

Ions of iron are the most promising object of the modeling, because Fe is an abundant ion that should not be completely stripped even at rather high energies, and there is a set of atomic parameters needed to calculate all the charge states [*Arnaud and Raymond*, 1992]. That is not to say that the charge-changing rates as given by *Arnaud and Raymond* [1992] for thermal ions are directly applicable to energetic Fe projectiles. However, many necessary atomic parameters can be extracted from that paper. Those parameters were used by *Kocharov et al.* [2000] for evaluating the energy-dependent ionization and recombination rates that are applicable also to non-thermal ions.

A charge state distribution of energetic ions is established by the competition of ionizing collisions and recombinations. There are two main contributors to the ionization – collisions with thermal electrons and collisions with thermal protons. The latter are significant for solar energetic ions, because the velocity of the ~1 MeV/nucleon ion is not less than the velocity of a thermal electron, and it is only the relative velocity of target and projectile that is important for a charge-changing collision. Proton-impact ionization was ignored in a number of the SEP charge-state calculations [*Luhn and Hovestadt*, 1987; *Ruffolo*, 1997; *Barghouty and Mewaldt*, 1999]. On the other hand, there was also a strong overestimate of proton-impact ionization, because of more than one order of magnitude underestimate of the threshold energy [*Kartavykh*, 1999; *Kartavykh and Ostryakov*, 1999], caused by a misinterpretation of the results of *Sidorovich and Nikolaev* [1988] (also *Nikolaev and Sidorovich* [1989]).

It is known from atomic physics that (1) the cross section for ionization by a high-energy proton well above the ionization threshold coincides with the electron cross section at equal velocities with respect to the atom/ion to be ionized [*McDaniel*, 1989]; and (2) the threshold energy for proton-impact ionization estimated from the energy-time indeterminacy relation is $E_{th} \approx \frac{1}{4}(M_p/m_e)I$ (per nucleon), where M_p/m_e is the proton-to-electron mass ratio, and I is the ionization potential of the corresponding electron. For instance, the proton-impact ionization threshold for Fe^{+11} is about 150 keV/nucleon. Proton-impact ionization was firstly incorporated into SEP calculations by *Kharchenko and Ostryakov* [1987], being applied to helium and oxygen. However, there are different theoretical estimates of the ionization cross section near threshold, depending on parameter range and theoretical assumptions [*Bethe*, 1930; *Bohr*, 1948; *Gryzinski*, 1965; *McGuire and Richard*, 1973; *Inokuti et al.*, 1978; *Knudsen et al.*, 1984; *Janev*, 1983; *Janev et al.*, 1994; *Barghouty*, 2000], whereas experimental data are fragmentary [*Goffe*, 1979; *Haugen et al.*, 1982; *Dietrich et al.*, 1992; *Sant'Anna et al.*, 1998]. No experimental data are available for highly ionized Fe ions colliding with protons. Different theoretical cross-sections for the proton-impact ionization of Fe ions were compared by *Kovaltsov et al.* [2001]. Disagreement in the near-threshold behavior is not small. The impact of the cross section uncertainty on the calculated SEP charge states is found to be noticeable but not crucial, especially under strongly non-equilibrium conditions.

An ion recombination rate comprises two terms – radiative recombination (the initially free electron emits excess energy in the form of electromagnetic radiation) and dielectronic recombination (the electron's energy is taken up by the excitation of another, bound electron). Each rate should be obtained by convolving the corresponding cross section with the ambient electron distribution as seen in the rest frame of the SEP ion. Hence, in the general case these rates depend on both the ambient electron temperature and the SEP ion energy. In the low energy (thermal) limit, the energy dependence disappears, and the recombination rate must coincide with the thermal recombination rate [*Arnaud and Raymond*, 1992]. For energetic Fe ions the recombination rates are given by *Kocharov et al.* [2000]. The assumption of thermal rates for SEPs is a poor approximation that gives unreasonable results. It is equally impossible to use a system of energy-dependent rates by *Rodríguez-Frías et al.* [2000, 2001], because it fails a test of consistency with the Arnaud-and-Raymond's results [*Kovaltsov et al.*, 2002; *Kocharov et al.*, 2002].

It is important that recombination in a hot plasma is much slower than recombination in neutral gases or solids. Studies of fast ions in plasmas indicate significantly higher ionization of the projectile in plasma than the ionization of the same projectile species in a cold neutral target [*Nardi and Zinamon*, 1982, *Peter and Meyer-ter-Vehn*, 1986]. In the plasma studies kinetic computations are performed and verified against experimental data (e.g., *Peter and Meyer-ter-Vahn* [1991], *Dietrich et al.* [1992]). A similar, kinetic approach should be adapted also for energetic ions in the plasma of the solar corona.

3. EQUILIBRIUM CHARGE OF ENERGETIC IONS

The mean equilibrium charge, $Q_{eq}(E)$, is the ionic charge averaged over the equilibrium distribution, $N_i^{eq}(E)$ which is defined as a solution of the charge balance equation for ions with a fixed energy:

$$(S_i + \alpha_i)N_i^{eq}(E) = S_{i-1}N_{i-1}^{eq}(E) + \alpha_{i+1}N_{i+1}^{eq}(E), \qquad (1)$$

where $i \equiv Q + 1$; Q is the ion's charge number (i.e., for Fe ions, $i = 1, ..., 27$); $S_i = S_i(T, E)$ is the temperature-energy-dependent ionization rate coefficient for transition from the ionization state i to the state $i + 1$; $\alpha_i = \alpha_i(T, E)$ is the recombination rate coefficient from the ionization state i to $i - 1$; S_i and α_i are in units of cm^3 s^{-1}, because they shall be multiplied by the electron number density, n, to get a process rate; E traditionally stands for kinetic energy per nucleon. *Kovaltsov et al.* [2001] solved equation (1) for Fe ions with three different versions of the proton impact cross section and obtained three markedly different estimates of $Q_{eq}(E)$ under the typical coronal conditions (one of the estimates is shown in Figure 1).

Rather than solve the complex system of charge-changing reactions, often in astrophysical applications a simple, semi-empirical expression for the mean charge dependence on energy is used,

$$Q_{eq}/Z = 1 - \exp[-V/(v_0 Z^{2/3})], \qquad (2)$$

where V and Z are the speed and nuclear charge of the projectile and $v_0 = e^2/\hbar$ is the Bohr velocity [*Barkas*, 1963; *Northcliffe*, 1963; *Betz et al.*, 1966]. The exponent in equation (2), which is based on the Thomas-Fermi model of the ion, assumes that a fast heavy ion penetrating through rarefied gases retains all of its electrons with orbital velocities exceeding the velocity of the ion. Expression (2) seems to describe fairly reasonably the equilibrium charge state of an ion as a function of energy as it traverses neutral, dense solids and gases [*Pierce and Blann*, 1968; *Betz*, 1972]. In contrast to the applications to cold neutral gases, however, the simple relation between mean charge and velocity in equation (2) fails in applications to plasmas. One can attempt to improve on equation (2) by accounting for some of the salient peculiarities of charge-changing cross sections in plasmas. At relatively low recombination rate and high contribution of protons to the ionization, the threshold velocity for ionization becomes close to $\frac{1}{2}$ of the orbital electron velocity $[E_{th} \approx \frac{1}{4}(M_p/m_e)I]$. Based on this, one can argue that a Fe ion retains all its electrons with orbital velocity greater than twice the ion velocity. In addition, a correction can be done for the thermal motion of ambient electrons [*Reames, Ng,*

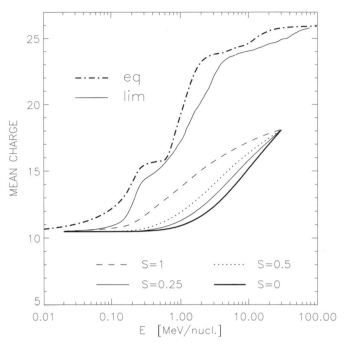

Figure 1. Nonequilibrium mean-charge of energetic Fe ions for regular acceleration at different dependencies of the acceleration rate on the energy (parameter S in equation (6)). Employed values of S are 0 (heavy solid curve), 0.25 (light solid curve), 0.5 (dotted curve), and 1 (dashed curve). The density × acceleration-time parameter is $n \times \tau_1/(10^{10}$ cm^{-3}s$) = 0.025, 0.045, 0.075, 0.2$ for $S = 0, 0.25, 0.5, 1$, respectively [*Kovaltsov et al.*, 2001]. As a reference, the dash-dotted line indicates a mean equilibrium charge at a fixed energy, $Q_{eq}(E)$. The dot-dot-dot-dashed line corresponds to the case $S = 0.25$ and $n \times \tau_1/(10^{10}$ cm^{-3}s$) = 2.5$. The latter is a limiting value near which the Coulomb energy-losses terminate the ion acceleration to below ~0.5 MeV/nucleon. During the acceleration the charge states situated above the dot-dot-dot-dashed line could not be attained. All calculations are done with the temperature $T = 1.26 \times 10^6$ K and the cross section for proton-impact ionization based on the Bates-Griffing relation [*Barghouty*, 2000; *Kovaltsov et al.*, 2001].

and Tylka, 1999]. Those modifications together result in a formula:

$$Q_{eq}/Z = 1 - \exp[-b(2\beta + \beta_{th})/Z^{2/3}]. \qquad (3)$$

where $\beta = V/c$; $b = \hbar c/e^2 = 137$ and $\beta_{th}^2 = 2k_B T/(m_e c^2)$. Equation (3) but with the empirical value $b = 110$ and without correction to the plasma temperature was suggested by *Tatischeff, Ramaty, and Kozlovsky*, [1998], based on kinetic calculations for C, N, O, and Ne, and with both correction included by *Kovaltsov et al.* [2001]. The charge-energy dependence in the form of equation (3) with $b = 110$–125 may roughly approximate the results of kinetic calculations by *Kovaltsov et al.* [2001], being still within the uncertainty range caused by the uncertainty in the proton-impact cross section.

Unfortunately, quantitative estimates show that the observed charge states of Fe are non-equilibrium [*Ruffolo*, 1997]. A numerical modeling reveals that during concurrent acceleration and charge-change at coronal temperatures, iron ions typically follow the rising trajectories on the charge-energy plane (Figure 1 and Section 4). These trajectories are situated below the mean equilibrium charge curve defined from the balance of ionization and recombination at fixed energy. The equilibrium charge curve for Fe ions might be attained only in the extreme case, when the acceleration was slowed down to below the Coulomb-loss limit, which would result in very few particles accelerated to >0.5 MeV/nucleon. On the other hand, during stopping the iron ions cross the equilibrium charge curve and run through a series of charge states above the mean equilibrium charge at current energy, because the Coulomb deceleration rate significantly exceeds the rate of ion recombination in a hot plasma [*Kocharov et al.*, 2001]. As a result, the variety of possible trajectories on the ion charge-energy plane turns out to be much wider than would be expected based on the equilibrium charge-state approximation. Perhaps, the main role of the equilibrium charge curve, $Q_{eq}(E)$, is to provide a reference for the SEP charge states found in observations and in numerical modeling.

4. MODELS OF CONCURRENT ACCELERATION AND CHARGE-EQUILIBRATION

Most of the models of concurrent acceleration and charge change are either the steady-state models of shock acceleration [*Kurganov and Ostryakov*, 1991; *Ostryakov and Stovpyuk*, 1999] or the leaky-box models of stochastic or shock acceleration [*Ostryakov and Stovpyuk*, 1997; *Barghouty and Mewaldt*, 1999]. Probably the most advanced leaky-box model of Fe-ion acceleration is the model by *Kocharov et al.* [2001]. The model incorporates both stochastic and regular acceleration, which are thought to be due to MHD turbulence and shocks, respectively:

$$\frac{\partial N_i}{\partial t} + \frac{\partial}{\partial E}\left(\frac{dE}{dt}N_i\right) - \frac{\partial^2}{\partial E^2}(D_E N_i) + \frac{N_i}{\tau_{esc}}$$
$$+ n\left[(S_i + \alpha_i)N_i - S_{i-1}N_{i-1} - \alpha_{i+1}N_{i+1}\right] = X_i\delta(t)\delta(E - E_o),$$

$$(4)$$

where $N_i(E)$ is the energy distribution of the ions of charge $Q \equiv i - 1$; n is the ambient electron number density; X_i is a source ion distribution; τ_{esc} is the particle escape (leaky-box) time; and the advection rate in energy space is of the form

$$\frac{dE}{dt} = \left(\frac{dE}{dt}\right)_d + \left(\frac{dE}{dt}\right)_{sh} - \left(\frac{dE}{dt}\right)_c, \qquad (5)$$

incorporating a systematic energy gain produced by stochastic acceleration, a regular acceleration rate that represents shock acceleration, and Coulomb energy losses, respectively. The regular acceleration rate, $(dE/dt)_{sh}$, is parameterized in the form:

$$\left(\frac{dE}{dt}\right)_{sh} = \frac{E_1}{\tau_1}\left(\frac{E}{E_1}\right)^S \qquad (6)$$

where τ_1 and S are model parameters; $E_1 = 1$ MeV/nucleon is normalization energy. The acceleration rate $(dE/dt)_d$ is similarly parameterized, with the same S but another characteristic time.

A Monte Carlo technique had been applied for finding the time-integrated solutions of equation (4). It was concluded that the acceleration type, either stochastic or regular, has a minor effect on the mean charge states of iron, $\langle Q \rangle (E)$, but the shapes of the charge-state distributions, $N_i(E)$, produced by regular acceleration and by stochastic acceleration are different. The magnitude of the mean ionic charge at a certain energy depends on the density × acceleration-time product, whereas the shape of the energy-charge profile $\langle Q \rangle (E)$ is ruled by the energy dependence of the acceleration rate (see parameter S in equation (6) and Figure 1). For this reason, *Kovaltsov et al.* [2001] emphasized the potential importance of careful measurements of charge-energy profiles, along with ion energy spectra, for the deciphering the acceleration mechanism responsible for the coronal production of SEPs.

The mean Fe-charge value $\langle Q \rangle$ (30 MeV/nucl.) = 18, inferred for the 1997 November 6 event [*Cohen et al.*, 1999], suggests that the product $n \times t_{0.1\text{-}30}$ is close to 0.8×10^{10}cm^{-3} s, where $t_{0.1\text{-}30}$ stands for the time it takes for the ion to be accelerated from 0.1 to 30 MeV/nucleon. This estimate is not sensitive to details of the acceleration model and remains nearly the same for $S = 0 - 0.5$ [*Kovaltsov et al.*, 2001].

5. CONSECUTIVE ACCELERATION AND CHARGE-CHANGE

Coronal and interplanetary transport after acceleration also can affect the ion charge-state distributions observed at 1 AU. Inside the accelerator, concurrent processes of ion acceleration and charge-change imply higher charge states for higher energy particles, because acceleration to a higher energy takes a longer time and a higher column density will be traversed. However, the column density traversed during coronal transport is not expected to increase with particle energy. For this reason, ion stripping after acceleration may result in non-rising charge-energy profiles of iron. Another transport effect may be adiabatic deceleration of particles in the interplanetary medium, which shifts the $\langle Q \rangle$ (E) curve to lower energies, so that highly-ionized Fe ions may appear at low energies, even above the equilibrium charge curve $Q_{eq}(E)$ [*Kartavykh et al.*, 2005]. Adiabatic deceleration during coronal transport may be also significant.

A numerical model incorporating stripping and adiabatic deceleration of Fe ions during coronal transport following shock acceleration was recently considered by *Lytova and Kocharov* [2005]. The model suggests acceleration of non-relativistic ions at a parallel shock wave propagating in a radial magnetic field through a spherical turbulent layer at the base of the solar wind (Figure 2). The general transport equation for fast charged particles under strong scattering conditions is the diffusion-convection equation [*Parker*, 1965; *Toptygin*, 1985]. When written for the ion distribution F_i as a function of kinetic energy per nucleon E, the equation is of the form:

$$\frac{\partial F_i}{\partial t} = \frac{1}{r^2} \frac{\partial}{\partial r}\left(r^2 \kappa \frac{\partial F}{\partial r}\right) - \frac{1}{r^2} \frac{\partial}{\partial r}\left(r^2 w F_i\right)$$

$$+ \frac{\partial}{\partial E} F_i\left[\frac{2E}{3r^2} \frac{\partial}{\partial r}(r^2 w) + \dot{E}_C\right] + \dot{F}_{CC}, \qquad (7)$$

where $F_i(E, r, t)$ is the number of ions per unit of E and unit of volume; $i = Q + 1$; $w(r, t)$ is the fluid speed of scattering centers; $\kappa(E, r, t)$ is the ion diffusion coefficient, which depends also on the turbulence spectrum; and $\dot{E}_C$ accounts for Coulomb energy-losses. The last term of equation (7) describes the charge-changing processes – ionization and recombination: $\dot{F}_{cc} \equiv n[S_{i-1}F_{i-1} - (S_i + \alpha_i)F_i + \alpha_{i+1}F_{i+1}]$. The velocity field in the solar frame, $w(r, t)$, comprises a uniform upstream flow of scattering centers, $w = W_1$, a steep increase within the shock front to $w = W_2$, and a far-downstream relaxation region, in which the velocity gradually returns to its upstream value, $W_2 \rightarrow W_1$. The initial diffusion coefficient, κ_1, is assumed to be small only inside a turbulent layer, $r_1 \leq r \leq r_2$, allowing the shock to accelerate particles within the layer, from heliocentric distance $r_1 = 1.2 R_\odot$ to $r_2 = 1.7 R_\odot$. Particles can freely escape at the outer boundary of the layer, r_e, which initially is $r_e = r_2$, but later on, after the shock has arrived at $r = r_2$, the layer starts to expand. The layer's outer boundary, r_e, moves in the shock downstream region either with the local speed of scattering centers, $w(r_e)$, or somewhat slower: $dr_e/dt = w(r_e) - \delta w$. For Figures 2 and 3 the later parameter is $\delta w = 30$ km/s, which allows particles to escape mainly via convection.

Figure 3 demonstrates how the ionic charge develops during downstream transport. In the basic case the shock-accelerated ions escape at the outer boundary of the model layer, $r = r_e$, but ions are registered twice – (i) at a point in the shock frame situated 10^3 km downstream of the shock and (ii) upon the particle arrival at the outer boundary $r = r_e$. Results of those two registrations are shown with points in Figure 3. It is seen that immediately after the acceleration (pluses) the energy-charge dependence is nearly linear on a log-linear scale, whereas after the coronal (downstream) transport to r_e (diamonds) a wide plateau is formed from ≈ 1.5 MeV/nucleon to 50 MeV/nucleon. In an additional simulation (histogram) Fe ions were registered in a far downstream region, 10^4 km from the shock, and immediately removed from the simulation, i.e., ions were allowed to escape at a fixed, downstream point of the shock frame (similar to an assumption of *Ostryakov and Stovpyuk*, [2003]). It is seen that the charge states of the far-downstream escape differ from the charge states of ions escaping into the interplanetary medium. The downstream-escape charge states are not directly related to the charge states in the interplanetary medium.

These authors employed an empirical density profile in the solar corona [*Guhathakurta et al.*, 1996] and also varied that profile by a factor of 0.2 to 5. The observed mean charge of ~ 30 MeV/nucleon iron [*Cohen et al.*, 1999] can be naturally explained by shock acceleration at heliocentric distances of $\sim 1.5 - 2 R_\odot$. However, uncertainties in the coronal density distribution and in the turbulent layer height allow perhaps

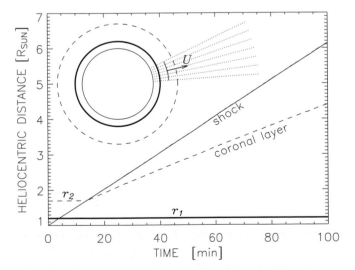

Figure 2. Coronal layer model of the shock acceleration region. The sketch in the insert illustrates the initial state of the turbulent spherical layer (r_1, r_2) at the base of solar wind. Inner and outer boundaries of the layer are respectively shown by the thick solid line and the dashed line, in both the height-time plot and the sketch. The shock height-time profile is plotted by a thin solid line ($U = 600$ km s^{-1}). The post-shock expansion of the coronal layer is shown for $\delta w = 30$ km/s (Model 2 of *Kocharov et al.* [2005]).

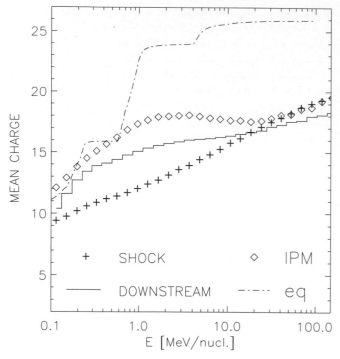

Figure 3. Mean charge states of Fe in ions escaping into the interplanetary medium (diamonds), in ions registered near the shock (plus signs), and in ions far downstream of the shock (histogram). Points correspond to the coronal layer model of shock acceleration (Figure 2). Histogram is for the same model but with the particle escape at a fixed distance (10^4 km) downstream of the shock. The mean equilibrium charge (dash-dotted line) is given as a reference. Strong adiabatic deceleration of shock-accelerated ions in the solar corona shifts the low-energy ions to the equilibrium charge curve, and even beyond it. All calculations are done with the temperature $T = 1.0 \times 10^6$ K and the Bohr's cross section for the proton-impact ionization [*Bohr*, 1948; *Luhn et al.*, 1984]. The turbulence spectrum index is $q = 1.5$.

a factor 1.5 uncertainty in the heliocentric distance estimate. Ions start to escape into the interplanetary medium upon shock arrival at the upper boundary of the acceleration region ($\sim 2-3\,R_\odot$) and continue to escape from beyond the shock when it travels farther from the Sun. At this extreme, all turbulence and hence all acceleration is concentrated in a coronal layer at the base of the solar wind. At the other extreme, the shock might drive a turbulent layer of proton-amplified waves in the upstream region and become a moving source of SEPs. The coronal layer model has been intentionally oversimplified to reveal underlying physical, and should be replaced in future with a more realistic model that accounts for a continual distribution of turbulence and particle re-acceleration in interplanetary space.

6. MIXING AND FRACTIONATION

As a result of poor counting statistics, charge-state data typically require a long integration time, whereas spectral and composition observations indicate that there may be different particle populations registered during different periods of a single SEP event [*Torsti et al.*, 2000; *Kocharov and Torsti*, 2002]. This implies that a combination of two or more theoretical models and a careful analysis of time profiles of a particular event may be required for deciphering physical processes behind the charge-energy profiles observed. The stripping models are related to the ion acceleration in the solar corona, which probably does not last more than ~1.5 hr, whereas charge-state observations typically integrate over much longer periods.

Along with ion stripping in the solar corona or independently of it, in the interplanetary medium energy-dependent charge states may be produced by a fractionation effect at the exponential turnover of the shock-accelerated ion spectra [*Klecker et al.*, 2001; *Tylka et al.*, 2001]. *Klecker et al.* [2001] argued that some of the SEP events show evidence for local acceleration at interplanetary shocks, where ion stripping is not expected but the energy-dependent charge states of Fe can be produced by the fractionation effect. For example, if the exponential turnover energy is proportional to the ion charge to mass ratio [*Ellison and Ramaty*, 1985], the total intensity of all Fe ions is in the form:

$$I(E) = \sum_Q I_0(Q) E^{-\gamma} \exp\left[-\frac{E}{E_0(Q)}\right], \quad E_0(Q) = \frac{Q}{A} E_{0p}. \quad (8)$$

This implies a concave charge-energy profile in the energy-range of the spectrum turnover [*Klecker et al.*, 2001].

7. CONCLUSIONS

Kinetic calculations for plasmas indicate that equilibrium mean charge states of energetic ions differ from those in neutral matter. The original Barkas's formula applied to plasma significantly underestimates the equilibrium mean charge of a heavy ion moving during a sufficiently long time with a constant speed. Furthermore, in plasma it seems impossible to maintain a constant speed for a long duration because of strong Coulomb losses and the high values of the final, equilibrium charge to be achieved. The latter is caused by the low rate of recombination in plasma. Energetic Fe ions in coronal plasma cannot naturally attain equilibrium charge. An equilibrium charge value, $Q_{eq}(E)$, either in numerical form or in the form of the modified Barkas's formula, may provide a useful reference for analysis but should not be directly used for fitting SEP ionic-charge data.

In the absence of universal charge-energy dependence for energetic ions in plasmas, a history (scenario) of the ion injection, acceleration, and transport becomes essential. Models of concurrent acceleration and charge-change (e.g., leaky-box models) produce the rising energy-charge profiles of iron, $\langle Q \rangle$ (E), which are situated below the equilibrium charge curve $Q_{eq}(E)$. The $\langle Q \rangle$ (E)-profile curvature depends on the shape of the energy dependence of the acceleration rate. This could be used for determining the energy dependence of the ion-acceleration rate, irrespective of the assumptions on the escape time. However, the energy-charge profiles of the consecutive acceleration and charge-equilibration (e.g., coronal shock acceleration followed by downstream transport) are significantly different from those for concurrent acceleration and charge-equilibration, despite the fact that the energy spectra may be the same.

Most of the previous computations of charge-consistent ion acceleration were performed in the framework of the escape-time approximation (leaky-box model), i.e., the energy change and the charge change were treated as concurrent processes. They resulted in a nearly logarithmic rise of the mean ionic charge of Fe with ion energy [*Ostryakov and Stovpyuk*, 1997], very similar to the plus signs in Figure 3. In contrast, during coronal transport after acceleration, a kind of plateau is formed in the range from a few MeV/nucleon to few tens MeV/nucleon (e.g., diamonds in the same figure). The flat pattern shown in Figure 3 is explained by stripping during the escape. A leaky-box model, with *concurrent* acceleration and escape, is a reasonable approximation for stochastic acceleration, whereas shock acceleration and downstream transport are mostly *consecutive* processes. Because of this difference, stochastic acceleration and shock acceleration (still without self-excited waves) result in significantly different energy-charge profiles of iron ions in the ~1–30 MeV/nucl. range. Note, that stochastic acceleration is not necessarily flare acceleration, but may also be induces by a CME-driven shock on a global coronal scale [*Kocharov et al.*, 2005]. On the other hand, the shock acceleration results of *Lytova and Kocharov* [2005] are valid only for acceleration in a coronal layer of external turbulence. In such a model, ions are accelerated in a shock wave and left behind the shock. Then they get ionized during the escape from shock-downstream corona to the interplanetary medium. Future modeling results for shock acceleration in co-moving layer of proton-amplified waves may be different.

The modeling work will be continued to include self-generated waves and more realistic geometries of coronal shocks and magnetic fields. Available and expected models of coronal acceleration (box models and traveling shocks), models of coronal and interplanetary transport, and models of interplanetary shock (re-)acceleration should allow one to fit the energy-dependent charge states of SEPs observed at 1 AU.

However, the scenario-sensitivity of the modeling results suggests that an integrated analysis of a particular solar-interplanetary event may be unavoidable. It is important to measure an accurate energy-charge profile of Fe ions in the 1–100 MeV/nucleon energy range with a time resolution of about one hour. Other particle and electromagnetic data, especially particle flux anisotropies and abundances, should also be employed. For instance, an empirical correlation has been found between high-energy charge of Fe and the abundance ratio Fe/O [*Labrador et al.*, 2003; *Dietrich and Tylka*, 2003; *Tylka et al.*, 2005]. Therefore a model that explains the increase in ionic charge of Fe with energy should also explain why Fe/O also attains enhanced values. Standard diffusive shock-acceleration mechanism employed by many authors (e.g., *Lytova and Kocharov* [2005]) naturally produces a universal power-law spectrum for all ion species, and hence should be modified to account for the spectral differences observed between different species. Employment of a full set of SEP data can reduce uncertainties in a choice of acceleration and transport models for the ionic charge computations.

Acknowledgments.   This work was supported by grant 106120 from the Academy of Finland. The author would like to thank Richard Mewaldt, who served as scientific editor to this paper, and two referees for their detailed comments. I am grateful to Gennady Kovaltsov and Allan Tylka for valuable discussions regarding this work.

REFERENCES

Arnaud, M., and J. Raymond, Iron ionization and recombination rates and ionization equilibrium, *Astrophys. J.*, 398, 394-406, 1992.

Barghouty, A.F., Robust estimates of hydrogen-impact ionization cross sections over a wide energy range, *Phys. Rev.*, A61(5), 052702(3 pp.), doi:10.1103/PhysRevA.61.052702, 2000.

Barghouty, A.F., and R.A. Mewaldt, Charge states of solar energetic iron: Nonequilibrium calculation with shock-induced acceleration, *Astrophys. J.*, 520, L127-L130, 1999.

Barghouty A.F., and R.A. Mewaldt, Simulation of charge-equilibration and acceleration of solar energetic ions, in *Acceleration and Transport of Energetic Particles Observed in the Heliosphere, AIP#528*, edited by R.A. Mewaldt, J.R. Jokipii, M.A. Lee, E. Möbius, and T.H. Zurbuchen, p. 71, AIP, Washington, D.C., 2000.

Barkas, W.H. *Nuclear Research Emulsions*, v.1, p. 371, Academic Press, New York, 1963.

Bethe, H., Theory of the passage of rapid corpuscular rays through matter, *Ann. Physik*, 5, 325-400, 1930.

Betz, H.-D., Charge states and charge-changing cross sections of fast heavy ions penetrating through gaseous and solid media, *Rev. Mod. Phys.*, 44, 465-539, doi: 10.1103/RevModPhys.44.465, 1972.

Betz H.D., G. Hortig, E. Leischner, Ch. Schmelzer, B. Stadler, and J. Weihrauch, The average charge of stripped heavy ions, *Phys. Lett.*, 22, 643-644, 1966.

Bohr, N., The penetration of atomic particles through matter, *Kgl. Danske Videnskabernes Selskab. Math.-fys. Medd.*, 18(No 8), 1948.

Cohen, C.M.S., A.C. Cummings, R.A. Leske, R.A. Mewaldt, E.C. Stone, B.L. Dougherty, M.E. Wiedenbeck, E.R. Christian, T.T. von Rosenvinge, Inferred charge states of high energy solar particles from the Solar Isotope Spectrometer on ACE, *Geophys. Res. Lett.*, 26, 149-152, 1999.

Dietrich K.-G., D.H.H. Hoffmann, E. Boggasch, J. Jacoby, H. Wahl, M. Elfers, C.R. Haas, V.P. Dubenkov, A.A. Golubev, Charge state of fast heavy ions in a hydrogen plasma, *Phys. Rev. Lett.* 69, 3623-3626, 1992.

Dietrich, W.F., and A. Tylka, Time-to-maximum studies and inferred ionic charge sates in the solar energetic particle events of 14 and 15 April 2001 *Proc. 28th Internat. Cosmic Ray Conf. (Tsukuba)* 3291-3294, 2003.

Ellison, D.C., and R. Ramaty, Shock acceleration of electrons and ions in solar flares, *Astrophys. J.*, 298, 400-408, 1985.

Ginzburg, V.L., and S.I. Syrovatskii *The Origin of Cosmic Rays (§7)*, 384 pp., Macmillan, New York, 1964.

Goffe, T.V., M.B. Shah, and H.B. Gilbody, One-electron capture and loss by fast multiply charged boron and carbon ions in H and H_2, *J. Phys. B* 12, 3763-3773, 1979.

Gryzinski, M., Classical theory of atomic collisions. I. Theory of inelastic collisions, *Phys. Rev.*, 138, 336-358, 1965.

Guhathakurta, M., T.E. Holzer, and R.M. MacQueen, The Large-scale density structure of the solar corona and the heliospheric current sheet, *Astrophys. J.*, 458, 817-831, 1996.

Haugen, H.K., L.H. Andersen, P. Hvelplund, H. Knudsen H., Ionization of helium by fast, multiply charged ions. Deviations from a q^2 scaling, *Phys. Rev. A* 26, 1950-1961, 1982.

Inokuti, M., Y. Itikawa, and J.E. Turner, Addenda: Inelastic collisions of fast charged particles with atoms and molecules-The Bethe theory revisited, *Rev. Mod. Phys.*, 50, 23-35, 1978.

Janev, R.K., General classical scaling of electron-loss cross sections in collisions of atoms with highly charged ions, *Phys. Rev.* A 28, 1810-1812, 1983.

Janev, R.K., G. Ivanovski, E.A. Solov'ev, Ionization of hydrogen atoms by multiply charged ions at low energies: The scaling law, *Phys. Rev.* A 49, R645-R648, doi:10.1103/PhysRevA.49.R645, 1994.

Kartavykh, Y.Y., Stochastic acceleration of heavy ions in solar flares: Coulomb losses and charge change, PhD Thesis, 123 pp., Pulkovo Observatory, St. Petersburg, Russia, 1999.

Kartavykh, Y.Y., and V.M. Ostryakov, Plasma diagnostics by the charge distributions of heavy ions in impulsive solar flares, *Proc. 26th Internat. Cosmic Ray Conf. (Salt Lake City)*, 6, 272, 1999.

Kartavykh, J.J., W. Dröge, V.M. Ostryakov, and G.A. Kovaltsov, Adiabatic deceleration effects on the formation of heavy ion charge spectra in interplanetary space, *Solar. Phys.*, 227, 123-135, 2005.

Kharchenko, A.A., and V.M. Ostryakov, On the charge state of solar energetic particles, *Proc 20th Internat. Cosmic Ray Conf. (Moscow)*, 3, 248-251, 1987.

Klecker, B., E. Möbius, M.A. Popecki, M.A. Lee, A.T. Bogdanov, On the energy dependence of ionic charge states in solar energetic particle events, in *Solar and Galactic Composition, AIP Conf. Proc.*, v. 598, edited by R.F. Wimmer-Schweingruber., pp. 317-321, AIP, Melville, New York, 2001.

Knudsen, H., L.H. Andersen, P. Hvelplund, G. Astner, H. Cederquist, H. Danared, L. Liljeby, K.-G. Rensfelt, An experimental investigation of double ionisation of helium atoms in collisions with fast, fully stripped ions, *J. Phys. B* 17, 3545-3564, 1984.

Kocharov, L., and J. Torsti, Hybrid solar energetic particle events observed on board SOHO, *Solar Phys.,*, 149-157, 2002.

Kocharov, L., G.A. Kovaltsov, J. Torsti, and V.M. Ostryakov, Evaluation of solar energetic Fe charge states: effect of proton-impact ionization, *Astron. Astrophys.*, 357, 716-724, 2000.

Kocharov, L., G.A. Kovaltsov, and J. Torsti, Dynamical cycles in charge and energy for iron ions accelerated in a hot plasma, *Astrophys. J.*, 556, 919-927, 2001.

Kocharov, L., G.A. Kovaltsov, and J. Torsti, Comment on 'Particle charge evolution during acceleration processes in solar flares' by M.D. Rodríguez-Frías, L. del Peral and J. Pérez-Peraza, *Journal of Physics G: Nuclear and Particle Physics*, 28(6), 1511-1514, 2002.

Kocharov, L., M. Lytova, R. Vainio, T. Laitinen, and J. Torsti, Modeling the shock aftermath source of energetic particles in the solar corona *Astrophys. J.*, 620, 1052-1068, 2005.

Korchak, A.A., Coulomb losses and the nuclear composition of the solar flare accelerated particles, *Solar Phys.*, 66, 149-158, 1980.

Korchak, A.A., and B.P. Filippov, On the formation of the charge spectrum and composition of particles accelerated in solar flares, *Sov. Astron.*, 23, 323-328, 1979.

Kovaltsov, G.A., A.F. Barghouty, L. Kocharov, V.M. Ostryakov, and J. Torsti, Charge-equilibration of Fe ions accelerated in a hot plasma, *Astron. Astrophys.*, 375, 1075-1081, 2001.

Kovaltsov, G.A., Y.Y. Kartavykh, L. Kocharov, V.M. Ostryakov, and J. Torsti, Comment on "Model of ionic charge states of impulsive solar energetic particles in solar flares" by M. Dolores Rodríguez-Frías, Luis del Peral, and Jorge Pérez-Peraza, *J. Geophys. Res.*, 107(A10), 1276, doi: 10.1029/2001JA009173, 2002.

Kurganov, I.G., and V.M. Ostryakov, Heavy particle acceleration at a shock front - charge-changing reactions, *Soviet Astron. Lett.*, 17, 77-84, 1991.

Labrador, A.W., R.A. Leske, R.A. Mewaldt, E.C. Stone, and T.T. von Rosenvinge, High energy ionic charge state composition in recent large solar energetic particle events, *Proc. 28th Internat. Cosmic Ray Conf. (Tsukuba)*, 3269-3272, 2003.

Leske, R. A., J.R. Cummings, R.A. Mewaldt, E.C. Stone, and T.T. von Rosenvinge, Measurements of the ionic charge states of solar energetic particles using the geomagnetic field, *Astrophys. J.*, 452, L149-L152, 1995.

Luhn, A., and D. Hovestadt, Calculation of the mean equilibrium charges of energetic ions after passing through a hot plasma, *Astrophys. J.*, 317, 852-857, 1987.

Luhn, A., B. Klecker, D. Hovestadt, M. Scholer, G. Gloeckler, F.M. Ipavich, C.Y. Fan, L.A. Fisk, Ionic charge states of N, Ne, Mg, SI and S in solar energetic particle events, *Adv. Space Res.*, 4(NO.2-3), 161-164, 1984.

Lytova, M., and L. Kocharov, Charge states of energetic solar ions from coronal shock acceleration, *Astrophys. J.*, 620, L55-L58, 2005.

Mason, G.M., J.E. Mazur, M.D. Looper, and R.A. Mewaldt, Charge state measurements of solar energetic particles observed with SAMPEX, *Astrophys. J.*, 452, 901-9111, 1995.

Mazur, J.E., G.M. Mason, M.D. Looper, R.A. Leske, and R.A. Mewaldt, Charge states of solar energetic particles using the geomagnetic cutoff technique: SAMPEX measurements in the 6 November 1997 solar particle event, *Geophys. Res. Lett.*, 26, 173-176, 1999.

McDaniel, E.W. *Atomic collisions*, p. 432, Wiley, New York, 1989.

McGuire, J.H., and P. Richard, Procedure for computing cross sections for single and multiple ionization of atoms in the binary-encounter approximation by the impact of heavy charged particles, *Phys. Rev.* A8, 1374-1384, 1973.

Möbius, E., Y. Litvinenko, H. Grünwaldt, M.R. Aellig, A. Bogdanov, F.M. Ipavich, P. Bochsler, M. Hilchenbach, D. Judge, B. Klecker, D. Morris, and W.K.H. Schmidt, Energy dependence of the ionic charge state distribution during the November 1997 solar energetic particle event, *Geophys. Res. Lett.*, 26, 145-148, 1999.

Möbius, E., M. Popecki, B. Klecker, L.M. Kistler, A. Bogdanov, A.B. Galvin, D. Heirtzler, D. Hovestadt, D. Morris, Ionic charge states of solar energetic particles from solar flare events during the current rise of solar activity as observed with ACE SEPICA, *Adv. Space Res.*, 29, 1501-1512, 2002.

Möbius, E., Y. Cao, M.A. Popecki, L.M. Kistler, H. Kucharek, D. Morris, and B. Klecker, Strong energy dependence of ionic charge states in impulsive solar events, *Proc. 28th Internat. Cosmic Ray Conf. (Tsukuba)*, 3273-3276, 2003.

Nardi, E., and Z. Zinamon, Charge state and slowing of fast ions in a plasma, *Phys. Rev. Lett.*, 49(NO.17), 1251-1254, 1982.

Nikolaev, V.S., and V.A. Sidorovich, The single and double ionization of helium atoms by fast nuclei and multicharged ions, *Nucl. Instrum. and Methods in Phys. Res.*, B 36, 239-248, 1989.

Northcliffe, L.C., Passage of heavy ions through matter, *Ann. Rev. Nucl. Sci.*, 13, 67-102, 1963.

Oetliker, M., D. Hovestadt, B. Klecker, M.R. Collier, G. Gloeckler, D.C. Hamilton, F.M. Ipavich, P. Bochsler, G.G. Managadze, The ionic charge of solar energetic particles with energies of 0.3-70 MeV per nucleon, *Astrophys. J.*, 477, 495-501, 1997.

Ostryakov, V.M., and M.F. Stovpyuk, Stochastic acceleration of heavy ions with variable charge, *Astron. Reports*, 41, 386-393, 1997.

Ostryakov, V.M., and M.F. Stovpyuk, Charge state distributions of iron in gradual solar energetic particle events, *Solar Phys.*, 189, 357-372, 1999.

Ostryakov, V.M., and M.F. Stovpyuk, Non-Homogeneous Charge-Consistent Model for Acceleration of Iron in the Solar Corona, *Solar Phys.*, 217, 281-299, 2003.

Ostryakov, V.M., Y.Y. Kartavykh, D. Ruffolo, G.A. Kovaltsov, and L. Kocharov, Charge state distributions of iron in impulsive solar flares: Importance of stripping effects, *J. Geophys. Res.*, 105(A12), 27315-27322, 2000.

Parker, E.N., The passage of energetic charged particles through interplanetary space, *Planet. Space Sci.*, 13, 9-49, 1965.

Peter, T., and J. Meyer-ter-Vehn, Influence of dielectronic recombination on fast heavy-ion charge states in plasma, *Phys. Rev. Lett.*, 57(NO.15), 1859-1862, 1986.

Peter, T., and J. Meyer-ter-Vehn, Energy loss of heavy ions in dense plasma. I - Linear and nonlinear Vlasov theory for the stopping power. II - Nonequilibrium charge states and stopping powers, *Phys. Rev.*, A43(4), 1998-2030, 1991.

Pierce, T.E., and M. Blann, Stopping powers and ranges of 5-90 MeV S^{32}, Cl^{35}, Br^{79}, and I^{127} ions in H_2, He, N_2, Ar, and Kr: A semiempirical stopping power theory for heavy ions in gases and solids, *Phys. Rev.*, 173, 390-405, 1968.

Popecki, M., E. Möbius, B. Klecker, A.B. Galvin, L.M. Kistler, A.T. Bogdanov, Ionic charge state measurements in solar energetic particle events, in *Acceleration and transport of energetic particles observed in the heliosphere, AIP Conf. Proc.*, vol. 528, *AIP* edited by R.A. Mewaldt, J.R. Jokipii, M.A. Lee, E. Möbius, and T.H. Zurbuchen, p. 63, AIP, Melville, New York, 2000.

Reames, D.V., C.K. Ng, and A.J. Tylka, Energy-dependent ionization states of shock-accelerated particles in the solar corona, *Geophys. Res. Lett.*, 26, 3585-3588, 1999.

Rodríguez-Frías, M.D., L. del Peral, and J. Pérez-Peraza, Particle charge evolution during acceleration processes in solar flares, *J. Phys. G.: Nucl. Part. Phys.*, 26, 259-265, 2000.

Rodríguez-Frías, M.D., L. del Peral, and J. Pérez-Peraza, Model of ionic charge states of impulsive solar energetic particles in solar flares, *J. Geophys. Res.*, 106, 15657-15664, 2001.

Ruffolo, D., Charge states of solar cosmic rays and constraints on acceleration times and coronal transport, *Astrophys. J.*, 481, L119-L122, 1997.

Sant'Anna, M.M., W.S. Melo, A.C. Santos, G.M. Sigaud, E.C. Montenegro, M.B. Shah, and W.F. Meyerhof, Absolute measurements of electron-loss cross sections of He^+ and C^{3+} with atomic hydrogen at intermediate velocities, *Phys. Rev.*, A 58, 1204-1211, 1998.

Sidorovich, V.A., and V.S. Nikolaev, The cross section scaling for ionization of hydrogen and helium by multicharged ions, *Lecture Notes in Physics* 294, *Proc. 3rd Workshop on High-Energy Ion-Atom Collisions* edited by D. Berényi and G. Hock, pp. 437-446, Springer-Verlag, Berlin, 1988.

Stovpyuk, M.F., Heavy ion acceleration in solar flares and heliosphere in the framework of charge-consistent model, PhD Thesis, 128 pp., St. Petersburg State Polytechnic University, St. Petersburg, Russia, 2000.

Tatischeff, V., R. Ramaty, and B. Kozlovsky, X-Rays from accelerated ion interactions, *Astrophys. J.*, 504, 874-888, 1998.

Torsti, J., P. Mäkelä, M. Teittinen, and J. Laivola, SOHO/Energetic and Relativistic Nucleon and Electron Experiment measurements of energetic H, He, O, and Fe fluxes during the 1997 November 6 solar event, *Astrophys. J.*, 544, 1169-1180, 2000.

Toptygin, I.N. *Cosmic rays in interplanetary magnetic fields (Geophysics and Astrophysics Monographs, No 27)*, 387 pp., D. Reidel, Dordrecht, 1985.

Tylka, A.J., P.R. Boberg, J.H. Adams, Jr., L.P. Beahm, W.F. Dietrich, and T. Kleis, The mean ionic charge state of solar energetic Fe ions above 200 MeV per nucleon, *Astrophys. J.*, 444, L109-L113, 1995.

Tylka, A.J., C.M.S. Cohen, W.F. Dietrich *et al.*, Evidence for remnant flare suprathermals in the source population of solar energetic particles in the 2000 Bastille Day event, *Astrophys. J.*, 558, L59-L63, 2001.

Tylka, A.J., C.M.S. Cohen, W.F. Dietrich *et al.*, Shock geometry, seed populations, and the origin of variable elemental composition at high energies in large gradual solar particle events, *Astrophys. J.*, 625, 474-495, 2005.

L. Kocharov, Department of Physics, University of Turku, FIN-20014 Turku, Finland. (kocharov@utu.fi)

Solar Energetic Particle Composition

Berndt Klecker

Max-Planck-Institut für extraterrestrische Physik, Garching, Germany

Over the last ~10 years, advanced instrumentation onboard several spacecraft (e.g., WIND, SAMPEX, SOHO, and ACE), extended our ability to explore energy spectra, elemental, isotopic, and ionic charge composition of solar energetic particle (SEP) events in a wide energy range from ~10 keV/nuc to ~100 MeV/nuc. Due to the much improved sensitivity of the instrumentation, spectral and compositional measurements are now available for a large range of particle intensities, i.e. not only for large SEP events generally associated with coronal mass ejections (CMEs) and coronal/interplanetary shocks, but also for small, flare associated events. Originally, these two types of events were classified as *gradual* and *impulsive*, based on the duration of the associated soft X-ray emission, but this two-class paradigm of SEP events was also useful to distinguish differences in the acceleration processes involved. However, in the last few years, the two-class paradigm has been challenged by several observations, as, for example, significant enrichment of ^{3}He and heavy ions and high Fe charge states of ~20 at high energies, were found in interplanetary shock related events. In this paper the recent observations on ion composition will be reviewed and their implications for our understanding of the different acceleration scenarios will be discussed.

1. INTRODUCTION

In their energy spectra, elemental, isotopic, and ionic charge composition solar energetic particles (SEP) carry fundamental information on the source region and their acceleration and propagation processes. High-energy particles originating at the Sun were first reported by *Forbush* [1946]. At that time there was little doubt that these particles were closely related to contemporary solar flares. Later it became clear that acceleration at coronal and interplanetary shocks is also an efficient mechanism for particle acceleration [e.g., *Bryant et al.*, 1962]. In the early 70s a new type of event was discovered that showed enhanced ^{3}He abundances with ^{3}He/^{4}He>1 [*Balasubrahmanyan and Serlemitsos*, 1974], while the corresponding ratio in the corona or solar

wind is $<5\times10^{-4}$. Such events were later found to also exhibit enhancements of heavy ions by an order of magnitude relative to coronal abundances [*Mason et al.*, 1986]. Based on these observations and on other characteristic differences, e.g., differences in the electron to proton ratio, in intensity-time profiles, and in the mean ionic charge of heavy ions, SEP events were classified as *impulsive* and *gradual*, following a classification of flares based on the length of soft X-ray emission [*Pallavicini et al.*, 1977]. In this scenario *impulsive* SEP events were related to flares and the *gradual* SEP events were related to coronal mass ejection (CME) driven coronal and interplanetary shocks [e.g., *Reames*, 1999; *Kallenrode*, 2003].

However, new results with advanced instrumentation from several missions (e.g., Wind, SAMPEX, SOHO, ACE) have shown that this picture was oversimplified. New composition and ionic charge measurements show that enrichments in ^{3}He are also common in interplanetary shock accelerated populations [*Desai et al.*, 2001], that enrichments in heavy ions are often observed in large events at high energies

Solar Eruptions and Energetic Particles
Geophysical Monograph Series 165
Copyright 2006 by the American Geophysical Union
10.1029/165GM15

[e.g., *Cohen et al.*, 1999; *Cane et al.*, 2003], and that high charge states of Fe are also observed in events usual classified as *gradual* [*Leske et al.*, 1995]. Therefore, the classification into two distinct types of events is presently in question and the relative contributions of flares and coronal/ interplanetary shocks are under debate.

In this paper we will summarize the new observations of SEP composition and their implications.

2. ^{3}HE- AND HEAVY ION-RICH EVENTS

The large enrichments of ^{3}He and heavy ions found in event-integrated abundances of ^{3}He- and heavy ion-rich events have been used as one of their defining characteristics as *impulsive* events. Although some of the characteristics (e.g., enrichment of ^{3}He relative to solar wind and coronal abundances, see paragraph 3.1) are observed to some extent also in large (*gradual*) events, several signatures of the ^{3}He- and heavy ion rich events are unique, suggesting a different acceleration process and warrant a classification as a separate class of event.

2.1. Time Dispersion

The high sensitivity of new instrumentation provides unprecedented statistics for small events. Plate 1 shows as an example the energy versus time profile for several events in August 1998 as observed with the ULEIS experiment onboard ACE [*Mason et al.*, 2000]. This type of display, plotting the arrival time of individual ions versus their energy, shows velocity dispersion that can be compared with scatter-free propagation along the interplanetary magnetic field, and allows the identification of individual injections at the Sun. This display also demonstrates that to correctly evaluate spectra and composition in these events, start and stop times for the averaging of data need to be energy dependent, as indicated by the 'boxes' enclosing individual injections. Plate 1 also shows that some events are cut off because of a loss of the magnetic connection to the acceleration site. These well defined injection profiles can also be used to infer large scale diffusion parameters of ions in the heliospheric magnetic field [*Giacalone et al.*, 2000; *Mazur et al.*, 2000].

2.2. Energy Spectra

In a survey of energy spectra of ions in ^{3}He-rich events in the mass range He to Fe and in the energy range 80 keV/nuc to 15 MeV/nuc *Mason et al.* [2002] found two classes of events. Class 1 events exhibit power laws that often steepen above ~1 MeV/nuc; in some cases the major species ^{3}He, ^{4}He, O, Fe have similar spectral slopes, while in other cases the ^{3}He slope below ~1 MeV/nuc is distinctly harder than the

others. Class 2 events show curved ^{3}He and Fe spectra at low energies, while ^{4}He has power law spectra. As a consequence of the different spectral shapes of ^{3}He and ^{4}He the ^{3}He/^{4}He-ratio in Class 2 events is strongly energy dependent; also the largest ^{3}He/^{4}He-ratios (at ~1 MeV/nuc) are observed in this class of events.

2.3. Elemental and Isotopic Abundances

It is known since the early measurements of heavy ion composition in ^{3}He-rich events that these events exhibit also an enrichment in heavy ions by an order of magnitude (for Fe), relative to coronal abundances [*Hurford et al.*, 1975; *Mason et al.*, 1986], with a systematic increase of the enrichment factor with mass. With advanced instrumentation onboard the Wind and ACE spacecraft the measurement of heavy ions in these events has been extended to trans-iron elements in the nuclear charge range $34 < Z < 82$ [*Reames*, 2000] and to Ultra-Heavy ions (UH, taken as 78-220 amu), showing that the enrichment factor continues to increase dramatically for high masses. Figure 1 [*Mason et al.*, 2004] shows the enhancement of elemental abundances relative to coronal abundances in the mass range 4 to ~220. Figure 1 shows that the enhancement factor is monotonically increasing to ~200 for high masses. It was also found by *Mason et al.* [2004] that the enhancement factor is ordered by mass per charge (M/Q).

In addition to information on the composition of heavy ions by particles escaping from the acceleration site in the corona into interplanetary space, γ-ray line observations provide information on the composition of heavy ions interacting with the ambient corona. *Murphy et al.* [1991] found that the composition of interacting heavy ions is similar to the abundances in ^{3}He-rich SEP events as observed in interplanetary space.

2.4. Ionic Charge States

Because of the large differences between the mean ionic charge of Fe in the MeV/nuc energy range in events identified as *gradual* ($Q_{Fe} \sim 14$) and *impulsive* ($Q_{Fe} \sim 20$) [*Gloeckler et al.*, 1976; *Hovestadt et al.*, 1981; *Klecker et al.*, 1984; and *Luhn et al.*, 1987), the ionic charge states were also used as a defining characteristic for this classification. However, new measurements of ionic charge states with instruments of improved sensitivity over a wide energy range on several spacecraft (SAMPEX, SOHO, ACE) have shown that this picture was oversimplified. In interplanetary (IP) - shock related events, the ionic charge state of heavy ions was observed to increase in many events with energy [*Oetliker et al.*, 1997 and references therein; *Möbius et al.*, 1999; *Leske et al.*, 2001a], with a large event to event

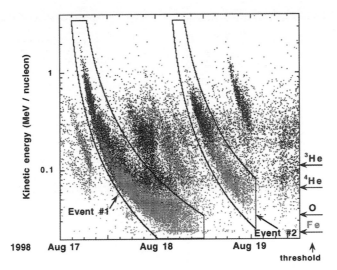

Plate 1. Ion velocity spectrograms for 1998 August 17–19.5 from *ACE*/ULEIS showing ion species ³He, ⁴He, O, and Fe Below 60 keV nucleon, Fe dominates because of the instrument efficiency and threshold effects. *Thin line*: Event "boxes" used for calculating fluences. *Right axis arrows*: Approximate instrument threshold for different ions (from *Mason et al.* [2000]).

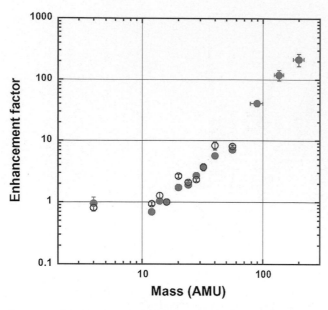

Figure 1. Filled circles: Enhancement factor for heavy ion abundances in ^{3}He-rich events, compared with gradual SEP abundances and solar system abundances (*Mason et al.*, 2004). Open circles: Values from *Reames* [1995].

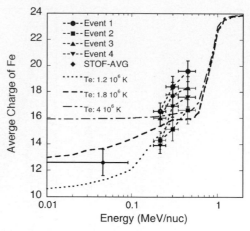

Figure 2. Mean ionic charge state of Fe as a function of energy for four impulsive events and energy dependence obtained for an equilibrium model, including charge stripping by electrons and protons [*Klecker et al.*, 2005]. The model calculation shows that at energies < 100 keV/nuc the ionic charge is predominantly determined by the temperature, whereas at energies above ~200 keV/nuc charge stripping is the dominant process.

variability. For impulsive events *Möbius et al.* [2003] reported a systematic increase of the mean ionic charge of Fe with energy by several charge units from ~14-16 at 180-250 keV/nuc to 16-20 at 350-550 keV/nuc. However, measurements with STOF/CELIAS/SOHO showed that in these events the ionic charge of Fe at energies below 100 keV/nuc was significantly smaller (12.5±0.9, *Klecker et al.* [2005]). The large increase of the mean ionic charge of iron in the energy range ~10-550 keV/nuc as observed with SOHO and ACE can only be explained in terms of impact ionization by protons and electrons in a dense environment in the low corona. Models including the effect of stochastic acceleration, coulomb energy loss, charge changing processes such as impact ionization and recombination with ambient electrons have been investigated for impulsive events by e.g., *Ostryakov et al.* [2000] and *Kartavykh et al.* [2001].

Figure 2 shows the mean ionic charge of Fe for 4 impulsive SEP events, together with the energy dependence obtained for the simplified case of equilibrium conditions in a charge stripping model [*Klecker et al.*, 2005]. The model calculations show that (1) at low energies below ~100 keV/nuc the mean ionic charge is essentially determined by the temperature, (2) at energies above ~200 keV/nuc charge stripping effects are dominating and result in a strong increase of the mean ionic charge of Fe at energies <1 MeV/nuc. Note that in non-steady state models, at any given energy, the mean ionic charge will be below the equilibrium values shown in Figure 2

(e.g., *Kocharov et al.*, 2000). However, Figure 2 shows that for 3 out of 4 events the increase of the charge state with energy is even steeper than for charge state equilibrium. To explain this apparent discrepancy energy loss by adiabatic deceleration of the energetic particle population during their transport from the corona to the observer has been invoked and successfully used by *Kartavykh et al.* [2005] to reproduce the energy dependence of the ionic charge of Fe in some of the impulsive events shown in Figure 2.

2.5. Discussion of Acceleration Processes

The basic observations any acceleration process of ^{3}He- and heavy ion rich events needs to explain are (1) short injection time scales, (2) enrichment of ^{3}He and heavy ions by a factor of up to 10^4 (for ^{3}He), ~10 (for Fe) and ~100 for Ultra-Heavy ions, (3) energy dependent ionic charge states at energies < 1 MeV/nuc, and (4) the spectral characteristics as summarized above. For a recent review of particle acceleration processes in solar flares see e.g., *Aschwanden* [2002]. In this section we will concentrate on some characteristics of ^{3}He- and heavy ion-rich events and possible mechanisms.

Since shock acceleration models typically produce power-law spectra (see also 3.3), they could be adopted to model the Class 1 spectra discussed in section 2.2. However, to explain the large enrichment of ^{3}He and heavy ions, two stage processes have been invoked. Such scenarios include selective heating by resonant wave-particle interactions as a first step [e.g., *Fisk*, 1978; *Temerin and Roth*, 1992; and *Zhang*, 1995), followed by a second step that could be a Fermi process (*Zhang*, 1995). However, for the event survey of *Mason et al.* [2002] no coronal shock was observed.

Curved spectra at low energies as observed for ^{3}He and Fe in Class 2 events can arise from stochastic acceleration processes by Alfvén turbulence. At low energies these processes have been shown to provide good fits to the data [*Möbius et al.*, 1982; *Mazur et al.*, 1992]. However, the spectra observed by *Mason et al.* [2002] are much harder at high energies (e.g., above ~1 (10) MeV/nuc for Fe (^{3}He), respectively) than obtained with a stochastic acceleration model [*Mason et al.*, 2002]. However, models based on cascading MHD turbulence (*Miller* [1998], and references therein) provide promising fits to the heavy ion spectra [*Mason et al.*, 2002], although in this model a different mechanism is needed for ^{3}He (see also Figure 3).

The energy dependent ionic charge states of heavy ions provide even more severe constraints on the acceleration models. Model calculations of the mean ionic charge as a function of energy in a plasma of temperature T, density N, including the effects of radiative and dielectronic recombination, and impact ionization by electrons, protons and He show that a large increase of the mean ionic charge with energy below 1 Mev/nuc can only be obtained for N x τ ~10^{10} - 10^{11} cm^{-3}s, where τ is the acceleration time [e.g., *Kocharov et al.*, 2000; *Ostryakov et al.*, 2000]. If we assume acceleration time scales of ~10-100s, this results in densities of 10^8-10^{10} cm^{-3}, corresponding to altitudes in the low corona below ~2 R$_S$. A model including stochastic acceleration, charge stripping, energy loss, spatial diffusion in the acceleration region, and spatial diffusion, convection, and adiabatic deceleration in interplanetary space was recently presented by *Kartavykh et al.* [2005]. They were able to reproduce for several events the energy dependence of the ionic charge composition above 200 keV/nuc. It will be interesting in the future to extend these models, to include also the heating process (i.e. the enrichment of ^{3}He and heavy ions), or to extend other models by including the effects of charge stripping and propagation near the Sun and in interplanetary space.

3. LARGE (GRADUAL) EVENTS

3.1. Event Integrated Abundances and Fractionation Effects

Event integrated elemental and isotopic SEP abundances have been related to photospheric, coronal, and solar wind abundances or have been used to infer solar system abundances that may not be accessible otherwise [e.g., *Meyer*, 1985]. When comparing SEP abundances with photospheric abundances (Figure 4) it has been realized for many years that both coronal and SEP abundances show a dependence on

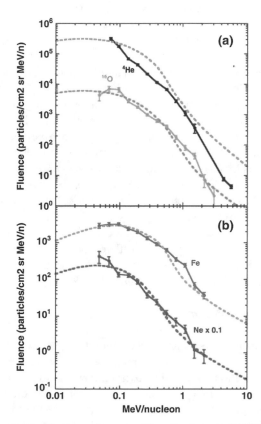

Figure 3. Comparison of spectra from the August 7, 1999 event with spectra calculated using a model of cascading MHD turbulence. (a) ^{4}He and O; (b) Ne (multiplied by 0.1) and Fe (adopted from *Mason et al.* [2002]).

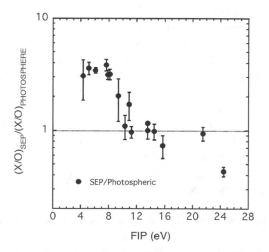

Figure 4. Ratio of SEP abundances in the energy range 5–12 MeV/nuc (for Z>2 ions), relative to photospheric abundances, as a function of first ionization potential. The increase of the abundance of low-FIP elements by a factor ~3–4 is clearly visible (*Reames* [1999], and references therein).

the first ionization potential (FIP) or first ionization time [*Hovestadt et al.*, 1974; *Meyer*, 1985; and *Geiss*, 1998), suggesting ion-neutral separation in the chromosphere as an important fractionation mechanism (see *Hénoux* [1998] for a recent review).

Furthermore, abundances in individual large SEP events show fractionation effects that monotonically depend on mass (M) and mass per charge (M/Q), usually approximated as a power law in M/Q [*Brenemann and Stone*, 1985]. This M/Q fractionation is observed for both elemental and isotopic abundances (e.g., *Leske et al.*, 1999) and the correlation between isotopic and elemental abundances in individual events has been used to infer the abundance of the coronal source [*Leske et al.*, 2001b].

However, in a comparison of elemental abundances in the mass range C to Ni in a large number of SEP events with the combined FIP- and M/Q dependent fractionation patterns, it was recently shown that the FIP effect observed in SEPs shows a large event-to-event variability [*Mewaldt et al.*, 2002; *Slocum et al.*, 2003], with significant differences from the FIP effect observed in both the slow and fast solar wind. This suggests that solar particles with energies > 5 MeV/nuc are accelerated out of a pool of coronal material other than the bulk solar wind observed at 1AU.

3.2. Ionic Charge States

The early measurements of ionic charge states of heavy ions about 25 years ago revealed for IP-shock related events an incomplete ionization of heavy ions in the mass range C to Fe at energies of ~1 MeV/nuc, with $Q_{Fe} \sim 10-14$, indicative of a source temperature of about $1.5-2 \times 10^6$K [*Gloeckler et al.*, 1976; *Hovestadt et al.*, 1981; and *Luhn et al.*, 1984]. However, these measurements were limited to a small energy range around 1 MeV/nuc. With advanced instrumentation onboard SAMPEX, SOHO and ACE, the energy range for ionic charge determination was significantly increased to 0.01–80 MeV/nuc (for Fe). These new measurements showed a significant increase of the mean ionic charge of Fe with energy for several gradual events, with a large event-to-event variability (Figure 5). At energies below ~100 keV/nuc, the mean ionic charge is mostly consistent with solar wind charge states [*Bogdanov et al.*, 2000, 2002; *Klecker et al.*, 2000; and *Bamert et al.*, 2002]. At somewhat higher energies (<1 MeV/nuc) the mean ionic charge of Fe shows a large event-to-event variability: in some events it is constant with energy [*Mazur et al.*, 1999], in other events it increases with energy by several charge units [*Möbius et al.*, 1999]. At even higher energies a mean ionic charge of ~18-20 was observed for Fe at ~40-80 MeV/nuc in several gradual events [*Leske et al.*, 1995, *Oetliker et al.*, 1997, *Leske et al.*, 2001a; *Labrador et al.*, 2003; and *Popecki et al.*, 2003].

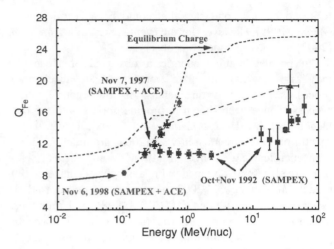

Figure 5. Mean ionic charge of Fe as a function of energy for 3 large *gradual* SEP events (data from *Mazur et al.* [1999]; *Möbius et al.* [1999]; *Popecki et al.* [2003]) and energy dependence obtained for equilibrium conditions with a charge stripping model [*Klecker et al.*, 2005].

A small increase of Q_{Fe} by 1 or 2 charge units at energies <1 MeV/nuc and somewhat larger increases as observed above ~10 MeV/nuc would be consistent with an M/Q-dependent exponential roll-over of the energy spectra at high energies [*Klecker et al.*, 2001]. However, a large increase of Q_{Fe} at energies <1 MeV/nuc by several charge units as sometimes observed in *gradual* events requires charge stripping in a dense environment as discussed for *impulsive* events in chapter 2.4. The large event-to-event variability could then be due to the combination of variable acceleration rates with non-equilibrium conditions for charge stripping [*Barghouty and Mewaldt*, 1999; *Ostryakov and Stovpyuk*, 1999; *Stovpyuk and Ostryakov*, 2001; and *Kovaltsov et al.*, 2001]. Another scenario that would lead to an increase of Q_{Fe} with energy in *gradual* events is the concept of two sources, e.g., a mixture of a coronal particle population with ions from a heavy ion rich 'flare' source, that would also have higher charge states, as discussed in section 3.5 below [*Tylka et al.*, 2001, 2005; and *Cane et al.*, 2003].

3.3. Energy Spectra

The energy spectra as observed in interplanetary space are the result of acceleration and propagation processes between the acceleration site and the observer. Power laws can often fit energy spectra in large events with exponential roll-over at high energies. These spectral forms are suggestive of shock acceleration: in the ideal case of an infinite and planar shock geometry and steady state conditions, the particle flux dJ/dE could be described by a power law: $dJ/dE \sim E^{-\gamma}$, where γ is related to the shock compression ratio (e.g., *Axford et al.*, 1978;

Blandford and Ostriker, 1978). However, because coronal and interplanetary shocks are not planar, and because only a limited time is available for acceleration, steady state will not be reached, in particular at high energies. Thus non steady-state conditions [*Forman and Webb*, 1985] and losses due to particles escaping upstream will result in a roll-over (e.g., *Ellison and Ramaty*, 1985) of the power law spectra at high energies. This spectral form is often observed and can be fitted by

$$dJ/dE \sim E^{-\gamma} \exp[-E/E_0] \tag{1}$$

where E_0 depends on mass M and ionic charge Q of the ions

$$E_0 = E_{0p}(Q/M)^\alpha \tag{2}$$

[*Tylka et al.*, 2000]. E_{0p} is the e-folding energy of protons and α is related to the power spectrum of the magnetic field fluctuations [*Cohen et al.*, 2003]. The e-folding energy E_{0p} and α show a large event-to-event variability, ranging for E_{0p} from ~10 MeV [*Tylka et al.*, 2000] to 800 MeV [*Lovell et al.*, 1998] and for α from ~0.8 to 2.7 [*Cohen et al.*, 2005].

The variability in the spectral form of heavy ions also results in a corresponding variability of elemental abundances with energy. If we use (1) and (2) for example to compute the abundances of Fe relative to O, then for α larger (smaller) than zero we will get a systematic decrease (increase) of the Fe/O ratio (e.g., Figure 6 of *Cohen et al.* [2003]). Other possibilities to explain systematic abundance variations with energy are the mixture of two (or more) sources as discussed in section 3.5.

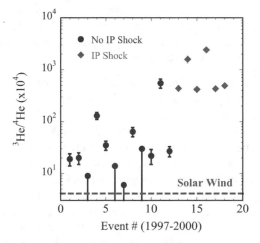

Figure 6. ^{3}He/^{4}He-ratio in the energy range 0.5-2.0 MeV/nuc in several large SEP events. Filled circles: 12 large events with no IP shock signature at 1AU [*Mason et al.*, 1999]; Diamonds: 6 events correlated with IP shocks and ^{3}He/^{4}He>100 times the solar wind [*Desai et al.*, 2001]; dashed line: solar wind abundance.

3.4. Observation of 'Flare' Material in Gradual Events

The large enrichments of ^{3}He and heavy ions, and high charge states of heavy ions of Q_{Fe}~20 found in event-integrated abundances of ^{3}He- and heavy ion-rich events have been used as one of their defining characteristics as *impulsive* events. However, these signatures have now been observed also in gradual events: *Mason et al.* [1999] reported ^{3}He abundances, significantly higher than solar wind abundances, in a survey of 12 large gradual events, with an average value of ^{3}He/^{4}He = 1.9 ± 0.2×10^{-3} in the energy range of 0.5-2.0 MeV/nuc. Even larger enrichments of ^{3}He with ^{3}He/^{4}He-ratios between 0.0014 and 0.24 have been found near the passage of 25 out of 56 IP shocks at 1 AU [*Desai et al.*, 2001]. The results of *Mason et al.* [1999] and *Desai et al.* [2001] are summarized in Figure 6. At higher energies (> 5 MeV/nuc) *Wiedenbeck et al.* [2000] reported for 3 large SEP events ^{3}He/^{4}He-ratios of about a factor of 10 above the average solar wind value, and a statistical survey using daily averages of ^{3}He and ^{4}He fluxes in the energy range 15–30 MeV/nuc for the years 1999 to 2002 showed an average value of ^{3}He/^{4}He = 0.015, also significantly higher than the ^{3}He abundance in the solar wind or corona [*Torsti et al.*, 2003].

Also, high abundances of Fe at energies of ~1 MeV/nuc [*Mason et al.*, 1999] and at energies of 12–60 MeV/nuc [*Cohen et al.*, 1999] and 25–80 MeV/nuc [*Cane et al.*, 2003] have been observed in many large events. Furthermore, high charge states of Q_{Fe}~20 at energies >10 MeV/nuc, another tracer for ions accelerated in *impulsive* events close to the Sun, are not uncommon (see section 3.2). Thus the question arises whether we have to reconsider our classification of SEP events into *impulsive* and *gradual*. This question is presently under debate and possible scenarios are discussed in the following section.

3.5. 'Flare' Material in Gradual Events – Scenarios

As a possible source of ^{3}He in large, IP-shock related events *Mason et al.* [1999] suggested remnant suprathermal particles from previous impulsive events, serving as seed particles for the injection at the IP-shock. In fact, *Wiedenbeck et al.* [2003] showed that during the time period November 1997 to May 2002 significant fluxes of ^{3}He in the energy range ~0.2–16.3 MeV/nuc were observed for at least 60% of the time.

In this scenario also high (and variable) heavy ion abundances (e.g., Fe/C) could be interpreted as a mixture of two sources: (1) suprathermal heavy ions from previous *impulsive* events and from *gradual* events. This suggestion is also supported by the finding that the interplanetary particle composition during quiet times shows enhancements in the ^{3}He/^{4}He and heavy ion composition: *Richardson et al.* [1990] found for Fe/O in the energy range ~1-10 MeV/nuc enhancements by a factor of ~10 over the coronal value and *Slocum et al.* [2002] reported enhancements of ^{3}He/^{4}He (4-15 MeV/nuc) and Fe/C (8-20 MeV/nuc) by two orders of magnitude and by a factor of ~8, respectively in the quiet-time population. Furthermore, there is now observational evidence that the IP-shock related heavy ion population at energies of ~1 MeV/nuc is not directly accelerated from a solar wind source: the elemental abundance in the IP-shock related particle population as observed in 72 events during 1997–2002 was poorly correlated with (slow) solar wind abundances, but positively correlated with the abundances upstream, suggesting the acceleration of suprathermals with a composition different from solar wind abundances [*Desai et al.*, 2003]. However, because of the much smaller enrichment factors for heavy ions than for ^{3}He in ^{3}He-rich events (~10 for Fe versus ~$10^2 - 10^4$ for ^{3}He), it is not clear whether there would be enough Fe for the high fluxes observed in IP-shock related events. Relating suprathermal iron densities at energies >0.01 MeV/nuc during quiet periods to the corresponding densities in Fe-rich events, *Mewaldt et al.* [2003] concluded that there may not be enough iron at quiet times to account for the overall enrichment of iron.

The concept of two sources with different composition, combined with an M/Q dependent roll-over of the spectra as described by equations (1) and (2) was successfully used by *Tylka et al.* [2001] to explain the hardening of the Fe spectra, the increase of the Fe/C-ratio and of Q_{Fe} at energies above ~10 MeV/nuc in the 2000, July 14 SEP gradual event. In a more general treatment of this problem *Tylka et al.* [2005] proposed as an explanation for the large variability of spectral, compositional, and ionic charge state features at high energies (i.e. above 10s of MeV/nuc) in large *gradual* SEP events a model where this variability arises from the interplay of two factors: shock geometry and the mixture of two seed populations with coronal/solar wind composition and 'flare' composition, i.e. a composition as observed in ^{3}He- and heavy ion-rich events. In this scenario the shock geometry plays an important role. It is, in particular, assumed that quasi-perpendicular shocks require a higher initial speed of the ions for effective injection and therefore preferentially accelerate suprathermal seed particles from flares, whereas quasi-parallel shocks generally draw their seed particles from the corona/solar wind suprathermals. In this model the shock geometry–via the injection threshold–determines which of the two components dominates and thus determines spectral shapes, heavy ion abundances and ionic charge states at high energies (see also *Tylka*, this volume).

In an alternative scenario, direct injection of the particles, accelerated in the CME related flare, has been proposed (e.g., *Klein and Trottet* [2001], and references therein;

Kocharov and Torsti [2002]; *Cane et al.*, [2003]). In this scenario, *gradual* events generally consist of two populations: (1) a population at low energies predominantly accelerated at the coronal/interplanetary shock, and (2) a high energy component (above ~10s of MeV/nuc), probably flare generated, with composition and charge states similar to *impulsive* events, possibly re-accelerated in the CME related shock. The relative intensity of the two components as observed at Earth will vary from event-to-event with the shock parameters (e.g., speed) and the flare size and location. However, at high energies the second component usually dominates, giving rise to the heavy ion enrichment and high charge states at high energies [*Cane et al.*, 2003].

4. SUMMARY

There is considerable progress over the last ~10 years in our knowledge of the elemental and isotopic composition, the energy spectra and ionic charge states in *gradual* and *impulsive* SEP events. However, the new results show that the classification into two distinct types of events was too simplistic. The important questions to be addressed in the future relate to the acceleration processes involved and include the following topics:

(1) There is a wealth of information on the abundances, energy spectra, and ionic charge states in *impulsive* events. The new observations of energy dependent ionic charge states indicate that the acceleration in these events takes place at low coronal altitudes (i.e. charge stripping and energy loss effects are important), and that interplanetary propagation effects are essential for understanding the measurements. However, models combining processes for the enrichment of ^{3}He and heavy ions with acceleration and charge stripping at the Sun and propagation in interplanetary space still need to be developed.

(2) The relative contribution of acceleration in coronal/interplanetary shocks and acceleration in the flare to the particle population in *gradual* events, in particular at high energies, is presently under active discussion. More work is needed here. For example, a verification by in-situ measurements and/or realistic (3D) simulations of the differences in the injection thresholds for quasi-parallel and quasi-perpendicular shocks will be important for models relying on these differences to explain the variability of spectra and abundances at high energies. The proposals suggesting direct injection of particles from the contemporary flare, in particular at high energies, will need to be complemented by physical models for the acceleration up to the observed energies of ~10s of MeV/nuc, including the process of ^{3}He- and heavy ion enrichment.

(3) Three dimensional simulations of CMEs and ICMEs, including particle injection and acceleration in the dynamically evolving magnetic fields, will also greatly improve our understanding of the complex acceleration processes near the Sun.

Significant progress can also be expected from future missions. STEREO (Solar Terrestrial Relations Observatory), to be launched in 2006, will provide a stereoscopic view of CMEs and interplanetary shocks, and compositional measurements from two vantage points, separated in solar longitude. The Solar Orbiter [*Marsden and Fleck*, 2003] with its perihelion at ~0.2AU, where propagation effects are significantly reduced, will allow unique studies of the sources of impulsive and gradual events, and of the acceleration in flares and coronal and interplanetary shocks close to the Sun [*Heber and Klecker*, 2005].

Acknowledgments.   We thank the conveners for the invitation to the Chapman Conference on Solar Energetic Plasmas and Particles and the organizers for their hospitality and perfect organization.

REFERENCES

Aschwanden, M.J., Particle acceleration and kinematics in solar flares, *Space Science Rev.*, 101, 1-227, 2002.

Axford, W.I., E. Leer, and G. Skadron, The acceleration of cosmic rays by shock waves, In: *15th Intern. Cosmic Ray Conf.*, Plovdiv, Bulgaria. Vol. 11, 132-137, 1978.

Balasubrahmanyan,V.K., and A.T. Serlemitsos, Solar energetic particle event with He-3/He-4 greater than 1, *Nature*, 252, 460-462, 1974.

Bamert, K., R.F. Wimmer-Schweingruber, R. Kallenbach, M. Hilchenbach, B. Klecker, A. Bogdanov, and P. Wurz, Origin of the May 1998 suprathermal particles: Solar and Heliospheric Observatory/Charge, Element, and Isotope Analysis System/(Highly) Suprathermal Time of Flight results, *J. Geophys. Res.*, 107 (A7), 1130, doi: 10.1029/2001 JA900173, 2002.

Barghouty A.F., and R.A. Mewaldt, Charge states of solar energetic iron: nonequilibrium calculation with shock-induced acceleration, *Astrophys. J.*, 520, L127-L130, 1999.

Blandford, R.D., and J.P. Ostriker, Particle acceleration by astrophysical shocks, *Astrophys. J.*, 221, L29-L32, 1978.

Bogdanov, A.T., B. Klecker, E. Möbius, M. Hilchenbach, L.M. Kistler, M.A. Popecki, and D. Hovestadt, Energy dependence of ion charge states in CME related solar energetic particle events observed with ACE/SEPICA and SOHO/STOF. In: *Proc. Acceleration and Transport of Energetic Particles in the Heliosphere*, ACE 2000 Symposium. (Eds.) R.A. Mewaldt, J.J.R. Jokipii, M.A. Lee, E. Möbius, and T.H. Zurbuchen. AIP 528, 143-146 (2000).

Bogdanov, A.T., B. Klecker, E. Möbius, M. Hilchenbach, M.A. Popecki, L.M. Kistler, and D. Hovestadt, Observations of heavy ion charge spectra in CME driven gradual solar energetic particle events, *Adv. Space Res.* 30 (1), 111-117, 2002.

Breneman, H.H., and E.C. Stone, Solar coronal and photospheric abundances from solar energetic particle measurements, *Astrophys. J. Lett.*, 299, L57-L61, 1985.

Bryant, D.A., T.L. Cline, U.D. Desai, and F.B. McDonald, Explorer 12 observations of solar cosmic rays and energetic storm particles after the solar flare of September 28, 1961, *J. Geophys. Res.*, 67 (16), 4983-5000, 1962.

Cane, H.V., T.T. von Rosenvinge, C.M.S. Cohen, and R.A. Mewaldt, Two components in major solar particle events, *Geophys. Res. Lett*, 30 (12), 8017, doi: 10.1029/2002 GL016580, 2003.

Cohen, C.M.S., R.A. Mewaldt, R.A. Leske, A.C. Cummings, E.C. Stone, M.E. Wiedenbeck, E.R. Christian, and T.T. von Rosenvinge, New observations of heavy-ion-rich solar particle events from ACE, *Geophys. Res. Lett.*, 26 (17), 2697-2700, 1999.

Cohen, C.M.S., R.A. Mewaldt, A.C. Cummings, R.A. Leske, E.C. Stone, T.T. von Rosenvinge, and M.E. Wiedenbeck, Variability of spectra in large solar energetic particle events, *Adv. Space Res.*, 32, 2649-2654, 2003.

Cohen, C.M.S., E.C. Stone, R.A. Mewaldt, R.A. Leske, A.C. Cummings, G.M. Mason, M.I. Desai, T.T. von Rosenvinge, and M.E. Wiedenbeck, Heavy ion abundances and spectra from the large solar energetic particle events of October–November 2003, *J. Geophys. Res.*, 110, A09S16, doi: 10.1029/2005JA011004, 2005.

Desai, M.I., G.M. Mason, J.R. Dwyer, J.E. Mazur, C.W. Smith, and R.M. Skoug, Acceleration of ^{3}He nuclei at interplanetary shocks, *Astrophys. J.*, 553, L89-L92, 2001.

Desai M.I., G.M. Mason, J.R. Dwyer, J.E. Mazur, R.E. Gold, S.M. Krimigis, C.W. Smith, and R.M. Skoug, Evidence for a suprathermal seed population of heavy ions accelerated by interplanetary shocks near 1 AU, *Astrophys. J.*, 588, 1149-1162, 2003.

Ellison, D.C., and R. Ramaty, Shock acceleration of electrons and ions in solar flares, *Astrophys. J.*, Part 1, 298, 400-408, 1985.

Fisk, L.A., He-3-rich flares - a possible explanation, *Astrophys. J.*, 224, 1048-1055, 1978.

Forbush, S.E., Three unusual cosmic-ray increases possibly due to charged particles from the Sun, *Phys. Rev.*, 70, 771-772, 1946.

Forman, M.A., and G.M. Webb, Acceleration of energetic particles, In: *Collisionless shocks in the heliosphere: a tutorial review.* Washington, DC, American Geophysical Union, 91-114, 1985.

Geiss, J., Constraints on the FIP mechanisms from Solar Wind abundance data, *Space Science Rev.*, 85, 241-252, 1998.

Giacalone, J., J.R. Jokipii, and J.E. Mazur, Small-scale gradients and large-scale diffusion of charged particles in the heliospheric magnetic field, *Astrophys. J.*, 532, L75-L78, 2000.

Gloeckler, G., R.K. Sciambi, C.Y. Fan, and D. Hovestadt, A direct measurement of the charge states of energetic iron emitted by the Sun, *Astrophys. J.*, 209, L93-L96, 1976.

Heber, B, and B. Klecker, Remote sensing of solar activity by energetic charged and neutral particles with Solar Orbiter, *Adv. Space Res.*, in press, 2005.

Hénoux J., FIP Fractionation: Theory, *Space Science Rev.*, 85, 215-226, 1998.

Hovestadt, D., Nuclear composition of solar cosmic rays, In: *Solar Wind Conference*, pp. 2-92, 1974.

Hovestadt D., H., Höfner, B. Klecker, M. Scholer, G. Gloeckler, F.M. Ipavich, C.Y. Fan, L.A. Fisk, and J.J. OGallagher, Direct observation of charge state abundances of energetic He, C, O, and Fe emitted in solar flares, *Adv. Space Res.* 1, 61-64, 1981.

Hurford, G.J., R.A. Mewaldt, E.C. Stone, and R.E. Vogt, Enrichment of heavy nuclei in He-3-rich flares, *Astrophys. J.*, 201, L95-L97, 1975.

Kallenrode, M.-B., Current views on impulsive and gradual solar energetic particle events, *J. Phys. G: Nucl. Part. Phys.*, 29, 965-981, 2003.

Kartavykh, Y.Y., V.M. Ostryakov, D. Ruffolo, E. Möbius, M.A. Popecki, Heavy ions from impulsive SEP events and constraints on the plasma temperature in the acceleration site, In: *Proc. 27th Intern. Cosmic Ray Conf.*, 8, 3091-3093, 2001.

Kartavykh, Y.Y., W. Dröge, V.M. Ostryakov, and G.A. Kovaltsov, Adiabatic deceleration effects on the formation of heavy ion charge spectra in interplanetary space, *Solar Phys.*, 227, 123-135, 2005.

Klecker B., D. Hovestadt, M. Scholer, G. Gloeckler, F.M. Ipavich, C.Y. Fan, and L.A. Fisk, Direct determination of the ionic charge distribution of helium and iron in ^{3}He-rich solar energetic particle events, *Astrophys. J.*, 281, 458-462, 1984.

Klecker, B., A.T. Bogdanov, M.A. Popecki, R.F. Wimmer-Schweingruber, E. Möbius, R. Schaerer, L.M. Kistler, A.B. Galvin, D. Heirtzler, D. Morris, D. Hovestadt, G. Gloeckler, Comparison of ionic charge states of energetic particles with solar wind charge states in CME related events, In: *Acceleration and Transport of Energetic Particles Observed in the Heliosphere*, AIP Conf. Proc., Vol 528, pp 135-138, 2000.

Klecker, B., E. Möbius, M.A. Popecki, M.A. Lee, and A.T. Bogdanov, On the energy dependence of ionic charge states in solar energetic particle events, In: *Solar and Galactic Composition*, AIP Conference Proceedings, Vol 598, pp 317-322, 2001.

Klecker, B., E. Möbius, M.A. Popecki, L.M. Kistler, H. Kucharek, and M. Hilchenbach, Observation of energy dependent ionic charge states in impulsive solar energetic particle events, *Adv. Space Res.*, in press, 2005.

Klein, K.-L., and G. Trottet, The origin of solar energetic particle events: coronal acceleration versus shock wave acceleration, *Space Science Rev.*, 95, 215-225, 2001.

Kocharov, L., G.A. Kovaltsov, J. Torsti, and V.M. Ostryakov, Evaluation of solar energetic Fe charge states: effect of proton-impact ionization, *Astron. Astrophys.*, 357, 716-724, 2000.

Kocharov, L., and J. Torsti, Hybrid solar energetic particle events observed on board SOHO, *Solar Physics*, 207, 149-157, 2002.

Kovaltsov, G.A., A.F. Barghouty, L. Kocharov, V.M. Ostryakov, and J. Torsti, Charge-equilibration of Fe ions accelerated in a hot plasma, *Astron. Astrophys.*, 375, 1075-1081, 2001.

Labrador, A., R.A. Leske, R.A. Mewaldt, E.C. Stone, and T.T. von Rosenvinge, High energy ionic charge state composition in recent large solar energetic particle events, In: *Proc. 28th Intern. Cosmic Ray Conf.*, Tsukuba, Japan, 6, pp. 3269-3272, 2003.

Leske, R.A., J.R. Cummings, R.A. Mewaldt, E.C. Stone, and T.T. von Rosenvinge, Measurements of the ionic charge states of solar energetic particles using the geomagnetic field, *Astrophys. J. Lett.*, 452, L149-152, 1995.

Leske, R.A., C.M.S. Cohen, A.C. Cummings, R.A. Mewaldt, and E.C. Stone, Unusual isotopic composition of solar energetic particles observed in the November 6, 1997 event, *Geophys. Res. Lett.*, 26 (2), 153-156, 1999.

Leske, R.A., R.A. Mewaldt, A.C. Cummings, E.C. Stone, and T.T. von Rosenvinge, The ionic charge state composition at high energies in large solar energetic particle events in solar cycle 23, In: *Joint SOHO/ACE workshop "Solar and Galactic Composition"*, AIP Conf. Proc. 598, pp. 171-176, 2001a.

Leske, R.A., R.A. Mewaldt, C.M.S. Cohen, E.R. Christian, A.C. Cummings, P.L. Slocum, E.C. Stone, T.T. von Rosenvinge, and M.E. Wiedenbeck, Isotopic abundances in the solar corona as inferred from ACE measurements of Solar Energetic Particles, In: *Joint SOHO/ACE workshop "Solar and Galactic Composition"*, AIP Conf. Proc., Vol 598, pp 127-132, 2001b.

Lovell, J.L., M.L. Duldig, and J.E Humble, An extended analysis of the September 1989 cosmic ray ground level enhancement, *J. Geophys. Res.*, 103 (A10), 23733-23742, 1998.

Luhn, A., B. Klecker, D. Hovestadt, M. Scholer, G. Gloeckler, F.M. Ipavich, C.Y. Fan, and L.A. Fisk, Ionic charge states of N, Ne, Mg, SI and S in solar energetic particle events, *Adv. Space Res.*, 4 (2-3), 161-164, 1984.

Luhn, A., B. Klecker, D. Hovestadt, and E. Möbius, The mean ionic charge of silicon in He-3-rich solar flares, *Astrophys. J.*, 317, 951-955, 1987.

Marsden, R.G., and B. Fleck, The Solar Orbiter mission, *Adv. Space Res.*, 32, 2699-2704, 2003.

Mason, G.M., D.V. Reames, T.T. von Rosenvinge, B. Klecker, and D. Hovestadt, The heavy-ion compositional signature in He-3-rich solar particle events, *Astrophys. J.*, 303, 849-860, 1986.

Mason, G.M., J.E. Mazur, and J.R. Dwyer, ^{3}He enhancements in large solar energetic particle events, *Astrophys. J.*, 525, L133-L136, 1999.

Mason, G.M., J.R., Dwyer, J.E. Mazur, New properties of ^{3}He-rich solar flares deduced from low-energy particle spectra, *Astrophys. J.*, 545, L157-L160, 2000.

Mason, G.M., M.E. Wiedenbeck, J.A. Miller, J.E. Mazur, E.R. Christian, C.M.S. Cohen, A.C. Cummings, J.R. Dwyer, R.E. Gold, S.M. Krimigis, R.A. Leske, R.A. Mewaldt, P.L. Slocum, E.C. Stone, and T.T. von Rosenvinge, Spectral properties of He and heavy ions in ^{3}He-rich solar flares, *Astrophys. J.*, 574, 1039-1058, 2002.

Mason, G.M., J.E. Mazur, J.R. Dwyer, J.R. Jokipii, R.E. Gold, and S.M. Krimigis, Abundances of heavy and ultra-heavy ions in ^{3}He-rich solar flares, *Astrophys. J.*, 606, 555-564, 2004.

Mazur, J.E., G.M. Mason, B. Klecker, and R.E. McGuire, The energy spectra of solar flare hydrogen, helium, oxygen, and iron - evidence for stochastic acceleration, *Astrophys. J.*, 401, 398-410, 1992.

Mazur, J.E., G.M. Mason, M.D. Looper, R.A. Leske, and R.A. Mewaldt, Charge states of solar energetic particles using the geomagnetic cutoff technique: SAMPEX measurements in the 6 November 1997 solar particle event, *Geophys. Res. Lett.*, 26 (2), 173-176, 1999.

Mazur, J.E., G.M. Mason, J.R. Dwyer, J. Giacalone, J.R. Jokipii, and E.C. Stone, Interplanetary magnetic field line mixing deduced from impulsive solar flare particles, *Astrophys. J.*, 532, L79-L82, 2000.

Meyer, J.-P., The baseline composition of solar energetic particles, *Astrophys. J. Suppl.*, 57, 151-171, 1985.

Mewaldt, R.A., C.M.S. Cohen, R.A. Leske, E.R. Christian, A.C. Cummings, E.C. Stone, T.T. von Rosenvinge, and M.E. Wiedenbeck, Fractionation of solar energetic particles and solar wind according to first ionization potential, *Adv. Space Res.*, 30, 79-84, 2002.

Mewaldt, R.A., C.M.S. Cohen, G.M. Mason, M.I. Desai, R.A. Leske, J.E. Mazur, E.C. Stone, T.T. von Rosenvinge, and M.E. Wiedenbeck, Impulsive Flare material: a seed population for large solar particle events?, In: *Proc. 28th Int. Cosmic Ray Conf.*, Tsukuba, Japan, Vol. 6, 3299 - 3302, 2003.

Miller, J.A., Particle acceleration in Impulsive Solar Flares, *Space Science Reviews*, 86, 79-105, 1998.

Möbius, E., M. Scholer, D. Hovestadt, B. Klecker, and G. Gloeckler, Comparison of helium and heavy ion spectra in He-3-rich solar flares with model calculations based on stochastic Fermi acceleration in Alfvén turbulence, *Astrophys. J.*, 259, 397-410, 1982.

Möbius, E., M.A. Popecki, B. Klecker, L.M. Kistler, A. Bogdanov, A.B. Galvin, D. Heirtzler, D. Hovestadt, E.J. Lund, D. Morris, and W.K.H. Schmidt, Energy dependence of the ionic charge state distribution during the November 1997 solar energetic particle event, *Geophys. Res. Lett.*, 26 (2), 145-148, 1999.

Möbius, E., Y. Cao, M.A. Popecki, L.M. Kistler, H. Kucharek, D. Morris, and B. Klecker, Strong Energy Dependence of Ionic Charge States in Impulsive Solar Events, In: *Proc. 28th Int. Cosmic Ray Conf.*, Vol. 6, 3273 - 3276, 2003.

Murphy, R.J., R. Ramaty, D.V. Reames, and B. Kozlovsky, Solar abundances from gamma-ray spectroscopy - comparisons with energetic particle, photospheric, and coronal abundances, *Astrophys. J.*, 371, 793-803, 1991.

Oetliker, M., B. Klecker, D. Hovestadt, G.M. Mason, J.E. Mazur, R.A. Leske, R.A. Mewaldt, J.B. Blake, and M.D.Looper, The ionic charge of solar energetic particles with energies of 0.3-70 Mev per nucleon, *Astrophys. J.*, 477, 495-501, 1997.

Ostryakov, V.M., and M.F. Stovpyuk, Charge state distributions of iron in gradual solar energetic particle events, *Solar Physics*, 189, 357-372, 1999.

Ostryakov, V.M., Y.Y. Kartavykh, D. Ruffolo, G.A. Kovaltsov, and L. Kocharov, Charge state distributions of iron in impulsive solar flares: Importance of stripping effects, *J. Geophys. Res.*, 105 (A12), 27315-27322, 2000.

Pallavicini, R., S. Serio, and G.S. Vaiana, A survey of soft X-ray limb flare images - The relation between their structure in the corona and other physical parameters, *Astrophys. J.*, 216, 108-122, 1977.

Popecki, M.A., J.E. Mazur, E. Möbius, B. Klecker, A. Bogdanov, G.M. Mason, and L.M. Kistler, Observation of energy-dependent charge states in solar energetic particle events, In: *Proc. 28th Intern. Cosmic Ray Conf.*, Tsukuba, Japan, Vol 6, 3283-3286, 2003.

Reames, D.V., Coronal abundances determined from energetic particles, *Adv. Space Res.*, 15 (7), 41-51, 1995.

Reames, D.V., Particle acceleration at the Sun and in the heliosphere, *Space Science Rev.*, 90, 413-491, 1999.

Reames, D.V., Abundances of trans-iron elements in solar energetic particle events, *Astrophys. J.*, 540, L111-L114, 2000.

Richardson, L.G., D.V. Reames, K.-P., Wenzel, and J. Rodriguez-Pacheco, Quiet-time properties of low-energy (<10 MeV/Nucleon) interplanetary ions during Solar Maximum and Solar Minimum, *Astrophys J. Letters*, 363, L9-L12, 1990.

Slocum, P.L., M.E. Wiedenbeck, E.R. Christian, C.M.S. Cohen, A.C. Cummings, R.A. Leske, R.A. Mewaldt, E.C. Stone, T.T. von Rosenvinge, Energetic particle composition at 1AU during periods of moderate interplanetary particle fluxes, *Adv. Space Res.* 30 (1), 97-104, 2002.

Slocum, P.L., E.C. Stone, R.A. Leske, E.R. Christian, C.M.S. Cohen, A.C. Cummings, M.I. Desai, J.R. Dwyer, G.M. Mason, J.E. Mazur, R.A. Mewaldt, T.T. von Rosenvinge, and M.E. Wiedenbeck, Elemental fractionation in small solar energetic particle events, *Astrophys. J.*, 594, 592-604, 2003.

Stovpyuk, M.F., and V.M. Ostryakov, Heavy ions from the 6 November 1997 SEP event: a charge consistent acceleration model, *Solar Phys.*, 198, 163-167, 2001.

Temerin, M., and I. Roth, The production of ^{3}He and heavy ion enrichments in ^{3}He-rich flares by electromagnetic hydrogen cyclotron waves, *Astrophys. J.*, 391, L105-L108, 1992.

Torsti, J., J. Laivola, and L. Kocharov, Common overabundance of ^{3}He in high-energy solar particles, *Astron. Astrophys.*, 408, L1-L4, 2003.

Tylka, A.J., P.R. Boberg, R.E. McGuire, C.K. Ng, and D.V. Reames, Temporal evolution in the spectra of gradual solar energetic particle events, In: *Acceleration and Transport of Energetic Particles Observed in the Heliosphere*, eds. R.A. Mewaldt, J.R. Jokipii, M.A. Lee, E. Möbius, and T.H. Zurbuchen (Melville: AIP), AIP Conf. Proc. 528, 147-152, 2000.

Tylka, A.J., C.M.S. Cohen, W.F. Dietrich, C.G. Maclennan, R.E. McGuire, C.K. Ng, and D.V. Reames, Evidence for remnant flare suprathermals in the source population of solar energetic particles in the 2000 Bastille Day event, *Astrophys. J.*, 558, L59-L63, 2001.

Tylka, A.J., C.M.S. Cohen, W.F. Dietrich, M.A. Lee, C.G. Maclennan, R.A. Mewaldt, C.K. Ng, and D.V. Reames, Shock geometry, seed populations, and the origin of variable elemental composition at high energies in large gradual solar particle events, *Astrophys. J.*, 625, 474-495, 2005.

Wiedenbeck, M.E., E.R. Christian, C.M.S. Cohen, A.C. Cummings, R.A. Leske, R.A. Mewaldt, P.L. Slocum, E.C. Stone, and T.T. von Rosenvinge, Enhanced abundances of ^{3}He in large solar energetic particle events, In: *Acceleration and Transport of Energetic Particles Observed in the Heliosphere*, AIP Conf. Proc., Vol 528, pp 107-110, 2000.

Wiedenbeck, M.E., G.M. Mason, E.R. Christian, C.M.S. Cohen, A.C. Cummings, J.R. Dwyer, R.E. Gold, S.M. Krimigis, R.A. Leske, J.E. Mazur, R.A. Mewaldt, P.L. Slocum, E.C. Stone, and T.T. von Rosenvinge, How common is energetic ^{3}He in the inner Heliosphere? In: *Solar Wind 10*, AIP Conf. Proc. Vol. 679, pp 652-655, 2003.

Zhang, T.X., Solar ^{3}He-rich events and ion acceleration in two stages, *Astrophys. J.*, 449, 916-929, 1995.

Radiative Diagnoses of Energetic Particles

A.L. MacKinnon

DACE, University of Glasgow, UK

We give an overview of the use of X- and γ-rays and radio radiation to diagnose energetic particle distributions at the Sun.

The bremsstrahlung hard X-ray spectrum yields a density-weighted mean electron energy distribution above 10 keV, via forward fitting or statistical regularization techniques. The thick target interpretation of the mean distribution minimizes fast electron numbers but still places severe demands on particle accelerators so we recall its implicit assumptions: diluteness of fast electrons with respect to the background plasma, monotonic slowing-down via binary collisions in a cold medium, distinct acceleration and interaction/radiation regions. Observations bearing on these assumptions are discussed. Recent insights from radio on 0.1-1 MeV electrons are reviewed.

Fast protons, α-particles and heavier ions in the MeV energy range and beyond all play a role in forming the γ-ray line component of the spectrum, between photon energies of about 0.5 and 8 MeV. Continuum radiation sometimes observed above 10 MeV bears witness to ions in the GeV energy range. The various signatures may be modelled to produce a picture of the ion energy distribution. Target chemical abundances and the distinct roles of protons and α particles are further complicating factors. γ-ray line shapes offer new diagnostic potential for tying down fast proton and α-particle distributions. Ions of energies <1 MeV/nucleon may be energetically significant in flares but remain elusive observationally.

We conclude with some comments on possible future lines of attack on diagnosis of fast particles.

1. INTRODUCTION

Particle acceleration seems to be a major, central component of the solar flare process, not just an inconsequential side-effect. We have very detailed information on flare-associated fast particles detected *in situ* in the interplanetary medium (IPM), but it seems likely that these particles have been accelerated en route to Earth, possibly in association with a CME [e.g., *Reames*, 1999, also several of the other

papers in this volume]. Moreover their interpretation at Earth is complicated by effects due to transport [e.g., *Dröge*, this volume]. To learn about particles accelerated in the flare primary energy release process, then, we must detect and interpret radiation from these particles. Such diagnostic studies are not just of interest for flare particle acceleration in its own right, but as an essential ingredient of unravelling the physics of the flare energy release process. Radio emission, both coherent and incoherent, electron-ion X-ray bremsstrahlung and nuclear γ-ray lines bear witness to the presence of energetic flare particles and we may deduce properties of their distribution functions in consequence. This review will look at the means by which X-ray, γ-ray and radio observations in particular can be made to tell us about the numbers and energy distributions of fast electrons

Solar Eruptions and Energetic Particles
Geophysical Monograph Series 165
Copyright 2006 by the American Geophysical Union
10.1029/165GM16

and ions at the actual flare site. It is aimed particularly at people who see statements of the numbers and energy distributions of fast particles in flares and would like some sense of the means by which these conclusions are reached. We also attempt to expose at least some of the - often unstated - caveats that go with such statements. With a view to the information most important for diagnosing the overall importance of fast particles in the flare process, we concentrate on spatially unresolved observations and the interpretation of spectra. Images are discussed primarily for their use in determining whether assumptions used in interpreting spatially unresolved spectra may be justified or not. We make some comments on particle acceleration mechanisms but the emphasis will be on deducing fast particle distributions rather than explaining their presence. We leave this task to Aschwanden, these proceedings; see also, e.g., *Vilmer and MacKinnon* [2003].

The standard interpretation of flare hard X-rays [HXR - *Brown*, 1971; *Hoyng et al.*, 1976; *Lin and Hudson*, 1976] leads to the conclusion that several tenths of all energy manifested in the flare is initially released as energetic electrons. We shall describe the means by which this conclusion is reached. The main ideas are well-rehearsed in the literature now, but theory still struggles to explain this finding, so the underlying assumptions deserve repeated and continuing scrutiny. The RHESSI mission [Ramaty High Energy Solar Spectroscopic Imager - *Lin et al.*, 2002] has unprecedented capabilities for imaging and spectroscopy that yield new insights on flare electrons and we shall mention some of its most ground-breaking findings. The situation regarding ions is slightly less clear. In some flares the energy content of ions in the MeV energy range appears to be comparable with that of deka-keV electrons. While HXR's appear to be a ubiquitous ingredient of the flare phenomenon, the γ-ray lines that testify to ion acceleration are not always detected, even from large flares [*Hudson et al.*, 2003].

Although certain ideas, e.g., the concept of a thick target source, are common to the interpretation of X-ray continuum and γ-ray lines, the differing character of the radiation mechanisms makes it sensible to deal with them separately. The next section discusses the interpretation of flare HXR's and the resulting picture of the role of electrons in flares. The notions of thick and thin target sources are now very familiar and may need little introduction, but X-ray observations place demands on particle acceleration theories that have been difficult to meet so we feel it important to revisit these ideas paying particular attention to the implicit assumptions, or at least those that seem particularly important to this author. Radio observations offer valuable corroboration of what has been learnt from HXR's, as well as a refined window on electrons in the >0.1 MeV energy range, and we briefly look at these in Section 3. Section 4 turns to interpretation of γ-rays and what

they tell us about flare fast ions. Section 5 makes very brief concluding comments.

2. FAST ELECTRONS AS REVEALED BY HXR'S

2.1. Information Content of HXR Spectrum

Flare electrons of energy E produce photons of energies $\varepsilon < E$ via electron-ion bremsstrahlung (*Korchak*, 1967 - models involving other radiation mechanisms must invoke populations of extremely relativistic electrons for which no other evidence exists: *McClements and Brown*, 1986). X-ray detectors give us spectra $I(\varepsilon, t)$ (in practice integrated over time intervals centred on times t), or at least photon count rates in several energy channels which can be modelled or deconvolved to yield $I(\varepsilon, t)$. Ignore the possibility of spatial information for the moment. Bremsstrahlung is not highly directional at photon energies of 10's of keV (the polar diagram for 100 keV electrons emitting 30 keV photons, for example, has a halfwidth of about 40°, getting broader for lower energy electrons - see e.g., *Massone et al.*, 2004), so we may integrate over electron directions. Let $I(\varepsilon, t)$ denote the photon flux differential in photon energy ε (photons cm^{-2}keV^{-1}s^{-1}) detected at Earth at time t. By writing down the bremsstrahlung emissivity at any point in the source region and integrating over the whole of its volume, we find that photon fluxes result from the source electron distribution thus [*Brown*, 1971; *Lin and Hudson*, 1976; *Tandberg-Hanssen and Emslie*, 1988; *Brown et al.*, 2003]:

$$I(\varepsilon,t) = \frac{1}{4\pi D^2} \int_\varepsilon^\infty F(E,t)v(E)\frac{\mathrm{d}\sigma}{\mathrm{d}\varepsilon}\mathrm{d}E \qquad (1)$$

Here $D = 1$ AU, $\mathrm{d}\sigma/\mathrm{d}\varepsilon$ is the bremsstrahlung cross-section differential in photon energy and $v(E)$ is the speed of an electron of energy E. $F(E, t)$, the angle-averaged, density-weighted, source integrated distribution of fast electrons has units of (electron energy)$^{-1}$(volume)$^{-1}$ (e.g. keV^{-1}cm^{-3}) and is given explicitly by

$$F(E,t) = \int n(\mathbf{r}) \int f(E,\theta,\phi,\mathbf{r},t)\mathrm{d}^2\Omega\mathrm{d}^3\mathbf{r} \qquad (2)$$

where $f(E, \theta, \phi, \mathbf{r}, t)$ is the distribution function at position $\mathbf{r}$, electron energy E and time t of electrons travelling in direction (θ, ϕ) (expressed in spherical polar coordinates), $n(\mathbf{r})$ is the ambient hydrogen density at position $\mathbf{r}$ and $\mathrm{d}^2\Omega = \sin\theta\,\mathrm{d}\theta\,\mathrm{d}\phi$.

The quantity F is some way removed from the quantities of greatest physical interest for understanding the flare (total number and energy content of fast electrons) but it is nonetheless of obvious, non-trivial interest, and can be related to more physical quantities given some assumptions, as we see below. Expressions for the bremsstrahlung cross-sections have been given e.g., by *Koch and Motz* [1959] and one can use them to deduce instantaneous electron populations e.g., by convolving parametric forms of the electron distribution with the cross-section and comparing the results with observations [e.g., *Kane and Anderson*, 1970; *Crosby et al.*, 1993; *Holman et al.*, 2003].

As an alternative to such forward fitting we may invert the observed X-ray spectrum to deduce F in a non-parametric way [*Brown*, 1971; *Brown et al.*, 2003]. Suppress the time-dependence of *F* for the moment, and follow *Brown* [1971] by specialising to the case of the non-relativistic Bethe-Heitler cross-section:

$$\frac{d\sigma}{d\varepsilon} = \frac{\sigma_0}{\varepsilon E} \ln\left[\frac{1-\sqrt{1-\frac{\varepsilon}{E}}}{1+\sqrt{1-\frac{\varepsilon}{E}}}\right] \qquad (3)$$

where σ_0 is a constant. With this cross-section, Eqn. (1) becomes, after a little reduction including an integration by parts:

$$\varepsilon \frac{dI}{d\varepsilon} \sim \int_\varepsilon^\infty \frac{F(E)}{\sqrt{E-\varepsilon}} dE \qquad (4)$$

This in turn is Abel's integral equation which can be solved [*Brown*, 1971; *Courant and Hilbert*, 1953] to give $F(E)$ in terms of the observed $I(\varepsilon)$:

$$F(E) \sim \frac{1}{E} \int_\varepsilon^\infty \frac{\varepsilon(3I(\varepsilon)+7\varepsilon I'(\varepsilon)+2\varepsilon^2 I''(\varepsilon))}{\sqrt{\varepsilon-E}} d\varepsilon \qquad (5)$$

With this set of assumptions and choice of cross-section we have an analytic solution for $F(E)$, showing explicitly the sense in which the source electron distribution may be deduced from measurements of the X-ray spectrum. In practice we are dealing with real, noisy data that may not be safely differentiated, at least not without preliminary smoothing [*Kontar and MacKinnon*, 2005]. One must employ a regularization method which simultaneously deconvolves and smoothes [e.g., *Craig and Brown*, 1986; *Johns and Lin*, 1992; *Thompson et al.*, 1992]; with such a numerical procedure in place, however, we can relax the assumptions and cross-section choices that make analytical solutions like eqn. (5) possible. We may use a more correct cross-section (as compiled in e.g., *Koch and Motz*, 1959, or given more recently by *Haug*, 1997), and in particular include the modest electron-electron contribution [*Haug*, 1998]. We may even relax the assumption of isotropy and determine how much the deduced electron distribution varies with differing assumptions about source angular electron distribution [*Massone et al.*, 2004].

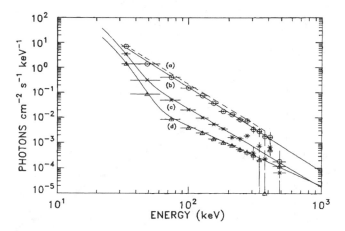

Figure 1. Photon spectra from a flare on 3 Feb 1982, from the HXRBS scintillation detector on SMM [from *Nitta et al.*, 1989]. The letters a, b, c, d denote a sequence of spectra obtained from the peak of the burst (spectrum *a*) through the decay phase. These data typify the best that can be done by way of photon energy resolution with scintillation detectors; compare with the detail of the spectra obtained with RHESSI (Figure 2).

Most missions before RHESSI employed scintillation detectors which have comparatively low photon energy resolution ($\Delta\varepsilon/\varepsilon$ typically 10%). Figure 1 shows data from *Nitta et al.* [1989], representative of the pre-RHESSI era. It is clear from the width of the energy bins and the number of data points that data like these would not justify a full, non-parametric inversion procedure (a numerical implementation of Eqn. (5), for example). Data from such detectors are dealt with by first fitting parametric forms. A power-law $I(\varepsilon) \sim \varepsilon^{-\gamma}$, with $\gamma > 0$, often provides an acceptable fit to the observed spectrum above 10-20 keV. Usually $2 \le \gamma \le 6$ [*Dennis*, 1988]. Then the mean source electron distribution is also a power-law in energy, $F(E) \sim E^{-\delta}$ [*Brown*, 1971], with $\delta = \gamma - 1/2$. One may also find that the data are more accurately described by a 'double power law' form, using different values of γ below and above some break energy. At other times, especially at lower photon energies, the spectrum may be better described by the form characteristic of an isothermal, Maxwellian plasma. These various spectral components may be seen, to varying degrees in the spectra shown in Figures 1 and 2.

The sequence of spectra in Figure 1 includes examples of both power-law and thermal forms. At the peak of the burst (labelled (a)) the spectrum is well described by a power-law in photon energy, through the full energy range of the instrument, from about 30 to 400 keV. As the overall intensity declines (spectra (b) - (d) in sequence) the spectrum increasingly resembles a sum of two distinct components.

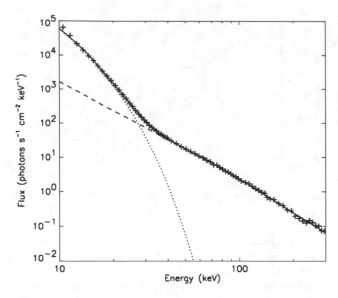

Figure 2. Photon spectrum (solid line) from a flare on 23 July 2002, from RHESSI [from *Lin et al.*, 2003]. The dashed and dotted lines show thermal and power-law components, respectively, of a fit to the overall spectrum.

The power-law form remains appropriate at higher photon energies but the spectrum below about 50 keV is increasingly well described, as the overall intensity declines, by the form representing an (iso)thermal plasma.

Although they illustrate these various forms of spectrum, the main reason for including these data here is to higlight the qualitative improvement in data quality yielded by RHESSI's germanium detectors. These have much finer energy resolution, FWHM of ~1 keV at low energies, increasing to ~5 keV at 5 MeV [*Lin et al.*, 2002]. Figure 2 gives an example of a RHESSI spectrum. Again one may fit parametric forms, and this particular spectrum has been shown with the thermal and power-law spectra whose sum gives a best fit to the data. With such data one can realistically aim to go further than parametric fitting of assumed functional forms, however. The data may be differentiated, with a little care, allowing study of the energy-dependence of the photon spectral index and in particular of departures from strict power-law form [*Kontar and MacKinnon*, 2005]. Regularized inversion of such data to recover the (source-averaged) electron distribution $F(E)$ in a non-parametric way also becomes possible [*Massone et al.*, 2003; *Piana et al.*, 2003; *Kontar et al.*, 2005a]. At the RHESSI Workshop series there was a comparative trial of the regularization strategies employed by several groups. Reassuringly, while reconstructions obtained using different methods show differences of detail, all give similar results for $F(E)$ [*Brown et al.*, 2006, submitted to Ap.J.].

2.2. Interpreting the Mean Electron Distribution

How do we relate $F(E, t)$ (or possibly its integral over non-negligible time intervals) to some quantity of physical interest? Which of the possible procedures to use is determined by the relative magnitudes of τ_{obs}, the time over which the spectrum has been integrated, and $\tau_E(\varepsilon)$, the time over which electrons of energy equal to the photon energy ε change their energies significantly [see also *Lin and Hudson*, 1976]. We may write

$$\tau_E(E) = \frac{E}{|\,dE/dt\,|} \quad (6)$$

We have written τ_E in a way that assumes electron evolution in energy is dominated by a systematic energy loss rate dE/dt but a similar definition can be given in terms of $\partial f/\partial t$ if e.g., electrons mostly diffuse in energy.

If $\tau_{obs} \ll \tau_E(\varepsilon)$, we may usefully assume that F represents the instantaneous distribution of electrons in a spatially homogeneous source of ambient density n_0 and total volume V. Then, clearly,

$$f_{thin}(E,t) = \frac{F}{n_0 V} \quad (7)$$

The subscript 'thin' denotes that this deduction of f has been made assuming a 'thin target' source, one in which electrons leave the source with their energies essentially unchanged (or at least in which electron energies change negligibly during the period over which the spectrum is obtained). This is a useful approach if we believe electrons are trapped in (coronal) regions where their lifetimes are much longer than the instrument time resolution [e.g., *Bai and Ramaty*, 1979; *Vilmer et al.*, 1982]. One result of the form (7) is that it can always be interpreted as an appropriate spatial average if the assumption of homogeneity is not satisfied, but its meaning is much less clear if τ_{obs} is not $\ll \tau_E(\varepsilon)$.

If τ_{obs} is not $\ll \tau_E(\varepsilon)$ we have to worry about how electron energies have evolved over the period of the observation. A limiting case of great utility is the *collisional thick target* introduced by *Brown* [1971]. This treatment assumes that the emitting electrons are accelerated rapidly in a region where they radiate negligibly, then released so that they stop completely in a distinct, HXR source region, via binary collisions with ambient particles [as described e.g., by *Spitzer*, 1956; *Trubnikov*, 1965]. This last assumption in turn requires that self-collisions of X-ray emitting electrons are rare, i.e. that the X-ray emitting population is dilute with respect to the background plasma, and indeed we usually assume that the ambient plasma thermal energy is $\ll \varepsilon$. Then $F(E)$ represents an average over the spatial and temporal lives of electrons in the source. Averaging f over the electron lifetimes, and using the monotonicity of energy loss rate to change variable to E (or v), the definition (2) of F implies

$$F(E) = \frac{1}{|dE/dN|} \int_E^\infty S(E) dE \qquad (8)$$

where dE/dN is the rate of change of electron energy with column depth:

$$\frac{dE}{dN} = \frac{1}{vn(\mathbf{r})} \frac{dE}{dt} \qquad (9)$$

and $S(E)$ is the rate of injection of electrons of energy E (again, angle-integrated and volume-integrated). S represents the product of the electron acceleration process. Details of the atmospheric density structure play no role in this result since the ratio of electron energy loss rate to bremsstrahlung emissivity depends only on electron energy. Even if we are unsure about applying it moment by moment through the flare, (8) will give correct results for S integrated over the event if we apply it to the event-integrated HXR flux I.

On the one hand a thick target produces the greatest bremsstrahlung yield, at all photon energies, from a given population of electrons. If we are happy that acceleration and HXR production regions may be separated, making the thick target assumption minimizes the number of electrons needed to explain a given observed $I(\varepsilon)$ [*MacKinnon and Brown*, 1989]. On the other hand it implies that the injected electron energy distribution $S(E)$ is steeper, and has a greater energy content, than appears to be the case from a thin-target interpretation of $F(E)$. In consequence the energy content of HXR emitting electrons (of energy 25 keV and above) has been found to be comparable to the total energy otherwise manifested in the flare [*Hoyng et al.*, 1976; *Lin and Hudson*, 1976; *Saint-Hilaire and Benz*, 2005]. Upwards of 10^{36} electrons per second are accelerated to energies of 25 keV and above, with a total energy content exceeding 10^{32} ergs in the largest flares. With steeply falling, power-law type energy distributions $E^{-\delta}$ ($\delta > 2$), energy content is dominated by the lowest electron energy E_0 for which the power-law form continues to hold. If the appropriate value of E_0 is in fact significantly less than 25 keV, electron energy content will be correspondingly greater and nonthermal electrons may easily become the single biggest component of the flare energy budget. E_0, clearly a quantity of great importance, is poorly determined from observations because the power-law type photon spectrum from accelerated electrons begins to merge with the ('thermal') spectral signature of the bulk of the flare hot plasma [see e.g., *Saint-Hilaire and Benz*, 2005] (cf. the low energy portion of Figure 2). Worse, if electrons thermalize in coronal regions where the

ambient temperature is $>10^7$K, their relaxation further masks any low-energy cutoff characteristic of the acceleration mechanism [*Galloway et al.*, 2005; and Section 2.4.4 below].

The large number and energy content of accelerated electrons has a couple of major consequences for the overall picture of flares. On the one hand, fast electrons must be a major means of energy transport in flares and the resulting, electron-heated picture of secondary flare phenomena [e.g., *Dennis and Schwartz*, 1989] is appealing enough to have attained the status of a 'paradigm'. On the other hand, flare particle acceleration theories have so far proven unequal to the challenge of accelerating electrons with the required efficiency [*Miller et al.*, 1997]. The sheer number of electrons apparently involved in understanding flare HXR's is particularly challenging. Thus there is continued interrogation of the assumptions of the thick target model.

2.3. HXR Images

Here we comment particularly on the use of HXR images to discuss the correctness of the ideas outlined above.

The phrase 'thick target' originally comes from radiation physics and refers simply to a source thick enough to completely stop the bombarding particles. In flare physics the 'thick target' has become identified with the chromospheric regions at the footpoints of the loops involved in the flare, although this need not be the case: high ambient density sources, collisionally thick for electrons up to 50 keV in energy, have been reported in the corona by *Veronig and Brown* [2005]. Even a low density source, e.g., a coronal magnetic bottle [*Takakura and Kai*, 1966; *Bai and Ramaty*, 1979], may be a thick target in the sense that no particles escape and the event-integrated X-ray spectrum from such a situation will be that predicted by a thick target calculation.

In the absence of trapping, electrons of 10s of keV energy, accelerated in the corona, will stop and produce most of their bremsstrahlung yield as they encounter increasing densities for the first time. Imaging observations of HXR bright points, coincident with footpoints of soft X-ray loops and with EUV and Hα brightenings, thus constitute strong support for a thick target interpretation of HXR's. Images with distinct bright points were first observed with the HXIS instrument on SMM [*Hoyng et al.*, 1981; *Machado et al.*, 1985; *MacKinnon et al.*, 1985], at photon energies up to 30 keV. Yohkoh's HXT confirmed and extended this finding, but also gave us images of 'above-the-loop' sources [*Masuda et al.*, 1994] believed to indicate the electron acceleration region, although it is not immediately obvious that the acceleration region, in the low-density corona, should be a seat of enhanced HXR emission [see *Fletcher*, 1995; *Fletcher and Martens*, 1998; *Conway et al.*, 1998].

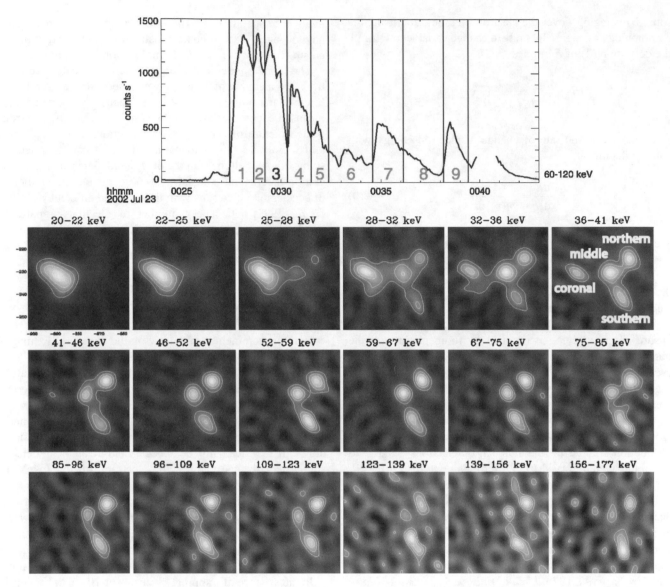

Figure 3. The top panel shows the spatially integrated HXR light curve for the flare of 23 July 2002. The bottom panels show X-ray spatial structure, in several energy channels as shown, for the period 3 of the top panel [from *Emslie et al.*, 2003].

RHESSI now gives us spatial resolution down to 2″, with good enough count statistics that images may be constructed in many, narrow photon energy channels. *Emslie et al.* [2003] show a set of images from one period of the X4.8 flare of 23 July 2002, reproduced in Figure 3. A pair of bright points may be identified with footpoints of loops. For the first time these features are resolved at photon energies into the 100 keV energy range. Improved resolution once again reveals a more complex situation, however. The two clear footpoints are accompanied by a third bright point of uncertain nature, complicating a simple loop interpretation of the observations. At low (10-30 keV) photon energies there is yet another spatially distinct source, apparently high in the corona. No doubt these various sources have a more complex tale to tell, as yet incompletely unravelled, of flare magnetic topology and the partitioning of energy between bulk heating and particle acceleration. Figure 3 illustrates how these various features appear in different energy ranges - at least in this event.

Such imaging spectroscopy will continue to influence our ideas of where electrons are accelerated and where they deposit energy, and thus of how best to interpret HXR spectra. We revisit some implications of images below (Section 2.4.1, and also in the concluding Section 5).

2.4. Thick Target Assumptions Interrogated

In this section we discuss further some of the physics usually assumed in thick target calculations and deductions of electron energy distributions, and some of the possible consequences if these assumptions do not in fact hold. First we note that even apparently modest development of the standard approach in the direction of increased realism may have substantial consequences. In particular, the nonuniform ionization of the chromosphere influences thick target emergent photon spectra because electron energy loss rates depend (in magnitude, not functional form) on the degree of ionization of the ambient plasma [Brown, 1973]. A dismaying consequence is that the inverse problem for the primary, accelerated electron energy distribution no longer has a unique solution [Brown et al., 1998]. Kontar et al. [2003] suggest that the HXR consequences of nonuniform atmospheric ionization may already have been detected, in departures of RHESSI spectra from power-law behaviour.

2.4.1. Separation of acceleration and interaction regions.
Conventional thick target interpretations of HXR's imply that a large fraction of all available (coronal) electrons must be accelerated and precipitated into the chromosphere. If instead the accelerator continues to act on electrons while they radiate, their numbers could in principle be greatly reduced. An example of such a situation would be particle acceleration in encounters with multiple energy release sites, as studied e.g., by Arzner and Vlahos [2004] or Turkmani et al. [2005]. Such models would have many appealing features and might occur naturally in a Self-Organized Critical (SOC) picture of flare energy release [Lu and Hamilton, 1991]. Spicer and Emslie [1988] proposed a related picture, not necessarily involving spatial fragmentation, in which HXR-emitting electrons would find themselves in a region of nonzero electric field.

One possible objection to these models comes from the observation of spatially well-separated footpoints, brightening simultaneously to within instrumental time resolution. MacKinnon et al. [1985] studied HXIS observations, in the 16-30 keV photon energy range, of flare footpoints separated by something like 10^4 km. They found that the footpoints brightened simultaneously to within 3s, rapidly enough to rule out synchronization by a signal propagating at the Alfvén speed. In Yohkoh HXT data Sakao et al. [1996] found double sources brightening simultaneously to within a fraction of a second. With RHESSI the picture, at least sometimes, seems less clear-cut: pairs of bright points appear to be well correlated but their positions change continuously during a flare, so the question of simultaneity becomes more elusive [e.g., Grigis and Benz, 2005a]. Synchronization of spatially separated locations in this way is difficult to reconcile with the stochastic progress, and indeed geometry of

self-organized flare models. Since the many, localized electric fields in these models are usually imagined in the corona [e.g., Arzner and Vlahos, 2004], we lose HXR production efficiency, reverting to the usual thick target picture, by allowing fast electrons to precipitate to chromospheric footpoints. Direct comparison with previous results, like those of Sakao et al. [1996], is further complicated by RHESSI's temporal resolution, routinely limited at present to the spacecraft's 4s spin period [as discussed in Hurford et al., 2002] but improvable in principle [Arzner, 2004].

2.4.2. Collisional losses.
Even if all the other assumptions of the collisional thick target are satisfied, electron energies will probably change due to other physical mechanisms. Of course any losses additional to binary collisions will only decrease the efficiency of the source region as an X-ray producer, and thus increase the number and energy content of electrons needed to explain a given observed HXR spectrum.

Electrons streaming away from the acceleration region will become anisotropic and thus unstable to generation of Langmuir or other electrostatic waves [Holman et al., 1982; McClements, 1987; Haydock et al., 2001]. Haydock et al. [2001] incorporate beam relaxation and velocity space plateau formation in a sort of marginal stability treatment (neglecting, however, 3-D aspects of the velocity distribution function) and find that the electron energy distribution implied by $I(\varepsilon)$ is unchanged from a purely collisional treatment, although the total number of electrons is greater.

In the chromosphere, collisional damping suppresses wave growth [Emslie and Smith, 1984]. Electrons that have become anisotropic enough to constitute a beam will still suffer the additional energy loss associated with the maintenance of a neutralising return current, first discussed in this context by Knight and Sturrock [1977]. Return current losses increase in proportion to primary beam current in such a way as to make the X-ray flux for large enough electron fluxes independent of the number of electrons [Emslie, 1981]. Return current modifications to deduced electron energy distribution and numbers have been discussed most recently by Zharkova and Gordovskyy [2005].

2.4.3. Diluteness.
Collisional slowing-down rates have always been calculated assuming that X-ray emitting electrons are 'dilute' with respect to the background plasma. Then, for instance, we may calculate collisional drift and diffusion coefficients once and for all ignoring collisions of energetic electrons with one another. The conventional treatment of return current losses also relies on being able to clearly separate 'beam' and 'background' electrons.

So-called thermal models for HXR production constitute the opposite extreme from the normal assumption of diluteness. Involving coronal volumes of $>10^8$K plasma, the continuing life of these models is motivated, once again, by a desire to reduce the number of fast electrons implied by observed HXR spectra. Historically, *Chubb et al.* [1966] considered the possibility of interpreting observed X-rays in terms of ~10^8K plasma, and *Brown* [1974] showed that non-isothermal emitting material with a suitable range of temperatures could in principle mimic observed, power-law X-ray spectra.

In the ideal case of a perfectly contained Maxwell-Boltzmann distribution, lifetimes of X-ray emitting electrons are effectively infinite and the number and energy content of electrons needed to explain typical HXR fluxes is much less than in the usual cold, thick target case. Electron collisional mean free paths in such a source are >> coronal loop lengths but ion-acoustic turbulence, driven as fast electrons try to stream into the surrounding, cooler plasma, may contain the X-ray emitting region [*Brown et al.*, 1979; *Smith and Lilliequist*, 1979]. The efficiency gain associated with such a source is less dramatic, however, because the source region expands (at the ion sound speed) and cools, and also loses energy because higher energy electrons are able to escape. While the escaping electrons are able to reconcile such models with observations of footpoints, the overall efficiency of this realistic form of thermal model may be no greater than that of nonthermal, thick target sources [*MacKinnon et al.*, 1985].

Some observations do suggest that X-ray emitting electrons are not always insignificant compared to their surroundings. For example, *Hudson et al.* [2001] describe a moving, coronal HXR source in which the pressure at least of the nonthermal electrons is comparable to the surrounding thermal plasma.

2.4.4. Cold target.   In the chromosphere, HXR emitting electrons will always be much more energetic than ambient, thermal particles. We call such a situation a 'cold target'. Energy loss and pitch-angle scattering rates may be calculated neglecting the energies of the field particles. *Emslie* [2003] points out, however, that likely coronal densities mean lower energy (~10 keV) electrons may stop entirely in the corona, in regions where their speeds are not >> the electron thermal speed v_e (Here $v_e = \left(\frac{kT}{m_e}\right)^{1/2}$, where T is the ambient temperature (K), k is Boltzmann's constant and m_e is the electron mass). Then their radiatively effective lifetimes are greater than in the case of a perfect cold target and we need fewer of them to account for the observed photon flux in this energy range. The total energy content of fast electrons is determined by the form of the energy distribution at the lowest energies so such revisions of the deduced form of the low energy electron distribution are potentially crucial. In the RHESSI images of the 23 July 2002 flare there is a softer sort of coronal HXR source [*Emslie et al.*, 2003], within which 10 keV electrons, for example, would not be in the cold target regime.

Galloway et al. [2005] make a first attempt at treating this situation via a linearized Fokker-Planck equation, incorporating velocity diffusion near the thermal speed. This allows them to treat the thermalisation of fast electrons in a way that automatically joins non-thermal and background components of the overall distribution. This work highlights the fact that electrons with speeds below about $3v_e$ thermalize, contributing to the X-ray spectral component that would be labeled 'thermal', so that attempts to determine from spectra a value of E_0 may be doomed to failure. Attempts to fit the resulting distribution to RHESSI data result in large total electron energy content, because the injected electron distribution appears to extend down in energy close to the ambient thermal speed. Time-dependence may alter the conclusions of this steady-state treatment, however. Moreover, accelerated electrons may sometimes be numerous enough to cast doubt on the assumption of diluteness that allows linearization of the Fokker-Planck equation. We await more complete modeling of this situation, and more detailed analysis of the RHESSI spectroscopically resolved images.

2.4.5. 'Dips' in deduced electron energy distributions. Some recent data may be inconsistent with the cold thick target model. *Piana et al.* [2003] show a reconstructed $F(E)$ for the period 00:30:00 - 00:32:20 during the flare of 23 July 2002, with an apparent minimum at about 50 keV. We saw above (8) that $F(E)$ in the cold thick target is essentially the cumulative distribution of the injected energy distribution $S(E)$; $F(E)$ should be monotonic decreasing in consequence and such a minimum appears irreconcilable with a cold thick target. This is a surprising result, and one that would have been impossible to obtain with data from any detector prior to RHESSI. Such mean energy distributions with dips are not commonplace, but similar results have been obtained in a few other flares [E Kontar, private communication; *Schwartz et al.*, 2003, *Kasparova et al.*, 2005].

In fact, the X-ray spectrum derived from the instrument count-rate spectrum may not correctly reflect the actual X-ray spectrum emitted by the flare fast electrons. At least two identifiable factors will lead to flattening in the right photon energy range to influence the inference of such dips. In large flares, RHESSI's spectral response becomes nonlinear due to 'pulse pile-up', the phenomenon of low energy photons arriving so rapidly that two or more are recorded as a single event in the instrument, as though they actually were a single photon of a higher energy [*Smith et al.*, 2002]. The resulting distortion of the spectrum is most pronounced in the same photon energy range that produces the 'dip' electron feature and may play a role in producing it, at least in large flares like the 23 July 2002 event [*Kontar et al.*, 2003].

Further, even the spectrum with all instrumental effects correctly accounted for may not be precisely that produced

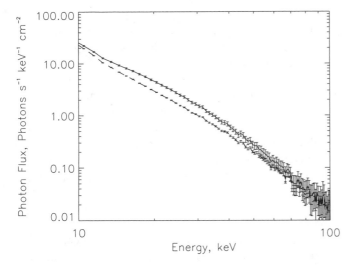

Figure 4. Photon spectra from the flare of 20 August 2002, before (solid) and after (dashed) deconvolution of a Compton backscattered spectral component [from *Kontar et al.*, 2005]. Note the change in the concavity of the spectrum in the 20-50 keV energy range, reducing the need for a 'dip' in the primary electron energy distribution $F(E)$.

maximum in the 30-50 keV energy range, lower energy photons being attenuated en route out of the photosphere by photoelectric absorption, and higher energy ones going too deep before being scattered into the line of sight [*Bai and Ramaty*, 1978]. Thus the total emergent spectrum may include a 'bulge' from backscattered photons in the right energy range to play a role in the results of *Piana et al.* [2003]; and other events as mentioned above. *Kontar et al.* [2005b] show how the Compton reflection Green's function of *Magdziarz and Zdziarski* [1995] may be used to deconvolve the Compton backscattered component from the total observed HXR spectrum, thus obtaining an estimate of the true primary spectrum. The only important assumption in this method is that the primary emitted radiation is isotropic. Figure 4 gives an example, the observed RHESSI spectrum from a flare of 20 August 2002, together with the primary spectrum deduced by *Kontar et al.* Note the slight upward 'bulge' in the observed spectrum, a flattening at around 20-30 keV, and its absence in the deduced primary spectrum. Because all electrons at energies $E > \varepsilon$ contribute to photon emission at photon energy ε, this flattening, taken at face value, demands a dip in the mean electron energy distribution $F(E)$. With the albedo contribution subtracted, however, the deduced primary spectrum is much closer to a single power law in ε, without the 'bulge' at 20-30 keV. Figure 5, again from *Kontar et al.* [2005b], shows $F(E)$, obtained via forward fitting from the RHESSI spectrum with and without the removal of the albedo contribution. The dip in $F(E)$

by the flare fast electrons. Photons emitted by electrons stopping in the high chromosphere may proceed directly to Earth (as assumed above in Section 2.1), or they may be emitted downward but suffer Compton scattering into the line of sight to Earth. The Compton backscattered component shows a

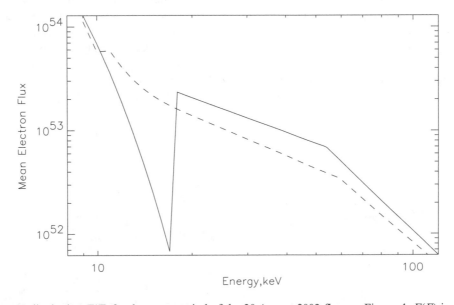

Figure 5. Mean electron distrbution $F(E)$ for the same period of the 20 August 2002 flare as Figure 4. $F(E)$ is obtained by forward fitting starting from the observed spectrum (solid lines) and from the primary spectrum (i.e. obtained by removing the albedo contribution from the observed spectrum). With the albedo contribution removed there is no need for a local minimum in $F(E)$. From *Kontar et al.* [2005]

demanded by the observed spectrum disappears when the albedo contribution is subtracted; what appears to the eye a modest change in the form of $I(\varepsilon)$ has much more dramatic consequences for the deduced $F(E)$. At the moment, then, it appears likely that proper inclusion of backscattered photons will suffice for understanding of this surprising spectral feature, rather than any radical revision of thick target orthodoxy. Conversely, photospheric albedo needs to be included for proper constraint of the source electron distribution in the 20-50 keV energy range.

3. RADIO EMISSION

3.1. Incoherent Gyrosynchrotron Radiation

The electrons that produce X-ray emission above ~10 keV also emit incoherent gyrosynchrotron radiation, typically in the frequency range from 1 to 10s of GHz. In consequence, flare impulsive phase hard X-rays and broadband microwave emission are usually found in close, but not perfect time coincidence; see the review of *Bastian et al.* [1998]. Flare microwave continuum may thus be used to corroborate deductions from X-rays. In practice there are several issues to be considered. The density of flare fast electrons is great enough to make self-absorption important. The emissivity and absorption coefficients are calculated as for synchrotron radiation [*Ginzburg and Syrovatskii*, 1965], but without taking the relativistic limit. The influence of the medium may be significant, in the form of Razin suppression. The viewing angle with respect to the magnetic field and the pitch-angle distribution of electrons are also important. *Ramaty* [1969] gives the full details of the necessary formalism for calculating emissivity and absorption coefficient as functions of frequency v.

In its simplest form, the observed spectrum $S(v)$ (erg cm^{-2}Hz^{-1}s^{-1}) has a single maximum at a frequency v_p [e.g.,

Dulk, 1985]; illustrated in Figure 6. Below v_p optical depth plays a crucial role in the form of S. An idealized, homogeneous slab source has the Rayleigh-Jeans form $S(v) \sim v^2$ for $v < v_p$; departures from this form indicate primarily source inhomogeneity. In view of all these factors, plus the importance of viewing angle with respect to the magnetic field, complete modelling of the microwave spectrum involves treating radiative transfer throughout a model for the spatial distribution of fast electrons [*Böhme et al.*, 1977]. Only then is convincing agreement between X-ray and microwave derived electron distributions obtained [*Klein et al.*, 1986].

For $v > v_p$, the source is optically thin. The form of $S(v)$ reflects the (line of sight integrated) electron energy distribution fairly directly and useful discussion of electron energy distribution is possible in advance of a full treatment of radiative transfer, etc. For instance, for an isotropic population of electrons with power-law energy dependence $E^{-\delta}$, $S(v) \sim v^{-\alpha}$ where

$$\delta = 1.11\alpha + 1.36 \qquad (10)$$

(*Dulk*, 1985 - this result replaces the well-known synchrotron relation $\delta = 2\alpha + 1$ in the mildly relativistic regime). The value of $S(v)$ for some v in this optically thin regime then fixes the number of electrons.

Recall that the 'synchrotron frequency' characteristic of emission by electrons of Lorentz factor γ is

$$v_s = 4.3 \times 10^6 \gamma^2 B (\text{Hz}) \qquad (11)$$

Observed peak frequencies are typically in the range 5-10 GHz [*Nita et al.*, 2004]. With fields of the order of 100 G, typical of the low corona, we see from Eq. (11) that electrons emitting in the optically thin regime, for instance at 10-20 GHz, will have energies of the order of 1 MeV. Flares with detectable X-rays in this energy range are not numerous. In almost ten years of operation, for example, the Solar

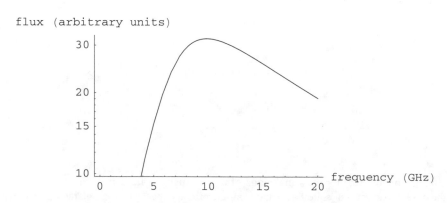

Figure 6. Idealized radio spectrum showing a maximum at a frequency v_p, here chosen to be about 10 GHz, and increasing and decreasing as powers of frequency v above and below v_p.

Maximum Mission's Gamma-Ray Spectrometer instrument detected just 185 flares with a large enough flux for determination of a photon spectral index above 300 keV - *Vestrand et al.* [1999]. The greater sensitivity of radio observations at 10s of GHz thus provides a most effective channel for studying electrons in the 100s of keV to MeV energy range.

In particular, there is HXR/γ-ray evidence in a handful of flares that the electron energy distribution hardens above a few 100 keV [e.g., *Peng*, 1995; *Marschhäuser et al.*, 1994; *Trottet et al.*, 1998; *Vilmer et al.*, 1999]. *Silva et al.* [2000] combined HXR and microwave observations, comparing spectral slope above v_p and HXR spectral index below 200 keV, to amass statistical evidence corroborating this finding in flares whose >300 keV electrons could not have been studied by X-rays.

Radio observations also suggest that electrons in the 100s of keV energy range evolve differently from those at lower energies. *Melnikov and Magun* [1998] studied more than 20 bursts in the frequency range between 8 and 50 GHz, commonly finding a decrease with time in the radio spectral index at $v > v_p$, i.e. a consistent hardening of the electron energy distribution in the MeV energy range. This is in marked contrast to the spectral behaviour in X-rays below about 200 keV, where the electron energy distribution normally hardens to burst peak and then softens again ['soft-hard-soft' behaviour, discussed in many papers, e.g., *Kane and Anderson*, 1971; *Hoyng et al.*, 1976; *Benz*, 1977; *Grigis and Benz*, 2005b]. This hardening at higher energies could conceivably be due to some acceleration process that gradually produces more and more higher energy electrons as the flare progresses, but there is also a fairly natural explanation in terms of trapping of higher energy electrons. The energy-dependence of collisional energy losses results in hardening of the electron distribution as time goes on [*Takakura and Kai*, 1966; *Bai and Ramaty*, 1979; *Vilmer et al.*, 1982], offering an explanation of the observed spectral behaviour. *Lee and Gary* [2000] take this a step further, modelling such trapping in sufficient detail to constrain the pitch-angle distribution of electrons resulting from the acceleration mechanism.

3.2. Coherent Radiation

Various narrowband, rapidly fluctuating, high brightness temperature emissions from 10s to 1000s of MHz additionally bear witness to the presence of energetic electrons throughout the corona. Flares are often, but not always accompanied by such emissions [*Benz et al.*, 2005]. Their quantitative interpretation is difficult because of the nonlinearity of the radiation mechanisms (even when these are known). Nonetheless they yield vital clues on the locations of fast electrons, the degree of fragmentation of the accelerator and many other questions [e.g., the recent review of *Benz*, 2003].

3.3. Mm and Sub-mm Radiation

Recent years have seen the concerted exploration of mm and sub-mm radiation from flares. Some flares at least display radiation at very high frequencies [e.g., 230 and 345 GHz - *Lüthi et al.*, 2004a; 212 GHz - *Trottet et al.*, 2002; 212 and 450 GHz - *Kaufmann et al.*, 2001; *Kaufmann et al.*, 2004]. Sometimes this high frequency radiation appears consistent with the extension upwards in frequency of the gyro-synchrotron spectrum described in Section 3.1 [*Trottet et al.*, 2002], albeit from an electron population extending in energy well into the MeV range. In other events a new component of the flare spectrum has been detected, with flux density increasing between 212 and 450 GHz and possibly on into the infrared [*Kaufmann et al.*, 2004]. Flare sub-mm radiation has been found to exhibit two components in temporal behaviour [*Trottet et al.*, 2002; *Lüthi et al.*, 2004b]: a rapidly fluctuating one during the impulsive phase and a more gradually varying component that continues through the flare gradual phase. The latter, resulting from a more spatially extended source [*Lüthi et al.*, 2004a], seems likely to result from thermal bremsstrahlung from hot flare plasma. The former component, coincident with flare X-rays and γ-rays, is more compact and still of less certain nature. It may result from synchrotron radiation of relativistic electrons, or even positrons produced by protons of energies in excess of 300 MeV (as in Section 4.4 below). Just as radio continuum at 10s of GHz gives a sensitive window on electrons in the 0.1-1 MeV energy range, it appears that the sub-mm wavelength range may offer a new, sensitive window on even higher energy particles.

4. γ-RAYS AND FAST IONS

Upward of 400 keV in photon energy we find the continuation of the electron bremsstrahlung spectrum but also line features resulting from various nuclear processes. These various, broad and narrow lines constitute the observable evidence for accelerated ions at the flare. Features of observed γ-rays, including many of the exciting new data from RHESSI and INTEGRAL, are also discussed in these proceedings by Share and Murphy. Some key findings from the SMM and Compton eras are reviewed by *Chupp* [1984] and by *Ryan and Hudson* [1995]. Here we concentrate particularly on inference of fast ion energy distributions and numbers.

Figure 7 shows the RHESSI γ-ray spectrum from the flare of 23 July, 2002, together with the various components used to model the spectrum [*Lin et al.*, 2003]: narrow, nuclear deexcitation lines due to collisions of fast protons and α particles with ambient nuclei; Doppler broadened lines due to collisions of heavier primary ions; the 2.223 neutron capture

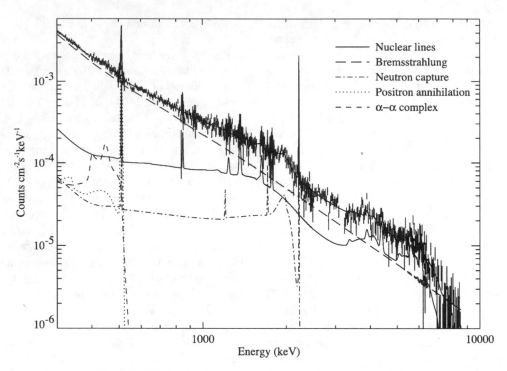

Figure 7. γ-ray spectrum obtained by RHESSI for the flare of 23 July 2002, plus the form of the major components used to fit the data [from *Lin et al.*, 2003].

line; the positron annihilation line at 0.511 keV; the lines at 0.429 and 0.478 MeV from nuclei of ^{7}Be and ^{7}Li, formed in fusion reactions of fast α's; and the bremsstrahlung continuum radiation of primary accelerated electrons, assumed here to have power-law form. These various features give us direct information on flare site fast ions in the 1-100 MeV energy range, as discussed further below. The only major component of the γ-ray spectrum not shown here is the continuum sometimes seen at high (>10 MeV) photon energies (Section 4.4).

Free neutrons are produced when ambient nuclei are bombarded by fast ions. If these neutrons thermalize before escaping from the Sun they may then capture on ambient protons, forming deuterium nuclei and emitting a very narrow line at 2.223 MeV in the process [most recently modelled by *Hua et al.*, 2002]. Neutrons escaping from the Sun may also be directly detected in space [*Chupp et al.*, 1987; *Ryan et al.*, 1993] or, at particularly high energies, with groud-based neutron monitors [e.g., *Debrunner et al.*, 1997; see also *Lopate*, this volume].

Positrons are another product of fast ion collisions, either via high-energy processes like $pp \rightarrow pn\pi^+$ (with a proton energy threshold of 270 MeV, and production of the positron via decay of the π^+), or via spallation or capture reactions that result in β-unstable radioactive nuclei. Thermalized positrons annihilate and give rise to the familiar line at 0.511 MeV. The width of the 0.511 MeV line, and the relative magnitudes of the line

itself and the three-photon continuum resulting from ortho-positronium formation, depend sensitively on the conditions in the region where positrons annihilate [see *Share et al.*, 2004]. Changes during a single flare in the relative intensities of the three-photon continuum and the two-photon annihilation line may reflect variations in the proportion of positrons produced via p - p collisions and pion decay, processes that have a proton energy threshold of ~300 MeV. The substantial change in line width reported during a single flare by *Share et al.* [2004] seems to indicate rapid changes in the location of positron annihilation, or rapid evolution of the structure of the flaring atmosphere although implications for ion energy distributions, if there are any, seem less clear at present.

4.1. Narrow Lines and Ion Energy Distributions

Narrow de-excitation lines, excited by both fast protons and α-particles, result from transitions between states of target nuclei (e.g., lines at 1.37 MeV from ^{24}Mg, 1.63 MeV from ^{20}Ne, 4.44 MeV from ^{12}C and 6.13 MeV from ^{16}O). They play an important role in deductions of fast ion energy distributions, but the situation is rather different from that of electrons and bremsstrahlung X-rays. Rather than a continuous photon spectrum, directly reflecting a continuous distribution of emitting particle energies, we have a set of weighted moments of the ion energy distributions.

Most observations of deexcitation lines prior to RHESSI yielded only an event-integrated line spectrum and for the moment we neglect temporal development. Denote by Φ_ε the fluence (event-integrated flux) of photons in a line at photon energy ε, from deexcitation of nuclei of species X. For such a fluence it is particularly appropriate to assume a thick target source. Then [e.g., *Ramaty*, 1986]

$$\Phi_\varepsilon = \Phi_{\varepsilon,p} + \Phi_{\varepsilon,\alpha} \qquad (12)$$

where $\Phi_{\varepsilon,p}$ and $\Phi_{\varepsilon,\alpha}$ are the contributions to the line at ε from excitation by fast protons and α-particles, respectively, and e.g.,

$$\Phi_{\varepsilon,p} = a_X \int_{E_{th,p}}^{\infty} \frac{\sigma_p(E)}{|dE_p / dN|} \Psi_p(E) dE \qquad (13)$$

where a_X is the abundance relative to hydrogen of species X, $\sigma_p(E)$ is the cross-section for production of the line at ε by protons of energy E, dE_p/dN is the rate of change of proton energy with hydrogen column depth, $E_{th,p}$ is the threshold energy for excitation of this particular line by protons, and $\psi_p(E)$ is the cumulative energy distribution of protons, i.e. the total number of protons injected at energies $\geq E$. $\Phi_{\varepsilon,\alpha}$ is given by the identical expression for α's. Contributions from spallation reactions should be added, in which nuclei of species X are produced via collisions with other species, but these are a complication of detail rather than principle.

Each line, then, gives us a weighted measure of the number of accelerated particles above its particular threshold energy, with proton and α contributions in general mixed together. Several lines together yield information on the energy distributions of protons and α's to the extent that their threshold energies, energy-dependences $\sigma(E)$ and sensitivity to protons and α's differ from one another. In fact the functions $\sigma(E)$ have a similar form and thresholds do not in general differ greatly, mostly lying in the 4-8 MeV range [*Ramaty et al.*, 1979; *Kozlovsky et al.*, 2002]. The 1.63 MeV line of ^{20}Ne, however, has a significantly lower energy threshold (2 MeV), for excitation by protons, than the other lines, so the ratio of $\Phi_{1.63}$ to the fluence in one or more of the other lines gives a measure of the steepness of the proton energy distribution, in the energy range between 2 and a few 10s of MeV [*Ramaty et al.*, 1996]. Its importance in this respect was pointed out by *Share and Murphy* [1995], after it had been found to be surprisingly strong in SMM-era data. The ratio $\Phi_{1.63}/\Phi_{6.13}$ involving the 6.13 MeV line of ^{16}O, for instance, has been particularly studied. Thresholds for excitation by α's vary much less, however, so the interpretation of such a ratio is significantly influenced by one's beliefs or assumptions about the relative numbers [*Ramaty et al.*, 1996] and energy distributions [*Toner and MacKinnon*, 2004] of protons and α's.

At present the key, observed quantity for discussing relative numbers of protons and α's is the fluence $\Phi_{0.4-0.5}$ in the 0.4-0.5 MeV energy range, from the two, broadened lines at 0.429 and 0.487 MeV [*Kozlovsky and Ramaty*, 1974]. Resulting from fusion reactions of fast α's with ambient ^{4}He, to give lines of ^{7}Li and ^{7}Be, these lines include no contribution from fast protons. With exciting α energy thresholds in the region of 10 MeV/nucleon, however, this feature carries no information on the bulk of the α's important in formation of the deexcitation lines. In the past it has been used, along with the assumption that protons and α's have the same energy distributions, to find large (0.5) values of α/p, the α to proton abundance ratio in the accelerated ion population [*Share and Murphy*, 1997]. An alternative interpretation of observed fluences is suggested by *Toner and MacKinnon* [2004], in which $\Phi_{0.4-0.5}$ fixes only the number of α's above 10 MeV/nucleon, and a range of proton and α spectral indices are consistent with the data. In all cases α's are found to have harder energy distributions than protons. On the one hand this finding might be consistent with MHD turbulent cascade acceleration mechanisms [*Miller and Roberts*, 1995]. On the other hand, proton and α distributions in the IPM do not generally differ greatly from one another [see e.g., the various examples in *Mewaldt et al.*, 2005] - although of course different processes may act on flare site and IPM populations.

In view of the role of the 1.63 MeV line, the ^{20}Ne abundance in the γ-ray source region is clearly a key quantity in deductions of fast ion energy distributions. There has always been some uncertainty over the appropriate value to use. Ideally, target elemental abundances are determined side by side with ion energy distributions in a process that fits all of the γ-ray spectrum [*Murphy et al.*, 1991]. An enhanced abundance of ^{20}Ne seems to be indicated, to an even greater degree than suggested by other solar and inner heliosphere indicators [*Meyer*, 1993]. In the absence of such a complete fit the assumed ^{20}Ne abundance assumes a crucial significance. *Share and Murphy* [1995] were able to demonstrate that an enhanced ^{20}Ne abundance could not on its own account for the observed strength of the 1.63 MeV line in 19 SMM flares, but it clearly needs to be taken into account in deducing fast ion energy distributions. Particularly worrying is the possibility that there is no single, correct value to use, and that the ^{20}Ne abundance in fact varies from one location to the next (as suggested by X-ray studies, e.g., *Schmelz et al.*, 1996).

Observations of stars other than the Sun have recently offered a new slant on this question. Chandra observations suggest that ^{20}Ne is commonly enhanced to as much as three times its photospheric abundance in active stars [*Drake et al.*, 2001]. Such a Ne abundance enhancement could resolve certain problems in helioseismology [*Drake and Testa*, 2005], offering a further indirect argument for its appropriateness to the Sun. A greater ^{20}Ne abundance would imply a harder fast proton (ion) distribution, with lower total energy. Repeating the approach of *Toner and MacKinnon* [2004] with a tripled ^{20}Ne to ^{16}O abundance ratio, we find that the parameter sets allowed by the observations would include cases with identical proton and α energy distributions, although harder and with less total energy than found previously.

Other α-only line features, at 0.339 MeV and around 1.02 MeV, have lower energy thresholds than the lines at 0.429 and 0.487 MeV. They have considerable potential for resolving the relative roles of protons and α's in forming the observed γ-ray spectrum, but remain poorly constrained by data [*Share and Murphy*, 1998].

The ions that dominate production of the 2.223 MeV neutron capture line tend to have higher energies than those responsible for the deexcitation lines [see *Lockwood et al.*, 1997]. The ratio of 2.223 MeV fluence to one or more deexcitation lines (or even all of the fluence in the 4-7 MeV range) thus gives another measure of ion energy power-law spectral index [e.g., *Ramaty*, 1986]. In principle the 2.223 MeV line might be used to resolve some of the ambiguity opened up by *Toner and MacKinnon* [2004], but in practice it also involves auxiliary assumptions that complicate its use. Assumptions about the ion primary pitch-angle distribution are particularly important.

Ultimately, the interest of these studies for flare physics is in constraining the importance of ion acceleration in the flare energy release process, and of ions in secondary energy transport in the flare. For example, *Murphy et al.* [1997] studied observations of the flare of 4 June 1991 made using the Oriented Scintillation Spectrometer Experiment (OSSE) on the Compton Gamma Ray Observatory. They deduced that there were $(6.7 \pm 1.2) \times 10^{32}$ protons present above 30 MeV, and that the total energy in all ions (protons, α's and heavier species - assuming a power-law form down to 1 MeV/ nucleon and a flat extension to zero energy) could be as great as 10^{33} ergs. In this instance, observed properties of ion acceleration seem to pose challenges as great as those routinely attending electron acceleration. The approach of *Toner and MacKinnon* [2004] would vary the distribution of ion energy between protons and heavier species, but total ion energy would remain large.

All of the above refers primarily to total (event-integrated) fluences. Few published studies have addressed temporal development of the γ-ray spectrum, or tried to derive light curves in individual lines. One exception is the OSSE/CGRO study by *Murphy et al.* [1997] of the flare of 4 June 1991. These authors found evidence for changes of *target* abundances during the flare, specifically for an increase in the ratio of low-FIP to high-FIP elements. These apparent changes in target abundances during a flare are certainly interesting in their own right, but represent a significant complication for attempts to study development of the fast ion distribution during a flare. The ratio $\Phi_{1.63}/\Phi_{6.13}$ should be less affected by this complication, however, since both ^{16}O and ^{20}Ne are high-FIP elements.

4.2. Shapes of Narrow Lines

Line widths reflect both the pitch-angle and energy distributions of the exciting particles. Everything else being equal, the heavier α particles produce a greater recoil of the emitting nuclei and hence broader lines than fast protons. Line shapes thus have a lot of potential for resolving the issues highlighted in Section 4.1, as well as for yielding unique information on ion pitch-angle distributions.

Line widths have been clearly resolved for the first time in the flare of 23 July 2002 with RHESSI [*Smith et al.*, 2003]. Redshifts fall off with the mass of the target species, as predicted, but are larger than expected for line production by vertically precipitating ions in a flare at this location (73° heliocentric angle). Likely explanations invoke magnetic field lines significantly tilted to the vertical, thus providing some insight into the magnetic geometry in which ions propagate. Directionality of high-energy (>10 MeV) continuum has previously been used similarly to probe average flare magnetic field geometry [*MacKinnon and Brown*, 1990].

INTEGRAL data for the 28 October 2003 flare resolve the 4.44 and 6.13 MeV lines of ^{12}C and ^{16}O respectively. *Kiener et al.* [2006] use detailed calculations of line shapes to interpret these observations. The line shapes place constraints additional to those provided by fluences on fast ion properties. There cannot be too many α-particles, for instance, relative to the number of protons, or the lines would be broader than observed, although such statements must depend also on the angular distribution of ions in the line production region, and their energy distribution. Thus one may delineate allowed regions of a parameter space for the fast ion distributions. *Kiener et al.*'s preferred values of fast α to proton ratio are <0.1, lower than found in other events, but they come with large error bars and also depend fairly strongly on assumed ion energy distribution. Nonetheless these results represent the fullest realisation so far of the diagnostic potential of line shapes combined with fluences.

4.3. Broad Lines

A deexcitation line from nuclear species X may result from collisions of fast, light ions (protons and α particles) with target nuclei of type X, or from collisions of fast nuclei of species X with ambient protons and α's. In the latter case the excited nuclei still have substantial velocities so the resulting lines are Doppler broadened, to the extent that they are not individually distinguishable. Broad lines formed in this way combine to form a single spectral feature roughly from 1-7 MeV. Although individual lines are not distinguished the shape of this feature does depend on the composition and energy distribution of heavier accelerated species. It has been successfully used to constrain heavy ion acceleration in at least some flares [e.g., *Ramaty et al.*, 1997; see also *Share and Murphy*, this volume].

4.4. Pion Decay Radiation

In the >10 MeV photon energy range, protons above 270 MeV energy can give rise to a very hard spectral component from direct decay of π^0 and from bremsstrahlung and annihilation in flight of π^+ decay positrons [*Murphy et al.*, 1987; first reported by *Forrest et al.*, 1985]. In principle the shape of the π^0 decay feature, a very broad line centred on 67 MeV, carries information on the ion energy distribution at the highest energies. An unambiguous determination of the highest photon energy produced would also tell us the highest ion energy produced. Typical detectors (usually BGO scintillators in this energy range) cover the 10-100 MeV photon energy range with just a handful of energy bins. A maximum around 70 MeV may be evident, but there are too few data points to comment on the width of this π^0 feature, particularly in the presence of the π^+ continuum plus any primary electron bremsstrahlung still present at these energies. Only on a couple of occasions has this high-energy continuum been observed by an instrument capable of detecting photons above 100 MeV [*Akimov et al.*, 1992; *Kanbach et al.*, 1993], whereas we know that the Sun can accelerate protons to maximum energies well into the GeV energy range, implying pion continuum extending well into the 100s of MeV photon energy range. Exploitation of the full diagnostic potential of the pion decay continuum thus remains beyond the capabilities of most existing instruments. A good example is given by the flare of 24 May 1990, well observed by the PHEBUS detector on GRANAT (sensitive up to 100 MeV). Detailed modelling of the high-energy continuum by *Vilmer et al.* [2003] could only delineate acceptable ranges of parameters, however. The analysis was limited by (1) insufficient resolution around the 67 MeV maximum of the π^0 feature, (2) ambiguity in model fitting between the π^+ continuum and primary electron bremsstrahlung, (3) inadequate energy coverage. We hope that future detectors will be able to detect this high-energy continuum, with finer energy resolution, on something like a routine basis, and look forward to new data on this feature, extending well into the 100s of MeV photon energy range, from the upcoming GLAST mission [http://glast.gsfc.nasa.gov; and *Ritz et al.*, 2004].

4.5. Low Energy Protons (<1 MeV)

As we have seen, the strong lines, all produced by ions of at least a few MeV/nucleon, can give us some information on the form of the ion energy distribution above a few MeV. We are left with a gulf of ignorance between these energies and the thermal (~keV) energies ions undoubtedly start with. *MacKinnon* [1989] suggested that radiative capture lines (e.g., at 2.37 MeV from $^{12}C(p, \gamma)^{13}N$) might be used to constrain the proton distribution at lower energies. The cross-sections for these reactions have resonances for incident proton energies in the 100s of keV range which dominate any fluence in these lines. The cross-sections are very small and upper limits to the line fluences of better than 10^{-5} photons.cm^{-2} would have to be attained. With COMPTEL, whose mode of operation ensured a low background, an upper limit in the region of 10^{-4} cm^{-2} was possible (R Suleiman, private communication). The 2.37 MeV line is probably too close to the 2.223 MeV neutron capture line to yield a useful constraint during flares, but might usefully constrain fast ion acceleration in cases where ions do not attain MeV energies. The next strongest of these lines is at 8.07 MeV, where it might benefit from the absence of other strong lines nearby. Only upper limits, still an order of magnitude or more short of the useful level, have been placed on their fluences [in the quiet Sun, using COMPTEL, by *McConnell et al.*, 1997; in flares using SMM/GRS by *Share et al.*, 2001].

Other, less direct diagnostics for lower energy ions have been suggested. Fast protons may gain an electron in charge exchange reactions, producing Doppler shifted emission in e.g., the Lyman α line at 121.5 nm, while they are still moving [*Canfield and Chang*, 1985]. This has possibly been detected in a stellar flare by *Woodgate et al.* [1992]. An anisotropic, bombarding proton distribution will induce line impact polarization in the Hα line [*Henoux et al.*, 1990] and there are possible observations [*Emslie et al.*, 2000]. A recent study [*Bianda et al.*, 2005] finds no evidence for any Hα linear polarization, however.

We have to conclude that ions in the 100s of keV energy range are still the 'Loch Ness Monster' of flare physics [*Dennis*, 1988], possibly of great importance but very poorly constrained by observations.

4.6. Thick Targets, Transport, etc.

As was the case for electrons, assumptions about ion transport underlying the above findings are always up for discussion. There seems to be a consensus that γ-ray emitting ions stop in the deep layers of the atmosphere, automatically a thick target. It is difficult to reconcile the column depths needed to stop line-emitting ions with observed flare timescales of a few s, unless ions stop in a dense region. Some observations exist, however, indicating a substantial fraction of line emission coming from the corona. Specifically, *Barat et al.* [1994] observed intense γ-ray line emission from the flare of 1 June 1991 with the PHEBUS instrument on GRANAT. The flare produced one of the largest fluences observed up until then, even although it was over the limb from the spacecraft's vantage point. Evidently the coronal, unocculted region of the flare was the seat of substantial deexcitation line production. The PHEBUS spectrometer did not have sufficient energy resolution to yield fluxes in individal narrow lines, but it was possible to separate the electron bremsstrahlung continuum and nuclear line contributions to the spectrum in several energy channels [*Ramaty et al.*, 1997]. From the broad shape of the nuclear line component, *Ramaty et al.* were able to show that the accelerated ions must have been emitting in a thin target region, i.e. a region in which their residence times were << their collisional slowing-down times.

Even partial containment (by simple magnetic convergence, by Alfvén turbulence or by some other mechanism) of a significant number of ions in the high-temperature corona would force a revision of line ratio diagnostics, because the collisional slowing-down rate is reduced when ion velocities are not >> ambient, thermal electron velocities [*MacKinnon and Toner*, 2003]. Were this to occur, we would have to deduce harder ion energy distributions with less total energy.

Hurford et al. [2003] obtained the first ever images in the 2.223 MeV line, of the 23 July 2002 flare. RHESSI's spatial resolution is only 35" at this photon energy but this still allows useful conclusions to be drawn. At this resolution the γ-ray source is an unresolved point source, arguing against any picture in which gamma-rays result from an extended region. As was the case for HXR's, such a finding is consistent with a cold, thick target interpretation of observed spectra. Remarkably, the γ-ray and X-ray source centroids are clearly spatially separated by 20". This finding argues for pictures of flare particle acceleration in which fast ions and electrons are produced on distinct field lines [see also the papers by *Share and Murphy* and by *Aschwanden*, this volume].

5. CONCLUSIONS

X-rays, γ-rays and radio observations give the only direct information on flare site fast electrons and ions. Ideally we would obtain spectra in these ranges, deduce the total number and energy distribution and then try to devise particle acceleration theories able to account for these. Further tests of these theories would result from information on the spatial distribution and temporal evolution of fast particles, similarly revealed by these radiations. As we have seen, however, things might not be so simple. The number of particles we believe are present depends on how long we think each one survives to radiate. Our estimates of fast particle numbers may be in error if (1) significant numbers of them are contained in the acceleration region, (2) they collide mostly with other fast particles, or (3) they move in a region where ambient electron

energies are not completely negligible in comparison to their own. Such possibilities might not attract our continuing attention were it not for the difficulty of accelerating 10^{35} electrons per second, as implied by the collisional, cold thick target interpretation. While we stated at the outset that we would leave the problems of acceleration to others, we cannot completely ignore such studies.

Where should we look for further progress on these questions, for reassurance that our methods of deduction of fast particle distributions are sound, or pointers on how they must be modified? Imaging observations must play a critical role. As we have already seen, observations of HXR footpoints appear to support a traditional, cold thick target approach to electron numbers. These footpoints, sites of HXR production no matter where the electrons have first been energized, need continued attention. Their areas are important: if they are too small, stability of the electron beam itself, or its associated return current is thrown into doubt. Electrons may collide frequently with other particles of similar energies. RHESSI should shed light on this question but UV observations offer another slant. TRACE, for example, routinely offers 1" spatial resolution and can provide image cadence as high as 1s. UV brightenings may function as proxies for the sites of electron precipitation, as discussed by *Fletcher and Warren* [2003]. Beam cross-sectional areas indicated by TRACE observations suggest that electron beams may not propagate stably, in which case ideas of electron numbers and energy distribution would have to be revised.

Additionally, imaging observations may clarify the feasibility of the magnetically connected region of the corona supplying enough electrons for HXR fluxes observed from footpoints. If we need more fast electrons than can be supplied by the available part of the atmosphere, clearly something is wrong with our interpretation of HXR emission. Of course it may be that the same electrons are recycled many times to the accelerator, e.g., in a beam-return current system, so that they need not all be present at once in the corona, but such a picture seems to demand that agents beyond just binary collisions (e.g., the electric field that maintains the return current) act on X-ray emitting electrons.

In the SMM era, *MacKinnon et al.* [1985] raised what were then seen as difficulties for the traditional thick target picture. From HXIS data they found that the HXR flux from footpoint pixels, in the 16-30 keV photon energy range, could be as low as 20% of the photon flux from the whole of the instrument's field of view. This finding was difficult to reconcile with the idea that accelerated electrons, precipitating to the footpoints, were responsible for most of the X-ray emission at these energies. At the low HXIS energies a significant fraction of the X-rays could be thermal. Such quantitative studies need to be repeated with RHESSI data. In Figure 3 it appears that footpoints only become distinct around 35 keV, and that most of the emission below this energy comes from the coronal, presumably thermal source. Does this mean that similar problems may arise from RHESSI? Or that electrons below about 35 keV are close enough to thermal speeds to place them in the regime studied by *Galloway et al.* [2005]? More quantitative studies are needed.

Arguments of this kind are used by *Sui et al.* [2005]. They combined temporal and imaging information in an attempt to identify the most likely value of electron minimum energy to use, mostly finding values in the region of 20 keV. They still found a fast electron total energy comparable to the peak energy content of the thermal plasma and certainly no conflict with the collisional thick target picture, beyond the general challenges of efficiency such findings pose to particle acceleration theories.

Observational indications of acceleration processes taking place in spatial coincidence with HXR footpoints would be particularly interesting, as would clear evidence that such processes do not take place there. Attempts have been made in the past to diagnose the presence of electric fields via observations of Stark effect polarization in lines of the hydrogen Paschen sequence [*Foukal and Behr*, 1995; *Foukal*, 1998]. Further such efforts, concentrating on the locations of HXR footpoints, would be very interesting whether they yield positive or negative results. Investigations of UV line shapes may sometimes diagnose the presence of turbulence [e.g., *McClements et al.*, 1991; *Erdelyi et al.*, 1998] or non-Maxwellian distributions [*Cranmer*, 1998] and the applicability of such techniques to the flaring chromosphere could profitably be investigated.

The Frequency Agile Solar Radiotelescope (FASR) will produce high-resolution, high-dynamic-range images across a wide frequency range, 0.1 to 24 GHz. It will certainly provide a further, crucial window on fast electrons in flares [e.g., *Bastian*, 2003]. In particular we may look to it to advance understanding of the relative roles of trapping and aceleration in producing the apparent hardening of the electron distribution at high energies (Section 3.1).

Turning particularly to fast ions, any observation that could pin down the most appropriate ^{20}Ne abundance will be valuable, since the 1.63 MeV line plays such a key role in representing the lowest proton energies revealed by deexcitation lines. Observations of the lines produced in α-Fe reactions could help to determine the the overall role of α's in forming the observed γ-ray spectrum.

ESA's Solar Orbiter mission, planned to reach within 0.3 AU of the Sun, will include a neutron detector, as may also NASA's proposed Solar Sentinels. Low-energy (<10 MeV) neutrons almost all decay before reaching spacecraft at 1 AU but will be detected in large numbers in the inner heliosphere, where these missions are expected to spend much of their time. Such observations will provide another constraint

on e.g. the relative numbers of protons and (comparatively neutron-prolific) α particles. Instruments optimized for detecting low-energy neutrons and operating in the inner heliosphere are under development [*Moser et al.*, 2004; *Bravar et al.*, 2005].

Observations of flare light curves in individual deexcitation lines would give a valuable insight to the evolution of flare ions, possibly testing further the thick target assumptions that go into the interpretation of flare-integrated spectra. The reasons for the apparent evolution of target abundances during flares need to be clarified, however. Are abundances actually changing in the regions where ions interact, are the locations of the interaction regions changing (just as the locations of HXR footpoints change), or is a changing ion energy distribution sampling a different range of heights in an atmosphere with height-dependent abundances?

The lowest energy flare fast particles dominate the energy content and have the greatest importance for understanding the flare phenomenon in its entirety. The highest energy particles produced in flares pose a different sort of challenge. We may hope for further insights into their energies, and indeed the frequency of their production, possibly from more detailed interpretation of sub-mm observations. The GLAST mission also should improve our knowledge of these highest energy ions, hopefully fulfilling some of the diagnostic potential of the pion decay spectral feature and further extending this high-energy frontier of solar physics.

At the outset of this contribution we noted the difficulty of learning about flare site particle acceleration from IPM particles. Nonetheless both IPM and flare site populations have useful things to say about one another. These questions are reviewed at length elsewhere [e.g., *von Rosenvinge*, this volume; *Reames*, 1999; *Ryan et al.*, 2000; *Kallenrode*, 2003] and we make only the briefest of comments. Impulsive particle events are believed to represent primary accelerated flare particles, while gradual events result from shock acceleration of a seed population, possibly energized in the events of the flare but certainly not an 'unprocessed' sample of flare site fast particles. Questions remain about the relationship between the populations at the flare and at 1 AU, of course. For instance the paucity of fast electrons in space compared to the large numbers deduced at the flare [*Ramaty et al.*, 1993] seems to require an explanation. *Ramaty et al.* [1993] also found ion distributions in space to be consistently harder than those deduced from γ-rays at the flare site. Confirming and understanding such findings, as well as clarifying the role of flare primary particles in the broader heliosphere, will benefit from refined deductions of flare site particle distributions.

Kontar et al. [2004] find a correlation between energetic electron energy distributions at the flare (from HXR's) and in space, but not the one that would be expected on the basis of the cold, thick target model for HXR production. What this result means is not yet clear, but it does at least make the point that, initial comments notwithstanding, IPM particles may indeed have something useful to say about flare site processes.

Acknowledgments. I am grateful to many colleagues in many places but particularly L. Fletcher, H. Hudson, K.-L. Klein, E. Kontar and M. Toner, for discussions on topics dealt with here. I thank R. Mewaldt for a helpful email exchange. The comments and suggestions of the referees were very helpful in many respects. Solar physics research in Glasgow is supported by a PPARC Rolling Grant. I thank the organisers for inviting me to this most interesting meeting, and for supporting my attendance.

REFERENCES

Akimov, V.V., V.G. Afanasev, A.S. Belousov *et al.*, High-energy gamma-rays recorded by the GAMMA-1 telescope from the 1991 Mar 26 and 1991 Jun 15 solar flares, *Sov. Astron. Letts.*, 18, 69-71, 1992.

Arzner, K., Visibility-based demodulation of RHESSI light curves, *Adv. Sp. Res.*, 34, 456-461, 2004.

Arzner, K., and L. Vlahos, Particle acceleration in multiple dissipation regions, *Astrophys. J.*, 605, L69-L72, 2004.

Bai, T., and R. Ramaty, Backscatter, anisotropy, and polarization of solar hard X-rays, *Astrophys. J.*, 219, 705-726, 1978.

Bai, T., and R. Ramaty, Hard X-ray time profiles and acceleration processes in large solar flares, *Astrophys. J.*, 227, 1072-1081, 1979.

Barat, C., G. Trottet, N. Vilmer, J.-P. Dezalay, R. Talon, R. Sunyaev, O. Terekhov, and A. Kuznetsov, Evidence for intense coronal prompt gamma-ray line emission from a solar flare, *Astrophys. J.*, 425, L109-L112, 1994.

Bastian, T.S., The Frequency Agile Solar Radiotelescope, *Adv. Sp. Res.*, 32, 2705-2714, 2003.

Bastian, T.S., A.O. Benz, and D.E. Gary, Radio emission from solar flares, *Ann. Rev. Ast. Ap.*, 36, 131-188, 1998.

Benz, A.O., Spectral features in solar hard X-ray and radio events and particle acceleration, *Astrophys. J.*, 211, 270-280, 1977.

Benz, A.O., Radio diagnostics of flare energy release, *LNP*, 612, 80-95, 2003.

Benz, A.O., P.C. Grigis, A. Csillaghy, Saint-Hilaire, Survey on Solar X-ray Flares and Associated Coherent Radio Emissions, *Solar Phys.* 226, 121-142, 2005.

Bianda, M., A.O. Benz, J.O. Stenflo, G. Küveler, and R. Ramelli, Absence of linear polarization in Hα emission of solar flares, *A&A*, 434, 1183-1189, 2005.

Böhme, A., F. Fuerstenberg, J. Hildebrandt, O. Saal, A. Krueger, P. Hoyng, and G.A. Stevens, A two-component model of impulsive microwave burst emission consistent with soft and hard X-rays, *Solar Phys.* 53, 139-151, 1977.

Bravar, U., P.J. Bruillard, E.O. Flueckiger, A.L. MacKinnon, J.R. Macri, P.C. Mallik, M.L. McConnell, M.R. Moser, and J.M. Ryan, Imaging solar neutrons below 10 MeV in the inner heliosphere, in *Optics for EUV, X-Ray, and Gamma-Ray Astronomy II, Proceedings of the SPIE, Vol. 5901*, edited by O. Citterio and S.L. O'Dell, p141-150, 2005.

Brown, J.C., The deduction of energy spectra of non-thermal electrons in flares from the observed dynamic spectra of hard X-ray Bbursts, *Solar Phys.* 18, 489, 1971.

Brown, J.C., Thick target X-ray bremsstrahlung from partially ionised targets in solar flares, *Solar Phys.* 28, 151-158, 1973.

Brown, J.C., On the thermal interpretation of hard X-ray bursts from solar flares, in *Coronal Disturbances: Proceedings from IAU Symposium No. 57*, edited by G.A. Newkirk, Reidel, Dordrecht, 1974.

Brown, J.C., A.G. Emslie, and E.P. Kontar, The determination and use of mean electron flux spectra in solar flares, *Astrophys. J.*, 595, L115-L117, 2003.

Brown, J.C., D.B. Melrose, and D.S. Spicer, Production of a collisionless conduction front by rapid coronal heating and its role in solar hard X-ray bursts, *Astrophys. J.*, 228 592-597, 1979.

Brown, J.C., G.K. McArthur, R.K. Barrett, S.W. McIntosh, and A.G. Emslie, Inversion of Thick Target Bremsstrahlung Spectra from Nonuniformly Ionised Plasmas, *Solar Phys.* 179, 379-404, 1998.

Canfield, R.C., and C.-R. Chang, Ly-alpha and H-alpha emission by superthermal proton beams, *Astrophys. J.*, 295, 275-284, 1985.

Chubb, T.A., R.W. Kreplin, and H. Friedmann, Observations of hard X-ray emission from solar flares, *J. Geophys. Res.*, 71, 3611, 1996.

Chupp, E.L., High-energy neutral radiations from the Sun, *Ann. Rev. Ast. Ap.* 22, 359-387, 1984.

Chupp, E.L., H. Debrunner, E. Flueckiger, D.J. Forrest, F. Golliez, G. Kanbach, W.T. Vestrand, J. Cooper, and G. Share, Solar neutron emissivity during the large flare on 1982 June 3, *Astrophys. J.*, 318, 913-925, 1987.

Conway, A.J., A.L. MacKinnon, J.C. Brown, and G. McArthur, Analytic description of collisionally evolving fast electrons, and solar loop-top hard X-ray sources, *A&A*, 331, 1103-1107, 1998.

Courant, R., and D. Hilbert, *Methods of Mathematical Physics*, Interscience Publishers, New York, 1953.

Craig, I.J.D., and J.C. Brown, *Inverse Problems in Astronomy*, Adam Hilger Ltd., Bristol and Boston, 1986.

Cranmer, S., Non-Maxwellian redistribution in solar coronal Ly-alpha emission, *Astrophys. J.*, 508, 925-939, 1998.

Crosby, N.B., M.J. Aschwanden, and B.R. Dennis, Frequency distributions and correlations of solar X-ray flare parameters, *Solar Phys.*, 143, 275-299, 1993.

Debrunner, H., J.A. Lockwood, C. Barat, R. Buetikofer, J.P. Dezalay, E. Flueckiger, A. Kuznetsov, J.M. Ryan, R. Sunyaev, O.V. Terekhov, G. Trottet, and N. Vilmer, Energetic neutrons, protons, and gamma rays during the 1990 May 24 solar cosmic-ray event, *Astrophys. J.*, 479, 997, 1997.

Dennis, B.R., Solar flare hard X-ray observations, *Solar Phys.*, 118, 49-94, 1988.

Dennis, B.R., and R.A. Schwartz, Solar flares - The impulsive phase, *Solar Phys.*, 121, 75-94, 1989.

Drake, J.J., and P. Testa, The 'solar model' problem solved by the abundance of neon in nearby stars, *Nature*, 436, 525-528, 2005.

Drake, J.J., N.S. Brickhouse, V. Kashyap, M.J. Laming, D.P. Huenemoerder, R. Smith, and B.J. Wargelin, Enhanced noble gases in the coronae of active stars, *Astrophys. J.*, 548, L81-L85, 2001.

Dulk, G.A., Radio emission from the sun and stars, *Ann. Rev. Ast. Ap.* 23, 169-224, 1985.

Emslie, A.G., On the importance of reverse current ohmic losses in electron-heated solar flare atmospheres, *Astrophys. J.*, 249, 817-820, 1981.

Emslie, A.G., The determination of the total injected power in solar flare electrons, *Astrophys. J.*, 595, L119-L121, 2003.

Emslie, A.G., and D.F. Smith, Microwave signature of thick-target electron beams in solar flares, *Astrophys. J.*, 279, 882-895, 1984.

Emslie, A.G., E.P. Kontar, S. Krucker, and R.P. Lin, RHESSI hard X-ray imaging spectroscopy of the large solar flare of 2002 July 23, *Astrophys. J.*, 595, L107-L110, 2003.

Emslie, A.G., J.A. Miller, E. Vogt, J.-C. Hénoux, and S. Sahal-Bréchot, Hα polarization during a well-observed solar flare: proton energetics and implications for particle acceleration processes, *Astrophys. J.*, 542, 513-520, 2000.

Erdelyi, R., J.G. Doyle, M.E. Perez, and K. Wilhelm, Center-to-limb line width measurements of solar chromospheric, transition region and coronal lines, *A&A*, 337, 287-293, 1998.

Fletcher, L., On the generation of loop-top impulsive hard X-ray sources, *A&A*, 303, L9-L12, 1995.

Fletcher, L., and P.C.H. Martens, A model for hard X-ray emission from the top of flaring loops, *Astrophys. J.*, 505, 418-431, 1998.

Fletcher, L., and H.P. Warren, The energy release process in solar flares; Constraints from TRACE observations, *LNP*, 612, 58-79, 2003.

Forrest, D.J., and W.T. Vestrand, E.L. Chupp, E. Rieger, J.F. Cooper, G.H. Share, Neutral pion production in solar flares, *ICRC 19*, Vol. 4, 146-149, 1985.

Foukal, P., Plasma electric field measurements as a diagnostic of neutral sheets in prominences, *ASP Conf Ser. 150: IAU COlloq. 167: New Perspectives on Solar Prominences*, p119, 1998.

Foukal, P., and B.B. Behr, Testing MHD models of prominences and flares with observations of solar plasma electric fields, *Solar Phys.*, 156, 293-314, 1995.

Galloway, R.K., A.L. MacKinnon, E.P. Kontar, and P. Helander, Fast electron slowing-down and diffusion in a high-temperature coronal X-ray source, *A&A*, 438, 1107-1114, 2005.

Ginzburg, V.I., and S.I. Syrovatskii, Cosmic magnetobremsstrahlung (synchrotron radiation), *Ann. Rev. Ast. Ap.*, 3, 297-350, 1965.

Grigis, P.C., and A.O. Benz, The evolution of reconnection along an arcade of magnetic loops, *Astrophys. J.*, 625, L143-L146, 2005a.

Grigis, P., and A.O. Benz, The spectral evolution of impulsive solar X-ray flares II. Comparison of observations with models, *A&A*, 434, 1173-1181, 2005b.

Haug, E, On the use of nonrelativistic bremsstrahlung cross sections in astrophysics, *A&A*, 326, 417-418, 1997.

Haug, E., Photon spectra of electron-electron bremsstrahlung, *Solar Phys.* 178, 341-351, 1998.

Haydock, E.L., J.C. Brown, A.J. Conway, and A.G. Emslie, The effect of wave generation on HXR bremsstrahlung spectra from flare thick-target beams, *Solar Phys.*, 203, 355-369, 2001.

Henoux, J.C., G. Chambe, D. Smith, D. Tamres, N. Feautrier, M. Rovira, and S. Sahal-Brechot, Impact line linear polarization as a diagnostic of 100 keV proton acceleration in solar flares, *Astrophys. J. (Supp.)*, 73, 303-311, 1990.

Holman, G.D., M.R. Kundu, and K. Papadopoulos, Electron pitch angle scattering and the impulsive phase microwave and hard X-ray emission from solar flares, *Astrophys. J.*, 257, 354-360, 1982.

Holman, G.D., L. Sui, R.A. Schwartz, and A.G. Emslie, Electron bremsstrahlung hard X-Ray spectra, electron distributions, and energetics in the 2002 July 23 solar flare, *Astrophys. J.*, 595, L97-L101, 2003.

Hoyng, P., J.C. Brown, and H.F. van Beek, High time resolution analysis of solar hard X-ray flares observed on board the ESRO TD-1A satellite, *Solar Phys.*, 48, 197-254, 1976.

Hoyng, P., A. Duijveman, M.E. Machado, D.M. Rust, Z. Svestka, A. Boelee, C. de Jager, K.T. Frost, H. Lafleur, G.M. Simnett, H.F. van Beek, and B.E. Woodgate, Origin and location of the hard X-ray emission in a two-ribbon flare, *Astrophys. J.*, 246 L155, 1981.

Hua, X.-M., B. Kozlovsky, R.E. Lingenfelter, R. Ramaty, and A. Stupp, Angular and energy-dependent neutron emission from solar flare magnetic loops, *Astrophys. J. (Supp.)*, 140, 563-579, 2002.

Hudson, H.S., R.P. Lin, and D.M. Smith, Gamma-ray flare occurrence patterns, *AGU Fall Meeting Abstracts*, A168, 2003.

Hudson, H.S., T. Kosugi, N.V. Nitta, and M. Shimojo, Hard X-radiation from a fast coronal ejection, *Astrophys. J.*, 561, L211-L214, 2001.

Hurford, G.J., E.J. Schmahl, and R.A. Schwartz et al., The RHESSI imaging concept, *Solar Phys.*, 210, 61-82, 2002.

Hurford, G.J., R.A. Schwartz, Krucker, R.P. Lin, D.M. Smith, and N. Vilmer, First gamma-ray images of a solar flare, *Astrophys. J.*, 595, L77-L80, 2003.

Johns, C.M., and R.P. Lin, The derivation of parent electron spectra from bremsstrahlung hard X-ray spectra, *Solar Phys.*, 137, 121-140, 1992.

Kanbach, G., D.L. Bertsch, and C.E. Fichtel et al., Detection of a long-duration solar gamma-ray flare on June 11, 1991 with EGRET on COMPTON-GRO, *A&AS*, 97, 349-353, 1993.

Kane, S.R., and K.A. Anderson, Spectral characteristics of impulsive solar flare X-rays $\geq$ 10 keV, *Astrophys. J.*, 162, 1003-1018, 1970.

Kasparova, J., M. Karlicky, R.A. Schwartz, and B.R. Dennis, X-ray and Hα emission of the 20 Aug 2002 flare, in *Solar Magnetic Phenomena, Proceedings of the 3rd Summerschool and Workshop held at the Solar Observatory Kanzelhöhe*, edited by A. Hanslmeier, A. Veronig, and M. Messerotti, Springer, Dordrecht, 2005.

Kaufmann, P., J.-P. Raulin, and E. Correia et al., Rapid submillimeter brightenings associated with a large solar flare, *Astrophys. J.*, 548, L95-L98, 2001.

Kaufmann, P., J.-P. Raulin, C.G.G. de Castro et al., A new solar burst spectral component emitting only in the Terahertz range, *Astrophys. J.*, 603, L121-L124, 2004.

Kiener, J., M. Gros, V. Tatischeff, and G. Weidenspointner, Properties of the energetic particle distributions during the October 28, 2003 solar flare from INTEGRAL/SPI observations, *A&A*, 445, 725-733, 2006.

Klein, K.-L., G. Trottet, and A. Magun, Microwave diagnostics of energetic electrons in flares, *Solar Phys.*, 104, 243-252, 1986.

Knight, J.W., and P.A. Sturrock, Reverse current in solar flares, *Astrophys. J.*, 218, 306-310, 1977.

Koch, H.W., and J.W. Motz, Bremsstrahlung cross-section formulas and related data, *Rev. Mod. Phys.*, 31, 920-955, 1959.

Kontar, E.P., S. Krucker, and R.P. Lin, Spectra of solar energetic electrons in flares and near Earth, *AGU Fall Meeting Abstracts*, A1130, 2004.

Kontar, E.P., and A.L. MacKinnon, Regularized energy-dependent solar flare hard X-ray spectral index, *Solar Phys.*, 227, 299-310, 2005.

Kontar, E.P., J.C. Brown, A.G. Emslie, R.A. Schwartz, D.M. Smith, and R.C. Alexander, An Explanation for Non-Power-Law Behavior in the Hard X-Ray Spectrum of the 2002 July 23 Solar Flare, *Astrophys. J.*, 595, L123-L126, 2003.

Kontar, E.P., A.G. Emslie, M. Piana, A.M. Massone, and J.C. Brown, Determination of Electron Flux Spectra in a Solar Flare with an Augmented Regularization Method: Application to RHESSI Data, *Solar Phys.* 226, 317-325, 2005a.

Kontar, E.P., A.L. MacKinnon, R.A. Schwartz, and J.C. Brown, Angle dependent Greens function correction for solar flare Compton backscattered X-rays *A&A*, in press, 2005b.

Korchak, A.A., Possible mechanisms for generating hard X-rays in solar flares, *Sov. Astron. - AJ*, 11, 258-263, 1967.

Kozlovsky, B., and R. Ramaty, 478-keV and 431-keV line emissions from alpha-alpha reactions, *Astrophys. J.*, 191, L43-L44, 1974.

Kozlovsky, B., R.J. Murphy, and R. Ramaty, Nuclear Deexcitation Gamma-Ray Lines from Accelerated Particle Interactions, *Astrophys. J. (Supp.)*, 141, 523-541, 2002.

Lee, J., and D.E. Gary, Solar microwave bursts and injection pitch-angle distribution of flare electrons, *Astrophys. J.*, 543, 457-471, 2000.

Lin, R.P., and H.S. Hudson, Non-thermal processes in large solar flares, *Solar Phys* 50, 153-178, 1976.

Lin, R.P., and B.R. Dennis, G.J. Hurford et al., The Reuven Ramaty High-Energy Solar Spectroscopic Imager (RHESSI), *Solar Phys.*, 210, 3-32, 2002.

Lin, R.P., S. Krucker, and G.J. Hurford et al., RHESSI observations of particle acceleration and energy release in an intense solar gamma-ray line flare, 595, L69-L76, 2003.

Lockwood, J., H. Debrunner, and J.M. Ryan, The relationship between solar flare gamma-ray emission and neutron production *Solar Phys.*,173, 151-176, 1997.

Lu, E.T., and R.J. Hamilton, Avalanches and the distribution of solar flares, *Astrophys. J.*, 380, L89-L92, 1991.

Lüthi, T., Lüdi, A. Magun, Determination of the location and effective angular size of solar flares with a 210 GHz multibeam radiometer, *A&A*, 420, 361-370, 2004a.

Lüthi, T., A. Magun, and M. Miller, First observation of a solar X-class flare in the sub-millimeter range with KOSMA, *A&A*, 415, 1123-1132, 2004b.

Machado, M.E., C.V. Sneibrun, and M.G. Rovira, Hard X-ray imaging evidence of non-thermal and thermal burst components, *Solar Phys.*, 99, 189-217, 1985.

MacKinnon, A.L., A potential diagnostic for low-energy, nonthermal protons in solar flares, *A&A*, 226, 284-287, 1989.

MacKinnon, A.L., and J.C. Brown, On the bremsstrahlung efficiency of nonthermal hard X-ray source models, *Solar Phys.* 122, 303-311, 1989.

MacKinnon, A.L., and J.C. Brown, Implications of the solar flare gamma-ray limb-brightening observations for particle acceleration and the flare magnetic environment II. Numerical results for a class of loop models, *A&A*, 232, 544-555, 1990.

MacKinnon, A.L., J.C. Brown, and J. Hayward, Quantitative analysis of hard X-ray 'footpoint' flares observed by the Solar Maximum Mission, *Solar Phys.* 99, 231-262, 1985.

MacKinnon, A.L., and M.P. Toner, Warm thick target solar gamma-ray source revisited, *A&A*, 409, 745-753, 2003.

Magdziarz, P., and A.A. Zdziarski, Angle-dependent Compton reflection of X-rays and gamma-rays, *MNRAS*, 837-848, 1995.

Marschhäuser, Rieger, E., G. Kanbach, Temporal evolution of bremsstrahlung-dominated gamma-ray spectra of solar flares, in *High-Energy Solar Phenomena - a New Era of Spacecraft Measurements. Proceedings of the Workshop Held in Waterville Valley, New Hampshire, March 1993* edited by J.M. Ryan and W.T. Vestrand, *AIP Conference Proceedings*, 294, 171, 1994.

Massone, A.M., M. Piana, A.J. Conway, and B. Eves, A regularization approach for the analysis of RHESSI X-ray spectra, *A&A*, 405, 325-330, 2003.

Massone, A.M., A.G. Emslie, E.P. Kontar, M. Piana, M. Prato, and J.C. Brown, Anisotropic bremsstrahlung emission and the form of regularized electron flux spectra in solar flares, *Astrophys. J.*, 613, 1233-1240, 2004.

Masuda, S., T. Kosugi, H. Hara, S. Tsuneta, and Y. Ogawara, A loop-top hard X-ray source in a compact solar flare as evidence for magnetic reconnection, *Nature*, 371, 495, 1994.

McClements, K.G., The quasi-linear relaxation and bremsstrahlung of thick target electron beams in solar flares, *A&A*, 175, 255-262, 1987.

McClements, K.G., and J.C. Brown, The inverse Compton interpretation of fast-time structures in solar microwave and hard X-ray bursts, *A&A*, 165, 235-243, 1986.

McClements, K.G., R.A. Harrison, and D. Alexander, The detection of wave activity in the solar corona using UV line spectra, *Solar Phys.*, 131, 41-48, 1991.

McConnell, M.L., K.W. Bennett, A.L. MacKinnon, R. Miller, G. Rank, J.M. Ryan, and V. Schönfelder, A Search for MeV Gamma-Ray Emission from the Quiet-Time Sun, *Proc. 25th ICRC, Durban (South Africa)*, 1, 13, 1997.

Melnikov, V.F., and A. Magun, Spectral flattening during solar radio bursts at cm-mm wavelengths and the dynamics of energetic electrons in a flare loop, *Solar Phys.*, 178, 153-171, 1998.

Mewaldt, R.A., C.M.S. Cohen, and A.W. Labrador et al., Proton, helium and electron spectra during the large solar particle events of October-November 2003,, A09S18, doi:1029/2005JA011038, 2005.

Meyer, J.-P., Elemental abundances in active regions, flares and interplanetary medium, *Adv. Sp. Res.*, 13, 377-390, 1993.

Miller, J.A., and D.A. Roberts, Stochastic proton acceleration by cascading Alfvén waves in impulsive solar flares, *Astrophys. J.*, 452, 912-932, 1995.

Miller, J.A., P.J. Cargill, A.G. Emslie, G.D. Holman, B.R. Dennis, T.N. LaRosa, R.M. Winglee, S.G. Benka, and S. Tsuneta, Critical issues for understanding particle acceleration in impulsive solar flares, *J. Geophys. Res.*, 102, 14631-14660, 1997.

Moser, M.R., E.O. Flückiger, J.M. Ryan, M.L. McConnell, and J.R. Macri, A Fast Neutron Imaging Telescope for Inner Heliosphere Missions, *35th COSPAR Scientific Assembly*, 2677, 2004.

Murphy, R.J., C.D. Dermer, and R. Ramaty, High-energy processes in solar flares, *Astrophys. J. (Supp.)*, 63, 721-748, 1987.

Murphy, R.J., R. Ramaty, D.V. Reames, and B. Kozlovsky, Solar abundances from gamma-ray spectroscopy - Comparisons with energetic particle, photospheric, and coronal abundances, *Astrophys. J.*, 371, 793-803, 1991.

Murphy, R.J., G.H. Share, J.E. Grove, W.N. Johnson, R.L. Kinzer, J.D. Kurfess, M.S. Strickman, and G.V. Jung, Accelerated particle composition and energetics and ambient abundances from gamma-ray spectroscopy of the 1991 June 4 solar flare, *Astrophys. J.*, 490, 883-900, 1997.

Nita, G.M., D.E. Gary, and J. Lee, Statistical study of two years of solar flare radio spectra obtained with the Owens Valley Solar Array, *Astrophys. J.*, 605, 528-545, 2004.

Nitta, N., A.L. Kiplinger, and K. Kai, The spatial, spectral, and temporal character of the hard X-ray flare of 1982 February 3, *Astrophys. J.*, 337, 1003-1016, 1989.

Piana, M., A.M. Massone, E.P. Kontar, A.G. Emslie, J.C. Brown, and R.A. Schwartz, Regularized electron flux spectra in the 2002 July 23 solar flare, *Astrophys. J.*, 595, L127-L130, 2003.

Peng, L., Directivity and its energy dependence in solar flare energetic emission, *Astrophys. J.*, 443, 855-862, 1995.

Ramaty, R., Gyrosynchrotron emission and absorption in a magnetoactive plasma, *Astrophys. J.*, 158, 753-770, 1969.

Ramaty, R., Nuclear processes in solar flares, in *Physics of the Sun (vol. 2)*, edited by P.A. Sturrock et al., p291-323, Kluwer, 1986.

Ramaty, R., B. Kozlovsky, and R.E. Lingenfelter, Nuclear gamma-rays from energetic particle interactions, *Astrophys. J. (Supp.)*, 40, 487-526, 1979.

Ramaty, R., N. Mandzhavidze, B. Kozlovsky, and J.G. Skibo, Acceleration in solar flares: Interacting particles versus interplanetary particles, *Adv. Sp. Res.*, 13, 275-284, 1993.

Ramaty, R., N. Mandzhavidze, and B. Kozlovsky, Solar Atmospheric Abundances from Gamma Ray Spectroscopy, in *High Energy Solar Physics*, p172, American Institute of Physics, 1996.

Ramaty, R., N. Mandzhavidze, C. Barat, and G. Trottet, The giant 1991 June 1 flare: evidence for gamma-ray production in the corona and accelerated heavy ion abundance enhancements from gamma-ray spectroscopy, *Astrophys. J.*, 479, 458-463, 1997.

Reames, D.V., Particle acceleration at the Sun and in the heliosphere, *Space Sci. Rev.*, 90, 413-491, 1999.

Ritz, S., P.F. Michelson, C. Meegan, and J. Grindlay, GLAST Mission, The Gamma-ray Large Area Space Telescope (GLAST) Mission, *BAAS* 205, 606, 2004.

Ryan, J.M., J.A. Lockwood, and H. Debrunner, Solar energetic particles, *Space Sci. Rev.*, 93, 35-53, 2000.

Ryan, J., and H.S. Hudson, High-energy particles in solar flares, *Ann. Rev. Ast. Ap.*, 33, 239-282, 1995.

Ryan, J.M., K. Bennett, H. Debrunner, D.J. Forrest, J. Lockwood, M. Loomis, M. McConnell, D. Morris, V. Schönfelder, B.N. Swanenburg, and W. Webber, Comptel measurements of solar flare neutrons, *Adv. Sp. Res.*, 13, 255-258, 1993.

Saint-Hilaire, P., and A.O. Benz, Thermal and non-thermal energies of solar flares, *A&A*, 435, 743-752, 2005.

Sakao, T., T. Kosugi, S. Masuda, K. Yaji, M. Inda-Koide, and K. Makishima, Characteristics of hard X-ray double sources in impulsive solar flares, *Adv. Sp. Res.*, 17, 67, 1996.

Schwartz, R.A., J. Kasparova, B.R. Dennis, and M. Karlicky, The Unusual Hard X-ray Spectrum of the Flare of 20 August 2002, *AGU Fall Meeting Abstracts*, A171, 2003.

Schmelz, J.T., J.L.R. Saba, D. Ghosh, K.T. Strong, Anomalous coronal neon abundances in quiescent solar active regions, *Astrophys. J.*, 508, 519-532, 1996.

Share, G.H., and R.J. Murphy, Gamma-ray measurements of flare-to-flare variations in ambient solar abundances, *Astrophys. J.*, 452, 933-943, 1995.

Share, G.H., and R.J. Murphy, Intensity and directionality of flare-accelerated α-particles at the Sun, *Astrophys. J.*, 485, 409-418, 1997.

Share, G.H., and R.J. Murphy, Accelerated and ambient helium abundances from gamma-ray line measurements of flares, *Astrophys. J.*, 508, 876-884, 1998.

Share, G.H., R.J. Murphy, and E.K. Newton, Limits on radiative capture γ-ray lines and implications for energy content in flare-accelerated protons, *Solar Phys.*, 201, 191-200, 2001.

Share, G.H., R.J. Murphy, D.M. Smith, R.A. Schwartz, and R.P. Lin, RHESSI e^+–e^- annihilation radiation observations: Implications for conditions in the flaring solar chromosphere, *Astrophys. J.*, 615, L169-L172, 2004.

Silva, A.V.R., H. Wang, and D.E. Gary, Correlation of hard X-ray and microwave spectral parameters, *Astrophys. J.*, 546, 1116-1123, 2000.

Smith, D.F., and C.G. Lilliequist, Confinement of hot, hard X-ray producing electrons in solar flares, *Astrophys. J.*, 232, 582-589, 1979.

Smith, D.M., The RHESSI spectrometer *Solar Phys.*, 210, 33-60, 2002.

Smith, D.M., G.H. Share, R.J. Murphy, R.A. Schwartz, A.Y. Shih, and R.P. Lin, High-resolution spectroscopy of gamma-ray lines from the X-class solar flare of 23 July 2002, *Astrophys. J.*, 595, L81-L84, 2003.

Spicer, D.S., and A.G. Emslie, A new quasi-thermal trap model for solar flare hard X-ray bursts - an electrostatic trap model, *Astrophys. J.*, 330, 997-1007, 1988.

Spitzer, L.J. *Physics of Fully Ionized Gases*, Interscience, New York, 1956.

Sui, L., G.D. Holman, B.R. Dennis, Determination of low-energy cutoffs and total energy of nonthermal electrons in a solar flare on 2002 April 15, *Astrophys. J.*, 626, 1102-1109, 2005.

Takakura, T., and K. Kai, Energy distribution of electrons producing microwave impulsive bursts and X-ray bursts from the Sun, *PASJ* 18, 57-76, 1966.

Tandberg-Hanssen, E., and A.G. Emslie, *The Physics of Solar Flares*, Cambridge University Press, Cambridge, 1988.

Thompson, A.M, I.J.D. Craig, J.C. Brown, and C. Fulber, Inference of non-thermal electron energy distributions from hard X-ray spectra, *A&A*, 265, 278-288, 1992.

Toner, M.P., and A.L. MacKinnon, Do protons and α-particles have the same energy distributions in solar flares? *Solar Phys.*, 223, 155-167, 2004.

Trottet, G., N. Vilmer, C. Barat, A.O. Benz, A. Magun, A. Kuznetsov, R. Sunyaev, O.A. Terekhov, multiwavelength analysis of an electron-dominated γ-ray event associated with a disk solar flare, *A&A*, 334, 1099-1111, 1998.

Trottet, G., J.-P. Raulin, P. Kaufmann, M. Siarkowski, K.-L. Klein, and D.E. Gary, First detection of the impulsive and extended phases of a solar radio burst above 200 GHz, *A&A*, 381, 694-702, 2002.

Trubnikov, B.A., Particle interactions in a fully ionized plasma, *Reviews of Plasma Physics*, 1, 105, 1965.

Turkmani, R., L. Vlahos, K. Galsgaard, P.J. Cargill, and H. Isliker, Particle acceleration in stressed coronal magnetic fields, *Astrophys. J.*, 620, L59-L62, 2005.

Veronig, A.M., J.C. and Brown, A coronal thick-target interpretation of two hard X-ray loop events, *Astrophys. J.*, 603, L117-L120, 2005.

Vestrand, W.T., G.H. Share, R.J. Murphy, D.J. Forrest, E. Rieger, E.L. Chupp, and G. Kanbach, The Solar Maximum Mission atlas of gamma-ray flares, *Astrophys. J. (Supp.)*, 120, 409-467, 1999.

Vilmer, N., and A.L. MacKinnon, What can be learned about competing acceleration models from multiwavelength observations? *LNP*, 612, 127-160, 2005.

Vilmer, N., S.R. Kane, and G. Trottet, Impulsive and gradual hard X-ray sources in a solar flare, *A&A*, 108, 306-313, 1982.

Vilmer, N., A.L. MacKinnon, G. Trottet, C. Barat, High energy particles during the large solar flare of 1990 May 24: X/γ ray observations, *A&A*, 412, 865-874, 2003.

Vilmer, N., G. Trottet, C. Barat, R.A. Schwartz, S. Enome, A. Kuznetsov, R. Sunyaev, and O. Terekhov, Hard X-ray and γ-ray observations of an electron dominated event associated with an occulted solar flare, *A&A*, 342, 575-582, 1999.

Woodgate, B.E., R.D. Robinson, K.G. Carpenter, S.P. Maran, and S.N. Shore, Detection of a proton beam during the impulsive phase of a stellar flare, *Astrophys. J.*, 397, L95-L98, 1992.

Zharkova, V.V., and M. Gordovskyy, The kinetic effects of electron beam precipitation and resulting hard X-ray intensity in solar flares, *A&A*, 432, 1033-1047, 2005.

A.L. MacKinnon, DACE, University of Glasgow, 11 Eldon Street, Glasgow G3 6NH, UK. (a.mackinnon@educ.gla.ac.uk)

Gamma Radiation From Flare-Accelerated Particles Impacting the Sun

Gerald H. Share[1]

Department of Astronomy, University of Maryland, College Park, MD 20742

Ronald J. Murphy

E.O. Hulburt Center for Space Research, Naval Research Laboratory Washington, DC 20375

We discuss how remote observations of gamma-ray lines and continua provide information on the population of electrons and ions that are accelerated in solar flares. The radiation from these interactions also provides information on the composition of the flaring chromosphere. We focus our discussion on recent *RHESSI* observations and archival observations made by *SMM* and *Yohkoh*.

1. INTRODUCTION

Eruptive solar events such as flares and shocks from coronal mass ejections (CMEs) accelerate electrons and ions to high energies. The solar energetic particles (SEPs) that reach interplanetary space have origins in both flares and CMEs (e.g. *Reames*, 1999) and there is debate over their relative importance (e.g. *Cane et al.*, 2003; *Tylka et al.*, 2005; *Li and Zank*, 2005). Timing (e.g. *Tylka et al.*, 2003) and composition studies support the idea that flares are primarily responsible for the 'impulsive' electron and ^{3}He-rich particle events in space and also contribute a seed population (*Mason, Mazur, and Dwyer*, 1999) for 'gradual' events that have their origin in CME-produced shocks. The processes that impulsively accelerate particles into interplanetary space along open magnetic field lines may also generate the ions and electrons along closed loops that interact in the chromosphere and photosphere.

In Figure 1 we show a cartoon depicting injection of accelerated particles (e.g. *Miller et al.*, 1997; *Aschwanden*, 2004) onto a closed magnetic loop, their transport, and their impact on the solar atmosphere. Calculations of the gamma-rays

and neutrons resulting from these interactions have generally assumed that they take place in a thick target atmosphere (e.g. *Ramaty et al.*, 1996) having photospheric or coronal compositions (e.g. *Anders and Grevesse*, 1989; *Reames*, 1995); however, there was one flare in which the emission was believed to arise in a thin target (*Ramaty et al.*, 1997). The resulting continuum and line γ radiation provides a means to remotely measure the characteristics of these trapped particles and relate them to the particles that escape from the same acceleration site at the Sun on open field lines. The flares of 2003 October 28 and November 2, and 2005 January 20 offer the best opportunity for making simultaneous measurements of these two particle populations. Other papers in this monograph will discuss the particles measured in space. In this paper we discuss the relationship between the >100 keV photon emissions and both the accelerated particles and the solar material in which they interact. We summarize some preliminary results derived from new *RHESSI* and archival *SMM* and *Yohkoh* observations.

2. WHAT PRODUCTS OF PARTICLE INTERACTIONS TELL US ABOUT FLARE-ACCELERATED ELECTRONS AND IONS

In this Section we briefly discuss the products of flare-accelerated electrons and ions that impact the Sun (Figure 1). Electrons produce a bremsstrahlung continuum that reflects their spectrum. For example an electron spectrum following

---

[1]Also at Naval Research Laboratory, Washington, DC 20375

Solar Eruptions and Energetic Particles
Geophysical Monograph Series 165
10.1029/165GM17

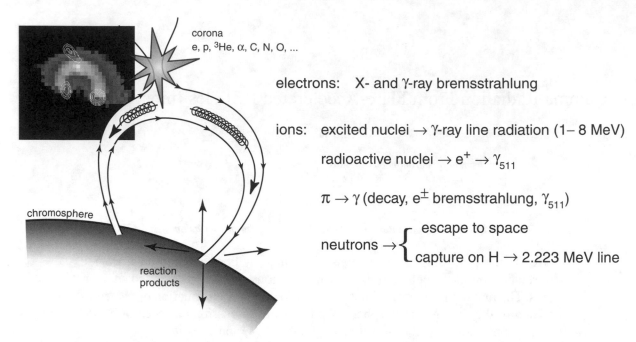

corona
e, p, ^{3}He, α, C, N, O, ...

electrons: X- and γ-ray bremsstrahlung

ions: excited nuclei $\rightarrow$ γ-ray line radiation (1– 8 MeV)

radioactive nuclei $\rightarrow$ e$^+$ $\rightarrow$ γ_{511}

$\pi \rightarrow \gamma$ (decay, e$^\pm$ bremsstrahlung, γ_{511})

neutrons $\rightarrow \left\{ \begin{array}{l} \text{escape to space} \\ \text{capture on H} \rightarrow \text{2.223 MeV line} \end{array} \right.$

chromosphere

reaction
products

Figure 1. Cartoon showing particle acceleration, magnetic loop transport, interaction in the solar atmosphere, and interaction products. *Yohkoh* image of limb flare is shown in upper left corner (*Masuda et al.*, 1994).

a power-law in energy with index β produces a photon spectrum with index $\sim\beta - 1$ from 0.3-1.0 MeV (*Ramaty et al.*, 1993). Protons and α-particles interact with ambient solar nuclei to produce neutrons, γ-ray lines from transitions of excited nuclei to lower-energy states, positrons from radioactive nuclei (*Ramaty, Kozlovsky, and Lingenfelter*, 1979), and other forms of radiation. If the incident particles have energies above a few hundred MeV/nucleon, π mesons are produced that decay with emission of electrons, positrons, neutrinos and γ rays (*Murphy, Dermer, and Ramaty*, 1987). The neutrons can escape from the Sun and be detected at Earth before they decay (e.g. *Chupp et al.*, 1982), decay at the Sun, or be captured on H or ^{3}He; capture on H in a quiet photosphere produces ^{2}H with emission of a 2.223-MeV γ-ray having a width expected to be $\lesssim 1$ eV, while capture on ^{3}He produces no γ radiation (*Hua and Lingenfelter*, 1987).

The γ-ray lines from ambient nuclei excited by interaction of accelerated protons and α particles are Doppler broadened by nuclear recoil (*Ramaty, Kozlovsky, and Lingenfelter*, 1979; *Kiener, de Séréville, and Tatischeff*, 2001). The shape of the lines is dependent on the angular distribution, spectrum, and assumed α/p ratio of the accelerated particles and the viewing angle of the observer. The width of the line can be characterized by its full-width at half maximum (FWHM) although the shape is not always Gaussian. Alpha-particle interactions produce broader lines than do protons with the

same energy/nucleon. The FWHM width of the 4.439 MeV line from de-excitation of ^{12}C is typically $\sim$100 keV. Much larger Doppler widths result from interactions of accelerated heavy ions when they de-excite after interacting with ambient H or He; e.g. the width of the ^{12}C line is $\sim$1 MeV FWHM. Positrons produced in radioactive decays, following the interactions, annihilate with electrons to produce a line at 511 keV having a width of a few keV and a continuum below that energy.

In Figure 2 we plot the γ-ray spectrum observed from 200 keV to 8.5 MeV by *RHESSI* during the decay phase of the 2003 October 28 flare. Earlier observations by *CORONAS-F* (*Veselovsky et al.*, 2004; *Kuznetsov et al.*, 2005) and by *INTEGRAL* (*Gros et al.*, 2004) revealed an $\sim$1-min interval dominated by electron bremsstrahlung up to $\sim$40 MeV followed by an $\sim$3-min interval in which strong nuclear emission was observed. The spectrum in the Figure reveals all the different components described above. The most distinct features are the 511-keV annihilation line and 2.223-MeV neutron capture line that rise above both the electron bremsstrahlung continuum and 'narrow' and 'broad' nuclear lines from proton-α and heavy-ion interactions, respectively.

2.1. Bremsstrahlung from Flare-Accelerated Electrons

The best fitting bremsstrahlung continuum shown in Figure 2 has a shape approximated by two power laws with

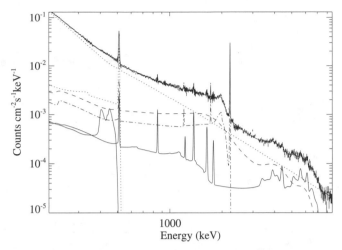

Figure 2. Fitted *RHESSI* count spectrum accumulated over the first 4 minutes of its observation of the 2003 Oct. 28 flare. Fits to the bremsstrahlung (dots), annihilation radiation (dots), and α-⁴He (dot-dot-dash), narrow (solid), broad (dashes) and 2.22 MeV (dashes-dots) nuclear line components are shown separately.

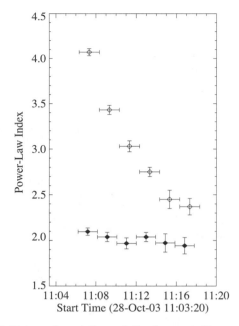

Figure 3. Temporal variation of the bremsstrahlung power-law indices below (open diamonds) and above (filled diamonds) the break energy during the 2003 October 28 flare.

spectral indices ~3.8 and 2.1 below and above an ~460 keV break energy, respectively. Such hardening was observed in several γ-ray flares (e.g. *Vilmer et al.*, 1999) and is larger than expected from relativistic effects (e.g. *McTiernan and Petrosian*, 1991). This suggests that the accelerated electron spectra may harden near ~1 MeV. The hardening in the October 28 flare may, in part, be due to a separate electron component from decay of high-energy pions whose neutral component appears to have been detected in high-energy γ-rays by *CORONAS-F* (*Veselovsky et al.*, 2004; *Kuznetsov et al.*, 2005). It will be interesting to compare the inferred electron spectrum at the Sun with that observed in space for this flare.

The bremsstrahlung spectrum hardened monotonically with time during the flare. This hardening is reflected in the decreasing power-law indices above and below the break energy that we plot in Figure 3; there was a similar hardening during the November 2 flare. It would be interesting to determine whether such hardening was observed in the electron spectrum observed in space. The high-energy electron spectrum can also vary rapidly but not monotonically in flares. This is illustrated in Figure 4 that displays variations in the single power-law index fit to the bremsstrahlung spectrum from the 1989 March 6 flare observed by *SMM*.

2.2. Nuclear Lines from Accelerated Particle Interactions

The 2003 October 28 and November 2 and 2005 January 20 flares offer a unique opportunity to compare accelerated ion spectra and composition at the Sun and in interplanetary space. Earlier studies (e.g. *Ramaty et al.*, 1993) used more

limited data but demonstrated that the relative numbers of accelerated protons at the Sun and in interplanetary space is highly variable and may be correlated with whether the flare has an impulsive or gradual time profile in soft X-rays. Other papers in this volume discuss SEP observations. We discuss observations of the characteristics of the accelerated particles at the Sun inferred from γ-ray observations.

2.2.1. Directionality of interacting protons and α-particles. The shapes of the 'narrow' nuclear lines provide information on

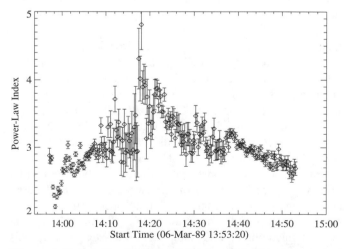

Figure 4. Temporal variation of the bremsstrahlung power-law index during the 1989 March 6 flare.

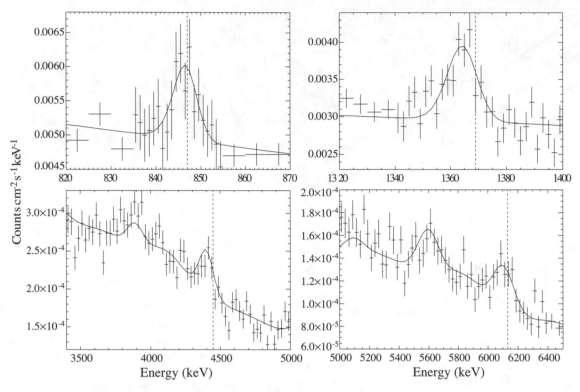

Figure 5. Fits to 4 lines observed by *RHESSI* during the 2003 October 28 flare. Instrumental Compton scattered and 1st escape peaks are visible at high energy. Top panels: ^{56}Fe and ^{24}Mg low FIP (First Ionization Potential) lines; Bottom panels: ^{12}C and ^{16}O high FIP lines. Dotted lines show the rest energies of the de-excitation lines.

the angular distribution of interacting protons and α-particles. We plot four of the lines used in the study of the 2003 October 28 flare in Figure 5. The lines are red-shifted from their laboratory energies (*Smith et al.*, 2004; *Gan*, 2005) by an amount that is consistent with both *INTEGRAL* observations (*Gros et al.*, 2004) and what is calculated for a broad downward-directed distribution of accelerated particles as found in a study of 19 *SMM* flares (*Share et al.*, 2002). We have simultaneously fit the energies of strong nuclear de-excitation lines in three *RHESSI* and 19 *SMM* flares to study their redshift as a function of heliocentric angle. In Figure 6 we plot the percentage line shift for these 22 flares relative to the 4.439 MeV laboratory energy of the ^{12}C line. *RHESSI* detected larger redshifts than expected for the heliocentric angles of the 2002 July 23 (*Smith et al.*, 2003) and 2003 November 2 flares (*Murphy and Share*, 2005). The *RHESSI* redshift at 70° appears to be barely consistent with the distribution of shifts observed by *SMM* in this preliminary analysis. Such large shifts suggest that the magnetic loops constraining the particles are tilted from the normal (*Smith et al.*, 2003). Such tilts are not uncommon; for example, *Bernasconi et al.* (1995) discuss measurements of flux tubes with an average inclination of ~14° relative to local vertical in the photosphere.

2.2.2. Spectrum of accelerated protons and α particles. Line-flux ratios provide an estimate of the energy spectrum of accelerated protons and α particles that impact the solar atmosphere (*e.g.*, *Murphy and Share*, 2005). This is possible because the cross sections for γ-ray line production can have significantly different energy dependences (*Kozlovsky, Murphy, and Ramaty*, 2002). As discussed below these determinations are dependent on various assumptions about the accelerated particles and ambient composition. Comparison of fluxes in the ^{20}Ne (1.63 MeV) line and in the ^{16}O (6.13 MeV) and ^{12}C (4.43 MeV) lines provides information on the spectrum between ~2 and 20 MeV nucleon^{-1}. To determine the spectrum from ~10–50 MeV nucleon^{-1} we typically compare fluxes in the 511-keV annihilation line produced following β^{+} radioactive decays with the ^{12}C and ^{16}O de-excitation lines. Comparison of the neutron-capture line with the ^{12}C and ^{16}O lines provides information on the spectrum from ~10-100 MeV nucleon^{-1}. Even though nuclear line spectrometers typically are only sensitive up to ~10-20 MeV they can detect evidence for protons up to hundreds of MeV nucleon^{-1} through detection of annihilation radiation from positrons following pion decay (see discussion in Section 4.2).

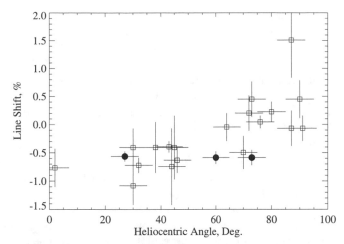

Figure 6. Measurements of the percent line shifts relative to the 4.43 MeV ^{12}C line vs. heliocentric angle made by *SMM* (squares) and *RHESSI* (solid circles, 2002 July 23: 72°; 2003 Oct. 28: 30°; Nov. 2: 60°).

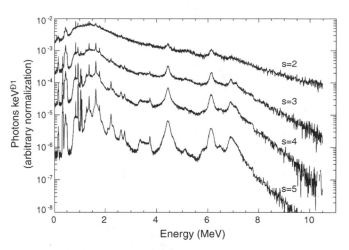

Figure 7. Calculated nuclear de-excitation spectra for four power-law indices. Text explains assumptions about accelerated-particle and ambient abundances.

Ramaty et al. (1995, 1996) evaluated the flare-averaged accelerated-particle spectra from 19 *SMM* flares (*Share and Murphy*, 1995). This determination was dependent on the assumed ambient and accelerated particle compositions and the simplification that all the accelerated particles have the same spectra. We assume a coronal ambient composition where the particles interact (*Reames*, 1995) and that the accelerated particles have an impulsive SEP composition. This latter composition (*Ramaty et al.*, 1996) is coronal (*Reames*, 1995) for C, N, Ne, Mg, Al, Si, S, Ca, and Fe relative to O but with Ne/O, Mg/O, Si/O and S/O increased by a factor of 3 and Fe/O increased by a factor of 10, α/O = 50, and α/p = 0.5. With these assumptions the average index for the 19 flares was ~4.3 between ~2 and 20 MeV nucleon^{-1} (*Share et al.*, 2002; *Murphy, and Share*, 2005).

From a preliminary analysis, under the same assumptions, we find that the spectra of the 2003 October 28 (in the decay phase) and November 2 flares in the same energy range were considerably harder than this average; the power-law indices were estimated to be 2.5 ± 0.4 and 2.5 ± 0.9, respectively, based on fluxes in the Ne, C, and O lines (*Share et al.*, 2004b). Measurement by *INTEGRAL* (*Kiener et al.*, 2005) of these same lines appears to confirm the hardness of the decay phase spectrum of the October 28 flare and indicates that the spectrum may have been even harder a few minutes earlier. Such a hard spectrum is also required to produce the observed 511 keV line flux (*Share et al.*, 2004a) and the π^o-decay γ-rays that were apparently observed by *CORONAS-F* just before the *RHESSI* observations began on October 28 (*Veselovsky et al.*, 2004; *Kuznetsov et al.*, 2005). *Gan* (2005) suggested that a steeper spectrum would be derived for an ambient Ne/O concentration of 0.15 in lieu of the value of

0.25 that was assumed (*Ramaty et al.*, 1996). The solar Ne/O ratio is uncertain with recent publications suggesting values from ~0.15 to 0.4 (*Drake and Testa*, 2005; *Bahcall et al.*, 2005; *Schmeltz et al.*, 2005; *Young*, 2005). Power-law spectral indices between about 3 and 4 were derived for the October 28 flare by comparing fluxes in the 2.223 MeV neutron capture line and the C and O de-excitation lines (*Share et al.*, 2004a; *Tatischeff, Kiener, and Gros*, (2005)); there are several other parameters that affect this measurement, however.

Perhaps the hardest spectrum of accelerated particles ever observed to impact the Sun occurred during the 2005 January 20 flare. It differs from the October 28 spectrum plotted in Figure 2 in that the narrow de-excitation lines are barely detectable, the annihilation line is much stronger relative to the 2.223-MeV neutron-capture line, and the bremsstrahlung and nuclear continua are more dominant. In Figure 7 we plot calculations of the nuclear de-excitation line spectra as a function of the spectral index of the accelerated particles. The nuclear continuum dwarfs the discrete lines for very hard spectra. The photon spectrum for this flare extended up to energies in excess of tens of MeV with evidence for pion-decay gamma rays based on *CORONAS-F* observations (*Priv. Comm. V. Kurt*, 2005). *RHESSI* observations throughout the flare show significant emission above 10 MeV over and above extrapolation of the hard X-ray bremsstrahlung. Such high-energy emission produces a significant amount of 511-keV line radiation from material around the *RHESSI* detectors. For this reason we cannot simply use the measured (511-keV)/(2.223-MeV) line ratio, without correction, to infer the spectrum of accelerated particles at the Sun. Work is in progress to determine this accelerated particle spectrum for comparison with the spectrum of solar energetic particles observed in space (*Mewaldt et al.*, 2005).

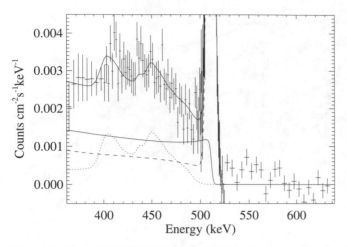

Figure 8. Fit to *RHESSI* 2003 October 28 spectrum revealing the α-^{4}He contribution (dots) to the total spectrum after bremsstrahlung and other nuclear contributions have been removed; fits to the annihilation line (dashes) and continuum (solid curve) are also plotted. From *Share et al.*, 2004; reproduced courtesy of the Univ. of Chicago Press.

2.2.3. Accelerated helium abundance.   The intensity of the α-^{4}He fusion lines (^{7}Be and ^{7}Li) in flares observed by the *SMM* spectrometer has been found to be more intense than expected for assumed accelerated α/p and ambient ^{4}He/H abundances of 0.1 (*Share and Murphy*, 1997). *Kozlovsky, Murphy, and Share* (2004) set a 1 σ lower limit of 0.35 to the α/p ratio in the 2002 July 23 flare observed by *RHESSI*, consistent with the *SMM* observations. Plotted in Figure 8 is the spectrum between 350 and 650 keV accumulated from 11:06-11:10 UT during the 2003 October 28 flare after subtracting bremsstrahlung and nuclear contributions. The dotted line shows the best fit to the α-^{4}He line shape for a downward isotropic distribution of accelerated particles. (In Section 4.2 we discuss the annihilation line and its continuum that are also shown in the Figure.) If we assume that the accelerated-particle power-law index is 2.5 and ambient ^{4}He/H = 0.1, we obtain preliminary accelerated α/p ratios of 0.65 (-0.3, $+0.15$) and 0.4 (-0.15, $+0.2$) for the 2003 October 28 and November 2 *RHESSI* flares, respectively (*Share et al.*, 2004b). Due to differences in the p and α cross sections, we derive lower α/p ratios for softer particle spectra. We note that if the protons and α-particles had different spectral indices this will affect the value of the accelerated α/p ratio derived from the gamma-ray line measurements (*Toner and Mackinnon* (2004)). *Gan* (2005) concluded that the α/p ratio in both flares probably did not exceed 0.1. He based his conclusion on comparative studies of the Ne/O and n-capture/C line flux ratios; this method is rather indirect because the results are also dependent on other factors. *Tatischeff et al.* (2005) used a similar method but did not attempt to draw any conclusions about the ratio. As we mentioned earlier it is possible to use the measured de-excitation line shape to estimate the α/p ratio (*Smith et al.*, 2003). Although this method is most sensitive to the ratio for softer accelerated particle spectra than observed in these two flares, *Kiener et al.* (2005) have used it in the INTEGRAL study of the October 28 flare.

Our estimate of the accelerated α/p ratio is inversely dependent on the ambient ^{4}He/H ratio, assumed to be 0.1. *Feldman, Landi, and Laming* (2005) recently measured a ^{4}He/H ratio of 0.122 ± 0.024 in high-temperature solar flare plasmas. *Mandzhavidze, Ramaty, and Kozlovsky* (1997) suggested a method to determine whether these high α-He line fluxes are due to an elevated α/p ratio and/or to an elevated ambient ^{4}He/H ratio. This requires comparison of the fluxes in lines produced by α-^{56}Fe (339 keV) and p-^{56}Fe (847 keV) reactions. There is evidence for a weak line at 339 keV in data obtained with moderate resolution NaI spectrometers, consistent with an elevated α/p ratio (*Share and Murphy*, 1998). Our preliminary assessment is that the spectra of the October 28 and November 2 flares may be too hard for the 339-keV line to be detected because its production cross section peaks at low energies.

Our finding that the accelerated α/p ratio inferred from the gamma-ray measurements generally appears closer to 0.5 than to 0.1 can be compared with ratios from 0.005 to as high as 0.3 derived from interplanetary particle measurements, for impulsive events (*Reames, Meyer*, and *von Rosenvinge*, 1994; *Kallenrode, Cliver, and Wibberenz*, 1992). However, these measurements were for particle energies of 4.4-6.4 MeV nucleon^{-1} that are just below the threshold for the observed α-^{4}He fusion lines. *Torsti et al.* (2002) reported observations up to 100 MeV nucl^{-1} where the α/p-ratio ranged between 0.15-0.5 for more than 10 hr during particle events on 1998 May 27 and 1999 December 28.

The key line features for understanding the accelerated ^{3}He abundance appear near 0.937, 1.040, and 1.08 MeV (*Mandzhavidze, Ramaty, and Kozlovsky*, 1997). *Share and Murphy* (1998) found evidence for an average ^{3}He/^{4}He ratio of 0.1 in flares observed by *SMM* and *Mandzhavidze, Ramaty, and Kozlosky*, (1999) additionally suggested that a ratio as high as 1 have occurred in some flares. Such high ^{3}He/^{4}He ratios are consistent with what is observed in smaller impulsive solar energetic particle events (e.g. *Reames, Meyer, and Von Rosenvinge*, 1994). *Mason et al.* (2002) report ratios for 14 impulsive particle events that varied between ~0.1 and 6.5. *RHESSI* can more easily resolve the ^{3}He lines from other nearby lines. Based only on the 937-keV line, we obtained preliminary 99% upper limits of 0.8 and 2.5 on the ^{3}He/^{4}He ratios in the October 28 and November 2 flares, respectively (*Share et al.*, 2004b).

2.2.4. Accelerated heavy ions.   From the fit to the overall *RHESSI* spectrum shown in Figure 2, we note that the broad nuclear lines (dashed curve) that are in part due to accelerated heavy ions impacting on ambient H can contribute up to about

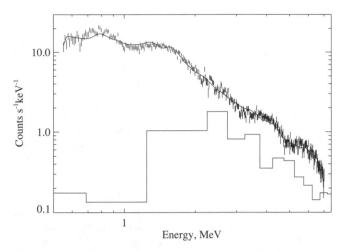

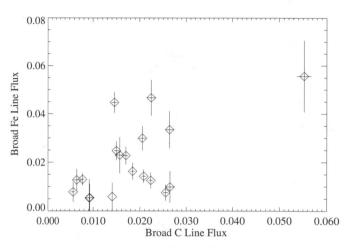

Figure 9. Broad nuclear-line component observed in the summed spectrum of 19 flares observed by *SMM*. The curve shows the calculated shape for accelerated particles with impulsive composition, $\alpha/p = 0.5$, power-law index 4.5. The broad peaks near 0.8, 4, and 6 MeV are attributed to Doppler-broadened de-excitation lines from accelerated ^{56}Fe, ^{14}C, and ^{16}O. The highly broadened peak between 1-2 MeV is partly due to Doppler-broadened lines from ^{56}Fe, ^{24}Mg, ^{20}Ne, and ^{28}Si. The histogram is discussed in the text.

Figure 10. Comparison of broad ^{56}Fe and ^{12}C line fluxes in 19 *SMM* flares.

25% of the total flux in certain energy ranges. We see evidence for the broad lines near 4 MeV and 6 MeV, attributable to flare-accelerated ^{12}C and ^{16}O. Unfortunately, there is an intense instrumental and nuclear continua below the 2.223-MeV line that may mask the presence of the broadened lines at lower energy, including the ^{56}Fe line near 847 keV observed in flares by *SMM*. *Share and Murphy* (1999) presented the broad-line spectrum from the sum of 19 flares observed by *SMM*; this spectrum is plotted in Figure 9 after the narrow lines and bremsstrahlung have been subtracted. Highly Doppler-broadened lines from accelerated ^{56}Fe, ^{12}C, and ^{16}O appear to be resolved. Analysis indicated an $\sim 5 \times$ excess in the abundance of accelerated ^{56}Fe relative to its coronal ambient abundance, similar to that found in impulsive SEPs. The shape of the spectrum is also in agreement with calculations (solid curve) for accelerated particles with an impulsive SEP composition, $\alpha/p = 0.5$ (see discussion in Section 2.2.3), power-law index 4.5, and downward isotropic distribution. Thus impulsively accelerated heavy ions at the Sun and in space may have similar compositions. We have preliminary evidence that the accelerated Fe concentration may be highly variable in flares. In Figure 10 we plot the broad ^{56}Fe line flux vs. the broad ^{12}C line flux observed in 19 *SMM* flares. We note that some of the variability may be due to flare-to-flare variations in the spectra of the accelerated particles.

Chadwick et al. (1999) have performed detailed calculations of the total γ-ray yield for various nuclei that can be used to determine the shape of the unresolved nuclear-line and continuum spectrum that have compromised our ability to study the accelerated heavy-ion component in flares. The histogram plotted in Figure 9 is a preliminary estimate of the shape of the unresolved nuclear line and continuum emitted from ^{27}Al.

3. COMPARISON OF ACCELERATED ELECTRONS AND IONS

3.1. Temporal Variations in Accelerated Ions and Electrons

Important information about the acceleration and transport of electrons and ions can be obtained by a comparison of the time histories observed in different energy bands of the hard X-ray and γ-ray spectra. *Chupp* (1990) discussed an early study of *SMM* data indicating that the peaks in individual bursts observed in the 4.1-6.4 MeV energy band (mostly due to ion interactions) were delayed between 2 s and 45 s from the corresponding maxima observed in the electron bremsstrahlung continuum near 300 keV (see also *Share and Murphy*, 2004c). This delay appears to be proportional to the rise time of the pulse. Some of these delays may be explained by transport effects in magnetic loops (*e.g.*, *Hulot et al.*, 1992; *Murphy and Share*, 2005).

Such delay analyses may not reveal the true complexity of the variation between the nuclear and electron emissions, however. In Figure 11 we show variations in the nuclear-line to bremsstrahlung ratio with time in two flares observed by *Yohkoh* (in the 4-7 MeV region) and one observed by *SMM* (from 0.3-8.5 MeV). The nuclear/bremsstrahlung ratio appears to increase with time during the 2001 August 25 flare, decrease with time during the 2001 April 15 flare,

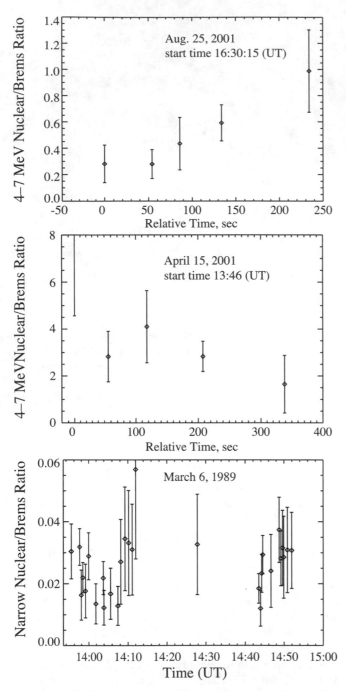

Figure 11. Nuclear to bremsstrahlung ratios vs. time observed by Yohkoh in the 2001 August 25 (top panel) and April 15 flares (middle panel) and by SMM in the 1989 March 6 flare (bottom panel).

and show variable behavior in the 1989 March 6 flare. These disparate variations are likely due to different phenomena such as transport in short and long loops, trapping, and distinct acceleration episodes. We are planning a systematic study of these variations in a large sample of flares.

3.2. Energy in Accelerated Electrons and Ions

Using the derived accelerated-particle spectra it is possible to compare the energies contained in flare-accelerated ions and electrons. *Ramaty and Mandzhavidze* (2000) performed this study for 19 flares observed by *SMM* and found that the ions and electrons contained comparable energies. *Emslie et al.* (2004) obtained a similar result for the 2002 July 23 flare; for the 2002 April 21 flare they placed an upper limit on the energy in ions that is just above that measured in electrons. We have made a preliminary estimate of the energies contained in protons in the 2003 October 28 and November 2 flares. We use both *RHESSI* and *INTEGRAL* (*Gros et al.*, 2004) data for the October 28 flare. The energy in protons is strongly dependent on both the spectrum and low-energy cutoff energy. The γ-ray line studies provide information on protons with energies $\gtrsim 3$ MeV. We can obtain an estimate of the energy contained in the protons by assuming that the spectrum extends without a break down to 1.0 MeV and is flat below that energy. For a power-law index of 3 we estimate that protons contained $\sim 1.5 \times 10^{31}$ ergs in the October 28 flare and $\sim 0.5 \times 10^{31}$ ergs in the November 2 flare. At present there are no estimates for the energy content in electrons for these flares.

4. ACCELERATED PARTICLES PROBE THE SOLAR ATMOSPHERE AND PHOTOSPHERE

Flare-accelerated protons and alpha particles and secondary neutrons and positrons all act as probes of the flaring chromosphere and photosphere. Gamma-ray emission from their interactions have revealed a dynamic chromosphere that exhibits the FIP effect seen in the corona and puzzling evidence for striking temperature and ionization changes.

4.1. FIP Effect at Chromospheric Densities

Fluxes of narrow lines from ambient heavy nuclei observed from 19 flares (*Share and Murphy*, 1995) have been used by *Ramaty et al.* (1995) to infer that there is a strong FIP effect (overabundance of low first-ionization potential elements relative to their composition in the photosphere) where the particles interact ($\gtrsim 10^{14}$ H cm^{-3}; see *e.g.*, *Murphy and Share*, 2005), suggesting that the ambient plasma has a coronal composition. *Laming* (2004) provided an explanation for the FIP effect under less dynamic conditions at lower densities ($\sim 10^{12}$ H cm^{-3}). *Share and Murphy* (1995) pointed out that the FIP ratio appears to vary from flare-to-flare. The average low-FIP (Fe, Mg, Si) to high-FIP (Ne, C, O) line ratio for 19 *SMM* flares was ~ 0.45. *Murphy et al.* (1997) showed evidence that the FIP ratio can increase with time in an extended flare observed by *CGRO*/OSSE. Evidence for

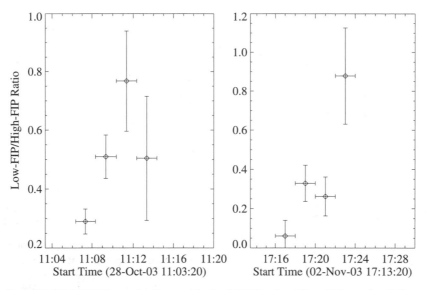

Figure 12. Variation in Low-FIP/High-FIP line ratio observed in the 2003 October 28 and November 2 flares observed by *RHESSI*.

increase in the ratio is also provided by *RHESSI* observations of the 2003 October 28 and November 2 flares (*Share et al.*, 2004b; *Shih et al.*, 2004) plotted in Figure 12. These observations suggest that the accelerated particles interact in ambient compositions that change with time. It would be of interest to determine whether chromospheric evaporation of this fractionated material could contribute significantly to the coronal material and provide seed material for subsequent acceleration into interplanetary space. We plan on studying the larger SMM data base for additional evidence of temporal changes in the ambient composition.

4.2. Positrons Probe the Chromosphere and Photosphere

Positrons are emitted in the radioactive decay of nuclei produced when flare-accelerated ions interact with the solar atmosphere (*Kozlovsky, Lingenfelter, and Ramaty*, 1987). The positrons annihilate directly and through formation of positronium. Annihilation of a positron with an electron yields either a line (2 photons) or continuum (3 photons) depending on the local density, temperature, and ionization state (*Crannell et al.*, 1976); thus the measured line-to-continuum ratio helps to determine the conditions of the ambient medium. Until the *RHESSI* 2002 July 23 flare observation (*Share et al.*, 2003), the 511-keV line had not been clearly resolved. The line had a width of 8.1 ± 1.1 keV which was consistent with both annihilation in an ionized medium 4–7×10^5 K and in a quiet atmosphere at 6000K. Detailed particle transport and interaction calculations are required to confirm our supposition that the latter location does not provide enough column depth for the production of radioactive nuclei and for the slowing down and annihilation of the positrons.

We have now studied annihilation radiation in four *RHESSI* flares. The 2003 October 28 flare produced the largest fluence of annihilation radiation observed to date (*Share et al.*, 2004a). It also displayed a remarkable change in width of the 511-keV line (see bottom panel of Figure 13). The width of the line changed from an average of ~6.5 keV to ~1 keV within about two minutes when there was no marked change in the bremsstrahlung, nuclear de-excitation line, and annihilation line fluxes. The change in line width is most clearly shown in Figure 14. There was a rapid change in the relative flux in the measured continuum below the line early in the flare (see third panel in Figure 13). This step-function increase below the line can be seen in Figure 8. We believe that the high continuum flux in the first 2-3 minutes is due to Compton scattering of the annihilation line under 5 to 10 g cm^{-2} of H. This depth is consistent with an e$^+$-origin from decay of π^+-mesons; γ- rays from π^o-meson decay were reported a few minutes earlier by *CORONAS-F* (*Veselovsky et al.*, 2004; *Kuznetsov et al.*, 2005). From fits to the line shapes we conclude that the broad line is likely to be due to annihilation in a $> 2 \times 10^5$ K medium and that the narrow line is produced at temperatures $< 10^4$ K, albeit in a highly ionized medium. These observations raise issues concerning the characteristics of the solar atmosphere during flares.

There have been two new developments since publication of the 2003 October 28 and November 2 annihilation radiation studies. *Murphy et al.* (2005) have developed an improved algorithm that calculates the 511-keV annihilation line spectrum and relative strength of the 3-γ continuum for a wide range of physical environments relevant to the flaring solar atmosphere. We are using these calculations to study the annihilation emission from all the flares observed by

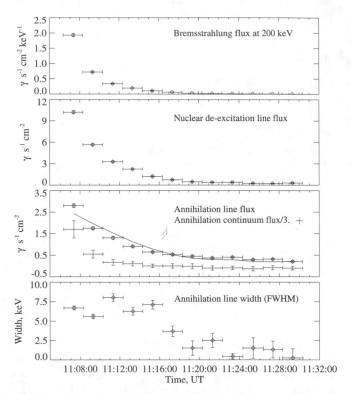

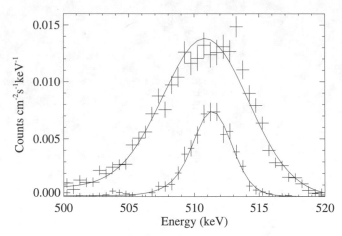

Figure 14. Count spectra of the solar 511-keV annihilation line (instrumentally broadened) derived by subtracting bremsstrahlung and nuclear contributions during the October 28 flare when the solar line was broad (11:06 - 11:16 UT) and narrow (11:18 - 11:30 UT). The solid curve is the best-fitting model that includes a Gaussian line and positronium continuum. From *Share et al.* (2004); reproduced courtesy of the Univ. of Chicago Press.

Figure 13. Time histories of the bremsstrahlung and total nuclear de-excitation line fluxes for the October 28 flare are plotted in the top two panels. Also shown are the time histories of the 511-keV annihilation line flux and width, as well as what we believe is the Compton scattered continuum below the line. From *Share et al.* (2004); reproduced courtesy of the Univ. of Chicago Press.

RHESSI. Recent studies indicate that the annihilation line shape is consistent with Gaussians over the entire October 28 flare, implying temperatures $> 2 \times 10^5$ K during the first 10 min and $< 10^4$ K ($> 20\%$ ionized) during the last 10 min. During both intervals, the inferred density is significantly $> 10^{14}$ H cm^{-3}. What is especially puzzling is the rapid, 2 minute, transition from broad to narrow line width (Figures 13 & 14). The line observed during the November 2 flare is also broad but it can be fit with the shape expected for annihilation in a quiet solar atmosphere at a temperature of 5000 K (*Murphy et al.*, 2005); there is uncertainty whether the continuum below the line is consistent with this origin, however.

The annihilation line studies, the hardness of the accelerated particle spectra, and the apparent detection of pion-decay radiation suggest that we are also studying sub-photospheric conditions with gamma rays. It is puzzling how the apparently high temperatures and ionization states can be supported at these densities. A new aspect is the detection of acoustic waves below the photosphere following the

October 28 and 29 flares (*Donea and Lindsey*, 2005). Our preliminary estimates indicate that there is sufficient energy in $\gtrsim 100$ MeV protons to account for the acoustic waves. If this association with high-energy protons is correct, acoustic waves should have been detected after the 2005 January 20 flare. The intense bremsstrahlung and nuclear continuum combined with weak de-excitation line emission in the January 20 flare resembles the electron-rich events suggested by *Rieger, Gan, and Marschhäuser* (1998). Could these be due to episodes with very hard proton spectra that produce pions? The hard bremsstrahlung then could result from pion-decay electrons and positrons. Chupp (private communication, 2005) is studying high-energy spectra from electron-rich episodes in the 1989 March 6 flare to determine whether the high-energy gamma-ray spectrum is consistent with pion-decay emission.

4.3. Search for Photospheric ^{3}He using Fast Neutrons

Neutrons produced in accelerated particle interactions with the solar atmosphere can escape the Sun, decay, or slow down and get captured by either H or ^{3}He. Capture on H produces a line at 2.223 MeV while capture on ^{3}He is radiationless. Both the angular distribution of accelerated particles and concentration of ^{3}He affect the time profile of the capture line. *Murphy et al.* (2003) have compared the time history of the line with calculations to determine both the angular distribution of accelerated particles and photospheric

³He concentration in the 2002 July 23 flare. The nuclear-line flux has been used as a surrogate for the acceleration time profile. A physically based transport model has been used in performing the analysis. We find that the accelerated particles suffer significant pitch-angle scattering in the corona, yielding a broad downward-directed distribution. Because pitch-angle scattering, spectral hardness, and ³He all affect the neutron-capture line timehistory and the nuclear line flux was relatively weak, the photospheric ³He abundance is not well constrained (³He/H = $(0.5\text{-}10.0) \times 10^{-5}$). *Tatischeff, Kiener, and Gros* (2005) obtained a similar result for the 2003 October 28 flare.

Acknowledgments.   We wish to thank David Smith, Albert Shih, Bob Lin, Richard Schwartz, Kim Tolbert, and the rest of the *RHESSI* team for the excellent performance of the instrument, its calibration, and for the analysis routines that have made this work possible. We also wish to thank the referees and editor for helping to clarify the manuscript and Benz Kozlovsky for discussions. This work was supported by NASA DPR 10049 and NNG04ED181, and by the Office of Naval Research.

REFERENCES

Anders, E., and N. Grevesse, Abundances of the elements - Meteoritic and solar, *Geochim. et Cosmochim. Acta*, 53, 197-214, 1989.

Aschwanden, M., *Physics of the Solar Corona, An Introduction*, 842 pp., Springer/ Praxis Publishing, Chichester, UK, 2004.

Bahcall, J.N., S. Basu, and A.M. Serenelli, What is the neon abundance of the Sun?, *Astrophys. J.*, 631, 1281-1285, 2005.

Bernasconi, P.N., C.U. Keller, H.P. Povel, and J.O. Stenflo, Direct measurements of flux tube inclinations in solar plages, *Astron. Astrophys.*, 302, 533-542, 1995.

Cane, H.V., T.T. von Rosenvinge, C.M.S. Cohen, and R.A. Mewaldt, Two components in major solar particle events, *Geophys. Res. Lett.*, 30, 8017, doi:10.1029/ 2002GL016580, 2003.

Chadwick, M.B., P.G. Young, S. Chiba, S.C. Frankle, *et al.*, Cross-section evaluations to 150 MeV for accelerator-driven systems and implementation in MCNPX, *Nucl. Sci. & Eng.*, 131, 293-328, 1999.

Chupp, E.L., Transient particle-acceleration associated with solar flares, *Science*, 250, 229-236, 1990.

Chupp, E.L., D.J. Forrest, J.M. Ryan, J. Heslin, *et al.*, A direct observation of solar neutrons following the 0118-UT flare on 1980 June 21, *Astrophys. J.*, 263, L95-L99, 1982.

Crannell, C.J., and G. Joyce, R. Ramaty, and C. Werntz, Formation of 0.511 MeV line in solar flares, *Astrophys. J.*, 210, 582-592, 1976.

Donea, A.-C., and C. Lindsey, Seismic emission from the solar flares of 2003 October 28 and 29, *Astrophys. J.*, in press, 2005.

Drake, J.D., and P. Testa, The 'solar model problem' solved by the abundance of neon in nearby stars, *Nature*, 436, 525-528, 2005.

Emslie, A.G., H. Kucharek, B.R. Dennis, N. Gopalswamy, *et al.*, Energy Partition in Two Solar Flare/CME Events, *J. Geophys. Res.*, 109, A10104, 2004.

Feldman, U., E. Landi, and J.M. Laming, Helium abundance in high-temperature solar flare plasmas, *Astrophys. J.*, 619, 1142-1152, 2005.

Gan, W.Q., Gamma-ray line analysis for the flares of 28 October and 2 November 2003, *Adv. Sp. Res*, 35, 1833-1838, 2005.

Gros, M., V. Tatischeff, J. Kiener, B. Cordier, *et al.*, INTEGRAL/SPI observation of the 2003 Oct 28 solar flare, *Proc. 5ᵗʰ INTEGRAL Workshop, Munich, Germany, ESA SP-552*, 669-676, 2004.

Hua, X.-M. and R.E. Lingenfelter, A determination of the (He-3)/H ratio in the solar photosphere from flare gamma-ray line observations, *Solar Physics*, 319, 555-566, 1987.

Hulot, E., N. Vilmer, E.L. Chupp, B.R. Dennis, and S.R. Kane, Gamma ray and X-ray time profiles expected from a trap-plus-precipitation model for the 7 June 1980 and 27 April 1981 solar flares, *Astron. Astrophys.*, 256, 273-285, 1992.

Kallenrode, M.-B, E.W. Cliver, and G. Wibberenz, Composition and azimuthal spread of solar energetic particles from impulsive and gradual flares, *Astrophys. J.*, 391, 370-379, 1992.

Kiener, J., N. de Séréville, and V. Tatischeff, Shape of the 4.438 MeV gamma-ray line of C-12 from proton and alpha-particle induced reactions on C-12 and O-16, *Phys. Rev. C*, 64, 025803, 2001.

Kiener, J., M. Gros, V. Tatischeff, and G. Weidenspointner, Properties of the energetic particle distributions during the October 28, 2003 solar flare from *INTEGRAL*/SPI observations, *Astron. Astrophys.*, in press, 2005.

Kozlovsky, B., R.E. Lingenfelter, and R. Ramaty, Positrons from accelerated particle interactions, *Astrophys. J.*, 316, 801-818, 1987.

Kozlovsky, B., R.J. Murphy, and R. Ramaty, Nuclear deexcitation gamma-ray lines from accelerated particle interactions, *Astrophys. J. (Supp.)*, 141, 523-541, 2002.

Kozlovsky, B., R.J. Murphy, and G.H. Share, Positron-emitter production in solar flares from ³He reactions, *Astrophys. J.*, 604, 892-899, 2004.

Kuznetsov, S.N., V.G. Kurt, B.Y. Yushkov, I.N. Myagkova, and K. Kudela, Gamma-ray and high-energy particle measurements of the solar flare of 28 October 2003 on board *CORONAS-F*, *Solar Physics, submitted*, 2005.

Laming, J.M., A unified picture of the first ionization potential and inverse first ionization potential effects, *Astrophys. J.*, 614, 1063-1072, 2004.

Li, G., and G.P. Zank, Mixed particle acceleration at CME-driven shock and flares, *Geophys. Res. Lett.*, 32, L02101, doi:10.1029/2004GL021250, 2005.

Mandzhavidze, N., R. Ramaty, and B. Kozlovsky, Solar atmospheric and solar flare accelerated helium abundances and gamma-ray spectroscopy, *Astrophys. J.*, 489, L99-L102, 1997.

Mandzhavidze, N., R. Ramaty, and B. Kozlovsky, Determination of subcoronal ⁴He and solar flare-accelerated ³He and ⁴He from gamma-ray spectroscopy, *Astrophys. J.*, 518, 918-925, 1999.

Mason, G.M., J.E. Mazur, and J.R. Dwyer, ³He enhancements in large solar energetic particle events, *Astrophys. J.*, 525, L133-L136, 1999.

Mason, G.M., M.E. Wiedenbeck, J.A. Miller, J.E. Mazur, *et al.*, Spectral properties of He and heavy ions in ³He-rich solar flares, *Astrophys. J.*, 574, 1039-1058, 2002.

Masuda, S, T. Kosugi, H. Hara, and Y. Ogawara, A loop top hard x-ray source in a compact solar-flareas evidence for magnetic reconnection, *Nature*, 371, 495-497, 1994.

McTiernan, J.M., and V. Petrosian, Center-to-limb variations of characteristics of solar flare hard x-ray and gamma-ray emission, *Astrophys. J.*, 379, 381-39, 1991.

Mewaldt, R., C., M.D. Looper, C.M.S. Cohen, G.M. Mason, *et al.*, Solar-particle energy spectra during the large events of October-November 2003 and January 2005, *Proc. 29ᵗʰ Int. Cosmic Ray, Pune, India, SH 1.2*, 101-104, 2005.

Miller, J.A., P.J. Cargill, A.G. Emslie, G.D. Holman, *et al.*, Critical issues for understanding particle acceleration in impulsive flares, *J. Geophys. Res.*, 102, 14631-14659, 1997.

Murphy, R.J., C.D. Dermer, and R. Ramaty, High-energy processes in solar flares, *Astrophys. J. (Supp.)*, 63, 721-748, 1987.

Murphy, R.J., B. Kozlovsky, J.G. Skibo, and G.H. Share, The Physics of Positron Annihilation in the Solar Atmosphere, *Astrophys. J. (Supp.)*, 161, -, 2005.

Murphy, R.J., and G.H. Share, What gamma-ray de-excitation lines reveal about solar flares, *Adv. Sp. Res.*, 35, 1825-1832, 2005.

Murphy, R.J., G.H. Share, J.E. Grove, W.N. Johnson, *et al.*, Accelerated particle composition and energetics and ambient abundances from gamma-ray spectroscopy of the 1991 June 4 solar flare, *Astrophys. J.*, 490, 883-900, 1997.

Murphy, R.J., G.H. Share, X.-M. Hua, R.P. Lin, *et al.*, Physical implications of *RHESSI* neutron-capture line measurements, *Astrophys. J.*, 595, L93-L97, 2003.

Ramaty, R., B. Kozlovsky, and R.E., Lingenfelter, Nuclear gamma-rays from energetic particle interactions, *Astrophys. J. (Supp.)*, 40, 487-526, 1979.

Ramaty, R. and N. Mandhavidze, Highly energetic physical processes and mechanisms for emission from astrophysical plasmas, *IAU Symposia*, 195, 123-132, 2000.

Ramaty, R., N. Mandzhavidze, B. Kozlovsky, and J.G. Skibo, Acceleration in solar flares – interacting particles versus interplanetary particles, *Adv. Sp. Res.*, (9), 275-284, 1993.

Ramaty, R., N. Mandzhavidze, B. Kozlovsky, and R.J. Murphy, Solar atmospheric abundances and energy content in flare-accelerated ions from gamma-ray spectroscopy, *Astrophys. J.*, 455, L193-L196, 1995.

Ramaty, R., N. Mandzhavidze, and B. Kozlovsky, Solar atmospheric abundances from gamma ray spectroscopy, *High Energy Solar Physics, ed. R. Ramaty, N. Mandzhavidze, and X-M Hua, AIP Conf. Proc.*, 374, 172-183, 1996.

Ramaty, R., N. Mandzhavidze, C. Barat, and G. Trottet, The giant 1991 June 1 flare: evidence for gamma-ray production in the corona and accelerated heavy ion abundance enhancements from gamma-ray spectroscopy, *Astrophys. J.*, 479, 458-463, 1997.

Reames, D.V., Coronal abundances determined from energetic particles, *Adv. Space Res.*, 15 (7), 41, 1995.

Reames, D.V., Particle acceleration at the Sun and in the heliosphere, *Space Sci. Rev.*, 90, 413-491, 1999.

Reames, D.V., Meyer, J.P., and von Rosenvinge, T.T., Energetic-particle abundances in impulsive solar-flare events, *Astrophys. J. (Supp.)*, 90, 649-667, 1994.

Rieger, E., W.Q. Gan, and H. Marschhäuser, Gamma-ray line versus continuum emission of electron-dominated episodes during solar flares, *Solar Physics*, 183, 123-132, 1998.

Schmelz, J.T., K. Nasraani, J.K. Roames, L.A. Lipper, and J.W. Garst, Neon lights up a controversy, the solar Ne/O abundance, *Astrophys. J.*, 634, L 197-L 200, 2005

Share, G.H., and R.J. Murphy, Gamma-ray measurements of flare-to-flare variations in ambient solar abundances, *Astrophys. J.*, 452, 933-943, 1995.

Share, G.H., and R.J. Murphy, Intensity and directionality of flare-accelerated alpha-particles at the Sun, *Astrophys. J.*, 485, 409-418, 1997.

Share, G.H., and R.J. Murphy, Accelerated and ambient He abundances from gamma-ray line measurements of flares, *Astrophys. J.*, 508, 876-884, 1998.

Share, G.H., and R.J. Murphy, Gamma-ray measurement of energetic heavy ions at the sun, *Proc. of the 26th Cosmic Ray Conference*, 6, 13-16, 1999.

Share, G.H., and R.J. Murphy, Solar Gamma-Ray Line Spectroscopy Physics of a Flaring Star, in Stars as Suns: Activity, *Evolution and Planets, IAU Conf. Series 219*, ed. A.K. Dupree and A.O. Benz, 133-144, 2004c.

Share, G.H., R.J. Murphy, J. Kiener, and N. de Sereville, Directionality of solar flare accelerated protons and alpha particles from gamma-ray line measurements, *Astrophys. J.*, 573, 464-470, 2002.

Share, G.H., R.J. Murphy, J.G. Skibo, D.M. Smith, *et al.*, High-resolution observation of the solar positron-electron annihilation line, *Astrophys. J.*, 595, L85-L88, 2003.

Share, G.H., R.J. Murphy, D.M. Smith, R.A. Schwartz, and R.P. Lin, *RHESSI* e+ - e− annihilation radiation observations: implications for conditions in the flaring solar chromosphere, *Astrophys. J.*, 615, L169-L172, 2004a.

Share, G.H., R.J. Murphy, D.M. Smith, R.P. Lin, and A.Y. Shih, Accelerated particle spectra and abundances in the 2003 October 28 and November 2 solar flares (abstract), *Fall AGU Mtg.*, SH13A-1135, 2004b.

Shih, A.Y., D.M. Smith, R.P. Lin, G.H. Share, *et al.*, *RHESSI* observations of gamma-ray lines from solar flares (abstract), *Fall AGU Mtg.*, SH24A-01, 2004.

Smith, D.M., G.H. Share, R.J. Murphy, R.A. Schwartz, *et al.* A.Y., High-resolution spectroscopy of gamma-ray lines from the x-class solar flare of 2002 July 23, *Astrophys. J.*, 595, L81-L84, 2003.

Smith, D.M., A.Y. Shih, R.J. Murphy, G.H. Share, R.A. Schwartz, and R.P. Lin, *RHESSI* spectroscopy of nuclear de-excitation lines in x-class flares (abstract), *AAS Mtg.*, 204, 02.01, 2004.

Tatischeff, V., J. Kiener, and M. Gros, Phyical implications of *INTEGRAL*/ SPI gamma-ray line measurements of the 2003 October 28 solar flare, Proceedings of the 5th Rencontres du Vietnam, "New Views on the Universe", Hanoi, Vietnam, in press (astro-ph/0501121)

Toner, M.P., and A.L. MacKinnon, Do fast protons and α particles have the same energy distributions in solar flares?, *Solar Physics*, 223, 155-168, 2004

Torsti, J., L. Kocharov, J. Laivola, S. Pohjolainen, *et al.*, Solar particle events with helium-over-hydrogen enhancement in the energy range up to 100 MeV nucl⁻¹, *Solar Physics*, 205, 123-147, 2002.

Tylka, A.J., C.M.S. Cohen, W.F. Dietrich, S. Krucker, *et al.*, Onsets and release times in solar particle events, *Proc. of the 28th Cosmic Ray Conference*, 7, 3305-3308, 2003.

Tylka, A.J., C.M.S. Cohen, W.F. Dietrich, M.A. Lee, *et al.*, Shock geometry, seed populations, and the origin of variable elemental composition at high energies in large gradual solar particle events, *Astrophys. J.*, 625, 474-495, 2005.

Veselovsky, I.S., M.I. Panasyuk, S.I. Avdyushin, G.A. Bazilevskaya, *et al.*, Solar and Heliospheric Phenomena in October November 2003: Causes and Effects, *Cosmic Res.*, 42, 435-488, 2004.

Vilmer, N., G. Trottetet, C. Barat, R.A. Schwartz, S. Enome, A. Kuznetsov, R. Sunzaev, and O. Terekhov, Hard X-ray and gamma-ray observations of an electron dominated event associated with an occulted solar flare, *Astron. Astrophys.*, 342, 575-582, 1999.

Young, P.R., The Ne/O abundance ratio in the quiet Sun, *Astron. Astrophys*, 444, L45-L48, 2005.

G.H. Share and R.J. Murphy, E.O. Hulburt Center for Space Research Naval Research Laboratory Washington, DC 20375. (gerald.share@nrl.navy.mil and murphy@ssd5.navy.mil)

Particle Acceleration in Solar Flares and Escape Into Interplanetary Space[1]

Markus J. Aschwanden

Lockheed Martin ATC, Solar and Astrophysics Laboratory, Palo Alto, California, USA.

We review the physics of particle acceleration and kinematics in solar flares under the particular aspect of their escape and propagation into interplanetary space. The topics include the magnetic topology in acceleration regions, the altitude of flare acceleration regions, evidence for bi-directional acceleration, the asymmetry of upward versus downward acceleration, and particle access to interplanetary space.

1. INTRODUCTION

High-energy particles associated with solar flares can be generated (1) either directly in the coronal flare site with subsequent escape into interplanetary space, or (2) they can be accelerated in CME (coronal mass ejection)-associated shocks that propagate through interplanetary space. It is often difficult to decide between the two different origins of accelerated particles, but this question is of fundamental importance to understand the origin of solar energetic particles (SEP) detected in space. Theoretically, the velocity dispersion of high-energy particles detected by a spacecraft in near-Earth orbit or in interplanetary space allows us to determine the distance to the acceleration or injection region, but in practice, the accuracy is generally insufficient to discriminate between sources in coronal flare sites or in CME locations near the Sun, say within a few solar radii. Furthermore, the time-of-flight measurement technique implies also the two assumptions of (1) a sharp onset of the injection profile and of (2) no significant scattering of the particles during propagation. In this review we restrict ourselves to the first

possibility, i.e., particle acceleration in coronal flare sites, and focus on the particular aspect of particle escape and propagation into interplanetary space. The reader has also to bear in mind that most of the following discussion is based on electrons as diagnosed from their copious hard X-ray and radio emission, while we are far less certain about the acceleration regions and magnetic topologies of ions, which can only be diagnosed in gamma-rays or in-situ.

2. MAGNETIC TOPOLOGY OF ACCELERATION REGION

The most crucial criterion whether high-energy particles can escape into interplanetary space is the magnetic topology in the acceleration region. Accelerated particles can escape if they have access to open field lines, otherwise they remain trapped on closed field lines.

There are three basic topologies of magnetic reconnection geometries between open and closed field lines: The pre-reconnection geometry consists of a pair of (1) open-open, (2) open-closed, or (3) closed-closed magnetic field lines (*Aschwanden*, 2002; 2004). If these pairs of pre-reconnection magnetic field lines are coplanar, we have a 2D model, as shown in Figure 1 (top row, thick dashed lines). The disjoint field lines are brought into contact with each other during the reconnection process (dotted lines in Figure 1, top row), and then relax into the post-reconnection configuration (shown with solid double lines in Figure 1 top row).

A standard 2D flare model of the dipolar or open-open type is the Carmichael-Sturrock-Hirayama-Kopp-Pneuman

[1] AGU Monograph of AGU Chapman Conference "Solar Energetic Plasmas and Particles", Turku, Finland, 2–6 August 2004 (eds. *N. Gopalswamy et al.*), in press, 2005.

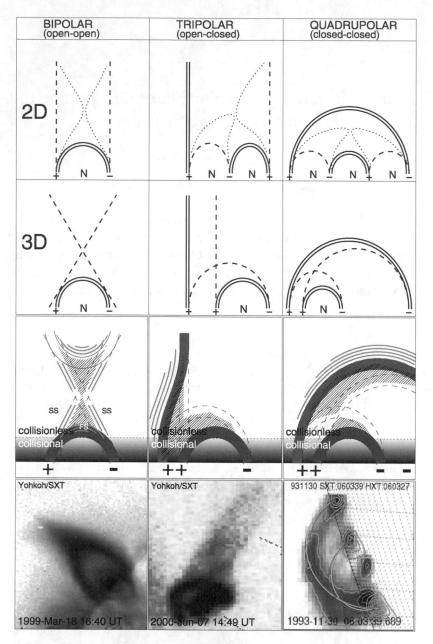

Figure 1. The topology of magnetic reconnection regions is classified into three classes: bipolar or open-open (left column), tripolar or open-closed (middle column), and quadrupolar or closed-closed field line reconnection (right column). The 2D versions are shown in the top row, with the pre-reconnection field lines marked with dashed lines, during reconnection with dotted lines, and post-reconnection field lines with double solid linestyle. The 3D versions are indicated in the second row, where the pre-reconnection field lines are not coplanar, but located behind each other. The third row indicates the acceleration regions (hatched), the relative densities (greyscale), and upward/sideward directed shocks (grey lines). The bottom row shows flare observations from Yohkoh/SXT that correspond to the three different reconnection topologies (adapted from Aschwanden 2002).

(CSHKP) reconnection model. It starts with a helmet-streamer configuration with two antiparallel magnetic field lines above the cusp of the streamer, where a Y-type reconnection geometry occurs in the cusp, as observed in the famous "candle-flame" flare of 1992-Feb-21 (*Tsuneta et al.,* 1992), which is similar to the 1999-Mar-18 flare shown in Figure 1 (bottom left). We see that the end product is one closed (postflare) loop (Figure 1, top left). The observations (Figure 1, bottom left) show only the lower part with a cusp and postflare loop, but in a vertically symmetric X-type geometry we would expect also an upward reconnected segment that escapes into interplanetary space.

The tripolar type involves three magnetic poles (Figure 1, top middle), but this term is not widely used in literature. This tripolar case has been disussed in the context of escape of impulsive energetic particles from flare sites (*Reames*, 2002), and as a way of avoiding the interplanetary magnetic flux buildup from CMEs (*Crooker et al.*, 2002), called "interchange reconnection" therein. Variants of this type of magnetic reconnection in tripolar geometries were also envisioned in the context of emerging-flux models (*Heyvaerts, Priest, and Rust*, 1977) and particularly after the discovery of soft X-ray plasma jets with Yohkoh (*Shibata et al.*, 1992). The observation of long straight soft X-ray jets (e.g., Figure 1, bottom middle) were taken as evidence of plasma flows along open field lines, a fact that constitutes a flare-like process between a closed and an open field line. The end product of tripolar (open-closed) reconnection is one closed post-reconnection (postflare) loop and one open field line (Figure 1, top middle), usually associated with a soft X-ray jet.

The quadrupolar type (Figure 1, top right) is also called interacting-loop model and has been theoretically modeled in terms of magnetic flux transfer between two current-carrying loops (*Melrose*, 1997). Classical examples have been observed with Yohkoh/SXT by *Hanaoka*, (1996), *Nishio et al.*, (1997), and modeled in terms of 3D quadrupolar geometries by *Aschwanden et al.*, (1999). The initial situation as well as the end product of quadrupolar reconnection are two closed loops, but the footpoint connectivities between opposite polarities are switched during reconnection. The outcomes are similar in 2D and 3D (Figure 1, second row), except that the footpoints and loops are not lined up in a single plane in 3D, but can have arbitrary shear angles between the pre-reconnection loops.

Observations usually do not make the pre-reconnection configuration visible, but display the post-reconnection field lines only, because they become filled with dense hot flare plasma by the chromospheric evaporation process, which is easily to detect in soft X-rays, as shown in the examples in Figure 1 (bottom row). For escape into interplanetary space, we need the involvement of open field lines, which is the case for bipolar and tripolar reconnection. In contrast, the quadrupolar reconnection does not involve open field lines, neither in the pre-reconnection nor in the post-reconnection configuration, and thus is likely to confine flare-generated high-energy particles.

3. ALTITUDE OF ACCELERATION REGIONS

The localization of the acceleration region in solar flares has been greatly helped by *Masuda's* discovery of above-the-looptop hard X-ray sources. *Masuda et al.*, (1994a) discovered an above-the-looptop hard X-ray source at energies of ≳20–50 keV with Yohkoh/HXT in about 10 flares (four examples are shown in Figure 2, left), besides the well-known (usually double-footpoint) chromospheric footpoint sources. Initially it was not clear how electrons can emit collisional bremsstrahlung in such low plasma densities as measured above flare loops, typically $n_e \lesssim 10^8$–10^9 cm^{-3}. A plausible explanation is collisional bremsstrahlung from trapped electrons, which are directly fed in from the accelerator in the cusp region beneath the reconnection point. The location of the coronal hard X-ray source was measured to be about 10" (7250 km) above the soft X-ray loop, and in slightly higher altitudes in images taken in the higher (≳50 keV) energy bands (*Masuda et al.*, 1994b). This location is fully consistent with the cusp geometry in bipolar reconnection models (Figure 1, left), and thus the coronal hard X-ray emission has to be emitted relatively close to the acceleration region associated with the reconnection point.

An independent method to determine the geometric location of the acceleration region are electron time-of-flight (TOF) measurements between the acceleration region and the collisional stopping region, which is the chromosphere in the thick-target model. In the simplest scenario, electrons are accelerated at some coronal altitude h_{acc} and propagate along the magnetic field lines down to the chromosphere, where they are stopped at height h_{stop} in the transition region or upper chromosphere. Because hard X-ray photons at different energies are caused by electrons with different kinetic energies or velocities, say v_1 and v_2, there will be an electron time-of-flight (TOF) difference $\Delta t = t_2 - t_1$ in the arrival time of electrons, if they are simultaneously injected and propagate over the same distance $L = h_{acc} - h_{stop}$,

$$t_2 - t_1 = \left(\frac{L}{v_2} - \frac{L}{v_1} \right).$$

Such time-of-flight differences could be measured from fast time structures in different hard X-ray spectral channels with the Burst and Transient Source Experiment (BATSE) Large Area Detectors (LAD) onboard the Compton Gamma Ray Observatory (CGRO) (*Aschwanden et al.*, 1995a,b; 1996a,b,c). Correcting for the curvature of the time-of-flight path, the pitch angle of the free-streaming electrons, and the twist of the magnetic field lines, a scaling law was found between the projected TOF distance (l') and the flare loop half length s (Figure 2, right),

$$L'/s = 1.43 \pm 0.30$$

This scaling law is also approximately the ratio between the ltitude of the acceleration region h_{acc} and the height of the flare loop h_{loop}. Since the investigated flare loops have

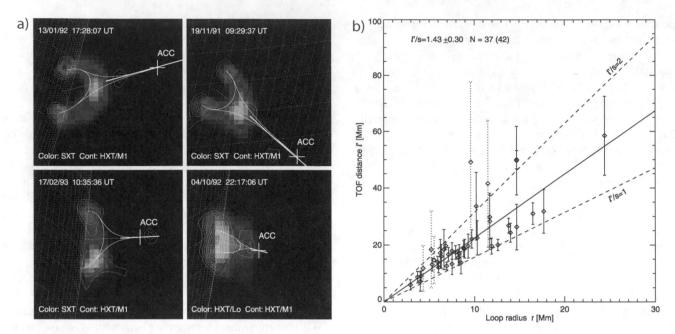

Figure 2. *Left:* The altitude of acceleration regions is inferred from electron time-of-flight distances based on energy-dependent time delays of 20-200 keV hard X-ray emission measured with BATSE/CGRO (crosses marked with ACC). Simultaneous direct detections of above-the-looptop hard X-ray sources (contours) and thermal hard X-ray emission (greyscale) from the low-energy channel of Yohkoh/HXT/Lo outline the cusp above the flare loop. *Right:* A scaling law of the TOF distance l' versus the loop radius r is determined from 42 flare events simultaneously observed with CGRO and Yohkoh. The average ratio of the TOF distance l' and loop half length s is indicated with a solid line, amounting to a mean of $l'/s = 1.43 \pm 0.30$. For comparison, a ratio of $l'/s = 1$ and $l'/s = 2$ is indicated with dashed lines.

a range of $h_{loop} \approx 3 - 20$ Mm, the height of the acceleration sites amounts to $h_{acc} \approx 5 - 35$ Mm, which is less than 5% of a solar radius.

Earlier stereoscopic multi-spacecraft measurements of occulted flares yielded maximum heights of $h \lesssim 30$ Mm for impulsive hard X-ray emission (*Kane et al.*, 1979; *Kane et al.*, 1982; *Hudson*, 1978; *Kane*, 1983). A record height of $\gtrsim 200$ Mm was determined for an occulted flare (1984-Feb-16, 09:12 UT) at hard X-ray energies of $\gtrsim 5$ keV, by combined stereoscopic measurements with the PVE, ICE and GOES spacecraft (*Kane et al.*, 1992). This most extreme case could be associated with thin-target emission from trapped electrons in a large-scale flare loop. Otherwise both the stereoscopic and TOF method yield a consistent upper limit of $h_{arr} \lesssim 0.05 \, R_\odot$.

4. BI-DIRECTIONAL ACCELERATION

During the flare of 1992-Sep-06, clear evidence for the acceleration of bi-directional electron beams has been observed in the radio dynamic spectra. A common frequency band was observed for positively (III) and negatively drifting (RS) radio bursts, which suggests that the start frequencies

are co-spatial with the acceleration region. Figure 3 shows a dynamic radio spectrum of the frequency range of 400-1500 MHz, together with a >25 keV hard X-ray profile. A total of 52 radio bursts were identified during the central 45 s around the flare peak, consisting of 30 narrow-band (<50 MHz) and 22 broad-band bursts, the latter containing 12 normal-drifting type III bursts and 20 reverse-slope (RS) bursts. The most interesting finding is that 10 structures consist of pairs of oppositely drifting (type III + RS) bursts, starting simultaneously at the same start frequency in upward and downward direction, suggesting a symmetric injection from a common accelerator. Although the start frequencies of all 52 bursts scatter over a relatively large frequency range of 400-1500 MHz, there is a concentration of one half of the bursts to start in a fairly narrow range of $f = 610 \pm 20$ MHz, which is only a variation of $\Delta f \approx 6\%$ in plasma frequency. This corresponds, using a hydrostatic density model ($n_e(h) \propto \exp(-h/\lambda)$) and the plasma frequency ($f \approx f_p \propto \sqrt{n_e}$), to a height difference of

$$\Delta h = \left(\frac{dh}{dn_e} \right) \left(\frac{dn_e}{df} \right) \Delta f = 2\lambda \frac{\Delta f}{f}$$

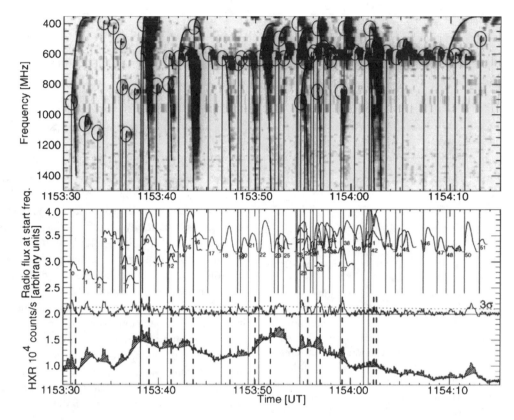

Figure 3. Evidence for bi-directional electron acceleration is shown during (45 s) the impulsive phase of the 1992-Sep-6 flare. The upper panel shows a radio spectrum recorded by ETH Zurich in the 400-1500 MHz frequency range. A total of 52 radio bursts are identified: the start frequencies (measured at the earliest detection time) are marked with circles and connected with a thin line to the hard X-ray data shown in the lower panel (CGRO/BATSE/DISCSC data with 64 ms resolution from the two most sunlit detectors). Drifting burst structures are traced out by thick lines; the precipitation times of reverse-drifting bursts are indicated by dashed lines. The time profiles of the start frequency of 52 radio bursts are displayed in the bottom panel (numbered with no. 0-51) to faciliate comparison with hard X-ray peaks. Fast pulses are shaded in the hard X-ray time profile (bottom profile), and a lower envelope is subtracted (second-lowest time profile) (*Aschwanden, Benz, and Schwartz*, 1993).

which is 12% of the density scale height, i.e. $\Delta h \approx 6000$ km for a standard coronal temperature of $T \approx 1$ MK. Thus we conclude that during the main flare phase the majority of electron beams are accelerated in a relatively compact region with a radius of a few 1000 km, while a smaller number of beams originates in a larger volume dispersed over a few 10,000 km. If we associate the acceleration sites with the cusps of reconnection regions, which have typical electron densities of $n_e \approx 10^9 - 10^{10}$ cm^{-3}, based on Yohkoh observations, we expect radio start frequencies of $f \approx 0.3 - 1.0$ GHz. The case shown in Figure 3, with a center frequency of $f = 0.6$ GHz, falls middle in this range. Therefore, if coronal conditions are favorable to detect plasma emission in this frequency range, the start frequencies $f_{SF}(n_e)$ of dual type III + RS radio bursts provide a direct density measurement n_e of the reconnection region. Moreover, if a suitable density model $n_e(h)$ is used, the scatter of start frequencies can be mapped

to a distribution of altitudes $h(n_e[f_{SF}])$ of elementary acceleration sites.

5. UPWARD VERSUS DOWNWARD ACCELERATION

The best evidence for high-energy particles that both escape into interplanetary space and produce flare emission in the lower corona probably comes from the generation of bi-directional electron beams, which are manifested in the form of simultaneously starting pairs of normal-drifting type III radio bursts and reverse-drifting type (RS) radio bursts (*Aschwanden et al.*, 1995c), and moreover coincide with synchronized hard X-ray pulses (*Aschwanden et al.*, 1995d). Type III bursts are generally interpreted as plasma emission resulting from electron beams along open field lines, while type J, U, or N bursts originate along closed field lines (Figure 4). However, although we might conceive the

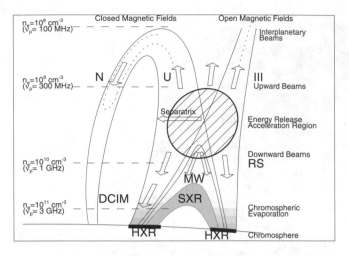

Figure 4. Radio burst types in the framework of the standard flare scenario: The acceleration region is located in the reconnection region above the soft X-ray-bright flare loop, accelerating electron beams in the upward direction (type III, U, N bursts) and in the downward direction (type RS, DCIM bursts). Downward moving electron beams precipitate to the chromosphere (producing hard X-ray emission and driving chromospheric evaporation), or remain transiently trapped, producing microwave (MW) emission. Soft X-ray loops become subsequently filled up, with increasing footpoint separation as the X-point rises. The insert shows a dynamic radio spectrum (ETH Zurich) of the 92-Sept-06, 1154 UT, flare, showing a separatrix between type III and type RS bursts at ≈600 MHz, probably associated with the acceleration region (*Aschwanden*, 1998).

acceleration region to be associated with a symmetric X-point, and thus might expect similar amounts of particle numbers to be accelerated in upward and downward direction, number estimates based on hard X-ray fluxes reveal a large asymmetry, with the upgoing propagating >20 keV electrons escaping into the interplanetary medium being 2-4 orders of magnitude lower than the downward accelerated electrons that produce hard X-rays (*Lin and Hudson*, 1976). There might be a number of reasons for this asymmetry: (1) the acceleration efficiency may be different, since both the electron density and the magnetic field are expected to be lower in the corona above an X-point than below; (2) many of the newly-reconnected magnetic field lines above the X-point may be closed rather than open (Figure 4, left side); (3) a lot of the upward-propagating low-energy electrons may lose their energy in the lower corona above the X-point before they have a chance to escape into the interplanetary medium, or (4) the electron beam formation and/or radiation propagation effects of radio type III emission are less favorable in upward direction. A semi-quantitative model of bi-directional electrons beams based on stochastic Langmuir- wave growth, three-wave interactions, and quasi-linear beam relaxation (e.g., *McLean and Labrum*, 1985) investigated the relative

occurrence rate of radio type III, U, and RS burst emission in the fundamental and harmonic mode and found that the upward/downward asymmetry could be explained by the decrease of the ratio of magnetic to thermal energy with altitude, which controls the heating rate and affects the beam relaxation process (*Robinson and Benz*, 2000).

6. ACCESS TO INTERPLANETARY SPACE

Given the dominance of closed-field regions over active regions, where virtually all flares are harbored, how can flare-accelerated electrons have access to interplanetary space at all? The getaway to interplanetary space seems to be facilitated on one hand by dynamic flare processes that open temporarily and locally the field, such as filament eruptions, coronal mass ejections, or some particular magnetic reconnection processes that involve open field lines, such as the magnetic breakout model (*Antiochos et al.*, 1999). On the other hand, open field regions that connect directly to the interplanetary space exist not only in coronal holes, but also to a substantial fraction in active regions. *Schrijver and DeRosa* (2003) found from potential-field extrapolations of the global magnetic field over the entire solar surface that the interplanetary magnetic field (IMF) originates typically in a dozen disjoint regions, around the solar cycle maximum. While active regions are often ignored as a source for the interplanetary magnetic field, *Schrijver and DeRosa* (2003) found that the fraction of the IMF that connects directly to magnetic plages of active regions increases from $\lesssim 10\%$ at cycle minimum up to 30–50% at cycle maximum, with even direct connections between sunspots and the heliosphere (Figure 5). Additional support for the magnetic connectivity comes also from solar wind investigators who tied open interplanetary fields to active region sources (*Neugebauer et al.*, 2002; *Liewer et al.*, 2004)

What is the relation of interplanetary electron beams to coronal radio type III bursts? There is not a simple one-to-one correspondence between interplanetary electron beams and solar type III electrons beams. *Lin* (1997) has undertaken an investigation on this connection and has concluded that a coronal source of interplanetary electron streams related to solar type III bursts exists. A more detailed analysis was undertaken by *Poquérusse et al.*, (1996), using the Nançay Artemis spectrograph to relate ground-based observations of solar type III radio bursts with Ulysses measurements of interplanetary type III radiation. *Poquérusse et al.*, (1996) find that to almost every interplanetary type III burst, whose spiral field line is connected to the visible disk or limb of the Sun, a group of solar type III bursts can be related. Figure 6 shows an example of such an Artemis-Ulysses reconstruction of the connection between solar and interplanetary type III bursts. Extending the slope of the leading edge of the

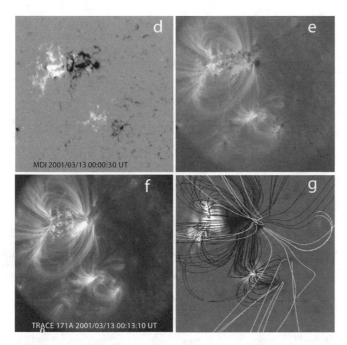

Figure 5. *Panel d:* MDI magnetogram; *Panel e:* MDI magnetogram overlayed with TRACE 171Å; *Panel f:* TRACE 171Å image of 2001-Mar-13, 00:13 UT; *Panel g:* Potential field extrapolation using a source-surface model. Closed field lines of active regions are indicated with black color, the open field lines that connect to interplanetary space with white (courtesy of *Schrijver and DeRosa*, 2003).

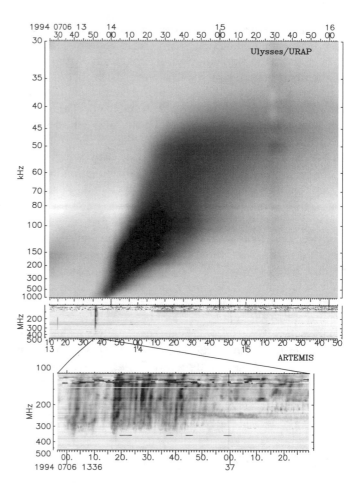

Figure 6. Artemis and Ulysses observations of a group of solar metric type III bursts and an interplanetary type III radio burst. The latter is represented with frequency decreasing upward as is common for solar type III bursts. The scale is chosen in such a way that its prolongation to higher frequencies gives the starting point of the interplanetary burst in the corona. It is seen that this point coincides with the low-frequency starting point of the solar type III group. The lowest panel shows the high time resolution dynamic spectrum of the type III group identifying it as a large group consisting of many members (*Poquérusse et al.*, 1996).

interplanetary type III emission across the measurement gap down to the higher frequencies of the coronal band, the origin of the interplanetary type III burst is found to coincide with the end of the coronal type III group. Almost all interplanetary type III bursts are found to be rooted in coronal type III bursts. Their electron beams originate in the solar corona and propagate thus stably all the way through the corona. Inversely, not every solar type III burst, even if strong, produces an interplanetary type III burst. To become an interplanetary electron beam requires additional conditions to be satisfied, presumably open magnetic fields and possibly other conditions as well. Moreover, type III bursts giving rise to interplanetary type III bursts and electron beams usually appear in groups. These groups are not resolved at low frequencies when they merge in interplanetary space. In the solar atmosphere, each of the bursts lasts up to several seconds only. These groups of electrons obviously merge into one broad beam due to velocity dispersion, when entering interplanetary space.

There are two sources of solar energetic particles in interplanetary space, either flare-related magnetic reconnection sites in the solar corona that are connected to interplanetary space via open field lines, or shock acceleration sites associated with super-Alfvénic CME fronts that propagate through

interplanetary space. Since the plasma in interplanetary space is collisionless, superthermal and high-energy particles can propagate through interplanetary space and form particle beams (e.g., electron beams or ion beams). The beam free energy is converted into Langmuir waves, and some Langmuir wave energy is converted to radio waves at the fundamental or harmonic local plasma frequency (e.g., *McLean and Labrum*, 1985). Thus, beam-driven type III-like radio bursts are common in interplanetary space (Figure 6), and occasionally there occur also type IV-like radio bursts (i.e., synchrotron emission caused by energetic electrons confined in a magnetic trap created behind an interplanetary shock wave). The spatial size of interplanetary radio bursts can be very large, since the extent of the radio source grows with distance from the Sun. However, interplanetary type III emission is not produced continuously along the propagation path of electron beams, but rather seem to occur in localized, unresolved regions of the interplanetary medium. There occur also interplanetary type III-like bursts, also called shock-associated (SA) events, believed to be electron beams that are produced by collisionless shocks associated with passing CMEs and propagate in the antisunward direction from the (type II-emitting) shock (*Cane et al.*, 1981). Thus, interplanetary radio bursts provide a rich diagnostic on the acceleration and propagation of energetic particles and shock waves. However, only radio bursts with plasma frequencies $\gtrsim 20$ MHz (above the Earth's ionospheric cutoff frequency) can be observed with ground-based radio telescopes, which extends only out to about 1-2 solar radii, while all interplanetary radio bursts further out have lower plasma frequencies and require space-based radio detectors.

More details on interplanetary particle beams and radio bursts can be found in reviews of *Lin* (1974, 1993, 2000), *Simnett* (1986), *Dulk* (1990), *Schwenn and Marsch* (1991a,b), *Aschwanden and Treumann* (1997), *Robinson* (1997), *Reames* (1999, 2000), *Cairns et al.*, (2000), *Reiner* (2001), *Bougeret* (2000), and *Gopalswamy* (2004). In this monograph, see also articles on related topics, such as on solar energetic particle (SEP) events (*Von Rosenvinge, Klecker, Kocharov, Mewaldt, Tylka, Droege*), ion acceleration by shocks (*Lee, Mason et al.*), flare gamma-rays (*Share*), flare hard X-rays (*Lin, MacKinnon*), and flare radio emission (*Klein, Mann, Gopalswamy*).

REFERENCES

Antiochos, S.K., C.R. DeVore, and J.A. Klimchuk, A model for solar coronal mass ejections, *ApJ*, 510, 485-493, 1999.

Aschwanden, M.J. What did Yohkoh and Compton change in our perception of particle acceleration in solar flares?, in *Observational Plasma Astrophysics: Five Years of Yohkoh and Beyond*, Proc. Yohkoh Conference, held in Yoyogi, Tokyo, Japan, 1996 Nov 6-8, (eds. T. Watanabe, T. Kosugi, and A.C. Sterling), Astrophysics and Space Science Library Vol. 229 (398 pages), Kluwer Academic Publishers, Dordrecht, The Netherlands, p.285-294, 1998.

Aschwanden, M.J., Particle Acceleration and Kinematics in Solar Flares. A Synthesis of recent observations and theoretical concepts, *Space Science Reviews*, 101, 1-227, 2002.

Aschwanden, M.J. *Physics of the Solar Corona. An Introduction*. 842 pp., Praxis Publishing Ltd., Chichester UK, and Springer, New York, 2004.

Aschwanden, M.J., A.O. Benz, B.R. Dennis, and R.A. Schwartz, Solar electron beams detected in hard X-rays and radio waves, *ApJ*, 455, 347-365, 1995c.

Aschwanden, M.J., A.O. Benz, and R.A. Schwartz, The timing of electron beam signatures in hard X-rays and radio: solar flare observations with BATSE/Compton Gamma-Ray Observatory and PHOENIX, *ApJ*, 417, 790-804, 1993.

Aschwanden, M.J., H.S. Hudson, T. Kosugi, and R.A. Schwartz, The scaling law between electron time-of-flight distances and loop lengths in solar flares, *ApJ*, 464, 985-998, 1996a.

Aschwanden, M.J., T. Kosugi, Y. Hanaoka, M. Nishio, and D.B. Melrose, Quadrupolar magnetic reconnection in Solar Flares: I.3D Geometry inferred from Yohkoh observations, *ApJ*, 526, 1026-1045, 1999.

Aschwanden, M.J., T. Kosugi, H.S. Hudson, M.J. Wills, and R.A. Schwartz, The scaling law between electron time-of-flight distances and loop lengths in solar flares, *ApJ*, 470, 1198i-1217, 1996b.

Aschwanden, M.J., M. Montello, B.R. Dennis, and A.O. Benz, Sequences of correlated hard X-ray and type III bursts during solar flares, *ApJ*, 440, 394-406, 1995d.

Aschwanden, M.J. and R.A. Schwartz, Accuracy, uncertainties, and delay distribution of electron time-of-flight measurements in solar flares, *ApJ*, 455, 699-714, 1995b.

Aschwanden, M.J., R.A. Schwartz, and D.M. Alt, Electron time-of-flight differences in solar flares, *ApJ*, 447, 923i-935, 1995a.

Aschwanden, M.J. and R.A. Treumann, Coronal and Interplanetary Particle Beams, in "Coronal Physics from Radio and Space Observations", Proc. of the CESRA Workshop held in Nouan-le-Fuzelier, France 3-7, June 1996, (ed. G. Trottet), Lecture Notes in Physics Vol. 483, Springer, Berlin, pp.108-134, 1997.

Aschwanden, M.J., M.J. Wills, H.S. Hudson, T. Kosugi, and R.A. Schwartz, Electron time-of-flight distances and flare loop geometries compared from CGRO and YOHKOH observations, *ApJ*, 468, 398-417, 1996c.

Bougeret, J.L., Solar Wind: Interplanetary Radio Bursts, in *Encyclopedia of Astronomy and Astrophysics*, Institute of Physics Publishing, Grove's Dictionaries, Inc., New York, (ed. P. Murdin), 2000.

Cairns, I.H., P.A. Robinson, and G.P. Zank, Progress on coronal, interplanetary, foreshock, and outer heliospheric radio emissions, *Publ. Astron. Soc. Australia*, 17, 22, 2000.

Cane, H.V., R.G. Stone, J. Fainberg, R.T. Stewart, J.L. Steinberg, and S. Hoang, Radio evidence for shock acceleration of electrons in the solar corona, *Journal of Geophysical Research*, 8/12, 1285-1288, 1981.

Crooker, N.U., J.T. Gosling, and S.W. Kahler, Reducing heliospheric magnetic flux from coronal mass ejections without disconnection, *Journal of Geophysical Research (Space Physics)*, 107/A2, SSH 3-1, CiteID 1028, DOI: 10.1029/2001JA000236, 2002.

Dulk, G.A., Interplanetary particle beams, *Solar Phys.*, 130, 139-150, 1990.

Gopalswamy, N., Interplanetary radio bursts, in "Solar and Space Weather radiophysics", (eds., D.E. Gary and C.U. Keller), Kluwer Academic Publishers, The Netherlands, pp.305-333, 2004.

Hanaoka, Y., Flares and plasma flow caused by interacting coronal loops, *Solar Phys.*, 165, 275-301, 1996.

Heyvaerts, J., E.R. Priest, and D.M. Rust, An emerging flux model for the solar flare phenomenon, *ApJ*, 216, 123-137, 1977.

Hudson, H.S., A purely coronal hard X-ray event, *ApJ*, 224, 235-240, 1978.

Kane, S.R., Spatial structure of high energy photon sources in solar flares, *Solar Phys.*, 86, 355-365, 1983.

Kane, S.R., K.A. Anderson, W.D. Evans, R.W. Klebesadel, and J. Laros, Observation of an impulsive solar X-ray burst from a coronal source, *ApJ*, 233, L151-L155, 1979.

Kane, S.R., F.E. Fenimore, R.W. Klebesadel, and J.G. Laros, Spatial structure of >100 keV X-Ray sources in solar flares, *ApJ*, 254, L53-L57, 1982.

Kane, S.R., J. McTiernan, J. Loran, E.E. Fenimore, R.W. Klebesadel, and J.G. Laros, Stereoscopic observations of a solar flare hard X-ray source in the high corona, *ApJ*, 390, 687-702, 1992.

Liewer, P.C., M. Neugebauer, and T. Zurbuchen, Characteristics of active-region sources of solar wind near solar maximum *Solar Phys.*, 223, 209-229, 2004.

Lin, R.P., Non-relativistic solar electrons, *Space Science Reviews*, 166, 189-256, 1974.

Lin, R.P. and H.S. Hudson, Non-thermal processes in large solar flares, *Solar Phys.*, 50, 153-178, 1976.

Lin, R.P., The relationship between energetic electrons interacting at the Sun and escaping to the interplanetary medium, *Adv. Space Res.*, 13/9, 265-273, 1993.

Lin, R.P., Observations of the 3D Distributions of Thermal to Near-Relativistic Electrons in the Interplanetary Medium by the Wind Spacecraft, *Lecture Notes in Physics.*, 483, 93-107, 1997.

Lin, R.P., Particle Acceleration in Solar Flares and Coronal Mass Ejections, in *Highly Energetic Physical Processes and Mechanisms for Emission from Astrophysical Plasmas*, IAU Colloquium 195, Proc. Conference, held at Montana State University, Bozeman, Montana, 1999 July 6-10, (eds. P.C.H. Martens, S. Tsuruta, and M.A. Weber), Astronomical Society of the Pacific, p.15-25, 2000.

Masuda, S., T. Kosugi, H. Hara, S. Tsuneta, and Y. Ogawara, A loop-top hard X-ray source in a compact solar flare as evidence for magnetic reconnection, *Nature* 371, No. 6497, 495-497, 1994a.

Masuda, S. 1994b, Hard X-ray sources and the primary energy release site in solar flares, 131 pp., Natl. Astronom. Obs. at Mitaka, Tokyo, 1994b.

McLean, D.J., and N.R. Labrum, (eds.), *Solar Radiophysics*, Cambridge University Press, Cambridge, 1985.

Melrose, D.B., A solar flare model based on magnetic reconnection between current-carrying loops, *ApJ*, 486, 521-533, 1997.

Neugebauer, M., P.C. Liewer, E.J. Smith, R.M. Skoug, and T.H. Zurbuchen, Sources of the solar wind at solar activity maximum, *Journal of Geophysical Research (Space Physics)*, 107, Issue A12, pp. SSH 13-1, CiteID 1488, DOI: 10.1029/2001JA000306, 2002.

Nishio, M., K. Yaji, T. Kosugi, H. Nakajima, T. Sakurai, Magnetic field configuration in impulsive solar flares inferred from coaligned microwave/X-ray images, *ApJ*, 489, 976-991, 1997.

Poqérusse, M., S. Hoang, J.L. Bougeret, and M. Moncuquet, Ulysses-ARTEMIS radio observation of energetic flare electrons, in *Solar Wind Eight*, Internat. Solar Wind Conference, held in Dana Point, California, June 1995, (eds. D. Winterhalter, J.T.

Gosling, S.R. Habbal, W.S. Kurth, and M. Neugebauer), AIP Press, New York, American Institute of Physics Conference Proceedings AIP CP-382, pp.62-65, 1996.

Reames, D.V., Particle Acceleration at the Sun and in the Heliosphere, *Space Science Rev.*, 90, 3/4, 413-491, 1999.

Reames, D.V., What we don't understand about ion acceleration in flares, in "High Energy Solar Physics - Anticipating HESSI", Astronomical Society of the Pacific Conference Series Vol. 206, Proc. HESSI Conference, held in College Park, Maryland, 1999 Oct 18-20, (eds. R. Ramaty and N. Mandzhavidze), ASP, San Francisco, California, pp.102-111, 2000.

Reames, D.V., Magnetic topology of impulsive and gradual solar energetic particle events, *ApJ*, 571, L63-L66, 2002.

Reiner, M.J., Kilometric Type III Radio Bursts, Electron Beams, and Interplanetary Density Structures, *Space Science Rev.*, 97, 129-139, 2001.

Robinson, P.A., Nonlinear wave collapse and strong turbulence, *Reviews of Modern Physics*, 69/2, 507-567, 1997.

Robinson, P.A., and A.O. Benz, Bidirectional type III solar radio bursts, *Solar Phys.*, 194, 345-369, 2000.

Schrijver, C.J., and M.L. DeRosa, Photospheric and heliospheric magnetic fields, *Solar Phys.*, 212, 165-200, 2003.

Schwenn, R., and E. Marsch, (eds.), *Physics of the Inner Heliosphere. I. Large-Scale Phenomena*, Physics and Chemistry in Space, Vol. 20, Space and Solar Physics, Springer Verlag, Berlin, 1991a.

Schwenn, R., and E. Marsch, (eds.), *Physics of the Inner Heliosphere. I. Particles, Waves, and Turbulence*, Physics and Chemistry in Space, Vol. 21, Space and Solar Physics, Springer Verlag, Berlin, 1991b.

Shibata, K., Y. Ishido, L.W. Acton, K.T. Strong, T. Hirayama, Y. Uchida, A.H. McAllister, R. Matsumoto, S. Tsuneta, T. Shimizu, H. Hara, T. Sakurai, K. Ichimoto, Y. Nishino, and Y. Ogawara, Observations of X-ray jets with the Yohkoh Soft X-Ray Telescope, *PASJ*, 44, L173-L180, 1992.

Simnett, G.M., Interplanetary phenomena and solar radio bursts, *Solar Phys.*, 104, 67-91, 1986.

Tsuneta, S., H. Hara, T. Shimizu, L.W. Acton, K.T. Strong, H.S. Hudson, and Y. Ogawara, Observations of a solar flare at the limb with the Yohkoh Soft X-Ray Telescope, *PASJ*, 44, L63-L70, 1992.

M.J. Aschwanden, Lockheed Martin ATC, Solar and Astrophysics Laboratory, Building 252, Org. ADBS, 3251, Hanover St., Palo Alto, CA 94304, USA. (aschwanden@lmsal.com)

Solar Energetic Electrons, X-rays, and Radio Bursts

R.P. Lin

Physics Department and Space Sciences Laboratory
University of California, Berkeley, CA

We review recent results on solar energetic electrons observed in situ near 1 AU, and on the radio and X-ray emission generated by these electrons. Near solar maximum, hundreds of solar impulsive electron events are detected per year by the Wind spacecraft. About ~10% are prompt events, where the electrons at all energies are injected from the Sun at the time of the solar type III radio burst, but in the rest the injection of tens of keV electrons is delayed by ~10 minutes. For prompt events with associated flare hard X-ray (HXR) emission, the electron spectrum measured at 1 AU shows a rough correlation with that of the HXR-producing electrons at the Sun, but the correlation does not fit a simple thick or thin target model. Recent analysis of several scatter-free events detected from ~0.4 to 300 keV show that they involve two separate injections, the low energy, ~0.4 to ~10 keV electron injection starts prior to the type III burst, while >13 keV electrons are injected ~10 minutes afterwards. The electron, radio and in situ plasma wave observations are all consistent with the type III radio emission being produced by~0.4 – 10 keV electrons. Finally, RHESSI has detected weak HXR bursts that may be a signature of the type III burst acceleration itself and not associated with flares.

INTRODUCTION

The acceleration of electrons is observed to accompany almost all transient releases of energy by the Sun, from the largest to the smallest. In large solar flares, HXR and gamma-ray continuum emission produced by bremsstrahlung collisions of energetic electrons with the solar atmosphere, have been observed up to >~0.1-1 GeV, with the accelerated >~20 keV electrons typically containing up to >~10-50% of the total energy (>~10^{32} ergs) released in the flare. HXR observations show that flares of all sizes, down to microflares (~10^{-6} the energy of the largest flares) that occur every few minutes, accelerate electrons (but to lower energies) [*Lin et al.*, 2003]. The frequency of flares increases

rapidly as the energy released decreases, suggesting the possibility that the total flare energy release, averaged over time, might be significant for the heating of the active corona.

In intense solar energetic particle (SEP) events, electrons up to tens of MeV have been directly detected by *in situ* space observations [*Datlowe*, 1971] near 1 AU. These intense SEP events are believed to result from acceleration by shocks driven by fast coronal mass ejections (CMEs). Smaller impulsive electron events are detected at ~1-100 keV energies (sometimes down to ~0.1 keV) hundreds of times a year near solar maximum [see *Lin*, 1985]. Such low energy electrons must originate high in the corona since energy losses to Coulomb collisions limit the amount of coronal material they can traverse.

As the accelerated electrons escape from the Sun, the faster ones run ahead of the slower ones, producing a bump-on-tail distribution that is unstable to the growth of Langmuir waves [see *Lin*, 1985 for review]. These waves then interact with the ambient plasma to produce radio emission at the plasma frequency or its harmonic. As the

Solar Eruptions and Energetic Particles
Geophysical Monograph Series 165
Copyright 2006 by the American Geophysical Union
10.1029/165GM19

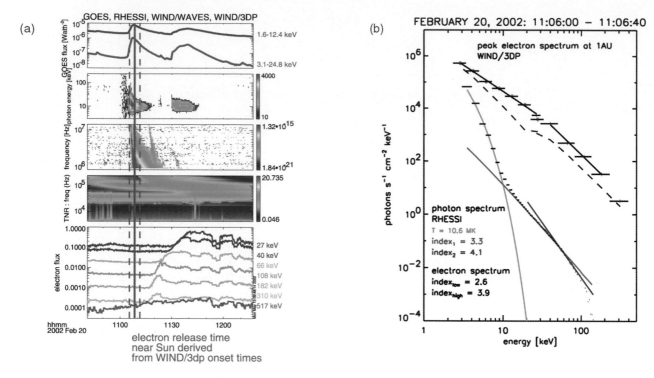

Plate 1. a) Example of a flare hard X-ray burst observed by RHESSI with corresponding solar type III radio burst and energetic electrons (and Langmuir waves) observed in situ by the Wind spacecraft [*Krucker and Lin*, 2002]. Top panel: GOES soft X-rays; second panel: Spectrogram of RHESSI X-rays from 3 to 250 keV; third and fourth panels: radio emission observed by the Wind WAVES instrument; fifth panel: electrons from ~20 to ~400 keV observed by Wind 3-DP instrument. **b)** Top trace: energy spectrum of the electrons observed by Wind 3-D P instrument; bottom trace: X-ray spectrum observed by RHESSI, fitted to a thermal spectral shape at low energies, and to a double power-law at high energies [*Krucker and Lin*, 2002].

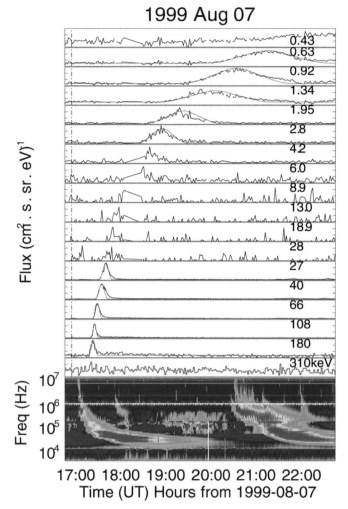

1999 Aug 07

Flux (cm^2 . s . sr . eV)$^{-1}$

0.43
0.63
0.92
1.34
1.95
2.8
4.2
6.0
8.9
13.0
18.9
28
27
40
66
108
180
310keV

Freq (Hz)

10^7
10^6
10^5
10^4

17:00 18:00 19:00 20:00 21:00 22:00
Time (UT) Hours from 1999-08-07

Plate 2. A scatter-free solar impulsive electron event observed by the Wind 3DP instrument at energies from ~0.4 to 180 keV (top two panels) and radio dynamic spectrum from the Wind WAVES instrument. A type III radio burst is observed starting at 1703 UT (dashed line in upper panels). Langmuir waves are detected (thin line at ~17 KHz) from ~1805 to 1920 UT, but are time-averaged in this plot (masking their bursty nature).

electrons travel outward to lower density regions, the radio emission goes to lower frequencies, leading to the characteristic fast drift solar type III radio burst. Impulsive electron events observed near 1 AU are generally associated with a solar type III radio burst observed down to the low frequencies characteristic of the interplanetary medium (IPM).

Low energy, ~ 0.01 to 1 MeV/nucleon ions are commonly observed to accompany impulsive electron events. These generally show enhancements of the isotope ^{3}He by ~10^2–10^4 over solar abundance (sometimes the ratio ^{3}He/^{4}He >1) and of heavy elements such as Fe. These electron/^{3}He-rich SEP events are called impulsive events because the associated flare soft X-ray burst usually is of short duration (<~10 min).

The relationship between the energetic particles at the Sun and the energetic particles observed near 1 AU is not understood. Here we review recent results obtained by comparing the electron measurements from solar wind to ~300 keV obtained by the 3D Plasma & Energetic Particles (3DP) instrument [*Lin et al.*, 1995] on the Wind spacecraft near 1 AU, with RHESSI HXR observations [*Lin et al.*, 2002] that provide detailed information on the energetic electron populations at the Sun, and with radio and plasma wave measurements from the Wind WAVES instrument.

PROMPT IMPULSIVE ELECTRON EVENTS AND HXR BURSTS

If the HXR-producing and escaping electrons come from a single acceleration at the Sun, a HXR burst should be detected with a near simultaneous type III burst starting at high frequencies. When the type III burst drifts down to near the local plasma frequency at 1 AU, the escaping electrons and the Langmuir waves they generate can be directly detected *in situ*. Plate 1a shows an example of a flare HXR burst detected by RHESSI, together with an associated type III radio burst observed by the Wind WAVES instrument from 14 MHz down to the local plasma frequency (~20 kHz) and an impulsive electron event observed by the Wind 3DP instrument up to ~300 keV. An analysis of the velocity dispersion of the onsets is consistent with a simultaneous injection of electrons at all energies, at the time indicated by the solid vertical line (dashed lines indicate uncertainty), followed by propagation over a 1.2 AU path length (about the Parker spiral field path to 1 AU). Thus, all the observations are consistent with a single acceleration at the Sun of both the HXR-producing and escaping electrons.

Plate 1b shows the flare X-ray spectrum (both thermal and HXR) observed by RHESSI, and the electron spectrum measured by the Wind 3DP instrument. Both spectra fit a double power-law with a relatively sharp downward break at a few tens of keV. Such double-power-law spectra are commonly observed for both flare hard X-ray bursts and impulsive event

electrons. The presence of a break in the hard X-ray spectrum implies a break in the X-ray producing electrons; in fact, because the bremmstrahlung cross-sections are quantitatively well known, precise high spectral resolution X-ray observations (such as those obtained by RHESSI) can be inverted to obtain the spectrum of the X-ray producing electrons [*Johns and Lin*, 1992].

Figure 1 shows a comparison of power-law exponents above the break for the electron spectra observed by Wind at 1 AU, with exponents for the HXR photon spectra observed by RHESSI, for ~15 events that have the timing consistent with a single prompt acceleration for both the X-ray producing and escaping electrons [*Kontar et al.*, 2005]. All these events are from magnetically well-connected flares. The data points show a rough linear correlation with larger electron exponents for larger HXR exponents (with the exception of the behind the limb event), suggesting that the electrons producing the HXRs at the Sun indeed are related to the electrons in these impulsive events observed in the IPM. The points should fall on the "Thick" target line if the escaping electrons directly sample the accelerated population (without any energy changes), and the accelerated electrons produce the HXRs as they lose all their energy to Coulomb collisions, i.e., if the acceleration occurs high in the corona and some of the electrons escape to the IPM while the rest are trapped in the solar atmosphere. The "Thin" target line would be for the case where the electrons produce the HXRs as they escape, but the collisions are too few to modify the spectrum. The RHESSI images typically show the HXRs come from footpoints where the ambient density is high - presumably the electrons are losing their energy to collisions, i.e., thick target. The points, however, do not lie on the "Thick" line or the "Thin" line, indicating that the relationship is more complex than these simple models; energy-dependent escape or propagation may be important. Note that the number of escaping electrons is typically only ~0.1-1% of the X-ray producing population [*Lin*, 1974].

DELAYED IMPULSIVE ELECTRON EVENTS

For most of the impulsive electron events observed at energies of tens of keV, the inferred injection of electrons back at the Sun appears to be delayed by ~10 minutes from the start of the type III radio burst [*Krucker et al.*, 1999; *Haggerty & Roelof*, 2002]. Many impulsive electron events extend down to below ~1 keV [*Lin et al.*, 1996] and some are detected even in the energy range ~0.1 to ~1 keV [*Gosling et al.*, 2003]. Recently, Wind 3DP observations (which covers from a few eV up to >~300 keV electrons) of three scatter-free impulsive electron events with delayed onset at >~ 38 keV [*Haggerty & Roelof*, 2002] were carefully analyzed to accurately determine

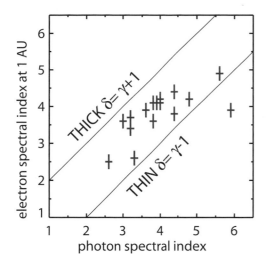

Figure 1. Comparison of power-law exponent γ for the hard X-ray spectrum at the peak of the burst measured by RHESSI with the power-law exponent δ for the electron spectrum measured at the time of maximum at each energy, both above the break energy [*Kontar et al.*, 2005].

the injection near the Sun [*Wang et al.*, 2005]. The event shown in Plate 2 extends from ~ 300 keV down to < ~0.4 keV.

The flux versus time profile at each energy shows a rapid, nearly symmetric rise and decay, indicating essentially scatter-free propagation from the Sun to 1 AU (since scattering would result in a slowly decaying tail in the time profile; see *Lin*, 1974).

Wang et al., (2005) applied a model where the injection time profile at the Sun is assumed to be triangular, with equal time for rise to the peak and decay back to zero. The injected electrons were assumed to travel ~1.2 AU, the Parker spiral field line length appropriate for the measured ~400 km/s solar wind. Model time profiles were calculated using the measured spectrum of the event and integrating over the width of each energy channel. The injection time and width were then adjusted to fit the observed profile in each energy channel. As can be seen (Figure 2), the injection profiles at energies above ~20 keV are similar, with comparable widths of ~10 minutes. The best-fit injection times are the same at all energies above ~20 keV, confirming that ~1.2 AU is appropriate for the path length. The onset of the injection for >20 keV electrons is clearly delayed by ~8 minutes relative to the type III burst injection.

A data gap and poor statistics in the ~8 to 20 keV measurements precluded accurate timing for those energies. The inferred injection profiles for ~0.4 to ~8 keV electrons show onsets starting prior to or at the type III burst injection. The peaks of the injection for ~0.4 to 8 keV electrons, however, are delayed relative to the injection peaks for > ~20 keV electrons, and the injection durations are much longer, ~50 to

~100 minutes. In this study, the other two other scatter-free events with delays at energies >~20 keV also show the same injection characteristics.

ELECTRONS AND SOLAR TYPE III RADIO EMISSION

Thus, the injection of electrons at energies of ~0.4 to ~10 keV starts early enough at the Sun that they could be the source of the solar type III radio burst. We note that a delay is needed for the fluxes of newly injected electrons to rise above the background plasma to produce the significant positive slopes in the electron reduced parallel velocity distribution function required for the growth of the Langmuir waves. These, in turn, scatter to produce the type III radio emission at the fundamental and harmonic of the plasma frequency [see *Lin*, 1990 for summary].

Previous *in situ* observations at 1 AU have shown that spiky bursts of Langmuir waves at the local plasma frequency, required for production of type III radio emission, are typically detected when ~1-10 keV electrons arrive in impulsive events [see *Lin*, 1985, *Ergun et al.*, 1997]. This is consistent with the prompt injection found in the above analysis, but leaves the question of why the delayed injection of higher energy (>~13 keV) electrons are generally not associated with type III bursts (to be addressed in a future paper). Their velocity dispersion also produces a bump-on-tail distribution, but no Langmuir waves are detected at that time. The delayed injection of >~13 keV electrons suggests a second acceleration, possibly due to the coronal counterpart of the flare waves detected by the SoHO EIT instrument [*Krucker et al.*, 1999], or to shock waves associated with a coronal mass ejection [*Haggerty and Roelof*, 2002], or to restructuring in the corona [*Maia & Pick*, 2004].

A large fraction, perhaps most, of type III bursts are not accompanied by flares. It should be noted that many HXR bursts do not have associated type III radio bursts - presumably the electrons are trapped and unable to escape. On the other hand, many type III bursts are not accompanied by HXR bursts - either the electrons are accelerated high in the corona where the ambient density and/or number of accelerated electrons are too low for detectable hard X-ray emission. In one period of ~15 minutes (Plate 3) the Wind WAVES instrument detected six solar type III bursts and RHESSI detected 12-15 keV HXR emission from all six [*Christe et al.*, 2005]. Only two of the events show any flare-like thermal emission. Perhaps this weak HXR emission is related to the type III electron acceleration process.

Further more detailed comparisons between RHESSI HXR/gamma-ray emission and impulsive electron events and radio/plasma wave emissions observed by Wind (and STEREO in the near future) in the IPM are needed to finally resolve the relationship between particles at the Sun and in the IPM.

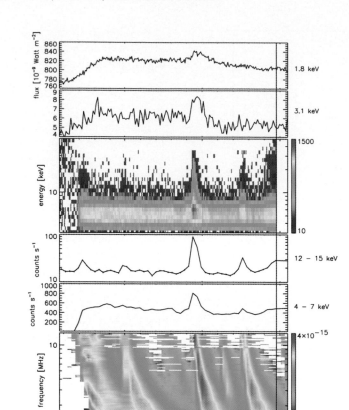

Plate 3. From the top, Panels 1 and 2: GOES soft X-ray on linear scale; Panel 3: RHESSI X-ray Spectrogram; Panel 4: RHESSI 12-15 keV (non-thermal) X-ray count rate; Panel 5: thermal (4-7 keV) count rate; Panel 6: Radio Spectrogram from Wind/WAVES. Increases in the 12-15 keV counts are seen for each type III solar radio burst [*Christe et al.*, 2005].

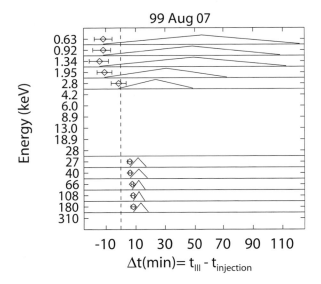

99 Aug 07

Figure 2. The inferred injection time profiles at the Sun (left triangles) and fit (smooth curves on right) to the time profiles observed by the Wind 3-DP instrument near 1 AU (note the expanded time scale of the left panel covers only 1 hour, while the right panel covers 5 hours). The error bars on the onsets are typically a few minutes [*Wang et al.*, 2005].

Acknowledgements. I'm pleased to acknowledge useful discussions with S. Krucker and L. Wang at Berkeley. This research was supported in part by NASA contract NAS5-98033-05/05 and NASA grant NNG05GH18G-03/08.

REFERENCES

Christe, S., S. Krucker, and R.P. Lin, *Ap. J. Lett.*, in prep., 2005.

Datlowe, D., Relativistic Electrons in Solar Particle Events, *Solar Phys.*, 17, 436-458, 1971.

Ergun, R., *et al.*, Wind spacecraft observations of solar impulsive electrno events associated with solar type III radio bursts, *Astrophys. J.*, 503, 435-445, 1998.

Gosling, J.T., R.M. Skoug, and D.J. McComas, Solar electron bursts at very low energies: Evidence for acceleration in the high corona?, *Geophys. Res. Lett.*, 30(13), 1697-1700, 2003.

Haggerty, D.K. and E.C. Roelof, Impulsive near-relativistic solar electron events: delayed injection with respect to solar electromagnetic emission, *Astrophys. J.*, 579, 841-853, 2002.

Johns, C.M., and R.P. Lin, The derivation of parent electron spectra from bremsstrahlung hard X-ray spectra, *Solar Phys.*, 137(1), 121, 1992.

Kontar. E., S. Krucker, and R.P. Lin, in prep. 2005.

Krucker, S., *et al.*, On the origin of impulsive electron events observed at 1 AU, *Astrophys. J.*, 519, 864-875, 1999.

Krucker, S. and R.P. Lin, Relative timing and spectra of solar flare hard X-ray sources, *Solar Phys.*, 210, 229-243, 2002.

Lin, R.P., *et al*, RHESSI observations of particle acceleration and energy release in an intense solar gamma-ray line flare, *Astrophys. J. Lett.*, 595, L69, 2003.

Lin, R.P., *et al.*, The Reuven Ramaty High-Energy Solar Spectroscopic Imager (RHESSI), *Solar Phys.*, 210, 3-32, 2002.

Lin, R.P., *et al.*, Observation of an impulsive solar electron event extending down to ~0.5 keV energy, *Geophys. Res. Lett.*, 23, 1211-1214, 1996.

Lin, R.P., *et al.*, A three-dimensional (3-D) plasma and energetic particle experiment for the Wind spacecraft of the ISTP/GGS mission, *Space Sci. Rev.*, 71, 125-153, 1995.

Lin, R.P., in *Basic Plasma Processes on the Sun*, edited by E.R. Priest and V. Krishnan, p.p. 467, International Astronomical Union, The Netherlands, 1990.

Lin, R.P., Energetic solar electrons in the interplanetary medium, *Solar Phys.*, 100, 537-561, 1985.

Lin, R.P., Non-relativistic solar electrons, *Solar Phys.*, 16, 189-256, 1974.

Maia, D., and M. Pick, Revisiting the origin of impulsive electron events: coronal magnetic restructuring, *Astrophys. J.*, 609, 1082, 2004.

Wang, L., R.P. Lin, S. Krucker, and J.T. Gosling, *Geophys. Res. Lett.*, accepted., 2005.

R.P. Lin, Physics Department and Space Sciences Laboratory, University of California, Berkeley, CA 94720-7450. (rlin@ssl.berkeley.edu)

Coronal Mass Ejections and Type II Radio Bursts

Nat Gopalswamy

Solar System Exploration Division, NASA Goddard Space Flight Center, Greenbelt, Maryland

The simultaneous availability of white light data on CMEs from the Solar and Heliospheric Observatory (SOHO) and radio data on shock waves from the Radio and Plasma Wave experiment on board the Wind spacecraft over the past decade have helped in making rapid progress in understanding the CME-driven shocks. I review some recent developments in the type II-CME relationship, focusing on the properties of CMEs as shock drivers and those of the medium supporting shock propagation. I also discuss the solar cycle variation of the type II bursts in comparison with other eruptive phenomena such as CMEs, flares, large solar energetic particle events, and shocks detected in situ. The hierarchical relationship found between the CME kinetic energy and wavelength range of type II radio bursts, non-existence of CMEless type II bursts, and the explanation of type II burst properties in terms of shock propagation with a realistic profile of the fast mode speed suggest that the underlying shocks are driven by CMEs, irrespective of the wavelength domain. Such a unified approach provides an elegant understanding of the entire type II phenomenon (coronal and interplanetary). The blast wave scenario remains an alternative hypothesis for type II bursts only over a small spatial domain (within one solar radius above the solar surface) that is not accessible to in situ observation. Therefore the existence of blast waves cannot be directly confirmed. CMEs, on the other hand, can be remote sensed from this domain.

1. INTRODUCTION

Since their initial discovery by *Payne-Scott et al.* (1947) and subsequent classification by *Wild and McCready* (1950), the type II solar radio bursts in the corona have been studied for more than half a century (*Kundu*, 1965; *Zheleznyakov*, 1969; *Nelson and Melrose*, 1985; *Aurass*, 1997; *Cane*, 2000; *Gopalswamy*, 2000; *Reiner*, 2000). The type II bursts are thought to be produced by electrons accelerated at MHD shock fronts by complex plasma processes (e.g., *Uchida*, 1960). In the interplanetary (IP) medium, these bursts were first detected by *Malitson et al.* (1973)

Solar Eruptions and Energetic Particles
Geophysical Monograph Series 165
10.1029/165GM20

using data from the IMP 6 mission. The IP shocks first detected by space missions were soon linked to coronal shocks inferred from metric type II bursts (*Pinter*, 1973). Voyager (*Boischot et al.*, 1980) and ISEE-3 (*Cane et al.*, 1982) spacecraft also observed IP type II bursts. *Payne-Scott et al.* (1947) clearly alluded to the relationship of the radio source to mass ejections. Soon after the discovery of white-light coronal mass ejections (CMEs) by the OSO-7 satellite, *Stewart et al.* (1974a,b) suggested that the metric type II bursts were due to CME-driven shocks. Despite the counter example reported by *Kosugi et al.* (1976) in which the CME and type II burst were temporally far apart to have a causal relationship, other observations pointed to a close CME-shock relationship: the above–average speed of CMEs associated with type II bursts [*Gosling et al.*, 1976; *Robinson*, 1985] and the near one-to-one correspondence between limb type II bursts and CMEs [*Munro et al.*, 1979]. The arguments against a close CME-type II relationship include: 1. The large number of CMEless metric type II bursts (*Sheeley et al.*, 1984; *Kahler et al.*, 1984) require a non-CME shock source (flare blast waves). 2. The projected heights of the type II sources were smaller than the corresponding CME leading edges, an observation thought to be inconsistent with the CME source (*Wagner and MacQueen*, 1983; *Gary et al.*, 1984; *Cane*, 1984; *Robinson and Stewart*, 1985; *Gopalswamy and Kundu*, 1992). 3. The disparity in speeds and directions of propagation of the CMEs and the associated shocks (*Gergely*, 1984) does not seem to support CME-driven mechanism.

The above controversy is mostly centered on metric type II bursts, which occur over a height range of ~1 Rs (solar radius) above the solar surface. IP type II bursts at frequencies below 2 MHz (occurring at heliocentric distances ≥10 Rs) were clearly CME-associated. Observations were seldom made in the 2-20 MHz range, which contributed to the independent treatment of metric and IP type II bursts. When the WAVES experiment (*Bougeret et al.*, 1995) on board Wind began observing type II bursts in the 1-14 MHz frequency range (see, e.g., *Kaiser et al.*, 1998; *Reiner et al.*, 1998a; *Gopalswamy et al.*, 2000b) the situation changed. The wavelength range corresponding to 1-14 MHz is decameter-hectometric or DH, for short. Simultaneous availability of coronagraph data from the Solar and Heliospheric Observatory (SOHO) mission, whose field of view (2-32 Rs) overlapped with the coronal/IP domain containing the 1-14 MHz plasma levels, enabled studies on the connection between CMEs and type II bursts. The DH type II bursts were also closely linked to CMEs that are faster and wider on the average (*Gopalswamy et al.*, 2001b). Wind/WAVES also has frequency coverage below 1 MHz down to 20 kHz, which, when combined with ground based observations, made it possible to study type II bursts over the entire Sun-Earth distance.

Studying a set of metric type II bursts without IP counterparts and another set of IP shocks detected in situ without metric type II bursts over the same time interval, *Gopalswamy et al.* (1998) concluded that the shocks inferred from metric type II bursts and the IP shocks were of different origin. The implication was that the metric type II bursts were of flare origin and the IP shocks were of CME origin. None of the metric type II bursts studied by *Gopalswamy et al.* (1998) had counterparts in the WAVES spectral domain (below 14 MHz). When type II bursts started appearing at DH wavelengths, it was found that some type II bursts continued beyond the outer corona into the IP medium (see Figure 7 of *Gopalswamy*, 2000). However, the discordance between the drift rates of metric and IP type II bursts continued to be present supporting the requirement for blast waves and CME-driven shocks present in the same eruptive event (*Cane*, 2000; *Reiner et al.*, 2001). In the meanwhile, the existence of CMEless type II bursts was brought into question.

The CMEless type II bursts may be an artifact stemming from the nature of the CME visibility function, which favored limb CMEs (*Cliver et al.*, 1999). *Gopalswamy et al.* (2001a) found that the CMEless type II bursts were indeed associated with EUV eruptions originating from close to the disk center (see also *Classen and Aurass*, 2002). Another development was the use of a realistic profile of the characteristic speed in the corona (*Krogulec et al.*, 1994) for interpreting type II burst spectra (*Mann et al.*, 1999; *Gopalswamy et al.*, 2001a), which can account for the drift rate discrepancy in terms of CME-driven shocks. Finally, the close relationship between CME kinetic energy and the wavelength range of type II bursts provides a unified view of the type II bursts as a CME-related phenomenon.

This chapter provides a global view of the type II radio bursts and their physical connection to CME-driven shocks, irrespective of the wavelength domain of occurrence. One of the results highlighted in this chapter is that the type II phenomenon can be explained by CME-driven shocks without resorting to the blast waves, thought to originate from the sites of associated flares. After a brief introduction to CMEs (section 2) and type II bursts (section 3), their interconnection is discussed and shown that the CME kinetic energy organizes the wavelength range of type II bursts (section 4). The solar cycle variation of type II bursts, CMEs, flares and solar energetic particles are presented in sections 5 and 6. A unified approach to the type II phenomenon, as dictated by the CME kinetic energy and the radial profile of the characteristic speed in the corona and IP medium, is presented in section 7. Finally the flare-type II relationship is discussed in the context of the unified approach (section 8) before providing the concluding remarks (section 9).

2. CORONAL MASS EJECTIONS

CMEs are large-scale magnetized plasma structures erupting from closed field regions such as active regions, filament regions, active region complexes and trans-equatorial interconnecting regions on the Sun (*Tousey*, 1973). Pre-eruption evolution of the closed field regions involving flux emergence, shearing motion or flux cancellation is thought to store free energy in magnetic fields. Release of this free energy often results in CMEs. Within the coronagraphic field of view, CMEs have speeds ranging from a few km/s to more than 2500 km/s (see e.g., *Gopalswamy*, 2004b and references therein), with an average value of ~450 km/s, which is slightly higher than the slow solar wind speed. The apparent angular width of CMEs ranges from a few degrees to more than 120 degrees, with an average value of ~47 deg (counting only CMEs with width less than 120 deg). The width of CMEs occurring close to the limb is likely to be the true width, whereas the width of CMEs occurring close to the disk center are severely affected by projection effects (*Gopalswamy et al.*, 2000b: *Burkepile et al.*, 2004). The total mass ejected ranges from a few times 10^{13} g to more than 10^{16} g with an average value of ~6.7×10^{14} g (*Vourlidas et al.*, 2002; *Gopalswamy*, 2004b). Accordingly, the kinetic energy of CMEs with angular width <120° ranges from ~10^{27} erg to ~10^{32} erg, with an average value of 5×10^{29} erg (see e.g., *Hundhausen*, 1997; *Vourlidas et al.*, 2002). Some very fast and wide CMEs can have kinetic energies exceeding 10^{33} erg, generally originating from large active regions [*Gopalswamy et al.*, 2005a].

CMEs occurring close to the disk center often appear to surround the occulting disk of the coronagraph and are known as halo CMEs (*Howard et al.*, 1982). Only ~3% of CMEs are observed as halo CMEs, which are faster (~1000 km/s) on the average (*Gopalswamy*, 2004b). When front-sided, these CMEs can directly impact Earth causing geo-magnetic storms, provided the magnetic field contained in the CMEs have a southward component. Such CMEs are known to be geoeffective. If the speed of the CMEs exceeds the local Alfven speed in the corona and interplanetary (IP) medium they can drive shocks, which can accelerate electrons and ions, generally known as solar energetic particles (SEPs). Such CMEs are sometimes referred to be SEPeffective. Accelerated electrons are inferred from the radio emission they produce, while the ions are detected after they propagate to particle detectors suitably located. The type II radio emission produced by CME-driven shocks is the primary subject matter of this chapter.

3. TYPE II RADIO BURSTS

Type II bursts are nonthermal radio emission originating from fast mode MHD shocks. The current paradigm for the generation of type II bursts is as follows: The shocks accelerate nonthermal electrons, which in turn produce radio emission at the fundamental and harmonic of the local plasma frequency via well-known plasma processes. In the dynamic spectra (intensity of radio emission displayed in the frequency-time plane) type II bursts appear as slanted features with the slope related to the speed of the shock and the density scale height in the medium. The spectral feature typically contains fundamental-harmonic components because radio emission occurs at the plasma frequency (fp) and its harmonic (2fp). Occasionally, emission is observed at the third harmonic (3fp) (*Zlotnik et al.*, 1998). The components can be further split into upper and lower bands, thought to be caused by the density structure in the shock (see e.g., *Nelson and Melrose*, 1985; *Vrsnak et al.*, 2001). Type II bursts occur at frequencies below ~150 MHz, although occasionally they are observed at higher frequencies (*Vrsnak et al.*, 1995; *Klein et al.*, 1999).

The high-frequency end of type II bursts corresponds to the radio emission close to the Sun, while the low frequency end corresponds to a location far away from the Sun where the radio intensity drops below the background. Type II bursts have been observed up to Earth orbit and beyond to a few AU.

3.1. Type II Burst Varieties

The appearance of type II bursts in various wavelength domains is shown in Figure 1: 1) bursts confined to the metric (m) domain; 2) bursts starting in the m domain but continuing into the DH domain; 3) bursts confined to the DH domain; 4) bursts starting in the DH domain and continuing into the kilometric (km) domain; 5) bursts having counterparts in all the wavelength domains, m-to-km; 6) bursts confined to the km domain. In the schematic picture, we have not shown the details of harmonic structure or band-splitting, which may or may not be present in all the events. In (5), the components in various spectral domains may and may not have direct continuity. The m variety (1) occurs typically above the ionospheric cutoff at ~20 MHz, observed from ground based radio telescopes. Radio emission from the Sun at longer decametric and km wavelengths cannot penetrate the terrestrial ionosphere, so spaceborne instruments observe varieties 2-6. Coronal densities similar to the ionospheric densities occur at a heliocentric distance of ~3 Rs, which is also considered to be the location of the source surface of the solar magnetic field. The ambient medium beyond the source surface is considered to be the IP space. Thus, the bursts at frequencies above the ionospheric cutoff are known as coronal (or m) type IIs, while the ones occurring at frequencies

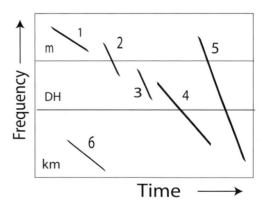

Figure 1. Schematic dynamic spectrum showing the commonly observed varieties of type II bursts confined to various wavelength ranges: metric (m), decameter-hectometric (DH), and kilometric (km). m type II bursts are generally observed by ground based radio telescopes; DH and km bursts need to be observed from space because of the ionospheric cutoff between DH and m domains.

below the cutoff are known as the IP (or DH, km) type II bursts (*Gopalswamy*, 2004c). Occasionally, one can observe population 2 (m-to-DH) bursts using ground based instruments at geographical locations where the ionospheric cutoff is well below the nominal 20 MHz (*Erickson*, 1997). One has to combine ground and space based observations to see the continuation from the m-domain to DH and km domains.

3.2. Type II Drift Rates

One of the recent findings is the universal relationship between the drift rate of type II bursts and the frequency of emission (*Vrsnak et al.*, 2001; *Aguilar-Rodriguez et al.*, 2005): on a log-log scale, the measured drift rate (df/dt) has an excellent correlation with the emission frequency (f) with a correlation coefficient of >0.9. Figure 2 shows plots of df/dt versus f for two sets of events: (1) type II bursts observed by Wind/WAVES and ISEE-3 (*Lengyel-Frey and Stone*, 1989) in various spectral domains, and (2) a set of m-to-km events from *Gopalswamy et al.* (2005b) for which measurement of drift rate was possible in the DH and km domains. The two sets were combined with metric type II data from *Mann et al.* (1996). Ideally one should have used the metric type II burst data in the same epoch as the DH and km type II bursts. Nevertheless, the trend is very clear that at higher frequencies the bursts have larger drift rates and the relation holds over six orders of magnitude in drift rate and three orders of magnitude in frequency. Thus one gets a power law relationship: df/dt ~ f$^\alpha$ where α ~2. Taken alone, the km type II bursts show a slightly steeper slope. In fact, *Aguilar-Rodriguez et al.* (2005) found an increase in the power law index from α ~1.4 in the m domain to α ~2.3 in the km domain. This variation is most likely due to changing shock speed. The close relationship between df/dt and f can be understood from the fact that the shock travels with a speed V emitting at successively lower frequencies determined by the local plasma density (n), which decreases with heliocentric distance (r) as r^{-2}: |df/dt| = V(df/dr) = V(f/2n)(dn/dr) = Vf^2. Here it is assumed that the emission occurs at the fundamental plasma frequency (f ~ n$^{1/2}$). This simple relationship is remarkably similar to the observed one, provided V is approximately constant and the range of speeds is not too wide. This is not a bad assumption in individual domains, but we do know that CMEs and shocks decrease in speed between the Sun and 1 AU (*Gopalswamy et al.*, 2000a) because of the drag force of the ambient medium acting on the CMEs (*Gopalswamy et al.*, 2001b; *Vrsnak*, 2001). Also V can increase or decrease in individual domains. For example, close to the Sun (m domain) most CMEs are likely to be accelerating, while decelerating in the IP medium. For purely km type IIs, the CMEs may be accelerating far into the IP medium (*Gopalswamy*, 2004a). In order to fully understand the df/dt − f relationship, one also needs to consider

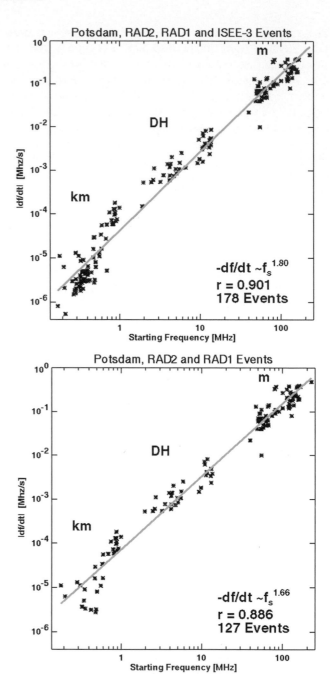

Figure 2. The drift rate and emission frequency of type II bursts. (top) m, DH, and km measurements are combined from various sources (ground based, WAVES, and ISEE-3). The data points in the km and DH domains may or may not correspond to the same burst. (bottom) WAVES type IIs with measurements in both in DH and km domains combined with m type IIs. The top and bottom plots use the same metric data from *Mann et al.* (1996).

the variation of the solar wind speed. Additional considerations include the deviation from the r^{-2} dependence for density in the inner corona (m domain). One of the important implications of this universal relationship is that the m type II bursts behave very similar to the DH and km type II bursts.

4. CME AND TYPE II BURST HIERARCHY

The likelihood of an interplanetary type II burst following a metric type II was found to be greatly increased if accompanied by strong, long-lasting H-alpha and soft X-ray flares (*Robinson et al.*, 1984). Such flares are known to be indicative of energetic CMEs. CMEs associated with DH type II bursts were generally faster and wider than those associated with metric type II bursts (*Gopalswamy et al.*, 2000b; *Lara et al.*, 2003). Furthermore, m type IIs followed by IP events (shocks and/or IP CMEs) were always accompanied by halo or partial halo CMEs (*Gopalswamy et al.*, 2001a), which are generally more energetic than other CMEs (*Gopalswamy*, 2004a). IP type II bursts below 1 MHz were known to be associated with energetic CMEs (*Cane et al.*, 1987), but it is not clear whether they belong to variety 4, 5 or 6 in Figure 1. The trend of more energetic CMEs resulting in longer wavelength type II bursts was revealed in a systematic study involving: i) m type IIs with no counterparts in the DH or km domains (same as variety 1 in Figure 1), ii) DH type IIs, irrespective of the presence of counterparts in the metric and km domains (varieties 2-5), iii) m-to-km type IIs (variety 5), and iv) purely km type IIs (variety 6). The purely m type II bursts were chosen such that the associated source regions were within 30 deg. of either limb (hence referred to as m-limb events). When the CME properties of these populations were compiled and compared, a systematic relationship was found (*Gopalswamy et al.*, 2005b): the CME speed, width, deceleration and the fraction of full halos (apparent width = 360 deg.) increased in the following order: m, DH, m-to-km (see Table 1). Since the width is proportional to the mass, faster and wider CMEs are more energetic. Thus, CMEs associated with m type IIs (population i) are the least energetic, while those associated with the m-to-km type IIs (population iii) are the

most energetic. CMEs associated with DH type IIs (population ii) are of intermediate kinetic energy. Table 1 shows that the CMEs associated with purely km type IIs do not quite fit into this hierarchy: while the CME speed and width are similar those of m type II bursts, the acceleration is of opposite sign (positive). These CMEs accelerate gradually and attain shock-driving capability only far into the IP medium when the speed becomes high enough to be super-Alfvenic (*Gopalswamy*, 2004b). The fraction of halo CMEs is also much larger than that of the m type II bursts (17.2% vs. 3.8%). The purely km population also explains the presence of IP shocks without metric type II bursts, as found in *Gopalswamy et al.* (1998). Table 1 also shows that all the CME populations associated with type II bursts are more energetic than the general population.

The link between CME kinetic energy and the wavelength extent of type II bursts has an important practical implication: it is possible to isolate the small number of energetic CMEs that are geoeffective and SEP-effective based on the observation of m-to-km type II bursts.

4.1. CME Height and Metric Type II Onset

The starting frequency of type II bursts indicates the distance from the eruption center at which the shock begins to accelerate electrons. The frequency of emission is proportional to the square-root of the electron density in the vicinity of the shock, so higher starting frequencies imply shock formation closer to the Sun. The starting frequency of type II bursts rarely exceeds ~150 MHz, although higher starting frequencies have been reported occasionally (*Vrsnak et al.*, 1995). Considering a set of 80 purely m-limb type II bursts with known emission mode (fundamental or harmonic), the average starting frequency was found to be 101 MHz [*Gopalswamy et al.*, 2005b]. The starting frequency of metric type II bursts with interplanetary counterparts was quite similar, if not higher (111 MHz). *Robinson et al.* [1984] who used a sample of only 16 metric type II bursts with IP counterparts found that ~78% of them had starting frequencies <45 MHz, compared to ~20% for all type II bursts. From this they concluded that MHD shocks which formed higher in the corona were more likely to produce IP type II bursts. We could not reproduce this result because the fraction of metric type II bursts with low starting frequencies (below 50 MHz) is rather small: ~33% for purely metric type IIs and ~17% for the m-to-km events. Their alternative suggestion that blast waves becoming shocks at large heights in the corona can escape into the IP medium is also not supported by the recent results.

Considering only limb events (to avoid projection effects) the CME leading edge was found to be at a heliocentric distance of ~2.2 Rs at the onset of purely metric type II bursts

Table 1. CME-type II burst hierarchical relationship compiled from Gopalswamy (2005b). Column 1 includes all the CMEs observed by SOHO from 1996 to the end of 2004. Other columns list properties of CMEs associated with type II bursts in various wavelength domains.

Property	All	m	DH	mkm	km
Speed (km/s)	487	610	1115	1490	539
Width (deg)	45	96	139	171	80
Halos (%)	3.3	3.8	45.2	71.4	17.2
Acceleration (m/s²)	−2	−3	−7	−11	+3

and virtually the same distance (2.3 Rs) for the m-to-km events (*Gopalswamy et al.*, 2005b). The similarity in CME leading edge heights (which always refer to the Sun center unless otherwise stated) of the m-to-km and purely metric populations reflects the similarity in starting frequencies of the two populations. In other words, type II bursts form roughly at the same heights irrespective of whether or not an IP type II burst follows. *Robinson et al.* (1984) had estimated that the shocks responsible for metric type II bursts form in the height range of $1.6 - 2$ Rs, which is not too different from (but slightly less than) the heights of CME leading edges obtained by *Gopalswamy et al.* (2005b). This remarkable similarity between the type II burst heights and the leading edges of CMEs associated with type II bursts indicate that the type II bursts are physically related to CMEs. The slightly smaller heights of type II bursts indicate that they may be originating from the flanks of the CME-driven shocks. A more important point is that a CME is present in the corona at the time of metric type II burst as a possible shock driver.

5. SOLAR CYCLE VARIATION

Solar-cycle variation of m and DH type II burst rates binned by Carrington Rotations (CRs) is shown in Plate 1 for the period from 1996 to the end of 2004. The number of CMEs and flares per CR (divided by 10 and 20, respectively to fit the scale) are also given for comparison. Only C, M, and X-class GOES flares have been included. There is a clear increase in the number of type II bursts from the solar minimum to maximum like any other indicator of solar activity. In particular, the number of metric type II bursts tracks the CME rate. The number of DH type II bursts also has a minimum-to-maximum variation, but the dependence on CME rate is less pronounced. This is because the DH type II bursts are associated with more energetic CMEs. Plate 1 also demonstrates that the type II bursts are a relatively rare phenomenon. Only 850 of the 9000+ CMEs (<10%) detected by SOHO between 1996 and 2004 were associated with m type II bursts. The fraction of CMEs associated with DH (2.5%) and m-to-km (0.8%) is much smaller. Cumulative distribution of CMEs as a function of speed (*Gopalswamy*, 2006) is consistent with these association rates because the number of CMEs fall rapidly at higher speeds needed for type II association (see also Table 1). On the other hand, the number of flares per CR (counting only C-, M-, and X-class flares) is higher by a factor of ~20 than the m type II burst rate.

Periods with large number of metric type II bursts with virtually no DH type II bursts have been reported (*Gopalswamy et al.*, 2004b) when the mean CME speeds are lower. The overall number of m type II bursts is also typically 4 times that of DH type II bursts. For both m and DH type II bursts, the first requirement is the presence of a CME (*Gopalswamy*

et al., 2005b). Then comes the speed, because the average speed of CMEs associated with DH type II bursts is almost twice the average speed of CMEs associated with m type II bursts. It was shown by *Gopalswamy et al.* (2003a) that the peaks in DH type II rate coincided with the peaks in the mean speed of CMEs.

Figure 3 shows the correlation between m and DH type II bursts with the CME occurrence rate and mean speed. As we noted before, the number of m type II bursts has the best correlation with the CME rate (r = 0.67). The type II occurrence rate is also reasonably correlated with the CME mean speed (r = 0.50). Note that there are no type II bursts when the CME mean speed is less than 200 km/s (see Figures 3a,b), which is close to the characteristic speed of the inner corona. If the type II bursts are not closely connected to CMEs, one would not obtain such a speed relationship.

6. TYPE II BURSTS AND SOLAR ENERGETIC PARTICLES

The close connection between metric type II bursts and solar energetic particles was recognized as early as 1971 by *Dodson and Hedemen*, who noted that "a type II burst was the only unusual aspect of a flare apparently associated with a proton enhancement" [quoted by *Svestka and Fritzova-Svestkova*, (1974)]. *Kahler et al.* (1978) found that "a mass-ejection event is a necessary condition for the occurrence of a prompt proton event", suggesting a physical link between CME-driven shocks and particle acceleration. The starting time of a metric type II burst marks the time when a shock is present nearest to the Sun ($\leq$2.2 Rs). At the time of SEP release near the Sun, the CME is at a height of ~2.7 Rs (*Kahler*, 1994; *Kahler et al.*, 2003). For SEP events with ground level enhancements (GLEs), the corresponding CME height is somewhat larger (4.5 Rs, *Gopalswamy et al.*, 2005d). The DH type II bursts originate in the same height range as the estimated release heights of SEPs, and it is not surprising that there is a 100% association between SEP events and DH type II bursts (*Gopalswamy*, 2003; *Cliver et al.*, 2004b). Type II bursts at frequencies below 2 MHz were also found to be associated with large SEP events (*Cane and Stone*, 1984). Combining all these, one can conclude that the same shock accelerates electrons to produce type II bursts, and ions detected in situ. The relative variation of IP shocks, large SEP events, and IP type II bursts is illustrated in Plate 2. The shocks were detected in situ by spacecraft in the solar wind. The large SEP events (events with proton intensity >10 pfu in the >10 MeV channel) were recorded by the GOES satellite. The occurrence rates of these events (binned over Carrington Rotations) are close to each other. The small differences can be attributed to the differing observability functions. For example, the SEP events are generally smaller in number

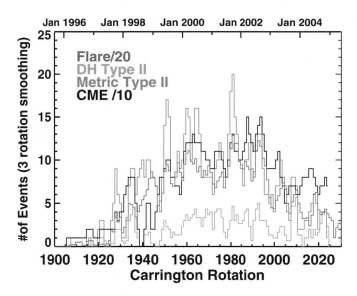

Plate 1. Solar cycle variation of m and DH type II bursts compared with the CME and flare rates (counting flares at and above C-class). All quantities are averaged over Carrington rotation (CR) periods. The plots have been made by smoothing over 3 rotations. The CME and flare rates are divided by 10 and 20, respectively to fit the scale. The UT scale is given at the top.

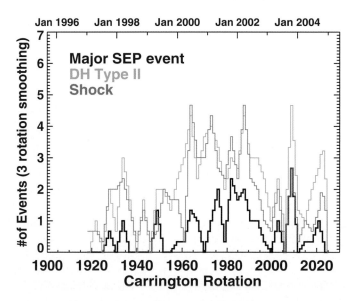

Plate 2. The occurrence rates of IP shocks, DH type II bursts and major SEP events. The major SEP events are defined as those which produce a 10-pfu proton event at Earth in the >10 MeV channel. Note that all the three events are closely related.

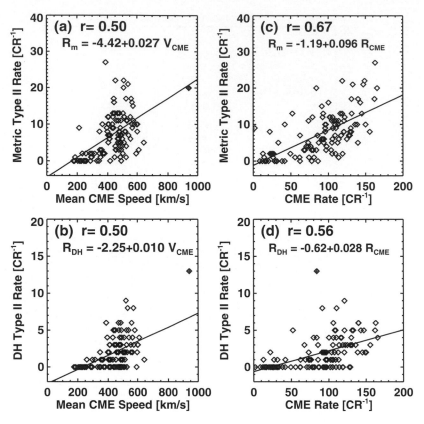

Figure 3. Correlation between the occurrence rates of m (R_m) and DH (R_{DH}) type II bursts with CME mean speed (V_{CME}, left) and CME occurrence rate (R_{CME}, right). All numbers are averaged over Carrington rotations (CRs). The correlation coefficients and the equations of the regression lines are noted on the plots.

because they are observed only when the SEP source is well-connected to the observer. The DH type II bursts are slightly larger in number because even backside eruptions can produce them, whereas the shocks from these eruptions may not arrive at Earth. The rate is also similar for fast and wide CMEs from the front-side western hemisphere of the Sun (*Gopalswamy et al.*, 2003b). This is also expected because the shocks DH type II bursts, and SEPs are associated with fast and wide CMEs. One of the important practical implications is that the DH type II bursts (promptly detected compared to SEPs and in situ shocks) can clearly isolate the small number of SEP-effective CMEs. As we noted above, it may not be possible to tell whether the associated shock will propagate far into the IP medium from the observation of metric type II burst alone. If the metric type II burst is associated with a fast and wide CME, then it is likely that a DH type II and an SEP event (if the CME is western) will follow.

6.1. CME Interaction and SEPs

Interaction between CMEs near the Sun was first identified from a long wavelength radio enhancement in the Wind/WAVES dynamic spectra (*Gopalswamy et al.*, 2001c) in association with two colliding CMEs within the field of view of SOHO/LASCO. Some active regions are copious producers of CMEs, so the corona above such regions is expected to be highly inhomogeneous due to preceding CMEs and their aftermath. When a shock passes through density (n) and/or magnetic field (B) inhomogeneities, the upstream Alfven speed (Va) will be modified according to $dVa/Va = dB/B - (1/2) dn/n$. If $dn/n > 2dB/B$, then $dVa/Va < 0$, which means an increase in the upstream density (above the quiet values) can lower Va and hence increase the Mach number of the shock. Stronger shocks accelerate more electrons resulting in enhanced radio emission. Other situations may arise depending on the signs of dB/B and dn/n and their relative magnitudes (*Lugaz et al.*, 2005). In the same way interacting CMEs affect the type II radio emission, the SEP events may also be affected (*Gopalswamy et al.*, 2002, 2004a). If a CME-driven shock propagates through a medium with density and magnetic field fluctuations, the shock strength will be modified. If the shock propagates through a preceding CME, trapping of particles in the closed loops of preceding CMEs can repeatedly return the particles back to the shock, thus enhancing the efficiency of acceleration (*Gopalswamy et al.*, 2004; *Kallenrode and Cliver*, 2001). A systematic survey of the source regions of large SEP events of cycle 23 has revealed that the SEP intensity is high when a CME-driven shock propagates into a preceding CME or its aftermath originating from the same solar source (*Gopalswamy et al.*, 2004a, 2005c). According to theoretical calculations, existence of preceding CMEs can greatly enhance the turbulence upstream of the shock, resulting in shorter acceleration times and higher SEP intensities (*Li and Zank*, 2005).

7. A UNIFIED APPROACH TO TYPE II BURSTS

Flare blast waves and CME-driven shocks have been considered as two possible sources of metric type II bursts, while the DH and longer wavelength bursts are due to CME-driven shocks. The primary observational support for the blast wave scenario are: (i) CMEless type II bursts and (ii) the discrepancy between the metric and IP type II bursts. In this section, we present evidence showing that these two may not hold anymore, further supporting the unified approach to the type II bursts in terms of CMEs.

7.1. CMEless Type II Bursts

The existence of CMEless metric type II bursts (*Sheeley et al.*, 1984; *Kahler et al.*, 1984) became questionable when the solar sources of such type II bursts were examined (*Cliver et al.*, 1999; *Gopalswamy et al.*, 2001a): most of the CMEless type II bursts originated from close to the disk center. Coronagraphs, by their very nature, are ill-positioned to detecting CMEs occurring close to the disk center (*Cliver et al.*, 1999), although such eruptions can be clearly seen in coronal images obtained in EUV and soft X-rays (*Gopalswamy et al.*, 2001a). This is especially true for purely metric type II bursts because they are associated with CMEs of just above average kinetic energy and weaker flares (see Section 8). The CME visibility function is such that about half of the CMEs associated with C-class flares occurring close to the disk center may not be detected by LASCO, whereas most of them will be detected if the associated flares occurred near the limb (*Yashiro et al.*, 2005). Furthermore, considering only type II bursts occurring close to the solar limb, there is nearly a 100% association with CMEs (*Gopalswamy and Hammar*, under preparation). Thus, the CMEless type II bursts clearly are an artifact of the visibility function of CMEs.

In a recent paper, *Classen and Aurass* (2002) suggested that metric type II bursts belong to three different classes originating from: 1. flare blast waves, 2. nose of CME-driven shocks, and 3. flanks of CME-driven shocks. However, the majority of m type II bursts interpreted with the blast wave scenario originated from close to the disk center. In fact, when we reexamined the 19 class 1 metric type II bursts (kindly provided by *T. Classen*), we found that all of them were associated with EUV eruptions with many having spatial extent much larger than that of the active region. Such a signature is indicative of CMEs, which might have been missed by LASCO because of the occulting disk. Some of them might have been missed because of multiple CMEs from the same region. One of the criteria used by *Classen and Aurass* (2002) to designate a type II event as a blast wave case is that the temporal separation between the type II burst and the associated CME must exceed 1 hr. Longer delay is expected for the disk-center CMEs because they have to expand significantly before appearing above the occulting disk. Thus the existence of CMEless type II bursts, which was considered to be the strongest support for blast waves, is in serious doubt. This issue can be settled once and for all by studying CMEs and type II bursts associated with a set of C-class flares detected by the two spacecraft of the STEREO mission. From an entirely different point of view, *Mancuso and Raymond* (2004) suggest that most of the type II bursts are consistent with a CME-driven shock scenario with the radio emission originating from the nose or flanks of the shock. They obtained the coronal density profile using synoptic UVCS observations in the corona (1.5-3.5 Rs) before the occurrence of 29 metric type II bursts and showed that the computed type II burst locations were consistent with CME-driven shocks.

7.2. Relation Between Metric and IP Type II Bursts

The discrepancy between the drift rates (or shock speeds) of metric type II bursts and IP type II bursts (see, e.g., *Cane*, 2000; *Reiner et al.*, 2001) has been thought to argue against the same shock causing the metric and IP type II bursts; it favors a blast wave for the metric type II burst and a CME-driven shock for the IP type II burst with both disturbances originating from the same eruption. The drift rate problem can be traced to the over-simplified radial variation of the characteristic speed. In fact, it will be shown that a realistic profile of the characteristic speed naturally explains the drift rate discrepancy and all the observed features of the type II bursts phenomenon.

In the mid 1980s, when the debate regarding the source of coronal type II bursts was underway, it was thought that the Alfven speed had a discontinuous jump from tens of km/s in the chromosphere to ~500 km/s in the corona (the thick solid curve in Figure 4, adapted from *Bougeret*, 1985). The classical definition of "fast CMEs" (400-500 km/s), stems from this characteristic speed (*Gosling et al.*, 1976, *Bougeret*, 1985; *Cliver et al.*, 1999). The thick solid curve in Figure 4 implies that only disturbances propagating with speeds

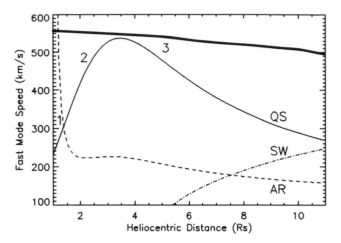

Figure 4. Fast mode speed in the quiet (QS, thin solid curve) and active region corona (AR, dashed line) compared with the solar wind speed (marked SW) adapted from *Gopalswamy et al.* (2001a). The thick solid curve is a sketch of the Alfven speed profile from *Bougeret* (1985). Region 1 corresponds to active region core where the fast-mode speed is high, so it is difficult to form shocks in this region. Metric type II bursts occur in region 2. IP type IIs occur in region 3. The Alfven speed is close to the fast mode speed since the sound speed is very small, so we use fast mode and Alfven speeds interchangeably.

exceeding ~500 km/s can drive shocks in the corona. But some metric type II bursts do occur in association with CMEs slower than 500 km/s (see, e.g., *Kundu et al.*, 1989; *Gopalswamy et al.*, 2001a). If the shocks are blast waves the CMEs are just accompanying events with no role in shock driving. The other possibility is that the local Alfven speed is not a constant value, but can be low enough for the slow CMEs to drive shocks. Recent studies indicate that this to be the case. A low and variable Alfven speed in the inner corona is evident even with simple models of the density and magnetic field in the quiet corona (see, e.g., *Krogulec et al.*, 1994). As shown in Figure 4 (thin solid curve marked QS), the fast mode speed starts from a low value (~200 km/s) near the coronal base, reaches a peak in the outer corona, and then slowly declines in the IP medium. The fast mode speed is similar to the constant value used in *Bougeret* (1985) only around the peak.

Mann et al. (1999) used such a speed profile (QS in Figure 4) to suggest that flare generated coronal shocks need to have speeds exceeding the peak value in order to penetrate into the IP space. Most of the type II bursts occur in the vicinity of active regions, so using the quiet Sun fast mode speed is not quite appropriate especially close to the active region. To remedy this, *Gopalswamy et al.* (2001a) introduced the active region component of the fast mode speed (the curve marked AR in Figure 4), which drops from a few thousand km/s at

the active region core to ~200 km/s at 2 Rs (see also *Mann et al.*, 2003). The resultant fast mode speed has a minimum in the inner corona (region 1 in Figure 4), rises to a peak in the outer corona (region 2), and then slowly declines in the IP space (region 3). It was noted in Figure 3 that type II bursts occur only when the CME mean speed is greater than 200 km/s, which is the minimum speed in Figure 4 that has to be exceeded for shock formation. Furthermore, when coronal shocks are driven by CMEs, the initial speed need not exceed the peak fast mode speed in order to penetrate into the IP space, since slower and accelerating CMEs are routinely observed beyond the height of the fast mode peak. CMEs typically accelerate in region 1, so they are likely to have higher speed in region 3 than in regions 1 and 2. This is an important factor that contributes to the speed discrepancy for shocks derived from type II bursts in regions 2 and 3. Another factor is the location of the radio source with respect to the radial direction. If the metric type II emission occurs at the flanks of the shock (where quasiperpendicularity is satisfied, see *Holman and Pesses*, 1983), then the drift rate is determined by the scale height along the locus of the quasi-perpendicular region, rather than along the density gradient. If the IP type II burst, on the other hand occurs at the nose of the shock, then one would also expect different drift rates for the metric and IP type II bursts. Emission from the nose and flanks of the same shock might explain the occasional events with simultaneous bands of emission in the metric and IP domains (*Raymond et al.*, 2000). A final possibility is that the flank and nose of the shock correspond to different sections of the fast mode profile in Figure 4. The flank is likely to a lower fast mode speed region compared to the nose.

The modified fast mode speed profile (AR+QS) in Figure 4 can account for various observed features of type II bursts from the corona and IP medium if the shock driver is a CME. 1. The high fast mode speed in the active region core does not allow shock formation there, thus providing a natural explanation for the relatively low starting frequency of type II bursts (~150 MHz). 2. At heliocentric distances <2 Rs where metric type II bursts form, the fast mode speed is relatively low, so it is easy to drive shocks. This explains the higher abundance of metric type II bursts compared to the IP type II bursts (see Plate 1). 3. The average speed of CMEs associated with type II bursts confined to the metric domain is ~600 km/s (see Table 1). This is about the peak fast mode speed (see Figure 4). These CMEs are super-Alfvenic only in region 2, and hence the corresponding type II bursts are confined to the metric domain. 4. Some of the energetic shocks can continue beyond the fast-mode peak, resulting in the m-to-DH and m-to-km type II bursts. CMEs associated with such type II bursts are of highest speed (>1200 km/s) and hence are capable of driving such shocks (see Table 1). 5. CMEs with intermediate speeds can drive shocks in the metric domain (where the fast mode speed is low), lose the shock in the outer corona (where the fast mode speed has its peak) and again drive a shock beyond the outer corona when the fast mode speed declines. A blast wave cannot do this because once it ceases to be a shock it is lost for ever since there is no driver behind it. 6. Accelerating, low-speed CMEs may produce shocks in the DH and/or km domains even though they do not drive shocks in the metric domain (*Reiner et al.*, 1998; *Gopalswamy*, 2004a). This again suggests that the initial speed of disturbances need not exceed the fast mode peak for shocks forming in the IP space. 7. Since the constraint on the speed of the shock driver is different in the inner and outer corona (dictated by the fast mode profile), drift rates derived from the type II radio bursts are expected to be different even for the same driver. 8. The fast mode speed profile in Figure 4 is based on simple density and magnetic field profiles. In reality, it may vary in the peak value and the location of the peak depending on the prevailing physical conditions in the corona. This allows for the possibility that a 250 km/s CME could be a fast CME, while a >1000 km/s CME could be a slow CME depending on the local fast mode speed. This way, the large number of fast and wide CMEs without type II bursts can be explained as a consequence of the high characteristic speed (*Sheeley et al.*, 1984; *Gopalswamy et al.*, 2001b). In summary, the possibilities resulting from the combination of CME and medium properties can result in all the known varieties of type II bursts shown in Figure 1, thus providing a simple possibility of explaining type II bursts in all spectral domains using CME-driven shocks.

The place of blast waves in the overall picture of type II phenomenon needs to be mentioned: the varieties 2-6 in Figure 1 are due to CME-driven shocks, while a fraction of events under variety 1 may be due to flare blast waves (*Vrsnak and Lulic*, 2000). This would imply that the CME-driven shock mechanism works for the entire Sun-Earth distance and for a narrow region of ~1 Rs from the solar surface an additional mechanism (blast waves) may operate. Unfortunately, there is no simple way to detect such blast waves because they do not propagate far from the Sun for in situ detection. Theoretical studies indicate that blast waves may not survive to reach the IP medium due to the refraction of the waves towards solar surface (*Vainio and Khan*, 2004). Since all the metric type II bursts are associated with super-Alfvenic CMEs (see point 3 in the previous paragraph), the blast waves and CMEs need to coexist near the Sun, which complicates the blast wave propagation. Interpretation of Moreton waves (in H-alpha, EUV) as blast waves is also problematic because CMEs accompany Moreton waves. Papers seeking to identify Moreton waves with blast waves (e.g., *Hudson et al.*, 2003) do not account for the accompanying CMEs, physically present in the spatial domain of the problem, and hence cannot be ignored.

8. FLARES AND TYPE II BURSTS

It must be pointed out that all metric type II bursts are associated with flares although only a small fraction of flares are associated with type II bursts (see *Cliver et al.*, 1999; see also Plate 1). Unlike disk CMEs, there is no problem in detecting flares associated with type II bursts (except for the behind-the-limb flares). Flares also fit into the overall hierarchical relationship between CMEs and type II bursts discussed in section 4. Table 2 shows the X-ray flare sizes corresponding to the m-limb, DH, and m-to-km type II bursts (*Gopalswamy et al.*, 2005b). Flares of size ≤B1.0 (GOES X-ray class) are listed as "other". The m-limb type II bursts are predominantly associated with C- and M-class flares (84%), while the vast majority of the m-to-km bursts are associated with M- and X-class flares (86%). For the DH type II bursts, the flare size is intermediate: the M and X class flares still constitute the majority (73%), but about a quarter of the flares are of C-class. Clearly the m-to-km type II bursts are associated with biggest flares while the m-limb type II bursts are associated with the smallest flares (see also *Robinson et al.*, 1984). Association of the m-to-km type II bursts with the most energetic CMEs and largest flares reminds us of the "Big-flare Syndrome" (*Kahler*, 1982).

There are ~20 times more flares and 10 times more CMEs than the number of metric type II bursts (see Plate 1) but only those flares accompanying CMEs are associated with type II bursts. Such a conclusion is consistent with the result obtained many years ago that metric type II bursts are associated only with eruptive flares (*Munro et al.*, 1979). Eruptive flares are so called because of the accompanying mass motion in the form of H-alpha ejecta, which we now know form the core of CMEs. Non-eruptive (or compact flares) are neither associated with CMEs nor with type II bursts.

The close connection between flare size and type II wavelength may appear consistent with the blast-wave scenario for metric type II bursts. But the required presence of CMEs complicates such an interpretation. X-ray observations have shown that the flare site is typically located close to the Sun (~10^4 km above the surface, see *Catalano and van Allen*, 1973), while the CME leading edge is at a much larger height when the flare starts. In the CSHKP model of an eruptive event (see *Anzer and Pneuman*, 1982 for example), the flare

Table 2. Fraction of soft X-ray flares associated with m-limb, DH, and m-to-km type II bursts. "other" denotes B-class and lower size flares.

	X	M	C	other
m-to-km	42%	44%	8%	6%
DH	25%	48%	23%	4%
m-limb	3%	40%	44%	13%

site is considered to be the reconnection site far below the CME leading edge. If a blast wave starts from the flare site, it has to propagate through the moving medium (CME material), which is not favorable for shock formation. Therefore, the blast wave speed with reference to the CME speed has to exceed the local characteristic speed to drive a shock, whereas the CME has to simply exceed the coronal Alfven speed to drive a shock. It is interesting to note that even X-class flares are not associated with type II bursts if they are non-eruptive.

9. SUMMARY AND CONCLUSIONS

There is a hierarchical relationship between CME kinetic energy and the wavelength range over which type II bursts occur: purely metric type II burst are associated with CMEs of low average speed (~600 km/s) while the m-tokm type II bursts are associated with much faster CMEs (average speed ~1500 km/s) with the DH type II bursts associated with CMEs of intermediate speed (~1100 km/s). The widths are also progressively higher as one goes from metric to m-to-km bursts, which implies a progressive increase in kinetic energy. This organization of type II bursts by CME kinetic energy lends support to the idea that the whole type II phenomenon can be explained by CME-driven shocks. The initial kinetic energy essentially decides how far a CME can drive a shock into the IP medium. The most energetic CMEs obviously can drive shocks far into the IP medium, so the shock produces radio emission at various distances from the Sun (and hence at various wavelengths). As expected, such energetic CMEs are also highly associated with SEP events. The DH and km type II bursts correspond to the spatial domain from 2-200 Rs and are associated with CMEs. It seems reasonable to extend the applicability of CME-driven shocks by another solar radius or so to include the m type II bursts in the unified approach to the type II phenomena. This is further supported by the non-existence of CMEless type II bursts. The universal relationship found between the drift rates of type II bursts in various wavelength domains is also consistent with such an interpretation. The consistency between type II source heights from radioheliographic observations and CME leading edges at the time of type II bursts also calls for a close association between type II bursts and CMEs. The discordant drift rates (or derived shock speeds) in the corona and IP medium, an argument often used for two different shock sources, can be readily explained when a realistic radial profile of the fast mode speed is used. There is undeniable relationship between flares and type II bursts, but the flares need to be eruptive (accompanied by CMEs). The presence of CMEs in eruptive flares implies that flare blast waves, if present, have to propagate through moving plasmas (i.e., CMEs with an average speed of at least 600 km/s), and hence less conducive for shock formation. The strongest

argument against the blast waves is that they have never been observed in the IP medium. Future in situ observations close to the Sun (such as from the Solar Orbiter and Solar Probe) may settle the issue of blast waves. The fact that the shock-driving ability of CMEs depends on their kinetic energy has an important practical utility: the type II bursts extending to the IP medium can isolate the small number of CMEs relevant for space weather.

Acknowledgments. Research supported by NASA/LWS and SR&T programs. I thank M.L. Kaiser for the Wind/WAVES type II catalog and G. Michalek, S. Yashiro, S. Petty for help with figures.

REFERENCES

Aguilar-Rodriguez, E., N. Gopalswamy, R.J. MacDowall, and M.L. Kaiser, A study of the drift rate of type II bursts at different wavelengths, Solar wind 11, in press, 2005.

Anzer, U., and G.W. Pneuman, Magnetic reconnection and coronal transients, *Solar Phys.*, 79, 129, 1982.

Aurass, H., Coronal mass ejections and type II radio bursts, in Coronal Physics from Radio and Space Observations, edited by G. Trottet, Springer, Berlin, p.135, 1997.

Bale, S.D., M.J. Reiner, J.-L. Bougeret, M.L. Kaiser, S. Krucker, D.E. Larson, R.P. Lin, The source region of an interplanetary type II radio burst. *Geophys. Res. Lett.*, 26, 1573-1576, 1999.

Boischot, A., A.C. Riddle, and J.B. Pearce, and J.W. Warwick, Shock waves and type II radio bursts in the interplanetary medium, *Solar Phys.*, 5, 397, 1980.

Bougeret, J.-L., Observations of shock formation and evolution in the solar atmosphere, in Collisionless shocks in the heliosphere: Reviews of current research, Washington, DC, American Geophysical Union, 1985, p. 13-32, 1985.

Bougeret, J.-L., *et al.*, Waves: The Radio and Plasma Wave Investigation on the Wind Spacecraft. *Space Sci. Rev.*, 71, 231-263, 1995.

Burlaga, L.F., K.W. Behannon, and L.W. Klein, Compound streams, magnetic clouds, and major geomagnetic storms, *J. Geophys. Res.*, 92, 5725-5734, 1987.

Burkepile, J.T., A.J. Hundhausen, A.L., Stanger, O.C. St. Cyr, and J.A. Seiden, Role of projection effects on solar coronal mass ejection properties: 1. A study of CMEs associated with limb activity, *J. Geophys. Res.*, 109, 3103, 2004.

Cane, H.V., The relationship between coronal transients, Type II bursts and interplanetary shocks, *Astron. Astrophys.*, 140, 205, 1984.

Cane, H.V., R.G. Stone, J. Fainberg, J.L. Steinberg, S. Hoang, Type II solar radio events observed in the interplanetary medium. I - General characteristics, *Solar Phys.*, 78, 187-198, 1982.

Cane, H.V. and R.G. Stone, Type II solar radio bursts, interplanetary shocks, and energetic particle events, *Atrophys. J.* 282, 339-344, 1984.

Cane, H.V., ISEE-3 observations of radio emission from coronal and interplanetary shocks. Radio Astronomy at Long Wavelengths, Geophysical Monograph 119, AGU, Washington DC, 147-153, 2000.

Cane, H.V., N.R. Sheeley, and R.A. Howard, Energetic interplanetary shocks, radio emission, and coronal mass ejections, *J. Geophys. Res.*, 92, 9869-9874, 1987.

Catalano, C.P., and J.A. Van Allen, Height distribution and directionality of 2-12 A X-ray flare emission in the solar atmosphere, *Astrophys. J.*, 185, 335-350, 1973.

Classen, H.T., and H. Aurass, On the association between type II radio bursts and CMEs, *Astron. Astrophys.*, 384, 1098-1106, 2002.

Cliver, E.W., D.F. Webb, and R.A. Howard, On the origin of solar metric type II bursts. *Solar Phys.*, 187, 89-114, 1999.

Cliver, E.W., N.V. Nitta, B.J. Thompson, and J. Zhang, Coronal Shocks of November 1997 Revisited: The CME Type II Timing Problem, *Solar Phys.*, 225, 105-139, 2004a.

Cliver, E.W., S.W. Kahler, and D.V. Reames, Coronal Shocks and Solar Energetic Proton Events, *Astrophys. J.*, 605, 902-910, 2004b.

Erickson, W.C., The Bruny Island radio spectrometer, *Pub. Astron. Soc.* Australia, 14, 278-282, 1997.

Gary, D.E., *et al.*, Type II bursts, shock waves, and coronal transients - The event of 1980 June 29, 0233 UT. *Astron. Astrophys.*, 134, 222-233, 1984.

Gergely, T., On the Relative Velocity of Coronal Transients and Type II Bursts, Proc. of the STIP Symposium, Edited by M.A. Shea, D.F. Smart, and S.M.P. McKenna-Lawlor, p.347, 1984.

Gopalswamy, N., Type II Solar Radio Bursts. Radio Astronomy at Long Wavelengths, Geophysical Monograph 119, AGU, Washington DC, 123-135, 2000.

Gopalswamy, N., Solar and geospace connections of energetic particle events. *Geophys. Res. Lett.*, 30, 1-4, 2003.

Gopalswamy, N., Recent advances in the long-wavelength radio physics of the Sun, Planetary Space Sci, 52, 1399-1413, 2004a.

Gopalswamy, N. A global picture of CMEs in the inner heliosphere, in "The Sun and the Heliosphere as an Integrated system", ASSL series, edited by G. Poletto and S. Suess, KLUWER/Boston, Chapter 8, p. 201, 2004b.

Gopalswamy, N., Interplanetary radio bursts, in Solar and Space Weather Radiophysics edited by D.E. Gary and C.O. Keller, Kluwer, Boston, chapter 15, p.305, 2004c.

Gopalswamy, N., Coronal Mass Ejections of Solar Cycle 23, J. Astrophy. Astr., in press, 2006.

Gopalswamy, N. and M.R. Kundu, Are coronal type II shocks piston driven? AIP Conference Proceedings # 264: Particle Acceleration in Cosmic Plasmas, ed. by G.P. Zank T.K. Gaisser, American Institute of Physics, New York, 257-260, 1992.

Gopalswamy, N., et al., Origin of coronal and interplanetary shocks - A new look with WIND spacecraft data. J. Geophys. Res., 307-316, 1998.

Gopalswamy, N., A. Lara, R.P. Lepping, M.L. Kaiser, D. Berdichevsky, and O.C. St. Cyr, Interplanetary acceleration of coronal mass ejections, Geophys. Res. Lett., 27, 145, 2000a.

Gopalswamy, N., M.L. Kaiser, B.J. Thompson, L. Burlaga, A. Szabo, A. Lara, A. Vourlidas, S. Yashiro, and J.-L. Bougeret, Radio-rich solar eruptive events, Geophys. Res. Lett., 27, 1427-1430, 2000b.

Gopalswamy, N., A. Lara, M.L. Kaiser, J.-L. Bougeret, Near-Sun and near-Earth manifestations of solar eruptions. J. Geophys. Res., 106, 25261-25278, 2001a.

Gopalswamy, N., S. Yashiro, M.L. Kaiser, R.A. Howard, J.-L. Bougeret, Characteristics of coronal mass ejections associated with long-wavelength type II radio bursts. J. Geophys. Res., 106, 29219-29230, 2001b.

Gopalswamy, N., S. Yashiro, M.L. Kaiser, R.A. Howard, J.-L. Bougeret, Radio Signatures of Coronal Mass Ejection Interaction: Coronal Mass Ejection Cannibalism? Astrophys. J., 548, L91-L94, 2001c.

Gopalswamy, N., S. Yashiro, G. Michalek, M.L. Kaiser, R.A. Howard, D.V. Reames, R. Leske, and T. von Rosenvinge, Interacting Coronal Mass Ejections and Solar Energetic Particles, Astrophys. J., 572, L103-L107, 2002.

Gopalswamy, N., A. Lara, S. Yashiro, S. Nunes, R.A. Howard, In: Solar variability as an input to the Earth's environment. Ed.: A. Wilson. ESA SP-535, Noordwijk: ESA Publications Division, p. 403, 2003a.

Gopalswamy, N., S. Yashiro, A. Lara, M.L. Kaiser, B.J. Thompson, P. Gallagher, Large solar energetic particle events of cycle 23: A global view Geophys. Res. Lett., 30 (12), SEP 3-1, CiteID 8015, 2003b.

Gopalswamy, N., S. Yashiro, S. Krucker, G. Stenborg, and R.A. Howard, Intensity variation of large solar energetic particle events associated with coronal mass ejections, J. Geophys. Res., 109, 12105, 2004a.

Gopalswamy, N., S. Nunes, S. Yashiro, R.A. Howard, Variability of solar eruptions during cycle 23, Ad. Space Res., 34 (2), 391-396, 2004b.

Gopalswamy, N., S. Yashiro, Y. Liu, G. Michalek, A. Vourlidas, M.L. Kaiser, and R.A. Howard, Coronal Mass Ejections and Other Extreme Characteristics of the 2003 October-November Solar Eruptions, JGR, 110, Issue A9, CiteID A09S15, 2005a.

Gopalswamy, N., E. Aguilar-Rodriguez, S. Yashiro, S. Nunes, M.L. Kaiser, and R.A. Howard, Type II Radio Bursts and Energetic Solar Eruptions, 110, Issue A12, CiteID A12S07, 2005b.

Gopalswamy, N., S. Yashiro, S. Krucker, and R.A. Howard, CME Interaction and the Intensity of Solar Energetic Particle Events, in Coronal and Stellar Mass Ejections, edited by K.P. Dere, J. Wang, and Y. Yan, International Astronomical Union, 367-372, 2005c.

Gopalswamy, N., H. Xie, S. Yashiro, I. Usoskin, Coronal Mass Ejections and Ground Level Enhancements. in Proceeding of 29th International Cosmic Ray Conference, Pune, 2005.

Gosling, J.T., E. Hildner, R.M. MacQueen, R.H. Munro, A.I. Poland, C.L. Ross, The speeds of coronal mass ejection events, Solar Phys., 48, 379, 1976.

Holman, G.D. and M.E. Pesses, Solar type II radio emission and the shock drift acceleration of electrons. Astrophys. J., 267, 837-843, 1983.

Howard, R.A., D.J. Michels, N.R. Sheeley, Jr., M.J. Koomen, The observation of a coronal transient directed at earth, Astrophys. J., 263, L101-L104, 1982.

Hudson, H.S., J.I. Khan, J.R. Lemen, N.V. Nitta, and Y. Uchida, Solar Phys., 212, 121, 2003.

Hundhausen, A.J., in Coronal Mass Ejections, edited by. N. Crooker, J.A Joselyn, and J. Feynman, AGU Monograph 99, p1, 1997.

Kahler, S.W., The role of the big flare syndrome in correlations of solar energetic proton fluxes and associated microwave burst parameters, J. Geophys. Res., 87, 3439-3448, 1982.

Kahler, S.W., Injection profiles of solar energetic particles as functions of coronal mass ejection heights, Astrophys. J., 428, 837-842, 1994.

Kahler, S.W., Solar Fast Wind Regions as Sources of Gradual 20 MeV Solar Energetic Particle Events, Proc. of the 28th International Cosmic Ray Conference, Edited by T. Kajita, Y. Asaoka, A. Kawachi, Y. Matsubara, and M. Sasaki, p. 3415, 2003.

Kahler, S.W., E. Hildner and M.A.I. van Hollebeke, Prompt solar proton events and coronal mass ejections, Solar Phys., 57, 429-443, 1978.

Kahler, S., N.R. Sheeley, Jr., R.A. Howard, D.J. Michels, M.J. Koomen, Solar Phys., 93, 133, 1984.

Kaiser, M.L., M.J. Reiner, N. Gopalswamy, R.A. Howard, O.C. St. Cyr, B.J. Thompson, and Bougeret, J.-L., Type II radio emissions in the frequency range from 1-14 MHz associated with the April 7, 1997 solar event, Geophysical Res. Lett., 25, 2501-2504, 1998.

Kallenrode, M.-B. and E.W. Cliver, Rogue SEP events: Modeling, Proc. of ICRC 2001, 3318-3321, 2001.

Klein, K.-L., J.I. Khan, N. Vilmer, J.-M. Delouis, and H. Aurass, X-ray and radio evidence on the origin of a coronal shock wave, Astron. Astrophys., 346, L53-L56, 1999.

Kosugi, T., Type II–IV radio bursts and compact and diffuse white-light clouds in the outer corona of December 14, 1971, Solar Phys., 48, 339 – 356, 1976.

Krogulec, M., Z.E. Musielak, S.T. Suess, S.F. Nerney, and R.L. Moore, Reflection of Alfven waves in the solar wind. J. Geophys. Res. 99, 23489-23501, 1994.

Kundu, M.R., Solar Radio Astronomy, Interscience Publishers, New York, 1965.

Kundu, M.R., N. Gopalswamy, S.M. White, P. Cargill, E.J. Schmahl, and E. Hildner, The radio signatures of a slow coronal mass ejection - Electron acceleration at slow-mode shocks?, Astrophys. J., 347, 505, 1989.

Lara, A., N. Gopalswamy, S. Nunes, G. Munoz, and S. Yashiro, A statistical study of CMEs associated with metric type II bursts. Geophys. Res. Lett., 30, No.12, SEP 4-1., 2003.

Lengyel-Frey, D., and R.G. Stone, JGR, 94, 159, 1989.

Li, G., and G.P. Zank, Multiple CMEs and large SEP events, 29th ICRC Conference, Pune, in press, 2005.

Lugaz, N., W.B. Manchester, and T.I. Gombosi, Numerical Simulation of the Interaction of Two Coronal Mass Ejections from Sun to Earth. Astrophys. J. 634, 651, 2005.

Malitson, H.H., J. Fainberg, and R.G. Stone, Observation of a Type II Solar Radio Burst to 37 Rsun. Astrophys. Lett. 14, 111, 1973.

Mancuso, S., and J.C. Raymond, Coronal transients and metric type II radio bursts. I. Effects of geometry, Astron. Astrophys., 413, 363-371, 2004.

Mann, G., A. Klassen, H.T. Classen, H. Aurass, D. Scholz, R.J. MacDowall, R.G. Stone, Catalogue of solar type II radio bursts observed from September 1990 to December 1993 and their statistical analysis, Astron. Astrophys. 119, 489-498, 1996.

Mann, G., A. Klassen, C. Estel, and B.J. Thompson, Coronal Transient Waves and Coronal Shock Waves. Proc. of 8th SOHO Workshop, Edited by J.-C. Vial and B. Kaldeich-Schmann. 477-481, 1999.

Mann, G., A. Klassen, H. Aurass, H.-T. Classen, Formation and development of shock waves in the solar corona and the near-Sun interplanetary space, Astron. Astrophys. 400, 329, 2003.

Munro, R.H., J.T. Gosling, E. Hildner, R.M. MacQueen, A.I. Poland, C.L. Ross, The association of coronal mass ejection transients with other forms of solar activity, Solar Phys., 61, 201-215, 1979.

Nelson, G.J. and D.B. Melrose, Type II bursts, in Solar radiophysics: Studies of emission from the sun at metre wavelengths (A87-13851 03-92). Cambridge and New York, Cambridge University Press, p. 333-359, 1985.

Payne-Scott, R., D.E. Yabsley, and J.G. Bolton, Nature, 160, 256, 1947.

Pinter, S., Close connexion between flare-generated coronal and interplanetary shock waves, Nature Phys. Sci., 243, 96-97, 1973.

Raymond, J.C., et al., SOHO and radio observations of a CME shock wave, Geophys. Res. Lett., 27, 1439, 2000.

Reiner, M.J., in Radio Astronomy at Long Wavelengths, Geophysical monograph Series, vol. 119, Edited by R.G. Stone, K.W. Weiler, M.L. Goldstein, and J.-L. Bougeret. Washington, DC: American Geophysical Union, p.137, 2000.

Reiner, M.J., M.L. Kaiser, J. Fainberg, J.-L. Bougeret, and R.G. Stone, Geophys. Res. Lett., 25, 2493, 1998a.

Reiner, M.J., M.L. Kaiser, Fainberg, and Stone, A new method for studying remote type II radio emissions from coronal mass ejection-driven shocks, JGR, 103, Issue A12, p29651, 1998b.

Reiner, M.J., M.L. Kaiser, N. Gopalswamy, H. Aurass, G. Mann, A. Vourlidas, and M. Maksimovic, 2001. Statistical analysis of coronal shock dynamics implied by radio and white-light observations. J. Geophys Res. 106, 25279-25290, 2001.

Robinson, R.D., R.T. Stewart, & H.V. Cane, Properties of metre-wavelength solar bursts associated with interplanetary type II emission, Solar Phys., 91, 159, 1984.

Robinson, R.D., R.T. Stewart, A positional comparison between coronal mass ejection events and solar type II bursts. Solar Phys., 97, 145-157, 1985.

Sheeley, N.R., R.A. Howard, D.J. Michels, R.D. Robinson, M.J. Koomen, and R.T. Stewart, Associations between coronal mass ejections and metric type II bursts. Astrophys. J., 279, 839-847, 1984.

Stewart, R.T., M. McCabe, M.J. Koomen, R.T. Hansen, G.A. Dulk, Observations of Coronal Disturbances from 1 to 9 R_sun. I: First Event of 1973, January 11, Solar Phys., 36, 203, 1974a.

Stewart, R.T., R.A. Howard, F. Hansen, T. Gergely, M. Kundu, Observations of Coronal Disturbances from 1 to 9 R_sun. I: First Event of 1973, January 11, Solar Phys., 36, 219, 1974b.

Svestka, Z. and L. Fritzová-Svestková, Type II Radio Bursts and Particle Acceleration, Solar Phys., 36, 417, 1974.

Tousey, R., The solar corona, Space Res., 13, 713, 1973.

Uchida, Y., On the Exciters of Type II and Type III Solar Radio Bursts, PASJ, 12, 376, 1960.

Vainio, R. and J. Khan, Solar Energetic Particle Acceleration in Refracting Coronal Shock Waves, *Astrophys. J.*, 600, 451, 2004.

Vourlidas, A., D. Buzasi, R.A. Howard, and E. Esfandiari, Solar variability: from core to outer frontiers, edited by. A. Wilson. ESA SP-506, vol. 1. Noordwijk: ESA Publications Division, p91, 2002.

Vrsnak, B., Deceleration of coronal mass ejections, *Solar Phys.*, 202, 173, 2001.

Vrsnak, B., and S. Lulic, Properties of metre-wavelength solar bursts associated with interplanetary type II emission, *Solar Phys.*, 197, 156, 2000.

Vrsnak, B., V. Ruzdjak, P. Zlobec, H. Aurass, Ignition of MHD shocks associated with solar flares. *Solar Phys.*, 158, 331, 1995.

Vrsnak, B., H. Aurass, J. Magdalenic, and N. Gopalswamy, Band-splitting of coronal and interplanetary type II bursts. I. Basic properties, *Astron. Astrophys.*, 377, 321-329, 2001.

Vrsnak, B., J. Magdalenic, H. Aurass, and G. Mann, Band-splitting of coronal and interplanetary type II bursts. II. Coronal magnetic field and Alfvén velocity, *Astron. Astrophys.*, 396, 673, 2002.

Wagner, W.J. and R.M. MacQueen, The excitation of type II radio bursts in the corona. *Astron. Astrophys.*, 120, 136, 1983.

Wild, J.P. and L.L. McCready, Observations of the spectrum of high-intensity solar radiation at metre wavelengths. I. The Apparatus and Spectral Types of Solar Burst Observed. *Aust. J. Sci. Res.*, A3, 387-398, 1950.

Yashiro, S., N. Gopalswamy, S. Akiyama, G. Michalek, and R.A. Howard, Visibility of coronal mass ejections as a function of flare location and intensity, *J. Geophys. Res.*, 110, No. A12, A12S05, doi:10.1029/2005JA011151, 2005.

Zheleznyakov, V.V., Radio emission of the Sun and Planets. Pergamon Press, New York, 1969.

Zlotnik, E. Ya, A. Klassen, K.-L. Klein, H. Aurass, G. Mann, Third harmonic plasma emission in solar type II radio bursts. *Astron. Astrophys.*, 331, 1087, 1998.

N. Gopalswamy, Solar System Exploration Division, Bldg 21, Room 260, Code 695, NASA Goddard Space Flight Center, Greenbelt, MD 20771, USA.

EIT Waves and Coronal Shock Waves

G. Mann

Astrophysikalisches Institut Potsdam, D-14482 Potsdam, Germany

EIT waves have been discovered by the EIT instrument aboard the SOHO space-craft as a global wave phenomenon in the low corona. Most of them are associated with solar type II radio bursts a signatures of shock waves in the high corona. Considering both phenomena to be caused by the same initial energy release (flare), they can be used as diagnostic tools for the magnetic field in the corona. This study leads to a local minimum and maximum of the Alfvén speed in the middle of the corona and at a radial distance of about 4 solar radii from the center of the Sun, respectively. Since shock waves are one source of energetic particles the occurrence of such a minimum and maximum has effects on the temporal evolution of solar energetic particle events.

1. INTRODUCTION

The observations of the **E**xtreme ultraviolet **I**maging **T**elescope (**EIT**) instrument [*Delaboudiniere et al.*, 1995] aboard the SOHO spacecraft reveal a new wave phenomenon on the Sun, the so-called *coronal transient* (or *EIT*) *waves* [*Moses et al.*, 1997; *Thompson et al.*, 1998]. In the EIT images, these waves appear as a diffuse bright rim (sometimes circularly) expanding around the flaring active region over one hemisphere of the Sun (see Plate 1).

Although the EIT waves remind of the well-known *Moreton waves* [*Moreton and Ramsey*, 1960] they are visible in EUV spectral lines (e.g., 195 Å of Fe XII) emitting by a $1.6 \cdot 10^6$ K hot corona, whereas the Moreton waves are seen in the H_α-images (see Figure 1 for example) coming from the 10^4 K hot chromosphere.

Few events show a sharp and bright wavefront somewhat reminscent to Moreton waves (e.g., *Thompson et al.* [2000]) in contrast to the diffuse ones. They are usually called "brow waves" [*Gopalswamy et al.*, 2000]. They are only observed very close to an active region and coincide with Moreton waves in space and time [*Khan and Aurass*, 2002]. These wave phenomena can also be seen in the X-ray light as

Solar Eruptions and Energetic Particles
Geophysical Monograph Series 165

reported by *Warmuth et al.* [2005]. The EIT front could also be produced by a sudden expansion of a part of the magnetic field lines [*Delannee and Aulanier*, 1999; *Delannee*, 2000], instead by a flare.

The solar event on May 12, 1997 illustrates the relation of EIT waves to other phenomena of the flareing corona (see *Mann et al.* [2001]). Plate 1 (top) shows a sequence of running difference EIT images at 195 Å. As seen, a disturbance initially generated by a flare in a small area on 04:17-04:35 UT is propagating like a circular wave over the Sun until 05:24 UT. The wave has a velocity ≈ 300 km/s [*Klassen et al.*, 2000]. At the bottom the corresponding dynamic radio spectrum of the same event is presented in the frequency range 40-800 MHz. Solar radio bursts already occurred as signatures of energetic electons on 04:40 UT. A solar type II radio burst started at 90 MHz on 04:54 UT (see Plate 1). It occurs as stripes of enhanced radio emission slowly drifting from high to low frequencies and should be regarded as the radio signature of a shock wave travelling through the corona [*Nelson and Melrose*, 1985; *Mann*, 1995a]. A radial shock velocity of ≈ 1000 km/s is deduced from the drift rate -0.06 MHz/s of the type II burst at 28 MHz (fundamental emission) [*Klassen et al.*, 2000]. The close temporal proximity between the EIT waves and the solar type II radio burst is seen in this particlular event (see Section 3 for further informations). During the same event a coronal mass ejection (CME) rose from the Sun on 05:15 UT and drove an interplanetary (IP) shock ahead itself. It was seen as an IP

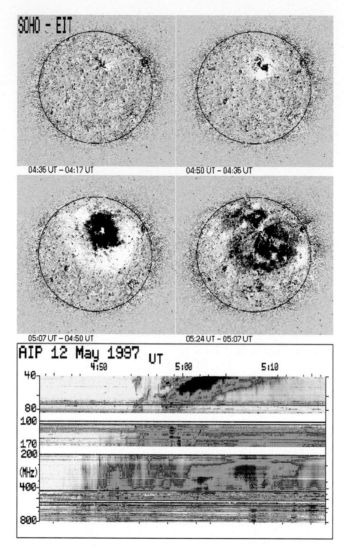

Plate 1. Running-difference images of the EIT instrument of the solar event on May 12, 1997 (top) and the corresponding dynamic radio spectrum in the range 40–800 MHz (bottom) recorded by the radiospectralpolarimeter of the Astrophysikalisches Institut Potsdam [*Mann et al.*, 1992] (see *Mann et al.* [1999a])

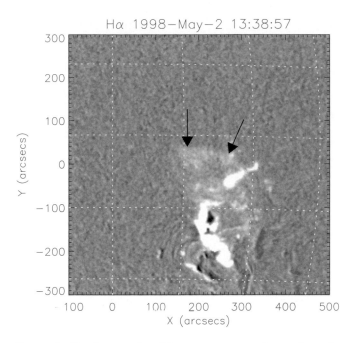

Figure 1. H$_\alpha$ image of a Moreton wave as observed at the Kanzelhöhe Solar Observatory (Austria).

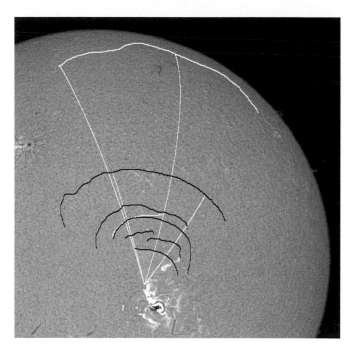

Figure 2. EIT (white) and Moreton (black) wave fronts at different times on the disc of the Sun for the event on May 2, 1998. The corresponding H$_\alpha$ image is underlying.

type II burst below 14 MHz by the WAVES instrument [*Bougeret et al.*, 1995] aboard the WIND spacecraft. A jump of the local plasma line recorded by WAVES at 1 AU on 01:00 UT on May 15 indicates the arrival of the IP shock resulting in the mean shock velocity of ≈600 km/s from the Sun up to 1 AU. The COSTEP instrument [*Müller-Mellin et al.*, 1995] aboard the SOHO spacecraft measured the onset of enhanced fluxes of electrons (0.25-0.7 MeV) and protons (4.3-7.8 MeV) on 05:12 UT and 07:40 UT [*Mann et al.*, 2001] during the same event, respectively. In summary, this particular example illustrates that EIT waves are in close temporal proximity with other (well-known) transient phenomena during solar flares (see further discussions in *Biesecker et al.* [2002]).

2. RELATIONSHIP TO MORETON WAVES

EIT waves have velocities in the range 150-400 km/s with a mean value at 270 km/s (see Figure 4) [*Mann et al.*, 1999a; *Klassen et al.*, 2000], whereas the Moreton waves seen in the H$_\alpha$ light show quite higher velocities of >400 km/s (sometimes ≈1000 km/s) [*Moreton*, 1960]. In order to clarify this difference, *Warmuth et al.* [2001] studied two solar events, in which both Moreton and EIT waves have been observed.

Let us focus to the event on May 2, 1998 (see Figure 2). Here, the wave fronts of the Moreton wave (black) and the

associated EIT wave (white) at different times are drawn on the underlying surface of the Sun. Note the different time resolution of the EIT instrument (≈12 minutes) [*Delaboudiniere et al.*, 1995] and the H$_\alpha$ telescope (≈1 minute) of the Kanzelhöhe Solar Observatory. The centre of these semi-circle wavefronts is located near an active region. Thus, this centre can be regarded as the location of the wave initiation. Combining the H$_\alpha$ and EIT data, the distance–time diagram, i.e., r(t), of the movement of the leading edge of this wave is presented in Figure 3. This diagram shows that the EIT wave represent the continuation of the Moreton wave in the corona. Taking the first derivative of r(t) one yields the velocity of the wave (see insert in Figure 3). One sees that the velocity drops from ≈1000 km/s down to ≈400 km/s in the first three minutes. The difference between the measured velocities of the Moreton- and EIT waves is caused by both the different time resolution of the EIT instrument and the H$_\alpha$ telescopes and the deceleration of the waves during their travelling in the solar atmosphere.

Thus, Moreton- and EIT waves are closely related to each other and caused by the same origin. That confirms the "sweeping-skirt" scenario [*Uchida et al.*, 1973; *Uchida*, 1974], i.e., the Moreton waves should be regarded as the imprint of the EIT waves on the chromosphere. (Further detailed discussions on this subject can be found in the papers by *Warmuth et al.*, [2004a,b].)

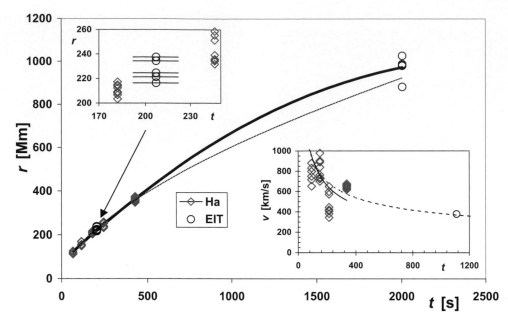

Figure 3. Radial distance versus time diagram of the H_α and EIT wave fronts. The radial distance is measured from the center of circles built up from the wave fronts. The first derivative of the function r(t) is presented in the insert. It gives the velocities of the wave fronts.

3. RELATIONSHIP TO TYPE II BURSTS

Moreton waves are accompanied by solar type II radio bursts [*Kai*, 1970]. *Uchida* [1968] regarded Moreton waves and type II burst related shock waves to be caused by the same origin (flare). Whereas the relationship between Moreton waves and type II bursts is well accepted [*Uchida*, 1968], *Klassen et al.* [2000] revealed the close relationship between EIT waves and type II radio bursts by a statistical analysis of 21 solar events with EIT waves. The results of this study is presented in Figure 4. The EIT wave speeds V_{EIT} are deduced from the running-difference EIT images at 195 Å. These values must be regarded as averaged ones over a 12 minute interval. The velocity of the type II burst sources are found from their drift rate in the dynamic radio spectrum.

It is generally accepted, that the radio emission takes place near the local electron plasma frequency $f_{pe} = (e^2 N_e/\pi m_e)^{1/2}$ (*e*, elementary charge; N_e, electron number density; m_e, electron mass) and/or its harmonics [*Melrose*, 1985]. Then, a relationship between the radial radio source velocity $V_{source,r}$ and the drift rate measured at the frequency f ($\approx f_{pe}$ for fundamental emission) can be derived to be

$$D_f = \frac{1}{2} \cdot \frac{1}{N_e} \frac{dN_e}{dR} \cdot V_{source,r} \qquad (1)$$

A onefold *Newkirk* [1961] model

$$N_e(r) = N_0 \cdot 10^{4.32 R_S / R} \qquad (2)$$

($N_0 = 4.2 \cdot 10^4$ cm^{-3}; R_S, solar radius; R, distance from the center of the Sun) has been adopted for a coronal density model for the solar corona. It corresponds to a barometric height formula with a temperature of $1.4 \cdot 10^6$ K [*Mann et al.*, 1999b]. Such a model agrees very well with white light scattering measurements by *Koutchmy* [1994] at quiet equatorial regions of the solar corona.

As various observations show, shock waves are established near but out of active regions in the solar corona [*Aurass et al.*, 1998; *Klassen et al.*, 1999; *Klein et al.*, 1999a; *Gopalswamy et al.*, 2000; *Khan and Aurass*, 2002]. Therefore, a onefold *Newkirk* [1961] model is an appropriate one for describing the radial density behaviour in the surrounding of type II radio burst sources. Therefore, it is adopted for deducing the radial velocity of the type II burst sources as presented in the histogram (top) of the Figure 5. The socomputed type II burst velocities must be regarded as a rough estimate because of several uncertainties. At first, the density can be higher in the source region as expected according to the onefold *Newkirk* [1961] model. At second, if the type II burst source is moving very obliquely towards the radial direction, there would be no drift in the dynamic radio

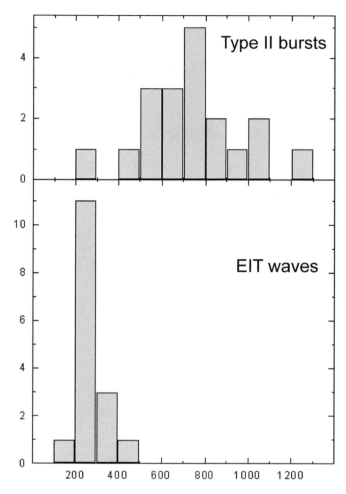

Figure 4. Distribution of the radial velocities of the solar type II radio burst sources (top) and the associated EIT waves (bottom) for the sample of events studied by *Mann et al.* [1999a].

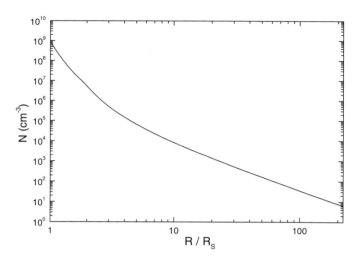

Figure 5. Radial behaviour of the electron number density according to the one-fold *Newkirk* [1961] model for R ≤ 1.8R_S and the heliospheric density model by *Mann et al.* [1999b] for R ≥ 1.8R_S (see also Sec. 5.1) (see *Mann et al.* [2003]).

spectrum. In conclusion, the type II burst source velocites presented in Figure 4 can be regarded as a low limit of the real velocity of them. Figure 4 shows the histogramms of the distribution of the velocities of the type II burst sources ($V_{II,r}$, top) and the EIT waves (V_{EIT}, bottom). It is seen, that the type II radio burst velocites are well above those of the EIT waves. The mean values of 270 km/s and 740 km/s are found for the velocities of the EIT waves and the associated solar type II radio burst sources, respectively. Once again, these values are rough estimates as discussed above.

As seen in Plate 1 EIT waves are globally travelling on the Sun. Thus, they are mainly propagating in quiet solar regions, i.e., outside active regions, although they are initially excited near an active region. Assuming a temperature T = 1.4·10^6 K as a typical value in the quiet solar corona, a sound speed $c_s = (\gamma k_B T / \tilde{\mu} m_p)^{1/2} = 180$ km/s (γ, ratio of specific heats;

k_B, Boltzmann's constant; $\tilde{\mu}$, mean molecular weight; m_p, proton mass) is obtained for $\gamma = 5/3$ and $\tilde{\mu} = 0.6$ (see *Priest* [1982]). Thus, the EIT wave speeds exceed well the sound speed. Consequently, the EIT waves should be regarded as *fast magnetosonic waves* (see *Priest* [1982]) in the low corona. Outside active regions the magnetic field is assumed to be radially directed, i.e., the EIT waves are propagating nearly perpendicular to the ambient magnetic field. Then, the EIT wave velocity V_{EIT} is related to the sound c_s and Alfvén speed v_A by $V_{EIT} = (v_E^2 + c_s^2)^{1/2}$. Here the Alfvén speed is usually defined by $v_A = B/(4\pi\tilde{\mu}Nm_p)^{1/2}$ (B, magnetic field strength; N, full particle number density) (see *Priest* [1982]). Note that the full particle number density N is related to the electron number density N_e by $N = 1.92N_e$ for $\tilde{\mu} = 0.6$. Thus, a mean Alfvén speed of 200 km/s can be found for $V_{EIT} = 270$ km/s and $c_s = 180$ km/s. It is generally assumed that the EUV light at 195 Å is emitted from the 1.6·10^6 K hot corona at a radial distance 1.01-1.08R_S from the center of the Sun [*Delabudiniere et al.*, 1995]. Thus, a magnetic field strength of 2-4 G can be deduced at the quiet photospheric level [*Mann et al.*, 1999a; *Warmuth and Mann*, 2005]. Here, the magnetic field has been maped from the low corona towards the photosphere by using the conservation of the magnetic flux, i.e., B·R^2 = const. That demonstrates that EIT waves can be used as a diagnostic tool (*coronal seismology*) for the low corona. (For further information on this subject it is refered to the papers by *Mann et al.* [2003] and *Warmuth and Mann* [2005].)

4. FORMATION AND DEVELOPMENT OF SHOCK WAVES IN THE CORONA

In the solar corona, shock waves can evolve from blast waves due to the initial energy release [*Uchida et al.*, 1973; *Vršnak et al.*, 1995] and/or driven by CMEs [*Stewart et al.*, 1974a,b; *Aurass*, 1997; *Lara et al.*, 2003]. Furthermore, loop oscillations can be the source of magnetohydrodynamic waves steepening into shock waves [*Hudson and Warmuth*, 2004]. Some of these shocks are able to continue into the interplanetary space. Coronal and IP shocks can be the source of type II radio radiation in the metric [*Uchida*, 1960] and deca-hectometric wave range [*Cane et al.*, 1981; *Reiner and Kaiser*, 1999], respectively. *Cane* [1983] suggested to differ between two kinds of shocks, the flare produced blast wave steepening into a shock during its journey through the inner corona and the CME driven shocks mainly appear in the outer corona and/or interplanetary space [*Wagner and MacQueen*, 1983; *Gopalswamy and Kundu*, 1992]. *Gopalswamy et al.* [1998] confirmed *Cane's* [1983] suggestion by investigating a large sample of solar events with type II radio bursts. On a similar result was reported by *Classen and Aurass* [2002] in the case of a particular solar event. But there are also events, in which several CMEs also cause different shocks in the corona [*Lara et al.*, 2003].

Yohkoh and SOHO images of the Sun and complementary spatial observations of the Nancay radio-heliograph showed, that solar type II radio burst sources mainly propagate non-radially away from active regions [*Gopalswamy et al.*, 1997, 2000; *Aurass et al.*, 1998; *Klassen et al.*, 1999; *Klein et al.*, 1999a]. Therefore, the knowledge of the global behaviour of the Alfvén and the sound speed as the characteristic speeds in a magnetoplasma is necessary for studying the formation and development of shock waves in the solar corona and interplanetary space. Since the Alfvén speed is depending on the magnetic field strength B and the full particle number density N, a density and magnetic field model is needed for doing that.

4.1. Density Model

As already mentioned, coronal shock waves usually appear near but outside active regions, so that a onefold *Newkirk* [1961] model can be adopted as an appropriate density model for the solar corona for this study (see also Sec. 3). That is confirmed in a recent study by *Warmuth and Mann* [2005], once more. They showed that the onefold *Newkirk* [1961] model agrees very well with recently performed measurements of the density in the corona up to a height of $0.5R_S$ above the photosphere.

But the *Newkirk* [1961] model can not describe the density behaviour in the interplanetary space. On the other hand,

Mann et al. [1999b] presented a heliospheric density model as a special solution of *Parker's* [1958] wind equation

$$\frac{v^2}{v_c^2} - \ln\left(\frac{v^2}{v_c^2}\right) = 4\ln\left(\frac{R}{r_c}\right) + 4\cdot\frac{r_c}{R} - 3 \qquad (3)$$

with $v_c = (k_B T / \tilde{\mu} m_p)^{1/2}$ as the critical velocity and $r_c = GM_\odot / 2v_c^2$ (G, gravitational constant; $M_\odot$, mass of the Sun) as the critical radius. Eq. (3) provides the radial solar wind speed v(R) as a solution. Then, the radial density behaviour N(R) can be obtained by means of the equation of continuity, i.e., $NvR^2 = C = \text{const}$ [*Mann et al.*, 1999b]. The constant C can be determined to be $C = 6.3 \cdot 10^{34}$ s^{-1} by in-situ measurements at 1 AU [*Schwenn*, 1990]. The resulting radial behaviour of the density N(R) agrees very well with observations from the corona up to 5 AU (see e.g., *Leblanc et al.* [1998]), if a global temperature of $1 \cdot 10^6$ K is used in Eq. (3). Requiring continuity, the onefold *Newkirk* [1961] model is adopted for regions with $R \leq 1.8R_S$, whereas that by *Mann et al.* [1999b] is used for regions $R \geq 1.8R_\odot$. The so-found density model is depicted in Figure 5.

4.2. Magnetic Field Model

The magnetic field of the corona is composed by that of an active region $\vec{B}_{ar}$ and of the quiet Sun $\vec{B}_{qS}$, i.e., $\vec{B} = \vec{B}_{ar} + \vec{B}_{qS}$. The magnetic field of an active region is modelled by a magnetic dipol, i.e.,

$$\vec{B}_{ar} = \frac{3(\vec{M}\cdot\vec{r})\vec{r}}{r^5} - \frac{\vec{M}}{r^3} \qquad (4)$$

with the magnetic moment $\vec{M}$. r denotes the distance from the center of the dipole located in a depth λ under the photosphere (see Figure 6).

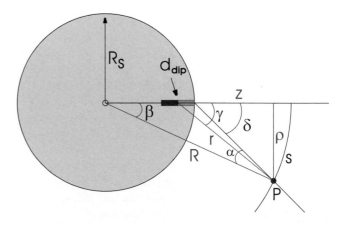

Figure 6. Scheme of reference frame used in Sec. 4 (see *Mann et al.* [2003]).

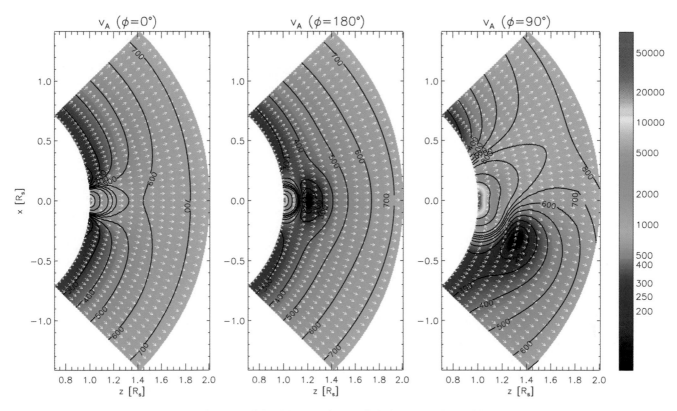

Plate 2. Coronal distribution of the Alfvén speed for the case of a parallel ($\phi = 0°$), antiparallel ($\phi = 180°$), and horizontal ($\phi = 90°$) direction of the magnetic dipole. The arrows indicate the direction of the magnetic field $\vec{B}$ composed by that of the dipole (see Eq. (4)) and the quiet Sun (see Eq. (5)). v_A is given in km/s. (see *Warmuth and Mann* [2005]).

The magnetic field of the quiet Sun is assumed to be radially directed, i.e.,

$$\vec{B}_{qS} = B_S \left(\frac{R_S}{R}\right)^2 \cdot \vec{e}_R \qquad (5)$$

with B_S as the magnetic field strength of the quiet Sun at the photosphere. The behaviour of the magnitude of magnetic field as given in Eq. (5) roughly agrees with the solar magnetic field model by *Banaszkiewicz et al.* [1998], especially near the ecliptic plane, i.e., above quiet equatorial regions.

The framework adopted in this paper with respect to this subject is depicted on Figure 6. In this study the magnetic dipole is considered to be directed either parallel ($\phi = 0°$) or anti-parallel ($\phi = 180°$) to the z-axis in order to model an unipolar active region. In this case the magnetic moment is related to the magnetic field strength B_0 at the z-axis on the photospheric level, i.e., $r = \lambda/2$, by $M = B_0 \lambda^3/16$. On the other hand, a bipolar active region can be modelled by a magnetic dipole inclined by $\phi = 90°$ to the z-axis. Then, the magnetic moment is given by $M = B_0 \lambda^3/8$.

4.3. Behaviour of the Alfvén Speed

In the previous subsections models of the density and the magnetic field in the corona has been introduced. With these models the Alfvén speed can be calculated at each point P in the corona (see. Figure 6). The result of this calculation is presented for the three different orientations ($\phi = 0°, 90°, 180°$) of the dipole with the parameters chosen to be $B_0 = 0.8$ kG, $B_S = 2.2$ G, and $\lambda = 0.1 R_S$ in Plate 2. In the case $\phi = 0°$, a minimum of the Alfvén speed appears at both sides of the active region in the low corona. Otherwise, such a minimum occurs above an active region in the middle of the corona, i.e., at an height of $\approx 0.5 R_S$, in the case of $\phi = 180°$. If the dipole is inclined by 90° with respect to the z-axis, i.e., in the case of an bi-polar active region, a minimum of the Alfvén speed only appears at one side of the active region in the middle of the corona. Note that the fast magnetosonic waves are refracted toward regions of low Alfvén speeds [*Uchida*, 1974].

In order to demonstrate the behaviour of the Alfvén speed from the corona up to the interplanetary space at 1 AU ($= 214 R_S$), its variation has been calculated along a straight line going from the center of the active region ($z = R_S$) with an inclination of $\delta = 45°$ (see Figure 6) with respect to the z-axis (see Figure 7). Such an inclination has been chosen since the type II burst sources mainly propagate obliquely away from active regions, as already mentioned. Figure 9 reveals, that the Alfvén speed has a global maximum around ≈ 740 km/s at $3.8 R_S$ [*Mann et al.*, 1999a; *Gopalswamy et al.*, 2001] as well as a local minimum in the middle of the corona

below $2 R_S$. The dotted line in Figure 7 shows the Alfvén speed resulting only from the magnetic field of the quiet Sun, i.e., according to Eq. (5) [*Mann et al.*, 1999a]. The influence of the magnetic field of an active region becomes neglible beyond a distance of $\approx 2.2 R_S$ as revealed in Figure 7. (For further discussion of this subject it is refered to the papers by *Mann et al.* [2003] and *Warmuth and Mann* [2005].)

4.4. Discussion

The occurrence of a local minimum and a global maximum of the Alfvén speed in the middle of the corona and the near-sun interplanetary space has obviously influence of the formation and development of shock waves and the associated type II radio bursts in the corona. In order to study that, a disturbance propagating along a straight path with an inclination of $\delta = 45°$ (see Figure 6 and 7) away from an active region is considered, since type II radio burst sources usually propagating non-radially away from an active region. The mean radial velocity V_r of a type II burst source is about 740 km/s [*Klassen et al.*, 2000], i.e., its full velocity V is found by $V = V_r/\cos \delta$ to be ≈ 1000 km/s. If such a disturbance is initially produced near the photosphere at $R = R_S$ and, subsequently, travelling along this path it enters regions with $V > v_A$ and becomes a shock wave. Then, it has a maximum Alfvén-Mach number of $M_A = 2.3$ and 4.4 at the local minimum of the Alfvén speed in the case of the parallel ($\phi = 0°$) and anti-parallel ($\phi = 180°$) dipole direction, respectively. Afterwards it reaches a region with increasing Alfvén speed (see Figure 7), i.e., V smaller than v_A, leading to the disappearance of the shock and the associated type II radio burst. That can explain that solar type II radio bursts usually occur below 150 MHz [*Mann et al.*, 2003] and disappear at frequencies above 14 MHz [*Gopalswamy et al.*, 2001]. Since the Alfvén speed is decreasing beyond the global maximum the considered disturbance can drive a bow shock ahead itself beyond a radial distance of $\approx 6 R_S$ (see Figure 7). If the disturbance has a velocity of 1000 km/s as assumed, it reach this distance after ≈ 1 hour, i.e., the on-set of the interplanetary shock as a source of SEP happens roughly after one hour of the initial energy release of the flare [*Mann et al.*, 2003]. It should be noted that the assumed model of a magnetic dipole for an active region is a simplified scenario, but it reflects the essential properties of an active region in a qualitative manner. Additionally, the basic parameters adopted in the previous study are changing from case to case in reality. But nevertheless the occurrence of a minimum and maximum of the Alfvén velocity in the middle of the corona and the near-Sun interplanetary space can be regarded as a model independent result. A quantitative study, in which way the variation of the basic parameters changes the presented results is given in the paper by *Warmuth and Mann* [2005].

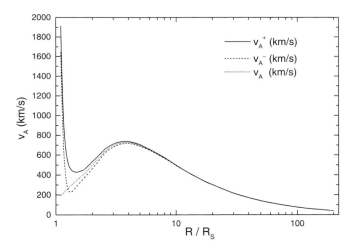

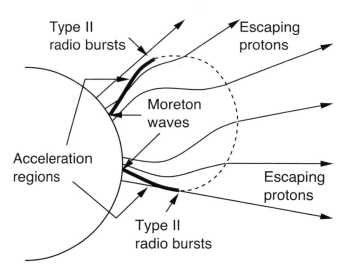

Figure 7. Radial behaviour of the Alfvén speed along a straight line with $\delta = 45°$ away from an active region (see Figure 6). The full and dashed line shows the Alfvén speed for the parallel ($\phi = 0°$) and antiparallel ($\phi = 180°$) case, respectively. The dotted line represents the behaviour of the Alfvén speed only due to the magnetic field of the quiet Sun (see Eq. (5)) (see *Mann et al.* [2003]).

Figure 8. Sketch of the propagation of Moreton waves and coronal shock waves as the source of solar type II radio bursts in the corona (see *Vainio and Khan* [2004]).

Krucker et al. [1999] and *Haggerty and Roelof* [2002] reported on a time delay between the on-set of the impulsive phase of the flare and the coronal release time of electrons with relativistic energies as obserevd by in-situ measurements at 1 AU. This time delay is about 10 minutes and more. If these particles are shock-accelerated it can be explained by the travelling time of a disturbance from the flare region to the region, where the Alfvén speed has its minimum, i.e., where the Alfvén-Mach number of the associated shock takes its maximum. It is well-known that high Alfvén-Mach number shocks are able to accelerate electrons very efficiently [*Kennel et al.*, 1985]. In addition, it should be noted that this delay can alternatively explained in the following manner: The magnetic field geometry is changed due to the flare in the corona and the interplanetary space, so that the magnetic connection between the site of the electron acceleration and the spacecraft occurs at a later time during the flare. Consequently, relativistic electrons produced in a later impulsive phase of the flare can reach the spacecraft along a magnetic field line after a time delay with respect to the start time of the flare. For example, that was the case in the particular solar event on June 2, 2002 [*Classen et al.*, 2003].

5. MORETON- AND EIT WAVES AS A SOURCE OF SOLAR ENERGETIC PARTICLES

Moreton and EIT waves are often associated with SEP events [*Bothmer et al.*, 1997; *Torsti et al.*, 1998; *Krucker et al.*, 1999; *Vainio and Khan*, 2004]. There are some observational

hints that the energetic particles seen as SEP events has been released at times when the EIT wave touchs the magnetic field line connecting the corona with the spacecraft (see Figure 8). As already mentioned in Section 4, highly energetic electrons are released from the corona up to 30 minutes after the type III radio burst onset. *Krucker et al.* [1999] suggested that this time delay can be caused by the travelling time needed by an EIT wave to reach an open magnetic field line establishing the connection with the spacecraft at which the energetic electrons are detected. That indicate that EIT waves may accelerate particles.

As already mentioned Moreton and EIT waves represent fast magnetosonic waves in the coronal plasma (see Sec. 3). Such waves are accompanied by both a density and magnetic field compression (see *Priest* [1982]). Figure 9 shows the intensity enhancement across a Moreton wave front seen in the H_α light in comparison to that in the He I line [*Vršnak et al.*, 2002]. The intensity of the H_α light looks like a *simple wave*, i.e., Moreton and EIT waves can be regarded as the manifestation of *simple magnetohydrodynamic waves* in the solar atmosphere [*Mann*, 1995b; *Vršnak and Lulić*, 2000]. In a low β plasma (β, ratio of the thermal to the magnetic pressure) as typically found in the corona, the velocity V of a simple fast magnetosonic wave is related to the density N_{comp} and magnetic field B_{comp} compression by $V = v_A \cdot N_{comp}^{1/2} = v_A \cdot B_{comp}^{1/2}$ [*Mann*, 1995b]. Thus, the part of the wave with a higher magnetic field compression is moving faster than that with a lower one. That leads to wave steepening. The time of wave steepening can be determined by

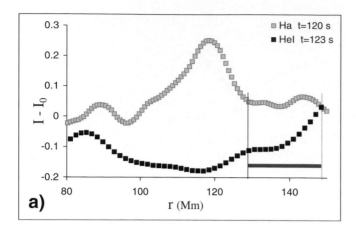

Figure 9. Intensity behaviour across a Moreton wave front seen in the H_α light in comparison to that of the He-I line (see *Vršnak et al.* [2002]).

$$t_s = \frac{L}{v_A} \cdot \frac{1}{\left(\sqrt{N_{comp}} - 1\right)} \qquad (6)$$

with L as the characteristic width of the wave pulse [*Mann*, 1995b]. Thus, a wave steepening time $t_s \approx 240$ s is obtained for reasonable values $L = 60 \cdot 10^3$ km, $v_A = 600$ km/s, and $N_{comp} = 2$ [*Vršnak et al.*, 2002].

Due to wave steepening the wave front appears as a moving magnetic mirror at which particles can be accelerated via *shock drift acceleration* [*Holman and Pesses*, 1983; *Ball and Melrose*, 2001]. The resulted distribution function of the accelerated particles has the shape of a *shifted loss-cone* one [*Leroy and Mangeney*, 1984; *Wu*, 1984; *Mann and Klassen*, 2005].

Thus, not only a shock wave but also a simple fast magnetosonic wave can be a particle accelerator. It should be emphasized, that a shock wave is a discontinuity with a transmitting mass flow. It is well defined in *magnetohydrodynamics* (MHD) by conservation laws leading to the well-known *Rankine-Hugoniot* relationships (see e.g., *Priest* [1982]. Consequently, it is independent from it exciting agent, e.g., either generated by a blast wave or driven by any piston [*Gopalswamy and Kundu*, 1992]. A shock wave is a dissipative structure. In contrast to that, a simple wave is a non-dissipative structure governed by the one-dimensional, non-linear MHD-equations. But a moving structure with a local magnetic field compression is only necessary for particle acceleration via shock drift acceleration, since it acts like a moving magnetic mirror. Therefore, also an EIT wave as a simple fast magnetosonic wave is able to accelerate particles like a shock wave.

6. SUMMARY

As illustrated Moreton and EIT waves are a common wave phenomenon on the Sun during flares. They occur in proximity to all of these other phenomena of the active Sun like solar type II radio bursts, CMEs, interplanetary shocks, and solar energetic particle events, for instance. EIT waves are generated due to the initial energy release of a flare. The Moreton waves are the imprint of the EIT waves on the chromosphere. Thus, they are closely related to each other. Since EIT waves are globally travelling waves on the Sun, they can be used as a diagnostic tool, i.e., in a sense of *coronal seismology*, for determining the quiet Sun magnetic field of 2-4 G at the photospheric level.

Modelling the magnetic field of an active region by a magnetic dipole located under the photosphere, the spatial behaviour of the Alfvén speed can be determined in the corona and interplanetary space. Of course, it is well-known that the spatial behaviour of the Alfvén speed is permanantly changing in the corona and interplanetary space, especially from event to event. But, nevertheless, the occurrence of a local minimum of the Alfvén speed in the middle of the corona and a global maximum of it at a distance of $\approx 4R_S$ from the center of the Sun in the near-Sun interplanetary space should be considered as a general result of these studies.

Also EIT waves as a manifestation of simple magnetosonic waves in the solar corona can act as a particle accelerator like shock waves, i.e., as a source of solar energetic particles, since they are accompanied by a magnetic field compression.

Acknowledgments. I am grateful to N. Gopalswamy for inviting me to give a talk on the subject of this paper at the AGU Chapman Conference *Solar Energetic Plasmas and Particles* held in Turku (Finland) on August 2-6, 2004. In particular, I would like to express my thanks to Drs. Klassen and Warmuth for their assistance and helpful comments for preparing this review paper.

REFERENCES

Aurass, H., Coronal mass ejections and type II radio bursts, in *Coronal Physics from Radio and Space observations*, edited by G. Trottet, *Lecture Notes in Physics* 483, pp. 135-253, Springer, Berlin, New York, 1997.

Aurass, H., A. Hofmann and H.-W. Urbarz, The 09 September 1989 γ-ray flare - multisite particle acceleration and shock-excited radio emission during quasi-perpendicular and quasi-parallel propagation, *Astron. Astrophys.*, 334, 289-298, 1998.

Banaszkiewicz, M., W.I. Axford, and J.F. McKenzie, An analytic solar magnetic field model, *Astron. Astrophys.*, 337, 940-944, 1998.

Ball, L. and D.B. Melrose, Shock drift acceleration of electrons, *Publ. Astron. Soc. Austr.*, 18, 361-378, 2001.

Biesecker, D.A., D.C. Myers, B.J. Thompson, D.M. Hammer, and A. Vourlidas, Solar phenomena associated with "EIT waves", *ApJ*, 569, 1009-1015, 2002.

Bothmer, V., A. Posner, H. Kunow, and 14 co-authors, Solar energetic particle events and coronal mass ejections: new insights from SOHO, *Proc. 31st ESLAB Symposium, ESA SP*-415, 207-212, 1997.

Bougeret, J.-L., M.L. Kaiser, P.J. Kellogg, and 9 co-authores, Waves: The Radio and Plasma Wave Investigation on the Wind Spacecraft, *Space Sc. Rev.*, 71, 231-263, 1995.

Cane, H.V., R.G. Stone, J.L. Fainberg, R.T. Stewart, J.L. Steinberg, and S. Hoang, Radio evidence of shock acceleration of electrons in the solar corona, *Geophys. Res. Lett.*, 8, 1285-1289, 1981.

Cane, H.V., Velocity profiles of interplanetary shocks, in *Solar Wind Five, NASA Conf. Publ.*, CP-2280, 703-708, 1983.

Chen, R.M., S.T. Wu, K. Shibata, and C. Fang, Evidence of EIT and Moreton waves in numerical simulations, *ApJ*, 572, L99-L102, 2002.

Classen, H.T., and H. Aurass, On the association between type II radio bursts and CMEs, *Astron. Astrophys.*, 384, 1098-1106, 2003.

Classen, H.T., G. Mann, A. Klassen, and H. Aurass, Relative timing of electron acceleration and injection at solar flares: A case study, *Astron. Astrophys.*, 409, 309-316, 2003.

Delaboudiniere, J.-P., G.E. Artzner, J. Brunaud and 25 co-authores, EIT: Extreme-Ultraviolet Imaging Telescope for the SOHO Mission, *Solar Phys.*, 162, 291-312, 1995.

Delannee, C., Another view of the EIT wave phenomenon, *ApJ.*, 545, 512-523, 2000.

Delannee, C. and G. Aulanier, CME associated with transequatorial loops and a bald patch flare, *Solar Phys.*, 190, 107-129, 1999.

Gilbert, H.R., T.E. Holzer, B.J. Thompson, and J.T. Burkepile, A comparison of CME-associated atmospheric waves observed in coronal (Fe XII 195 Å) and chromospheric (He I 10830 Å) lines, *ApJ*, 607, 540-553, 2004.

Gopalswamy, N., and M.R. Kundu, Are coronal shocks piston driven, in *Particle Acceleration in Cosmic Plasmas,* edited by G.P. Zank and T.K. Gaisser, American Inst. of Physics, New York, pp. 257-260, 1992.

Gopalswamy, N., M.L. Kaiser, R.P. Lepping, S.W. Kahler, K. Ogilvie, D. Berdichevsky, T. Kondo, T. Isobe, and M. Akioka, Origin of coronal and interplanetary shocks - A new look with WIND spacecraft data, *J. Geophys. Res.*, 103, 307-322, 1998.

Gopalswamy, N., and B.J. Thompson, Early life of cornal mass ejections, *J. Atmosph. and Sol.-Terr. Phys.*, 62, 1457-25, 268, 2000.

Gopalswamy, N., M.L. Kaiser, J. Sato, and M. Pick, Shock waves and EUV transient during a flare, in *High Energy Solar Physics*, edited by R. Ramaty and N. Mandzhavidze, *PASP Conf. Ser.* 206, pp. 351-358, 2000.

Gopalswamy, N., A. Lara, M.L. Kaiser, and J.-L. Bougeret, Near-Sun and near-Earth manifestations of solar eruptions, *J. Geophys. Res.*, 106, 25,261-25,268, 2001.

Haggerty, D.K. and E.C. Roelof, Impulsive nearly-relativistic electron events - delayed injection with respect to solar electromagnetic emission, *ApJ*, 579, 841-853, 2002.

Holman, G.D. and M.E. Pesses, Solar type II radio emission and shock drift acceleration, *ApJ*, 267, 837-843, 1983.

Hudson, H.S. and A. Warmuth, Coronal loop oscillations and flare shock waves, *ApJ*, 614, L85-L88, 2004.

Kahler, S.W., Injection heights of solar eneregtic particles as functions of coronal mass ejection heights, *ApJ.*, 428, 837-846, 1994.

Kai, K., Expanding arch structure of a solar radio outburst, *Solar Phys.*, 11, 456-462, 1970.

Kennel, C.F., J.P. Edmiston, and T. Hada, A Quarter Century of Collisionless Shock Research, in *Collisionless Shocks in the Heliosphere: A Tutorial Review*, edited by R.G. Stone and B.T. Tsurutani, pp. 1-36, *Geophysical Monograph*, 34, AGU, Washington D.C., 1985.

Khan, J.I. and A. Aurass, X-ray observations of a large-scale solar coronal shock wave, *Astron. Astrophys.*, 383, 1018-, 2002.

Klassen, A., H. Aurass, K.-L. Klein, A. Hofmann, and G. Mann, Radio evidence on shock wave formation in the solar corona, *Astron. Astrophys.*, 343, 287-299, 1999.

Klassen, A., H. Aurass, G. Mann, and B.J. Thompson, Catalogue of the 1997 SOHO-EIT coronal transient waves and associated type II radio burst spectra, *Astron. Astrophys.*, suppl., 141, 357-371, 2000.

Klein, K.-L., J.I. Khan, N. Vilmer, J.-M. Delouis, and H. Aurass, X-ray and radio evidence on the origin of a coronal shock wave, *Astron. Astrophys.*, 346, L53-L57, 1999a.

Klein, K.-L., E.L. Chupp, G. Trottet, A. Magun, P.P. Dunphy, E. Rieger, and S. Urpo, Flare-associated energetic particles in the corona and at 1 AU, *Astron. Astrophys.*, 348, 271-285, 1999b.

Koutchmy, S., Coronal physics from eclipse observations, *Adv. space Res.*, 14, 29-39, 1994.

Krucker, S., D.E. Larson, R.P. Lin, and B.J. Thompson, On the origin of impulsive electron events observed at 1 AU, *Apj.*, 519, 864-875, 1999.

Lara, A., N. Gopalswamy, S. Nunes, G. Munoz, and S. Yashiro, A statistical study of CMEs associated metric type II radio bursts, *Geophys. Res. Lett.*, 30, 12-16, 2003.

Leblanc, Y., G.A. Dulk, and J.-L. bougeret, Tracing the electron density from the corona to 1 AU, *Solar Phys.*, 183, 165-172, 1998.

Leroy, M.M. and A. Mangeney, A theory of energetization of solar wind electrons by Earth's bow shock, *Ann. Geophys.*, 2, 449-456, 1984.

Mann, G., Theory and observations of coronal shock waves, in *Coronal Magnetic Energy Release*, edited by A.O. Benz and A. Krüger, *Lecture Notes in Physics* 444, pp. 183-200, Springer, Berlin, New York, 1995a.

Mann, G., On simple magnetohydrodynamic waves, *J. Plasma Phys.*, 53, 109-121, 1995b.

Mann, G., H. Aurass, W. Voigt and J. Paschke, Preliminar observations of solar type II radio bursts with the new radiospectrograph in Tremsdorf, *Proc. 1st SOHO Workshop*, ESA SP-348, 129-133, 1992.

Mann, G., H. Aurass, A. Klassen, C. Estel, and B.J. Thompson, Cornal transient waves and coronal shock waves, *Proc. 8th SOHO Workshop*, ESA SP-446, 477-481, 1999a.

Mann, G., F. Jansen, R.J. MacDowall, M.L. Kaiser, and R.G. Stone, A heliospheric density model and type III radio bursts, *Astron. Astrophys.*, 348, 614-620, 1999b.

Mann, G., A. Klassen, H. Aurass, H.-T. Classen, V. Bothmer, and M.J. Reiner, EIT waves, coronal shock waves, and solar energetic particle events, *Proc. Planetary Radio Emission V* 445-450, 2001.

Mann, G., A. Klassen, H. Aurass, and H.T. Classen, Formation and development of shock waves in the solar corona and the near-Sun interplanetary space, *Astron. Astrophys.*, 400, 329-336, 2003.

Mann, G. and A. Klassen, Electron beams generated by shock waves in the solar corona, *Astron. Astrophys.*, 441, 319-326, 2005.

Melrose, D.B., Plasma emission mechanisms, in *Solar Radio Physics*, edited by D.J. McLean and N.R. Labrum, pp. 177-210, Cambridge Univ. Press, Cambridge, 1985.

Moreton, G.E., H$_\alpha$ Observations of Flare-Initiated Disturbances with Velocites ≈ 1000 km/s, *ApJ*, 65, 494-502, 1960.

Moreton, G.E. and H.E. Ramsey, Recent Observations of Dynamical Phenomena Associated with Solar Flares, *Publ. Astron. Soc. Pacific*, 72, 357-366, 1960.

Moses, D., F. Clette, J.-P. Delaboudiniere and 32 co-authors, EIT Observations of the Extreme Ultraviolet Sun, *Solar Phys*, 175, 571-599, 1997.

Müller-Mellin, H. Kunow, V. Fleissner, and 16 co-authores, COSTEP – Comprehensive Suprathermal and Energetic Particle Analyser, *Solar Phys.*, 162, 483-504, 1995.

Nelson, G.S. and D. Melrose, Type II bursts, in *Solar Radio Physics*, edited by D.J. McLean and N.R. Labrum, pp. 333-360, Cambridge Univ. Press, Cambridge, 1985.

Newkirk, G., The solar corona in the active regions and the thermal origin of the slowly varying component of solar radiation, *ApJ*, 133, 983-992, 1961.

Parker, E.N., Dynamics of the interplanetary gas and magnetic field, *ApJ*, 1128, 664-667, 1958.

Priest, E.R., *Solar Magnetohydrodynamics,* 73 pp., D. Reidel Publ. Comp., Dordrecht, 1982.

Reames, D.V., L.M. Barbier and C.K. Ng, The spatial distribution of particles accelerated by coronal mass ejection-driven shocks, *ApJ*, 466, 473-483, 1996.

Reiner, M.J. and M.L. Kaiser, High-frequency type II radio emissions associated with shocks driven by coronal mass ejections, *J. Geophys. Res.*, 104, 16,979-16,988, 1999.

Schwenn, R., Large-Scale Structure of the Interplanetary Medium, in *Physics of the Inner Heliosphere,* edited by R. Schwenn and E. Marsch, pp. 99-182, Springer Verlag, Berlin, 1990.

Stewart, R.T., M.K. McCabe, M.J. Koomen, F. Hansen, and G.A. Dulk, Observations from coronal disturbances from 1 to 9 R$_\odot$ - I: First event of 1973 January 11, *Solar Phys.*, 36, 203-218, 1974a.

Stewart, R.T., R.A. Howard, F. Hansen, T. Gergeley, and M. Kundu, Observations from coronal disturbances from 1 to 9 R$_\odot$ - II: Second event of 1973 January 11, *Solar Phys.*, 36, 219-228, 1974b.

Thompson, B.J., S.P. Plunkett, J.B. Gurman, J.S. Newmark, O.C. Cyr, St., and D.J. Michels, SOHO/EIT observations of an Earth-directed coronal mass ejection on May 12, 1997, *Geophys. Res. Lett.*, 25, 2465-2469, 1998.

Torsti, J. A. Anttila, L. Kocharov, and 7 co-authors, Energetic (~1 to 50 MeV) protons associated with Earth-directed coronal mass ejections, *Geophys. Res. Lett.*, 25, 2525-2529, 1998.

Uchida, Y., On the exciters of type II and type III solar radio bursts, *Publ. Astron. Soc. Japan*, 12, 376-383, 1960. Uchida, Y., Propagation of hydrodynamic disturbances in the solar corona and Moreton's wave phenomena, *Solar Phys.*, 4, 30-36, 1968.

Uchida, Y., Behaviour of flare-produced coronal MHD wavefront and the occurrence of type II radio bursts, *Solar Phys.*, 39, 431-438, 1974.

Uchida, Y., M.D. Altschuler and G. Newkirk, Flare-produced coronal MHD-fast-mode wavefronts and Moreton's wave phenomenon, *Solar Phys.*, 28, 495-504, 1973.

Vainio, R. and J.I. Khan, Solar energetic particle acceleration in refracting coronal shock waves, *ApJ*, 600, 451-457, 2004.

Vršnak, B. and S. Lulić, Ignition of MHD shocks associated with solar flares, *Solar Phys.*, 158, 331-344, 1995.

Vršnak, B. and S. Lulić, Formation of coronal MHD shock waves-II: The pressure pulse mechanism, *Solar Phys.*, 196, 157-171, 2000.

Vršnak, B., A. Warmuth, Brajša, and A. Hanslmeier, Flare waves observed in Helium I 10830 Å: A link between H$_\alpha$ Moreton and EIT waves, *Astron. Astrophys.*, 394, 299-310, 2002.

Wagner, W.J. and E.M. MacQueen, The excitation of type II radio bursts in the corona, *Astron. Astrophys.*, 120, 136-138, 1983.

Warmuth, A., B. Vršnak, H. Aurass, and A. Hanslmeier, Evolution of two EIT/H$_\alpha$ Moreton waves, *ApJ*, 25, L105-L109, 2001.

Warmuth, A., Vršnak, Magdalenić, A. Hanslmeier and W. Otruba, A multiwavelength study of solar flare waves: I. Observations and basic properties, *Astron. Astrophys.*, 418, 1101-1115, 2004a.

Warmuth, A., Vršnak, Magdalenić, A. Hanslmeier and W. Otruba, A multiwavelength study of solar flare waves: II. Perturbation characteristics and physical interpretation, *Astron. Astrophys.*, 418, 1117-1129, 2004b.

Warmuth, A. and G. Mann, A model of the Alfvén speed in the solar corona, *Astron. Astrophys.*, 435, 1123-1135, 2005.

Wu, C.S., A fast Fermi process: energetic electrons accelerated by a nearly perpendicular bow shock, *J. Geophys. Res.*, 89, 8857-8862.

G. Mann, Astrophysikalisches Institut Potsdam, D-14482 Potsdam, Germany. (GMann@aip.de)

Radio Bursts and Solar Energetic Particle Events

K.-L. Klein

Observatoire de Paris, LESIA - CNRS UMR 8109, F-92195 Meudon

Radio bursts in the solar atmosphere and interplanetary space come from energetic non-maxwellian electron populations. Since the radio emission is a non collisional process, it traces electrons also in dilute plasmas, where hard X-rays and gamma-rays are undetectable. Radio waves are therefore a probe of energetic electrons from the low corona to the interplanetary medium. This review outlines how radio signatures can be used to infer acceleration sites in flares and to visualize the magnetic structures which guide particles from the low corona to interplanetary space. Electron acceleration at coronal shock waves is then discussed. These are seen to accelerate less electrons than the impulsive acceleration processes operating in the low corona. The ability to determine onset times of SEP events to within a few minutes has recently revived the comparison with radio bursts as tracers of acceleration processes in the corona. Remarkable discrepancies in the timing of the escaping and interacting particles were ascribed to different acceleration processes of these particle populations (i.e. CME shock acceleration of the escaping particles), transport processes (delayed particle transport in interplanetary space) or a combination of time-extended acceleration in the magnetically stressed corona and transport to space along several diverging magnetic flux tubes. The different interpretations are confronted and discussed. It is concluded that particle acceleration at the Sun is not adequately described by a bimodal picture of "flare acceleration" of very short duration in very small volumes on the one hand and CME shock acceleration over large volumes on the other hand.

1. INTRODUCTION

Transient radio emission is created in the solar atmosphere by non-maxwellian populations of electrons at energies from a few keV to several MeV. Most flare-related radio emission at frequencies above a few GHz (wavelengths $\lesssim 10$ cm), a wavelength range often referred to as microwaves, is gyrosynchrotron radiation from mildly relativistic electrons (energies 100 keV to several MeV). At longer wavelengths the emission processes are mostly coherent, due to unstable electron distributions such as beams or loss cone

Solar Eruptions and Energetic Particles
Geophysical Monograph Series 165
Copyright 2006 by the American Geophysical Union
10.1029/165GM22

distributions, which excite and amplify waves in the coronal plasma which are then converted to electromagnetic waves near the local electron plasma frequency or its harmonic. The coherent emission processes are globally referred to as plasma emission. The sources are at places where the observing frequency is the ambient plasma frequency or its harmonic. Typical heliocentric distances of emission as inferred from imaging observations and models of coronal electron density are 1.1 R$_\odot$ near 300 MHz, 1.4 R$_\odot$ near 150 MHz, 6 R$_\odot$ near 1 MHz, 1 AU near 20 kHz. Unlike hard X-rays and gamma-rays, the radio signatures are not produced by collisional processes, and are therefore not restricted to the low corona and chromosphere. Radio waves are a sensitive tracer of energetic electrons from the low corona to far into the interplanetary medium, a property which can be used to infer acceleration regions and to track the propagation of electrons through the corona and interplanetary space.

This paper attempts to review the specific information brought by radio observations on the origin and propagation of energetic particles in the corona and their relationship with SEP events. It is important to bear in mind that radio waves come from energetic electrons, so that any comparison with ions in space is indirect. However, radio data provide information on acceleration regions which need not be restricted to electrons, especially since observations of hard X-rays (HXR), gamma-rays and radio waves show that interacting electrons and ions are closely related in the solar atmosphere (*Vilmer and MacKinnon*, 2003, and references therein). Radio bursts are also unique tracers of the magnetic structures which guide particles through the corona. This paper starts with a short overview of radio evidence on sites and time scales of electron acceleration and electron transport to the high corona and interplanetary space in impulsive flares. This information is compared with particle measurements in impulsive SEP events. Section 3 addresses radio evidence for the acceleration of electrons at travelling coronal shock waves. Recent studies of the onset timing of SEP events and the inferences on the timing of the first released particles with respect to electromagnetic signatures of coronal acceleration ("flare acceleration") are discussed in Sect. 4, where different interpretations of delayed releases, including CME shock acceleration, interplanetary transport and time-extended acceleration in the magnetically stressed corona, are confronted with the data.

2. RADIO EVIDENCE ON ELECTRON ACCELERATION AND PROPAGATION DURING IMPULSIVE FLARES

Microwave imaging and decimeter wave spectroscopy suggest that typical electron acceleration sites during impulsive flares are located in the low corona. A frequently encountered configuration in maps at wavelengths around 1 cm points to interacting loops (*Nishio et al.*, 1997; *Hanaoka*, 1999), corresponding to scenarios based on magnetic reconnection. No imaging observations are presently available at wavelengths between 20 cm and 70 cm (frequency range 450 – 1400 MHz), where spectrographic observations of type III bursts (plasma emission) reveal the origin of electron beams at ~10 keV which travel upward and downward through the solar atmosphere. From their frequency spectrum, ambient electron densities in the range of 10^9–10^{10} cm^{-3} are inferred in the acceleration region. This work is discussed elsewhere in this volume (*Aschwanden*, 2005). The acceleration time scales of ions up to a few MeV and of electrons up to 100 keV as inferred from hard X-ray (HXR) and gamma-ray bursts are <1 s (*Vilmer and MacKinnon*, 2003). Time scales and certain

spectral features of radio emissions have been interpreted as evidence for electron acceleration during bursty reconnection (*Kliem et al.*, 2000) or at the shocks produced by outflows from reconnection regions (*Aurass and Mann*, 2004).

The ambient electron densities of the electron acceleration region and the typical time scales can be compared with the requirements from energy-dependent charge states of ions in impulsive SEP events (*Möbius et al.*, 2003). Their interpretation by collisional stripping implies that the product of electron density and residence time in the acceleration region $n_e \tau > 10^{10}$ cm^{-3} s (*Kocharov et al.*, 2001; *Klecker et al.*, 2005). This requirement can be fit by the upper limit of the density inferred from the radio observations and by the upper limit of the acceleration time scales derived from HXR and gamma-rays. However, in a study of some impulsive SEP events and their related radio emission, *Klein and Posner* (2005) conclude that the type III emitting electrons and, by inference, the escaping protons (deka-MeV energies) which are released together with the type III emission within the error bars of some minutes, were accelerated in a more dilute plasma ($n_e \sim 10^8$ cm^{-3}). This acceleration region is reminiscent of the "high coronal flares" discussed by *Lin* (1985) and *Cliver and Kahler* (1991), also for the sake of reconciling the measurements of interacting and escaping particles. The combined analysis of the charge states of impulsive SEP events and the timing and starting frequency of the associated type III radio emission could clarify the question if interacting and escaping particles stem from the same acceleration region in impulsive SEP events or not.

Since type III radio sources at different frequencies trace the local plasma frequency corresponding to the observing frequency, imaging of these sources traces the magnetic structures that guide electrons from the low to the high corona and interplanetary space. It is readily seen from the comparison of decimetric and decametric-to-hectometric type III emission (*Poquérusse et al.*, 1996) that apparently single type III bursts in interplanetary space are the low-frequency counterparts of groups of often numerous metric type III bursts in the corona (*Aschwanden*, 2005, Figure 6). *Pick and Ji* (1986) showed that during most meter wave type III groups multiple sources are observed. This means that electron beams are injected into different flux tubes during a flare and therefore propagate along different diverging paths in the corona. The overall angular range covered by these flux tubes frequently reaches 25° (heliocentric). Following *Buttighoffer* (1998), a typical flux tube has a width around 7,000 km (10 arc seconds) at the Sun, and around 3×10^6 km at 1 AU. A given spacecraft may intercept successively all, several, or none of these flux tubes. The successive passage at a spacecraft of filled and empty flux tubes of this size is shown to create characteristic flux dropouts in the time histories of ions (energies ~20 keV/amu – 5 MeV/amu) in impulsive SEP events (*Mazur et al.*, 2000). Although this phenomenon can be created by the meandering of interplanetary field lines of an individual flux tube (*Giacalone et al.*, 2000), the radio observations show that particles propagate in several distinct flux tubes since their acceleration in the corona.

3. ELECTRON ACCELERATION AT TRAVELLING CORONAL SHOCK WAVES: TYPE II RADIO BURSTS

Type II radio bursts are the most prominent indicators of travelling shocks in the corona. In the dynamic spectrogram of Figure 1 (dark shading means bright emission), the type II emission consists of two slowly drifting, approximately parallel bands. Similar to type III bursts, the radio emission is plasma emission. In type II bursts it is generated by electrons accelerated at the coronal shock. The two bands are radio waves near the fundamental and the harmonic of the local plasma frequency, their drift towards lower frequencies shows the progression of the shock from high to low densities. Using standard models of coronal electron density, the drift rates translate into speeds of some hundreds to about 2000 km s^{-1}. Detailed discussions of type II radio emission are given by *Nelson and Melrose* (1985), *Mann et al.* (1995) and *Knock and Cairns* (2005). The Mach number of coronal shocks is difficult to evaluate, since the coronal magnetic field is poorly known. *Mann et al.* (1995) suggest that type II

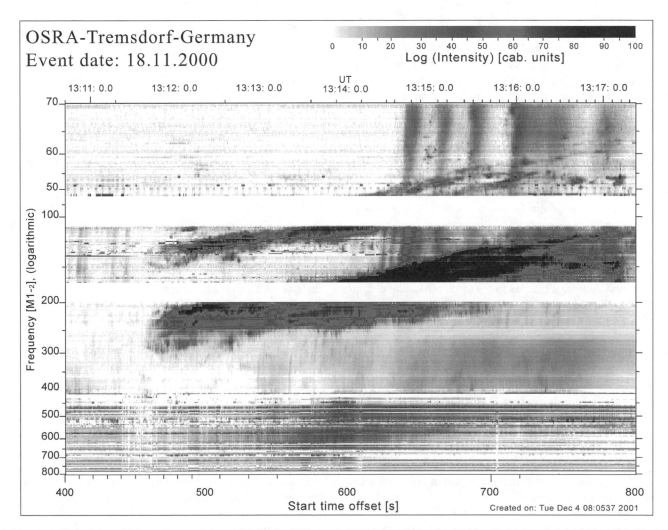

Figure 1. Dynamic spectrum of a type II burst (Tremsdorf Observatory, Potsdam Astrophysical Institute): The brightest features (dark in the figure) are two lanes that drift slowly from high frequencies (~300 MHz, 1312 UT) to low frequencies. The group of more rapidly drifting short bursts between 1315 and 1317 UT are type III bursts ascribed to electron beams which propagate within open magnetic flux tubes at speeds ~0.2*c*. White horizontal bands and other horizontal lines are the frequency bands of terrestrial emitters. From *Mann et al.* (2003).

shocks are weakly supercritical quasi-parallel shocks, with typical Alfvenic Mach numbers $\gtrsim 1.5$, from the spectral properties and analogies with shocks in interplanetary space. *Bale et al.* (1999) analyzed a quasi-perpendicular interplanetary shock with kilometric type II emission whose Alfvenic Mach number was 13.

3.1. Type II Burst Shocks in the Corona as Electron Accelerators

Because of the coherent character of the emission it is not possible to derive the number and energy of the radiating electrons directly from the spectrum of a type II burst. However, in some events short bursts emanating from the slowly drifting type II lane and drifting rapidly to lower or higher frequencies (or both) are observed. They are ascribed to electron beams accelerated at the shock, and their frequency drift rates were used by *Cairns and Robinson* (1987) and *Mann and Klassen* (2002) to infer the speed of the emitting electrons. The latter authors found systematically lower drift rates than in type III bursts, on average by a factor 1.9. With a standard coronal density model, the observed drift rates translate into an average speed of $0.1c$ for the electron beams accelerated by a type II shock, corresponding to an energy of a few keV, with a range consistent with the values of $(0.05–0.5)c$ found by *Cairns and Robinson* (1987). The speed could of course be higher if the electrons travelled along non radial magnetic field lines, e.g. in coronal loops. Electrons in the foreshock of the interplanetary shock observed in situ by *Bale et al.* (1999) extended to more than 100 keV (Bale, pers. comm.).

Hard X-ray emission is the most reliable probe of the energetics of non thermal electrons in the solar atmosphere. Since metric type II bursts most often occur with other types of radio emission (Figure 1), it is difficult to decide if a HXR signature of the shock-accelerated electrons exists or not. A unique possibility to record HXR emission from electrons accelerated at a coronal shock are events where the flaring active region is occulted by the solar limb, and only the type II source is in the field of view of Earth-orbiting X-ray detectors. In such an event, analysed by *Klein et al.* (2003), neither GOES nor CGRO/BATSE detected any hard X-ray emission at the time of the type II burst. The upper limits of the electron fluxes generated by the type II shock at energies ~20 keV were found two orders of magnitude smaller than the fluxes of electrons accelerated during the flare in the active region on the backside of the Sun that was in the field of view of the *Ulysses* HXR detector. The processes of electron acceleration in flaring active regions, whatever their nature, hence produce far more energetic electrons than travelling shocks in the corona.

Although the upper limits of the numbers of electrons accelerated at the type II shock are found to be smaller than

the numbers required for a large HXR burst, they are comparable to the electron numbers inferred by *Lin et al.* (1982) for large solar energetic electron events. Therefore the above estimation of electrons from shock acceleration in type II bursts does not formally exclude that coronal shocks are the accelerators of electron events in space. But since only upper limits could be derived even with the most sensitive solar HXR observations ever obtained, while copious amounts of electrons are accelerated by flare-related processes in the lower corona, it is hard to escape the conclusion that these latter processes are a major source, or even the principal source, of the large electron events in space.

3.2. CME Interaction and Electron Acceleration in the High Corona: A Means to Enhance Shock Acceleration?

Can shock acceleration be made more efficient in specific configurations? *Gopalswamy et al.* (2001b, 2002) identified a peculiar radio signature that they ascribed to the interaction of a fast CME with a slower previous one. The term "interaction" is purely phenomenological. It means that the *projected* trajectories of the fronts of the CMEs intersect each other, and that the surface activity associated with the two CMEs occurs in the same solar region. Figure 2.a illustrates one case: the initial group of type III bursts is followed by a faint narrow band of type II emission. Near 1810 UT the type II emission is superposed by a much brighter and broader emission (labeled "Enhancement") that lasts about 30 min and contains some narrow-band elements with a similar drift rate as the previous type II burst. As seen in Figure 2b, the radio enhancement starts at the time when (in projection) the front of the primary CME ("CME2" in the figure) travels through the brightest, hence densest, part ("core") of the slower preceding CME (CME1).

Gopalswamy and coworkers briefly mention magnetic reconnection between the CMEs, but focus on a different explanation of the radio emission: they surmise that a shock generated by the primary CME travels through the preceding CME, i.e. through a medium of higher density than the undisturbed corona. The density enhancement creates the jump to higher frequencies in the radio emission of the "Enhancement" with respect to the preceding type II burst. These authors argue that higher density implies that the shock has higher Mach number than if it travelled through the undisturbed corona, and therefore becomes a more efficient particle accelerator. The radio evidence provides little support to this idea: while a type II burst that can be attributed to the shock of the main CME is clearly seen before the interaction in Figure 2, none exists in the example of *Gopalswamy et al.* (2002), where two slower CMEs interact. But even if a shock exists, it is not obvious that it will be stronger in the enhanced density plasma of the previous CME than in the undisturbed

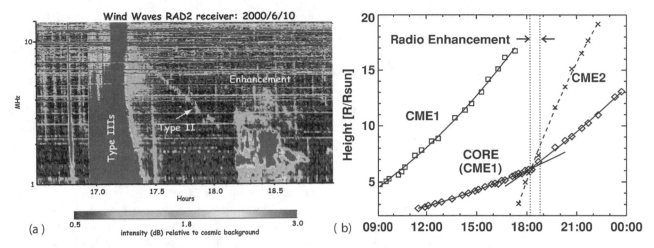

Figure 2. (a) Dynamic spectrum (Wind/WAVES) showing a type III burst, a type II burst and a bright emission ("Enhancement") attributed to CME interaction. Bright emission is colored in red in the original. Horizontal lines are instrumental artefacts. (b) Trajectories in the plane of the sky of the preceding CME (CME1) and the fast primary CME (CME2). From *Gopalswamy et al.* (2001b).

corona, because the shock speed must be evaluated in the frame of the moving plasma of the previous CME. In the extreme example of a one-dimensional case, the Mach number $M_{Ai} = \frac{\Delta V}{c_{Ai}}$, where $\Delta V = V_2 - V_1$ is the difference between the speeds of the two CMEs, c_{Ai} the Alfven speed in the previous CME. In the absence of the previous CME, the Mach number would be $M_{Ao} = \frac{V_2}{c_{Ao}}$, where the index o designates parameters of the undisturbed corona. The ratio of the Alfvenic Mach numbers within the previous CME and in its absence hence is

$$\frac{M_{Ai}}{M_{Ao}} \simeq \frac{\Delta V}{V_2} \frac{B_o}{B_i} \frac{v_{pi}}{v_{po}} \simeq 1.5 \frac{B_o}{B_i}. \qquad (1)$$

v_{pi}, v_{po} are the plasma frequencies inside the preceding CME and in the undisturbed corona, respectively. The CME speeds are used here as proxies of the shock speeds. In the case of Figure 2, $\Delta V = 550$ km s^{-1}, $V_2 = 660$ km s^{-1}, $\frac{v_{pi}}{v_{po}} \simeq \frac{3.2}{1.8} \simeq 1.8$, as read from the high-frequency limit of the radio emission in the ambient corona (type II burst) and within the preceding CME (Enhancement). Hence the increase of the Mach number due to the density enhancement is strongly reduced by the decrease of shock speed in the preceding CME. It would be further reduced if the previous CME had a high magnetic field. Although more complex scenarios than this simple overtaking of one-dimensional structures are likely, the example illustrates that it is questionable if the preceding CME is an appropriate region for

the formation of a shock or for the enhancement of the Mach number of a previously existing shock. It is to be investigated if magnetic reconnection, which is a prerequisite for the penetration of the fast CME into the preceding one, is an alternative. The particle acceleration should then start when the front of the fast CME reaches the rear of the slower preceding one, as observed in the cases where the episode of acceleration can be timed with the DH radio emission (e.g., Figure 2).

The means by which possible CME interaction could enhance SEP fluxes, as argued by *Gopalswamy et al.* (2004) on statistical grounds, is not clear in the light of the associated radio emission. E.g., it is an open question if particles readily escape from the acceleration region to 1 AU (*Kahler*, 2003).

4. RADIO EVIDENCE ON THE TIMING AND LOCATION OF SEP RELEASE TO INTERPLANETARY SPACE

The timing and location of energetic particle release at the Sun is the key to identify relationships between SEP events and dynamical processes in the corona. Time profiles of SEP measured far from the source are smeared out by transport in the turbulent interplanetary magnetic field (*Dröge*, 2000). The closest approach to the Sun so far was made by the *Helios* spacecraft. Figure 3 (*Kallenrode and Wibberenz*, 1991) compares the time histories of protons (27–37 MeV) and electrons (300–800 keV) measured at 0.4 AU from the Sun with X-ray, microwave (8.8 GHz) and hectometric

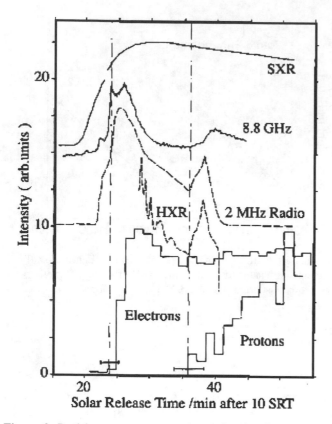

Figure 3. Particle measurements and associated radio and X-ray emission (HXR: hard X-rays, SXR: soft X-rays) in the corona during an SEP event (1980 June 08). The time profiles of electrons (300–800 keV) and protons (27–37 MeV) were observed by the Helios spacecraft at a heliocentric distance of 0.4 AU. The abscissa ("solar release time") gives the time when the particles (photons, electrons, protons) were released at the Sun. From *Kallenrode and Wibberenz* (1991).

(2 MHz) radio emission of electrons in the solar atmosphere. The abscissa gives the time when the particles were released at the Sun, computed by subtracting ~500 s from the photon arrival time at Earth and the travel time of the particles along the nominal Parker spiral to *Helios*. As usual, the HXR bremsstrahlung and the microwave gyrosynchrotron emission have similar time profiles, pointing to a common acceleration of the radiating electrons in the low corona. The 2 MHz emissions are type III bursts from electron beams escaping from the corona, seen at this frequency at several solar radii from Sun center. The radio and HXR time profiles show two parts: the most prominent impulsive phase emission, followed at 1036 SRT by a new rise at all frequencies that demonstrates a new episode of electron acceleration in the corona. The SEP time profiles show that the electron flux rises with the first radio and HXR emission, reaches its peak a few minutes afterwards and then decays. Protons start a few minutes later,

together with the new acceleration episode shown by the HXR and radio waves. It is possible that the electrons undergo a second rise then, but the plotted time profiles are count rates, and the flattening of the electron profile after 1036 SRT might also be due to contamination by protons. But there are clearly two different phases of the SEP event, an initial one that is electron rich, and a second one which is proton-rich. Both phases occur with distinct episodes of electron acceleration in the corona, shown by the microwave and HXR emission, and their escape to the high corona and interplanetary space, traced by the type III bursts. The common onset times of signatures of escaping particles and interacting electrons suggest a close physical relationship between the SEP and particle acceleration in the solar atmosphere. If this is true, the observation implies that the acceleration processes producing the radiating electrons also provide escaping electrons and protons, with variable abundance ratios.

Unfortunately, only few particle observations were reported from *Helios* in the inner heliosphere, so that for systematic studies one has to rely presently on measurements at 1 AU, where transport effects are expected to have a stronger influence on the particle profiles. The only promising comparison is that between coronal acceleration signatures and the release time of the *earliest* arriving particles from the Sun at the spacecraft, inferred through the energy (velocity) dispersion of the arrival times. Numerical investigations by *Kallenrode and Wibberenz* (1990) and *Lintunen and Vainio* (2004) show that from measurements at 1 AU the solar release time of SEP can only be determined to within an uncertainty of several minutes at best (cf. also *Klein et al.*, 2005). This limitation remains a major obstacle to firm conclusions at the present time.

Studies of the relative timing of SEP events (electrons and ions) have shown that frequently the inferred solar release time of the first particles detected by spacecraft lags behind the *start* of the associated flare, be it measured at soft X-rays or Hα, which are both thermal signatures of flare energy release, or by radio emission or other signatures of the acceleration of electrons in the corona (*Carmichael*, 1962; *Lin*, 1974; *Cliver et al.*, 1982; *Lockwood et al.*, 1990; *Krucker et al.*, 1999; *Klein et al.*, 1999; *Laitinen et al.*, 2000; *Haggerty and Roelof*, 2002; *Tylka et al.*, 2003). In recent systematic studies of electrons at energies above 25 keV (referred to as "near-relativistic" in the following), *Krucker et al.* (1999) and *Haggerty and Roelof* (2002) inferred delays ranging from 0 to several tens of minutes between the first electromagnetic signatures of non thermal electrons in the solar atmosphere and the start of the release of particles detected near the Earth. A variety of reasons may explain such delays, as will be discussed in the following: Escaping particles may be accelerated during the same events as interacting particles, but at different places and by different processes. If the SEP are accelerated at shock waves driven by fast CMEs, the delayed SEP onsets can be ascribed to the time

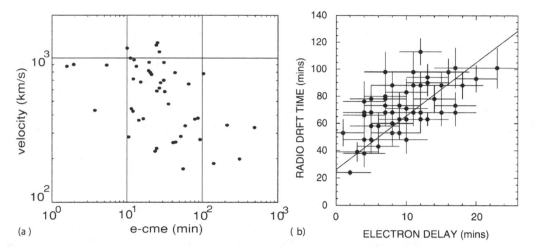

Figure 4. Correlations of electron release delays: (a) speed of the CME vs. delay between the release time of near-relativistic electrons after the launch of the associated CME (*Simnett et al.*, 2002), (b) travel time of type III emission from the Sun to the spacecraft vs. electron release delay (*Cane*, 2003).

the shock needs to become an efficient accelerator (Sect. 4.1). The arrival of particles at 1 AU can also be delayed by scattering at magnetic field irregularities in the interplanetary medium (Sect. 4.2). Finally, the electromagnetic signatures in Figure 3 illustrate that particle acceleration in the corona may consist of several time-extended episodes. If, as discussed in Sect. 2, particles propagate from solar active regions to interplanetary space along multiple flux tubes, a complex time relationship between interacting and escaping particles seen by a single spacecraft can be expected simply because the spacecraft can only be connected with at most one of these flux tubes at a time (Sect. 4.3).

4.1. SEP Onset Delays and Shock Acceleration in the High Corona

Simnett et al. (2002) showed that the intensities of the near-relativistic electron events of *Haggerty and Roelof* (2002) correlate with the CME speeds in the plane of the sky. They considered this as evidence for acceleration by the shock presumably driven by the CME. They evaluated the release delay of near-relativistic electrons with respect to the launch times of the associated CMEs, estimated as the time when the trajectory of the CME front in the plane of the sky extrapolates back to the solar limb. Their plot of the velocity of the CME vs. the release delay is shown in Figure 4a. The trend towards anticorrelation is interpreted through a characteristic height of acceleration by the CME shock at a few solar radii, which is reached the sooner, the faster the CME. This characteristic release height could be understood if the fast magnetosonic speed had a maximum in the corona, at typical heliocentric

distances between 3 and 4 $R_\odot$, as argued by *Mann et al.* (1999) and *Gopalswamy et al.* (2001a) based on models of the coronal magnetic field. A CME of a given speed will more likely drive a shock wave at heights above this critical value. *Kahler* (1994) used the idea that the onset of SEP events is related to the height of the CME front to explain the onset times of proton events from some MeV to some GeV: from a plot of the particle time profile vs. projected height of the CME front he concluded that relativistic protons are accelerated at lower heights (2.5–15 $R_\odot$) than protons at lower energies. *Krucker and Lin* (2000) came to a similar conclusion from a study of the onset times of proton events at energies up to a few MeV. These studies show that delayed SEP acceleration can be understood if the SEP are accelerated by the CME shock. The close temporal correspondence even of delayed SEP onsets with fresh accelerations of radiating electrons illustrated by Figure 3, however, does not readily fit in such a scenario.

Delayed SEP releases were also observed in some flare/CME events from the eastern solar hemisphere. Particles accelerated in these flaring active regions cannot directly reach the Earth along the undisturbed interplanetary magnetic field. *Torsti et al.* (1999) and *Krucker et al.* (1999) studied such SEP events, which were accompanied by large-scale propagating disturbances seen in EUV images to travel out from active regions ("EIT waves"), and are often interpreted as the trace in the low corona of travelling magneto-acoustic waves or shocks (*Biesecker et al.*, 2002; *Warmuth et al.*, 2004). *Krucker et al.* (1999) modeled the propagation of a large-scale disturbance from a flare in the eastern hemisphere. They showed that while the EIT wave itself seems to be too slow to explain the solar release of particles detected near the Earth by shock-acceleration in the low corona, the release time can be understood if the particles are accelerated when a wave with an appropriate 3D shape intercepts the Earth-connected interplanetary magnetic field line in the high corona, at several tens of heliocentric degrees from the flaring active region and at heliocentric distances of several solar radii. This appears consistent with the preferred statistical association of SEP events in poorly connected flares with type II bursts (*Hucke et al.*, 1992), i.e. signatures of travelling shocks in the corona which are often associated with EIT waves (*Klassen et al.*, 2000; *Khan and Aurass*, 2002).

What is the relationship of coronal shocks and CMEs with EIT "waves"? While the EUV features are most often interpreted as travelling waves, the observation of stationary fronts and dynamic loops in some events prompted for alternative ideas. E.g., *Delannée and Aulanier* (1999) interpreted the brightening in EIT images as a signature of the CME legs, and the enclosed dimming as evidence for plasma outflow. If this interpretation applies, the acceleration of the escaping particles can also be due to the interaction between the magnetic field of the CME with the ambient field in the corona, far from the flaring active region. The scenario has not yet been investigated in detail. But *Maia et al.* (1999) showed that the lateral expansion of a CME (1997 Nov 06) can be understood as the successive interaction of magnetic loops at increasing distance from the parent active region, accompanied by radio signatures of accelerated electrons at increasing distances from the flare. *Klein et al.* (2001) suggested that during the 2000 Jul 14 event moving radio sources ascribed to magnetically confined populations of non thermal electrons could reveal magnetic reconnection far from the flaring active region. Numerical simulations by *Chen et al.* (2005) might reconcile the different interpretations. They suggest that some EUV brightenings that show up as EIT "waves" are indeed compression regions at the interface between loops which are opened or stretched during a CME, and neighboring field lines. But they also predict a shock whose trace in the low corona propagates much faster than the EIT "wave" and would therefore be able to explain SEP production with a flare in the eastern hemisphere within a scenario similar to the one of *Krucker et al.* (1999). The understanding of the lateral evolution of CMEs and its relation to shock waves and magnetic reconnection is a key to further progress in our understanding of the acceleration and release of SEP.

even if SEP are accelerated together with the radiating particles, their interplanetary transport may delay the arrival at 1 AU. *Cane* (2003) compared the release delays of near-relativistic electrons and the time when the low-frequency edge of the associated type III emission reached the local electron plasma frequency at the *Wind* spacecraft. She considers the time interval between the start of the high-frequency type III emission in the corona and the arrival of the type III emission at the local electron plasma frequency as the travel time of the emitting electron beams. She calls this time interval the radio drift time. It can be used as a proxy of the electron travel time provided that (i) the electrons emitting the type III burst do not change speed along their trajectory, (ii) type III bursts whose low frequency reaches the electron plasma frequency at the spacecraft are emitted near the spacecraft, and are not electromagnetic radiation that comes from remote places. In this case the radio drift time is a measure of the travel time of electrons at energies between about 5 and 10 keV (*Buttighoffer*, 1998, and references therein). Hypothesis (i) is plausible, but caution is necessary for hypothesis (ii), since *Hoang et al.* (1997) showed that interplanetary type III bursts have emission cones several tens of degrees wide and can be observed even when the emitting electron beams are not intercepted. Irrespective of the validity of these assumptions, Figure 1 of *Cane and Erickson* (2003) shows that the assessment of the arrival time of the type III emission at the spacecraft may be ambiguous in the case of long radio delays.

In Figure 4b (*Cane*, 2003) the radio drift time is plotted vs. the release delay of electrons >25 keV for the events of *Haggerty and Roelof* (2002), i.e. the same events as in Figure 4a. Here the electron release delay is the time interval between the inferred release of near-relativistic electrons detected at ACE and the start of the type III emission in the corona. There is again a correlation, like in Figure 4a. *Cane* (2003) interprets it as evidence that the type III emitting low energy electrons and the electrons above 25 keV measured in situ belong to the same population, and that the delayed arrival times have a common cause, namely a prolonged travel time in interplanetary space due to particle scattering. *Cane and Erickson* (2003) also showed that the radio drift time tends to increase with increasing distance of the parent flare from the spacecraft-connected nominal Parker spiral, a finding they consider as evidence for cross-field transport of the protons and electrons detected at the spacecraft.

4.2. SEP Onset Delays and Interplanetary Particle Propagation

Kallenrode and Wibberenz (1990), *Cane* (2003), *Cane and Erickson* (2003), and *Lintunen and Vainio* (2004) argued that

4.3. SEP Onset Delays and Complex Magnetic Configurations in the Corona

The explanations invoking delayed shock acceleration of the SEP in the high corona or delayed particle propagation in

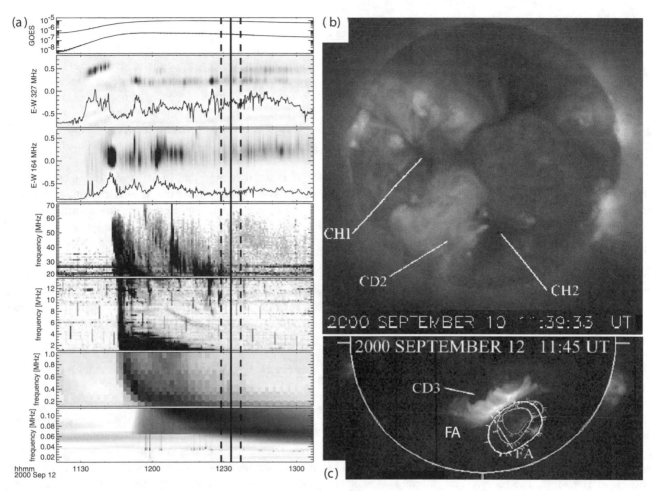

Figure 5. (a) Time histories of electromagnetic emissions around the time of near-relativistic electron release on 12 September 2000. The interval of electron release is delimited by vertical dashed lines, after correction for the different travel times of electrons and photons. From *Klein et al.* (2005). (b), (c) Yohkoh/SXT images of the Sun before (b) and after (c) the associated filament eruption (*Vršnak et al.*, 2003). The three closed curves in (c) are the half-maximum iso-intensity contours of the type IV radio emission at 12:34 UT at 327, 237 and 164 MHz (white shading increases with decreasing frequency).

the interplanetary medium both attribute the delayed SEP onsets only to the escaping particles. But the time-coincident rises of the SEP profiles and the HXR and radio profiles in Figure 3 suggest a much closer relationship between the SEP, including the protons whose onset is delayed, and particle acceleration processes related to a flare. Rather than focussing on the start time, we now discuss with more detail the time evolution of radio emission and its relationship with delayed releases of SEP.

The timing of 20 electron events from ACE (*Haggerty and Roelof*, 2002) was compared with radio emission by *Maia and Pick* (2004), and 40 electron events observed by the *Wind* spacecraft were similarly analysed by *Klein et al.* (2005). It was found in these studies that most, if not all, events where delays are observed - in any case, those events where these delays amount to 10 min or more - are associated with broad

CMEs, large flares and complex, long-lasting ≳ 1 hr) radio emission which is a tracer of time-extended electron acceleration in the corona.

The event with the most conspicuous onset delay of *Klein et al.* (2005) is shown in Figure 5a: the time interval where the first near-relativistic electrons reaching the *Wind* spacecraft are observed to be released is delimited by dashed vertical lines. It is compared with time profiles of the electromagnetic emission, from soft X-rays (top) over metric (second to fourth panel from top) to the decametric-to-kilometric radio emission (three bottom panels). In the second and third panel from top, the time history of the whole Sun flux density at 327 MHz (91 cm) and 164 MHz (1.83 m) is overlaid on a grey-scale plot of the 1D brightness (after integration over the north-south direction; dark shading means bright emission), which gives an idea of the spatial evolution of the sources. The timing of the various signatures of radiating and escaping electrons comprises at least three distinct phases during this event: (1) The first signature of electron acceleration in the corona is seen at 327 and 164 MHz, with no counterpart in the spectrum ≤70 MHz. The well-defined low-frequency cutoff of the radio spectrum suggests that the first electrons accelerated in the corona radiate in closed magnetic structures. The structures evolve dynamically, as illustrated in Figure 5a by the westward (i.e. upward in the figure) motion of the 327 MHz source between ~1132 and 1142 UT ("moving type IV burst"). (2) Some minutes later, the bright decametric-to-kilometric type III burst (~30–0.04 MHz) signals the escape of low-energy electrons to interplanetary space. Following *Vršnak et al.* (2003), the type III emission appears when the Hα flare ribbons, which trace the footpoints of the loop arcade FA in the *Yohkoh*/SXT image of Figure 5c, attain the region CD2 of weak soft X-ray emission shown in Figure 5b. These authors suggest that the electron beams are accelerated when the magnetic reconnection producing the arcade FA involves adjacent open field lines rooted in CD2. The positions of the interplanetary type III sources, inferred from the signal modulation by the spacecraft spin, outline a flux tube that extends far south of the ecliptic plane, as expected if it is rooted in CD2, so that it is plausible that no electrons were detected at that time by *Wind* (*Klein et al.*, 2005). (3) The first near-relativistic electrons detected at *Wind* are released about 40 min later. At this time the metric radio emission brightens again, as shown by the whole Sun flux at 327 MHz and by the 1D brightness at 164 MHz. The renewed brightening, as well as structural changes in the metric radio sources (i.e. the appearance of a new source at 327 MHz) reveal a new episode of electron acceleration in the corona, just in time to account for the release of the near-relativistic electrons detected at *Wind*. As seen in Figure 5c, the radio source (isocontours at half maximum at three frequencies) is located on top of the loop arcade FA. The relationship with the formation of a loop arcade suggests that the radio emitting electrons, and by inference from the common timing the electrons detected by *Wind*, are accelerated during reconnection of the stressed magnetic fields in the post-eruptive corona.

Besides these signatures of electron acceleration in different coronal structures, shock-accelerated electrons are also observed in Figure 5, as shown by the type II burst between 1200 and 1230 UT in the range (10–4) MHz (Figure 5a). The type II burst starts well before the inferred release of the first near-relativistic electrons, however, and does not seem to be related to their acceleration.

The different starting times of radio emissions in different frequency ranges during this and many other events demonstrate that delays of electron signatures are not a specific property of the electrons detected in situ. These delays result from different successive episodes of acceleration of electrons in the corona and their injection into different closed and open magnetic structures. These findings are not restricted to electrons, since evidence that delayed coronal acceleration processes of protons occur well behind the CME front and its shock at the time of delayed SEP releases is readily apparent in Figure 3 and has also been reported for relativistic protons (*Akimov et al.*, 1996; *Klein et al.*, 1999).

It is worth noting that in a few of the electron events of *Klein et al.* (2005) that were associated with fast and broad CMEs the electrons detected at the spacecraft were released at the very beginning of the flare-related radio emission. This is consistent with the idea that in these cases the spacecraft happened to be connected with the flux tube along which the first accelerated electrons escaped to interplanetary space. The finding is harder to reconcile with the idea that shocks accelerate these electrons above a characteristic coronal height, since then a delay would be inevitable.

5. SUMMARY AND CONCLUSIONS

The solar corona provides different regions where particles are accelerated to high energies. Radio emissions from non thermal electrons give tools to infer acceleration sites and to trace magnetic structures that guide charged particles through the corona and into interplanetary space. Hence, while the primary emission comes from electrons, radio waves also provide some clues to the understanding of ions in SEP events. The main results addressed in this paper are the following:

- Typical acceleration regions in impulsive flares are located in the low corona, with ambient electron densities $\sim(10^9–10^{10})$ cm^{-3}. Circumstantial evidence suggests that magnetic reconnection plays a key role.
- Energetic electrons (particles) are injected into several magnetic flux tubes in an active region and then propagate to interplanetary space along multiple paths, even in many relatively simple impulsive flares.
- During large flare/CME events, electron (particle) acceleration in the corona can continue over several tens of minutes or even several hours. The radio observations show that accelerated particles are released into different coronal flux tubes in the course of such events. Since a given spacecraft is connected to at most one of these flux tubes at a time, it is probable that it will miss part of the accelerated particles, and more or less often the first accelerated particles. Therefore onset time delays of SEP events are expected in single spacecraft observations, even if the SEP are accelerated together with radiating

particles in a flaring active region, and even if the propagation is scatter free.

- Traveling shock waves in the corona accelerate electrons seen at radio waves, but less than the acceleration processes which produce typical impulsive hard X-ray bursts. There is specific radio evidence for electron acceleration at interacting CMEs. It is still to be investigated if this acceleration reveals shock waves or magnetic reconnection.

The main result from the radio-SEP comparison is in the present author's view a much closer interrelationship between flares and CMEs as sources of SEP than is generally assumed. The detailed event analyses do not exclude shock acceleration of SEP, but they emphasize various possible coronal acceleration regions and do not support a clear cut distinction between flare-related and CME shock-related SEP events. Even in simple impulsive flares the origin of the SEP is not definitely settled, despite the closely related timing of interacting and escaping particles. Besides producing shocks, CMEs seem to provide small-scale acceleration sites over extended volumes in the corona. The large-scale restructuring of the coronal magnetic fields in CMEs likely plays a key role in particle transport. In that case the width of the CME, rather than its speed, should affect the SEP intensities.

Dual-spacecraft observations of CMEs and energetic particles with STEREO will further elucidate the role that CMEs and flares play in SEP events. The reduction of transport effects due to particle measurements in the inner heliosphere, where *Helios* showed us the potential of particle measurements under conditions of short travel distances, will be a further qualitative progress aimed at by the ESA *Solar Orbiter* concept. The full power of these new missions can only be exploited if they are combined with the observation of radiative signatures of particles in the corona. Radio observations, preferentially with wideband imaging as proposed by the FASR project, are a necessary complementary tool for understanding the origin of SEP events.

Acknowledgments. Radio observations of the Sun are provided by a complementary set of observatories around the world. The author gratefully acknowledges the long lasting efforts of those who run these observatories and make the data available. B. Vršnak is thanked for providing original figures, M.P. Issartel for her precious help with redrawing Figure 3, G. Aulanier, W. Dröge, S. Krucker, R.P. Lin, A. Posner and G. Trottet for many helpful discussions, and the referees for their careful reading and constructive criticism.

REFERENCES

Akimov, V.V., P. Ambrož, A.V. Belov, A. Berlicki, I.M. Chertok, M. Karlický, V.G. Kurt, N.G. Leikov, Y.E. Litvinenko, A. Magun, A. Minko-Wasiluk, B. Rompolt, and B.V. Somov (1996), Evidence for prolonged acceleration based on a detailed analysis of the long-duration solar gamma-ray flare of june 15, 1991, *Solar Phys.*, 166, 107.

Aschwanden, M.J. (2005), Particle acceleration in solar flares and escape into interplanetary space, in *Solar Energetic Plasmas and Particles*, edited by N. Gopalswamy, AGU Monograph, this volume.

Aurass, H., and G. Mann (2004), Radio observation of electron acceleration at solar flare reconnection outflow termination shocks, *ApJ*, 615, 526-530.

Bale, S.D., M.J. Reiner, J.-L. Bougeret, M.L. Kaiser, S. Krucker, D.E. Larson, and R.P. Lin (1999), The source region of an interplanetary type II radio burst, *GRL*, 26, 1573-1576.

Biesecker, D.A., D.C. Myers, B.J. Thompson, D.M. Hammer, and A. Vourlidas (2002), solar phenomena associated with "EIT waves", *ApJ*, 569, 1009-1015.

Buttighoffer, A. (1998), Solar electron beams associated with radio type III bursts: propagation channels observed by Ulysses between 1 and 4 AU, *A&A*, 335, 295-302.

Cairns, I.H., and R.D. Robinson (1987), Herringbone bursts associated with type II solar radio emission, *Solar Phys.*, 111, 365-383.

Cane, H.V. (2003), Near-relativistic solar electrons and type III radio bursts, *ApJ*, 598, 1403-1408.

Cane, H.V., and W.C. Erickson (2003), Energetic particle propagation in the inner heliosphere as deduced from low-frequency (<100 kHz) observations of type III radio bursts, *JGR*, 108, SSH 8-1-8-12 (doi:10.1029/ 2002JA009,488).

Carmichael, H. (1962), High-energy solar-particle events, *Space Sci. Rev.*, 1, 28.

Chen, P.F., C. Fang, and K. Shibata (2005), A full view of EIT waves, *ApJ*, 622, 1202-1210.

Cliver, E., and S. Kahler (1991), High coronal flares and impulsive acceleration of solar energetic particles, *ApJ*, 366, L91-L94.

Cliver, E.W., S.W. Kahler, M.A. Shea, and D.F. Smart (1982), Injection onsets of 2 GeV protons, 1 MeV electrons, and 100 keV electrons in solar cosmic ray flares, *ApJ*, 260, 362-370.

Delannée, C., and G. Aulanier (1999), CME associated with transequatorial loops and a bald patch Flare, *Solar Phys.*, 190, 107-129.

Dröge, W. (2000), Particle scattering by magnetic fields, *Space Science Reviews*, 93, 121-151.

Giacalone, J., J.R. Jokipii, and J.E. Mazur (2000), Small-scale gradients and large-scale diffusion of charged particles in the heliospheric magnetic field, *ApJ*, 532, L75-L78.

Gopalswamy, N., A. Lara, M.L. Kaiser, and J.-L. Bougeret (2001a), Near-Sun and near-Earth manifestations of solar eruptions, *JGR*, 106, 25,261-25,278.

Gopalswamy, N., S. Yashiro, M.L. Kaiser, R.A. Howard, and J.-L. Bougeret (2001b), Radio signatures of coronal mass ejection interaction: coronal mass ejection cannibalism?, *ApJ*, 548, L91-L94.

Gopalswamy, N., S. Yashiro, M.L. Kaiser, R.A. Howard, and J.-L. Bougeret (2002), Interplanetary radio emission due to interaction between two coronal mass ejections, *GRL*, 29, 106-1.

Gopalswamy, N., S. Yashiro, S. Krucker, G. Stenborg, and R.A. Howard (2004), Intensity variation of large solar energetic particle events associated with coronal mass ejections, *JGR*, 109, A12,105 (doi: 10.1029/ 2004JA01,602).

Haggerty, D.K., and E.C. Roelof (2002), Impulsive near-relativistic solar electron events: delayed injection with respect to solar electromagnetic emission, *ApJ*, 579, 841-853.

Hanaoka, Y. (1999), High-energy electrons in double-loop flares, *PASJ*, 51, 483-496.

Hoang, S., M. Poquérusse, and J.-L. Bougeret (1997), The directivity of solar kilometric type III bursts: Ulysses-Artemis observations in and out of the ecliptic plane, *Solar Phys.*, 172, 307-316.

Hucke, S., M.B. Kallenrode, and G. Wibberenz (1992), Interplanetary type III radiobursts and relativistic electrons, *Solar Phys.*, 142, 143-155.

Kahler, S. (1994), Injection profiles of solar energetic particles as functions of coronal mass ejection heights, *ApJ*, 428, 837-842.

Kahler, S.W. (2003), Energetic particle acceleration by coronal mass ejections, *Adv. Space Res.*, 32, 2587-2596.

Kallenrode, M.B., and G. Wibberenz (1990), Influence of interplanetary propagation on particle onsets, in *Proc. 21st ICRC*, vol. 5, pp. 229-232.

Kallenrode, M.B., and G. Wibberenz (1991), Particle injection following solar flares on 1980 may 28 and june 8 - evidence for different injection time histories in impulsive and gradual events?, *ApJ*, 376, 787-796.

Khan, J.I., and H. Aurass (2002), X-ray observations of a large-scale solar coronal shock wave, *A&A*, 383, 1018-1031.

Klassen, A., H. Aurass, G. Mann, and B.J. Thompson (2000), Catalogue of the 1997 SOHO-EIT coronal transient waves and associated type II radio burst spectra, *A&A Supp.*, 141, 357-369.

Klecker, B., E. Moebius, M.A. Popecki, M. Hilchenbach, L.M. Kistler, and H. Kucharek (2005), Observation of energy-dependent ionic charge states in impulsive solar energetic particle events, *Adv. Space Res., in press*.

Klein, K.-L., and A. Posner (2005), The onset of solar energetic particle events: prompt release of deka-MeV protons and associated coronal activity, *A&A*, 438, 1029-1042 (doi: 10.1051/0004-6361:20042,607).

Klein, K.-L., E.L. Chupp, G. Trottet, A. Magun, P.P. Dunphy, E. Rieger, and S. Urpo (1999), Flare-associated energetic particles in the corona and at 1 AU, *A&A*, 348, 271-285.

Klein, K.-L., G. Trottet, P. Lantos, and J.-P. Delaboudinière (2001), Coronal electron acceleration and relativistic proton production during the 14 July 2000 flare and CME, *A&A*, 373, 1073-1082.

Klein, K.-L., R.A. Schwartz, J.M. McTiernan, G. Trottet, A. Klassen, and A. Lecacheux (2003), An upper limit of the number and energy of electrons accelerated at an extended coronal shock wave, *A&A*, 409, 317-324.

Klein, K.-L., S. Krucker, G. Trottet, and S. Hoang (2005), Coronal phenomena at the onset of solar energetic electron events, *A&A*, 431, 1047-1060.

Kliem, B., M. Karlický, and A.O. Benz (2000), Solar flare radio pulsations as a signature of dynamic magnetic reconnection, *A&A*, 360, 715-728.

Knock, S.A., and I.H. Cairns (2005), Type II radio emission predictions: Sources of coronal and interplanetary spectral structure, *JGR (Space Physics)*, 110, A1101 (doi: 10.1029/ 2004JA010,452).

Kocharov, L.G., G.A. Kovaltsov, and J. Torsti (2001), Dynamical cycles in charge and energy for iron ions accelerated in a hot plasma, *ApJ*, 556, 919-927.

Krucker, S., and R.P. Lin (2000), Two Classes of Solar Proton Events Derived from Onset Time Analysis, *ApJ*, 542, L61-L64.

Krucker, S., D.E. Larson, R.P. Lin, and B.J. Thompson (1999), On the origin of impulsive electron events observed at 1 AU, *ApJ*, 519, 864-875.

Laitinen, T., K.-L. Klein, L. Kocharov, J. Torsti, G. Trottet, V. Bothmer, M.L. Kaiser, G. Rank, and M.J. Reiner (2000), Solar energetic particle event and radio bursts associated with the 1996 july 9 flare and coronal mass ejection, *A&A*, 360, 729-741.

Lin, R.P. (1974), Non-relativistic solar electrons, *Space Sci. Rev.*, 16, 189-256.

Lin, R.P. (1985), Energetic solar electrons in the interplanetary medium, *Solar Phys.*, 100, 537-561.

Lin, R.P., R.A. Mewaldt, and M.A.I. van Hollebeke (1982), The energy spectrum of 20 keV-20 MeV electrons accelerated in large solar flares, *ApJ*, 253, 949-962.

Lintunen, J., and R. Vainio (2004), Solar energetic particle event onsets as analyzed from simulated data, *A&A*, 420, 343-350.

Lockwood, J.A., H. Debrunner, and E.O. Flueckiger (1990), Indications for diffusive coronal shock acceleration of protons in selected solar cosmic ray events, *JGR*, 95, 4187-4201.

Möbius, E., Y. Cao, M.A. Popecki, L.M. Kistler, H. Kucharek, D. Morris, and B. Klecker (2003), Strong energy dependence of ionic charge states in impulsive solar events, in *Proc. 28th ICRC*, pp. 3273-3276.

Maia, D., and M. Pick (2004), Revisiting the origin of impulsive electron events: coronal magnetic restructuring, *ApJ*, 609, 1082-1097.

Maia, D., A. Vourlidas, M. Pick, R. Howard, R. Schwenn, and A. Magalhães (1999), Radio signatures of a fast coronal mass ejection development on november 6, 1997, *JGR*, 104, 12,507-12,514.

Mann, G., and A. Klassen (2002), Shock accelerated electron beams, in *Solar Variability: From Core to Outer Frontiers*, ESA SP-506, p. 245.

Mann, G., T. Classen, and H. Aurass (1995), Characteristics of coronal shock waves and solar type II radio bursts., *A&A*, 295, 775-781.

Mann, G., H.T. Classen, U. Motschmann, H. Kunow, and W. Dröge (1999), High energetic electrons accelerated by a coronal shock wave, *Astrophys. Spa. Sci.*, 264, 489-496.

Mann, G., A. Klassen, H. Aurass, and H.-T. Classen (2003), Formation and development of shock waves in the solar corona and the near-Sun interplanetary space, *A&A*, 400, 329-336.

Mazur, J.E., G.M. Mason, J.R. Dwyer, J. Giacalone, J.R. Jokipii, and E.C. Stone (2000), Interplanetary magnetic field line mixing deduced from impulsive solar flare particles, *ApJ*, 532, L79-L82.

Nelson, G.J., and D.B. Melrose (1985), Type II bursts, in *Solar Radiophysics: Studies of Emission from the Sun at Metre Wavelengths*, edited by D. McLean and N. Labrum, pp. 333-359, Cambridge University Press.

Nishio, M., K. Yaji, T. Kosugi, H. Nakajima, and T. Sakurai (1997), Magnetic field configuration in impulsive solar flares inferred from coaligned microwave/X-ray images, *ApJ*, 489, 976-991.

Pick, M., and S. Ji (1986), Type III burst sources and electron beam injection, *Solar Phys.*, 107, 159-165.

Poquérusse, M., S. Hoang, J.-L. Bougeret, and M. Moncuquet (1996), Ulysses-ARTEMIS radio observation of energetic flare electrons, in *Solar Wind Eight*, edited by D. Winterhalter, J. Gosling, S. Habbal, W. Kurth, and M. Neugebauer, *Am. Inst. Phys.*, pp. 62-65.

Simnett, G.M., E.C. Roelof, and D.K. Haggerty (2002), The acceleration and release of near-relativistic electrons by coronal mass ejections, *ApJ*, 579, 854-862.

Torsti, J., L. Kocharov, M. Teittinen, and B. Thompson (1999), Injection of ≳ 10 MeV protons in association with a coronal Moreton wave, *ApJ*, 510, 460-465.

Tylka, A.J., C.M.S. Cohen, W.F. Dietrich, S. Krucker, R.E. McGuire, R.A. Mewaldt, C.K. Ng, D.V. Reames, and G.H. Share (2003), Onsets and release times in solar particle events, in *Proc. 28th ICRC*, pp. 3305-3308.

Vilmer, N., and A. MacKinnon (2003), What can be learned about competing acceleration models from multiwavelength observations?, in *Energy conversion and particle acceleration in the solar corona, Lecture Notes in Physics*, vol. 612, edited by K.-L. Klein, pp. 127-160, Springer.

Vršnak, B., A. Warmuth, D. Maričić, W. Otruba, and V. Ruždjak (2003), Interaction of an erupting filament with the ambient magnetoplasma and escape of electron beams, *Solar Phys.*, 217, 187-198.

Warmuth, A., B. Vršnak, J. Magdalenić, A. Hanslmeier, and W. Otruba (2004), A multiwavelength study of solar flare waves. II. Perturbation characteristics and physical interpretation, *A&A*, 418, 1117-1129.

Diffusive Ion Acceleration by CME-Driven Shocks

Martin A. Lee

*Space Science Center, Institute for the Study of Earth, Oceans and Space,
University of New Hampshire, Durham, New Hampshire*

The acceleration of solar energetic ions in "gradual" events at coronal/ interplanetary shocks is reviewed including: a brief observational motivation for a shock origin of gradual events, the historical development of the theory of shock acceleration as applied to solar energetic particles, the basic features of the process of diffusive shock acceleration (first-order Fermi acceleration at the shock, wave excitation by the energetic protons, magnetic focusing and upstream escape of the ions), challenges to developing a successful theory, the basic equations of ion transport and acceleration, and current theoretical research. Finally, outstanding issues which must be addressed by future work are listed including: ion injection into the process of diffusive shock acceleration, careful evaluation of the wave intensity upstream of the shock, acceleration and injection at quasi-perpendicular shocks, long-time behavior of gradual events, the inclusion of wave propagation velocities in the effective shock compression ratio, and a rigorous treatment of the time-dependence.

1. INTRODUCTION

One of the most exciting discoveries in space during the last three decades has been the close association between energetic particles and shocks in the heliosphere. That association includes energetic storm particle (ESP) events at interplanetary traveling shocks, "diffuse" ions at Earth's bow shock and other planetary bow shocks, the "corotating" ion events at the forward and reverse shocks bounding corotating interaction regions in the solar wind, and the anomalous cosmic rays at the solar wind termination shock. Clearly shocks are efficient accelerators of energetic ions in the tenuous collisionless plasma of the heliosphere.

Our understanding of ion shock acceleration has evolved from early rudimentary ideas on the possibility of a collisionless shock (*Gold*, 1959) and the interaction of charged particles with laminar shocks (*Hudson*, 1965; *Sarris and VanAllen*, 1974), to the well-developed theory of diffusive shock acceleration (DSA) (*Axford et al.*, 1978; *Krymsky*,

Solar Eruptions and Energetic Particles
Geophysical Monograph Series 165
Copyright 2006 by the American Geophysical Union
10.1029/165GM23

1977; *Blandford and Ostriker*, 1978; *Bell*, 1978). DSA has been applied to all the ion/shock associations listed above as well as to the acceleration of galactic cosmic rays at supernova shocks. In general the mechanism, which readily produces power-law energy spectra as observed for cosmic rays, has been successful in these applications.

However, the acceleration of solar energetic particles (SEPs) by shock waves is somewhat more controversial. Although large "gradual" SEP events often include an ESP component concurrent with the passage of an interplanetary shock, the origin of the particles which arrive early in the event, usually those with the highest energies, is unclear. The uncertainty is due largely to the absence of direct observations close to the Sun where these particles are accelerated. In support of a shock origin is a high correlation of gradual events with CMEs and their driven shocks, the broad distribution of gradual events about optimal magnetic connection to the site of the flare/CME, and their generally coronal elemental and charge-state composition. In support of a flare or magnetic-reconnection origin of these particles is the very prompt onset of many events, which is generally considered to be a challenge for DSA which must accommodate the acceleration time and the shock formation time in an environment where the Alfvén speed V_A can exceed 10^3 km/s.

In addition, many gradual events exhibit compositional signatures of flare-site plasma.

Also providing support for a flare origin of all SEPs, is the unambiguous origin of the small "impulsive" events at flares. They exhibit extreme compositional fractionation relative to the corona. They are generally not associated with CMEs and interplanetary shocks, and not assumed to originate at CME-driven shocks. Impulsive events are usually thought to be accelerated by the electric field associated with magnetic reconnection (*Litvinenko*, 1996) or by turbulence generated at the reconnection site (*Miller and Roberts*, 1995; *Emslie et al.*, 2004), but may also be energized at reconnection-driven shocks (*Tsuneta and Naito*, 1998).

Nevertheless, based on simplicity the most appealing origin of a gradual event is a CME-driven coronal/interplanetary shock. The challenge to shock advocates is to account for the prompt onsets and the various elemental and charge-state compositional variations observed within and between most events. An excellent review of gradual and impulsive SEP events is provided by *Reames* (1999).

2. SHOCK ORIGIN OF GRADUAL EVENTS

Figure 1 is a schematic diagram of ion acceleration at an evolving coronal/interplanetary shock driven by a CME into the solar wind, Region 1. The dots denote the ions which are predominantly in Region 2, a sheath of enhanced turbulence upstream of the shock, and in Region 3, a sheath of enhanced turbulence between the shock and the CME. The ions are

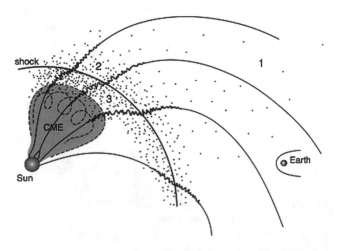

Figure 1. Schematic snapshot of an evolving coronal/interplanetary shock driven by a CME. Accelerated ions are denoted by dots. Magnetic field lines are shown, with wiggles denoting magnetic fluctuations. The spatial domain accessible to the ions is divided into solar wind (1), a proton-excited turbulent sheath upstream of the shock (2), and the turbulent shock-heated solar wind downstream of the shock (3).

injected out of the solar wind and/or the ambient suprathermal/energetic ion population into the process of DSA. The accelerating upstream ions stream relative to the solar wind at a speed greater than V_A, which excites hydromagnetic waves propagating antisunward relative to the solar wind. The waves are transmitted through the shock to create a turbulent sheath on both sides of the shock in which ions are strongly scattered and coupled to the shock compression, and thereby accelerated by the first-order Fermi process. Particles at the upstream edge of Region 2 are extracted from the turbulent sheath by the magnetic mirror force and stream nearly scatter-free to Earth orbit and beyond. These particles are the first to arrive at Earth orbit in gradual events. After the shock passes Earth orbit and/or weakens significantly, the particles diffuse to fill the inner heliosphere and decay as they adiabatically cool in the diverging solar wind (*Reames et al.*, 1996).

An important feature of DSA is that particles are accelerated not only by the first-order Fermi process, but also by shock-drift acceleration in which they drift (by virtue of the curvature and gradient of the average magnetic field **B** at the shock) parallel to the motional electric field of the solar wind as viewed in the frame of the shock (*Jokipii*, 1982). The contributions of the two acceleration mechanisms to the energy gains of the particles are frame dependent. Since in the deHoffman-Teller frame the motional electric field vanishes, the shock-drift energization also vanishes. The complementary situation occurs for perpendicular shocks for which the deHoffman-Teller frame is not physically accessible and the particle energy gain is completely due to shock-drift.

The process of DSA is deceptively simple. In fact there are numerous challenges to developing a predictive theory for gradual SEP events. To start with, there are complexities due to the configuration of these events and incomplete observations: (i) The events involve complicated temporal and spatial variation due to the evolution of the shock and magnetic connection to the shock nose where the shock is strongest; (ii) The spherical geometry of the heliosphere is crucial to describe the escape of ions from the shock sheath by magnetic mirroring; (iii) The higher-energy particles in gradual events are accelerated close to the Sun where shock parameters are not well known and observations are indirect.

In addition, there are required modifications to the standard application of DSA: (iv) The transition from the scatter-dominated ion transport in the turbulent sheath (Region 2) to the nearly scatter-free ion transport in the solar wind (Region 1) poses a challenge for theory since the traditional transport equation (*Parker*, 1965) for DSA is not valid for the escaping ions in Region 1; (v) Ion fractionation and escape is controlled sensitively by the frequency-dependence of the excited wave intensity in the turbulent sheath, which must be calculated carefully; (vi) Shock obliquity appears to play an important role in gradual event morphology (*Tylka et al.*, 2005), yet the structure of quasi-perpendicular shocks is not well understood; (vii) The seed populations and injection rates for DSA in gradual events are not well understood, yet they are crucial to the predicted elemental and charge-state composition of gradual events.

The emphasis in this paper is on ion acceleration at shocks. Electrons are also accelerated by shocks. Electrons are accelerated to 10s of keV at Earth's bow shock, and accelerated electrons at coronal shocks are responsible for the type II radio bursts which signal the presence and location of the shock. Nevertheless, electrons are much less efficiently accelerated than ions at most heliospheric shocks and they do not attain comparable energies. The reason is presumably that high-frequency waves resonant with electrons are not excited with sufficient growth rates to couple electrons effectively to the shock compression.

3. BASIC EQUATIONS

The basic particle transport equation is the focused transport equation, quoted here for radial magnetic field (*Roelof*, 1969; *Skilling*, 1971; *Earl*, 1976, 1981; *Isenberg*, 1997)

$$\frac{\partial f}{\partial t} + (V + v\mu)\frac{\partial f}{\partial r} - \frac{(1-\mu^2)}{r}Vv\frac{\partial f}{\partial v} + \frac{(1-\mu^2)}{r}(v + \mu V)\frac{\partial f}{\partial \mu} = \frac{\partial}{\partial \mu}\left[(1-\mu^2)D_\mu\frac{\partial f}{\partial \mu}\right] + Q \quad (1)$$

where $f(v, \mu, r, t)$ is the nonrelativistic velocity-space distribution function averaged over gyrophase, μ is the cosine of particle pitch-angle, D_μ is the pitch-angle diffusion coefficient, Q is a source, and particle speed v and μ are measured in the frame of the solar wind with speed V. Terms 2-5 describe convection, adiabatic deceleration, magnetic focusing, and pitch-angle diffusion. Equation (1) neglects the velocity of the scattering fluctuations, typically the Alfvén speed.

If $D_\mu >> V/r$, scattering is efficient and the particle distribution is nearly isotropic. Then equation (1) may be integrated over μ and generalized to yield

$$\frac{\partial f_0}{\partial t} + (\mathbf{V} + \mathbf{V}_D)\cdot\nabla f_0 - \nabla\cdot\mathbf{\kappa}\cdot\nabla f_0 - \frac{1}{3}\nabla\cdot\mathbf{V}p\frac{\partial f_0}{\partial p} - \frac{1}{p^2}\frac{\partial}{\partial p}\left(p^2 D_p\frac{\partial f_0}{\partial p}\right) = Q \quad (2)$$

where $f_0(p, \mathbf{r}, t)$ is the omnidirectional distribution function (averaged over pitch-angle and gyrophase), p is momentum

magnitude, D_p is the diffusion coefficient in p-space, $\mathbf{V}_D = pvc(3q)^{-1}\nabla\times(B^{-2}\mathbf{B})$ is the drift velocity, q is particle charge, $\mathbf{V}$ is the plasma or solar wind velocity, and equation (2) assumes $v >> V$. Equation (2) stems from the work of many including *Davis* (1956), *Parker and Tidman* (1958), *Parker* (1965), *Gleeson and Axford* (1967) and *Jokipii and Levy* (1977). The component of the spatial diffusion tensor $\mathbf{\kappa}$ parallel to $\mathbf{B}$, $\kappa_{\parallel}$, may be derived from the μ-integration of equation (1) as (*Earl*, 1974)

$$\kappa_{\parallel} = \frac{v^2}{8}\int_{-1}^{1}d\mu\frac{(1-\mu^2)}{D_\mu} \quad (3)$$

The components of $\mathbf{\kappa}$ perpendicular to $\mathbf{B}$, $\kappa_{\perp}$, are generally small and negligible for SEP transport; a possible exception is the region close to a nearly perpendicular shock (*Dwyer et al.*, 1997).

The pitch-angle diffusion coefficient for ions, D_μ, is given in terms of the hydromagnetic wave intensities $I_\pm(k,r,t)[\langle|\delta\mathbf{B}|^2\rangle = \sum_{+,-}\int_{-\infty}^{\infty}I_\pm(k)dk]$ by quasilinear theory as (*Lee*, 1983; *Gordon et al.*, 1999)

$$D_\mu = \frac{\pi}{2}\frac{\Omega^2}{B_0^2}\frac{1}{|\mu|v}\sum_\pm I_\pm \quad (k = \Omega\mu^{-1}v^{-1}) \quad (4)$$

where $\Omega (=qB_0/mc)$ is the ion cyclotron frequency, $(\pm)$ refer to propagation direction, and we restrict ourselves to nonrelativistic ion speeds $v >> \omega/k$, and wave propagation parallel to $\mathbf{B}$. Nonlinearity is apparent since the wave intensity adjacent to the shock is enhanced by the flux of protons upstream of the shock. Thus, the wave kinetic equation

$$\frac{\partial I_\pm}{\partial t} + (\mathbf{V} + \mathbf{V}_g)\cdot\nabla I_\pm = 2\gamma_\pm I_\pm \quad (5)$$

must be considered together with the ion transport equations, where $\mathbf{V}_g$ is the wave group velocity in the plasma frame, inhomogeneity of the background plasma is neglected, and the wave growth rate $\gamma_\pm$ is given by

$$\gamma_\pm = \pm 2\pi^3|k|^{-1}V_A c^{-2}q^2 m^{-1}\int_0^\infty dv v^2\cdot \int_{-1}^{1}d\mu(1-\mu^2)\delta(\mu - \Omega k^{-1}v^{-1})\partial f_p/\partial\mu \quad (6)$$

Here we make the traditional assumption that protons dominate wave excitation; the minor ions may be treated as "test particles." Equations (1)-(6) taken together are the starting point for the theoretical description of the shock origin of gradual events; generally D_p in equation (2) may be neglected, but see *Schlickeiser et al.* (1983).

Since the wave excitation described by equations (5) and (6) is dominated by lower-energy protons confined near the shock, the "lifetime" of the upstream waves is sufficiently short that inhomogeneity of the solar wind may generally be neglected as in equation (5). However, equation (5) is more precisely formulated in terms of wave action, which allows the wavevectors to refract and the waves to do work on the solar wind (*Ng et al.*, 1994; 2003).

4. HISTORICAL PERSPECTIVE AND CURRENT WORK

Early theoretical work on shock acceleration of SEPs proceeded in two directions. Firstly, following the development of the theory of diffusive shock acceleration based on equation (2) (*Axford et al.*, 1978; *Krymsky*, 1977; *Blandford and Ostriker*, 1978; *Bell*, 1978), there were applications of the theory to SEPs by *Achterberg and Norman* (1980), *Lee and Fisk* (1982) and *Lee and Ryan* (1986). These were simplified in both geometry and the form of the diffusion coefficient. There were also applications of the theory to energetic storm particle (ESP) enhancements observed at Earth orbit (*Forman*, 1981; *Lee*, 1983; *Gordon et al.*, 1999). An ESP event is actually one phase of a gradual event, which occurs if the shock still accelerates ions when it passes Earth. With the planar geometry appropriate for ESP events, *Lee* (1983) was able to include in the theory wave excitation described by equations (5) and (6). These models provided a reasonable description of the ion (and wave) enhancements near the shock where equation (2) is valid. Research on the behavior of shock-accelerated SEPs based on equation (2) continues to provide insights for more complex configurations: *Vainio et al.* (2000) demonstrate that a single shock with turbulence confined to a sheath upstream and downstream of the shock can produce both particles which escape into interplanetary space and particles which interact in the solar atmosphere. *Kocharov et al.* (2005) and *Lytova and Kocharov* (2005) invoke a turbulent layer at the base of the solar wind to trap shock-accelerated protons and release them gradually as a delayed component as observed in some events.

Secondly, there have been many applications of equation (1) to the nearly scatter-free transport of SEPs in interplanetary space early in an event when ion anisotropy may be large (*Heras et al.*, 1992, 1995; *Kallenrode*, 1993; *Ruffolo*, 1995; *Torsti et al.*, 1996; *Kallenrode and Wibberenz*, 1997; *Lario et al.*, 1998; *Anttila et al.*, 1998; *Kocharov et al.*, 1999; *Anttila and Sahla*, 2000; *Laitinen et al.*, 2000). These particles constitute an important phase of the event for space weather forecasting and usually include the most energetic

particles. These models include the shock acceleration heuristically as a source term Q, which is chosen with a power-law energy spectrum appropriate to a moving shock which is weaker on the flanks. They provide a reasonable description of the early phase of gradual events.

Current work has attempted to combine the advantages of these two past research directions in order to accommodate more realistic geometry, wave excitation, and the transition from scatter-dominated to nearly scatter-free ion transport with increasing distance upstream of the shock. *Ng* and coworkers (*Ng et al.*, 1999) have combined equations (1), (4), (5) and (6) to describe the upstream propagation of all ion species including the wave excitation essential to the turbulent sheath adjacent to the shock. Although this approach cannot describe the acceleration process, it does predict the upstream fractionation of different ion species; for the event of 20 April 1998 they find excellent agreement with observed abundance ratios (*Tylka et al.*, 1999). *Zank et al.* (2000) used a "hybrid" approach to calculate the proton time profiles expected in gradual events. They combined the shocked plasma flow from hydrodynamic numerical simulations, the upstream ion/wave configuration from *Gordon et al.* (1999) assuming a free-escape boundary at a prescribed position upstream of the shock, and a numerical calculation of the ion distribution downstream of the shock. In spite of this patchwork approach, the predicted time profiles are very good.

Lee (2005a) combined equations (4)-(6) with the 2-stream moments ($F_+ = \int_0^1 d\mu f$ and $F_- = \int_{-1}^0 d\mu f$) of equation (1) to accommodate large streaming anisotropy in the theory of diffusive shock acceleration combined with wave excitation. Although this model is effectively stationary, assumes a simple geometry, and neglects adiabatic deceleration and drift of ions, it is the first to describe analytically the extraction of ions from the turbulent sheath adjacent to the shock by magnetic focusing and the resulting cutoff in the power-law energy spectrum.

An interesting feature of this model is that it accounts qualitatively for the observed "streaming limit" (*Reames*, 1990; *Ng and Reames*, 1994) in gradual events. The escaping anisotropic protons are predicted to have an intensity with a "hard" energy spectrum, with a weak dependence on time prior to the onset of the ESP event associated with shock arrival, and independent of the proton injection rate at the shock. Using a different approach, *Vainio* (2003) also obtains an expression for the streaming limit.

The major weakness of the model appears to be the form of the wave intensity, which is $\propto k^{-3}$ throughout most of the turbulent sheath upstream of the shock. In addition to being in poor agreement with observations of wave spectra at several events (*Tsurutani et al.*, 1983; *Viñas et al.*, 1984), this form leads to extreme upstream fractionation between ion species with different A/Q, unreasonably hard energy spectra of the minor ions, and unrealistically "sharp" high-energy cutoffs for the minor ions. The k^{-3}-spectrum appears to arise from the approximation used in solving equation (6) that the growth rate is controlled by the lowest energy protons which can resonate with a particular wave.

Kota et al. (2003) have generalized equation (1) to describe transport within an arbitrary magnetic flux tube with variable $V(s)$ and $B(s)$, where s is arclength along the flux tube. In this approach the shock acceleration is included in the focused transport equation, as pioneered analytically by *Kirk and Schneider* (1987). For reasonable specified D_μ, numerical solution of the focused transport equation yields reasonable energetic ion distribution functions. *Ng and Reames* (2004) are taking a similar approach but including a generalized wave kinetic equation (5) to describe wave excitation by the energetic protons. Although these approaches must be modified to incorporate shock obliquity, they address all the key features of the shock acceleration and ion transport.

5. FUTURE PROSPECTS

The current models of SEP transport and acceleration in gradual events are able to account generally for the observed features of these events. However, there are a number of outstanding issues which must still be addressed:

(1) As mentioned in Section 2, the calculation of $I(k,z)$ requires care. Neglecting the ambient wave intensity, *Lee* (2005a) obtained $I \propto k^{-3}$ throughout most of the turbulent sheath, which suppresses large A/Q ion species relative to protons with increasing distance from the shock in contrast with observations (e.g. *Tylka et al.*, 1999). However, *Gordon et al.* (1999) noted that $I \sim k^{-2}$ for large k. *Lee* (2005b) found that I evolves in k-space from $I \propto k^{\beta-6}$ at small k, the form which obtains at the shock where β is the standard power-law index of the ion distributions at the shock [$\beta \equiv 3X(X-1)^{-1}$, where X is the shock compression ratio] (*Lee*, 2005a), to $I \propto k^{-2}$ at large k with a transition wavenumber which depends on z. This form of $I(k,z)$ with the addition of the ambient wave intensity needs to be incorporated into the ion transport and acceleration. The particular form, $I \propto k^{-2}$, is special since it leads to no ion fractionation according to A/Q.

(2) The current theories generally consider ion injection into the DSA process from the solar wind overtaken by the shock. However, recent composition measurements in gradual events imply that the ambient suprathermal/energetic ion population is a major seed population, which must be included (*Mason et al.*, 1999; *Desai et al.*, 2003; *Tylka et al.*, 2005). Within the theory of DSA this can be accomplished by including an omnidirectional distribution function $f_{i,\infty}(v)$ which is advected into the shock. In addition there still exists

the traditional problem of determining the fraction of each solar wind species which is injected at the shock. Observations imply that this fraction is $\sim 10^{-3} - 10^{-2}$ (*Gordon et al.*, 1999).

(3) The predicted dependence of the energetic ions and excited waves on θ, the upstream angle between **B** and the shock normal, is not well known for quasi-perpendicular shocks. The θ-dependence due to the streaming instability and $\kappa_\parallel$, as outlined in Section 3, is understood, but, as $\theta \to \pi/2$, $\kappa_\perp$ becomes important. Not only is $\kappa_\perp$, which in the solar wind is dominated by random walk of magnetic field lines, not well understood, but also the possible instabilities at a perpendicular shock due to the large ion spatial gradients perpendicular to **B** have not been explored. These may contribute to increasing $\kappa_\perp$. In addition, as discussed in issue (2), the injection rates out of the solar wind, and the speed threshold of $f_{i,\infty}(v)$ above which full participation in DSA can be assumed, clearly depend on θ. To make matters more complicated, a given event most probably involves a sequence of values of θ (*Tylka et al.*, 2005).

(4) One criticism often raised concerning the DSA origin of gradual/ESP events is the often large discrepancy between the observed power-law spectral index β_{ob} and that predicted by the observed compression ratio, β_{th}. This discrepancy was originally highlighted by *Van Nes et al.* (1984) in a statistical study of 75 ESP events. Except for the inherent difficulties in determining β_{ob} for the very irregular time profiles of energetic particles observed at quasi-perpendicular shocks, the agreement may not be so bad. For weak shocks the spectral index β_0 for the energetic ion seed population from previous gradual events or impulsive events can be smaller than β_{th}; in this case the shock essentially preserves β_0 in the acceleration process. For shocks for which the expected speed spectral index is β_{th}, it must be remembered that X actually refers to the average-wave-frame compression ratio, which is that which the particles sense. Accordingly the magnitude of the normal component of the upstream plasma velocity, V_u, is decreased by $V_A \cos\theta$ by virtue of the dominance of the proton-excited waves. Downstream of the shock, *Tan et al.* (1988) showed that the average wave frame at several ESP events propagates sunward relative to the solar wind at a speed $\sim V_A$. These modifications to V_u and V_d reduce X and increase the predicted value of β, particularly for the generally weak interplanetary traveling shocks relevant for ESP events. With these important corrections the agreement between β_{ob} and β_{th} should improve. Incidentally, an important finding of the Van *Nes et al.* study is that the peak proton flux decreases with increasing θ, an indication that the effective injection rate decreases with increasing θ.

(5) Ion transport downstream of the shock presents theoretical challenges, particularly late in an event. Early in an

event the downstream plasma is expected to be turbulent as upstream excited waves are transmitted downstream. A reasonable assumption is then that $\kappa = 0$ downstream. Later in the event, however, the turbulence decays as the wind expands and κ increases. The energetic ions then diffuse to nearly uniform intensity in the inner heliosphere, and decay as they cool adiabatically and diffuse to fill ever larger volumes. Treating the transition from scatter-dominated transport adjacent to the shock and/or at early times (required for shock acceleration) to nearly scatter-free transport far from the shock and/or late in the event is challenging. As a step in the right direction *Zank et al.* (2000) allow ions to diffuse from the downstream region back across the shock to the upstream region.

(6) A final challenge is to include time-dependence in the coupled solution of the ion acceleration/transport equations (1) or (2) and the wave kinetic equation (5). *Ng et al.* (2003) include time-dependence in their treatment of ion transport and wave excitation, but the acceleration is presumed to occur instantaneously. *Lee* (2005a) employs an adiabatic approximation which presumes that the acceleration time is small compared with the evolutionary timescale of the shock. Clearly these approaches are not valid close to the Sun where the evolutionary timescale is small or at high energies comparable to the cutoff energy for which the acceleration time is large.

Acknowledgments. The author wishes to thank the organizers of the Chapman Conference in Turku, Finland, for a very enjoyable informative conference. He also wishes to thank the AGU Monograph editors, Nat Gopalswamy and Dick Mewaldt, for their patience in awaiting this review paper, and two anonymous referees for valuable suggestions to improve the original version of this typescript. This work was supported, in part, by the NASA Sun-Earth Connection Theory Program grants NAG 5-11797 and NNG05GL40G, and by the DoD MURI grants to the University of Michigan and the University of California at Berkeley (subgrants to the University of New Hampshire).

REFERENCES

Achterberg, A., and C.A. Norman, Particle acceleration by shock waves in solar flares, *Astron. Astrophys.*, 89, 353, 1980.

Anttila, A., L.G. Kocharov, J. Torsti, and R. Vainio, Long-duration high-energy proton events observed by GOES in October 1989, *Ann. Geophysicae*, 16, 921-930, 1998.

Anttila, A., and T. Sahla, ERNE observations of energetic particles associated with Earth-directed coronal mass ejections in April and May, 1997, *Ann. Geophysicae*, 18, 1373-1381, 2000.

Axford, W.I., E. Leer, and G. Skadron, The acceleration of cosmic rays by shock waves, *Proc. Int. Conf. Cosmic Rays 15th*, 11, 132-137, 1978.

Bell, A.R., The acceleration of cosmic rays in shock fronts, *Mon. Not. R. astr. Soc.*, 182, 147, 1978.

Blandford, R.D., and J.P. Ostriker, Particle acceleration by astrophysical shocks, *Astrophys. J.*, 221, L29-L32, 1978.

Davis, L., Jr., Modified Fermi mechanism for the acceleration of cosmic rays, *Phys. Rev.*, 101, 351, 1956.

Desai, M.I., G.M. Mason, J.R. Dwyer, J.E. Mazur, R.E. Gold, S.M. Krimigis, C.W. Smith, and R.M. Skoug, Evidence for a suprathermal seed population of heavy ions accelerated by interplanetary shocks near 1 AU, *Astrophys. J.*, 588, 1149-1162, 2003.

Dwyer, J.R., G.M. Mason, J.E. Mazur, J.R. Jokipii, T.T. von Rosenvinge, R.P. Lepping, Perpendicular transport of low-energy corotating interaction region-associated nuclei, *Astrophys. J.*, 490, L115, 1997.

Earl, J.A., The diffusive idealization of charged particle transport in random magnetic fields, *Astrophys. J.*, 193, 231, 1974.

Earl, J.A., The effect of adiabatic focusing upon charged-particle propagation in random magnetic fields, *Astrophys. J.*, 205, 900, 1976.

Earl, J.A., Analytical description of charged particle transport along arbitrary guiding field configurations, *Astrophys. J.*, 251, 739-755, 1981.

Emslie, A.G., J.A. Miller, and J.C. Brown, An explanation for the different locations of electron and ion acceleration in solar flares, *Astrophys. J.*, 602, L69-L72, 2004.

Forman, M.A., Acceleration theory for 5-40 keV ions at interplanetary shocks, *Adv. Space Res.*, 1 (3), 97, 1981.

Gleeson, L.J., and W.I. Axford, Cosmic rays in the interplanetary medium, *Astrophys. J.*, 149, L115, 1967.

Gold, T., Plasma and magnetic fields in the solar system, *J. Geophys. Res.*, 64, 1665-1674, 1959.

Gordon, B.E., M.A. Lee, E. Möbius, and K.H. Trattner, Coupled hydromagnetic wave excitation and ion acceleration at interplanetary traveling shocks and Earth's bow shock revisited, *J. Geophys. Res.*, 104, 28,263-28,277, 1999.

Heras, A.M., B. Sanahuja, Z.K. Smith, T. Detman, and M. Dryer, The influence of the large-scale interplanetary shock structure on a low-energy particle event, *Astrophys. J.*, 391, 359-369, 1992.

Heras, A.M., B. Sanahuja, D. Lario, Z.K. Smith, T. Detman, and M. Dryer, Three low-energy particle events: modeling the influence of the parent interplanetary shock, *Astrophys. J.*, 445, 497-508, 1995.

Hudson, P.D., Reflection of charged particles by plasma shocks, *Mon. Not. R. Astr. Soc.*, 131, 23-49, 1965.

Isenberg, P.A., A hemispherical model of anisotropic interstellar pickup ions, *J. Geophys. Res.*, 102, 4719-4724, 1997.

Jokipii, J.R., and E.H. Levy, Effects of particle drifts on the solar modulation of galactic cosmic rays, *Astrophys. J.*, 213, L85-L88, 1977.

Jokipii, J.R., Particle drift, diffusion, and acceleration at shocks, *Astrophys. J.*, 255, 716-720, 1982.

Kallenrode, M.B., Particle propagation in the inner heliosphere, *J. Geophys. Res.*, 98, 19,037-19,047, 1993.

Kallenrode, M.-B., and G. Wibberenz, Propagation of particles injected from interplanetary shocks: A black box model and its consequences for acceleration theory and data interpretation, *J. Geophys. Res.*, 102, 22,311-22,334, 1997.

Kirk, J.G., and P. Schneider, On the acceleration of charged particles at relativistic shock fronts, *Astrophys. J.*, 315, 425-433, 1987.

Kocharov, L., J. Torsti, T. Laitinen, and M. Teittinen, Post-impulsive-phase acceleration in a wide range of solar longitudes, *Solar Phys.*, 190, 295-307, 1999.

Kocharov, L., M. Lytova, R. Vainio, T. Laitinen, and J. Torsti, Modeling the shock aftermath source of energetic particles in the solar corona, *Astrophys. J.*, 620, 1052-1068, 2005.

Kota, J., W.B. Manchester, J.R. Jokipii, D.L. de Zeeuw, and T.I. Gombosi, Acceleration and transport of solar energetic particles: modeling CME-driven shocks, *Proc. 28th Int. Cosmic-Ray Conf. (Tsukuba)*, 6, 3529, 2003.

Krymsky, G.F., A regular mechanism for the acceleration of charged particles on the front of a shock wave, *Dokl. Akad. Nauk SSSR*, 234, 1306, 1977.

Laitinen, T., K.-L. Klein, L. Kocharov, J. Torsti, G. Trottet, V. Bothmer, M.L. Kaiser, G. Rank, and M.J. Reiner, Solar energetic particle event and radio bursts associated with the 1996 July 9 flare and coronal mass ejection, *Astron. Astrophys.*, 360, 729-741, 2000.

Lario, D., B. Sanahuja, and A.M. Heras, Energetic particle events: Efficiency of interplanetary shocks as 50 keV < E < 100 MeV proton accelerators, *Astrophys. J.*, 509, 415-434, 1998.

Lee, M.A., and L.A. Fisk, Shock acceleration of energetic particles in the heliosphere, *Space Sci. Rev.*, 32, 205-228, 1982.

Lee, M.A., Coupled hydromagnetic wave excitation and ion acceleration at interplanetary traveling shocks, *J. Geophys. Res.*, 88, 6109-6119, 1983.

Lee, M.A., and J.M. Ryan, Time-dependent coronal shock acceleration of energetic solar flare particles, *Astrophys. J.*, 303, 829-842, 1986.

Lee, M.A., Coupled hydromagnetic wave excitation and ion acceleration at an evolving coronal/interplanetary shock, *Astrophys. J. Suppl.*, 158, 38-67, 2005a.

Lee, M.A., Generation of turbulence at shocks, in *The Physics of Collisionless Shocks*, edited by G. Li, G.P. Zank, and C.T. Russell, AIP Press, Melville, NY, pp. 240, 2005b.

Litvinenko, Y.E., Particle acceleration in reconnecting current sheets with a nonzero magnetic field, *Astrophys. J.*, 462, 997-1004, 1996.

Lytova, M., and L. Kocharov, Charge states of energetic solar ions from coronal shock acceleration, *Astrophys. J.*, 620, L55-L58, 2005.

Mason, G.M., J.E. Mazur, and J.R. Dwyer, He enhancements in large solar energetic particle events, *Astrophys. J.*, 525, L133-L136, 1999.

Miller, J.A., and D.A. Roberts, Stochastic proton acceleration by cascading Alfvén waves in impulsive solar flares, *Astrophys. J.*, 452, 912-932, 1995.

Ng, C.K., and D.V. Reames, Focused interplanetary transport of ~1 MeV solar energetic protons through self-generated Alfven waves, *Astrophys. J.*, 424, 1032-1048, 1994.

Ng, C.K., D.V. Reames, and A.J. Tylka, Effect of proton-amplified waves on the evolution of solar energetic particle composition in gradual events, *Geophys. Res. Lett.*, 26, 2145-2148, 1999.

Ng, C.K., D.V. Reames, and A.J. Tylka, Modeling shock-accelerated solar energetic particles coupled to interplanetary Alfvén waves, *Astrophys. J.*, 591, 461-485, 2003.

Ng, C.K., and D.V. Reames, Acceleration and transport of shock-accelerated energetic ions, *Eos*, 85(17), JA380, 2004.

Parker, E.N., and D.A. Tidman, Suprathermal particles, *Phys. Rev.*, 111, 1206, 1958.

Parker, E.N., The passage of energetic charged particles through interplanetary space, *Planet. Space Sci.*, 13, 9, 1965.

Reames, D.V., Acceleration of energetic particles by shock waves from large solar flares, *Astrophys. J.*, 358, L63-L67, 1990.

Reames, D.V., L.M. Barbier, and C.K. Ng, The spatial distribution of particles accelerated by coronal mass ejection-driven shocks, *Astrophys. J.*, 466, 473-486, 1996.

Reames, D.V., Particle acceleration at the Sun and in the heliosphere, *Space Sci. Rev.*, 90, 413-491, 1999.

Roelof, E.C., Propagation of solar cosmic rays in the interplanetary magnetic field, in *Lectures in High Energy Astrophysics*, edited by H. Ogelmann and J.R. Wayland, NASA Spec. Publ. SP-199, p. 111, 1969.

Ruffolo, D., Effect of adiabatic deceleration on the focused transport of solar cosmic rays, *Astrophys. J.*, 442, 861-874, 1995.

Sarris, E.T., and J.A. Van Allen, Effects of interplanetary shock waves on energetic charged particles, *J. Geophys. Res.*, 79, 4157-4173, 1974.

Schlickeiser, R., A. Campeanu, and I. Lerche, Stochastic particle acceleration at parallel astrophysical shock waves, *Astron. Astrophys.*, 276, 614, 1993.

Skilling, J., Cosmic rays in the Galaxy: convection or diffusion?, *Astrophys. J.*, 170, 265, 1971.

Tan, L.C., G.M. Mason, G. Gloeckler, and F.M. Ipavich, Downstream energetic proton and alpha particles during quasi-parallel interplanetary shock events, *J. Geophys. Res.*, 93, 7225-7243, 1988.

Torsti, J., L.G. Kocharov, R. Vainio, A. Anttila, and G.A. Kovaltsov, The 1990 May 24 solar cosmic-ray event, *Solar Phys.*, 166, 135-158, 1996.

Tsuneta, S., and T. Naito, Fermi acceleration at the fast shock in a solar flare and the impulsive loop-top hard X-ray source, *Astrophys. J.*, 495, L67-L70, 1998.

Tsurutani, B.T., E.J. Smith, and D.E. Jones, Waves observed upstream of interplanetary shocks, *J. Geophys. Res.*, 88, 5645-5656, 1983.

Tylka, A.J., D.V. Reames, and C.K. Ng, Observations of systematic temporal evolution in elemental composition during gradual solar energetic particle events, *Geophys. Res. Lett.*, 26, 2141-2144, 1999.

Tylka, A.J., C.M.S. Cohen, W.F. Dietrich, M.A. Lee, C.G. Maclennan, R.A. Mewaldt, C.K. Ng, and D.V. Reames, Shock geometry, seed populations, and the origin of variable elemental composition at high energies in large gradual solar particle events, *Astrophys. J.*, 625, 474-495, 2005.

Vainio, R., On the generation of Alfvén waves by solar energetic particles, *Astron. Astrophys.*, 406, 735-740, 2003.

Vainio, R., L. Kocharov, and T. Laitinen, Interplanetary and interacting protons accelerated in a parallel shock wave, *Astrophys. J.*, 528, 1015-1025, 2000.

Van Nes, P., R. Reinhard, T.R. Sanderson, K.-P. Wenzel, and R.D. Zwickl, The energy spectrum of 35- to 1600-keV protons associated with interplanetary shocks, *J. Geophys. Res.*, 89, 2122-2132, 1984.

Viñas, A.F., M.L. Goldstein, and M.H. Acuña, Spectral analysis of magnetohydrodynamic fluctuations near interplanetary shocks, *J. Geophys. Res.*, 89, 3762-3774, 1984.

Zank, G.P., W.K.M. Rice, and C.C. Wu, Particle acceleration and coronal mass ejection driven shocks: A theoretical model, *J. Geophys. Res.*, 105, 25,079-25,095, 2000.

Acceleration of SEPs: Role of CME-Associated Shocks and Turbulence

R. Vainio

Department of Physical Sciences, University of Helsinki, Finland

We review recent efforts to understand the connection between coronal mass ejections, their associated shock waves and solar energetic particle (SEP) events, and the underlying plasma turbulence that presumably enables the collisionless SEPs to feel the compressions of the shocked bulk flow and to stay in the vicinity of the shocks long enough to be accelerated to high energies. By combining several recent theoretical results, we discuss possible scenarios to explain observations of CME- and shock-associated SEP events. It is emphasized that while self-generated turbulence has to be an important ingredient of the transport and acceleration models for large SEP events, smaller events can not rely on this scenario to provide the turbulent conditions necessary for the diffusive shock acceleration mechanism. Thus, test-particle modeling still plays an important role towards our understanding of diffusive SEP acceleration in CME-associated coronal shock waves.

1. INTRODUCTION

Coronal mass ejections (CMEs) [*Tousey*, 1973, *Gosling et al.*, 1974], together with solar flares [*Carrington*, 1859, *Hodgson*, 1859], are the main sources of solar energetic particle (SEP) events [*Forbush et al.*, 1949]. CME-related SEP events [*Kahler et al.*, 1978] are currently believed to be accelerated in coronal shock waves associated with the mass motions of the CMEs [*Reames*, 1999, gives a recent review]. The most popular acceleration mechanism considered in connection with shocks is diffusive shock acceleration (DSA) [*Axford et al.*, 1977, *Krymsky*, 1977, *Bell*, 1978, *Blandford and Ostriker*, 1977]. In this mechanism, particles gain energy by crossing the compressive shock front many times. This requires efficient particle scattering off magnetic irregularities in the vicinity of the shock front. Thus, in the heart of the connection between CMEs, shocks and SEPs lies coronal hydromagnetic turbulence that can couple the collisionless SEPs to the background bulk plasma and let them be energized at the expense of the kinetic energy of the compressed flow ahead of the ejecta.

Propagating coronal shock waves have been studied by indirect observations for decades. Type II radio bursts [*Wild et al.*, 1963] are believed to be emission from the shock front at the local plasma frequency and its first harmonic. Using coronal electron density models their drifting frequencies can, therefore, be translated to radial distances of the shock front. The bursts start from metric wave lengths corresponding to low-coronal densities and continue up to dekametric [*Wild et al.*, 1963] and sometimes even to kilometric wavelengths [*Cane et al.*, 1982], the latter corresponding to interplanetary shocks driven by the associated coronal mass ejections.

Metric type II radio bursts may also be associated with Moreton waves [*Moreton*, 1960], which are chromospheric wave fronts following flares and propagating away from the associated flare site at about 1000 km s^{-1}. They have been identified as chromospheric footprints of coronal hydromagnetic disturbances, i.e., shock waves [*Uchida*, 1968]. EIT waves [*Thompson et al.*, 1999] are similar disturbances in the low corona, associated with flares and/or CMEs. Both Moreton waves and EIT waves are also associated with SEP events. Finally, the first direct imaging observations of coronal shock waves associated with flares and CMEs have

Solar Eruptions and Energetic Particles
Geophysical Monograph Series 165
10.1029/165GM24

recently been obtained in soft X-rays [*Hudson et al.*, 2003] and in white light [*Sheeley et al.*, 2000, *Vourlidas et al.*, 2003], respectively.

The observed speeds of coronal mass ejections vary from less than 100 to more than 2000 km s^{-1}. The values correspond to the plane-of-sky speed of the leading edge of the visible CME. The inferred speeds of the type II radio bursts are in the range from a few hundred to a couple of thousand kilometers per second. Coronal mass ejections related to SEP events are typically faster than about 500 km s^{-1}, although examples of slower CMEs related to SEP events exist [*Kocharov et al.*, 2001]. CMEs can be divided into two classes, impulsive and gradual, according to their acceleration [*Sheeley et al.*, 1999] and those associated with SEP events usually belong to the impulsive class [*Kocharov et al.*, 2001]. Impulsive CMEs are typically associated with flares and Moreton waves [*Sheeley et al.*, 1999] and SEP-associated impulsive CMEs with (metric) type II radio bursts, as well [*Kocharov et al.*, 2001].

The sound speed in the corona is usually well below 200 km s^{-1}, so most CMEs and practically all SEP-producing CMEs are supersonic. The Alfvén speed, v_A, in the quiet regions is about 200–300 km s^{-1} at the coronal base and increases to a local maximum of about 600–800 km s^{-1} at a few solar radii above the surface [*Mann et al.*, 2003]. After this maximum, the Alfvén speed again decreases outwards. In coronal holes [*Laitinen et al.*, 2003] and active regions [*Mann et al.*, 2003] the Alfvén speed may increase to several thousand km s^{-1}. As the shock strength is mostly determined by its Mach number, fast-mode shock waves in these regions should not be very strong. Conversely in the dense coronal streamers v_A should be smaller than the values in the quiet corona, especially near the streamer cusps and/or current sheets where the magnetic field is also low. Thus, shocks with largest compression ratios in corona should be found in such regions. The typical fast-magnetosonic Mach numbers of CME-related shock waves vary from just above unity for low-to-intermediate speed CMEs propagating through the quiet solar corona to about ten for fast CMEs propagating through regions of low magnetic fields.

Coronal shock waves related to CME mass motions may either be driven waves, e.g., ahead of a super-Alfvénic CME, or freely propagating, i.e., refracting, waves. The latter are typically regarded as flare blast waves [*Uchida*, 1968], but they may result from mass motions, as well. Note that such mass motions do not have to be supermagnetosonic to create shocks: submagnetosonic mass motions form large-amplitude compressive MHD waves, which propagate away from the disturbance and subsequently steepen to form shocks. As soon as the dimensions of the propagating wave are larger than the source dimensions, the wave will appear as a freely propagating shock, like in the case of blast waves.

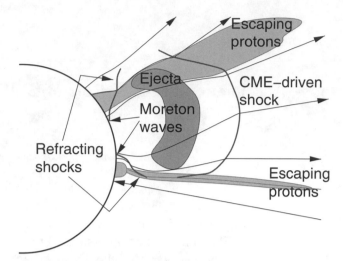

Figure 1. Some plausible locations of shocks connected to CMEs. Bow shocks driven by CMEs as well as freely propagating, refracting shocks initiated by a thermal energy release (flare) or the initial rapid transverse expansion of the ejecta are indicated. Both kinds of shock waves will strengthen in regions of low coronal Alfvén speed, such as streamers and current sheets.

An example of a piston capable of producing a refracting shock is provided by the initial lateral expansion of CMEs that tends to slow down very low in the corona. Fig. 1 provides a cartoon of different kinds of coronal shocks and their environments.

The direction of propagation of the coronal shock waves with respect to the magnetic field also varies. While the CME-bow shock in the outer corona should be a quasi-parallel shock, its flanks may have an oblique geometry, especially if the flanks interact with denser material with slower signal speeds (see Fig. 1). The propagation direction in the low corona is much more variable: the bow shock in the low corona may encounter horizontal fields and be quasi-perpendicular, but the refracting shocks in the low corona, believed to be associated with the Moreton waves, may even have reverse geometry: the point of intersection of the magnetic field lines and the shock front may also move inwards (Fig. 1). This results in some special effects for DSA in these shocks [*Vainio and Khan*, 2004].

In the following, we will review the basic theory of particle transport in the corona and solar wind (Section 2.1), some recent results on turbulence transport in the expanding plasma flow (Section 2.2), the theory particle acceleration at coronal shocks (Sections 3.1–3.2), and some recent modeling efforts (Section 3.3). We emphasize the possibility to combine observations and modeling from the different communities involved to learn more about the subjects. Finally, a brief summary of the paper is provided (Section 4).

2. PARTICLE TRANSPORT AND TURBULENCE EVOLUTION

2.1. Energetic Particle Transport in Solar-Wind Turbulence

The magnetic fluctuations of the solar wind cause particle diffusion in phase space. First of all, particles are diffusing in pitch angle, which causes the particle distributions to spread diffusively along the mean magnetic field [*Jokipii*, 1966, *Earl*, 1974]. The mean free path of this collision-like process is denoted by $\lambda_\parallel$, called the parallel mean free path, and related to a diffusion coefficient parallel to the mean field by $\kappa_\parallel = \frac{1}{3} v \lambda_\parallel$, where v is the particle speed. Another diffusion process is related to the spreading of the particle distribution in the direction perpendicular to the mean magnetic field [*Jokipii*, 1966, *Forman et al.*, 1974]. The resulting perpendicular diffusion coefficient $\kappa_\perp$ of the energetic particles is related to (i) random walk of the field lines themselves and (ii) the particles' ability to cross field lines. In this paper, we will concentrate on the efforts of modeling the parallel mean free path in the inner heliosphere and on its effects on particle acceleration in CMEs.

The mean free path $\lambda_\parallel$ of energetic particles in turbulent magnetic fields is defined by [*Jokipii*, 1966, *Earl*, 1974]

$$\lambda_\parallel(p) := \frac{3v}{8} \int_{-1}^{+1} \frac{(1-\mu^2)^2}{D_{\mu\mu}(\mu, p)} d\mu, \qquad (1)$$

where μ and p are the pitch-angle cosine and momentum of the particle, and $D_{\mu\mu}$ is the pitch-angle diffusion coefficient due to scattering off magnetic irregularities. The latter is determined by the properties of turbulent fluctuations of the medium.

Quasilinear theory can be used to calculate $D_{\mu\mu}$ from the intensity $I(\omega, k)$ of magnetic-field fluctuations in the solar wind [*Jokipii*, 1966, *Schlickeiser*, 1989]. Here, ω and k are the frequency and wavenumber of the fluctuations as measured in the solar wind frame. This theory results in a resonant interaction, where fluctuations with

$$\omega - k_\parallel v \mu = n\Omega \qquad (2)$$

contribute to the scattering. Here, $\Omega = qB/\gamma mc$ is the relativistic gyrofrequency of the energetic particle, q and γ are its charge and Lorentz factor, and n is an integer number. Fluctuation geometry (i.e., direction of wave vector k with respect to magnetic field B) affects the scattering rates. Fluctuations propagating parallel to the mean field $(k \parallel B)$ are most effective in scattering energetic particles, and those propagating perpendicular to it have very little contribution to pitch-angle scattering.

Observations imply that the solar-wind turbulence could be represented with two components [*Bieber et al.*, 1996]:

SLAB: ~20% of slab-mode fluctuations with $(k \parallel B)$ and $\delta B_{SLAB} \perp B$;
2D: ~80% of two-dimensional (2D) fluctuations with $k \perp B \perp \delta B_{2D}$ and $\delta B_{2D} \perp k$.

In this representation, the contribution of 2D-fluctuations to the scattering can be neglected, and the slab-component fully determines the pitch-angle diffusion coefficient. We, thus, get the classical expression

$$D_{\mu\mu} = \frac{1}{2}(1-\mu^2)\pi\Omega\frac{|k_\parallel| I_{SLAB}(k_\parallel)}{B^2}; \quad k_\parallel = -\frac{\Omega}{v\mu}, \qquad (3)$$

assuming that the wave frequencies are negligible compared to $k_\parallel v\mu$. This assumption breaks down close to $\mu = 0$, where the magnetostatic approximation ($\omega = 0$) yields $D_{\mu\mu} \propto |\mu|^{q-1}$ and, thus, infinite mean free paths for fluctuation spectra $I_{SLAB} \propto k_\parallel^{-q}$ steeper than $q = 2$. In the inertial range of solar-wind turbulence the spectral index is below 2, but at scales smaller than the ion inertial length the power spectrum typically steepens to $q > 3$. Taking account of the finite complex wave frequencies, however, produces resonance broadening giving finite scattering rates also near $\mu = 0$ [*Schlickeiser and Achatz*, 1993]. Also other effects such as turbulence dynamics [*Bieber et al.*, 1996] produce resonance broadening to give finite values of the mean free path. At high particle rigidities, i.e., $R \equiv pc/q \gtrsim 100$ MV in the solar wind, resonance broadening effects can be simulated by assuming static fluctuations with the inertial-range spectrum $I_{SLAB} \propto k_\parallel^{-q}$ ($q < 2$) extending to infinity [*Dröge*, 2003].

Solar energetic particle observations imply that $\lambda_\parallel > 0.1$ AU for protons above 1 MeV energies at 1 AU. Typically, the mean free path scales with the rigidity as $\lambda_\parallel \propto R^{2-q}$ with $q \sim 1.6$–1.8 for $R \gtrsim 100$ MV [*Dröge*, 2000]. Although the rigidity dependence of the mean free path seems rather stable, there is a lot of scatter in the magnitude of $\lambda_\parallel$ from event to event, but it seems that this can be explained by the variable scattering conditions in the solar wind [*Dröge*, 2003].

2.2. Slab-mode Turbulence Evolution in the Solar Wind

Slab-turbulence in the solar wind is often modeled as Alfvén waves. The power spectrum $P_+(r, f)$ of outward-propagating Alfvén waves on a steady-state background wind evolves as [*Tu et al.*, 1984]

$$\frac{\partial P_+}{\partial t} + \frac{1}{A}\frac{\partial}{\partial r}\left(A\frac{(u+v_A)^2}{v_A}P_+\right) + \frac{\partial F_+}{\partial f} = -\gamma P_+, \qquad (4)$$

where $A \propto 1/B_r$ is the flux-tube cross sectional area, $f = (u+v_A) k_r / 2\pi$ is the wave frequency in the inertial frame, u is the solar wind speed, $v_A = B_r / \sqrt{4\pi\rho}$ is the radial component of the Alfvén speed, $F_+(f, r)$ is the flux of Alfvén waves in the frequency space, related to non-linear wave-wave interactions, and γ is the damping rate of the Alfvén waves, which becomes large close to the local ion-cyclotron frequency. WKB-wave transport is recovered by setting $F = 0$. Motivated by the observations with $q \sim 5/3$ in the solar wind, the flux function is often taken to be consistent with the Kolmogorov phenomenology of turbulence, i.e., [Tu, 1987, 1988]

$$F_+ = 2\pi C^2 \frac{v_A}{u+v_A} \frac{f^{5/2} P_+^{3/2}}{B}, \qquad (5)$$

where $C^2 \propto \sqrt{P_-(f')/P_+(f)}$ is a dimensionless parameter with $P_-(f)$ being the power spectrum of inward-propagating Alfvén waves evaluated at $f' = |u - v_A| f/(u + v_A)$, because the wave-wave interactions occur at equal wavelengths of the interacting waves. Thus, a self-consistent description of the solar-wind Alfvén waves requires another equation for the inward waves, and the coupled set of non-linear partial differential equations is not analytically solvable. Qualitative modeling is, however, possible by regarding C^2 as a model parameter, as described below.

Coronal heating on open field lines by damping of high-frequency Alfvén–ion-cyclotron waves has been modeled by employing a fixed functional form of $C^2(r)$ and solving the wave-transport equation together with the hydrodynamic equations describing the solar-wind expansion [Tu, 1987, Hu et al., 1999]. We have derived an approximate analytical expression for the power spectrum in a given solar-wind stream, which is valid for a power-law wave spectrum emitted from the solar surface [Vainio et al., 2003]. It enables the evaluation of the SEP mean free path as a function of radial distance from the Sun. If the emitted spectrum is of the form

$$P_+(r_\odot, f) = \varepsilon_P \frac{B_\odot^2}{f}, \quad f_0 < f < f_1 \qquad (6)$$

the solar-wind spectrum can be approximated as

$$P_+(r, f) = \frac{P_{WKB}(r, f)}{1 + [f/f_c(r)]^{2/3}}, \quad f > f_0 \qquad (7)$$

where

$$P_{WKB}(r, f) = \varepsilon_P \frac{B_\odot^2}{f} \frac{M_{A\odot}(M_{A\odot}+1)^2}{M_A(M_A+1)^2} \qquad (8)$$

is the WKB behavior of the Alfvén waves, $M_A(r) = u/v_A$ is the local Alfvénic Mach number of the flow, subscript '$\odot$' denotes values at the solar surface $r = r_\odot$, and

$$\frac{1}{f_c} = 2\pi \varepsilon_p^{1/2} \int_{r_\odot}^{r} C^2 \frac{v_{A\odot}}{v_A} \frac{(M_{A\odot}+1)}{[M_A+1]^3} \frac{M_A^{1/2}}{M_{A\odot}^{1/2}} \frac{dr}{v_A} \qquad (9)$$

is the break-point time scale, where the spectral form of the slab-mode fluctuations changes from f^{-1} to $f^{-5/3}$ behavior. This broken power-law form corresponds to the observations of the Alfvén-wave power spectrum in the solar wind in a wide range of radial distances and frequencies [Horbury et al., 1996]. By adopting a spatial dependence for $C^2(r)$, we can compute the spectrum of Alfvén waves throughout the considered solar-wind stream [Vainio et al., 2003].

Instead of choosing a spatial dependence of C^2, we will here illustrate the effects of the turbulent cascade on the particle mean free path by considering a few functional forms for $f_c(r)$ explicitly. Neglecting the (presumably small) difference between the radial and the parallel direction, we can write $fP(r, f) = |k_\parallel| [I(k_\parallel, r) + I(-k_\parallel, r)]$. Assuming a vanishing net polarization, the spectral form (7) yields a mean free path of [Vainio et al., 2003]

$$\lambda_\parallel = \frac{2r_L}{\pi} \left[1 + \frac{27}{7} \left(\frac{u + v_A}{2\pi f_c r_L} \right)^{2/3} \right] \left(\frac{B}{\delta B_{WKB}} \right)^2, \qquad (10)$$

where $r_L = R/B$ is the Larmor radius of the particle and $\delta B^2_{WKB} = fP_{WKB}$.

Fig. 2 gives $\lambda_\parallel$ for four *ad-hoc* models:
MODEL 1: $f_c = f_{c0} \equiv 10^{-5}$ Hz.
MODEL 2: $f_c = f_{c1}(r) \equiv f_{c0}[\mathrm{AU}/(r-r_\odot)]$.
MODEL 3: $f_c = f_{c1}(r)[1 + (10r_\odot/r)^5]$
MODEL 4: $f_c = \infty$
$\varepsilon_P = 0.01$ is used in each model, which produces a reasonable mean free path at 1 AU for models 1–3. The obtained spectral shape in models 1–3 is consistent also with the low-frequency ($f \sim 10^{-4} - 0.1$ Hz) magnetic-field measurements at 1 AU. Note that model 1 corresponds to WKB-transport of a broken power-law spectrum and model 4 to WKB-transport of an f^{-1} spectrum. The latter is clearly inconsistent with values of $\lambda_\parallel$ at 1 AU.

Models 1–3 are all consistent with observed values of $\lambda_\parallel$ at 1 AU. Models 2–3 are most consistent with the observed spatial dependence of $\lambda_\parallel$ between 0.3 AU and 1 AU, corresponding to an approximately spatially constant value of the radial mean free path $\lambda_{rr} = \lambda_\parallel \cos^2\psi$ [Kallenrode, 1993], where ψ is the spiral angle of the Parker field. Since we do not have *in-situ* SEP measurements below $65r_\odot$, we can not say much about the mean free path in solar corona below, say, $10r_\odot$. We can try to constrain this parameter by the inferred delay

3. DIFFUSIVE PARTICLE ACCELERATION IN CME-DRIVEN SHOCK WAVES

3.1. Constraints from Turbulence for Test-Particle Acceleration in Coronal Shocks

In a turbulent medium, SEP transport becomes diffusive if pitch-angle scattering is strong enough to keep the particle distribution quasi-isotropic. Consider particle transport in the vicinity of a planar shock, propagating at an angle of θ_n with respect to the upstream magnetic field, $\boldsymbol{B}_1 = B_1 \boldsymbol{b}_1$. Neglecting perpendicular diffusion the diffusion coefficient in the direction of shock normal $\boldsymbol{e}_x$ is given by

$$\kappa = \frac{\lambda_\parallel v \cos^2\theta_n}{3}, \tag{11}$$

where $\cos\theta_n = |\boldsymbol{e}_x \cdot \boldsymbol{b}_1|$. If the turbulent upstream region extends to infinity, no particles will escape ahead of the shock. Thus, assume that there is a free-escape boundary at a fixed distance, L_1, ahead of the shock. In a steady state, the particle distribution is given by [*Vainio et al.*, 2000]

$$f(x,p) = \begin{cases} f_0(p)\dfrac{g(x,p)-g_1(p)}{1-g_1(p)}, & -L_1 < x < 0 \\ f_0(p), & x > 0 \end{cases}$$

$$f_0(p) = \frac{\sigma Q_0}{4\pi u_{1x} p_0^3}\left(\frac{p_0}{p}\right)^\sigma \exp\left\{-\sigma\int_{p_0}^p \frac{dp'}{e^{\eta_1(p')}-1}\right\}$$

$$\sigma = \frac{3r_{sc}}{r_{sc}-1}; \quad r_{sc} = \frac{u_{1x}}{u_{2x}} \tag{12}$$

$$g(x,p) = e^{-\eta(x,p)}; \quad g_1(p) = g(-L_1,p)$$

$$\eta(x,p) \equiv \int_x^0 \frac{u_{1x}dx'}{\kappa(x',p)},$$

where Q_0 is the injection rate of particles with $p = p_0$ to the acceleration process, $u_{1x[2x]}$ is the upstream [downstream] speed of the scattering centers in the shock-normal direction, and r_{sc} is the scattering center compression ratio. Standard results for an infinite upstream region are recovered by setting $L_1 \to \infty$.

The flux of particles escaping upstream is

$$\mathcal{F} = 4\pi p^2 \kappa \left.\frac{\partial f}{\partial x}\right|_{x=-L_1} = \frac{4\pi u_{1x}p^2 f_0(p)}{e^{\eta(p)}-1} \tag{13}$$

The observed SEP energy spectra are typically power-laws with an exponential cut-off at high energies. Thus, SEP events require a large and weakly p-dependent η_1 if they are

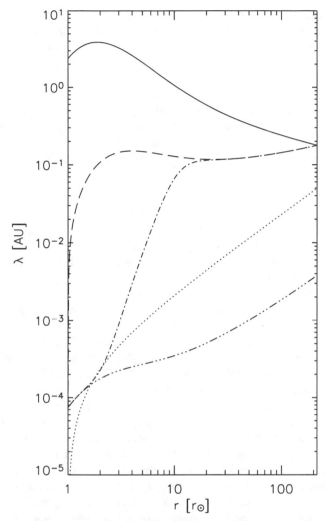

Figure 2. Mean free path of 10-MeV protons in the solar wind. The curves give $\lambda_\parallel$ for turbulence models 1 (solid), 2 (dashed), 3 (dot-dashed), and 4 (triple-dot-dashed), and the value of $\lambda_\parallel$ required for the production of 10-MeV protons by DSA ($u_{1x} = 1000$ km s^{-1}, $\theta_n = 70°$, and $r_{sc} = 4$) (dotted curve).

times of the particles with respect to their release at the Sun [*Torsti et al.*, 1998, *Krucker and Lin*, 2000]. Models 1–3 pass these tests with flying colors, as well [*Lintunen and Vainio*, 2004]. Thus, SEP transport does not constrain the coronal values of $\lambda_\parallel$ too severely.

We can also try to use turbulence measurements to deduce, how f_c evolves. Observations in fast solar wind streams, where the turbulence is less evolved and the spatial dependence of f_c is possible to deduce, suggest that there is an active cascade operating in the solar wind producing a $f_c \propto r^{-1}$ dependence [*Horbury et al.*, 1996] favoring, again, models 2–3 over models 1 and 4.

to be modeled with steady-state DSA in an outward-propagating shock. As external turbulence models, like the ones reviewed above, tend to produce very momentum dependent values of η_1, steady-state DSA can be used probably only in connection with self-generated turbulence in large SEP events (see below). However, as the onset of SEP events is usually observed to have some delay with respect to the electromagnetic emissions from the Sun [*Cliver et al.*, 1982, *Kahler et al.*, 2003], particle acceleration and release may actually be consecutive processes. We will come back to this possibility, as well.

Even if $\eta_1 \gg 1$ at all momenta, the power law spectrum is cut off at some high momentum p_c limited by the acceleration time:

$$\tau \equiv \frac{p_c}{\dot{p}_c} = \frac{3r_{sc}}{r_{sc}-1}\frac{\kappa(p_c)}{u_{1x}^2} = \frac{\sigma\kappa(p_c)}{u_{1x}^2} \sim \frac{\Delta x}{u_{1x}}, \qquad (14)$$

where the last equation results from equating the acceleration time scale with the dynamic time scale of the shock. Here we have also neglected a similar contribution to τ from downstream [*Drury*, 1983], assuming that the diffusion coefficient there is much smaller than in the ambient medium. The more oblique the quicker is the shock in accelerating energetic particles, since $\dot{p} \propto 1/\cos^2\theta_n$. If perpendicular transport can be neglected, we can easily estimate the available acceleration time (see Fig. 3) and calculate the required parallel mean free path to accelerate the particles to $p = p_c$ in a given time:

$$\frac{p_c}{\dot{p}_c} \sim \frac{\Delta s\cos\theta_n}{u_{1x}} \Rightarrow \qquad (15)$$

$$\lambda_\parallel(p_c) \sim \frac{3u_{1x}\Delta s}{\sigma v_c\cos\theta_n} = \frac{u_3}{\sigma\beta_c\cos\theta_n}\frac{\Delta s}{100} \qquad (16)$$

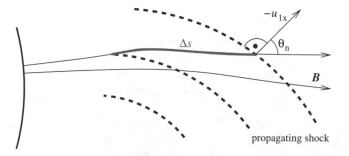

Figure 3. Propagation of an oblique coronal shock. If particles can not cross field lines, the acceleration time can be estimated as $\tau \sim \Delta s \cos \theta_n/u_{1x}$. The shock front in this sketch is drawn at three successive times and the magnetic field lines are drawn before they are intersected by the shock front; the tangential magnetic field component is expected to increase behind the shock front.

where $u_3 = u_{1x}/(10^3 \text{ km s}^{-1})$ and $\beta_c = v_c/c$, and s is the distance measured along the magnetic field. In practice, the neglect of perpendicular transport means that the turbulence has to be weak $(\delta B/B \ll 1)$ and the shock normal angle has to be limited to, say, $\theta_n < 80°$.[1]

A diverging mean magnetic field causes similar values for $\lambda_\parallel(p_c)$ because of adiabatic focusing and deceleration [*Vainio et al.*, 2000]. Of these two effects, focusing is more restrictive causing particles to stream along field lines away from the Sun at speed [*Kocharov et al.*, 1996, *Vainio et al.*, 2000]

$$U = \frac{\kappa_\parallel}{L_B} \sim \frac{2\lambda_\parallel}{3r}v, \quad L_B^{-1} \equiv -\frac{1}{B}\frac{\partial B}{\partial s} \qquad (17)$$

and if U exceeds the projected shock speed along the magnetic field, $\dot{s}_{sh}$, in the solar frame, particles will escape from an outward-propagating shock.

We have plotted the required value of $\lambda_\parallel(p_c)$ for $E_c = 10$ MeV, $u_3 = 1$, $\theta_n = 70°$, $r_{sc} = 4$, and $\Delta s = r - r_\odot$ in Fig. 2. As can be seen, only model 3 can simultaneously fulfill the requirements set by observations of SEP transport in the IP medium and the presumed test-particle DSA in an oblique shock in the solar corona. This model would predict acceleration of protons in CME-driven shocks up to about 10 MeV at heights of about a solar radius above the surface and the release of ions to the interplanetary medium as the shock leaves the corona to the less turbulent (at resonant scales) IP medium. We note also that for models 1–2 the computed coronal mean free path is so large that the modeling of particle transport by spatial diffusion is no-longer valid. It is, however, very unlikely that fully kinetic modeling would considerably enhance the particle acceleration efficiency of non-perpendicular shocks in these models.

3.2. Effect of Shocks and SEPs on Turbulence

As CME-driven shocks are super-fast-magnetosonic, they overtake the ambient turbulence, process it, and leave it behind in an excited state. Neglecting the microphysics at the shock front and considering the compression at the fluid level, only, the effect of shocks on turbulence is three-fold [*Vainio and Schlickeiser*, 1998, 1999]: (i) shocks compress the normal component of the wave vectors of upstream fluctuations; (ii) shocks amplify the intensity of the fluctuations at a given frequency; and (iii) shocks modify the cross-helicity state

---

[1] For quasi-perpendicular shocks, diffusion along the shock normal is determined by perpendicular transport. In this case, the available acceleration time is no longer proportional to $\cos\theta_n$, and Eq. (16) becomes invalid.

of the fluctuations enabling stochastic acceleration of particles downstream. The intensity of downstream Alfvén-waves at low-Mach-number ($u_1/v_A \sim 2$) quasi-parallel shocks may be more than 100 times the upstream intensity in a low-beta plasma, such as the solar corona [*Vainio and Schlickeiser*, 1999]. Thus, a slow-to-intermediate speed CME may create very turbulent upstream conditions for an overtaking fast CME. Therefore, interacting CMEs should be more efficient in accelerating SEPs than isolated ones. Observations, indeed, confirm this prediction [*Gopalswamy et al.*, 2002, 2004].

In large SEP events, streaming instabilities ahead of CME-driven shocks can amplify the ambient MHD waves, i.e., $-\gamma$ in Eq. (4) becomes large just ahead of the shock. As a result of the instability particles themselves can create the upstream turbulence ahead of the shock, and the requirement of a short ambient coronal mean free path can be relaxed. One can then consider a simplified turbulence transport model *without* nonlinear cascading, i.e., $F_+ = 0$ in Eq. (4). Then, the ambient mean free path is rather large throughout the inner heliosphere, like in our model 1. This is more or less the standard model for the acceleration of ions in large SEP events [*Ng et al.*, 2003, *Lee*, 2005] and it is covered in other chapters of this book.

Here we want to emphasize, however, that a model with a large ambient mean free path does *not* apply to all SEP events that appear to be accelerated by CME-driven shocks. It can be shown [*Vainio*, 2003] that in the case of a large ambient mean free path, a certain number of accelerated particles needs to be released in each flux tube before the coronal Alfvén waves have grown appreciably to allow rapid particle acceleration. For a fixed number of injected particles, the local time-integrated wave-growth rate is proportional to the Alfvén speed, so the wave-growth is most significant at the position of the global maximum of $v_A(r)$ [*Vainio*, 2003]. Numerically,

$$\frac{dN}{dp \, d\Omega_\odot} = \frac{10^{33}}{p} \sqrt{\frac{n_\odot}{2 \times 10^8 \, \mathrm{cm}^{-3}}} \frac{v_{A\odot}}{v_{A,\max}} \qquad (18)$$

gives the number of particles per unit momentum and unit solid angle at the solar surface that needs to be injected before the coronal Alfvén waves resonant with particles of momentum p have grown substantially. This number can be translated into a typical time-of-maximum proton intensity in the solar wind using the well-known Green's function for radial diffusion. Assuming an impulsive injection and a spatially constant radial mean free path in the solar wind, we get at $r = 1$ AU

$$\left.\frac{dJ}{dE}\right|_{\max} = \frac{1}{4\pi} \left.\frac{dN}{d^3r \, dp}\right|_{\max} \approx \frac{7.4 \times 10^{-2}}{r^3} \frac{dN}{dp \, d\Omega_\odot}$$

$$\approx \left(\frac{\mathrm{MeV}}{E}\right)^{1/2} \frac{15}{\mathrm{cm}^2 \, \mathrm{sr} \, \mathrm{s} \, \mathrm{MeV}} \qquad (19)$$

Note that this number does not depend on the mean free path at 1 AU, as long as it is smaller than $\sim$0.3 AU to allow the use of diffusion approximation. Thus, the time-of-maximum intensity of 10-MeV protons has to (promptly) exceed $\sim$5 $(\mathrm{cm}^2 \, \mathrm{sr} \, \mathrm{s} \, \mathrm{MeV})^{-1}$ before we can expect that the resonant Alfvén waves have grown in the corona during the event. In the largest SEP events exhibiting streaming-limited intensities [*Reames and Ng*, 1998], proton intensities exceed the threshold for significant wave growth at energies up to a few hundred MeV. Thus, the standard model may well be applied to these SEP events (perhaps excluding the earliest phases of particle acceleration). There are, however, a large number of CME-related events that *never* reach the threshold intensity even at MeV energies (e.g., the April 7 and May 12, 1997, events [*Torsti et al.*, 1998]). If these events are accelerated by (non-perpendicular) coronal shocks via the diffusive mechanism, turbulence conditions similar to our model 3 with a short coronal mean free path combined with a large one in the interplanetary medium are required.

3.3. Modeling Test-Particle Acceleration in Turbulent Corona

Finally, as the validity of test-particle calculations was established in the case of small CME-related SEP events, we will take a brief look at two recent test-particle modeling results on particle acceleration in coronal shocks.

As demonstrated above, steady-state DSA does not correspond to a power-law injection of particles towards the upstream region, unless the upstream turbulence is arranged in a special manner. This is not plausible unless the turbulence is self-generated [*Vainio*, 2003]. However, if the acceleration takes place in a refracting shock wave near the coronal base [*Vainio and Khan*, 2004], the observer is actually connected to the downstream region of the shock (see Fig. 1). Thus, the escaping spectrum is of the same form as that at the shock. Using values of λ given by wave intensities used in coronal cyclotron-heating models [*Laitinen et al.*, 2003], we can estimate that the cut-off energy in the accelerated particles in a refracting shock is some tens of MeV [*Vainio and Khan*, 2004], as often observed in small CME-related SEP events. In this model, the solar-frame speed of the upstream scattering centers (high-frequency Alfvén waves) is upward. Note, however, that if a refracting shock has a large value of r_{sc} near the solar surface, the solar-frame convection speed in the downstream region may be downward. For such a shock the thickness of the downstream region can not be too large; otherwise most of the particles get convected down to the solar surface and precipitate.

When considering acceleration in CME bow shocks, we need time-dependent modeling, where the acceleration of particles occurs low in the corona under turbulent conditions

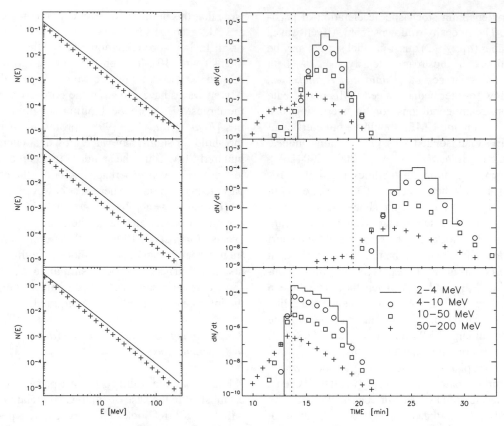

Figure 4. Energy spectra (left column) and escape rate profiles (right column) of protons escaping a turbulent coronal layer through a stationary, open boundary situated at $1.7r_\odot$ (Rows 1 and 3) or at $2r_\odot$ (Row 2). The inner boundary at $1.2r_\odot$ is closed. A shock propagates through the layer at the speed of 600 km s^{-1}. Low-energy protons are injected to the shock either impulsively at the heliocentric distance of $1.25r_\odot$ (Rows 1 and 2) or uniformly through the volume of the spherical layer (Row 3). The steady-state DSA spectrum is plotted as straight lines in the left column and the shock arrival time at the outer boundary is indicated with a vertical dotted line in the right column. [*Kocharov et al.*, 2005]

and as the shock wave moves outward to a region with a rapidly increasing mean free path, the particles may escape from the shock. This kind of modeling requires numerical solutions of the diffusive particle transport equation near the shock. We have performed test-particle simulations in a simplified model [*Kocharov et al.*, 2005] resembling the scattering conditions of the turbulence model 3 presented in section 2.2 of this paper (although more turbulent conditions near the solar surface were assumed). An example of the model results is given in Fig. 4. A shock with $r_{sc} = 3$ propagates through the layer at the speed of 600 km s^{-1}. The Alfvén speed in the layer is 200 km s^{-1}, and the upstream diffusion coefficient $\kappa_1 = \kappa_0(E/E_0)^{0.25}$ with $\kappa_0 = 2.5 \times 10^5$ km^2 s^{-1}. Protons are injected with an energy of $E_0 = 0.3$ MeV at the shock to the acceleration process either impulsively or uniformly in volume throughout the turbulent layer. As can be seen, the model produces reasonable energy spectra for protons as well as injection-time profiles that resemble

observations with delays of some ten minutes with respect to the start of the CME lift-off. The diffusion coefficient used in these calculations are extremely small, which leads to hard power-law spectra prevailing up to relativistic energies but increasing the mean free path in the turbulent layer will produce spectra with cut-off energies at tens of MeVs, consistent with many small SEP events [*Kocharov et al.*, 2005].

4. SUMMARY

We have reviewed the basic theory of SEP transport in the solar wind and some recent modeling efforts to understand the role of coronal turbulence in the acceleration of SEPs in coronal shocks. The main results of the studies can be summarized as follows:

(1) Solar-wind turbulence models can be used to calculate the evolution of the slab-component of turbulence and,

thus, the mean free path of SEPs in the solar corona and solar wind.

(2) Observations of solar-wind turbulence spectra and SEP event onset times are consistent with coronal turbulence models, which produce short SEP mean free path in the corona. This allows SEP acceleration by the DSA mechanism close to the solar surface.

(3) Self-generated turbulence has been shown to be a plausible mechanism to produce very short mean free paths ahead of the CME shocks in large SEP events, which may have power-law spectra extending to relativistic energies. However, small CME-related SEP events (10-MeV proton fluxes below ~5 (cm^2 s sr MeV)$^{-1}$ at 1 AU) can not rely on this mechanism to produce the upstream turbulence.

(4) Both refracting and driven shocks can load a turbulent corona with reasonable SEP spectra, which leak in to the interplanetary medium in the aftermath of CMEs.

Our study did not take into account particle transport perpendicular to the mean magnetic field. Thus, we could not address particle acceleration in perpendicular coronal shock waves. This shortcoming does not mean that we would regard such shocks as unimportant for the CME–shock–turbulence–SEP relation. While the validity of Eq. (16) breaks down at $\theta_n \rightarrow 90°$, it implies that nearly perpendicular shocks could be very efficient particle accelerators. The problem with nearly perpendicular shocks in the classical picture of scattering is, however, that the projected fluid speed along the mean magnetic field becomes large and it becomes increasingly difficult to draw particles from the thermal population to the acceleration process. However, particle transport across the mean field in magnetic fluctuations with finite amplitude differs from classical scattering, as particles cross the mean magnetic field mainly by following the actual turbulent field lines. Recent numerical simulations [*Giacalone*, 2005], indeed, indicate that particle injection in perpendicular shocks propagating into irregular magnetic fields is much more efficient than the classical diffusive treatment indicates. Thus, coronal turbulence models with a small amount of fluctuations at resonant scales but with a higher amount at lower wave numbers may also be able to explain particle acceleration to high energies without introducing self-generated waves.

REFERENCES

Axford, W.I., E. Leer, and G. Skadron, The acceleration of cosmic rays by shock waves, *Proc. 15th Int. Cosmic Ray Conf. (Plovdiv)*, 11, 132, 1977.

Bell, A.R., The acceleration of cosmic rays in shock fronts. I, *Mon. Not. R. Astron. Soc.*, 182, 147-156, 1978.

Bieber, J.W., W. Wanner, and W.H. Matthaeus, Dominant two-dimensional solar wind turbulence with implications for cosmic ray transport, *J. Geophys. Res.*, 101, 2511-2522, 1996.

Blandford, R.D., and J.P. Ostriker, Particle acceleration by astrophysical shocks, *Astrophys. J.*, 221, L29-L32, 1978.

Cane, H.V., R.G. Stone, J. Fainberg, J.L. Steinberg, and S. Hoang, Type II solar radio events observed in the interplanetary medium. I - General characteristics, *Solar Phys.*, 78, 187-198, 1982.

Carrington, R.C., Description of a Singular Appearance seen in the Sun on September 1, 1859, *Mon. Not. R. Astron. Soc.*, 20, 13-15, 1859.

Cliver, E.W., S.W. Cliver, M.A. Shea, and D.F. Smart, Injection Onsets of ~2 GeV Protons, ~1 MeV Electrons, and ~100 keV electrons in Solar Cosmic Ray Flares, *Astrophys. J.*, 260, 362-370, 1982.

Drury, L.O'C., An Introduction to the Theory of Diffusive Shock Acceleration of Energetic Particles in Tenuous Plasmas, *Rep. Progr. Phys.*, 46, 973-1027, 1983.

Dröge, W., Particle scattering by magnetic fields, *Space Sci. Rev.*, 93, 121-151, 2000.

Dröge, W., Solar particle transport in a dynamical quasi-linear theory, *Astrophys. J.*, 589, 1027-1039, 2003.

Earl, J.A., The Diffusive Idealization of Charged-Particle Transport in Random Magnetic Fields, *Astrophys. J.*, 193, 231-242, 1974.

Forbush, S.E., P.S. Gill, and M.S. Vallarta, On the mechanism of sudden increases of cosmic radiation associated with solar flares, *Rev. Mod. Phys.*, 21, 44-48, 1949.

Forman, M.A., J.R. Jokipii, and A.J. Owens, Cosmic-Ray Streaming Perpendicular to the Mean Magnetic Field, *Astrophys. J.*, 192, 535-540, 1974.

Giacalone, J., Particle Acceleration at Shocks Moving through an Irregular Magnetic Field, *Astrophys. J.*, 624, 765-772, 2005.

Gopalswamy, N., *et al.*, Interacting coronal mass ejections and solar energetic particles, *Astrophys. J.*, 572, L103-L107, 2002.

Gopalswamy, N., S. Yashiro, S. Krucker, G. Stenborg, and R.A. Howard, Intensity variations of large solar energetic particle events associated with coronal mass ejections, *J. Geophys. Res.*, 109, A12105, 2004.

Gosling, J.T., E. Hildner, R.M. MacQueen, R.H. Munro, A.I. Poland, and C.L. Ross, Mass ejections from the sun - A view from SKYLAB, *J. Geophys. Res.*, 79, 4581-4587, 1974.

Hodgson, R., On a curious Appearance seen in the Sun, *Mon. Not. R. Astron. Soc.*, 20, 15-16, 1859.

Horbury, T.S., A. Balogh, R.J. Forsyth, and E.J. Smith, The rate of turbulent evolution over the Sun's poles, *Astron. Astrophys.*, 316, 333-341, 1996.

Hu, Y.Q., S.R. Habbal, and X. Li, On the cascade processes of Alfvén waves in the fast solar wind, *J. Geophys. Res.*, 104, 24819-24834, 1999.

Hudson, H.S., J.I. Khan, J.R. Lemen, N.V. Nitta, and Y. Uchida, Soft X-ray observation of a large-scale coronal wave and its exciter, *Solar Phys.*, 212, 121-149, 2003.

Jokipii, J.R., Cosmic-Ray Propagation. I. Charged Particles in a Random Magnetic Field, *Astrophys. J.*, 146, 480-487, 1966.

Kahler, S.W., E. Hildner, and M.A. I. van Hollebeke, Prompt solar proton events and coronal mass ejections, *Solar Phys.*, 57, 429-443, 1978.

Kahler, S.W., G.M. Simnett, and M.J. Reiner, Onsets of Solar Cycle 23 Ground Level Events as Probes of Solar Energetic Particle Injections at the Sun, *Proc. 28th Int. Cosmic Ray Conf. (Tsukuba)*, 3415-3418, 2003.

Kallenrode, M.-B., Particle propagation in the inner heliosphere, *J. Geophys. Res.*, 98, 19037-19047, 1993.

Kocharov, L.G., J. Torsti, R. Vainio, and G.A. Kovaltsov, Propagation of solar cosmic rays: diffusion versus focused diffusion, *Solar Phys.*, 165, 205-208, 1996.

Kocharov, L., J. Torsti, O.C. St. Cyr and T. Huhtanen, A relation between dynamics of coronal mass ejections and production of solar energetic particles *Astron. Astrophys.*, 370, 1064-1070, 2001.

Kocharov, L., M. Lytova, R. Vainio, T. Laitinen, and J. Torsti, Modeling the shock aftermath source of energetic particles in solar corona, *Astrophys. J.*, 620, 1052-1068, 2005.

Krucker, S., and R.P. Lin, Two classes of solar proton events derived from onset time analysis, *Astrophys. J.*, 542, L61-L64, 2000.

Krymsky, G.F., A regular mechanism for accelerating charged particles at the shock front, *Dokl. Akad. Nauk SSSR*, 243, 1306-1308, 1977.

Laitinen, T., H. Fichtner, and R. Vainio, Toward a self-consistent treatment of the cyclotron wave heating and acceleration of the solar wind plasma, *J. Geophys. Res.*, 108, 1081, 2003.

Lee, M.A., Coupled Hydromagnetic Wave Excitation and Ion Acceleration at an Evolving Coronal/Interplanetary Shock, *Astrophys. J. (Supp.)*, 158, 38-67, 2005.

Lintunen, J., and R. Vainio, Solar energetic particle event onset as analyzed from simulated data, *Astron. Astrophys*, 420, 343-350, 2004.

Mann, G., A. Klassen, H. Aurass, and H.-T. Classen, Formation and development of shock waves in the solar corona and the near-Sun interplanetary space, *Astron. Astrophys.*, 400, 329-336, 2003.

Moreton, G.E., Observations of flare-initiated disturbances with velocities ~1000 km/sec, *Astron. J.*, 65, 494-495, 1960.

Ng, C.K., D.V. Reames, and A.J. Tylka, Modeling shock-accelerated solar energetic particles coupled to interplanetary Alfvén waves, Astrophys. J., 591, 461-485, 2003.

Reames, D.V., Particle acceleration at the Sun and in the heliosphere, *Space Sci. Rev.*, 90, 413-491, 1999.

Reames, D.V., and C.K. Ng, Streaming-limited Intensities of Solar Energetic Particles, *Astrophys. J.*, 504, 1002-1005, 1998.

Schlickeiser, R., Cosmic-ray transport and acceleration. I - Derivation of the kinetic equation and application to cosmic rays in static cold media, *Astrophys. J.*, 336, 243-263, 1989.

Schlickeiser, R., and U. Achatz, Cosmic-ray particle transport in weakly turbulent plasmas. Part 1. Theory, *J. Plasma Phys.*, 49, 63-77, 1993.

Sheeley, N.R., J.H. Walters, Y.-M. Wang, and R.A. Howard, Continuous tracking of coronal outflows: Two kinds of coronal mass ejections, *J. Geophys. Res.*, 104, 24739-24768, 1999.

Sheeley, N.R., W.N. Hakala, and Y.-M. Wang, Detection of coronal mass ejection associated shock waves in the outer corona, *J. Geophys. Res.*, 105, 5081-5092, 2000.

Thompson, B. J., *et al.*, SOHO/EIT observations of the 1997 April 7 coronal transient: possible evidence of coronal Moreton waves, *Astrophys. J.*, 517, L151-L154, 1999.

Torsti, J., *et al.*, Energetic (~1 to 50 MeV) protons associated with Earth-directed coronal mass ejections, *Geophys. Res. Lett.*, 25, 2525-2528, 1998.

Tousey, R., The solar corona, *Space Res.*, 13, 713-730, 1973.

Tu, C.-Y., Z.-Y. Pu, and F.-S. Wei, The power spectrum of interplanetary Alfvénic fluctuations. Derivation of the governing equation and its solution, *J. Geophys. Res.*, 89, 9695-9702, 1984.

Tu, C.-Y., A solar wind model with the power spectrum of Alfvénic fluctuations, *Solar Phys.*, 109, 149-186, 1987.

Tu, C.-Y., The damping of interplanetary Alfvénic fluctuations and the heating of the solar wind, *J. Geophys. Res.*, 93, 7-20, 1988.

Uchida, Y., Propagation of hydromagnetic disturbances in the solar corona and Moreton's wave phenomenon, *Solar Phys.*, 4, 30-44, 1968

Vainio, R., On the generation of Alfvén waves by solar energetic particles, *Astron. Astrophys.*, 406, 735-740, 2003.

Vainio, R., and J.I. Khan, Solar energetic particle acceleration in refracting coronal shock waves, *Astrophys. J*, 600, 451-457, 2004.

Vainio, R., and R. Schlickeiser, Alfvén wave transmission and particle acceleration at parallel shock waves, *Astron. Astrophys.*, 331, 793-799, 1998.

Vainio, R., and R. Schlickeiser, Self-consistent Alfvén-wave transmission and test-particle acceleration at parallel shocks, *Astron. Astrophys.*, 343, 303-311, 1999.

Vainio, R., L. Kocharov, and T. Laitinen, Interplanetary and interacting protons accelerated in a parallel shock wave, *Astrophys. J.*, 528, 1015-1025, 2000.

Vainio, R., T. Laitinen, and H. Fichtner, A simple analytical expression for the power spectrum of cascading Alfvén waves in the solar wind, *Astron. Astrophys.*, 407, 713-723, 2003.

Vourlidas, A., S.T. Wu, A.H. Wang, P. Subramanian, and R.A. Howard, Direct detection of a coronal mass ejection-associated shock in large angle and spectrometric coronagraph experiment white-light images, *Astrophys. J.*, 598, 1392-1402, 2003.

Wild, J.P., S.F. Smerd, and A.A. Weiss, Solar bursts, *Ann. Rev. Astron. Astrophys.*, 1, 291-366, 1963.

R. Vainio, Department of Physical Sciences, P.O.B. 64, FI-00014 University of Helsinki, Finland. (rami.vainio@helsinki.fi)

Spectral and Compositional Characteristics of Gradual and Impulsive Solar Energetic Particle Events

Allan J. Tylka

E. O. Hulburt Center for Space Research, Code 7652, Naval Research Laboratory, Washington, DC, USA

Martin A. Lee

*Space Science Center and Institute for the Study of Earth, Oceans, and Space,
University of New Hampshire, Durham, NH, USA*

The spectral and compositional characteristics of solar energetic particles (SEPs) reflect a number of factors, including the nature of the acceleration process and the accessible seed population. We review recent discoveries about these issues and discuss current hypotheses about the relationship between flares and CME-driven shocks in producing SEPs above a few tens of MeV/nucleon in large, gradual events. We also briefly review recent results on spectra and composition in impulsive events. We particularly emphasize how significant correlations among spectral and compositional characteristics serve to constrain models of SEP production.

1. INTRODUCTION

The classification of SEP events into two categories, "gradual" and "impulsive" [*Reames*, 1995a, 1999], is reviewed elsewhere in these Proceedings. In this paper, we will continue the traditional usage, in which these terms are short-hand for the likely acceleration site and mechanism. Specifically, "impulsive" events are those in which the particle acceleration occurs at sites associated with flares, probably through resonant wave-particle interactions arising from magnetic reconnection. These resonant interactions are the generally-favored candidate for the mechanism that produces the very large compositional distortions that are seen in these events. "Gradual" events, on the other hand, are those in which the acceleration occurs primarily at shocks, through first-order Fermi and shock-drift acceleration. Shocks accelerate particles from the appropriate seed population in a more or less even-handed fashion. The composition in gradual events therefore reflects that of the seed population, although – as we shall see – not necessarily in a simple way.

Much of the discussion in the SEP community in recent years has focused on the fact that the compositional distinctions between these two event types become blurred when we look at energies above the few MeV/nucleon where the categories were originally developed. It is therefore worthwhile to remind ourselves of recent evidence for two distinct SEP acceleration mechanisms. Figure 1 presents a nine-year survey of energetic ions at ~3 MeV/nucleon from the Low-Energy Matrix Telescope (LEMT) on *Wind* [*Reames and Ng*, 2004]. The two panels show different views of the same data, namely the correlation between eight-hour averaged Fe and O intensities. In the left panel, the symbol size represents the observed $^3He/^4He$ ratio at ~3 MeV/nucleon. In the right panel, the symbol size gives the observed ratio of ultraheavy ions in the Sn-Ba group (atomic numbers Z = 50-56) relative to oxygen.

Except at low intensities, where anomalous cosmic-ray (ACR) oxygen sometimes obscures the picture, the observations generally fall along two loci, corresponding to Fe/O~1 and Fe/O~0.1. The most striking feature of Figure 1 is the clear preference for the very large ^3He and ultraheavy enhancements to occur along the Fe/O~1 locus. All of these enhancements reflect a fundamental feature of the impulsive

Solar Eruptions and Energetic Particles
Geophysical Monograph Series 165
10.1029/165GM25

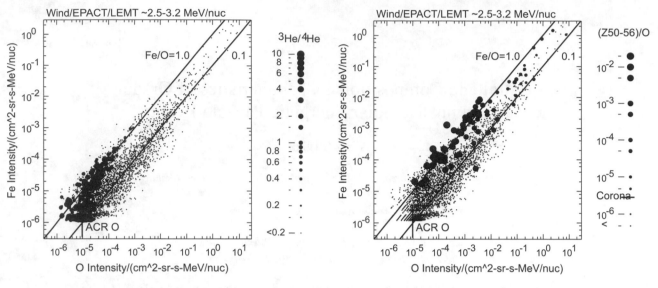

Figure 1. Correlation between 8-hour-averaged Fe and O intensities at ~3 MeV/nucleon from a ~9-year survey by the Low Energy Matrix Telescope (LEMT) on *Wind*. The symbol size indicates the ^{3}He/^{4}He ratio (left panel) or the $(50 \leq Z \leq 56)$/O ratio (right panel), as indicated by the legends. The intensities lie along two loci, at Fe/O~1 and at Fe/O~0.1, which characterize impulsive and gradual SEP events at these energies. The large ^{3}He and ultraheavy enhancements are found in impulsive events. Note that LEMT sacrificed resolution for collecting power, enabling it to make the first detailed survey of SEPs from the upper two-thirds of the periodic table [*Reames*, 2000] but allowing reliable ^{3}He/^{4}He measurements only when the ratio exceeds ~20%. (See *Reames and Ng* [2004] for a color version of this figure.)

acceleration process, which selectively energizes some species over others. Also noteworthy in Figure 1 is that Z > 2 intensities in impulsive events at a few MeV/nucleon rarely exceed ~0.01 ions/cm²-sr-s-MeV/nucleon.

2. GRADUAL SEP EVENTS

Before turning to spectral and compositional variation in gradual events, it is worthwhile to see *when* and *where* high-energy SEPs are produced. Figure 2 presents timing results for the two largest ground-level events (GLEs) so far in Solar Cycle 23. The top panels show the time profile of soft x-rays and gamma-rays, shifted by 8 minutes to account for travel time from the Sun. Vertical lines mark the reported onsets of metric type II emission [*Gopalswamy et al.*, 2005] and the first CME image from *SOHO*/LASCO [*Yashiro et al.*, 2004], also corrected for light-travel time. The bottom panels show the corresponding injection profile *at the Sun* for ~GeV protons, as deduced from time-intensity profiles and angular anisotropies from the "Spaceship Earth" world-wide network of neutron monitors [*Bieber et al.*, 2004; *Sáiz et al.*, 2005]. The observations have been modeled with a detailed numerical treatment of interplanetary transport that determines not only the time at which particles first departed the neighborhood of the Sun, but also the temporal structure of this injection process.

According to this modeling, in both events, particle emission into interplanetary space was delayed with respect to the event's impulsive phase, as represented by the gamma rays. The injection of ~GeV protons began and peaked when the leading-edge of the CME was only a few R_S above the solar surface. If we are to discover how shocks might produce these very high-energy particles, this region of the corona is where we must understand the shock's characteristics.

2.1. Spectral and Compositional Variability

Figure 3a shows results from a recent survey of spectral and compositional variability in gradual SEP events. The figure shows a correlation between key characteristics of the 42 largest SEP events of 1997-2004. (The detailed event list, including the parameters used in generating Figure 3a, is given in *Tylka et al.*, 2005.) All of these events were associated with large, fast (>800 km/s) CMEs. The sole event selection criterion was a total *GOES* proton fluence exceeding 2×10^5 cm⁻²-sr⁻¹ above 30 MeV. This energy biases the measure of size towards particles accelerated near the Sun; the fluence threshold selects events that are at least ~20-times larger than the biggest, undisputed impulsive event in this time period [*Leske et al.*, 2003; *Reames and Ng*, 2004].

The vertical axis in Figure 3a shows event-integrated Fe/O ratios from the Solar Isotope Spectrometer (SIS) on *ACE* at 30-40 MeV/nucleon, normalized to the nominal coronal value of 0.134 [*Reames*, 1995b]. These Fe/O ratios range over nearly

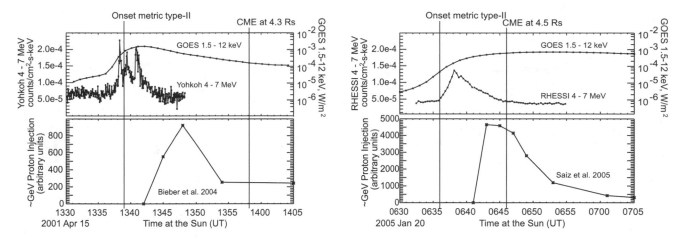

Figure 2. Time profiles in the two largest ground level events (GLEs) of Cycle 23, on 2001 April 15 (left panels) and 2005 January 20 (right panels). The top panels show gamma-rays at 4-7 MeV (left axis) and soft x-rays (right axis). The bottom panels show the injection profile of ~GeV protons at the Sun, as deduced from modeling the response of the Spaceship Earth neutron-monitor network [*Bieber et al.*, 2004; *Sáiz et al.*, 2005]. All times have been corrected for propagation from the Sun.

three orders of magnitude. The horizontal axis is a measure of the event's spectral rollover. Specifically, the event-integrated oxygen spectrum was fit to two independent power-laws, $E^{-\gamma_1}$ covering 3-10 MeV/nucleon (from *Wind*/LEMT) and $E^{-\gamma_2}$ covering 30-100 MeV/nucleon (from *ACE*/SIS). (Oxygen was chosen for this purpose since it is the best-measured heavy-ion species and its charge-to-mass ratio varies little from event to event.) Although these spectra are not necessarily broken power-laws, the difference between these two indices, $(\gamma_2 - \gamma_1)$, provides a measure of spectral steepening. Spectra that are nearly power-laws over this energy range have $\gamma_2 - \gamma_1 \sim 0$, while $\gamma_2 - \gamma_1 \gg 0$ indicates a spectrum that rolls over more or less exponentially as energy increases. Finally, symbol size in Figure 3a indicates the >30 MeV proton fluence, which also spans about three orders of magnitude.

The event selection in Figure 3a was based on proton fluence alone, without reference to heavy-ion characteristics. The plot therefore presents an unbiased survey of heavy-ion characteristics among the Cycle's large events. In particular, note that: (1) there are no events in the lower-left, where spectra would be power-laws ($\gamma_2 - \gamma_1 \sim 0$) while Fe/O is strongly suppressed; (2) there are no events in the upper-right, where spectra rollover steeply ($\gamma_2 - \gamma_1 \gg 0$) and Fe/O is enhanced; and (3) strong Fe enrichments are absent, or at least rare, among events with the largest proton fluences. The anti-correlation in Figure 3a (correlation coefficient = −0.86, corresponding to a random probability $<10^{-5}$) strongly indicates that there are common factors behind the variability in high-energy Fe/O and spectral shape. These factors also affect, at least to some extent, event size.

For a subset of the events in Figure 3a, we also have measurements of $<Q_{Fe}>$, the mean ionic charge of Fe above

~25 MeV/nucleon. These results are plotted in Figure 3b versus the corresponding Fe/O value at 30-40 MeV/nucleon. Figure 3b shows clear correlation between Fe/O and $<Q_{Fe}>$ at high energies: we do not see high $<Q_{Fe}>$ when Fe/O is suppressed; nor do we see low $<Q_{Fe}>$ when Fe/O is significantly enhanced. Taken together, the two panels of Figures 3 suggest that various facets of high-energy SEP variability – spectral shape, Fe/O, charge states, and event size – are closely inter-twined. The apparent correlations among these observables are high hurdles for SEP models, and dealing with these various features in a piece-meal fashion is unlikely to be adequate.

Figures 4 offer a closer look at two events, 2002 April 21 and 2002 August 24, that contribute to the extremes of variation in Figure 3a. As explained elsewhere [*Tylka et al.*, 2005, 2006], these two events were chosen for detailed comparisons because they arose from ostensibly similar CMEs and flares. The August event was a GLE, but the April event had a total proton fluence above 30 MeV that was ~14 times larger. Both events erupted from near the west limb, so that the high-energy particles observed at Earth were produced primarily near the Sun. From ~0.5 to ~10 MeV/nucleon, Fe/O ratios are nearly identical in the two events. *But as energy increases further, the events diverge, so that Fe/O differs by nearly two orders of magnitude at the highest energy.*

Figure 4 also shows the event-averaged oxygen spectra for these two events. In the April event, the spectrum is fit to the familiar *Ellison and Ramaty* [1985] functional form, $F(E) \sim E^{-\gamma} exp(-E/E_0)$, where E is the energy in MeV/nucleon and γ and E_0 are fit parameters. This simple form gives a remarkably good description of all the ion spectra in this event. The e-folding energy, E_0, differs from species to species, with iron having a smaller value than oxygen, thereby leading to suppressed Fe/O

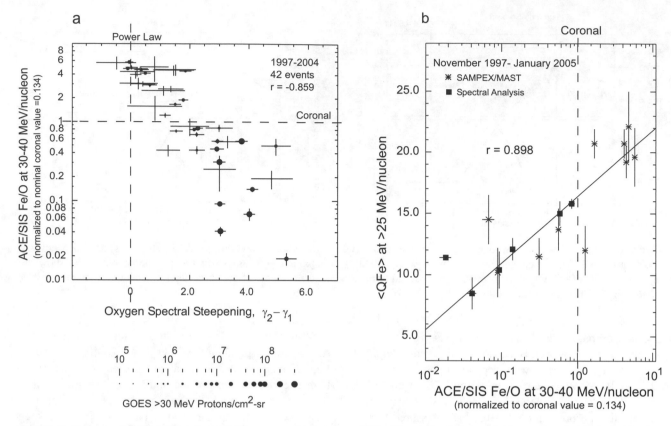

Figure 3. Two aspects of correlated variability among high-energy SEPs. Figure 3a (left panel) shows the correlation of event-integrated Fe/O at 30-40 MeV/nucleon vs. spectral steepening of the oxygen spectrum in the largest SEP events of 1997-2004, as described in the text. The weighted correlation coefficient for this plot is r = −0.859 for 42 events. The Fe-rich events tend to be more like power-laws. The symbol size indicates the event-integrated fluence of >30 MeV protons, as given in the legend at the bottom (See *Tylka et al.* [2005] for a color version of this figure.) Figure 3b (right panel) shows measurements of the mean ionic charge of iron ($<Q_{Fe}>$) from *SAMPEX* (asterisks) at ≳25 MeV/nucleon [*Leske et al.*, 2001; *Labrador et al.*, 2003, 2005] and from spectral analyses [*Tylka et al.*, 2000, 2001, 2006] at comparable energies (filled squares) plotted versus *ACE*/SIS Fe/O at 30-40 MeV/nucleon. Two events have been measured by both methods, which agree to within uncertainties. In order to avoid double counting, the correlation fit omitted the spectral analyses from these two events. The correlation fit also omitted the outlying datapoint on the far left. The weighted correlation coefficient in this panel is r = 0.898 for 13 events.

at high energies. In fact, by starting with the measured oxygen charge state from *SAMPEX*, linear scaling between the fitted e-folding energies and the charge-to-mass (Q/A) ratios [*Ellison and Ramaty*, 1985; *Klecker et al.*, 2003; *Tylka et al.*, 2000, 2001] yields reasonable inferred charge states for other species, all of which are consistent with a common source temperature of 1.44 ± 0.10 MK. (See *Tylka et al.* [2006] for further details.)

On the other hand, in the August event, the exponential rollovers that govern high energies in the April event are absent. Instead, one power-law rolls smoothly into a second, steeper power-law, which describes the SEP spectra out to the highest measureable energies. As illustrated in Figure 4, spectra in the August event can be fit to the *Band et al.* [1993] function, a purely empirical form, identical to the

Ellison and Ramaty function at low energies, but crafted to roll into another power-law with no discontinuity in either the function or its first derivative. The power-law indices at high energies differ from species to species, with the iron spectrum being harder than the oxygen spectrum, causing Fe/O to rise with energy.

The distinctive spectral morphologies of these two events, as illustrated by the oxygen spectra in Figure 4, are also seen in other species, including protons and carbon; the spectral distinction therefore cannot be explained by Q/A-dependent effects alone. Spectral differences are also apparent in time-dependent spectra (as opposed to event-averaged), even at the start of the event (see *Tylka et al.*, 2005, Figure 7), suggesting that the differences do indeed reflect inherent characteristics of the accelerator while near the Sun. The two events in

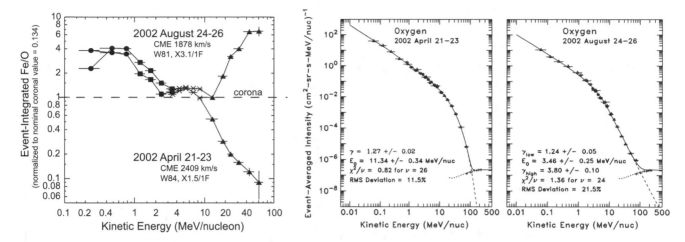

Figure 4. Left panel: event-integrated Fe/O (normalized to the nominal coronal value 0.134 [*Reames*, 1995b]) versus energy for the solar energetic particle events of 2002 April 21 and 2002 August 24. In order of increasing energy, the data come from the Ultra Low-Energy Isotope Spectrometer [ULEIS; *Mason et al.*, 1998; circles] on the Advanced Composition Explorer (*ACE*), the Electron Proton, and Alpha Monitor [EPAM; *Gold et al.*, 1999; squares] on *ACE*, LEMT [*von Rosenvinge et al.*, 1995; crosses] on *Wind*, and the Solar Isotope Spectrometer [SIS; *Stone et al.*, 1998; triangles] on *ACE*. Right panel: event-averaged oxygen spectra in these two events. The curves through the datapoints are fits to the *Ellison and Ramaty* [1985] functional form (for the April event) and the double power-law *Band et al.* [1993] functional form (for the August event). Fit parameters, reduced χ^2, and rms deviation from the fit are noted in each panel. The small crosses in the lower right corners are contemporaneous Galactic cosmic ray (GCR) measurements from *ACE*. Short dashes show estimated GCRs, long dashes are the extrapolated SEP fits, and solid curves show the sum of fitted SEPs and GCRs. Details about the data selection, fit procedures, and systematic uncertainties are given in *Tylka et al.* [2006].

Figure 4 also likely differ in their ionic charge states, as discussed in *Tylka et al.* [2006].

One further comment should be noted before we turn to modeling efforts. The relative simplicity of the 2002 April 21 event – with its clear exponential rollovers, apparently ordered by Q/A, and comparatively low charge states, consistent with a common source temperature for all species – is rare. Among the events catalogued in *Tylka et al.* [2005], only ~10% clearly exhibit this behavior. Thus, conditions that produce this distinctive event type arise only infrequently near the Sun[1]. Nevertheless, this particular event type must also be accounted for in our models. (As discussed below, in the shock-geometry hypothesis, this event type provides the baseline from which more complicated events are generated through parameter variation.)

2.2. The Shock Geometry Hypothesis

Tylka et al. [2005] put forth a hypothesis that addresses the correlated spectral and compositional variability in SEPs above a few tens of MeV/nucleon. The hypothesis invokes CME-driven shocks as the dominant accelerators in all cases, but points to two further factors whose interplay drive the variability. These factors are (1) evolution in θ_{Bn}, the angle between the shock-normal **n** and the upstream magnetic field **B**, as the CME-driven shock moves outward from the Sun; and (2) a compound seed population, typically comprising at least suprathermals from the corona (or solar wind) and suprathermals from flares.

Figure 5 presents results from two recent theoretical studies that bolster the notion that θ_{Bn} should contribute to SEP spectral variability. Figure 5a (from *Giacalone*, 2005) shows shock-accelerated proton spectra, as derived from time-dependent numerical simulations for different mean values of θ_{Bn}. The spectra rollover more steeply at the highest energies when θ_{Bn} is small, thus implying that if a shock takes on a range of θ_{Bn} values, the high energies will be dominated by particles produced when the shock was quasi-perpendicular. *Lee* [2005] derived qualitatively similar spectra by considering the explicit θ_{Bn} dependence of the diffusion coefficient that governs escape from the shock region.

---

[1] For traveling interplanetary shocks near Earth, the situation is just the opposite: events in which Fe/O increases with energy comprise just a few percent of the total number, while events in which Fe/O decreases with energy are much more common [*Desai et al.*, 2004; *Tylka et al.*, 2005]. Whatever conditions are responsibile for the variability in Figure 4, their relative probablilities are reversed at the Sun and at 1 AU.

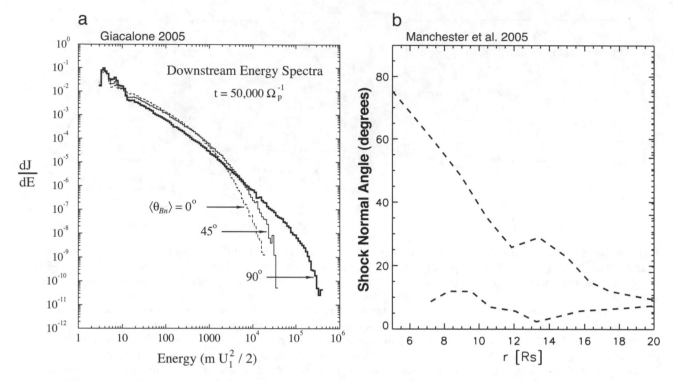

Figure 5. Results from two recent theoretical investigations with implications for the role of the shock-normal angle θ_{Bn} in SEP events. Figure 5a (left panel) shows calculations of the differential proton energy spectrum from time-dependent numerical calculations [*Giacalone*, 2005] for three different mean values of θ_{Bn}. Energy is given in terms of the proton mass m and U_1, the upstream in-flow speed in the shock rest frame. For $U_1 = 2000$ km/s, this energy scale corresponds to ~20 keV, and the maximum energy in the $\langle\theta_{Bn}\rangle = 90°$ spectrum is ~6 GeV. The total elapsed time in this calculation is 50,000 Ω_p^{-1}, where Ω_p is the proton cyclotron frequency. This elapsed time corresponds to ~20 seconds for a typical solar magnetic field of 0.25 G at ~3 R_S [*Gopalswamy et al.*, 2001; *Mann et al.*, 2003]. Figure 5b (right panel) shows evolution of θ_{Bn} along two magnetic flux lines, as derived from a numerical MHD simulation of a CME-driven shock moving through the corona at ~1000 km/s [*Manchester et al.*, 2005]. Taken together, these two figures suggest that evolution in θ_{Bn} should be a significant factor in SEP spectral variability at high energies.

Figure 5b (from *Manchester et al.*, 2005) shows results from a three-dimensional MHD simulation of a CME-driven shock moving through the corona. The figure shows the change in θ_{Bn} along two field lines at the same solar longitude but at different solar latitudes. Due to the non-radial expansion of the CME (which has been observed at low altitudes; see, for example, *Zhang et al.*, 2001), the shock initially broadsides one field line and is therefore nearly perpendicular. As the shock expands and envelopes this field line, θ_{Bn} decreases, falling to 10° by 20 R_S. However, on the other field line, the shock normal remains nearly parallel to the magnetic field along this whole distance. The shock-normal angles for both field lines reach values near 45° at 1 AU because of the spiral nature of the interplanetary magnetic field. Of course, this later evolution is generally not relevant for high-energy SEPs [*Zank et al.*, 2000, Figure 7].

Figure 5b is not precisely what we need to address SEP observations. For one thing, the simulations do not trace the

shock from ~2-3 R_S, where SEP production apparently starts, at least in some events (see Figure 2). The calculations also employ a solar-minimum coronal model, far simpler than coronal configurations at solar maximum, when most SEPs are produced. Nevertheless these simulations, combined with injection altitudes inferred from the SEP timing studies and the extensive body of theoretical work represented in Figure 5a, make evolution in θ_{Bn} a strong and natural candidate for explaining SEP spectral variability.

But, as we have already discussed, high-energy spectral shape and composition are tied together. How can θ_{Bn} affect SEP composition? Figure 6 (from *Tylka et al.*, 2005) sketches the potential connection. The seed population for shock-accelerated SEPs comprises both suprathermals from the corona (or solar wind, depending upon the location at which the acceleration occurs) and from flares. As noted in Figure 6, these two components have distinctive compositional characteristics. The flare ions in the seed population could be

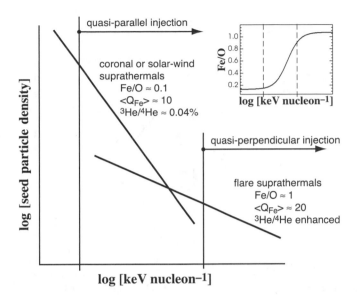

Figure 6. A schematic representation of the suprathermal seed population for shock-accelerated solar energetic particles, comprising both coronal (or solar-wind) and flare-accelerated ions. The flare suprathermals are more likely to be apparent in quasi-perpendicular shocks, for which the injection threshold is higher. The inset (upper right) shows how the Fe/O ratio in the seed population changes with energy. As the geometry evolves as the shock moves out from the Sun (generally, from quasi-perpendicular toward quasi-parallel), the nature of the accessible seed population would also change. [from *Tylka et al.*, 2005]

either remnants from previous activity [*Mason et al.*, 1999; *Tylka et al.*, 2001] or come from the associated flare activity [*Reames*, 2002; *Li and Zank*, 2005] if open field lines connect the flare site to the shock front. Surveys of interplanetary $^{3}He/^{4}He$ [*Richardson et al.*, 1990; *Laivola et al.*, 2003; *Torsti et al.*, 2003; *Wiedenbeck et al.*, 2003, 2005], compared to the average solar-wind value [*Gloeckler and Geiss*, 1998], suggest that flare suprathermals become more important at higher seed energies.

Under most conditions, efficient acceleration requires a higher initial particle speed at quasi-perpendicular shocks than at quasi-parallel shocks [*Forman and Webb*, 1985; *Jokipii*, 1987; *Webb et al.*, 1995; *Zank et al.*, 2004; but see *Giacalone*, 2005 for a dissenting view.] Thus, as also sketched in Figure 6, flare suprathermals are more likely to be the seed particles for quasi-perpendicular shocks. But as the shock moves outward and θ_{Bn} decreases, the spectra soften, while at the same time the injection threshold is lowered, so that the coronal component increasingly dominates the accessible seed population. The net effect of this evolution is to allow the unique characteristics of flare-suprathermals to be preferentially reflected among higher-energy SEPs, causing Fe/O (as well as $<Q_{Fe}>$ and $^{3}He/^{4}He$) to increase with energy.

Correlations among these quantities (Figure 3b above; also *Cane et al.*, 2003) thus arise naturally: they are "built into" the flare-component of the seed population.

Of course, the relative sizes of the coronal and flare components in the seed population can vary from event to event. Moreover, as suggested by Figure 5b, there can also be field lines along which the quasi-perpendicular phase is absent. In those cases, we might expect to see exponential rollovers within the energy range of our instruments, as well as charge states characteristic of the corona and/or solar-wind.

Tylka and Lee [2006] have presented a simple analytical implementation of this scenario. They generalize the *Ellison and Ramaty* [1985] functional form and write the differential spectrum F_i of ion species i as

$$F_i (E, \theta_{Bn}) = C_i E^{-\gamma} exp (-E/E_{0i}), \qquad (1)$$

where

$$E_{0i} = E_0[Q_i/A_i][\sec \theta_{Bn}]^{2/(2\gamma-1)}. \qquad (2)$$

In this equation, C_i is a normalization coefficient proportional to the abundance of species i in the seed population; Q_i/A_i is the charge-to-mass ratio of ion species i; and E_0 is a free parameter with dimensions of energy. Neglecting transport and non-equilibrium effects, the power-law index γ is determined by the compression ratio and is therefore explicitly taken to be the same for all species. The e-folding energy E_{0i}, on the other hand, reflects limits of the acceleration process, as imposed by escape from the shock region. Dependence on species and θ_{Bn} therefore resides in E_{0i}. The explicit dependence on θ_{Bn} comes from *Lee* [2005].

To approximate evolution in θ_{Bn}, spectra of this form are averaged over $\xi = \cos \theta_{Bn}$. The flare component of the seed population is accessible to the shock at all ξ. However, to represent the injection bias sketched in Figure 6, a heuristic weighting factor of ξ is applied to the coronal component, so that its contribution becomes increasingly suppressed as the shock approaches perpendicular. To complete the calculations, nominal relative abundances [*Reames*, 1995b] and ionic charge-state distributions are assumed for both components of the seed population. (See *Tylka and Lee*, 2006 for further details.)

Figures 7a and 7b show two calculations from this model, both with the same values of γ and E_0. In both calculations, the seed populations are also the same, with the flare component providing 5% of the oxygen ions and a concomitantly larger proportion of the Fe ions, as dictated by the different Fe/O ratios in the coronal and flare components. The two calculations differ only in the averaging range for θ_{Bn}. In the so-called "quasi-perpendicular" case, the averaging covers the full range, $0 \le \theta_{Bn} \le 90°$. In the "quasi-parallel" case, on

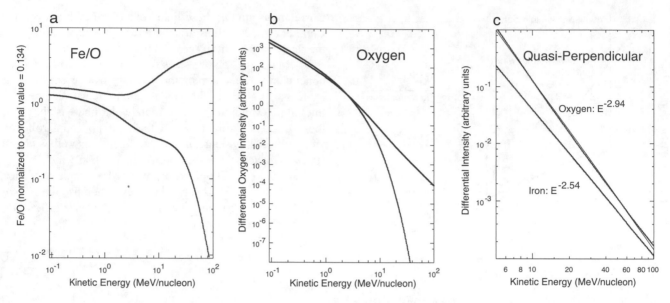

Figure 7. Model calculations from *Tylka and Lee* [2006] for (a) the Fe/O ratio (normalized to the nominal coronal value) versus energy and (b) the differential oxygen spectrum. The model parameters are the same in both cases except that one is the "quasi-perpendicular" case, in which the spectra are averaged over the full range of $0 \leq \theta_{Bn} \leq 90°$; the other is the "quasi-parallel" case in which $0 \leq \theta_{Bn} \leq 60°$. (See text for details.) Panel (c) offers a closer look at the oxygen and iron spectra from the quasi-perpendicular case at 5-100 MeV/nucleon. The lines are power-law fits to the model calculations. The fitted power-law indices for oxygen and iron are 2.94 and 2.54, respectively.

the other hand, the average runs over only $0 \leq \theta_{Bn} \leq 60°$, since the advantages of quasi-perpendicular geometry (in terms of the particles' rate of energy gain) become significant only when θ_{Bn} exceeds ~60° [*Jokipii*, 1987; *Webb et al.*, 1995]. In these calculations, the quasi-perpendicular case produces Fe/O that rises with energy and a roughly double-power-law oxygen spectrum. Omitting the quasi-perpendicular range of angles produces Fe/O that falls with energy and an oxygen spectrum that rolls over roughly exponentially. Except for the Fe/O below ~1 MeV/nucleon, these results are qualitatively similar to that of the two events shown in Figure 4.

Figure 7c offers a closer look at the iron and oxygen spectra in the quasi-perpendicular case at energies of 5-100 MeV/nucleon, where the rise in Fe/O with energy is often observed. The model calculations for both spectra are roughly power-laws over this restricted range. However, the power-law indices are different for the two species, with iron harder than oxygen, just as we see in the data [*Tylka and Dietrich*, 1999; *Tylka et al.*, 2002, 2006; *Lehtinen et al.*, 2005]. The magnitude of the spectral difference is also comparable to what we find in the data. Different power-laws for different species has been touted as an ostensible challenge for explaining these data in terms of shocks. But this calculation shows that species-dependence can arise naturally in the context of shock acceleration, even though the calculation

explicitly started with the same power-law index for all species.

Tylka and Lee [2006] show that the model, in spite of its simplicity, accounts for observed features of the high-energy SEP data in quantitative ways that no other model has matched to date. By varying parameters, the model reproduces the whole morphological range of spectral shapes and energy-dependent Fe/O. These same calculations also yield energy-dependent ^{3}He/^{4}He and $<Q_{Fe}>$ similar to those that have been reported in the data[2]. The shock-geometry model provides a quantitative explanation for the *Breneman and Stone* [1985] fractionation effect, a fundamental feature of SEP phenomenology that has been known for 20 years but has heretofore been unexplained. The shock-geometry hypothesis also has implications for other features of the data, including event size, Fe-richness in GLEs, the shapes of high-energy time profiles (a particular concern for *Cane et al.*, 2003), and the SEP event distributions in source longitude and CME speed.

---

[2] Elsewhere in these Proceedings, *Kocharov* reviews efforts to understand energy-dependent $<Q_{Fe}>$ in terms of stripping during the acceleration process. But it is presently unclear whether such models can account for the enhanced Fe/O ratios that accompany high $<Q_{Fe}>$.

However, the *Tylka and Lee* [2006] model can only be regarded as a promising starting point for further investigations. For example, the analytical implementation incoporates only evolution in θ_{Bn}. But other key factors, such as the compression ratio, seed particle densities, and ambient plasma conditions, also change as the shock moves outward. It will therefore be important to embed the model within a realistic CME-shock simulation, coupled with a rigorous treatment of particle transport, such as *Ng et al.* [2003]. These factors are probably particularly important for understanding observations below ~1 MeV/nucleon. A more sophisticated implementation will also make it possible to test the model with time-dependent intensities, spectra, and abundance ratios, rather than just event-integrated quantities.

2.3. A Direct Flare Component at High Energies

It is generally agreed that CME-driven shocks dominate in producing SEPs below ~10 MeV/nucleon. It has been argued above that higher-energy SEPs with flare-like composition are also produced by a shock, operating on a seed population that contains suprathermals from flare activity, thereby leading to the ostensible "blurring" of the compositional distinctions between impulsive and gradual events. However, an alternative hypothesis would be two distinct acceleration mechanisms operative in the same event, with the shock producing the majority of the lower-energy particles while the flare alone directly generates most of the higher-energy particles. The observational support for this direct-flare hypothesis is reviewed by *Cane et al.* in these Proceedings. *Tylka et al.* [2005] discuss some of the observational challenges for this alternative hypothesis. *Tylka and Lee* [2006] also show that the *average* high-energy Fe/O enhancement among Fe-rich events disfavors a direct flare component.

The word "flare" in this context is often used broadly, to refer to any particle acceleration via reconnection, regardless of when or where it occurs. In particular, "flare" need not refer to the main electromagnetic energy release in the "impulsive" phase of an event. Instead, "flare" can also refer to the post-eruption reconnection that occurs on bigger, lower-density loops beneath the CME. *Cane et al.* [2002] have proposed these processes as a significant source of interplanetary SEPs. Other authors [e.g., *Klein and Posner*, 2005; *Miroshnichenko et al.*, 2005] have a similar idea, although in their case, the reconnection may be part of a global restructuring of the corona after the CME launch and hence need not be at the launch site. Such models might naturally be consistent with the timing results shown in Figure 2. But further theoretical development is required before we can know what expectations these models bring for SEP spectra and composition, and how those expectations compare to the data. Moreover, as pointed out by *Cliver et al.* [2004],

post-eruption reconnection is a common feature of all CMEs. Accordingly, if these ideas are correct, one might wonder why slow CMEs do not also produce SEPs.

2.4. Smoothed Shocks and Compression Regions

As we have seen, one of the challenges in the high-energy SEP data is the fact that in many events, iron has a harder spectrum than oxygen, so that Fe/O increases with energy. *Eichler* [1979] first pointed out that such behavior could arise in the case of a so-called smoothed shock, in which there is gradual change in plasma conditions over a finite spatial region, rather than an abrupt discontinuity. Scattering mean free paths generally increase with rigidity. Iron ions, because of their higher A/Q, would therefore sample a larger compression ratio than oxygen. This behavior should be most dramatic when Fe has a particularly low charge-state, leading to anti-correlation between Fe/O enhancements and $<Q_{Fe}>$. Similar spectral differences might also arise at the compressive acceleration regions recently considered by *Jokipii et al.* [2003].

But as already shown in Figure 3b, these expectations are *precisely the opposite* of what we see in the data: the strongest Fe/O enhancements are seen in events in which the A/Q values of Fe and O are nearly the same; and when Fe does have a significantly higher A/Q than oxygen, Fe/O is strongly suppressed at high energies. In order to make this hypothesis work, it would apparently be necessary that the smoothed shocks and/or compressive regions arise preferentially in conjunction with high density regions, where additional stripping could occur after, or perhaps during, the acceleration processes.

In addition, protons have the smallest A/Q of all. In the smoothed-shock and compressive-region scenarios, we should therefore expect protons to have a *softer* energy spectrum (expressed in terms of MeV/nucleon) than heavier ions. To our knowledge, this behavior has never been observed.

2.5. Near-Shock Scattering Conditions

Cohen et al. [2003] proposed that event-to-event variability in energy-dependent Fe/O could be understood in terms of near-shock scattering conditions. In particular, they expressed this notion as a generalization of the *Ellison and Ramaty* form, in which the e-folding energy need not be directly proportional to Q/A, but rather scales as an arbitrary power, $E_{0i} = E_0(Q_i/A_i)^{\delta}$. (They also suggested that E_{0i} need not be an e-folding energy *per se*, but could also describe the energy at which one power law steepens into another, such as in the *Band*-function fit shown in Figure 4.) As shown in Figure 8, by choosing negative (or positive) values for δ, Fe/O can be made to increase (or decrease) with energy.

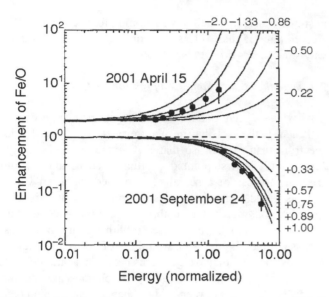

Figure 8. Model calculations for the energy dependence of Fe/O observed by *ACE*/SIS in two events, based on modifications to the rigidity-dependence of scattering conditions near the shock. The numbers labeling the curves indicate the value of δ, where particle spectra's e-folding energies scales as $(Q/A)^\delta$. See text for further discussion. [from *Cohen et al.*, 2003]

But this hypothesis hits a snag when we remember that SEP Fe ions arise from a *distribution* of charge states. In the case of $\delta < 0$, low-charge Fe ions would have a higher e-folding energy than high-charge Fe ions and therefore a harder spectrum at high energies. This model with $\delta < 0$ would therefore predict that $\langle Q_{Fe} \rangle$ should *decrease* with energy. This behavior has never been observed[3]. This prediction directly contradicts, for example, observations of the 2001 April 15 GLE in which both Fe/O [*Tylka et al.*, 2002] and $\langle Q_{Fe} \rangle$ [*Dietrich and Tylka*, 2003] increase with energy. Like the smoothed-shock hypothesis, using $\delta < 0$ is also problematical in that it would cause protons to have a softer spectrum than heavier ions.

---

[3] One might wonder whether $\langle Q_{Fe} \rangle$ decreasing with increasing energy is even physically reasonable. All processes by which charged particles are accelerated boil down to the actions of electric fields, even if it is more convenient to describe the processes in terms of diffusion or scattering from magnetic irregularities. Provided that all ions are subjected to the same environment, an electric field operating on a group of ions with the same mass, but different charges, will always push the larger-Q ions to the highest energies, causing $\langle Q_{Fe} \rangle$ to increase with energy.

3. IMPULSIVE SEP EVENTS

Mason et al. [2002] surveyed spectral shapes in 14 impulsive events with substantial $^3\text{He}/^4\text{He}$ enrichments (>4%) at 385 keV/nucleon. The survey examined the spectra of ^3He, ^4He, and heavy ions (C through Fe) at ~80 keV/nucleon to ~15 MeV/nucleon. (A notable absence in the survey is protons, whose distinctive charge-to-mass ratio might shed further light on the acceleration process.) *Mason et al.* classified the spectral morphologies into three types: (1) events in which ^3He and Fe are rounded below ~1 MeV/nucleon, in a manner consistent with a stochastic acceleration model, at least at the lower energies; (2) events in which the spectra of all species except ^3He are consistent with a common single power-law; (3) events in which all species, including ^3He, exhibit similar double power-laws, with the spectral break at ~1 MeV/nucleon. *Mason et al.* further speculated that these three event types may correspond to successive steps in a multi-stage acceleration process, in which the first type "gives us our closest look" into the basic mechanisms for the ^3He (and Fe) enrichment, while the other event types "represent further stages of acceleration that modify the spectra."

But *Reames and Ng* [2004] have presented observations challenging the notion of a multi-stage acceleration process in impulsive events. They examined spectral characteristics at ~3-10 MeV/nucleon in impulsive events with strong enhancements among ultraheavy ions with atomic number $Z > 34$. These ultraheavy enhancements grow with mass and attain values comparable to those of $^3\text{He}/^4\text{He}$. As shown in Figure 9, the magnitudes of the ultraheavy enhancements are correlated with the power-law index of Fe, so that enhancements are largest when the Fe spectrum is steepest. This correlation argues against two *independent* processes, with one causing the enhancements while the other produces the observed spectra. Furthermore, the largest ultraheavy enhancements were observed in the events with the smallest Fe fluence and the lowest x-ray intensities. *Reames and Ng* [2004] concluded that "small [flare] events with low energy input can produce only steep spectra of the dominant species but accelerate rare [ultraheavy] elements with great efficiency, probably by selective absorption of resonant waves in the flare plasma. With increased energy input, enhancements diminish as [ultraheavy] ions are depleted and spectra of the dominant species harden."

4. SUMMARY

In both gradual and impulsive events, there are correlations among spectral and compositional characteristics. These correlations are powerful constraints for our models.

In the case of gradual events, the shock-geometry hypothesis offers a promising synthesis of the rich phenomenology

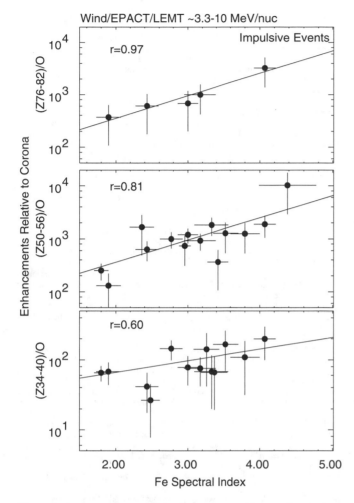

Figure 9. Enhancements of ultraheavy ions with atomic numbers Z = 76-82 (top), Z = 50-56 (middle), and Z = 34-40 (bottom) at ~3-10 MeV/nucleon in individual impulsive SEP events, plotted against the event's spectral index for Fe at those same energies. The results are correlated, so that enhancements are larger when the Fe spectrum is steeper. [from *Reames and Ng*, 2004]

revealed to us by the new instruments. This synthesis addresses not only the new data on high-energy spectral shapes and energy-dependent composition, but also long-standing puzzles on event-to-event variability, such as the origin of the Breneman & Stone fractionation effect. A great deal of additional work is needed to test this hypothesis and to put it on a firmer theoretical and observational basis. Are there adequate numbers of flare suprathermals in the corona to provide the seeds for the observed high-energy fluences in Fe-rich events? Do scattering conditions in the corona allow for a higher injection-threshold at quasi-perpendicular shocks? Can we find direct imaging or spectroscopic confirmation for our inferences about coronal shock geometry in individual events? In spite of these questions, one can nevertheless be

cautiously optimistic that we have identified key factors behind SEP spectral and compositional variability at high energies in most, if not all, gradual events.

In the case of impulsive events, the new phenomenology is no less rich. But at present, we have no comparable synthesis. In fact, one can argue that our most pressing need is a better understanding of impulsive events. After all, shock theory has given us definite ideas about how shock characteristics, such as speed and shock-normal angle, should be reflected in the energetic particles; the challenge now is to refine those ideas by implementing them in more sophisticated ways that better represent dynamic coronal and interplanetary conditions. But for impulsive events we have no comparable semi-quantitative framework for understanding how flare parameters, such as size, duration, and magnetic topology, drive the observed spectra and composition.

Acknowledgments.   We thank E.W. Cliver, C.M.S. Cohen, W.F. Dietrich, R.J. Murphy, C.K. Ng, D.V. Reames, and G.H. Share for informative and helpful comments on this manuscript. AJT was supported by NASA DPR NNH05AB581 and the Office of Naval Research. MAL was supported by NSF Grant ATM-0091527, NASA NNG05GL40G, and DoD MURI grants to the University of Michigan and the University of California at Berkeley.

REFERENCES

Band, D., *et al.*, BATSE observations of gamma-ray burst spectra. I. Spectral diversity, *ApJ*, 413, 281-292, 1993.

Bieber, J.W., *et al.*, Spaceship Earth observations of the Easter 2001 solar particle event, *ApJ*, 601, L103-L106, 2004.

Breneman, H.H. and E.C. Stone, Solar coronal and photospheric abundances from solar energetic particle measurements, *ApJ*, 299, L57-L61, 1985.

Cane, H.V., W.C. Erickson, and N.P. Presage, Solar flares, type III radio bursts, coronal mass ejections, and energetic particles, *J. Geophys. Res.*, 107, (A10), 1315, doi:10.1029/2001JA000320, 2002.

Cane, H.V., T.T. von Rosenvinge, C.M.S. Cohen, and R.A. Mewaldt, Two components in major solar particle events, *Geophys. Res. Lett.*, 30, 8017, doi:10.1028/2002GL016580, 2003.

Cliver, E.W., S.W. Kahler, and D.V. Reames, Coronal shocks and solar energetic proton events, *ApJ*, 605, 902-910, 2004.

Cohen, C.M.S., *et al.*, Variability of spectra in large solar energetic particle events, *Adv. Space Res.*, 32, (12), 2649-2654, 2003.

Desai, M.I., *et al.*, Spectral properties of heavy ions associated with the passage of interplanetary shocks at 1 AU, *ApJ*, 611, 1156-1174, 2004.

Dietrich, W.F. and A.J. Tylka, Time-to-maximum studies and inferred ionic charge states in the solar energetic particle events of 14 and 15 April 2001, *Proc. 28th Internat. Cosmic Ray Conference* (Tsukuba), 6, 3291-3294, 2003.

Eichler, D., Particle acceleration in collisionless shocks: regulated injection and high efficiency, *ApJ*, 229, 419-423, 1979.

Ellison, D. and R. Ramaty, Shock acceleration of electrons and ions in solar flares, *ApJ*, 298, 400-408, 1985.

Forman, M.A. and G. Webb, Acceleration of energetic particles, in *Collisionless Shocks in the Heliosphere: A Tutorial Review*, eds. R.G. Stone and B.T. Tsurutani, *Geophys. Monongr. 35*, 91-114, 1985.

Giacalone, J., Particle acceleration at shocks moving through an irregular magnetic field, *ApJ*, 624, 767-772, 2005.

Gloeckler, G. and J. Geiss, Measurement of the abundance of helium-3 in the sun and in the local interstellar cloud with SWICS on *Ulysses*, *Space Science Rev.*, 84, 275-284, 1998.

Gold, R.E., *et al.*, Electron, Proton, and Alpha Monitor on the Advanced Composition Explorer Spacecraft, *Space Sci. Rev.*, 86, 541-562, 1998.

Gopalswamy, N., A. Lara, M.L. Kaiser, and J.-L. Bougeret, Near-Sun and near-Earth manifestations of solar eruptions, *JGR*, 106, 25,261-25,277, 2001.

Gopalswamy, N., H. Xie, S. Yashiro, and I. Usoskin, Coronal mass ejections and ground level enhancements, *Proc. 29th Internat. Cosmic Ray Conference* (Pune), 1, 169-173, 2005.

Jokipii, J.R., Rate of energy gain and maximum energy in diffusive shock acceleration, *ApJ*, 313, 842-846, 1987.

Jokipii, J.R., J. Giacalone, and J. Kóta, Diffusive compression acceleration of charged particles, *Proc. 28th Internat. Cosmic Ray Conf.*, (Tsukuba), 6, 3685-3689, 2003.

Klecker, B., *et al.*, On the energy dependence of ionic charge states, *Proc. 28th Internat. Cosmic Ray Conf.* (Tsukuba), 6, 3277-3280, 2003.

Klein, K.-L. and A. Posner, The onset of solar energetic particle events: prompt release of deka-MeV protons and associated coronal activity, *A&A*, 438, 1029-1042, 2005.

Labrador, A.W., R.A. Leske, R.A. Mewaldt, E.C. Stone, and T.T. von Rosenvinge, High energy ionic charge state composition in recent large solar energetic particle events, *Proc. 28th Int. Cosmic-Ray Conf.* (Tsukuba), 6, 3269-3272, 2003.

Labrador, A.W., R.A. Leske, R.A. Mewaldt, E.C. Stone, and T.T. von Rosenvinge, High energy ionic charge state composition in the October/November 2003 and January 20, 2005 SEP events, *Proc. 29th Int. Cosmic-Ray Conf.* (Pune), 1, 99-103, 2005.

Laivola, J., J. Torsti, J, and L. Kocharov, A statistical study of ^{3}He enhancements in the high energy solar particles, *Proc. 28th Internat. Cosmic Ray Conf.* (Tsukuba), 6, 3233-3236, 2003.

Lee, M.A., Coupled hydrodynamic wave excitation and ion acceleration at an evolving coronal/interplanetary shock, *ApJS*, 158, 38-67, 2005.

Lehtinen, I., J. Torsti, and P. Mäkelä, Heavy elements in solar particle events during the Solar Cycle 23: *SOHO*/ERNE measurements, *Proc. 29th Int. Cosmic-Ray Conf.* (Pune), 1, 63-67, 2005.

Leske, R.A., R.A. Mewaldt, A.C. Cummings, E.C. Stone, and T.T. von Rosenvinge, The ionic charge state composition at high energies in large solar energetic particle events in Solar Cycle 23, *AIP Conference Proc.* 598, ed. R.F. Wimmer-Schweingruber, 171-176, 2001.

Leske, R.A., *et al.*, The unusual solar particle events of August 2002, *Proc. 28th Int. Cosmic-Ray Conf.* (Tsukuba), 6, 3253-3256, 2003.

Li, G. and G.P. Zank, Mixed particle acceleration at CME-driven shocks and flares, *GRL*, 32, L02101, doi:1029/2004GL021250, 2005.

Manchester, W.B., *et al.*, Coronal mass ejection shock and sheath structures relevant to particle acceleration, *ApJ*, 622, 1225-1239, 2005.

Mann, G., A. Klassen, H. Aurass, and H.-T. Classen, Formation and development of shock waves in the solar corona and the near-Sun interplanetary space, *A&A*, 400, 326-336, 2003.

Mason, G.M., *et al.*, The Ultra-Low-Energy Isotope Spectrometer (ULEIS) for the *ACE* Spacecraft, *Space Sci. Rev.*, 86, 409-448, 1998.

Mason, G.M., J.E. Mazur, and J.R. Dwyer, ^{3}He enhancements in large solar energetic particle events, *ApJ*, 525, L133-L136, 1999.

Mason, G.M. *et al.*, Spectral properties of He and heavy ions in ^{3}He-rich solar flares, *ApJ*, 86, 1039-1058, 2002.

Miroshnichenko, L.I., K.-L. Klein, G. Trottet, P. Lantos, E.V. Vashenyuk, and Yu.V. Balabin, Electron acceleration and relativistic nucleon production in the 2003 October 28 solar event, *Adv. Space Res.*, 35, 1864-1870, 2005.

Ng. C.K., D.V. Reames, and A.J. Tylka, Modeling shock-accelerated solar energetic particles coupled to interplanetary Alfvén waves, *ApJ*, 591, 461-485, 2003.

Reames, D.V., Solar energetic particles: a paradigm shift, *Rev. of Geophys.*, Supplement: U.S. National Report to the International Union of Geodesy and Geophysics, 1991-1994, Part 1: Contributions in Space Science, 585, 1995a.

Reames, D.V., Coronal abundances determined from energetic particles, *Adv. Space Res.*, 15,(7), 41-51, 1995b.

Reames, D.V., Particle acceleration at the Sun and in the heliosphere, *Space Sci. Rev.*, 90, 413-491, 1999.

Reames, D.V., Abundances of trans-iron elements in solar energetic particle events, *ApJ*, 540, L111-L114, 2000.

Reames, D.V., Magnetic topology of impulsive and gradual solar energetic particle events, *ApJ*, 571, L63-L66, 2002.

Reames, D.V. and C.K. Ng, Heavy-element abundances in solar energetic particle events, *ApJ*, 610, 510-522, 2004.

Richardson, I.G., D.V. Reames, K.-P. Wenzel, and J. Rodriguez-Pacheco, Quiet-time properties of low-energy (<10 MeV per nucleon) interplanetary ions during solar maximum and solar minimum, *ApJ*, 363, L9-L12, 1990.

Sáiz, A., *et al.*, Relativistic particle injection and interplanetary transport during the January 20, 2005 ground level enhancement, *Proc. 29th Internat. Cosmic Ray Conference* (Pune), 1, 229-233, 2005.

Stone, E.C., *et al.*, The Solar Isotope Spectrometer for the Advanced Composition Explorer, *Space Sci. Rev.*, 86, 357-408, 1998.

Torsti, J., J. Laivola, and L. Kocharov, Common overabundance of ^{3}He in high-energy solar particles, *A&A*, 408, L1-L4, 2003.

Tylka, A.J. and W.F. Dietrich, IMP-8 observations of the spectra, composition, and variability of solar heavy ions at high energies relevant to manned space missions, *Radiation Measurements*, 30, 345-359, 1999.

Tylka, A.J. and M.A. Lee, A model for spectral and compositional variability at high energies in large, gradual solar particle events, *ApJ*, 2006, in press.

Tylka, A.J., P.R. Boberg, R.E. McGuire, C.K. Ng, and D.V. Reames, Temporal evolution in the spectra of gradual solar energetic particles, *AIP Conference Proc.* 528, eds. R.A. Mewaldt *et al.*, 147-152, 2000.

Tylka, A.J., C.M.S. Cohen, W.F. Dietrich, C.G. Maclennan, R.E. McGuire, C.K. Ng, and D.V. Reames, Evidence for remnant flare suprathermals in the source population of solar energetic particles in the 2000 Bastille Day event, *ApJ*, 581, L59-L63, 2001.

Tylka, A.J., P.R. Boberg, C.M.S. Cohen, W.F. Dietrich, M.A. Lee, C.G. Maclennan, G.M. Mason, C.K. Ng, and D.V. Reames, Flare and shock-accelerated energetic particles in the solar events of 2001 April 14 and 15, *ApJ*, 581, L119-L123, 2002.

Tylka, A.J., C.M.S. Cohen, W.F. Dietrich, M.A. Lee, C.G. Maclennan, R.A. Mewaldt, C.K. Ng, and D.V. Reames, Shock geometry, seed populations, and the origin of variable elemental composition at high energies in large gradual solar particle events, *ApJ*, 625, 474-495, 2005.

Tylka, A.J., C.M.S. Cohen, W.F. Dietrich, M.A. Lee, C.G. Maclennan, R.A. Mewaldt, C.K. Ng, and D.V. Reames, Comparative study of ion characteristics in the large gradual solar energetic particle events of 2002 April 21 and 2002 August 24, *ApJS*, 2006, in press.

von Rosenvinge, T.T., *et al.*, The Energetic Particles: Acceleration, Composition, and Transport (EPACT) investigation on the *Wind* spacecraft, *Space Sci. Rev.*, 71, 155-206, 1995.

Webb, G.M., G.P. Zank, C.M. Ko, and D.J. Donohue, Multidimensional Green's functions and the statistics of diffusive shock acceleration, *ApJ*, 453, 178-206, 1995.

Wiedenbeck, M.E. *et al.*, How common is energetic ^{3}He in the inner heliosphere?, *AIP Conference Proc 679*, eds. M. Velli *et al.*, 652-655, 2003.

Wiedenbeck, M.E. *et al.*, The time variation of energetic ^{3}He in interplanetary space from 1997 to 2005, *Proc. 29th Internat. Cosmic Ray Conference* (Pune), 1, 117-120, 2005.

Yashiro, S., *et al.*, A catalog of white light coronal mass ejections observed by the *SOHO* spacecraft, *J. Geophys. Res.*, 109, A07105, doi:10.1029/2003JA010282, 2004.

Zank, G.P., W.K.M. Rice, and C.C. Wu, Particle acceleration and coronal mass ejection driven shocks: a theoretical model *J. Geophys. Res.*, 105, 25,079-25,095, 2000.

Zank, G.P., *et al.*, Perpendicular diffusion coefficient for charged particles of arbitrary energy, *J. Geophys. Res.*, 109, A04107, doi:10.1029/ 2003JA010301, 2004.

Zhang, J., K.P. Dere, R.A. Howard, M.R. Kundu, and S.M. White, On the temporal relationship between coronal mass ejections and flares, *ApJ*, 559, 452-462, 2001.

Allan J. Tylka, E. O. Hulburt Center for Space Research, Code 7652, Naval Research Laboratory, Washington, DC 20375, USA. (allan.tylka@nrl.navy.mil)

Martin A. Lee, Space Science Center and Institute for the Study of Earth, Oceans, and Space, University of New Hampshire, Durham, NH 03824-3525, USA. (marty.lee@unh.edu)

Observations of Energetic Storm Particles: An Overview

Christina M.S. Cohen

California Institute of Technology, Pasadena, California

An increase in the intensity of energetic particles associated with the passage of an interplanetary shock has long been referred to as an 'energetic storm particle (ESP) event.' Such increases have been observed since the 1960's and the particles were generally thought to be either locally accelerated or trapped in the vicinity of the shock and transported with it. This general overview will describe the initial observations and theories of ESP events as well as the measurements that have led to our current understanding of the creation of these events. The relevance of ESP events to space weather as well as furthering our understanding of particle acceleration by shocks is also discussed.

1. INTRODUCTION

Energetic storm particle (ESP) events have been studied for over 30 years. This overview attempts to describe the general evolution of the observations and theories of ESP events from their first detection to our current understanding. Naturally, such an overview is not exhaustive and cannot present all the relevant measurements or past theories. However, hopefully, it provides a useful context and background for the current studies, relevance, and conventional wisdom of ESP events.

1.1. Early Observations

Increases in energetic particle intensities near the vicinity of a shock were first reported by *Bryant et al.* [1962]. Observations on 30 September 1961 from the cosmic ray instrument onboard Explorer 12 showed substantial increases in the 2-15 MeV proton intensities just after the sudden commencement of a geomagnetic storm. It was this association with a geomagnetic storm and the corresponding Forbush decrease that led *Byant et al.* to dub the ions 'energetic storm particles' (ESPs). The authors found the ESPs to have a softer spectrum than solar energetic particles (SEPs)

Solar Eruptions and Energetic Particles
Geophysical Monograph Series 165
Copyright 2006 by the American Geophysical Union
10.1029/165GM26

and made the suggestion that they were solar protons trapped in a magnetic cloud region.

Several years later, *Rao et al.* [1967] reported on a survey of ESP events using cosmic ray detectors on the Pioneer 6 and 7 spacecraft. They found a one-to-one correspondence between the ESP events and Forbush decreases, as well as measurable anisotropies, which were not bi-directional, throughout the events. The authors argued against the ESPs being trapped in a magnetic cloud, primarily on the basis of the lack of bi-directional anisotropies, and suggested that interplanetary acceleration by the shock associated with the Forbush decrease was the cause of the enhanced intensities.

Kahler [1969] disagreed after performing his own survey of ESP events associated with magnetic sudden commencements using data from Explorers 12, 33, 34, and 35, Mariner 4, OGOs 1 and 3, Pioneer 6, and IMPs 3 and 4. The author argued that the association of ESP events with SEP events suggested that both particle populations were generated at the Sun (n.b. this was during the time when SEP events were thought to originate solely from solar flares). Previously *Lin et al.* [1968] had suggested that SEP events were comprised of 3 components: the diffusive component which could populate a wide range of magnetic field lines as it moved through the interplanetary medium, followed by the core and halo components which were emitted over a narrow range of longitudes (the halo being made of higher energy particles and thus having a larger spatial extent than the core). *Kahler* argued that the latter two components were always accompanied by an interplanetary plasma cloud. Such a cloud could drag magnetic field lines outward, distorting them such that appropriately positioned spacecraft would intercept the field lines populated with halo protons and subsequently observe and increase in particle intensity, i.e., an ESP event.

1.2. Early Theories

By the early 1970's the dominant theories suggested ESP events were a result of 1) acceleration by bouncing between the Earth's bow shock and a traveling interplanetary shock [*Axford and Reid*, 1963], 2) diffusive shock acceleration at quasi-parallel interplanetary shocks [e.g., *Jokipii*, 1966; *Scholer and Morfill*, 1975], 3) shock drift acceleration at quasi-perpendicular interplanetary shocks [e.g., *Hudson*, 1965], 4) sweeping by traveling magnetic structures [*Palmer*, 1972], or 5) trapping behind a moving interplanetary shock. Idea 1 was rejected when observations indicated that the distance between the bow shock and the position of the spacecraft during some ESP events was approximately an order of magnitude larger than the gyroradius of a typical ESP and not connected along nominal magnetic field lines indicating that particles were unlikely to be efficiently bouncing back and forth and gaining significant energy [*Rao et al.*, 1967]. The theory of magnetic sweeping (item 4, above) was discounted when observations indicated that the increase in particle intensity often occurred behind the shock and not exclusively ahead of the shock as would be expected by the theory [*Sarris and van Allen*, 1974]. Further, simulations were able to produce only ESP events significantly smaller than typically observed and even then required large diffusion coefficients and a totally reflecting shock, aspects not typical in interplanetary space [*Palmer*, 1972; *Rao et al.*, 1967]. Theory 5, trapping behind the shock, was not significantly pursued as an option as scientists focused heavily on shock-related acceleration (ideas 2 and 3, above). It was, however, preferred by *Simnett and Holt* [1970] as the explanation for an event observed by several spacecraft at different radial and longitudinal locations.

2. CATEGORIZATION

Sarris and van Allen [1974] proposed a two category system for ESP events based on the observed particle intensities: Spike events and classic ESP events. The characteristics are given in Table 1. The primary observational difference was the size of the event and the theoretical distinction was believed to be the acceleration mechanism. In general, this categorization is still used today, however the more detailed measurements available from instruments such as the Electron Proton Alpha Monitor (EPAM) on the ACE spacecraft have lead to the defining of three additional categories to describe more complicated events [e.g., see *Lario et al.*,

Table 1. Proton Characteristics of Spike and Classic ESP Events (adapted from *Sarris and van Allen* [1974])

Characteristic	Spike Event	Classic Event
Duration	5-20 minutes	Several hours
Arrival	Within 5-10 minutes of shock	Ahead or behind shock
Maximum Energy	~5 MeV	~20 MeV
Acceleration Mechanism	Shock Drift	Diffusive Shock
Shock orientation	Quasi perpendicular	Quasi parallel

2003 and *Ho et al.*, 2005]. Figures 1 and 2 show examples of spike and classic ESP events (as identified by *Ho et al.* [2005]) using EPAM ion data.

2.1. Spike Events

Spike events are short in duration (5-10 minutes) and arrive within minutes of the shock. Observed increases in proton intensities rarely exceed 5 MeV. All of these characteristics are apparent in the event shown in Figure 1. It was proposed by *Sarris and van Allen* that these events are a result of shock drift acceleration at quasi perpendicular shocks.

The authors argued that the short duration of these events is chiefly a consequence of the difficulty in maintaining a perpendicular (or quasi-perpendicular) shock for extended periods of time. As effective acceleration is likely to occur

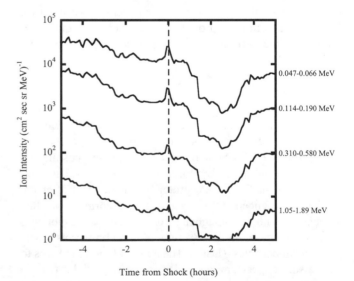

Figure 1. Example of a spike ESP event (2002 day 149), showing ion intensities from the EPAM LEMS120 telescope as a function of time from the shock passage (at 15:04 UT as identified by the MAG and SWEPAM instruments). Energy ranges for each trace are given to the right of the figure.

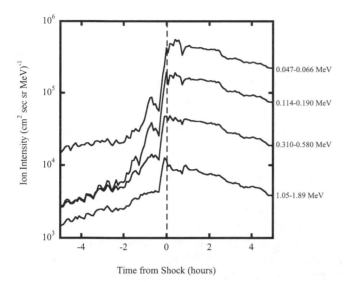

Figure 2. Example of a classic ESP event (2002 day 238), showing ion intensities from the EPAM LEMS120 telescope as a function of time from the shock passage (at 06:21 UT as identified by the MAG and SWEPAM instruments). Energy ranges for each trace are given to the right of the figure.

only over a small angular regime (80-90 degrees), magnetic field fluctuations easily cause the shock orientation to move out of that range.

The limited time that a shock remains perpendicular also affects the typical maximum energy to which the particles are accelerated. High energy ions require longer times for acceleration and it was suggested that the orientation of a shock will typically be stable only long enough to produce particles with maximum energies around a few MeV.

These suggestions were supported by a series of spike events observed on 20 February 1968. Four distinct increases in the proton intensity were observed over the course of 3 hours. It was noticed that each event occurred when the orientation of the magnetic field was roughly 80-90 degrees [see Figure 7 of *Sarris and van Allen*, 1974] with no increases observed when the field angle was smaller. Although only the first spike event was associated with a shock passage, the authors suggested that the subsequent events were 'remnants' of previous occasions when the shock was perpendicularly oriented. As these remnants diffused away from the shock they decreased in magnitude but remained large enough to be observed.

2.2. Classic ESP Events

The durations of classic ESP (or simply ESP) events are typically longer than those of spike events according to *Sarris and van Allen*. ESP events can last for several hours

and might arrive ahead or behind the shock. The maximum energy at which proton increases are observed is higher than for spike events, extending to ~20 MeV. These characteristics can be seen in the event in Figure 2. In contrast to the spike event in Figure 1, the lowest energy ion intensities do not peak at the shock passage but slightly afterwards. The >1 MeV ion intensities for this classic event show a strong ESP increase while the spike event is barely evident at these energies in Figure 1. It should be noted that the maximum energy to which particles are being locally accelerated in an ESP event is still a topic of some discussion. *Lario and Decker* [2002] have presented evidence suggesting the large ESP event of 20 October 1989 (where increases were seen at energies >500 MeV) was not the result of local shock acceleration (as presumed by *Reames* [1999b] and others) but was rather an earlier accelerated population trapped by a magnetic structure ahead of the shock and traveling with it. Regardless, the suggested source of the classic ESP events is diffusive shock acceleration at oblique or quasi-parallel shocks. As these events are generally bigger (in duration, energy extent, and intensity) there are more studies of classic ESP events, particularly in terms of composition and spectra.

The first study of heavy ion composition in such events was by *Klecker et al.* [1981] involving measurements of He, C, O, and Fe. The composition data were obtained using the Ultra-Low-Energy-Wide-Angle-Telescope (ULEWAT) and the Ultra-Low-Energy-Charge-Analyzer (ULECA) on the ISEE-3 spacecraft. Additionally, charge state measurements were available from the Ultra-Low-Energy-Z-E-Q-Analyzer (ULEZEQ) on the same spacecraft. The authors found that during the ESP event on 28-29 September 1978 the heavy ion spectra steepened by an amount that depended on the particle's mass/charge (A/Q) ratio. This resulted in Fe/He and Fe/O abundance ratios that decreased during the event when measured at a constant energy/nucleon value. *Klecker et al.* suggested that the A/Q effects were a result of diffusive shock acceleration involving rigidity-dependent mean-free path lengths.

More recent work on heavy ion composition in classic ESP events was reported by *Desai et al.* [2003]. This survey of 56 interplanetary shocks and the associated energetic particle increases revealed 25 cases of substantial amounts of ^{3}He being accelerated. The ^{3}He/^{4}He ratios varied between factors of 3 and 600 over the solar wind value. This, along with the overall heavy ion composition in these events, suggested that the solar wind was unlikely to be the sole seed population for these accelerated ions. The authors suggested the results could be best understood as the result of a A/Q-dependent acceleration of a seed population that itself was composed of multiple sources, including remnant flare material enriched in ^{3}He.

2.3. Anisotropy

Measurements of the anisotropy of 35-1600 keV proton increases during shock passages helped to seal the associations between spike events and shock drift acceleration and between ESP events and diffusive shock acceleration. Presenting two examples from a survey of ~40 events, *Wenzel et al.* [1985] illustrated how the anisotropies downstream of a classic ESP event and a spike event differed substantially in the manner predicted by diffusive shock acceleration and shock drift acceleration respectively.

Proton data from the low-energy proton experiment on ISEE-3 revealed isotropic distributions for several hours during a large particle intensity increase associated with the passing of a shock on 5 April 1979. In contrast, data obtained during a short-duration spike coincident with a shock passage on 9 March 1979 showed pitch-angle distributions that were peaked at 90 degrees from the magnetic field direction immediately downstream of the shock. The plasma and magnetic field measurements from instruments on ISEE 3 indicated that the shock on 5 April 1979 was quasi-parallel with significant upstream wave activity, while the shock on 9 March 1979 was quasi-perpendicular and had no wave activity upstream.

The particle, plasma, and field observations during the April event are consistent with the diffusive shock acceleration theory of *Lee* [1983]. His theory predicts upstream wave activity, downstream particle isotropy, and softening of the spectra as the shock approaches; all signatures seen in the ISEE 3 data from the April event but not in the March event. Analysis of acceleration by quasi-perpendicular shocks by *Decker* [1983] resulted in particle anisotropies and intensity increases that matched those observed in the 9 March 1979 event indicating this event was likely the result of shock-drift acceleration.

In the larger survey of ~40 shock passages (examined regardless of the presence of an associated particle increase), it was found that the particle anisotropy downstream increased (as well as the occurrence of 90 degree peaked distributions) as the angle between the shock and the magnetic field, θ_{Bn}, increased [*Sanderson et al.*, 1983]. It was thus concluded that quasi-perpendicular shocks ($\theta_{Bn} > {\sim}60°$) were a sign of shock-drift acceleration and quasi-parallel shocks ($\theta_{Bn} < {\sim}50°$) indicated diffusive shock acceleration. The authors observed more quasi-perpendicular shocks than quasi-parallel ones, but found the larger particle increases were associated with quasi-parallel shocks. This was consistent with the observations of *van Nes et al.* [1984] for shocks with $30° < \theta_{Bn} < 60°$ and suggested that diffusive shock acceleration was more efficient at accelerating 35-56 keV protons than shock-drift acceleration.

3. CURRENT UNDERSTANDING AND APPLICATION

3.1. Cause of ESP Events

In 1995, *Cane* [1995] suggested that SEP and classic ESP events were basically the same things - enhanced particle intensities as the result of shock acceleration. As a shock moves from near the Sun toward 1 AU, it may accelerate particles along the way, some of which will escape upstream of the shock to be observed by spacecraft as an SEP event. If a spacecraft is located such that it intercepts a passing shock and the shock is strong enough to still be accelerating particles, an ESP event will be observed.

Using data from the Helios 1 and 2 and IMP-8 spacecraft, *Cane* was able to examine how the particle intensity time profiles in a given event differed when observed at different longitudes. Cases where the nose of the shock (the strongest part of the shock) passed over the spacecraft were most likely to involve an ESP event. It was also found that SEP events with an eastern origin (relative to the spacecraft) had typical time profiles: a gradual rise peaking at or after the passage of a shock (when one was observed). Western events also had a common profile that differed from eastern events: a fast rise in the particle intensities followed by an exponential decay. Thus, the observed evolution of the particle intensities (and the presence/absence of ESP events) depended on the properties of the shock (strength/speed), the spacecraft's magnetic connection to the shock, and how both of these aspects evolved as the shock moved outwards from the Sun.

This view was further supported by *Kallenrode* [1996] in a survey of 351 interplanetary shocks observed by the two Helios spacecraft and the associated increases in the 5 MeV proton intensities. *Kallenrode* attempted to distinguish between particle intensity increases resulting from acceleration close to the Sun (and then propagating to the spacecraft) and those resulting from shock acceleration at or near the spacecraft. In cases where an ESP event was observed, this distinction was relatively straightforward. In other events where the increase was moderate, gradual, and not well peaked at the shock passage the distinction was more subjective. The local or interplanetary component of the events (i.e., typically the ESP event) was found to be larger, both absolutely and relative to the solar component, for observers closer to the flare normal, i.e., the nose of the shock. The size of the ESP event diminished when observed along the flanks of the shock.

The acceptance of a common acceleration mechanism for ESP and SEP events has sparked interest in ESP events as a potential way of understanding puzzling observations of SEP events. Contrary to expectations, initial results from Solar Isotope Spectrometer (SIS) on the ACE spacecraft found large SEP events with compositional signatures expected only in small, flare-related SEP events [*Cohen et al.*, 1999]; specifically, large enhancements of heavy ions (e.g., Fe/O) above 10 MeV/nucleon. Further measurements by instruments on ACE and Wind have confirmed these observations and there have been several attempts to find possible explanations [e.g., see *Mason et al.*, 1999; *Tylka et al.*, 2001; *Mewaldt et al.*, 2003; *Cohen et al.*, 2003; *Cane et al.*, 2003; *Tylka et al.*, 2005].

One suggested scenario involves the orientation of the shock as it accelerates particles near the Sun [*Tylka et al.*, 2005]. This idea proposes that perpendicular shocks are more likely to produce enrichments of heavy ions than parallel shocks due to the higher injection energy required. This causes the accelerated ions to be drawn primarily from the suprathermal population, where presumably there is a larger contribution from flare SEP remnants, at perpendicular shocks. Since it is currently not possible to obtain the required measurements (near the Sun) to confirm or disprove this hypothesis, studying ESP events, where the shock orientation and particle composition can both be directly measured, may prove useful.

According to *Tylka et al's* idea, one might expect to see enhancements in heavy ions in spike events where shock acceleration at quasi-perpendicular shocks is occurring. On 23 May 2002 SIS observed roughly order of magnitude increases in heavy ion intensities (He-Fe) around 14 MeV/nucleon (Figure 3) in association with a shock

passage at 10:15 UT. Fits to the data from the magnetometer (MAG) and solar wind plasma (SWEPAM) instruments on ACE indicate the shock θ_{Bn} was 84° (http://www-ssg.sr.unh.edu/mag/ace/ACElists/obs_list.htm). Although abundance ratios for N/O and Mg/O did not change significantly, the Fe/O ratio decreased by a factor of ~5 during the event (Figure 4) in contrast to expectations. However, it is hard to classify this event as a spike event in that the duration was atypically long (~2 hours) and it extended to unusually high energies. Additionally, other events with increased Fe/O ratios at quasi-perpendicular shocks have been observed [e.g., see Figure 4 of *Tylka et al.*, 2005] and low Fe/O ratios have been observed for shocks with a large range of θ_{Bn} values (including quasi-perpendicular). Clearly more study is warranted.

3.2. Space Weather

With the continual presence of humans in space and more sensitive scientific equipment being stationed there, temporal changes in the radiation environment (especially near Earth) is currently a strong concern. While it has been established that proton intensities in SEP events are generally governed by streaming limits [*Reames and Ng*, 1998], which can be theoretically calculated and predicted, these limits do not apply to ESP events. Large ESP events can result in dramatic, short-term increases in the particle intensities by more than

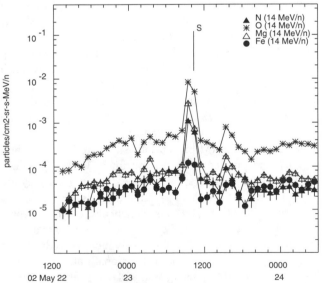

Figure 3. Heavy ion intensities (at 14 MeV/nucleon) from the SIS instrument as a function of time showing an ESP event. The time of shock passage is indicated by the 'S' and line at the center of the plot (courtesy of *T.T. von Rosenvinge*).

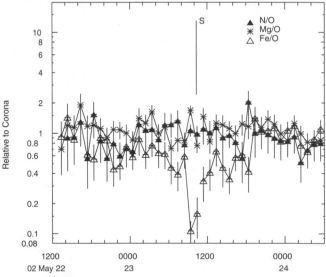

Figure 4. Elemental ratios at 14 MeV/nucleon as a function of time derived from the intensities shown in the previous figure, normalized by values of *Reames* [1999a]. The time of shock passage is indicated by the 'S' and line at the center of the plot (courtesy of *T.T. von Rosenvinge*).

an order of magnitude (particularly at lower energies, see Figure 5 or Figure 1 from *Cohen et al.* [2001]) creating a potentially unexpected, significant radiation hazard (see also, *Reames* [1999b]). Unfortunately, it is difficult to predict the size of an ESP event from the associated shock parameters. In fact, in a survey of 168 shock passages *Lario et al.* [2003] found that many shocks (65 or 113 out of 168 shocks for ions at 47-68 keV or 1.9-4.8 MeV, respectively) were not accompanied by an ESP event. *Kallenrode* [1996] also found that for a given shock speed, the corresponding particle increase at 5 MeV varied by up to 6 orders of magnitude. Similarly, *van Nes* [1984] found a ~2-3 order of magnitude spread in 35 keV proton intensities for a given shock strength.

Even the spectral characteristics of ESP events are not easily predicted from shock parameters. Although diffusive shock acceleration models, such as that of *Lee* [1983], suggest that the spectral index of the accelerated ion population should be a simple function of the shock compression ratio, this is not seen in the observations. *Desai et al.* [2004] found

the spectral index of oxygen in ESP events was not well correlated with the measured shock compression ratio. Thus predicting the size of an ESP event and the maximum energy to which it will be seen is unreliable when based on measured shock properties. Further, the shock properties themselves are not currently well predicted.

A less desirable, but more obtainable, alternative to forecasting ESP events is nowcasting. This involves reporting in situ measurements of shocks and ESP events in real time from spacecraft far enough upstream of the Earth to provide some degree of warning. The ACE spacecraft is currently orbiting the L1 Lagrangian point and can provide 30-60 minutes of warning for near-Earth operations (e.g., space shuttle flights, Earth orbiting satellites). This is illustrated in Figure 5 where an ESP event is seen first by SIS on ACE and then subsequently by the energetic particle detector on the GOES-8 spacecraft orbiting Earth. The similarity of the intensity time profiles is remarkable and leaves little doubt that the two spacecraft were observing the same shock, which was probably still accelerating particles. Comparison of the time profiles indicates ACE observed the shock and related ESP event ~45 minutes before GOES. Other examples are presented and discussed in *Cohen et al.* [2001].

Work is currently in progress to develop an automated shock identification routine using the real-time magnetic field and solar wind data from the MAG and SWEPAM instruments, respectively, on ACE. These data will be combined with real-time proton data from 50 to 5000 keV, and at 10 and 30 MeV from the EPAM and SIS instruments, respectively, in order to assess the presence and strength of ESP events at ACE. The ACE real-time data are typically available within 5 minutes of acquisition, thus a quick automated evaluation of ESP events could provide warnings of incoming particle-accelerating shocks 30-60 minutes before they reach sensitive equipment/personnel. This could be enough time to place sensors into protected configurations, cancel or end extra-vehicular activities of astronauts, and re-evaluate launch conditions for rockets/shuttles.

The real-time EPAM proton intensities are also currently being used to predict shock arrival times [*Vandegriff et al.*, 2005]. As these ions escape from the shock region, they are detected upstream and form a 'ramp' towards the shock passage, which is evaluated by a neural network. The results provide a 24 hour advance prediction of the shock arrival time with an uncertainty of 8.9 hours, which is reduced to 4.6 hours for a 12 hour advance prediction. Similar work done by *Posner et al.* [2004], using Wind/STICS suprathermal ion measurements, indicated 5-72 hours of advanced warning was possible. Although these techniques do not evaluate the size of the ESP event or its potential hazard they do provide longer advanced warning of incoming strong shocks than nowcasting efforts.

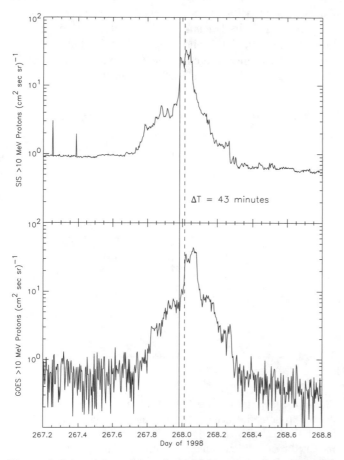

Figure 5. Integral intensities of >10 MeV protons from SIS/ACE and GOES-8 for an energetic storm particle event.

4. SUMMARY

Over the course of >30 years the causes of shock-associated particle increases, called energetic storm particle events, have been debated and data examined until the scientific community formed a general consensus. The larger, classical ESP events are a result of diffusive shock acceleration occurring as a shock passes the spacecraft; generally the same process that creates most large SEP events. The locally-accelerated particle population is usually isotropic and reflects a rigidity-dependent process acting on the upstream suprathermal seed population. Spike events are the short-lived results of shock drift acceleration at quasi-perpendicular shocks. These events exhibit strong anisotropies and are typically a low-energy phenomena.

The importance of understanding the causes and characteristics of ESP events is both fundamental and practical. Studying ESP events is currently the most useful way to examine interplanetary shock acceleration in general as the parameters of the shock can be measured directly, something currently not possible for SEP acceleration close to the Sun (where most >10 MeV/nucleon ions are accelerated). In a practical sense, ESP events can be a direct radiation hazard for space-based assets. Understanding their creation is key to predicting them, something not presently possible but highly desirable.

The statement by *Gombosi et al.* [1979] is unfortunately still applicable over 25 years later: 'Further progress in deeper understanding on interplanetary shock associated energetic particle events may heavily depend on new simultaneous, spatially separated spacecraft observations with comprehensive, high-resolution instrumentation and data analysis.' Hopefully, new results from the STEREO mission in 2006 will fit the bill.

Acknowledgments.   This work was supported by NASA at the California Institute of Technology (under grant NAG5-6912). I thank the EPAM, MAG, and SWEPAM teams and the ACE Science Center for making their data available online. GOES data were obtained through the SEC/NOAA website. I thank *R.A. Leske* for carefully reading this manuscript and his helpful comments, *T.T. von Rosenvinge* for providing Figures 3 and 4, and the organizers for putting together a stimulating and enjoyable conference.

REFERENCES

Axford, W.I., and G.C. Reid, Increases in Intensity of Solar Cosmic Rays before Sudden Commencements of Geomagnetic Storms, *Journal of Geophysical Research*, 68, 1793-1803, 1963.

Bryant, D.A., T.L. Cline, U.D. Desai, and F.B. McDonald, Explorer 12 Observations of Solar Cosmic Rays and Energetic Storm Particles after the Solar Flare of September 28, 1961, *Journal of Geophysical Research*, 67, 4983-5000, 1962.

Cane, H.V., The Structure and Evolution of Interplanetary Shocks and the Relevance for Particle Acceleration, *Nuclear Physics B - Proceedings Supplements*, 39 (1), 35-44, 1995.

Cane, H.V., T.T. von Rosenvinge, C.M.S. Cohen, and R.A. Mewaldt, Two Components in Major Solar Particle Events, *Geophysical Research Letters*, 30, SEP 5-1, doi: 10.1029/2002GL016580, 2003.

Cohen, C.M.S., R.A. Mewaldt, A.C. Cummings, R.A. Leske, E.C. Stone, P.L. Slocum, M.E. Wiedenbeck, E.R. Christian, and T.T. von Rosenvinge, Forecasting the Arrival of Shock-Accelerated Solar Energetic Particles at Earth, *Journal of Geophysical Research*, 106, 20979-20984, 2001.

Cohen, C.M.S., R.A. Mewaldt, A.C. Cummings, R.A. Leske, E.C. Stone, T.T. von Rosenvinge, and M.E. Wiedenbeck, Variability of Spectra in Large Solar Energetic Particle Events, *Advances in Space Research*, 32, 2649-2654, 2003.

Cohen, C.M.S., R.A. Mewaldt, R.A. Leske, A.C. Cummings, E.C. Stone, M.E. Wiedenbeck, E.R. Christian, and T.T. von Rosenvinge, New Observations of Heavy-Ion-Rich Solar Particle Events from ACE, *Geophysical Research Letters*, 26, 2697-2700, 1999.

Decker, R.B., Formation of Shock-Spike Events at Quasi-Perpendicular Shocks, *Journal of Geophysical Research*, 88, 9959-9973, 1983.

Desai, M.I., G.M. Mason, J.R. Dwyer, J.E. Mazur, C.W. Smith, and R.M. Skoug, Evidence for a Suprathermal Seed Population of Heavy Ions Accelerated by Interplanetary Shocks near 1 AU, *Astrophysical Journal*, 588, 1149-1162, 2003.

Desai, M.I., G.M. Mason, M.E. Wiedenbeck, C.M.S. Cohen, J.E. Mazur, J.R. Dwyer, R.E. Gold, S.M. Krimigis, Q. Hu, C.W. Smith, and R.M. Skoug, Spectral Properties of Heavy Ions Associated with the Passage of Interplanetary Shocks at 1 AU, *Astrophysical Journal*, 611, 1156-1174, 2004.

Gombosi, T., K. Kecskemety, and S. Pinter, On the Connection of Interplanetary Shock Wave Parameters and Energetic Storm Particle Events, *Geophysical Research Letters*, 6, 313-316, 1979.

Ho, G.C., D. Lario, R.B. Decker, M.I. Desai, Q. Hu, C.W. Smith, and R.M. Skoug, Transient Shocks and Associated Energetic Particle Distributions Observed by ACE during Solar Cycle 23, *Journal of Geophysical Research*, submitted.

Hudson, P.D., Reflection of Charged Particles by Plasma Shocks, *Monthly Notices of the Royal Astronomical Society*, 131, 23-50, 1965.

Jokipii, J.R., Pitch-Angle Dependence of First-Order Fermi Acceleration at Shock Fronts, *Astrophysical Journal*, 145, 616-622, 1966.

Kahler, S.W., A Comparison of Energetic Storm Protons to Halo Protons, *Solar Physics*, 8, 166-185, 1969.

Kallenrode, M.-B., A Statistical Survey of 5-MeV Proton Events at Transient Interplanetary Shocks, *Journal of Geophysical Research*, 101, 24393-24410, 1996.

Klecker, B., M. Scholer, D. Hovestadt, G. Gloeckler, and F.M. Ipavich, Spectral and Compositional Variations of Low Energy Ions During an Energetic Storm Particle Event, *Astrophysical Journal*, 251, 393-401, 1981.

Lario, D., and R.B. Decker, The Energetic Storm Particle Event of October 20, 1989, *Geophysical Research Letters*, 29, 1393, 10.1029/ 2001GL014017, 2002.

Lario, D., G.C. Ho, R.B. Decker, E.C. Roelof, M.I. Desai, and C.W. Smith, ACE Observations of Energetic Particles Associated with Transient Interplanetary Shocks, in *AIP Conf. Proc.* 679: *Solar Wind Ten*, pp. 640-643, 2003.

Lee, M.A., Coupled Hydromagnetic Wave Excitation and Ion Acceleration at Interplanetary Traveling Shocks, *Journal of Geophysical Research*, 88, 6109-6119, 1983.

Lin, R.P., S.W. Kahler, and E.C. Roelof, Solar Flare Injection and Propagation of Low-Energy Protons and Electrons in the Event of 7-9 July, 1966, *Solar Physics*, 4, 338-360, 1968.

Mason, G.M., C.M.S. Cohen, A.C. Cummings, J.R. Dwyer, R.E. Gold, S.M. Krimigis, R.A. Leske, J.E. Mazur, R.A. Mewaldt, E. Mobius, M. Popecki, E.C. Stone, T.T. von Rosenvinge, and M.E. Wiedenbeck, Particle Acceleration and Sources in the November 1997 Solar Energetic Particle Events, *Geophysical Research Letters*, 26, 141-144, 1999.

Mewaldt, R.A., C.M.S. Cohen, G.M. Mason, M.I. Desai, R.A. Leske, J.E. Mazur, E.C. Stone, T.T. von Rosenvinge, and M.E. Wiedenbeck, Impulsive Flare Material: A Seed Population for Large Solar Particle Events?, in *28th International Cosmic Ray Conference*, pp. 3229-3232, Tsukuba, Japan, 2003.

Palmer, I.D., Shock Wave Effects in Solar Cosmic Ray Events, *Solar Physics*, 27, 466-477, 1972.

Posner, A., N.A. Schwadron, D.J. McComas, E.C. Roelof, and A.B. Galvin, Suprathermal ions ahead of Interplanetary Shocks: New Observations and Critical Instrumentation Required for Future Space Weather Monitoring, *Space Weather Journal*, 2, S100004, 10.1029/2004SW000079, 2004.

Reames, D.V., Particle Acceleration at the Sun and in the Heliosphere, *Space Science Review*, 90, 413-491, 1999a.

Reames, D.V., Solar Energetic Particles: Is There Time to Hide?, *Radiation Measurements*, 30, 297-308, 1999b.

Reames, D.V., and C.K. Ng, Streaming-limited Intensities of Solar Energetic Particles, *Astrophysical Journal*, 504, 1002-1005, 1998.

Rao, U.R., K.G. McCracken, and R.P. Bukata, Cosmic-Ray Propagation Processes, *Journal of Geophysical Research*, 72(17), 4325-4341, 1967.

Sanderson, T.R., R. Reinhard, and K.-P. Wenzel, Anisotropies of 35-56 keV Ions Associated with Interplanetary Shocks, in *18th International Cosmic Ray Conference*, pp. 156-159, Bangalore, India, 1983.

Sarris, E.T., and J.A. van Allen, Effects of Interplanetary Shock Waves on Energetic Charged Particles, *Journal of Geophysical Research*, 79, 4157-4173, 1974.

Scholer, M., and G.E. Morfill, Simulation of Solar Flare Particle Interaction with Interplanetary Shock Waves, *Solar Physics*, 45, 227-240, 1975.

Simnett, G.M., and S.S. Holt, Long Term Storage of Relativistic Particles in the Solar Corona, *Solar Physics*, 16, 208-223, 1970.

Tylka, A.J., C.M.S. Cohen, W.F. Dietrich, M.A. Lee, C.G. Maclennan, R.A. Mewaldt, C.K. Ng, and D.V. Reames, Shock Geometry, Seed Populations, and the Origin of Variable Elemental Composition at High Energies in Large Gradual Solar Particle Events, *Astrophysical Journal*, 625, 474-495, 2005.

Tylka, A.J., C.M.S. Cohen, W.F. Dietrich, C.G. Maclennan, R.E. McGuire, C.K. Ng, and D.V. Reames, Evidence for Remnant Flare Suprathermals in the Source Population of Solar Energetic Particles in the 2000 Bastille Day Event, *Astrophysical Journal*, 558, L59-L63, 2001.

van Nes, P., R. Reinhard, T.R. Sanderson, K.-P. Wenzel, and R.D. Zwickl, The Energy Spectrum of 35- to 1600-keV Protons Associated with Interplanetary Shocks, *Journal of Geophysical Research*, 89, 2122-2132, 1984.

Vandegriff, J.K., K. Wagstaff, G. Ho, and J. Plauger, Forcasting space weather: predicting energetic store particle events using neural networks, *Advances in Space Research*, in press, 2005.

Wenzel, K.-P., R. Reinhard, T.R. Sanderson, and E.T. Sarris, Characteristics of Energetic Particle Events Associated with Interplanetary Shocks, *Journal of Geophysical Research*, 90, 12-18, 1985.

Christina M.S. Cohen, California Institute of Technology, MC 220-47, Pasadena, CA 91125 USA

Fifty Years of Ground Level Solar Particle Event Observations

C. Lopate

Space Science Center, University of New Hampshire, Durham, New Hampshire, USA

For the past fifty years ground based installations have been monitoring charged particle fluxes associated with Solar Particle Events (SPEs). While rare there have been a significant number of SPEs with associated charged particles fluxes extending to rigidities high enough to penetrate the Earth's magnetic fields and be observed as Ground Level Events (GLEs). Such events, during which SPE particle fluxes reach their highest energies – often on short time scales, constrain models of the processes involved in SPEs. This review will discuss the observations of these GLEs and their impact on the changing view of SPEs.

1. INTRODUCTION

From the time of historic flight of Victor Hess in 1912 scientists have been using ground based measurement instruments to study radiation from space. Ground level measurements of particle radiation from space rely on the fact that high-energy particles incident on the upper atmosphere will interact with the atmospheric gases creating a cascade of secondary particles (Figure 1). An air shower has three main channels through which energy is transmitted, identified as the electromagnetic (γ, e$^-$, and e$^+$), mesonic, or hard (π and μ), and nucleonic, or soft (p and n) channels. Primary particles with enough energy will create cascades that can reach the surface of the Earth.

The first widely used type of instrument to study cosmic radiation was the ionization chamber. An ionization chamber ultimately derives its response from the mesonic component of air showers. Muon detectors have a generally high (>4 GV) rigidity threshold and thus are relatively insensitive to Solar particle events (SPEs). Some of the larger SPEs will; however, generate particles of high enough energy to be observed with ground-based instrumentation. A trio of ground level enhancements (GLEs) seen in 1942 (Figure 2) was proposed by *Forbush* to be of Solar origin [*Forbush*, 1946]. Forbush's observation began the study of SPEs using ground-based measurements.

The idea of using a monitor for the study of secondary neutrons created during air showers arose from the work of Simpson in the late 1940's who studied the latitude dependence of the nucleonic component of air showers and compared this with the mesonic component. Simpson found that the latitude dependence of the nucleonic component showed variability on the order of ~300-400% where an ionization chamber would show only ~10-15% variability [*Simpson*, 1948]. Monitors to study the nucleonic component would be able to access the primary charged particle spectrum at significantly lower energies than would ionization chambers. There were a number of investigators who developed detectors to study the secondary neutron production in air showers [e.g., *Tongiorgi*, 1949; *Cocconi*, 1950; *Simpson et al.*, 1953]. During the International Geophysical Year (1957-1958), Simpson's neutron monitor (IGY-NM) was designated the standard neutron detector for the study of variations in the cosmic radiation (Figure 3). This neutron monitor was easy and inexpensive to build and relatively simple to maintain, allowing many scientists the opportunity to contribute to the study of space astro-particle physics. The basic idea behind the IGY-NM [see also *Hatton*, 1971] is a five-fold process: 1) secondary air shower neutrons are slowed in paraffin, which also acts to keep background nucleons from entering the instrument, 2) these neutrons interact in lead surrounding the counter creating evaporation neutrons, 3) the evaporation neutrons are slowed by paraffin until they enter the detector, 4) these slow neutrons are absorbed by boron in the detector through the process:

$$10B + n \rightarrow 11B^* \rightarrow 7Li + 4He + Q \quad (1)$$

Solar Eruptions and Energetic Particles
Geophysical Monograph Series 165
Copyright 2006 by the American Geophysical Union
10.1029/165GM27

283

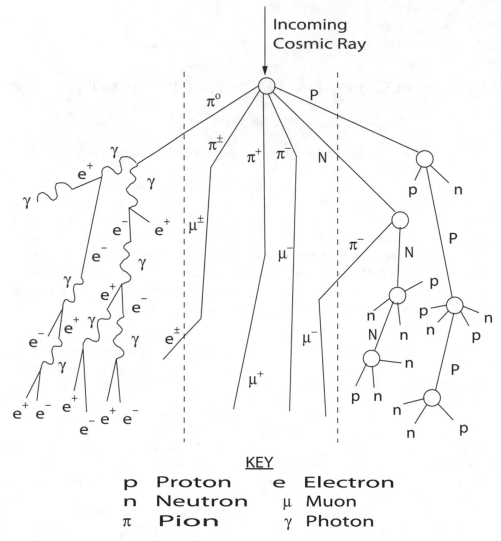

Figure 1. Air shower caused by high-energy particles incident on Earth's upper atmosphere. The primary particles are noted as P and N. Energy transfer in the air shower is through three main channels, the electromagnetic (γ, e⁻ and e⁺), mesonic, or hard (π and μ), and the nucleonic, or soft (p and n).

and 5) the 2.311 MeV α-particle (4He) emitted from the decay of the excited boron atom creates an ionization pulse which is registered. In practice, the neutron monitor counts all ionization pulses above a preset minimum threshold. The design allows for little background, typically 1-2 background counts per minute for an IGY-NM tube. An IGY-NM will respond to primary particles of incident rigidity above ~0.2 GV at the magnetic poles and above ~15 GV at the magnetic equator. A single IGY-NM tube at sea level on the equator will register ~0.2 counts/second. During the International Quiet Sun Year, 1964, the larger supermonitor (NM64) designed by *Carmichael* [1964] came into popular use. It's design is basically the same as the NM-IGY, but it's larger size gives an increase in count rate of a factor ~35 per tube.

While the neutron monitor was primarily used for the study of Galactic cosmic radiation, its increased sensitivity to lower energy primaries also allowed its use for the increased study of SPEs. The first major study of the high-energy particle production and propagation of SPEs were made of the large GLE observed on 23 February 1956. *Meyer, Parker* and *Simpson* studied the correlations between optical and radio observations of the flare and the subsequent ground level enhancement observed at Earth (Figure 4). Their work [*Meyer, Parker and Simpson*, 1956] included a model of the interplanetary medium and estimates for the effects of interplanetary magnetic fields on the propagation of high-energy charged particles from the Sun to the Earth, and was the first comprehensive study of Solar energetic particles (SEPs) in

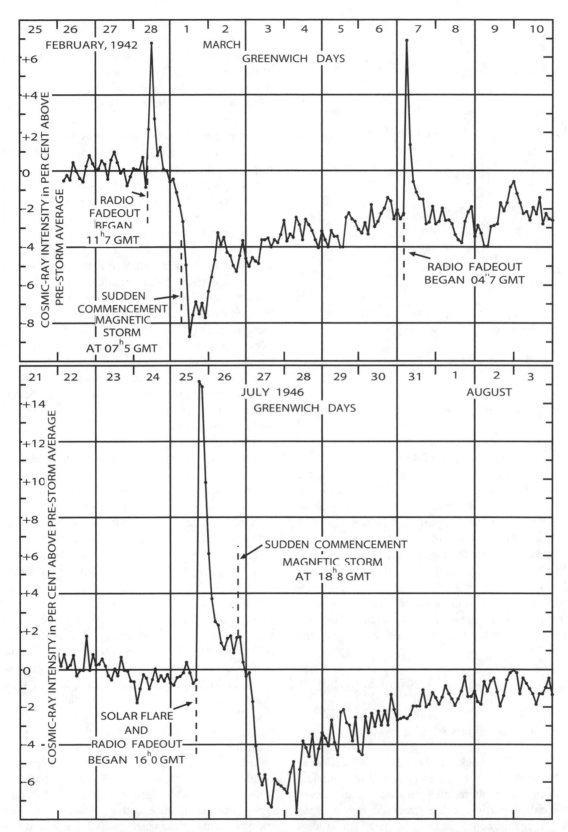

Figure 2. Three unusual increases in cosmic ray intensity at Cheltenham, Maryland during Solar flares and radio fadeouts [*Forbush*, 1946]. The first time ground level enhancements were identified directly with Solar phenomena.

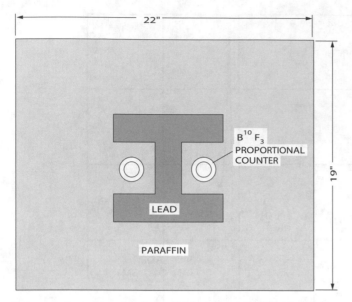

Figure 3. A cross-sectional view of the basic 'pile' used in the first neutron monitors developed by John A. Simpson [*Simpson et al.*, 1953].

an effort to understand all the physical processes involved in the SPE.

The processes in SPEs that accelerate charged particles usually do so only to moderate energies of a few 10's of MeV. Occasionally there are large events where particles are accelerated to multi-GeV energies. Because they are ultra-relativistic, it is difficult for space-based devices to accurately identify these high-energy particles. However, it is at just these energies where ground-based instruments begin to measure particles. Thus ground-based observatories for the measurement of high-energy charged particles are a useful adjunct to the generally more detailed space-based measurements. It is through the use of these ground-based observatories that we get most measurements of the extreme of particle acceleration and thus constrain the processes that occur during SPEs.

2. USING GROUND-BASED INSTRUMENTS FOR STUDYING SPEs

2.1. Response to Primary Protons

One of the major drawbacks to the use of ground-based measurements for the study of SPEs is that, due to nuclear interactions in the atmosphere, all direct information about the type of incident particle being observed is lost. Thus ground-based measurements cannot be used in studies involving elemental, isotopic and electronic charge states of

SEPs. Studies of charged SEPs using ground-based measurements have generally made the assumption that the incident particles are all protons. Some recent studies [*Tylka et al.*, 1999; *Lopate*, 2001] have shown that even assuming the hardest reasonable Fe spectrum during the 29 September 1989 SPE (an event with one of the largest Fe fluences as measured in space), the Fe contribution to the total Climax neutron monitor response was ~7.5%; thus the assumption that the charged particle flux observed in a GLE is entirely composed of protons is reasonable.

When studying charged particle induced ground level enhancements it is important to understand the effect of the Earth's magnetic field on particles transported through the heliosphere. Charged particles incident on the Earth's upper atmosphere are ordered by their rigidity, P ($= pc/Ze$), where p is the particle momentum, c is the speed of light, Z is the electric charge units of the incident particle, and e is the fundamental charge. The minimum rigidity of an incident charged particle necessary to produce a measurable signal at the ground is called the vertical cutoff rigidity, P_c. As mentioned previously, the Earth's magnetic field creates vertical cutoff rigidities less than ~0.2 GV at the magnetic poles increasing to ~15 GV at the magnetic equator. Thus the network of ground-based observatories, by being situated at different magnetic latitudes, can be used to determine the spectrum of the incident particle flux.

However, knowing the vertical cutoff rigidity is not enough to uniquely determine an incident particle's rigidity. One must also take into account the fact that a charged particle moving with a given velocity vector in interplanetary space will have that vector altered as it enters the Earth's magnetosphere. Adding even more complication, the difference between these two velocity vectors will be rigidity dependent. The angle of incidence of a particle entering the Earth's magnetosphere that will lead to that particle being vertically incident on the atmosphere above a given position is called its asymptotic direction [*McCracken*, 1962; *McCracken et al.*, 1968]. Figure 5 shows the asymptotic directions for the neutron monitor at Jungfraujoch, situated at 46.5° N and 8.0° E [*Flückiger*, 1993]. As can be seen in Figure 5, the deflection at the lowest incident particle rigidity is greatest. There is ~310° deflection for an incident particle with 4.8 GV rigidity decreasing to ~45° for an incident particle with 20 GV rigidity. We should note that the calculation of these asymptotic directions depends crucially on the exact topology of the Earth's magnetic field. The directions shown in Figure 5 are calculated for a quiet-time steady-state magnetic field. It is often the case that, during SPEs large enough to generate GLEs, the Earth's magnetic field is so disturbed from the quiet-time configuration, that without direct simultaneous measurements of the magnetic field, any derived asymptotic direction would be unreliable. Thus for each GLE and for

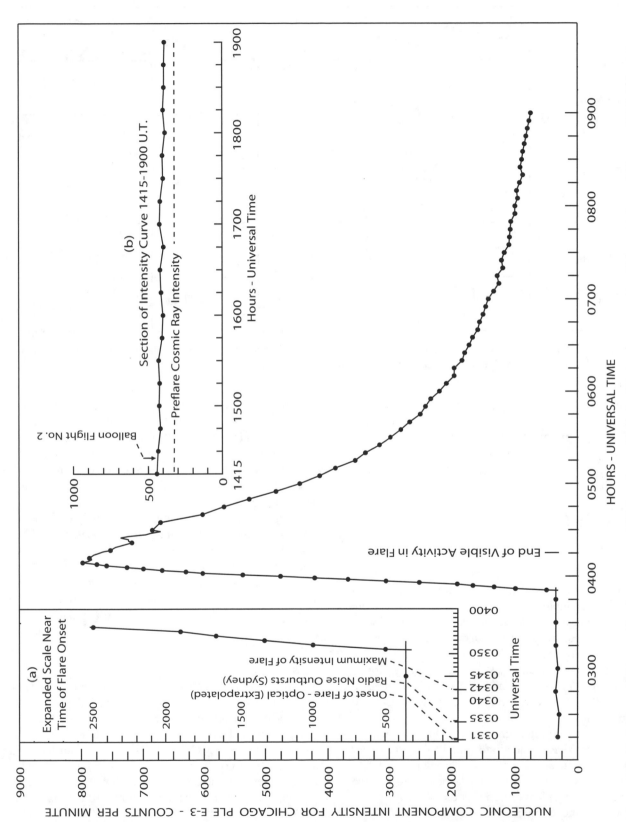

Figure 4. The 23 February 1956 ground level enhancement (GLE) as observed with the Chicago neutron monitor [*Meyer, Parker and Simpson*, 1956]. This study was the first attempt at a comprehensive explanation for the propagation of Solar energetic particles through interplanetary space from the Sun to the Earth.

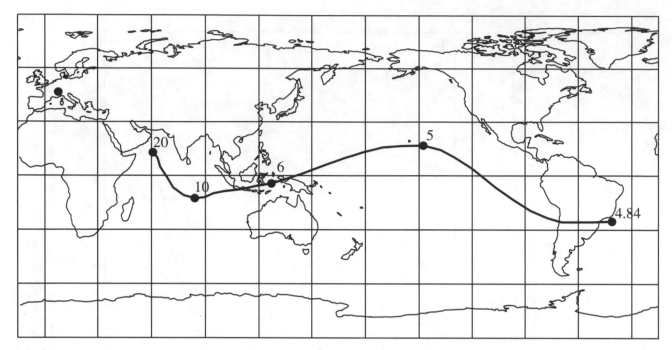

Figure 5. The asymptotic directions of cosmic rays observed entering vertically into the atmosphere at the Jungfraujoch neutron monitor [*Flückiger*, 1993]. Longitude and latitude lines are drawn every 30 degrees.

each ground based observatory it is necessary to calculate the asymptotic directions if one hopes to derive an accurate interplanetary flux from ground based observations.

When studying a GLE it is possible to relate the ground-based measurements to the absolute flux of charged particles creating the GLE as:

$$\Delta N(z,t) = \int_{P_c}^{\infty} S_p(P,z)\, F_p(\chi,t)\, J_p(P,t)\, dP \qquad (2)$$

where t is time, z is the atmospheric depth at which the measuring instrument sits, P is the rigidity of the incident protons, χ is the asymptotic separation between the incident protons and the measurement site, $\Delta N(z, t)$ is the increase in count rate of the measuring instrument, $S_p(P, z)$ is the proton specific yield function, $F_p(\chi, t)$ is the pitch angle distribution of the incident proton flux, and $J_p(P, t)$ is the primary particle rigidity spectrum. The general method used when trying to derive absolute fluxes from the ground level measurement is to assume a proton flux spectrum, J_p, and pitch angle distribution, F_p, then calculate an expected ground level enhancement by integrating the right side of equation (2) and fitting the calculation to the actual measurement, ΔN. Through an iterative process of judicious guessing of J_p and F_p one can determine a range and error for incident proton flux distribution.

Central to the above process is the calculation of the proton specific yield function, S_p. There have been three main methods used to calculate S_p, through theoretical derivation, through parameterization of data taken with a set of monitors, and through Monte Carlo analysis. Each of these has advantages and disadvantages and a comparison of various studies gives a good indication of the error implicit in any extrapolation from ground level measurements to absolute fluxes in space.

The biggest drawback to using a first-principles theoretical derivation is that the sheer number of processes, and their range in energy, involved in an air shower started by a primary proton is enormous, so exact closed form solutions for S_p are not possible. In practice, people have assumed some reasonable form. The depth dependent Dorman function [*Dorman*, 1970] has been one of the most popular:

$$\Delta N(z) = N_0(1 - \exp(-\alpha(z)P_c^{1-\kappa(z)})) \qquad (3)$$

where α and κ are fitted parameters that depend only on the atmospheric depth of the instrument, and N_0 depends on the exact type of measurement device. The derivative of equation (3) can be equated to the right side of equation (2) and if one uses periods when only Galactic cosmic rays contribute to an instrument response it is possible to determine J_p and F_p through alternate measurements so one can derive S_p. However, using this method, the simplified functional forms for ΔN make it difficult to separate out much of the physics in the air shower process.

When using a set of ground level measurements to derive a parameterized fit for the specific yield we give up all hope of describing the physical process involved in the air shower in exchange for a good functional fit for S_p. Since the analysis directly uses the instrument responses to derive S_p, these fits can give very good results. For neutron monitor studies the standardization to IGY-NM and NM64 have made parameterization methods popular because there are many identical instruments located at various positions around the Earth, so complete latitude surveys are available. Most of the parameterization fits for S_p are done with sea level neutron monitors or with balloon instrumentation at about 10 km altitude [*Nagashima et al.*, 1989] with interpolation used for instruments between these atmospheric depths. It is also important to understand that, since J_p varies in time, the instrument response is time-dependent so that any parameterized fit is best only for periods of activity nearly identical to that when the survey measurements were taken – often a drawback for GLE studies.

Because air shower processes are computationally difficult to program Monte Carlo analyses used to determine S_p are a more recent advents. It is only since the early 1980's that these methods have become popular. The advantage of using Monte Carlo's is that as computer processing has become better it has been possible to include more of the physical processes in the air shower. The main drawback is to insure that there are no computational effects masking or distorting the results. Despite the apparent advantage to using Monte Carlo codes different codes have led to very different response functions.

Figure 6 [*Clem and Dorman*, 2000] shows a comparison of S_p derived for sea-level NM64 from four different studies. The studies of *Lockwood et al.* [1974] and *Stoker* [1981]

were done using parameterization methods, while those of *Debrunner* [1982] and *Clem* [1999] were done using Monte Carlo analyses. As can be seen in Figure 6, although the derived form for S_p is similar in all four studies, there is more than an order magnitude difference observed at the low energy. If we use these forms of S_p for SPE studies, since the proton spectrum is sharply falling, the variations in S_p at low energy can make significant differences in calculated absolute flux or anisotropy. Despite the inherent error in using ground-based measurements in SPE studies, their usefulness in understanding the high-energy behavior should not be underestimated.

2.2. Response to Primary Neutrons

Ground level instruments also have response to neutrons produced in SPEs and which travel directly to the Earth. Neutrons of high enough energy will reach Earth before decaying and will have enough energy remaining to be observed with ground-based instrumentation. There are now two types of ground-based instruments that can observe these direct Solar neutrons: neutron monitors and neutron telescopes [*Muraki et al.*, 1992; *Matsubara et al.*, 1993]. Neutron telescopes are generally made from proportional counters (gas composition of 90% Ar and 10% CH_4) and plastic scintillator (C_9H_{10}). Neutrons enter the plastic scintillator and create recoil protons through the charge exchange processes $n + p \rightarrow p + n$ and $n + C \rightarrow p + X$. The plastic scintillator can then measure the ionization energy of the recoil protons. The proportional counters are used as pointing telescopes to identify Solar neutron events, and also as vetoes for charged particles that enter the instrument.

The measurement of direct Solar neutrons through neutron monitors differs from incident proton measurements through the fact that neutrons are not affected by the Earth's magnetic field. Since neutrons move directly from the Sun to the Earth only instruments that are on the day side of the Earth will be able to see direct neutrons. Also the atmospheric affects of neutrons and their response in the two types of neutron monitors (IGY-NM and NM64) are very different. When studying Solar neutron events the detector response is related to the initial neutrons through:

$$\Delta N(z,\theta,t) = R^{-2} \int_{t_{min}}^{\infty} \mu(t - t_s)\, Q(E_n)\, P(E_n)$$
$$\times S_n(E_n,z,\theta)\,(dE_n / dt_s)\, dt_s \qquad (4)$$

where t is time, t_s is the transit time from the Sun to the Earth for a neutron, E_n is the neutron energy defined by t_s, z is the atmospheric depth at which the measuring instrument sits, θ is the angular distance of the instrument to the

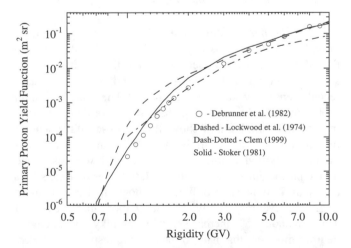

Figure 6. A comparison between four derivations of the proton specific yield function [*Clem and Dorman*, 2000].

sub-Solar point, P is the survival probability for a neutron of energy E_n reaching Earth, μ is the normalized time profile of the neutron emission at the Sun, Q is the time-integrated Solar emissivity, t_{min} is the high-energy cutoff of the neutron emissivity spectrum Q, dE_n/dt_s is the neutron time-energy dispersion relation [*Lingenfelter and Ramaty*, 1967] and S_n is the yield function of the instrument.

3. GROUND-BASED STUDIES OF SPEs

3.1. A Short History of GLE Investigations

Forbush [1946] made the first identification of GLEs with Solar events. Not much more was done in his initial papers than make the identification, but that was the key step toward studying and understanding the physics associated with GLEs.

The GLE observed on 23 February 1956 was used to take the next major step. *Meyer, Parker and Simpson* [1956] studied this event with an emphasis on the propagation of the energetic particles from the Sun to the Earth. They concluded that there must be some type of magnetic structure in the heliosphere that affected the particle transport. The idea of using high energy charged particles as probes of the interplanetary magnetic field has continued. Scattering off of magnetic irregularities beyond Earth was hypothesized as early as 1958 [*Simpson*, 1958]. However, it was still difficult to reconcile the different responses observed nearly simultaneously at ground stations around the world. After *Parker's* [1965] description of the Solar wind, the embedded magnetic field and particle transport in interplanetary space, the idea of magnetic connection between source and observer was established, and scattering off the magnetic field was generally assumed. *McCracken et al.* [1968] added significantly to this by formalizing the idea of magnetic connection between the interplanetary fields and the geomagnetic fields. This allowed for a more satisfactory explanation of why different ground station responded at different times and at different levels. While the models for the magnetic fields, both interplanetary [e.g., *Jokipii and Kopriva*, 1979] and geomagnetic [e.g., *Tsyganenko*, 1989] have advanced over the years, the idea of using these ultra-high energy particles as probes of the fields continues [e.g., *Humble et al.*, 1991; *Cramp et al.*, 1997].

In addition to studying how magnetic connection affects ground level observations, Solar wind theory has been applied to particle transport of the high energy particles. The first efforts put forth considered only the standard diffusion-convection models. Particle scattering in random magnetic fields was applied to both Galactic and Solar particles [e.g., *Jokipii*, 1966; *Hasselmann and Wibberenz*, 1968]. The success of these models was limited; they calculated long particle mean free paths (~0.5 AU) for Solar particles of energies

above 1 GeV and thus the underlying diffusive approximation was challenged. As models developed, focused transport was included [e.g., *Earl*, 1976]. This change allowed for models to physically address the non-diffusive regime necessary, especially for the highest energy particles. While the mean free paths for GeV energy particles were still large, better fits to the time-intensity profiles were made. As numerical modeling has become efficient still more complex models have been developed. Modern models include, for instance streaming and pitch-angle scattering [e.g., *Dröge*, 2000] or convection and adiabatic deceleration [e.g., *Ruffolo*, 1995]. These models derive slightly smaller particle mean free paths (~0.3 AU) and, when applied to the highest energy particles, can fit the time-intensity profiles for multiple neutron monitors.

Early particle acceleration models focused on second order Fermi acceleration or stochastic acceleration as the main mechanism [e.g., *Forman, Ramaty and Zweibel*, 1982]. When comparing earlier models to data, the in-situ plasma parameters were unknown, and so what we now know as unphysical situations were used to fit GLE spectra. *Ellison and Ramaty* [1985] extended early models with a semi-empirical method and showed that they could fit GLE spectra up to several GeV in energy with reasonable parameters. More recent models [e.g., *Dulk*, 2000] can fit both relativistic electrons and protons with a single shock acceleration model. Other non-shock acceleration models, using either electric field (current sheet) drift or magnetic reconnection [e.g., *Reiner et al.*, 2000] have also been applied to GLE observations.

Since it is now known that all these processes occur within a single event new models connecting these various processes are being tested. In order to demonstrate some of the types of analyses done we have chosen to show in more detail three GLE studies.

3.2. Studies of Proton Induced GLEs

3.2.1. The bastille day event, 14 july 2000. One of the most famous GLEs observed was the 14 July 2000 event – the Bastille Day Event. Because of the large enhancements seen with ground-based instruments and because it was a recent event many studies were reported. The study by *Duldig* [2001] was one that used about half of the worldwide neutron monitor network. Figure 7 shows the responses of the neutron monitors that were used in the study. By using neutron monitors that reported high-time resolution data it was possible to derive proton anisotropies and absolute fluxes on time scales as fast as 5 minutes. Also, Figure 7 shows the large variation in ground level response with a nearly 400% increase seen at Thule to about 25% increase seen at Lomniky Stit. There are neutron monitors that had no measurable response and were not included in Figure 7, though they were included in Duldig's analysis. Figure 8 shows the derived absolute proton

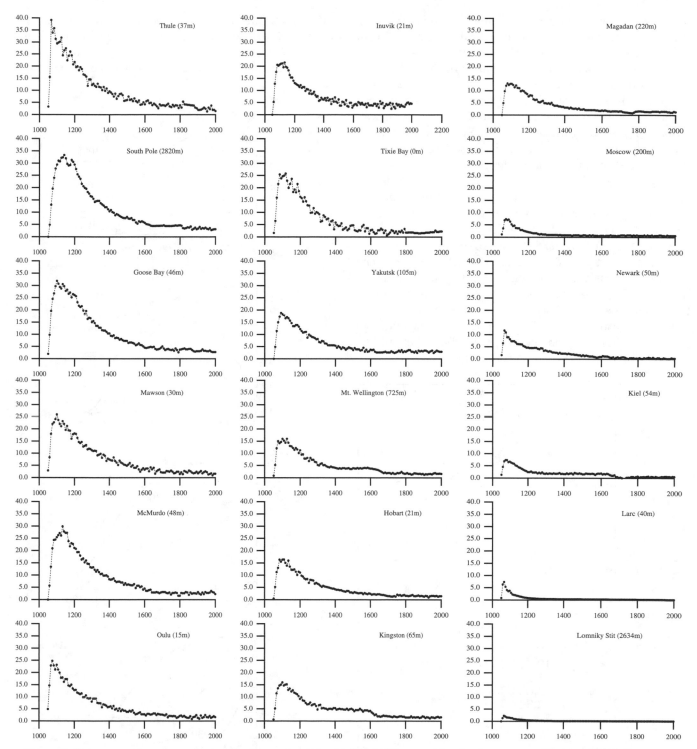

Figure 7. Neutron monitor responses used in the study of the 14 July 2000 ground level enhancements [*Duldig*, 2001]. Rates are percentage increases normalized to the Apatity neutron monitor (not shown).

fluxes at four times during the event. The rigidity spectrum shows a typical rollover, but is fairly hard at the beginning of the event, extending to ~10GV, and softening as the event progresses. This behavior is very typical of the high-energy behavior of these large SPEs. Figure 9 shows the proton anisotropy as a function of time. The strong anisotropy seen in the early times of this event were presumably due to scattering of high-energy protons near the Sun, then direct propagation to the Earth, not unusual in these large events. The strong anisotropy at the end of the event consists mostly of lower energy particles (the spectrum has softened considerably) and may be due to an enhanced acceleration period. The high energy particle spectra were fit using two different models [*Bieber et al.*, 2001] and it was shown that to the limits of the flux errors, a model of particle propagation using convection and adiabatic diffusion [*Ruffolo*, 1995] and one using focusing, streaming and pitch-angle scattering [*Dröge*, 2000] gave indistinguishable results which fit data from nine neutron monitors.

There are two theories currently in use to explain the high energy particles observed in GLEs, as typified by the 14 July 2000 event. One model has a shock driven by a coronal mass ejection (CME) forming quickly after the CME liftoff [e.g., *Kahler, Hildner and Van Hollebeke*, 1978; *Kahler et al.*, 1984; *Reames*, 1999] Since this shock would be moving

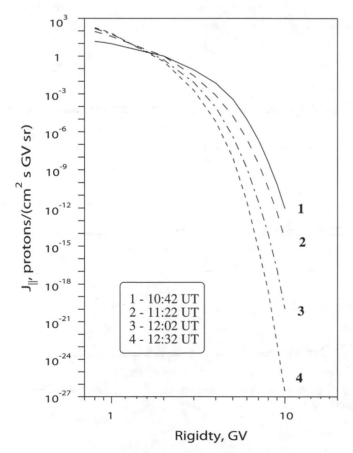

Figure 8. Derived interplanetary proton spectra made from neutron monitor response during the 14 July 2000 [*Duldig*, 2001].

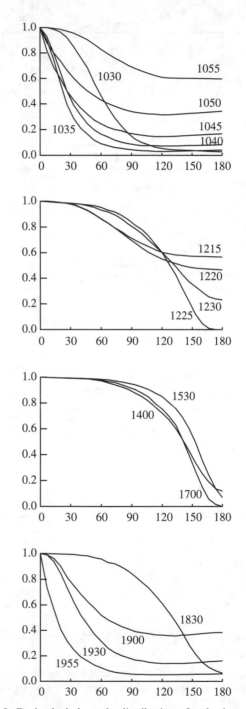

Figure 9. Derived pitch angle distributions for the interplanetary proton flux during the 14 July 2000 Solar particle event [*Duldig*, 2001]. Anisotropies response is normalized and angles are in degrees.

upward through the corona into increasingly weaker magnetic fields it would become less efficient at particle acceleration as time progressed, leading to the softening spectrum, and also as the event progressed, the shock front would extend to cover more interplanetary magnetic field lines causing the pitch-angle distribution at Earth to become more isotropic as the event progresses. This model has a major hurdle to overcome – in order for the timing to be correct, the CME driven shock must not only form, but also accelerate protons to 10 GV in a matter of minutes and at a height of 2-4 Solar radii, a process that some consider unlikely. The other model to explain this behavior is that the protons are accelerated during the (sometimes extended) magnetic reconnection of the flare loop and which propagates downward into the strong photospheric and chromospheric magnetic fields [e.g., *Svestka, Martin and Kopp*, 1980; *Litvinenko and Somov*, 1995]. Escape from the acceleration region could be accomplished by propagation along open field lines that would thread the region [e.g., *Klimchuk*, 1996]. There is some disagreement as to whether shock acceleration occurs during the reconnection with arguments in favor [e.g., *Dulk et al.*, 2000] and others against [e.g., *Reiner et al.*, 2000]. In either case, the softening of the proton spectrum is due to loss of particles into the photosphere and the early anisotropy is caused because the acceleration is localized to a few magnetic field lines on the Sun. This model; however, cannot easily explain the long period of nearly isotropic flux a few hours into the event and lasting nearly five hours. A possible explanation of this last observation is that the protons late in the event are mirroring off a structure outside the Earth's orbit and returning isotropically.

3.2.2. The 29 september 1989 event. The second study we wish to examine here is by *Lovell, Duldig and Humble* [1998] of the large 29 September 1989 GLE. The neutron monitor data they used to study the event is shown in Figure 10. Their detailed analysis of these 8 neutron monitor responses was compared with space-based measurements of the same SPE. In Figure 11 the results of their spectral analysis are compared with data from IMP-8, GOES-6 and GOES-7. All of the data is fit by an acceleration spectrum (basically a power-law spectrum modified by a exponential rollover at high energy) derived by *Ellison and Ramaty* [1985] to explain energetic proton spectra from Solar flares. This shows the importance of using both the high-energy data derived from ground-based measurements in conjunction with space-based measurements. With only space-based measurements it would be impossible to determine where the high-energy roll over in the acceleration process occurred; with only the ground-based measurements, the large error at high energy would not allow for so accurate absolute flux determination. Indeed *Bieber and Evenson* [1991] looked at the 19 October 1989

GLE using only neutron monitor observations using five different analysis techniques. All five techniques fit the high-energy spectrum to within one order of magnitude, but the extrapolations to low energies gave absolute fluxes differing by five orders of magnitude. Clearly without both low and high energy measurements it is impossible to derive an accurate proton spectrum.

While this report has so far been focusing on neutron monitor measurements on the ground, there are muon telescopes that also make very useful ground-based measurements. Muon telescopes have access to much higher differential energy measurements than do neutron monitors, thus can extend the spectral measurements of SPEs to even higher

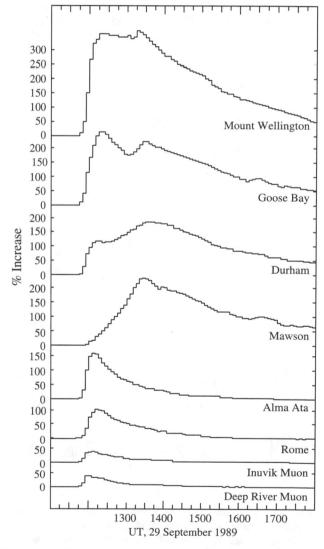

Figure 10. Neutron monitor responses used to study the 29 September 1989 Solar particle event [*Lovell, Duldig and Humble*, 1998]. Each rate is normalized to that neutron monitor's mean rate one hour prior to the ground level increases.

energy [*Falcone et al.*, 2003]. Some muon telescopes also have the ability to determine the incident particle direction without having to resort to statistical inter-comparisons between multiple instruments.

3.3. Studies of Neutron Induced GLEs

Events with direct Solar neutron components observable at Earth are very rare, due mainly to the low cross-section for energetic neutron production and their loss due to scattering and decay in transit from the Sun. The first reported observation of Solar neutrons was from the Gamma Ray Spectrometer on the Solar Maximum Mission satellite, in association with the flare of 21 June 1980 [*Chupp et al.*, 1982]. It was not until the GLE on 3 June 1982 when Solar neutrons were observed with ground based instruments [*Debrunner et al.*, 1983; *Efimov, Kocharov and Kudela*,

1983]. Prior to the advent of neutron telescopes (described above) in the early 1990's, there had been only six GLEs with associated response from direct Solar neutrons. In the present Solar cycle (23), using the specialized neutron telescopes, there have been another five GLEs identified with associated direct Solar neutrons.

3.3.1. The 24 may 1990 event.   Because of the large number of monitors which observed the 24 May 1990 GLE, both the direct Solar neutron and subsequent proton responses, and the unambiguous differentiation of these two phases of the GLE, this is considered a classic example of a direct Solar neutron event. Neutron monitor response during the early phase of the event is shown in Figure 12 [*Pyle, Shea and Smart*, 1991]. The flare time is marked with triangles; the neutron monitors show a response about 9 minutes after the flare and then a second increase about 20 minutes after the flare (marked by a dashed line). This analysis shows that first signal, due to direct Solar neutrons, is ordered not by cutoff rigidity (Inuvik with the lowest cutoff rigidity has the smallest response) but by atmospheric depth (Climax with the smallest atmospheric depth, 782 gm, has the greatest response); the proton response, starting around 2105 UT is ordered by cutoff rigidity as expected for charged particles. Another analysis of

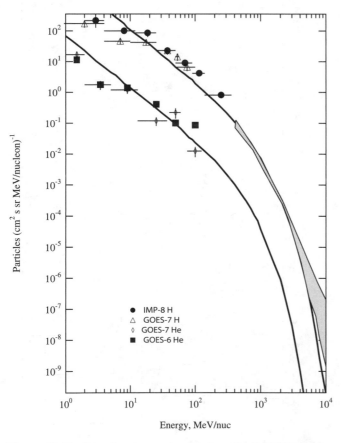

Figure 11. The interplanetary proton spectrum for the 29 September 1989 Solar particle event derived from neutron monitor measurements [*Lovell, Duldig and Humble*, 1998]. The hashed area shows the range of error for the ground-level data. Proton and alpha particle fluxes from IMP 8, GOES 6 and GOES 7 at the peak of the event are also presented. Also shown are best fits of all the data to the functional form derived in *Ellison and Ramaty* [1985].

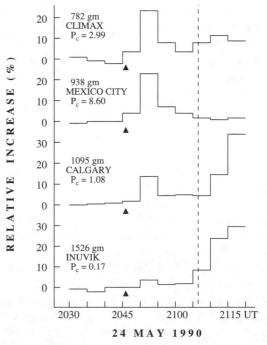

Figure 12. The neutron monitor response during the first phases of the 24 May 1990 ground level events [*Pyle, Shea* and *Smart*, 1991]. The flare time is marked with a triangle (about 20:46 UT) and the proton enhancement time is marked with a dashed line. The peaks at approximately 20:54 UT are due to direct Solar neutrons.

the 24 May 1990 event used an assumed power-law in energy $Q(E) = 10^{28}E^{-3}$ in equation (4) above [*Muraki and Shibata*, 1996]. With this spectrum they were able to fit the various neutron monitor responses. This spectrum is consistent with values derived from other Solar neutron events. For instance, the total fluence of Solar neutrons above 50 MeV during the 21 June 1980 event was calculated to be $\sim 3 \times 10^{28}$ neutrons/sr [*Chupp et al.*, 1982] and the derived Solar neutron spectrum in the 3 June 1982 GLE was found to have a power law spectral index of between -2 and -3 [*Debrunner et al.*, 1983; *Efimov, Kocharov and Kudela*, 1983]. A time-line for the neutron emission during the 24 May 1990 event indicates that the emission starts about 2-3 minutes after the associated flare x-ray emission starts, indicating it is associated with ions accelerated prior to the CME shock formation, thus likely associated with a shock formed during the magnetic reconnection phase of the event [*Ryan, Lockwood and Debrunner*, 2000]. This is typical for Solar neutron emission events. It has recently been noted that ground-level enhancements due to direct Solar neutrons are not ordered by either the longitude or magnitude of the Solar flare associated with the emission [*Watanabe*, 2005], thus models of Solar neutron production may need to account for a majority of the high energy neutron production occurring radially upward from the Solar surface. Comparisons between ground level enhancements due to direct Solar neutrons and associated Solar gamma-ray measurements can be used to expand the deduced neutron spectrum to higher energies than are available to the gamma-ray observations alone.

4. CONCLUSION

Over the past 50 years ground based instruments have been used to study Solar particle events. In their heyday, the 1950's-1960's, ground based studies opened the way to our understanding of space particle physics. Since the advent of space instrumentation ground-based instruments have been the most useful tools for studying the high-energy portion of the SEP spectrum. Despite the relative rarity of ground level enhancements, about 70 observed due to protons and about 11 observed due to neutrons, the ground-based measurements are crucial. There are still unresolved problems in our understanding of Solar particle events; for instance, where and how does the acceleration of all the particles occur, what are the key parameters of the particle transport, can particle storage or reflection occur? Because GLEs represent particles at the very extreme end of the Solar particle spectrum, they test the models in their limits; necessary tests for a complete understanding of Solar particle events.

We hope that the scientific community will support the efforts of workers using ground-based instrumentation in the future. More accurate computational work and a continued effort on latitude surveys will be crucial to future work. Also, as models become more sophisticated it will be necessary to have more accurate real-time maps of the Earth's magnetic field, especially during periods when the magnetosphere is disturbed. Despite our great strides in understanding the processes involved in SPEs there is still much work to be done.

Acknowledgments. We would like to thank the American Geophysical Union for the invitation to this Chapman Conference on Solar Energetic Plasmas and Particles. We would also like to thank the University of Turku and its Space Research Laboratory for their assistance and hospitality. The papers cited in this work are representative, not exhaustive, of all the work done using ground observations of Solar particle events over the last fifty years. This work was supported by NSF grant ATM-9912341.

REFERENCES

Bieber, J.W., W. Dröge, P. Evenson, R. Pyle, D. Ruffolo, U. Pinsook, P. Tooprakai, M. Rujiwarodom, T. Khumlumlert, and S. Krucker, Relativistic solar protons on Bastille Day 2000, *Proc. 21st Int'l. Cosmic Ray Conf. (Hamburg), Invited, Rapporteur, and Highlight Papers*, 281, 2001.

Bieber, J.W., and P. Evenson, Determination of energy spectra for the large solar particle events of 1989, *Proc. 22nd Int'l. Cosmic Ray Conf. (Dublin)*, 3, 129, 1991.

Carmichael, H., Cosmic Rays, *IQSY Instruction Manual No. 7*, Deep River, Canada, (IQSY Secretariat, London), 1964.

Chupp, E.L., D.J. Forrest, J.M. Ryan, J. Heslin, C. Reppin, K. Pinkau, G. Kanbach, E. Rieger, and G.H. Share, A direct observation of solar neutrons following the 0118 UT flare on 1980 June 21, *Astrophys. J.*, 263, L95, 1982.

Clem, J.M., Atmospheric yield functions and the response to secondary particles of neutron monitors", *Proc. 26th Int'l. Cosmic Ray Conf. (Salt Lake City)*, 7, 317, 1999.

Clem, J.M., and L.I. Dorman, Neutron monitor response functions, *Space Sci. Rev.*, 93, 335, 2000.

Cocconi, G., V.C. Tongiorgi, and M. Widgoff, Cascades of nuclear disintegrations induced by the cosmic radiation, *Phys. Rev.*, 79, 768, 1950.

Cramp, J.L., M.L. Duldig, E.O. Flückiger, J.E. Humble, M.A. Shea and D.F. Smart, The October 22, 1989, solar cosmic ray enhancement: an analysis of the anisotropy and spectral characteristics, *J. Geophys. Res.*, 102, 24237, 1997.

Debrunner, H., J.A. Lockwood, and E. Flückiger, *8th Europe Cosmic Ray Symp. (Rome)*, unpublished, 1982.

Debrunner, H., E. Flückiger, E.L. Chupp, and D.J. Forrest, The solar cosmic ray neutron event on June 3, 1983, *Proc. 18th Int'l. Cosmic Ray Conf. (Bangalore)*, 4, 75, 1983.

Dorman, L., Coupling and barometer coefficients for measurements of cosmic ray variations at altitudes of 260-400 mb, *Acta Phys. Acad. Sci. Hung.*, 29, 223, 1970.

Dröge, W., The rigidity dependence of solar particle scattering mean free paths, *Astrophys. J.*, 537, 1073, 2000.

Duldig, M.L., Fine time resolution analysis of the 14 July 2000 GLE, *Proc. 27th Int'l. Cosmic Ray Conf. (Hamburg)*, 8, 3364, 2001.

Dulk, G.A., Y. Leblanc, T.S. Bastion, and J.L. Bougeret, Acceleration of electrons at type II shock fronts and production of shock-accelerated type III bursts, *J. Geophys. Res.*, 105, 27343, 2000.

Earl, J.A., The effect of adiabatic focusing upon charged particle propagation in random magnetic fields, *Astrophys. J.*, 205, 900, 1976.

Ellison, D.C., and R. Ramaty, Shock acceleration of electrons and ions in solar flares, *Astrophys. J.*, 298, 400, 1985.

Efimov, Y.E., G.E. Kocharov, and K. Kudela, On the solar neutrons observation on high mountain neutron monitors, *Proc. 18th Int'l. Cosmic Ray Conf. (Bangalore)*, 10, 276, 1983.

Falcone, A., *et al.*, Observation of GeV solar energetic particles from the 1997 November 6 event using Milagro, *Astrophys. J.*, 588, 557, 2003.

Flückiger, E.O., Private communication to H. Debrunner, 1993.

Flückiger, E.O., and E. Kobel, Aspects of combining models of the Earth's internal and external magnetic filed, *J. Geomagn. Geoelectr.*, 42, 1123, 1990.

Forbush, S.E., Three recent cosmic-ray increases possibly due to charged particles from the Sun, *Phys. Rev.*, 70, 771, 1946.

Forman, M.A., R. Ramaty, and E.G. Zweibel, The acceleration and propagation of solar flare energetic particles, *NASA Tech. Memorandum*, (Goddard Space Flight Center, Greenbelt, MD), No. 83989, 1982.

Hasselmann, K., and G. Wibberenz, Scattering of charged particles by random electromagnetic fields, *Zeitschrift fur Geophysik*, 34, 353, 1968.

Hatton, C.J., The neutron monitor, in *Progress in Elementary Particle and Cosmic Ray Physics*, edited by J.G. Wilson and S.A. Wouthuysen, North-Holland Pub'l. Co., Amsterdam-London, 1971.

Humble, J.E., M.L. Duldig, D.F. Smart, and M.A. Shea, Detection of 0.5-15 GeV solar protons on September 29, 1989, at Australian stations, *Geophys. Res. Lett.*, 18, 737, 1991.

Jokipii, J.R., Cosmic-ray propagation. I. Charged particles in a random magnetic field, *Astrophys. J.*, 146, 480, 1966.

Jokipii, J.R., and D.A. Kopriva, Effects of particle drift on the transport of cosmic rays. III. Numerical models of galactic cosmic-ray modulation, *Astrophys. J.*, 234, 384, 1979.

Kahler, S.W., E. Hildner, and M.A.I. Van Hollebeke, Prompt solar proton events and coronal mass ejections, *Sol. Phys.*, 57, 429, 1978.

Kahler, S.W., N.R. Sheeley, Jr., R.A. Howard, M.J. Koomen, D.J. Michels, R.E. McGuire, T.T. von Rosenvinge, and D.V. Reames, Associations between coronal mass ejections and solar energetic proton events, *J. Geophys. Res.*, 89, 9683, 1984.

Klimchuk, J.A., Post-eruptive arcades and 3-D magnetic reconnection, in *Magnetic Reconnection in the Solar Atmosphere*, edited by R.D. Bentley and J.T. Mariska, *ASP Conf. Ser.*, 111, 319, 1996.

Lingenfelter, R.E., and R. Ramaty, High energy reactions in solar flares, in *High energy reactions in astrophysics*, Benjamin Press, edited by Shen, New York, 1967.

Litvinenko, Y.E., and B.V. Somov, Relativistic acceleration of protons in reconnecting current sheets of solar flares, *Solar Phys.*, 158, 317, 1995.

Lockwood, J., W. Webber, and L. Hsieh, Solar flare proton rigidity spectra deduced from cosmic ray neutron monitor observation, *J. Geophys. Res.*, 79, 4149, 1974.

Lopate, C., Climax neutron monitor response to incident iron ions: An application to the 29 Sept. 1989 ground level event, *Proc. 27th Int'l. Cosmic Ray Conf. (Hamburg)*, 8, 3398, 2001.

Lovell, J.L., M.L. Duldig and J.E. Humble, An extended analysis of the September 1989 cosmic ray ground level enhancement, *J. Geophys. Res.*, 103, 23733, 1998.

Matsubara, Y., *et al.*, New Bolivia solar neutron telescope, *Proc. 23rd Int'l. Cosmic Ray Conf. (Calgary)*, 3, 139, 1993.

McCracken, K.G., The cosmic-ray flare effect. 1. Some new methods of analysis, *J. Geophys. Res.*, 67, 423, 1962.

McCracken, K.G., U.R. Rao, B.C. Fowler, M.A. Shea, and D.F. Smart, The trajectories of cosmic rays in a high degree simulation of the geomagnetic field, *Tech. Rep. 77, Mass. Inst. of Tech.*, (Cambridge, MA), 1968.

Meyer, P., E.N. Parker, and J.A. Simpson, Solar cosmic rays of February 1956 and their propagation through interplanetary space, *Phys. Rev.*, 104, 768, 1956.

Muraki, Y., K. Murakami, M. Miyazuki, K. Mitsui, S. Shibata, S. Sukakibara, T. Sakai, T. Takahashi, T. Yamada, and K. Yamaguchi, Observation of solar neutrons associated with the large flare on 1991 June 4, *Astrophys. J.*, 400, L75, 1992.

Muraki, Y. and S. Shibata, Solar neutrons on May 24th, 1990, *AIP Conf. Proc.*, edited by R. Ramaty, N. Mandzhavidze, and X.-M. Hua, 374, 256, 1996.

Nagashima, K., S. Sakakibara, and K. Murakami, Response and yield functions of neutron monitor, galactic cosmic ray spectrum and its solar modulation derived from all the available world-wide surveys, *Nuovo Cimento*, 12C, 173, 1989.

Parker, E.N., The passage of energetic charged particles through interplanetary space, *Plan. Space Sci.*, 13, 9, 1965.

Pyle, K.R., M.A. Shea, and D.F. Smart, Solar flare generated neutrons observed by neutron monitors on 24 May 1990, *Proc. 22nd Int'l. Cosmic Ray Conf. (Dublin)*, 3, 57, 1991.

Reames, D.V., Particle acceleration at the sun and in the heliosphere, *Space Sci. Rev.*, 90, 413, 1999.

Reiner, M.J., J.M. Karlicky, K. Jiricka, H. Aurass, G. Mann, and M.L. Kaiser, On the solar origin of complex type III-like radio bursts observed at and below 1 MHz, *Astrophys. J.*, 530, 1049, 2000.

Ruffolo, D., Effect of adiabatic deceleration on the focused transport of solar cosmic rays, *Astrophys. J.*, 442, 861, 1995.

Ryan, J.M., J.A. Lockwood, and H. Debrunner, Solar energetic particles, *Space Sci. Rev.*, 93, 35, 2000.

Simpson, J.A., The latitude dependence of neutron densities in the atmosphere as a function of altitude, *Phys. Rev.*, 73, 1389, 1948.

Simpson, J.A., Solar flare cosmic rays and their propagation, *Il Nuovo Cimento, Ser. X, No. 2 Suppl.*, 8, 133, 1958.

Simpson, J.A., W. Fonger, and S.B. Treiman, Cosmic radiation intensity-time variations and their origin, I. Neutron intensity variation method and meteorological factors, *Phys. Rev.*, 90, 934, 1953.

Stoker, P.H., Primary spectral variations of cosmic rays above 1 GeV, *Proc. 17th Int'l. Cosmic Ray Conf. (Paris)*, 4, 193, 1981.

Svestka, Z., S.F. Martin, and R.A. Kopp, Particle acceleration in the process of eruptive opening and reconnection of magnetic fields, in *Solar and Interplanetary Dynamics*, edited by M. Dryer and E. Tandberg-Hanson, D. Reidel, Norwall, MA, 1980.

Tongiorgi, V.C., Neutrons in the extensive air showers of the cosmic radiation, *Phys. Rev.*, 75, 1532, 1949.

Tsyganenko, N.A., A magnetospheric magnetic field model with a warped tail current sheet, *Plan. Space Sci.*, 37, 5, 1989.

Tylka, A.J., W.F. Dietrich, C. Lopate, and D.V. Reames, High-energy solar Fe ions in the 29 September 1989 ground level event, *Proc. 26th Int'l. Cosmic Ray Conf. (Salt Lake City)*, 6, 67, 1999.

Watanabe, K., Solar neutron events associated with large solar flares in solar cycle 23, PhD thesis, Department of Physics, Nagoya University, Nagoya, Japan, 2005.

The Propagation of Solar Energetic Particles in the Interplanetary Medium

Wolfgang Dröge

Bartol Research Institute, University of Delaware, Newark, Delaware

The dynamics of solar particle events provide a direct link to the understanding of wave-particle interactions, and to the nature of solar wind fluctuations. Depending on their energy, the often simultaneously observed electrons, protons and ions interact with different wavenumber ranges of the fluctuations, and are sensitive to various aspects of the dynamical nature of the solar wind turbulence. In general, the evolution of particle events is also sensitive to the spatial variation of the transport parameters between the Sun and a few AU. Together with in-situ plasma and magnetic field observations this information can be used to extrapolate the properties of transport parameters into the more distant Heliosphere. Recent developments in the theory of parallel transport of energetic particles, and examples for the modeling of solar particle events and the derivation of transport parameters are considered. A dynamical quasi-linear theory is presented which gives special emphasis to the geometry and dynamic nature of the fluctuations, and which is able to provide particle mean free paths solely from observed plasma parameters, in good agreement with those derived by the modeling. Possibilities to apply the above results to the study of other energetic particle populations in the Heliosphere are discussed.

1. INTRODUCTION

A correct quantitative treatment of the interactions between cosmic rays and turbulent magnetic fields is one of the fundamental problems of modern astrophysics, and the study of energetic solar particle propagation offers a unique possibility to test model predictions with in-situ measurements. Considerable progress has been achieved in recent years towards a better understanding of the nature of the solar wind turbulence, and to overcome some of the deficiencies of the first, pioneering scattering theories (standard QLT) of *Jokipii* [1966] and *Hasselmann and*

Wibberenz [1968] that could not be reconciled with observations. New approaches to the theory which take into account the dissipation range, the dynamical character and the three-dimensional geometry of the turbulence [e.g., *Bieber et al.*, 1994]. have been shown to give better explanations for various aspects of the observations. Adopting the wave picture for the magnetic field fluctuations, it was shown that the effects of wave propagation, thermal wave damping and resonance broadening [e.g., *Schlickeiser*, 1993] can lead to quite similar results [for a recent review see *Dröge*, 2000a]. Both "wave" and "turbulence" approaches employ a resonance broadening of the scattering process which can be described by a single parameter related to wave damping in the first case, and the decorrelation of the fluctuations in the second. This parameter can be adjusted so that the theories can approximately reproduce the observed rigidity dependence [*Dröge*, 2000b] of the particle's scattering mean free path. To bring the (still too small)

Solar Eruptions and Energetic Particles
Geophysical Monograph Series 165
Copyright 2006 by the American Geophysical Union
10.1029/165GM28

absolute levels of the mean free paths into agreement with the observations *Bieber et al.* [1994] suggested a composite model for the fluctuations which consists of ≈20% slab and 80% 2-D fluctuations, the latter contributing little or not at all to particle scattering.

The possibility that an energetically significant fraction of the magnetic fluctuations could reside in a nearly two-dimensional (2D) geometry, in which both the wave vectors and the vectors of the fluctuating component are perpendicular to the average magnetic field was shown by *Matthaeus et al.* 1990] and *Zank and Matthaeus* [1992]. *Bieber et al.* [1996] demonstrated that the ratio of transverse spectral powers $P_{yy}(v)/P_{xx}(v)$ can provide an estimate of the percentage of slab and 2D fluctuations. Based on an analysis of Helios magnetic field data with low time resolution they found that, averaged over a large sample, the fluctuations in the inertial range are best described by a 85% 2D component. *Leamon et al.* [1998] extended the above method to higher frequencies, and found a 54% 2D component in the dissipation range. A determination of the slab/2D fraction based on the spectral ratio method was carried out in detail for an individual particle event by *Dröge* [2003]. This study revealed slab fractions of 20% and 40% in the inertial and dissipation ranges, respectively, and provided mean free paths which were in good agreement with the ones obtained from simultaneous particle observations.

In the present paper, we will first give a brief overview on the phenomenological description of the interplanetary transport of solar particles. Section 3 reviews the concept of dynamical scattering and resonance broadening in cosmic ray transport, and discusses the general background of the pitch angle scattering problem. Section 4 demonstrates how transport coefficients can be calculated with the theory outlined before from measured magnetic field fluctuations. Section 5 discusses the implications of the results presented in this work and summarizes our conclusions.

2. TRANSPORT MODELS

The interplanetary magnetic field can be approximately viewed as a smooth average field, represented by an Archimedian spiral, with superimposed irregularities. If the assumption is made that the distribution function $f(r, p, t)$ for energetic charged particles with momentum p at a position r at a time t is always nearly isotropic, their propagation in the Heliosphere can be described by a general Parker transport equation

$$\frac{\partial f}{\partial t} - \nabla \cdot (K \cdot \nabla f) + V_s \cdot \nabla f$$
$$- \frac{1}{3}(\nabla \cdot V_s)p\frac{\partial f}{\partial p} = Q \qquad (1)$$

which includes the effects of diffusion, drift, convection and adiabatic energy losses. Here V_S is the solar wind velocity, Q represents the sources of the energetic particles, and

$$K = \begin{pmatrix} \kappa_\perp & \kappa_A & 0 \\ -\kappa_A & \kappa_\perp & 0 \\ 0 & 0 & \kappa_\parallel \end{pmatrix} \qquad (2)$$

is the heliospheric diffusion tensor. Its diagonal elements describe diffusion perpendicular and parallel to the mean magnetic field, while the off-diagonal elements describe drift effects. A radial diffusion coefficient can be defined as $\kappa_r = \kappa_\parallel \cos^2\phi + \kappa_\perp \sin^2\phi$ where ϕ is the angle between the radial direction and the magnetic field. As another convenient measure for the scattering strength we introduce the mean free path λ which is related to the diffusion coefficient by $\kappa = 1/3 \, v\lambda$, where v is the particle speed.

Perpendicular diffusion plays an important role in the modulation of galactic cosmic rays and particle acceleration at shock waves. It is still under debate whether this effect is also of importance for the propagation of solar particles in the inner Heliosphere. Recent progress in numerical simulations and in theoretical work on perpendicular diffusion [*Giacalone and Jokipii*, 1999; *Matthaeus et al.*, 2003] suggests that $\kappa_\perp$ is closely related to the value of $\kappa_\parallel$, as well as to the geometry of the solar wind turbulence. The modeling of solar particle events can give valuable insight on both of these quantities, and thus also on general aspects of transport and acceleration processes in the Heliosphere.

In the treatment of solar particle propagation in the inner Heliosphere presented in this work perpendicular diffusion and drift effects will be neglected. If we assume that the non-radial gradients in the distribution function of the particles are small the scattering can approximately be considered as isotropic spatial diffusion and the transport equation can be written in the form [*Parker*, 1963]:

$$\frac{\partial N}{\partial t} + \frac{1}{r^2}\frac{\partial}{\partial r}\left(r^2 V_S N - r^2 \kappa_r \frac{\partial N}{\partial r}\right)$$
$$- \frac{2V_S}{3r}\frac{\partial}{\partial E}(\alpha E N) = Q(r, E, t) \qquad (3)$$

Here $N(r, E, t) = 4\pi (p^2/v) f(p)$ is the differential number density, and $\alpha = (E + 2E_0)/(E + E_0)$, with E and E_0 being the kinetic and rest energy of the particles, respectively. It is assumed that the solar wind velocity is constant and has only a radial component. In modeling observed time profiles of solar particle events with solutions of the transport equation, in order to derive the above transport coefficients, it is important to not only fit the isotropic part of the distribution function but also make use of the information contained in its angular dependence. An anisotropy can be defined as

$$A = \frac{3S}{vN} \quad (4)$$

where $S = CV_S N - \kappa_r \partial N/\partial r$ denotes the streaming of the particles and C is the Compton-Getting factor [cf., *Ng and Gleeson*, 1975]. Equation (3) has the advantage that numerical solutions can be obtained in a straight forward manner and solar particle events can be modeled including the effects of convection and adiabatic deceleration in the solar wind which are important for low energy (<1 MeV/n) ions. For some special cases even analytic solutions exist. The disadvantage of the spatial diffusion model is that it does not give very accurate predictions if the mean free path is large (greater than ~10% of the distance to the source), in particular during the rise phase of the event when also the anisotropy is large.

A better treatment in cases where the scattering is relatively weak is provided by the model of focussed transport which describes the particle scattering on a more elementary level and allows a direct comparison with predictions of wave-particle interaction theories. In this model the motion of the particles consists of two components, adiabatic motion along the smooth field and pitch angle scattering off the irregularities. The quantitative treatment of the evolution of the particle's phase space density $f(z, \mu, t)$ can be cast into the form

$$\frac{\partial f}{\partial t} + \mu v \frac{\partial f}{\partial z} + \frac{1-\mu^2}{2L} v \frac{\partial f}{\partial \mu} - \frac{\partial}{\partial \mu}\left(D_{\mu\mu}(\mu)\frac{\partial f}{\partial \mu}\right) = q(z,\mu,t) \quad (5)$$

where z is the distance along the magnetic field line, $\mu = \cos\theta$ the particle pitch angle cosine. The particle speed v remains constant. The systematic forces are characterized by $L(z) = B(z)/(-\partial B/\partial z)$, the focussing length in the diverging magnetic field B. Analytical solutions of Equation (5) are not known, numerical treatments based on finite-differences schemes were developed by *Ng and Wong* [1979], *Schlüter* [1985] and *Ruffolo* [1991]. Extensions of the model which incorporate the effects of adiabatic energy losses and convection in the solar wind into Equation (5) were presented by *Ruffolo* [1995] and, utilizing a Monte-Carlo method, by *Kocharov et al.* [1998].

If the scattering is sufficiently strong, $f(z, \mu, t)$ adjusts rapidly to a nearly isotropic distribution, and the solutions of Equation (5) become similar to those of the spatial diffusion model. The mean free path $\lambda_\parallel$ which relates the pitch angle scattering rate to the spatial diffusion parallel to the ambient magnetic field is given by [*Hasselmann and Wibberenz*, 1978]

$$\lambda_\parallel = \frac{3v}{8}\int_{-1}^{+1} d\mu \frac{(1-\mu^2)^2}{D_{\mu\mu}(\mu)} \quad (6)$$

The mean free path has proven to be a convenient parameter to characterize the varying degrees of scattering from one solar particle event to another, even when it adopts values close to or larger than the observers's distance from the Sun and the transport process cannot be considered as spatial diffusion. The transport of particles away from the coronal acceleration site and their subsequent injection at the beginning of the interplanetary magnetic field line connected to the observer can be expressed in terms of a Reid-Axford profile [*Reid*, 1964]

$$q(z,t) = \frac{C}{t}\exp\left\{-\frac{\tau_c}{t} - \frac{t}{\tau_L}\right\} \quad (7)$$

where the rise and decay time scales τ_c and τ_L originally were envisioned to represent coronal diffusion from a solar flare and escape onto the connecting field line, respectively. We use this functional form to conveniently parameterize the source function in Equations (3) and (5), but the actual transport does not need to be coronal diffusion. In fact, Equation (7) can be regarded as a phenomenological representation of a generic injection process in which there is a fast rise to maximum, followed by a monotonic decay indicating injection close to the Sun over a finite amount of time. It can thus describe not only coronal diffusion, but also enhanced scattering close to the Sun as well as particle injection from an acceleration process in the higher corona such as a coronal shock [*Lee and Ryan*, 1986; *Kocharov et al.*, 2005].

Solutions of the transport equations (3) and (5) have been widely used to model observed intensity and anisotropy time profiles. Successful fits to observed particle data often require a radial dependence of the mean free path, which is usually described by a power law, $\lambda_r \propto r^b$. The proper choice of b is important in order to determine the correct local value of λ at the point of observation. In many events values of b close to 0 are found to be sufficient for a good fit, indicating that there is no systematic radial dependence of λ_r between the Sun and $r \sim 3$ AU. However, the modeling is quite sensitive to local variations of the mean free path due to propagating disturbances such as shocks, coronal mass ejections (CMEs) and magnetic bottle configurations which

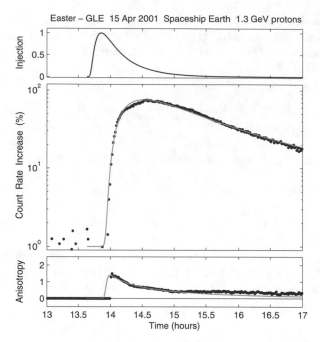

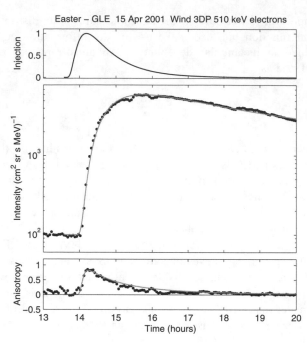

Figure 1. Fits to the *spaceship Earth* neutron monitor time-intensity and anisotropy profiles observed on 2001 Apr 15, assuming a constant $\lambda_r = 0.17$ AU. At a radial distance of 1 AU this value corresponds to $\lambda_\parallel = 0.34$ AU.

Figure 2. Fits to the time-intensity and anisotropy profiles of 510 keV electrons observed on *Wind* on 2001 April 15, assuming $\lambda_r = 0.14$ AU ($\lambda_\parallel = 0.26$ AU at Earth).

can have a significant influence on the particle propagation [cf., *Bieber et al.*, 2002].

As an example we show in Figures 1 and 2 the time history of ~1.3 GeV protons and ~510 keV electrons, respectively, observed on 2001 April 15. The particle event was associated with a solar flare located at S20 W85 which produced an X14.4 soft X-ray event, Type III radio emission and a CME. The GeV protons in this event were observed by the *Spaceship Earth* neutron monitor network, and a modeling of their transport was described in detail in *Bieber et al.* [2004]. Figure 1 also shows fits to the time-intensity and anisotropy profiles of the energetic protons from which a parallel mean free path of 0.34 AU at Earth was derived. A fit to 510 keV electrons measured with the *Wind* 3DP instrument [*Lin et al.*, 1995], derived assuming a constant $\lambda_r = 0.14$ AU ($\lambda_\parallel = 0.24$ AU at Earth) is shown in Figure 2. Note that the simultaneous fits to the intensity and anisotropy profiles also provide a fairly accurate determination of the injection profile close to the Sun (shown in the upper panels of the figures). Figure 3 shows results from a recent survey of near-earth mean free paths $\lambda_\parallel$ which was based on the modeling of selected solar events where electron and proton data over a large energy range were available [*Dröge*, 2000b].

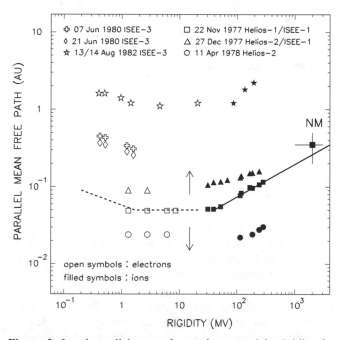

Figure 3. Local parallel mean free path vs. particle rigidity for selected solar particle events. The form of the rigidity dependence as indicated by the curve seems to be consistent with observations from any given event. Only the absolute height of the curve varies. Adapted from *Dröge* (2000b).

3. THEORY

Cosmic ray pitch angle scattering is caused by irregularities of the magnetic field which violate the conservation of the first adiabatic invariant. When the irregularities superposed on the average field are sufficiently small, a pitch angle diffusion coefficient $D_{\mu\mu}(\mu)$ can be obtained in closed form by calculating first-order corrections to the particle's orbit in the average field, and ensemble-averaging over the statistical properties of the turbulence [Jokipii, 1966]. Small irregularities mean that the changes of the particle's pitch angle during a single gyration are small and many gyrations are required to change the pitch angle considerably, implying that the particles are scattered predominantly by irregularities which are in resonance with the particle gyration. As a cumulative result of many small random changes in its pitch angle the particle experiences a macroscopic change in direction, leading to spatial diffusion along the field line. Because of the appearance of $D_{\mu\mu}(\mu)$ in the denominator of the integrand in the expression for $\lambda_\parallel$ (cf., Equation 6), the behaviour of the scattering rate at pitch angles where it is small has a disproportionate influence upon the mean free path. It can therefore be expected that $\lambda_\parallel$ is regulated by the scattering rate at pitch angles where scattering is slowest, generally at pitch angles close to 90°.

The statistical properties of the fluctuations (δB), superimposed onto an average magnetic field ($B = B_0$), are conveniently characterized by their two-point, two time correlation function. If the additional assumption is made that the fluctuations are homogeneous and stationary, the correlation functions depend only on the spatial and time lags $R_{ij}(\zeta, \tau) = \langle \delta B_i (r, t) \, \delta B_j (r + \zeta, t + \tau) \rangle$ where $\langle \rangle$ denotes the average over an ensemble of microscopic realizations of the fluctuations [e.g., Batchelor, 1970]. Alternatively, the fluctuations can be as well described by power spectra $P_{ij}(k, \omega)$ which are obtained by taking the Fourier transforms of the correlation functions with respect to their spatial and time coordinates.

Yet another possibility, which is suited to isolate the dynamical aspects of the correlation and which we will pursue here, is to perform the Fourier transforms only with respect to the spatial coordinates but not to the time coordinate. In the following, we will consider the slab component for which mixed spectral functions can be cast into the form [cf., Dröge, 2000a]

$$P_{ij}(\boldsymbol{k}, \tau) = P_{ij}(k) \exp^{i\omega_j^r(k)\tau - \Gamma_j(k)^\tau} \delta(k_x)\delta(k_y) \qquad (8)$$

where we use helical ($i, j = R, L; \; k = k_z = k_\parallel$) coordinates which allow the spectral functions to be related to the spectral

densities $I_{R,L}^{+,-}(k)$ of forward (+) and backward (−) propagating waves with right-hand (R) or left-hand (L) polarization, and to the helicity of the fluctuations. Here $\omega_{R,L}^r(k)$ is the real part of the dispersion relation in the wave picture, and $\Gamma_{R,L}(k)$ describes wave damping effects due to interactions with the warm ($T > 0$) background plasma, or decorrelation effects in the turbulence picture, in which $\omega_{R,L}^r(k) = 0$.

The pitch angle scattering coefficient can now be expressed as

$$D_{\mu\mu}(\mu) = \frac{\Omega^2(1 - \mu^2)}{2B_0^2} \int_{-\infty}^{+\infty} dk$$

$$\times \left\{ \frac{\Gamma_R(k)}{\Gamma_R^2(k) + (k\mu v - \omega_R^r(k) - \Omega)^2} P_{RR}(k) \right.$$

$$\left. + \frac{\Gamma_L(k)}{\Gamma_L^2(k) + (k\mu v - \omega_L^r(k) + \Omega)^2} P_{LL}(k) \right\} \qquad (9)$$

The occurrence of the function $\Gamma(k)$ in the spectral density (8) leads to a resonance broadening in the particles' interaction with the fluctuations, and the delta functions of standard QLT, describing the sharp resonance of the particles with fluctuations of a certain wave number are replaced by Breit-Wigner type resonance functions (the expressions preceding the power spectra P_{RR} and P_{LL}, respectively, in Eq. 9). As a result, particles with a given μv (in particular, $\mu = 0$) can now be scattered by fluctuations within a finite range of wave numbers. A number of different functional forms for $\Gamma(k)$ have been suggested. Achatz et al. [1993] found that

$$\Gamma(k) \simeq k v_{th,p} \exp(-\Omega_p / (\beta^{1/6} k V_A)) \qquad (10)$$

can describe the damping of ion cyclotron waves, where $v_{th,p}$ is the thermal background proton velocity, V_A the Alfvén speed, and $\beta = v_{th,p}^2 / v_A^2$ the plasma-beta. In the turbulence picture of the fluctuations, Bieber et al. [1994] consider a damping model with

$$\Gamma(k) = \alpha |k| V_A \qquad (11)$$

where the parameter α, which might be estimated from turbulence theory, allows to adjust the strength of the dynamical effects, ranging from $\alpha = 0$ (magnetostatic limit) to $\alpha \geq 1$ (strongly dynamical). By expressing $v_{th,p}$ in Equation (10) in terms of β and v_A we find that in the limit of large wave numbers $(k \gg \Omega_p/(\beta^{1/6}V_A))$ resonance broadening effects due to thermal damping and decorrelation are of similar form, i.e., $\Gamma(k) = \delta |k| V_A$, with $\delta = \alpha$ for the turbulence picture, and $\delta = \sqrt{\beta}$ for the wave picture. The exponential cut-off in Equation (10) towards smaller wave numbers reflects the fact that Alfvén waves are very mildly damped at low frequencies, and that the resonance function for thermal damping quickly approaches a delta-function, as in the original QLT. In this work we will use the simplified form of $\Gamma(k)$ quoted above for the description of thermal damping effects.

Under the assumption that the fluctuation spectra can be described by a single power law in k (i.e., neglecting the dissipation range), and imposing further simplifications, an analytical solution of Equation (9) can be obtained [Dröge, 2003] which is a fairly good approximation for the ion mean free paths in the case that the resonance broadening is sufficiently strong ($\alpha \sim 0.3$ or larger). However, for electrons with energies typical for solar events and for the case of weak resonance broadening the dissipation range has to be adequately incorporated, and numerical methods to calculate $D_{\mu\mu}(\mu)$ have to be applied.

In the following we will neglect helicity effects and assume that the spectral densities $I_{R,L}^{+,-}(k)$ are equal and of the form

$$I(k) = I_5 \frac{\left[1 + (k_i/k)^m\right]^{-q/m} k_5^{-q}}{\left[1 + (k/k_d)^n\right]^{(q_d - q)/n}} \qquad (12)$$

where k_5 is the wavenumber in units of 10^{-5} km^{-1}, $I_5 = I(k_5 = 1)$, q and q_d are the spectral indices in the inertial and dissipation range, respectively, and k_d is the wave number where the dissipation range sets in. The transition from the energy-containing range into the inertial range of the fluctuations, where the spectrum is assumed to be flat, is marked by k_i which is equivalent to the inverse of the correlation length. The sharpness of the transition from one power law exponent to another is modeled by the parameters m and n, respectively. The normalization is such that $I(k)$ represents the one-sided spectrum of one perpendicular component of the fluctuation, i.e., $\delta B_x^2 = \int_0^\infty dk I(k)$.

For comparison with earlier work we have calculated $D_{\mu\mu}(\mu)$ adopting the model spectrum of Bieber et al. [1994] who considered a turbulence model with the following parameters: a one-sided power spectral density at k_5 in one of the perpendicular components of $5 \cdot 10^4$ (nT)^2 km, B = 4.12 nT, $V_A = 33$ km/s, $k_d = 0.02$ km^{-1}, $q_d = 3$, and $\delta = 0.3$. Figure 4 shows the resulting pitch angle scattering coefficient normalized to particle speed for 30 MV protons and electrons as a function of the particle pitch angle cosine. Above $\mu \sim 0.01$ the scattering rate is similar in both the dynamical model and standard QLT. Below that value, the electrons are in resonance with the dissipation range of the turbulence, and the scattering is strongly suppressed relative to the dissipationless

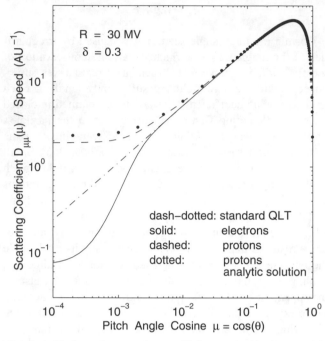

Figure 4. Pitch angle scattering coefficient normalized to particle speed for 30 MV protons and electrons as a function of the particle pitch angle cosine. The results shown here were derived for a resonance broadening parameter $\delta = 0.3$ and the model power spectrum described in the text. Solid and dashed curves were calculated numerically. For comparison, the analytical approximation and the prediction of standard QLT are also shown. Adapted from Dröge (2003).

theory. Below $\mu \sim 5 \cdot 10^{-4}$ resonance broadening sets in and leads to a finite scattering rate at $\mu = 0$, thereby avoiding the divergent mean free paths caused by the dissipation range in magnetostatic models [cf., Bieber et al., 1988]. For 30 MV protons, however, the normalized scattering rate at small pitch angles is greatly enhanced compared to the electrons, because they have a much lower gyrofrequency at the same rigidity.

The introduction of resonance functions into the calculation of $D_{\mu\mu}(\mu)$ suggests replacing the resonance condition in the magnetostatic theory, $k_{res} = \Omega/(\mu v)$, by an effective resonant wave number

$$k_{res} \sim \frac{\Omega}{\sqrt{(\mu v \pm v_{ph})^2 + \delta^2 V_A^2}} \qquad (13)$$

which for $\delta > 0$ is always finite, indicating a non-zero scattering rate through 90°. Here we have included the effects of wave propagation which are described by the wave's phase speed $v_{ph} = \omega / k$. Even for $\delta = 0$ wave propagation effects can lead to finite scattering for dissipation range spectra, as was shown by *Schlickeiser* [1988] and *Achatz et al.* [1993], although these models do not reproduce the observed rigidity dependence of the particles (cf., Figure 3) well. A linear dispersion relation which is valid for low-frequency Alfven waves, i.e., $v_{ph} = v_A$, can be easily incorporated into the analytical and numerical treatment of (2). However, even in the case of a cold ($\beta = 0$) plasma v_{ph} becomes dependent on the wave number in the dissipation range, and including finite-β and wave damping effects would introduce further complications in the computation of the dispersion relation and thus of the scattering coefficient. We will therefore try the hypothesis that, if plasma waves constitute the slab fraction of the fluctuations, damping rather than propagation is their leading order dynamical effect. Further we assume that the dispersion relation, or wave propagation effects in general are not essential for particle scattering through 90°, i.e., we set $\omega(k) = 0$.

In the following, we explore the effects of varying k_d, q_d, and δ, on the the mean free path $\lambda_\parallel$ which relates the pitch angle scattering rate to the spatial diffusion parallel to the ambient magnetic field. We adopt the suggestions of *Bieber et al.* [1994] and assume that only 20% of the above model spectrum represents the slab portion and contributes to particle scattering, and that the 80% in the 2D-component makes essentially no contribution. The latter assumption has also been confirmed by analytical calculations [*Shalchi and Schlickeiser*, 2004] which show that the 2D component is negligible for typical plasma conditions, except for particles of very low rigidities and if the slab fraction is below 10%. The results are presented in Figure 5. The top panel of the figure shows mean free paths calculated numerically by means of Equations (6) and (9) for a fluctuation spectrum assuming $\delta = 1$, an onset of the dissipation range at 0.02 km^{-1}, and for dissipation range spectral indices varying within the range of observations. The curves resulting from the combined effects of the dissipation range and resonance broadening are in good agreement with the shape of the rigidity dependence shown in Figure 3. As can be seen from the figure, at rigidities above 100 MV standard QLT is a good approximation for the particle mean free paths. A closer inspection shows that because of the still finite width of the resonance function at smaller resonant wave numbers, which are mostly in the inertial range, the curve lies somewhat below the standard QLT prediction. If we had used the exact expression (10) for the thermal damping rate instead of the approximation for large k, the curves in this range would be almost identical. The difference is so small that the simplification seems to be justified.

Below ~100 MV the mean free path is significantly larger for electrons, but smaller for protons compared to standard QLT. Dissipation ranges with steeper spectra lead to larger

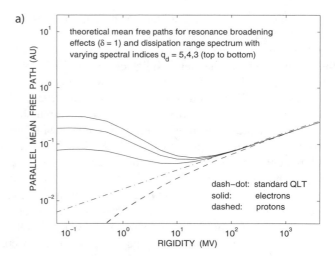

a)

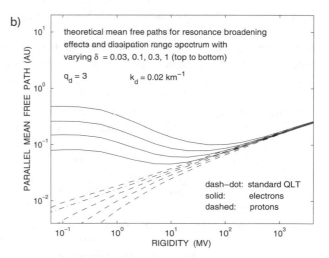

b)

Figure 5. Theoretical mean free paths for resonance broadening and a power spectrum with spectral index of 5/3 in the inertial range, and varying values of the spectral indices in the dissipation range (top), and varying values of the resonance broadening parameter (bottom). Adapted from *Dröge* (2003).

mean free paths for electrons below ~30 MV even for strong resonance broadening. The mean free paths for protons are almost not affected by the above variations. The bottom part of Figure 5 shows mean free paths calculated for k_d and q_d kept constant at 0.02 km^{-1} and 3, respectively, and δ varying from 0.03 to 1. In this case also the protons exhibit a variation, and for $\delta = 0.03$. have larger mean free paths compared to the prediction from standard QLT.

4. COMPARISON WITH OBSERVATIONS

Ideal conditions to test the model outlined above, and the simplifications made, would be as follows: (i) observations with directional information of electrons and ions over a large range in rigidity including the measurement of electrons and ions at the same rigidity, e.g., electrons at several MeV and ions at ~keV energies, (ii) a large-scale structure of the interplanetary magnetic field close to the nominal Archimedean spiral, (iii) the absence of interplanetary disturbances such as CMES and shocks, (iv) a constant level of the magnetic fluctuations (approximately stationary turbulence). As one might guess, the above conditions are rarely fulfilled. Particle events with electron fluxes at 10 MV or above which are large enough to derive meaningful values of λ are scarce, whereas the propagation of ions in that range, due to their low speeds is often affected by varying conditions in the solar wind and also, or even totally dominated by the effects of coronal mass ejections and interplanetary shocks. Powerful events which generate high energy particles frequently occur during periods of enhanced solar activity, with a high probability that preceding events have released shocks and created increased levels of turbulence in the solar wind.

A particle event on 1996 July 9 for which high resolution magnetic field data, plasma data, and electron and proton data at various rigidities were available and conditions (ii)-(iv) were approximately valid was analyzed by *Dröge* [2003]. Using a method suggested by *Bieber et al.* [1996] to estimate of the percentage of slab and 2D fluctuations from the ratio of transverse spectral powers $P_{yy}(\nu)/P_{xx}(\nu)$ it was possible to calculate mean free paths which were in good agreement with the particle observations. A similar result was obtained in another event (2000 July 14) where GeV protons were observed by *assuming* the apparently typical 20% value for the slab fraction because the above method was not applicable due to an irregular magnetic field configuration. Data from these events will be shown later in Figure 9, and discussed together with new observations.

Here we will continue with the analysis of the 2001 April 15 particle event and investigate whether the dynamical quasi-linear theory, together with the available solar wind plasma parameters, can reproduce the transport coefficients derived in the modeling. An overview of *Wind* MFI [*Lepping et al.*, 1995] observations of interplanetary magnetic field data from April 15, 00:00 UT to April 16, 24:00 UT is given in Figure 6. For most of the time all components of the magnetic field, as well as its absolute value, show large fluctuations, indicating a high level of solar wind turbulence. During the period shown the solar wind speed, observed with the *Wind* SWE instrument [*Ogilvie et al.*, 1995] decreased from 600 to ~400 km/s. At the onset of the particle event it had a value of $V_S = 500$ km/s which we have adopted for the calculations made in this work. The bottom panel of the figure shows that the azimuth angle of the magnetic field vector exhibits large deviations from its nominal value of ~40° for the above solar wind speed.

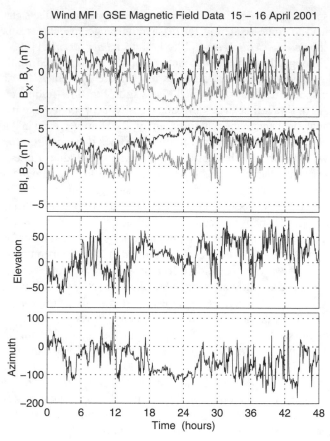

Figure 6. *Wind* MFI (*Lepping et al.*, 1995) magnetic field data for the period 2001 Apr 15-16.

The work of *Dröge* [2003] suggests that the spectral ratio test of *Bieber et al.* [1996] can be used to derive the slab/2D ratio for low to moderate levels of the magnetic fluctuations, and if the angles between the solar wind velocity and the magnetic field are within approximately 30 to 60 degrees. For an estimate of the slab component in the 2001 April 15 event, where the above conditions are violated, we will in the following *assume* that the magnetic field component perpendicular to the ecliptic (B_Z in GSE coordinates) is representative for the scattering strength, and that the slab fractions are 20% and 40%, respectively, in the inertial and dissipation range, in accordance with the statistical average values. A power spectrum for the magnetic fluctuations was computed from high resolution (3 vectors/s) *Wind* MFI observations for the time period 01/04/15 12:00 UT to 01/04/16 24:00 UT, covering approximately a period of one and a half days after the onset of the event. As suggested by *Dröge* [2003] the fluctuations during such a period are probably most representative for the scattering between the Sun and 1 AU, under the

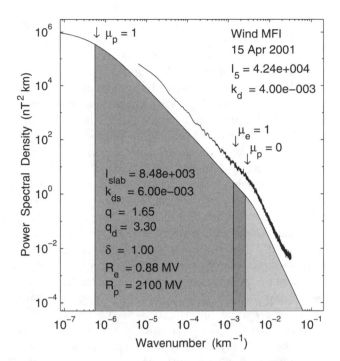

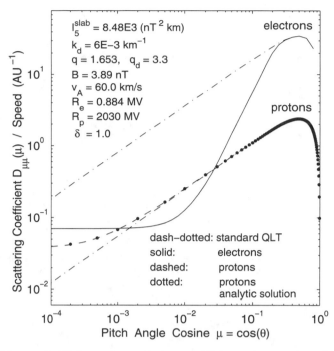

Figure 7. Upper curve: power spectra of the magnetic field component perpendicular to the ecliptic (in the GSE system) for the period 01/04/15 12:00 UT to 01/04/16 24:00 UT. Lower curve: estimated spectrum for the slab component. The shaded area below the curve indicates for the case of strong resonance broadening ($\delta = 1$) the range of the spectrum the 510 keV electron (light grey) and 1.3 GeV protons (medium grey) are effectively in resonance with (dark grey where they overlap).

Figure 8. Pitch angle scattering coefficients (assumed to be symmetric in μ) normalized to particle speed for 510 keV electrons and 1.3 GeV protons. The results shown here were derived for a resonance broadening parameter $\delta = 1.0$ and the plasma parameters obtained for the 2001 April 15 particle event.

hypothesis of stationary fluctuations. The spectrum is shown in Figure 7, together with the slab component which was estimated under the above assumptions. The ranges of the spectrum 510 keV electrons and 1.3 GeV protons are effectively in resonance with (cf., Equation 13) are indicated by the shaded areas below the curve. As can be seen from the figure, the protons are in resonance more or less only with the inertial range of the spectrum, whereas the resonant wavenumbers of the electrons are mostly in the dissipation range.

Pitch angle scattering coefficients, computed from the slab power spectrum and averages of plasma parameters for the above period, i.e., B = 3.89 nT, n = 2.0 cm^{-3}, T = 120 000 K, and for $\delta = 1$ are shown in Figure 8 for illustration. According to their respective resonance ranges, the scattering coefficients for electrons and protons have quite different shapes. We find that the scattering of high energy protons can be approximated well with dissipationless standard QLT. The scattering of electrons, on the other hand, is strongly reduced below $\sim\mu = 0.1$. The resulting mean free paths for the thermal damping model with the measured $\beta = 0.55$ are shown in

Figure 9 (upper panel, solid line). The agreement with the neutron monitor observations is very good, and the difference with the electron observations is less than a factor of 2. The prediction for the electrons can even be improved if we adopt the turbulence model and, to account for the high level of the fluctuations ($\delta B / B_0 \sim 0.6$), assume a large decorrelation parameter of $\alpha = 1.2$ (dashed line). We note that, in spite of their deviating scattering coefficients, the mean free paths of the electrons and protons considered here are of the same order. However, their pitch angle distributions are quite different.

The middle panel of Figure 9 shows observations and predictions for the 1996 July 9 event where the slab fraction could be determined with the spectral ratio method, the lower panel shows data for the 2000 July 14 event where also rather the characteristic value had to be assumed because of a high turbulence level. It can be seen from the figure that the observed rigidity dependence as well as the absolute values are reproduced well by the dynamical quasi-linear theory.

5. CONCLUSIONS

The recent results summarized in this work indicate that we finally might have arrived at a satisfactory quantitative

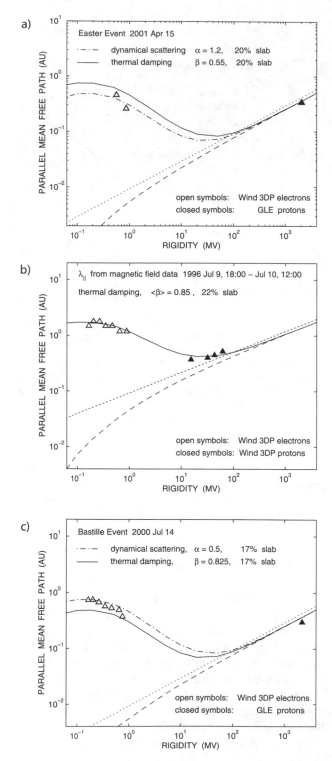

theory of solar energetic particle transport in the inner Heliosphere. Out to 2 or 3 AU, and close to the ecliptic plane, diffusion parallel to the average magnetic field due to pitch-angle scattering is the dominating process, with the addition of convection and adiabatic deceleration at low energies (ions below ~1 MeV/n). For the above scenario it appears that a dynamical quasi-linear theory (DQLT) which takes into account particle interaction with the slab component of the fluctuations and resonance broadening due to dynamic or wave damping effects is a good approximation to particle scattering at all pitch angles. Resonance broadening dominates the scattering through $\mu = 0$ and probably makes the need for including other effects such as non-linear corrections, mirroring, wave propagation and details of the dispersion relation, and the contribution of the 2D-component to particle scattering obsolete. The good agreement between the observations presented here and the above approach with basically no free parameters - within a factor ~2 overall uncertainty - seems to support the hypothesis that DQLT is the leading-order theory for parallel transport. If other effects, in the above scenario, would be of similar or greater importance this should be visible. It would be interesting to find solar particle events with larger deviations from DQLT predictions, or other observations of heliospheric transport processes which are in contradiction to DQLT. If systematic discrepancies with observations were detected for certain cases, DQLT predictions could be used to estimate the importance of other effects mentioned above.

The agreement between the predictions of DQLT and particle observations for a (admittedly, still limited) number of solar events studied in detail is promising. Resonance broadening is related to observable plasma parameters B, n, and T, so some prediction of local scattering conditions should be possible. However, a major remaining obstacle for a true prediction of transport parameters in the solar wind remains our lack of knowledge of the exact decomposition of the fluctuations, which is difficult to obtain from single spacecraft measurements, and the question how representative near-Earth measurements are for scattering conditions experienced by the particles in the inner Heliosphere. A reasonably accurate prediction of particle transport parameters from single-spacecraft observations, which we would define as being within a factor of ~2 compared to the observed values, should be expected for particle events during which the magnetic field obeys a regular structure and large disturbances in the solar wind are absent.

The dynamical quasi-linear theory for parallel transport asymptotically approaches standard QLT results at high rigidities, suggesting that in this range the simpler theory can be used to describe particle transport parallel to the mean

Figure 9. Mean free paths obtained from particle observations, and predictions from the dynamical quasi-linear theory for 3 solar particle events. Solid and dash-dotted lines show prediction for electrons in the thermal damping and dynamical scattering model, respectively. Dashed lines show prediction for ions (similar for both models), dotted lines show the standard QLT result.

field once the geometry of the fluctuations in the inertial range is properly taken into account. However, at low rigidities the differences introduced by the combined effects of the dissipation range and resonance broadening are quite dramatic. The dynamical theory predicts a much more efficient scattering through a 90° pitch angle and a reduced $\lambda_\parallel$ compared to standard QLT for ions, but large values of $\lambda_\parallel$ for electrons. As can be seen from Figure 9, standard QLT even with the correct slab fraction would have underestimated mean free paths of low energy electrons by almost two orders of magnitude.

Consequences of this prediction by DQLT could be tested for particle processes at low energies for which parallel transport is of importance. These processes comprise the scattering of pick-up ions and injection of particles into a shock acceleration process, e.g., at shocks driven by coronal mass ejections and shocks associated with co-rotating interaction regions. Dynamical scattering might in fact help to justify the assumed small upstream values of $\lambda_\parallel$ which are necessary to make shock acceleration work for ions, but it would be difficult to explain how electrons could be efficiently accelerated if mean free paths were organized such as derived from the work presented here, e.g., as shown in Figure 9. The ubiquitous low energy solar electrons, because of their large mean free paths, may be used as probes to study the properties of interplanetary turbulence in regions not accessible to spacecraft observations. Should more refined methods to analyse solar wind turbulence, and routine plasma observations from spacecraft located at various radial distances from the Sun become available in the future, major improvements in the forecast of solar particle events and a better understanding of particle transport and energization in the Heliosphere might be possible.

Acknowledgments. I thank the organizers of the Chapman Conference on Solar Energetic Plasmas and Particles for inviting me to present this paper, and the European Space Organisation for financial support. Thanks are also due to R.P. Lin and S. Krucker for providing *Wind* 3DP particle data, and to R.P. Lepping and A. Szabo for providing *Wind* MFI magnetic field data. The comments of two anonymous referees certainly helped to improve the manuscript.

REFERENCES

Achatz, U., W. Dröge, R. Schlickeiser, and G. Wibberenz, Interplanetary transport of solar electrons and protons - effect of dissipative processes in the magnetic field power spectrum, *J. Geophys. Res.*, 98, 13261-13280, 1993.

Batchelor, G.K. *Theory of Homogenous Turbulence*, Cambridge University Press, 1970.

Bieber, J.W., C.W. Smith, and W.H. Matthaeus, Cosmic ray pitch-angle scattering in isotropic turbulence, *Astrophys. J.*, 334, 470-475, 1988.

Bieber, J.W., W.H. Matthaeus, C.W. Smith, W. Wanner, M.-B. Kallenrode, and G. Wibberenz, Proton and electron mean free paths: the Palmer consensus revisited, *Astrophys. J.*, 420, 294-306, 1994.

Bieber, J.W., W. Wanner, and W.M. Matthaeus, Dominant two-dimensional solar wind turbulence with implications for cosmic ray transport, *J. Geophys. Res.*, 101, 2511-2522, 1996.

Bieber, J.W., W. Dröge, P.A. Evenson, R. Pyle, D. Ruffolo, U. Pinsook, P. Tooprakai, M. Rujiwarodom, T. Khumlumlert, and S. Krucker, Energetic particle observations during the 2000 July 14 solar event, *Astrophys. J.*, 567, 622-634, 2002.

Bieber, J.W., P.A. Evenson, W. Dröge, R. Pyle, D. Ruffolo, D.M. Rujiwarodom, P. Tooprakai, and T. Khumlumlert, Spaceship Earth observations of the Easter 2001 solar particle event, *Astrophys. J.*, 601, L103-L106, 2004.

Dröge, W., Particle scattering by magnetic fields, *Space Sci. Rev.* 93, 121-151, 2000a.

Dröge, W., The rigidity dependence of solar particle scattering mean free paths, *Astrophys. J.*, 537, 1073-1079, 2000b.

Dröge, W., Solar particle transport in a dynamical quasi-linear theory, *Astrophys. J.*, 589, 1027-1039, 2003.

Giacalone, J., and J.R. Jokipii, The transport of cosmic rays across a turbulent magnetic field, *Astrophys. J.*, 520, 204-214, 1999.

Hasselmann, K., and G. Wibberenz, Scattering of charged particles by random electromagnetic fields, *Z. Geophys.* 34, 353-388, 1968.

Jokipii, J.R., Cosmic-ray propagation. I. Charged particles in a random magnetic field, *Astrophys. J.*, 146, 480-487, 1966.

Kocharov, L., R. Vainio, G.A. Kovaltsov, and J. Torsti, Adiabatic deceleration of solar energetic particles as deduced from Monte Carlo simulations of interplanetary transport, *Solar Phys.* 182, 195-215, 1998.

Kocharov, L., M. Lytova, R. Vainio, T. Laitinen, and J. Torsti, Modeling the shock aftermath source of energetic particles in the solar corona, *Astrophys. J.*, 620, 1052-1068, 1999.

Leamon, R.J., C.W. Smith, N.F. Ness, W.H. Matthaeus, and H.K. Wong, Observational constraints on the dynamics of the interplanetary magnetic field dissipation range, *J. Geophys. Res.*, 103, 4775-4778, 1998.

Lee, M.A., and J.M. Ryan, Time-dependent coronal shock acceleration of energetic solar flare particles, *Astrophys. J.*, 303, 829-842, 1986.

Lepping, R.P., *et al.*, The Wind magnetic field investigation, *Space Sci. Rev.* 71, 207-229, 1995.

Lin, R.P., *et al.*, A three-dimensional plasma and energetic particle investigation for the Wind spacecraft, *Space Sci. Rev.* 71, 125-153, 1995.

Matthaeus, W.H., M.L. Goldstein, and D.A. Roberts, Evidence for the presence of quasi-two-dimensional nearly incompressible fluctuations in the solar wind, *J. Geophys. Res.*, 95, 20673-20683, 1990.

Matthaeus, W.H., G. Qin, J.W. Bieber, and G.P. Zank, Nonlinear collisionless perpendicular diffusion of charged particles, *Astrophys. J.*, 590, L53-L56, 2003.

Ng, C.K., and L.J. Gleeson, Propagation of Solar-flare Cosmic Rays Along Corotating Interplanetary Flux-tubes, *Solar Phys.* 43, 475-511, 1975.

Ng, C.K., and K.Y. Wong, Solar particle propagation under the influence of pitch-angle diffusion and collimation in the interplanetary magnetic field, *Proc. 16th Internat. Cosmic Ray Conf. (Kyoto)* 5, 252-257, 1979.

Ogilvie, K.W., *et al.*, SWE, a comprehensive plasma instrument for the Wind spacecraft, *Space Sci. Rev.* 71, 55-77, 1995.

Parker, E.N., *Interplanetary Dynamical Processes*, Wiley and Sons, New York, 1963.

Reid, G.C., A diffusive model for the initial phase of a solar proton event, *J. Geophys. Res.*, 69, 2659-2667, 1964.

Roelof, E.C., Propagation of solar cosmic rays in the interplanetary magnetic field, in *Lectures in High Energy Astrophysics*, edited by H. Ögelmann, H., and J.R. Wayland, pp. 111-135, NASA SP-199, 1969.

Ruffolo, D., Interplanetary transport of decay protons from solar flare neutrons, *Astrophys. J.*, 382, 688-698, 1991.

Ruffolo, D., Effect of adiabatic deceleration on the focused transport of solar cosmic rays, *Astrophys. J.*, 442, 861-874, 1995.

Shalchi, A., and R. Schlickeiser, The parallel mean free path of heliospheric cosmic rays in composite slab/two-dimensional geometry: I. the damping model of dynamical turbulence, *Astrophys. J.*, 604, 861-873, 2004.

Schlickeiser, R., On the interplanetary transport of cosmic rays, *J. Geophys. Res.*, 93, 2725-2729, 1988.

Schlickeiser, R., and U. Achatz, Cosmic-ray particle transport in weakly turbulent plasmas - 1. Theory, *J. Plasma. Phys.* 49(1), 63-77, 1993.

Schlüter, W., Die numerische Behandlung der Ausbreitung solarer flareinduzierter Teilchen in den magnetischen Feldern des interplanetaren Raumes, Ph.D. Thesis, 148 pp., University of Kiel, 1985. (In German)

Zank, G.P., and W.M. Matthaeus, Waves and turbulence in the solar wind, *J. Geophys. Res.*, 97, 17189-17194, 1992.

W. Dröge, Bartol Research Institute, University of Delaware, Newark, DE 19716. (droege@bartol.udel.edu)

Radial and Latitudinal Variations of the Energetic Particle Response to ICMEs

David Lario

The Johns Hopkins University, Applied Physics Laboratory, Laurel, Maryland, USA

We present energetic particle observations during events in which the interplanetary counterpart of a single coronal mass ejection (ICME) has been observed by spacecraft widely separated in radial distance and latitude. The effects that ICMEs have on energetic particle intensities depend on (1) the ability of ICMEs to drive strong shocks able to accelerate energetic particles, (2) the existence of an intra-ICME energetic particle population, and (3) the capability of the ICMEs to confine (or exclude) energetic particles. These three factors depend upon (1) the energy of the particles, (2) the solar wind medium in which the ICMEs propagate, (3) the mechanisms of energetic particle transport within and around the ICMEs, and (4) the magnetic topology of the ICMEs. As ICMEs expand, ICME-driven shocks weaken and intra-ICME magnetic field intensities decrease. These changes, together with the fortuitous occurrence of solar energetic particle events during the transit time of the ICMEs to large distances are the main causes for the different energetic particle signatures observed during the passage of the same ICME at two distant heliospheric locations.

1. INTRODUCTION

The interplanetary counterparts of coronal mass ejections (CMEs) are known as interplanetary CMEs (ICMEs). Solar wind and magnetic field signatures used to identify ICMEs are described elsewhere [e.g., *Gosling*, 2000]. Energetic particle (EP) signatures associated with the passage of ICMEs in the ecliptic plane and at 1 AU from the Sun have been widely studied [e.g., *Richardson*, 1997, and references therein]. These signatures include cosmic ray (CR) intensity depressions [*Cane*, 2000, and references therein], bidirectional ~1 MeV ion flows (BIFs) [*Marsden et al.*, 1987], bidirectional CR flows [*Richardson et al.*, 2000]. When ICMEs are preceded by fast shocks, low-energy (<20 MeV) proton intensities may peak around the time of the shock passage and decrease when the spacecraft enters into the ICMEs

Solar Eruptions and Energetic Particles
Geophysical Monograph Series 165
10.1029/165GM29

[*Cane et al.*, 1988]. Occasionally, during the passage of an ICME, it is possible to observe unusual particle flows due to a fresh injection of solar energetic particles (SEPs) from the Sun [*Richardson and Cane*, 1996].

Although many ICMEs show these EP signatures, not all of these signatures are present in every single ICME. An obvious requirement to observe these EP signatures is the presence of elevated EP fluxes. Cosmic rays are always present and *Cane et al.* [1997] found that the 75% of ICMEs observed at heliocentric radial distances $R \leq 1$ AU (identified using solar wind proton temperature depressions) showed decreases of $\geq 4\%$ in the >60 MeV cosmic ray flux [*Cane et al.*, 1997, Table 3]. However, there are ICMEs at $R = 1$ AU not accompanied by low-energy (<25 MeV) proton intensity depressions [*Cane et al.*, 1996, Table 1] and ICMEs at $R = 1$ AU that do not contain BIFs [*Marsden et al.*, 1987]. *Kahler and Reames* [1991] suggested that the most typical situation for those ICMEs able to drive strong shocks at 1 AU is a sharp decline in the ~25 MeV proton intensities near the leading edge of the ICME but not a corresponding increase at its trailing edge. Statistical analysis of the passage of ICMEs at 1 AU (identified using different solar wind signatures) show that BIFs are observed in ~80% of cases but are absent in the other ~20% of cases, because of either ion fluxes are too low to determine the flow or because there are no BIFs [*Richardson and Reames*, 1993]. In addition, BIFs are not always observed throughout the passage of the ICME, but may appear intermittently and not coincide exactly with the passage of the ICME [*Richardson and Reames*, 1993].

The solar wind disturbances generated by an ICME are determined by the ambient solar wind through which the ICME propagates [*Gosling et al.*, 1995]. As an ICME expands and propagates out from the Sun, it may undergo significant distortion, the shock driven ahead of a fast ICME weakens and the magnetic field magnitude inside the ICME decreases [*Riley et al.*, 2003]. These changes should be reflected in the EP response to the passage of ICMEs. For example, *Cane et al.* [1994, and references therein] showed that CR decreases at R = 1 AU are due to both the passage of ICMEs and of the enhanced magnetic field turbulence formed in the compressed medium between the shocks and the ICMEs. *Cane et al.* [1994] suggested that at large heliocentric distances, the effect of ICMEs on CR may become sufficiently reduced (as the ICME expands and the magnetic field magnitude decreases) and that only the turbulence in the compressed medium is important for producing CR decreases.

The observation of a single ICME by two spacecraft located at different helioradii allows us to study the radial evolution of the EP signatures associated with the passage of ICMEs. Radial alignments of spacecraft are rare and few cases have been reported in which the same ICME has been observed by two well separated spacecraft. The identification of ICMEs at large heliocentric distances and their association with ICMEs observed earlier, closer to the Sun, is difficult because ICMEs tend to be observed during periods of intense levels of solar activity when multiple transient events occur closely spaced in time and they may interact and merge as they move out through the heliosphere [*Burlaga et al.*, 2002]. In this paper we analyze cases in which the same ICME has been apparently observed by well-separated spacecraft in radial distance and latitude. We compare the effect that the passage of the same ICME at two different heliospheric locations produces on EP intensities. Other significant cases of individual ICMEs observed by spacecraft separated by more than 1.5 AU but not covered in this paper include the events described by *Skoug et al.* [2000], *Paularena et al.* [2001] and *Lario et al.* [2001]. Multi-spacecraft observations of ICMEs at distances R < 1 AU have been addressed elsewhere [e.g., *Burlaga et al.*, 1987; *Cane et al.*, 1994, 1997].

2. DATA SOURCES

Energetic particle, solar wind and magnetic field data used in this paper were provided by the National Space Science Data Center (NSSDC) through the web site nssdc.gsfc.nasa.gov. Energetic particle data from the Helios-1 and Helios-2 spacecraft (henceforth He-1 and He-2, respectively) were measured by the University of Kiel experiments [*Kunow et al.*, 1977], whereas magnetic field and solar wind data were measured by the magnetometer [*Neubauer et al.*, 1977] and plasma experiments [*Rosenbauer et al.*, 1977]. Observations at R = 1 AU were made by the Interplanetary Monitoring Platform (IMP-8), the Geosynchronous Observation Environmental Satellite (GOES-8), and the Advanced Composition Explorer (ACE). We use <13 MeV proton data from the Charged Particle Measurement Experiment (CPME) onboard IMP-8 [*Sarris et al.*, 1976] and <4.8 MeV ion data from the Electron Proton Alpha Particle Monitor (EPAM) onboard ACE [*Gold et al.*, 1998]. We also use >60 MeV particle fluxes inferred from the anti-coincidence guard of the Goddard Medium Energy (GME) experiment [*McGuire et al.*, 1978] onboard IMP-8, and >8 MeV proton intensities from the Energetic Particle Sensor (EPS) onboard GOES-8 [*Sauer*, 1993]. Magnetic field data at 1 AU are obtained from the MAG experiment onboard ACE [*Smith et al.*, 1998].

Observations at high heliographic latitudes were made by the Ulysses spacecraft. We use <4.8 MeV ion data from the HI-SCALE instrument [*Lanzerotti et al.*, 1992] and >8 MeV proton intensities from the COSPIN instruments [*Simpson et al.*, 1992]. Magnetic field data were collected by the VHM-FGM experiment [*Balogh et al.*, 1992]. Finally, observations from the outer heliosphere were made by the Voyager-2 spacecraft (henceforth V2). Energetic particle data come from the Low Energy Charged Particle (LECP) experiment [*Krimigis et al.*, 1977], solar wind data from the Plasma Science (PLS) experiment [*Bridge et al.*, 1977], and magnetic field data from the magnetometer onboard V2 [*Behannon et al.*, 1977].

3. IN-ECLIPTIC OBSERVATIONS

The observation of a single ICME by different spacecraft located in the ecliptic plane depends upon the longitudinal separation between the spacecraft and the longitudinal extent of the ICME. From single-spacecraft studies of large, flare-associated events, *Richardson and Cane* [1993] estimated that ICMEs in the ecliptic plane at R = 1 AU extend up to ~100° in longitude. However, multi-spacecraft studies of ICMEs at distances R ≤ 1 AU and associated with less energetic events indicate that these ICMEs typically do not extend much more than ~50° in longitude, suggesting that ICMEs with energetic solar events may extend over larger angular distances [*Cane et al.*, 1997]. Therefore, two spacecraft must be relatively close in heliolongitude to see the same ICME. The following examples show two cases in which the same ICME was observed by two spacecraft in the ecliptic plane and separated by more than 1 AU.

3.1. March 1978

Figure 1 shows energetic particle, solar wind and magnetic field data from He-1 (a), He-2 (b), V2 (c) and IMP-8 (d) from day 58 to day 74 of 1978. Figure 1e shows the location of these spacecraft on day 65. Heliospheric coordinates of each spacecraft are given in the caption of the figure, where R is the helioradius and Ψ the heliographic inertial longitude. The vertical solid lines in panels a-d indicate the passage of shocks (as identified by *Volkmer and Neubauer* [1985]; *Borrini et al.* [1982]; and *Burlaga et al.* [1984]). The gray vertical bars identify the passage of ICMEs. A first ICME (henceforth ICME-1) was observed by He-1 on day 61 (details of the ICME identification can be found in Figure 2 of *Cane et al.* [1997]). Figure 1a shows depressions of the EP intensities in both the >51 MeV/n ion and 4-13 MeV proton channels at the entry of He-1 into ICME-1. *Cane et al.* [1997] computed an 8% decrease in the >60 MeV/n particle fluxes during the passage of ICME-1 over He-1. *Richardson* [1994] measured BIFs in the post-shock region but for a period of less than 4 hours. Another period of 8.5 hours with BIFs was observed by He-1 from 62/1800 UT to 63/0230 UT but already outside ICME-1 [*Richardson*, 1994, Table 1].

Owing to the small longitudinal separation between V2 and He-1, it is proper to ask whether ICME-1 was also observed by V2. *Burlaga and Behannon* [1982] identified a magnetic cloud that moved past V2 from 0200 UT on day 66 (i.e., 66/0200 UT) to 0200 UT on day 68 (i.e., 68/0200 UT), whereas *Wang and Richardson* [2004] determined that the start and stop times of this ICME at V-2 were 66/0100 UT and 68/2100 UT, respectively. The gray vertical bar in Figure 1c is based on this latter identification. The distance between the location of V2 on day 67 (R = 2.51 AU, Ψ = 19.0°) and the location of He-1 on day 61 (R = 0.87 AU, Ψ = 30.4°) was ΔR = 1.64 AU in the radial direction and $\Delta\Psi$ = 11.4° in the longitudinal direction. Assuming that both spacecraft intercepted the same region of the ICME, we deduce an averaged radial transit speed of 520 km s^{-1} which seems reasonable considering the measured solar wind speed, the speed of the ICME-driven shock at He-1 (611 km s^{-1} according to *Cane et al.* [1997]), and the possible expansion of the ICME as it moves from 0.87 AU to 2.51 AU. Note also the similarity between the directions of the magnetic field rotation (from south to north) observed by both spacecraft.

He-2, separated 33° in longitude with respect to He-1, observed a weak shock at 60/0414 UT followed by a period of 11.5 hours of BIFs from 61/0600 UT to 61/1730 UT with plasma signatures typical of ICME [*Richardson*, 1994, Table 2]. We have indicated this period by the first vertical

gray bar in Figure 1b. Assuming that both Helios spacecraft observed the same ICME, He-2 only grazed a flank of ICME-1. IMP-8 separated 55° in longitude with respect to He-1 did not observe this ICME.

The EP signatures associated with the passage of ICME-1 over He-1 and He-2 ranged from intensity depressions at He-1 to a gradual decay of 4-13 MeV proton intensities without any significant change at the entry (exit) of He-2 into (from) the ICME. In contrast to He-2, He-1 presumably intercepted ICME-1 near its central part and penetrated well inside the ICME (note the complete rotation of magnetic field in Figure 1a), suggesting that EP depressions are more pronounced at the center of the ICME than at its edges. *Cane et al.* [1994] also showed that the closer the spacecraft intercepts the center of the ICME the greater the >60 MeV/n proton intensity depression. Another possibility for the lack of particle depression at He-2 is that the SEP event observed by He-2 with onset on day 60 was able to fill ICME-1 with SEPs and hence the continuous decay of proton intensities. However, if He-1 and He-2 intercepted the same ICME and the magnetic topology of this structure was such that particles propagating within the ICME were able to fill the whole ICME, such a filling of SEPs should also be seen by He-1. Only an intensity enhancement was observed by He-1 in the middle of day 61 and inside ICME-1.

The entry of V2 into ICME-1 (Figure 1c) showed a <17 MeV proton intensity decrease with respect to those measured on day 65 just after the passage of an interplanetary shock. The 3-17 MeV proton intensities (and also the 2.1-3.5 MeV proton channel not shown here) showed a gradual increase on day 66 already inside the ICME. The occurrence of an intense solar flare at 65/1159 UT at N26E20 as seen from the Earth [*Cane et al.*, 1994] resulted in a large SEP event as observed from He-1, He-2 and IMP-8. It is possible that this intensity increase seen by V2 within ICME-1 resulted from the injection of SEPs from this solar event. The peak intensity observed by V-2 on day 68 within ICME-1 resembles that observed by He-1 on day 61 suggesting that it may be the same intra-ICME particle population first observed by He-1 but convected to 2.5 AU within the ICME. Particle intensities increased at the exit of V2 from ICME-1 indicating that the arrival at V2 of SEPs injected from the Sun on day 65 was modulated by the ICME. The >210 MeV proton intensities observed by V2 during the time interval shown in Figure 1 seem to be anti-correlated with the magnetic field magnitude as suggested by *Burlaga et al.* [1985]. These intensities showed a ~5% decrease at the entry of the spacecraft into the ICME with a small peak at the end of day 67 suggesting the existence of some internal structure within the ICME.

The site of the flare at 65/1159 UT was E20 as seen from IMP-8, W02 as seen from He-2, W36 as seen from He-1

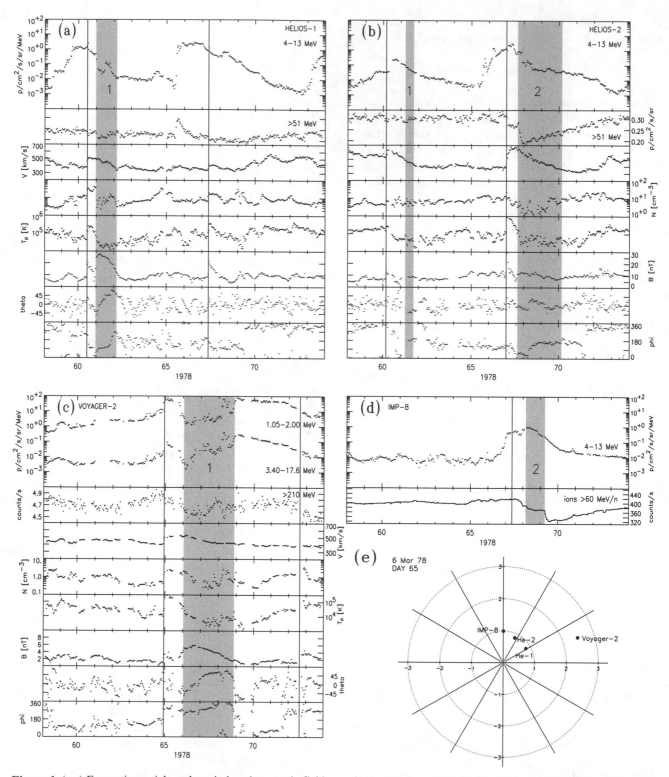

Figure 1. (a-c) Energetic particle, solar wind, and magnetic field magnitude data from He-1 (a), He-2 (b), and V2 (c). Magnetic field directions are in the RTN spacecraft centered coordinate system for V2 and in the spacecraft-centered solar ecliptic coordinate system for the Helios spacecraft. (d) Energetic particle intensities measured by IMP-8. (e) Location of the spacecraft on day 65 of 1978. He-1 was at R = 0.85 AU and Ψ = 34.1°; He-2 at R = 0.86 AU and Ψ = 67.2°; IMP-8 at R = 0.99 AU and Ψ = 89.6°; and V2 at R = 2.49 AU and Ψ = 18.5° Gray vertical bars and solid vertical lines in panels a-d indicate the passage of ICMEs and shocks, respectively.

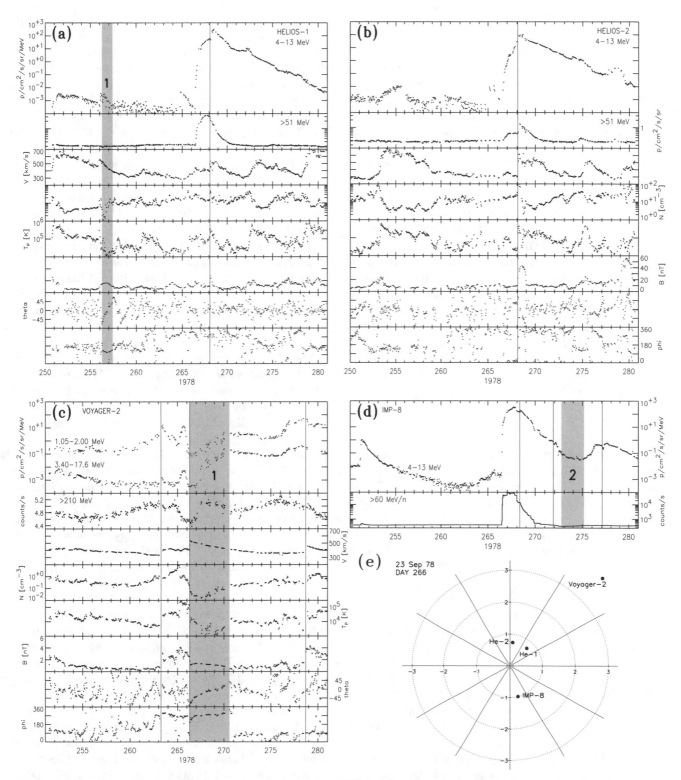

Figure 2. The same as Figure 1 but for the events in September 1978. Panel e shows the spacecraft locations on day 266 of 1978. He-1 was at R = 0.76 AU and Ψ = 45.5°; He-2 at R = 0.74 AU and Ψ = 83.0°; IMP-8 at R = 1.00 AU and Ψ = 284.3°; and V2 at R = 3.93 AU and Ψ = 44.3°.

and W51 as seen from V2. IMP-8 and He-2 clearly observed the ICME that originated at the Sun in temporal association with this flare (indicated with the number 2 in Figures 1b and 1d). However, He-1 only observed the shock driven by this ICME but not the ICME [*Cane et al.*, 1994]. The passage of the ICME over He-2 and IMP-8 produced large cosmic ray depressions. *Cane et al.* [1994] estimated that the size of the >60 MeV/n particle intensity decrease was 33% at He-2 and 21% at IMP-8. An additional decrease was observed in the >60 MeV/n ion intensities at IMP-8 on day 69 (but not in the 4-13 MeV proton intensities). The solar wind structure related to this additional decrease cannot be identified because of limited data available [*Cane et al.*, 1994]. Whereas 4-13 MeV proton intensities decreased when He-2 entered into the ICME, IMP-8 did not observe any significant change in the decaying proton intensities. BIFs were clearly observed throughout the passage of the ICME over He-2 [*Richardson*, 1994] but only intermittently at IMP-8 [*Richardson and Reames*, 1993]. *Cane et al.* [1994] suggested that IMP-8 just grazed an edge of this ICME whereas He-2 penetrated deeply into the ICME, indicating that the effects of this ICME on both low-energy and CR intensities were more noticeable near the center of the ICME than at its edges (cf. Figure 11 of *Cane et al.* [1994]).

3.2. September 1978

Figure 2 shows energetic particle, solar wind and magnetic field data from He-1 (a), He-2 (b), V2 (c) and IMP-8 (d) from day 250 to 281 of 1978. Figure 2e shows the location of these spacecraft on day 266. The origin of the major SEP event with onset on day 266 was associated with a solar flare at 266/1000 UT at W50 as seen from IMP-8 [*Reames et al.*, 1996]. The shock from this event was seen at 268/0229 UT by He-1, at 268/0133 UT by He-2, at 268/0705 UT by IMP-8, and presumably [*Reames et al.*, 1996] also at 278/1600 UT by V2 although this latter association is uncertain due to both the larger helioradii of V2 and the intense level of activity at that time [*Richardson et al.*, 1990]. None of these spacecraft observed ICME signatures immediately following the shock passage.

During the time interval shown in Figure 2, an ICME (indicated by the number 1 in Figure 2a, henceforth ICME-1) was observed by He-1 from 256/0600 UT to 257/1000 UT. *Cane et al.* [1997] computed a decrease of 10% in the >60 MeV/n particle fluxes during the passage of ICME-1 over He-1. *Richardson* [1994], however, did not identify any period of BIFs. The passage of ICME-1 was preceded (days 251-255) by a high-speed solar wind stream (observed again by He-1 on days 278-283). The 4-13 MeV proton intensities showed a weak enhancement in association with the passage of this high-speed stream. A new 4-13 MeV proton intensity

enhancement was observed prior to the passage of ICME-1 although no clear shock was observed preceding this ICME at He-1 [*Volkmer and Neubauer*, 1985]. The passage of this ICME was characterized by a gradual decay of the 4-13 MeV proton intensities reaching values close to the instrumental background level (Figure 2a).

V2 observed a magnetic cloud from 266/0700 UT to 271/1400 UT [*Burlaga and Behannon*, 1982]. This ICME was preceded by a structure bounded by a forward and reverse shock on days 263 and 266 respectively, that was identified by *Burlaga and Behannon* [1982] as a corotating interaction region (CIR). The distance between the location of V2 on day 268 (R = 3.95 AU, Ψ = 44.5°) and the location of He-1 on day 257 (R = 0.84 AU, Ψ = 35.9°) was ΔR = 3.11 AU in the radial direction and only $\Delta\Psi$ = 8.6° in the longitudinal direction. Assuming that both spacecraft observed the same ICME, we deduce an averaged radial transit speed of ~485 km s^{-1} to propagate from He-1 to V2, which is reasonable considering the measured solar wind speed at both spacecraft. Note also the similarity of the magnetic field rotations (from south to north) observed by both spacecraft.

The EP signatures associated with the entry of V2 into ICME-1 were an intensity decrease with respect to those measured in the preceding CIR, but followed by a gradual increase on day 267. This intensity increase was also observed in all V2/LECP proton energy channels from 60 keV to 30 MeV with indications suggestive of velocity dispersion. This increase was most likely due to SEPs injected during the major event on day 266 and able to propagate within the ICME-1. The >210 MeV proton intensities at V2 decreased already on day 265 in association with the passage of an enhanced magnetic field structure formed in front of ICME-1 and quickly recovered on day 267 as the magnetic field intensity decreased already inside the ICME. He-2 (37° westward of He-1 on day 257) and IMP-8 (120° eastward of He-1 on day 257) did not observe any plasma signatures that suggest the passage of ICME-1.

An additional ICME was observed by IMP-8 from 272/1700 UT to 275/0400 UT identified with the number 2 in Figure 2d [*Wang and Richardson*, 2004]. The passage of ICME-2 over the Earth produced a large Forbush decrease [*Belov et al.*, 2001]. The 4-13 MeV proton intensities at IMP-8 also decreased (Figure 2d), and intermittent periods of BIFs were observed within ICME-2 [*Richardson and Reames*, 1993]. *Wang and Richardson* [2004] associated ICME-2 with another ICME observed by V2 from 283/0100 UT to 286/1200 UT (not shown here). This association implies that this ICME was at least 115° wide when it arrived at V2 (IMP-8 was at Ψ = 291.2° on day 273, and V2 at Ψ = 45.8° on day 284). However, He-1 (separated 123° in longitude with respect to IMP-8 on day 273) did not observe any ICME signature. There is no evidence from data at R ≤ 1 AU that

ICME-2 extended at least 115° when it arrived at R = 3.9 AU. The time interval between days 265 and 280 was also a period of intense level of solar activity [*Richardson et al.*, 1990]. Therefore, it is possible that multiple CMEs were ejected in different directions and consequently, V2 and IMP-8 observed different ICMEs.

4. OUT-OF-ECLIPTIC OBSERVATIONS

Energetic particle signatures associated with the passage of ICMEs over the Ulysses spacecraft at high heliographic latitudes have been analyzed in a number of different works [see references included in the work of *Lario et al.*, 2004]. Clear differences have been observed between ICMEs propagating within high-speed solar wind streams and ICMEs propagating within slow solar wind streams. Particle signatures associated with ICMEs in slow solar wind streams range from intensity depressions [*Malandraki et al.*, 2003] to EP enhancements observed within the ICME and due to injection of SEPs by unrelated solar events [*Armstrong et al.*, 1994; *Malandraki et al.*, 2001]. By contrast, EP signatures at high latitudes and when Ulysses was immersed in high-speed solar wind flows generally showed low-energy particle intensity enhancements and CR depressions in association with the passage of the ICME even when no fresh injection of SEPs occurred during the transit time of the ICME to high latitudes [*Bothmer et al.*, 1995; *Lario et al.*, 2004]. Here we present three examples of ICMEs observed at both high and low latitudes.

4.1. February 1994

The first case of an ICME observed in the ecliptic at 1 AU and by Ulysses at high latitudes occurred in February 1994

[*Gosling et al.*, 1995]. Figure 3 shows EP data measured by IMP-8 and Ulysses from day 50 to 64 of 1994. Ulysses was at R = 3.5 AU, 12° west of Earth, at a heliographic latitude of Λ = 54°S and immersed in high-speed (>750 km s⁻¹) solar wind. Gray vertical bars and solid vertical lines indicate the passage of ICMEs and shocks, respectively, as identified by *Gosling et al.* [1995]. Differences in the solar wind disturbances generated by this ICME at low and high latitudes were attributed to the different speeds prevailing in the ambient solar wind ahead of the ICME. The part of the ICME propagating in the ecliptic plane was able to drive a strong shock that locally accelerated protons to very high (>100 MeV) energies [*Humble et al.*, 1995]. The part of the ICME at high latitudes had typical signatures of an over-expanding ICME preceded and followed by weak forward and reverse shocks [*Gosling et al.*, 1995] that were not able to locally accelerate ions above 4 MeV.

IMP-8 observations show that EP intensities decreased with respect to those observed at the time of the shock passage. By contrast, Ulysses observations showed low-energy (<4 MeV) ion intensities enhanced throughout the passage of the ICME [*Bothmer et al.*, 1995], whereas 250-2000 MeV proton intensities showed a ~7% decrease [*Bothmer et al.*, 1997]. As discussed by *Bothmer et al.* [1995] and *Lario et al.* [2004] the absence of an intense shock-accelerated particle population preceding the ICME is essential to see EP intensity enhancements during the passage of ICMEs. Additional low-energy ion intensity enhancements were observed within the ICME by IMP-8 on day 53 and by Ulysses at the end of day 58, that presumably were due to new injections of SEPs by unrelated solar events [*Pick et al.*, 1995].

A similar example of an ion intensity enhancement observed during the passage of an over-expanding ICME by Ulysses at R = 1.9 AU, Λ = 79°N (also immersed in a high-speed solar

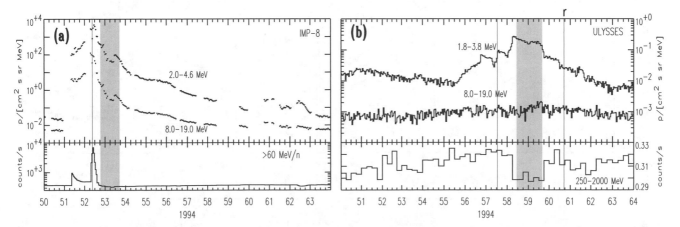

Figure 3. Energetic particle observations from IMP-8 in the ecliptic (a) and from Ulysses at high heliographic latitudes (b). Gray vertical bars indicate the passage of ICMEs as identified by *Gosling et al.* [1995]. Vertical solid lines indicate the shock passages (r indicates reverse shock).

wind stream and preceded and followed by weak forward and reverse shocks) occurred on days 271-272 of 2001 [*Lario et al.*, 2004]. In this case the intensity increase was observed at both <300 keV electron and <60 MeV proton intensities and no new injection of SEPs occurred during its passage. However, the association with an ICME observed at low latitudes was not obvious [*Lario et al.*, 2004].

4.2. November 2001

An extreme case of ICME observation by two spacecraft widely separated in heliolatitude occurred in November 2001 during the solar maximum northern polar passage of Ulysses [*Reisenfeld et al.*, 2003a, b]. Figure 4 shows EP data from day 308 to day 343 of 2001 collected by Ulysses at high heliographic latitudes ($\Lambda > 70°$) and R ranging from 2.20 to 2.42 AU (a), and by ACE and GOES-8 in the ecliptic plane at R = 1 AU (b). We have added in the bottom panel of Figure 4b neutron monitor data from the Climax station (cutoff rigidity 3 GV). During the time interval shown in Figure 4, Ulysses remained immersed in high-speed solar wind flow (>700 km s^{-1}) and observed three ICMEs numbered from 1 to 3 in Figure 4a. Ion intensities below 8 MeV increased close to the entry of Ulysses into these ICMEs [*Lario et al.*, 2004]. By contrast, 250-2000 MeV protons, mostly of galactic origin, showed clear depressions. *Lario et al.* [2004] interpreted the low-energy particle enhancements observed at the entry of Ulysses into these high-heliolatitude ICMEs as due to (1) the lack of an intense shock-accelerated population propagating outside the ICMEs, (2) the effects that local magnetic field structures have on particle transport within and around

the ICMEs, and (3) the confinement of low-energy ions by both the ICMEs and associated magnetic field disturbances.

Reisenfeld et al. [2003a, b] argued that the first and third ICMEs in Figure 4a were also observed in the ecliptic plane at 1 AU on days 310-313 and 328-330, respectively (numbered 1 and 3 in Figure 4b). These two ICMEs were able to drive strong shocks in the ecliptic plane, with >100 MeV proton intensities peaking at the arrival of the shocks at 1 AU (second panel of Figure 4b). However, the shocks driven by the same ICMEs at high latitudes were only able to locally accelerate ions below ~5 MeV [*Lario et al.*, 2004]. The entry of ACE into these two ICMEs was accompanied by a decrease of the particle intensities with respect to those measured at the time of the shocks. Whereas at 1 AU and in the ecliptic plane, the time-intensity profiles of the SEP events peaked around the arrival of the shocks, the highest intensities at high latitudes were observed during the passage of the ICMEs. Low-energy (<4.8 MeV) ion intensities at ACE (thick trace in the top panel of Figure 4b) showed also small increases within ICMEs −1 and −3 that do not have an obvious association with new injections of SEPs from the Sun and that resemble those observed by Ulysses. These elevated intra-ICME particle intensities resulted from either EPs already contained within the ICMEs since their liftoff time at the Sun or EPs that diffused into the ICMEs as they propagated out from the Sun. The observation of EP enhancements at Ulysses in association with the passages of these two ICMEs may be explained by assuming that intra-ICME particles remained efficiently confined within the ICMEs and associated field structures, and particle intensities outside the ICMEs decreased faster than the intra-ICME population as the ICMEs propagated to Ulysses.

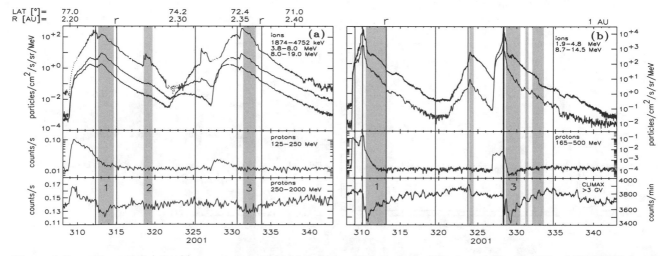

Figure 4. Energetic particle intensities measured at (a) high heliographic latitudes by Ulysses and (b) at 1 AU by ACE, GOES-8 and the Climax neutron monitor station. Gray vertical bars indicate the passage of ICMEs as identified by *Reisenfeld et al.* [2003a, b] and by *Cane and Richardson* [2003]. Vertical solid lines indicate the shock passages (r indicates reverse shocks).

5. OUTER HELIOSPHERE

As ICMEs propagate toward the outer heliosphere, they can interact with CIRs and other ICMEs to produce merged interaction regions (MIRs) [*Burlaga et al.*, 1986]. The association between MIRs and ICMEs previously observed in the inner heliosphere is not unequivocal because of both the multiple transient events that usually occur in short time intervals and the distortion, interaction and merging undergone by the ICMEs as they propagate to large distances [*Burlaga et al.*, 2002].

The passages of MIRs over spacecraft in the outer heliosphere are usually associated with variations in the EP intensities. The EP response to the passage of MIRs displays significant structure that is markedly different from event to event. The factors that determine the EP response depend on the existence of a shock preceding the MIR able to locally accelerate particles, the magnetic field structures formed around and within the MIR able to confine and modulate the transport of EPs, and the processes of deceleration undergone by the EPs propagating to the outer heliosphere [*Decker and Krimigis*, 1993, 2003].

Figure 5 shows the consequences of the extreme October-November 2003 solar events in the outer heliosphere as observed by V2 at R = 73.2 AU and Λ = 25°S. A large MIR associated with a fast solar wind stream moved past V2 during more than ~40 days [*Burlaga et al.*, 2005]. CR intensities decreased ~10 days (dashed line **a** in Figure 5) after the passage of a shock (solid vertical line in Figure 5) and reached a minimum ~10 days later (dashed line **b** in Figure 5). Magnetic field magnitude (shown in the work of *Burlaga et al.* [2005]) was enhanced after the shock passage, gradually increasing from day 128 (line **a**) to day 138 (line **b**). *Burlaga et al.* [2005] identified the region formed between the shock and the line **a** as a "sheath-like" region that was followed by a region of fast solar wind, high magnetic field and CR depression that resembles the sequence found in the two-step Forbush decreases observed at 1 AU [*Cane*, 2000]. Low-energy ions, however, showed intensity increases during the passage of the MIR. Elevated <80 keV ion intensities were observed between lines **a** and **b** in Figure 5, suggesting that these particles were confined in the high magnetic field region. The 0.5-1.5 MeV proton intensities increased for ~20 days before the shock and peaked twice at the time of the sheath-like region and within the high magnetic field region.

Similar examples of MIR effects on EP intensities were shown by *Decker and Krimigis* [1993, 2003]. Although there are no two events alike, the general characteristics are that low-energy (<500 keV) ion intensity enhancements are observed several days behind the shock and presumably confined by magnetic field structures. By contrast, higher

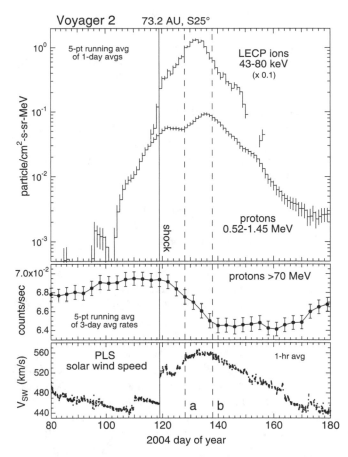

Figure 5. Energetic particle and solar wind data measured by V2 during the passage of the MIR associated with the October/November 2003 events. Solid vertical line indicates the passage of a shock and dashed vertical lines the passage of magnetic field discontinuities as identified by *Burlaga et al.* [2005].

energies (~1-20 MeV) exhibit gradual increases (if any) before the arrival of the shock with localized peaks close to the shock passage. Such time-intensity profiles were observed by V2 during the passage of the MIRs in May 1991, September 1991, January 2001 and October 2001 [*Decker and Krimigis*, 1993, 2003]. A possible interpretation of the ion intensity behavior at different energies is that low-energy (<500 keV) ions are part of a particle population accelerated earlier in the SEP events that generated the MIR. These low-energy particles propagate toward the outer heliosphere along the highly twisted magnetic field undergoing adiabatic deceleration, and thus are able to be caught by the traveling MIR and remain confined by the magnetic structures formed in the MIR. Higher-energy particles observed before and during the shock passage may consist of an ambient proton population magnetically reflected by and swept ahead of the MIR.

6. CONCLUSIONS AND SUMMARY

The examples shown in sections 3 and 4 of individual ICMEs observed by spacecraft well separated in radial distance and latitude allow us to conclude that:

-Differences in the characteristics of an ICME propagating in different solar wind regimes (e.g., at high and low latitudes for the ICMEs analyzed in section 4) lead to distinct EP signatures. Whereas at low latitudes ICMEs drive strong shocks able to accelerate particles, at high latitudes and in fast solar wind flows shocks are weak and particle intensities are higher inside than outside the ICMEs.

-CR depressions at large helioradii are mainly associated with increases in magnetic field magnitude and not with the ICME itself (e.g., Figures 1c and 2c; see also *Burlaga et al.* [1985]).

-Evolution of ICMEs as they propagate further out in the heliosphere lead to changes in the EP signatures. If the internal magnetic field magnitude diminishes, decreases in CR intensities are less pronounced and EP intensities show less structure during the passage of the ICME.

-Observations of the same ICME at different heliolongitudes (e.g., ICMEs −1 and −2 in Figure 1) show that decreases in the EP intensities are more pronounced near the center of the ICME than at its edges [see also *Cane et al.*, 1994].

-The fortuitous occurrence of SEP events during the transit time of ICMEs to large helioradii may fill the ICME with SEPs (e.g., ICME-1 in Figure 2 and the ICME in Figure 3).

-Our ability to observe EP intensity enhancements depends upon whether there is an intense shock-accelerated population able to mask the intra-ICME population and upon whether EPs remain confined within the ICMEs.

In the outer heliosphere, the highest low-energy ion intensities are observed in association with the passage of MIRs where EPs remain confined and convected by magnetic field structures.

Although all these conclusions are drawn from a few selected events, statistical studies of the EP response to the passage of ICMEs at different helioradii and heliolatitudes are required before establishing general dependences of the evolution of the ICME energetic particle signatures.

Acknowledgments. I gratefully acknowledge the assistance of *R.B. Decker* in the preparation of this article, NSDDC for providing the data used in it and *I.G. Richardson* for his comments on the manuscript. The work described in this paper was partially supported by NASA research grant NAG5-6113.

REFERENCES

Armstrong, T.P., *et al.*, Observation by Ulysses of hot (~270 keV) coronal particles at 32° south heliolatitude and 4.6 AU, *Geophys. Res. Lett.*, 21, 1747-1750, 1994.

Balogh, A., *et al.*, The magnetic field investigation of the Ulysses mission: Instrumentation and preliminary results, *Astron. Astrophys. Suppl. Ser.*, 92, 221-236, 1992.

Behannon, K.W., *et al.*, Magnetic field experiment for Voyagers 1 and 2, *Space Sci. Rev.*, 21, 235-257, 1977.

Belov, A.V., *et al.*, Pitch-angle features in cosmic rays in advance of severe magnetic storms: Neutron monitor observations, *Conf. Pap. Int. Cosmic Ray Conf. 27th*, 3507-3510, 2001.

Borrini, G., J.T. Gosling, S.J. Bame, and W.C. Feldman, An analysis of shock wave disturbances observed at 1 AU from 1971 through 1978, *J. of Geophys. Res.*, 87, 4365-4373, 1982.

Bothmer, V., *et al.*, The Ulysses south polar pass: Transient fluxes of energetic ions, *Geophys. Res. Lett.*, 22, 3369-3372, 1995.

Bothmer, V., *et al.*, The effects of coronal mass ejections on galactic cosmic rays in the high latitude heliosphere: Observations from Ulysses' first orbit, *Conf. Pap. Int. Cosmic Ray Conf. 25th*, 1, 333-336, 1997.

Bridge, H.S., *et al.*, The plasma experiment on the 1977 Voyager mission, *Space Sci. Rev.*, 21, 259-287, 1977.

Burlaga, L.F., and K.W. Behannon, Magnetic clouds: Voyager observations between 2 and 4 AU, *Solar Phys.*, 81, 181-192, 1982.

Burlaga, L.F., L.W. Klein, R.P. Lepping, and K.W. Behannon, Large-scale interplanetary magnetic fields: Voyager 1 and 2 observations between 1 AU and 9.5 AU, *J. of Geophys. Res.*, 89, 10659-10668, 1984.

Burlaga, L.F., *et al.*, Cosmic ray modulation and turbulent regions near 11 AU, *J. of Geophys. Res.*, 90, 12027-12039, 1985.

Burlaga, L.F., *et al.*, Formation of a compound stream between 0.85 AU and 6.2 AU and its effects on solar energetic particles and cosmic rays, *J. of Geophys. Res.*, 91, 13331-13340, 1986.

Burlaga, L.F., *et al.*, Compound streams, magnetic clouds and major geomagnetic storms, *J. of Geophys. Res.*, 92, 5725, 1987.

Burlaga, L.F., *et al.*, Successive CMEs and complex ejecta, *J. of Geophys. Res.*, 107, 1266, doi:10.1029/2001JA000255, 2002.

Burlaga, L.F., *et al.*, Voyager 2 observations related to the October-November 2003 solar events, *Geophys. Res Lett.*, 32, L03S05, doi:10.1029/2004GL021480, 2005.

Cane, H.V., Coronal mass ejections and Forbush decreases, *Space Sci. Rev.*, 93, 55-77, 2000.

Cane, H.V., and I.G. Richardson, Interplanetary coronal mass ejections in the near-earth solar wind during 1996-2002, *J. of Geophys. Res.*, 108(A4), 1156, doi:10.1029/2002JA009817, 2003.

Cane, H.V., *et al.*, The role of interplanetary shocks in the longitude distribution of solar energetic particles, *J. of Geophys. Res.*, 93, 9555-9567, 1988.

Cane, H.V., *et al.*, Cosmic ray decreases and shock structure: A multispacecraft study, *J. of Geophys. Res.*, 99, 21429-21441, 1994.

Cane, H.V., I.G. Richardson, and T.T. von Rosenvinge, Cosmic ray decreases: 1964-1994, *J. of Geophys. Res.*, 101, 21561-21572, 1996.

Cane, H.V., I.G. Richardson, and G. Wibberenz, Helios 1 and 2 observations of particle decreases, ejecta, and magnetic clouds, *J. of Geophys. Res.*, 102, 7075-7086, 1997.

Decker, R.B., and S.M. Krimigis, Two unusual shock events observed at the Voyagers in 1991, comparison, *Conf. Pap. Int. Cosmic Ray Conf. 23rd*, 3, 310-313, 1993.

Decker, R.B., and S.M. Krimigis, Voyager observations of low-energy ions during solar cycle 23, *Adv. Space Res.*, 32, 597-602, 2003.

Gold, R.E., *et al.*, Electron proton and alpha monitor on the advanced composition explorer spacecraft, *Space Sci. Rev.*, 86, 541-562, 1998.

Gosling, J.T., Coronal mass ejections, *in 26th Int. Cosmic Ray Conf.*, edited by B.L. Dingus, D.B. Kieda and M.H. Salamon, pp. 59-79, *AIP Conf.* Proc. 516, 2000.

Gosling, J.T., *et al.*, A CME-driven solar wind disturbance observed at both low and high heliographic latitudes, *Geophys. Res. Lett.*, 22, 1753-1756, 1995.

Humble, J.E., *et al.*, Observation of GV protons associated with an interplanetary shock, *Conf. Pap. Int. Cosmic Ray Conf. 24th*, 4, 900-903, 1995.

Kahler, S.W., and D.V. Reames, Probing the magnetic topologies of magnetic clouds by means of solar energetic particles *J. of Geophys. Res.*, 96, 9419-9424, 1991.

Krimigis, S.M., *et al.*, The Low Energy Charged Particle /LECP/ experiment on the Voyager spacecraft, *Space Sci. Rev.*, 21, 329-354, 1977.

Kunow, H., *et al.*, Cosmic ray measurements on board Helios-1 from Dec 1974 to Sep 1975. Quiet time spectra, radial gradients and solar events, *J. of Geophysics*, 42, 615-631, 1977.

Lanzerotti, L.J., *et al.*, Heliosphere instrument for spectra, composition and anisotropy at low energies, *Astron. Astrophys. Suppl. Ser.*, 92, 365-400, 1992.

Lario, D., *et al.*, Joint Ulysses and ACE observations of a magnetic cloud and the associated solar energetic particle event, *Space Sci. Rev.*, 97, 277-280, 2001.

Lario, D., *et al.*, Low-energy particle response to CMEs during the Ulysses solar maximum northern polar passage, *J. of Geophys. Res.*, 109, A01107, doi:10.1029/2003JA010071, 2004.

Marsden, R.G., *et al.*, ISEE-3 observations of low-energy proton bidirectional events and their relation to isolated magnetic structures, *J. of Geophys. Res.*, 92, 11009-11019, 1987.

Malandraki, O.E., *et al.*, Tracing the magnetic topology of coronal mass ejection events by Ulysses/HI-SCALE energetic particle observations in and out of the ecliptic, *Space Sci. Rev.*, 97, 263-268, 2001.

Malandraki, O.E., E.T. Sarris, and G. Tsiropoula, Magnetic topology of coronal mass ejections out of the ecliptic: Ulysses/ HI-SCALE energetic particle observations, *Ann. Geophysicae*, 21, 1249-1256, 2003.

McGuire, R.E., T.T. von Rosenvinge, and F. B. McDonald, The composition of solar energetic particles, *Astrophys. J.*, 301, 938-961, 1996.

Neubauer, F.M., *et al.*, Initial results from the Helios-1 search-coil magnetometer, *J. of Geophysics*, 42, 599-614, 1977.

Paularena, K.I., C. Wang, R. von Steiger, and B. Heber., An ICME observed by Voyager-2 and by Ulysses 5 AU, *Geophys. Res. Lett.*, 28, 2755-2758, 2001.

Pick, M., *et al.*, The propagation of sub-MeV electrons to latitudes above 50°S, *Geophys. Res. Lett.*, 22, 3373-3376, 1995.

Reames, D.V., L.M. Barbier, and C.K. Ng, The spatial distribution of particles accelerated by coronal mass ejection-driven shocks, *Astrophys. J.*, 466, 473-486, 1996.

Reisenfeld, D.B., *et al.*, CMEs at high northern latitudes during solar maximum: Ulysses and SOHO correlated observations, in *Solar Wind Ten*, edited by M. Velli *et al.*, *AIP Conf. Proc.*, 679, 210-213, 2003a.

Reisenfeld, D.B., *et al.*, Properties of high-latitude CME-driven disturbances during Ulysses northern polar passage, *Geophys. Res. Lett.*, 30(19), 8031, doi:10.1029 2003GL017155, 2003b.

Richardson, I.G., A survey of bidirectional ~1 MeV ion flows during the Helios-1 and Helios-2 missions: Observations from the Goddard Space Flight Center instruments, *Astrophys. J.*, 420, 926-942, 1994.

Richardson, I.G., Using energetic particles to probe the magnetic topology of ejecta, *in Coronal Mass Ejections,* edited by N. Crooker, J.A. Joselyn and J. Feynman, pp. 189-196, Geophysical Monograph 99, American Geophysical Union, 1997.

Richardson, I.G., and H.V. Cane, Signatures of shock drivers in the solar wind and their dependence on the solar source region, *J. of Geophys. Res.*, 98, 15295-15304, 1993.

Richardson, I.G., and D.V. Reames, Bidirectional ~1 MeV amu-1 ion intervals in 1973-1991 observed by the Goddard Space Flight Center instruments on IMP-8 and ISEE-3/ICE, *Astrophys. J. Suppl. Ser.*, 85, 411-432, 1993.

Richardson, I.G., and H.V. Cane, Particle flows observed in ejecta during event onsets and their implication for the magnetic topology, *J. of Geophys. Res.*, 101, 27521-27532, 1996.

Richardson, I.G., H.V. Cane, and T.T. von Rosenvinge, MeV ion anisotropies at interplanetary shocks: Observations from the ISEE-3/ICE medium energy cosmic ray experiment, *Conf. Pap. Int. Cosmic Ray Conf. 21st*, 5, 333-336, 1990.

Richardson, I.G., *et al.*, Bidirectional particle flows at cosmic ray and lower (~1 MeV) energies and their association with interplanetary coronal mass ejections/ejecta, *J. of Geophys. Res.*, 105, 12597-12592, 2000.

Riley, P., *et al.*, Using an MHD simulation to interpret the global context of a coronal mass ejection observed by two spacecraft, *J. of Geophys. Res.*, 108, doi:10.1029/2002JA009760, 2003.

Rosenbauer, H., *et al.*, A survey on initial results of the HELIOS plasma experiment, *J. of Geophysics*, 42, 561-580, 1977.

Sauer, H.H., *et al.*, GOES observations of energetic protons E>685 MeV: Description and data comparison, *Conf. Pap. Int. Cosmic Ray Conf. 23rd*, 3, 250-253, 1993.

Sarris, E.T., S.M. Krimigis, and T.P. Armstrong, Observations of magnetospheric bursts of high-energy protons and electrons at approximately 35 earth radii with IMP-7, *J. of Geophys. Res.*, 81, 2341-2355, 1976.

Simpson, J.A., *et al.*, The Ulysses cosmic ray and solar particle investigation, *Astron. Astrophys. Suppl. Ser.*, 92, 401, 1992.

Skoug, R.M., *et al.*, Radial variation of solar wind electrons inside a magnetic cloud observed at 1 and 5 AU, *J. of Geophys. Res.*, 105, 27269-27275, 2000.

Smith, C.W., *et al.*, The ACE magnetic field experiment, *Space Sci. Rev.*, 86, 613-632, 1998.

Volkmer, P.M., and F.M. Neubauer, Statistical properties of fast shock waves in the solar wind between 0.3 and 1 AU: Helios-1, 2 observations, *Ann. Geophysicae*, 3, 1-12, 1985.

Wang, C., and J.D. Richardson, Interplanetary coronal mass ejections observed by Voyager 2 between 1 and 30 AU, *J. of Geophys. Res.*, 109, A06104, doi:10.1029/2004JA010379, 2004.

D. Lario, The Johns Hopkins University, Applied Physics Laboratory, 11100 Johns Hopkins Rd., Laurel, MD 20723, USA

KET Ulysses Observations of SEP in and out of the Ecliptic

Alexei Struminsky

Space Research Institute, Russian Academy of Sciences, Moscow, Russia

Bernd Heber

Fachbereich Physik, Universität Osnabrück, Osnabrück, Germany

Observations of eight solar energetic particle (SEP) events in the polar helio-spheric regions and six events near the Jupiter orbit and their implications for the current paradigm of particle acceleration and propagation in the heliosphere are discussed. These events are easily identified with well-known episodes of the solar activity in 1992, 1997, 2000-2001 and 2003 and the corresponding SEP events near the Earth. Absolute intensities, fluences and propagation times of >30 MeV solar protons obtained from data of the Kiel Electron Telescope aboard the Ulysses spacecraft are compared with results of the GOES proton detector near the Earth. Analyzing the diversity and similarity of SEP events observed in different points of the heliosphere we conclude that the Sun should be a source of SEPs prolonged in space and time, much more complex than assumed before. Contrary, the global process of particle propagation in the heliosphere – along and cross-field diffusion – appears amazingly constant during the first 2-3 days since the particle release before the arrival of solar wind disturbances to the spacecraft. During the decay phase, when spatial gradients are weak or non-existing in the heliosphere (the reservoir effect), the events are different from each other, possibly reflecting the total number of particles injected into the heliosphere and the rate of particle escape from the storage region.

1. INTRODUCTION

The current views on acceleration of solar cosmic rays and their propagation in interplanetary space were formed based on observations in the ecliptic plane at a distance of 1 AU or less (see *Kallenrode*, 2003 and references therein). A variety of the SEP intensity time-profiles observed at the Earth imply complex behavior of the injection function into interplanetary space from the acceleration site, i.e., the

source of SEP might be extended in space and time. In order to explain the observed longitudinal distribution of SEP events coronal propagation and/or acceleration at Coronal Mass Ejection (CME) driven shock waves consequently, a prolonged and spatially extended injection into the inter-planetary space was introduced. It is adopted that SEPs pre-dominantly propagate along interplanetary magnetic field lines (IMF), so their flux across field lines should be depressed by $D_\perp/D_\parallel \sim 10^{-4}$ to 10^{-2}, the ratio of the cross and along field diffusion coefficients (see *Zhang et al.*, 2003b and references therein).

From Jovian electron measurements it became evident, that the diffusion coefficients perpendicular to the mean interplanetary magnetic field in radial D_r and meridional D_z direction are different (*Conlon*, 1978, *Ferrando et al.*, 1993,

Solar Eruptions and Energetic Particles
Geophysical Monograph Series 165
10.1029/165GM30

Ferreira et al., 2001). The radial diffusion coefficient D_{rr} at fixed heliographic latitude θ is given by the expression:

$$D_{rr} = D_\parallel \cos^2 \Phi + D_\perp \sin^2 \Phi,$$

where Φ is the spiral angle given by $\tan \Phi = (\Omega r \cdot \sin\Theta)/V_{SW}$. Close to the Sun D_{rr} is dominated by $D_\parallel$ while beyond 5 AU $D_\perp$ contributes significantly, although $D_\perp$ is only a few percent of $D_\parallel$. Therefore simultaneous observations from several distant points in the heliosphere are necessary to investigate temporal and spatial profiles of the source.

Studies of SEP events at large heliocentric distance in the ecliptic plane started with first observations of SEP events aboard Pioneer 10-11 (see *Hamilton*, 1981 and references therein) and Voyager 1-2 (*Decker et al.*, 1999; *Lanzerotti et al.*, 2001) followed two and three solar cycles later by the Ulysses (*Lim et al.*, 1996 and *Lario et al.*, 2000) and Cassini (*Lario et al.*, 2004b) observations. Simultaneous measurements of SEP intensities in the polar regions of the heliosphere aboard Ulysses (*McKibben et al.*, 2003) and at different distances in the ecliptic plane (near Earth and aboard the Cassini spacecraft) provide the first three-dimensional patterns of SEP events in the heliosphere.

One of the scientific goals of the Ulysses mission to high heliolatitudes for the Cosmic ray and Solar Particle Investigation (COSPIN) instrument is to determine the role of coronal magnetic fields for the storage and propagation of SEP, and the importance of emission of energetic particles from regions other than solar flares (*Simpson et al.*, 1992). Since the source might not be extended similarly in latitude as in longitude and the theoretical ratio of the cross and along field diffusion coefficients is rather small, it has been questioned before the Ulysses mission whether SEP would be detectable at polar latitudes in the heliosphere.

Below we present observations of the largest SEP events in the 1990's and 2000's from distant spacecraft and Earth. In order to put these observations into context, previous studies at Earth and at different heliospheric distances are reviewed and the question raised as to what extent Ulysses observations of SEP provide a challenge to the current views on SEP acceleration and propagation?

2. REVIEW OF PREVIOUS RESULTS

2.1. Radial and Azimuthal Distribution of SEPs

The azimuthal propagation of solar protons in the energy range of 10-30 MeV was investigated by *McKibben* (1972) by using observations on board the Earth satellite IMP 4 and the deep-space probes Pioneer-6 and 7 widely separated at ~1 AU from the Earth. It was found that during some events the decay phase could be divided into two phases. The first phase is characterized by strong azimuthal gradients of the proton intensity and exponential time decay constant in the range ~10-20 hours. During the second phase, when azimuthal gradients are weak or not present (the reservoir effect), the intensity decrease is characterized by time constants of more than 40 hours. Two decay constants observed during the decay phase of SEP events might be taken as an indication of two fundamentally different processes of particle propagation in the heliosphere.

Hamilton (1981) summarized the Pioneer 10, 11 observations at large heliospheric distances in terms of diffusion-convection propagation along the magnetic field line. These observations have confirmed that the separation of the connection longitude of the observing spacecraft from the flare site is important to at least 10 AU for detection of SEP particles. The model predicted a decrease of the time to maximum intensity (by ~39% at 5 AU) and an increase of the decay rate due to both convection and energy loss. These results should be applicable to the propagation of energetic particles along any magnetic line in the heliosphere.

The original work, which describes different multi-spacecraft observations in terms of shock wave acceleration and magnetic connection between the source and observer, is *Reames et al.* (1996). In their analysis the authors considered observations of particle time profiles of several SEP events at radial distances ≤1 AU by IMP 8 and Helios 1, 2 and concluded that the general features of the shock paradigm are well supported by the multi-spacecraft observations. However, data from Voyager 2 at ~4 AU were only presented for one event.

Reames et al. (1996) underlined that the lack of longitude gradients late in an event (the reservoir effect) is in striking contrast with the initial longitude dependence. They consider the reservoir as a region downstream of the shock, where particles are quasi-trapped or trapped in CME loops. The longitude independence results, partly from magnetic field lines drawn out by the Coronal Mass Ejection (CME) that coupled widely separated regions. The coupling may be aided by cross-field diffusion of the particles, especially in the turbulent region between the shock and the CME. Besides the upstream injection of particles from a new fast CME-driven shock near the Sun may fill old loops of previous CME's creating the upstream effect (*Reames*, 1999).

2.2. Ulysses and Cassini at Large Heliocentric Distances

Lim et al. (1996) analyzed energetic particle, plasma and magnetometer data from Ulysses obtained in October-November 1992. They described the interplanetary conditions as a combination of Corotating Interaction Region (CIR) and several shock waves and discussed arrival of flare particles, particles within CIRs and particles transported by a coronal mass ejection, but didn't find definite arguments in favor of some particular mechanism.

Later observations of <19 MeV protons at 1 and 5 AU aboard Wind and Ulysses during rising phase of the 23rd solar cycle were discussed by *Lario et al.* (2000) in terms of the SEP acceleration by CME driven shock waves and magnetic connection between the accelerator and observer. However, this association has appeared to be not so obvious, the longitudinal dependence of the SEP events at Ulysses is absent and, in contrast to 1 AU, the classification of SEP event profile as West-, East-, or Central meridian type is not useful. The absence of the longitudinal dependence is in contrast to the previous results of Pioneer 10, 11 and their interpretation by the standard diffusion model (*Hamilton*, 1981) as well as the spatial distribution of particles accelerated by CME driven shocks predicted by the shock wave acceleration model (*Reames et al.*, 1996). *Lario* and co-authors suggested the existence of a special topology of the interplanetary magnetic field (IMF) created by traveling CME-driven shocks and magnetic structures, which may explain the lack of longitudinal dependence as well as broadening and smoothening of the flux profiles at 5 AU.

Simultaneous observations of high-energy (>25 MeV) ion intensities aboard Cassini at 4-7 AU and near the Earth in 2000-2001 once again showed that the most intense SEP events were generally observed at both locations (Earth and Cassini) regardless of their longitude separation and were exclusively associated with SEP injected close to the Sun (*Lario et al.*, 2004b). The authors found that existing magnetic field structures between the Sun and Cassini appear to be important for the SEP observations. Such structures may prevent an undisturbed propagation of energetic particles through the interplanetary medium. The inability of the particles to cross these magnetic field structures may cause the deficit of energetic particles in front of these structures and leads to a trapped particle population behind them.

2.3. Ulysses at Polar Heliospheric Latitudes

McKibben et al., (2003) reviewed the COSPIN observations during the solar maximum south polar pass (August-December 2000) and fast latitude scan. The authors found that for almost all large SEP events significant particle increases both near Earth and at Ulysses were observed. Therefore their detection is independent of the spacecraft positions relative to the location of the flare and CME. The SEP events at highest latitudes are characterized by 1) a delay of the event onsets, 2) a longer rise time towards maximum and 3) significantly smaller maximum intensities when compared to measurements near the Earth. The Ulysses observations at high latitudes confirmed the reservoir effect.

The proton intensities observed near Earth by IMP8 and Ulysses are nearly the same after 3-4 days and decay with the same rate for the rest of the event (*McKibben et al.*, 2003). Note, the Cassini spacecraft was moving in the ecliptic plane from 4 AU to 7 AU and measured the proton intensities (*Lario et al.*, 2004b). The Ulysses and Cassini data for 2000-2001 are not yet analyzed together.

The particle release time from the solar source derived from the high latitude measurements is between 100 and 350 min later than the release time derived from in-ecliptic measurements (*Dalla et al.*, 2003a). They showed that the parameter, which best orders this characteristic, is the difference in latitude between the associated flare and the spacecraft. If shock acceleration is of some importance it is expected that the delay and the inverse speed of the corresponding interplanetary shock should be correlated. This was, however, not observed. According to *Dalla et al.* (2003b) the model of SEP acceleration by coronal mass ejections driven shocks does not account for the Ulysses observations.

During the rising phase of a solar particle event an efficient cross-field diffusion close to the observer can be excluded, since the particle flow direction during the rising phase of the events is essentially field aligned (*Sanderson et al.*, 2003). They found no evidence for any net flow across the field lines for the events, which occurred in the high-latitude high-speed solar wind (the north polar pass in August-December 2001).

One day after the event onset *Zhang et al.* (2003a, 2003b) found evidence for cross-field particle transport by analyzing the anisotropy measurements of 40-90 MeV protons during the July 14, 2000 event. They interpreted these anisotropies as a sign for cross-field diffusion. When fitting the intensity time profiles *Zhang et al.*, (2003a) obtained a ratio of the cross-field and along-field diffusion coefficient of about 1%, a number somewhat larger but consistent with values derived from Jovian electron and galactic cosmic ray studies between 1 and 5 AU (*Ferreira et al.*, 2001; *Burger et al.*, 2000). In contrast, *Zhang et al.*, (2003b) determine a ratio of 25% when analyzing the anisotropy measurements only.

3. KET/ULYSSES OBSERVATIONS

3.1. Instrumentation and Data Selection

The COsmic and Solar Particle INvestigation Kiel Electron Telescope (COSPIN/KET) on-board Ulysses measures protons and α-particles in the energy range from ~4 to >2000 MeV/n and electrons in the range from ~2 to >300 MeV in different energy channels (*Simpson et al.*, 1992). The experimental results discussed below are restricted to the 38-125 MeV protons from COSPIN/KET and the 40-80 MeV protons from the Energetic Particle Sensor (EPS) on board

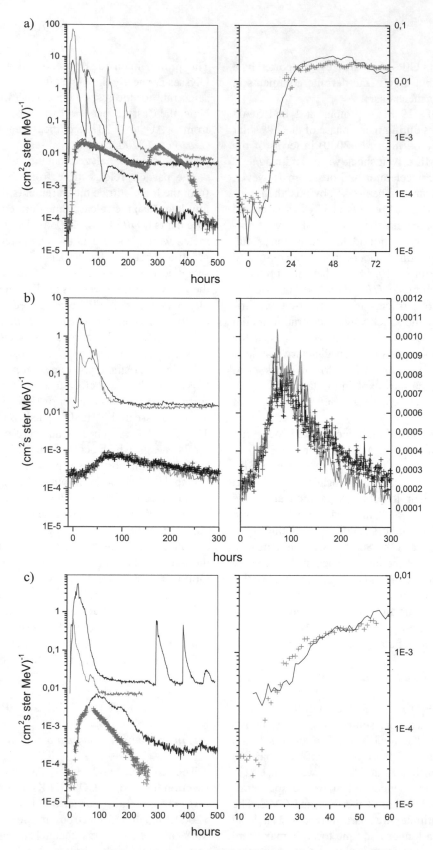

Figure 1. Hourly averaged 38-82 MeV and 38-125 MeV proton intensities measured by GOES EPS close to Earth and by Ulysses KET. The following six solar SEP events are shown: a) October 30, 1992 (black) and October 28, 2003 (gray), b) November 6, 1997 (black) and August 24, 1998 (gray), and c) June 25, 1992 (gray) and April 20, 1998 (black). The time is chosen that way that hour 0 corresponds to the X-ray onset of each event. The right panels display the Ulysses time profiles during the event onset.

the GOES satellite. We use the Solar X-Ray Imager (SXI) Subsystem aboard GOES to obtain information about solar flare activity (http://rsd.gsfc.nasa.gov/goes/text/goes. databook.html). *Ryan et al.* (2000) have suggested using Occam's razor, any model that describes the behavior of lower energy particles in space should be extended gracefully to higher energies to explain ground level enhancements (GLE) of solar cosmic rays. We choose the most powerful and energetic SEP events observed by Ulysses during the two last solar maxima, which are easily identified with well-known episodes of the solar activity and the corresponding

SEP events near the Earth, to investigate their properties in different points of the heliosphere:

- Six SEP events were observed by Ulysses near the Jupiter orbit at ~5 AU close to the ecliptic plane (Figure 1) and were associated with GLE events registered by neutron monitors at Earth (http://helios.izmiran.rssi.ru/cosray/main.htm).

- Eight SEP events registered at heliographic latitude >60° have an enhancement above background in the 125-250 MeV energy channel. Three and five of these events occurred while Ulysses was in the southern and northern hemisphere, respectively (Figure 2).

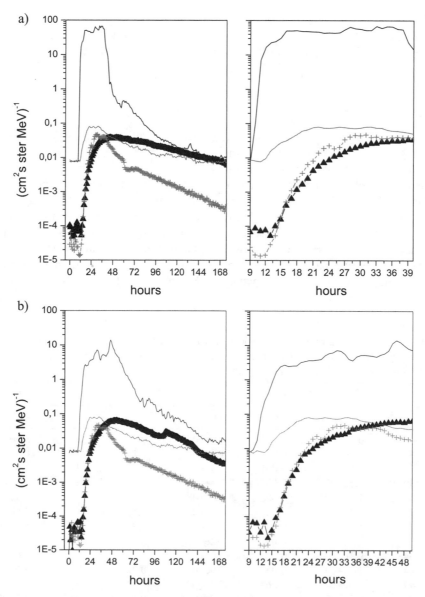

Figure 2. Hourly averaged 38-82 MeV proton intensities measured by GOES EPS (lines) close to Earth. Symbols correspond to 38-125 MeV Ulysses KET proton intensities at polar latitudes. The onset for the events of July 14, 2000 and September 24, 2001 in comparison with the September 12, 2000 event is shown in the right panels (adopted from *Struminsky et al.*, 2006). A time shift for the September 12, 2000 event has been applied to get a simultaneous proton onset at Ulysses.

Table 1. Possible parent solar flares and locations of Ulysses for the considered events

Date, day of the year	Day of GLE	X-ray event			Ulysses position		
		Onset	Imp	Coord	R_U	Lat	Long
25 Jun 92, 177	177	19:51	X3.9	N09W69	5.33	S12.8	E114
30 Oct 92, 304	307	16:59	X1.7	S26W63	5.17	S19.5	W120
06 Nov 97, 310	310	11:49	X9.4	S18W63	5.31	N2	W115
20 Apr 98, 110	122, 126	09:38	M1.4	S43W90	5.39	S6.3	E53
24 Aug 98, 237	237	21:50	X1.0	N30E07	5.34	S12.7	E166
28 Oct 03, 301	301, 302, 306	09:53	X17.2	S16E08	5.17	N6	W120
14 Jul 00, 196	196	10:03	X5.7	N22W07	3.2	S62	E115.7
12 Sep 00, 256		11:31	M1.0	S12W18	2.7	S71	E162.7
08 Nov 00, 313		22:42	M7.4	N10W77	2.4	S80	E182.9
15 Aug 01, 227		??	??	??	1.63	N63	W34.4
24 Sep 01, 267		09:32	X2.6	S18E27	1.8	N72	W31.1
04 Nov 01, 308	308	16:03	X1.0	N06W18	2.1	N78	W63.3
22 Nov 01, 326		20:18	M7.1	S25W67	2.31	N74	W60.1
26 Dec 01, 360	360	04:32	M3.8	N08W54	2.5	N67	W39

We consider a single event at Ulysses and a number of GOES events, which might contribute particles to the Ulysses event. For example, particular enhancements of >38-125 MeV protons associated with the GLE events of 2 November 1992, 2 and 6 May 1998, 29 October and 2 November 2003 are not distinguishable. The parent X-ray flares as identified for the GOES SEP events are considered as candidates for parent flares of the Ulysses events. Table 1 contains some information on the first parent X-ray flares and the Ulysses location, days of GLE events, which

occurred during the considered time interval. Many research groups reported various aspects of the events presented in Table 1.

We compare intensity-time profiles, T_U and T_G- delay times of protons since parent X-Ray flares, F_U and F_G fluences, J_U and J_G maximum intensities observed respectively from Ulysses and GOES during different events. Table 2 summarizes this information. Fluences above background were calculated for time periods presented in left panels of Figures 1 and 2.

Table 2. Comparison of SEP events observed by Ulysses (38-125 MeV) and GOES (38-82 MeV), where T_U and T_G – time delay of proton arrival relatively to the X-ray onset in hours, T_R – the exponential time decay constant, F_U and F_G – fluences [$cm^{-2}ster^{-1}MeV^{-1}$], J_U and J_G – maximum intensities [$s^{-1}cm^{-2}ster^{-1}MeV^{-1}$], $a+1=ln(J_G/J_U)/ln(R_U)$

Date	Ulysses						GOES			a+1
	T_U	T_R	F_U	J_U	$F_U R_U^2$	$J_U R_U^3$	T_G	F_G	J_G	
25 Jun 92	16	45	6.7e+02	0.003	1.9e+04	0.48	1	3.5e+04	0.7	3.25
30 Oct 92	11	56	7.4e+03	0.03	1.9e+05	4.1	1	6.7e+05	7.8	3.38
06 Nov 97	24	-	2.6e+02	0.001	7.3e+03	0.15	1	1.8e+05	3	4.85
20 Apr 98	16	50	2.5e+03	0.007	7.3e+04	1.1	1	3.8e+05	5.4	3.95
24 Aug 98	24	-	1.9e+02	0.001	5.4e+03	0.16	1	3.4e+04	0.5	3.71
28 Oct 03	12	74	1.4e+04	0.03	3.7e+05	4.1	1	4.0e+06	74	4.75
14 Jul 00	4	59	1.2e+04	0.04	1.2e+05	1.3	1	5.1e+06	68	6.40
12 Sep 00	2	39	4.4e+03	0.05	3.2e+04	0.97	1	1.6e+04	0.08	0.47
08 Nov 00	5	64	9.1e+03	0.02	5.2e+04	0.29	1	3.0e+06	51	8.96
16 Aug 01	-	44	1.1e+04	0.05	2.9e+04	0.22		7.3e+04	1.9	7.44
24 Sep 01	5	47	1.6e+04	0.06	5.2e+04	0.34	1	9.9e+05	14	5.49
04 Nov 01	4	26	8.3e+03	0.03	3.7e+04	0.27	1	3.3e+06	85	10.7
22 Nov 01	9	28	5.0e+03	0.03	2.7e+04	0.37	2	7.8e+05	11	7.05
26 Dec 01	4	50	2.8e+03	0.02	1.8e+04	0.32	2	8.0e+04	3.1	5.50

Figure 3 illustrates the procedure for fluence calculations. The GOES data suggest multiple injections, which blend together into one profile at Ulysses in October 1992 and October 2003. In both cases the plot of fluences shows clear steps corresponding to the GLE events. The event of 2 November 1992 nearly doubles the GOES fluence, but its impact for the Ulysses observations is minor. The later events added about one quarter to the total fluence at GOES and Ulysses in October-November 2003, but the second maximum at Ulysses nearly doubles it. Therefore the maximum uncertainty of fluence ratio is a factor of 2. We will discuss the nature of this second maximum below.

Ulysses fluences and maximum intensities are normalized to 1 AU by multiplying them by the square and cube of the Ulysses heliocentric distance, R_U^2 and R_U^3, respectively. This normalization corresponds to a case of delta-like injection and isotropic diffusion in 3D space. Although the last model cannot describe the particle transport, we use these normalization factors as a limit for the intensities we would expect when extrapolating the Ulysses intensities to 1 AU.

Since the flux tube section may increase more rapidly than the square of radius the maximum intensity should vary with radial distances $\sim r^{-(a+1)}$, where $2 < a < 3$ (*Hamilton*, 1977). The last column of Table 2 shows values of (a+1) calculated using the maximum intensities observed by GOES and Ulysses.

As in *Struminsky et al.* (2006) we consider two extreme cases of particle propagation from the Sun to an observer. First, a localized source at the Sun and two magnetic field lines, of which one is very well connected to the source and the other not, so the intensity at the second point should correlate with the first one and be depressed by a factor $D_\perp/D_\parallel$. Second, two magnetic field lines in the heliosphere might be connected to different parts of the extended source, so there might not be any correlation between the two measurements if the result of cross-field diffusion is minor in comparison with a direct injection into both magnetic field lines. Below we will refer to both these extremes and show that the observations possibly reflect their mixture. Note, a similarity of intensity-time profiles observed after the onset in different points in the heliosphere (or events) during some time interval requires the same combination of injection rate from the source and propagation conditions in the interplanetary medium.

In a first order approximation the ratio of fluences between a spacecraft close to Earth and Ulysses reflects an efficiency of particle transport to different points in the heliosphere, assuming a well magnetic connection between Earth and the acceleration site. Calculating fluences we account for all SEPs registered by the detector (diffusing along and across the interplanetary magnetic field, trapped, shock accelerated). Therefore we do not claim that the fluences reflect the number of injected particles into a given magnetic field line during the event. The best correspondence between the particle fluence and the number of injected particles is expected to be during time periods with no interplanetary disturbances and when the event duration is short compared to the rotation period of the Sun.

The second phase of a SEP event begins after 75-100 hours of the SEP event onset and is characterized by an exponential decay (*McKibben*, 1972). The corresponding time constants

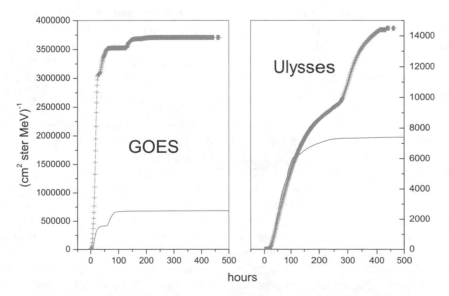

Figure 3. Proton fluences registered within 38-82 MeV and 38-125 MeV by GOES EPS close to Earth and by Ulysses KET during the October 30, 1992 (black line) and October 28, 2003 (gray crosses) SEP events. The time is chosen as in Figure 1.

T_R have been calculated for next 100 hours of the events and are presented in Table 2.

3.2. Onset and Rising to Maximum at ~5 AU in the Ecliptic

Protons with energies of about 50 MeV travel ~1 AU/h. The enhancement above background at Earth orbit was detected during the first or second hour since the X-ray onset (the delay is about 1 hour in all cases). Since the length of the magnetic field line is >10 AU for a nominal solar wind speed of ~400 km/s, these protons are expected to arrive more than 10 h after the flare onset at the position of Ulysses. From the analysis it is evident that the events can be divided into two groups, which are characterized by an expected (11-12 hours) and a delayed (24 hours) arrival.

Figures 1a and 1b show the nominal and delayed case, respectively. The two intermediate cases are displayed in Figure 1c. These two events, however, are similar to the nominal events during the first 50 hours, if Ulysses intensities are multiplied by a factor of 10. The normalized intensities (both Ulysses and GOES) of the June 1992 event and the original intensities of the October 1992 event are compared in the left panel of Figure 4. The event of 20 April has a larger and delayed maximum (Figure 1c). From Figure 4 (right panel) it is evident that the event of 20 April 1998 as observed near the Earth is comparable to the event of 30 October 1992 without any normalization. Possibly, in a case of April 1998 we saw a weak source in the beginning and much more intensive and prolonged later, which were merged together both at Ulysses and GOES.

Three nominal in-ecliptic events (excepting the event of October 2003) do not contradict the predictions of *Hamilton* (1977) that $\sim r^{-(a+1)}$, where $2 < a < 3$, and the assumption that the maximum intensity at 1 AU in the Ulysses magnetic field tube has been equal to that near the Earth. The near Earth intensity of protons on 28-29 October 2003 was about one order higher than within the Ulysses flux tube at 1 AU.

Although the Ulysses time profiles look similar in logarithmic scale for the events in October 1992, October 2003, June 1992 and April 1998 (left panels, Figure 1a, c) the absolute intensities are only the same within a factor of 2-3. These differences are larger than experimental errors, but they are very small in comparison with the observed variations at 1 AU. In contrast, the time profiles of the 'delayed' events are nearly the same in linear scale as shown in Figure 1b, i.e., are equal within the experimental errors during the whole period. However, the situation gets more complex when interplanetary disturbances, e.g., generated by previous (easterly) solar events, alter the particle propagation to Ulysses and with it the observed time histories.

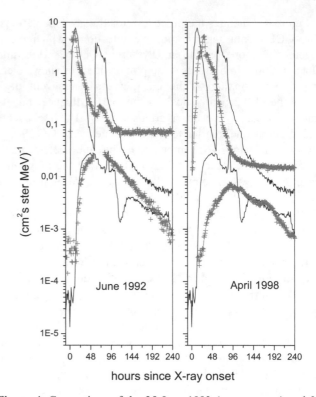

Figure 4. Comparison of the 25 June 1992 (gray crosses) and 20 April 1998 (gray crosses) SEP events with the 30 October (black line) 1992 SEP event. Note, both proton intensities (GOES and Ulysses) for the 25 June event are multiplied by 10.

The proton intensity in October 2003 relative to the intensity measured in October 1992 is displayed in Figure 5 and can be understood when taking into account IMF structures. The lower panel of Figure 5 shows the magnetic field strength as measured by the FGM/VHM instrument (*Balogh et al.*, 1992). The intensity of the October 2003 event is depressed by a factor 4-5 between regions of enhanced magnetic field during ~5 hours in the beginning of the SEP event. Another possible explanation of the depression is that there is a small temporal shift between the profiles, i.e., the X-ray onsets don't correspond to the moments of particle release into the heliosphere.

Taking into account the IMF conditions when interpreting the Ulysses measurements we do no find any obvious dependence of the time delay on the relative position of the parent flare with respect to Ulysses. In the case of the 30 October 1992 and 6 November 1997 events time delays and shapes of time profiles are different for practically the same position. On the other hand the time profiles of the 25 June 1992 and 30 October 1992 events (left panel, Figure 4) are similar during the first two days, although the Earth and Ulysses are 120° apart. Therefore, we conclude that 1) the flare position

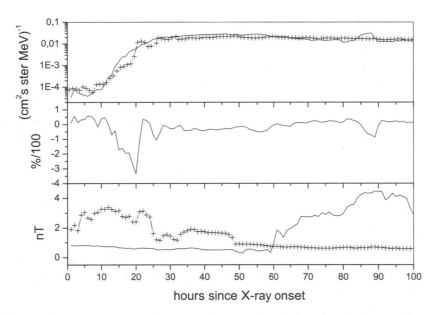

Figure 5. Proton intensity variations during the Ulysses SEP event of October 28, 2003 relatively to October 30, 1992. The upper and lower panels show respectively the proton intensity and the IMF strength as measured by Ulysses for the events of October 30, 1992 (line) and October 28, 2003 (crosses).

does not determine the point of the proton release into the interplanetary magnetic field and 2) cross-field diffusion does not explain the time profiles, at least during beginning of the events.

3.3. Decay Phase at ~5 AU in the Ecliptic

An exponential decay after 75-100 hours is the general tendency observed in most events, which is the main characteristic of the second decay phase (*McKibben*, 1972) and the reservoir effect. Only during the smaller in-ecliptic events we did not observe this exponential decay. Because a longitudinal dependence is not observed at 5 AU we conclude tentatively that corotation is of minor importance for the interpretation of the Ulysses events during the decay phase.

Figure 4 shows the importance of the total number of protons injected into the heliosphere for the second decay phase. The event of 20 April 1998 at the Earth is similar to that of 30 October 1992, as are their decay phases observed at Ulysses. The normalized time profiles observed during the event of 25 June 1992 are similar to those of 30 October 1992 at GOES during about first two days (the total number of particles is about 10 times less than in the event of October 1992), but those at Ulysses equal on a time-scale of 10 days.

The effect of later events in October-November 1992 and 2003, on the Ulysses time profiles is minor. In 1992 the first injection occurred on day 304 and the second on day 307, but the time interval between injections in 2003 was less

than 1 day (Figure 1a), however we do not see instant changes in time profiles. The additional injection of particles may increase the time decay constant T_R (Table 2), its value is minimal for the single event of 25 June 1992, but maximal for the quadruple event of October-November 2003. Integrating the exponential decay from the same moment to infinity and assuming that 100% is the value of the integral for 25 June 1992 (multiplied by 10) we get 24.4% additional protons from the event of 2 November 1992 and 64.4% additional protons from the later events of October-November 2003.

In some events this exponential decay is distorted by sharp increases and decreases of proton intensity on a time scale of one day or less, which correlate with IMF disturbances. For example, the intensity is enlarged by a factor of two at ~86 hour for the event of October 1992 (see Figures 5 and 6). The value of these modulation effects looks reasonable in comparison with Forbush decreases and pre-increases. We present examples of similar observations at high heliographic latitudes in the next section (Figure 7).

An extreme case of the modulation occurred in October-November 2003, when the largest decay constant and the reverse gradient of SEP were observed (Figure 1a). *Lim et al.* (1996) interpreted the interplanetary conditions at Ulysses in October-November 1992 as a combination of CIR and shock waves. From Figure 6 it is clear that a similar pattern has been observed in October-November 2003, but the second maximum of protons is larger and better pronounced.

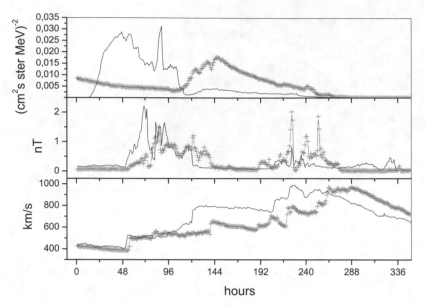

Figure 6. Hourly averaged proton intensities and solar wind parameters measured by the Ulysses instruments after 17 UT on 30 October 1992 (lines) and 10 UT on 4 November 2003 (gray crosses).

Additional proton injection into interplanetary space in November 2003 (Figure 1a) is the only difference.

3.4. Comparison of SEP Events in and out of the Ecliptic

In this section we summarize the KET/Ulysses observations of SEP events at high heliolatitudes presented in *Struminsky et al.* (2006) and compare them with characteristics of SEP events at ~5 AU in the ecliptic.

Figure 2 shows examples of solar proton time profiles observed during the large events of 14 July 2000 and 24 September 2001, which are compared with the small event of 12 September 2000. Symbols and lines correspond to Ulysses and GOES observations, respectively. Ulysses profiles have

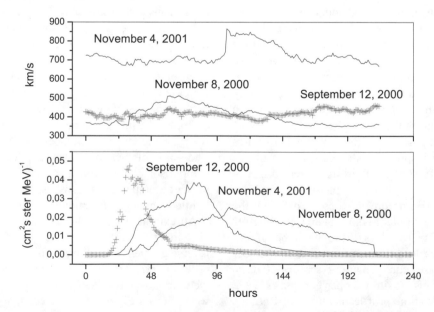

Figure 7. Hourly averaged 38-125 MeV proton intensities and solar wind speed for the events of September 12 and November 8, 2000 and November 4, 2001.

been shifted in time so that the onset of both events is the same at Ulysses. The left and right panels show the measurements during the first 7 and 1½ days of the event.

The time delay between the proton onset at polar latitudes and the X-ray onset of the possible parent flare (T_U, Table 2) looks arbitrary in comparison with cases considered in the previous section. The delay is ordered best by Ulysses latitude and suggests a later release from the source (*Dalla et al.*, 2003a).

Struminsky et al. (2006) concluded (1) that during the rising phase, the first 30 hours, the time histories for the high latitude events at Ulysses are similar, (2) that the maximum intensities and fluences at Ulysses are the same within a factor of ~3 and ~6, respectively, and (3) that the proton intensities and fluences are only weakly dependent on the spacecraft longitude and the flare location. These three facts exclude cross-field diffusion of SEP as the dominant particle transport from the ecliptic to high latitude regions. It rather requires a direct particle injection at high latitudes. Values of the power-law index of radial dependence presented in Table 2 for high latitude events definitely show that the magnetic field line did not connect Ulysses and GOES. It is important to note that the northern and southern hemisphere events show very similar time profiles, in spite of the solar wind conditions being different during the southern and northern polar cap passages.

In what follows the 12 September 2000 event has been chosen as a reference because magnetic field and solar wind measurements at Ulysses indicate only a few disturbances.

It is important that all large events observed in the ecliptic at 1 AU have a distinct high latitude partner. The intensity time profiles at high latitudes are only slightly modified by transient disturbances, in contrast to measurements in the ecliptic. We see the deficit of energetic particles in front of these structures and a trapped particle population behind them (Figure 7). This leads to a larger uncertainty when calculating the particle fluences.

The decay phase of the out-of-ecliptic time profile is better correlated to the in-ecliptic profiles at 1 AU than during the onset times. Such a behavior is expected if an effective interplanetary cross-field transport would operate (*Struminsky et al.*, 2006).

Because it is not possible to compare one and the same event at high heliolatitudes and at 5 AU in the ecliptic, we compare different events, as listed in Tables 1 and 2. While one event was observed by Ulysses in 2000/2001 at high latitudes the other one was observed at 5 AU in 1992, 1997, 1998 or 2003 close to the ecliptic. As an example, the June 1992 and October 1992 events are compared with the September 2000 and July 2000 high latitude events in Figure 8. The first and second pair of events corresponds to a small and a large event near Earth, respectively. Although not expected it is remarkable that the events with nearly the same fluences have also nearly the same time scale, independent of being observed at large heliocentric distances or at high heliospheric latitudes.

The fact that the exponential time decay constants during the late phase for both in-ecliptic and out-ecliptic events, as summarized in Table 2, are similar to those reported by

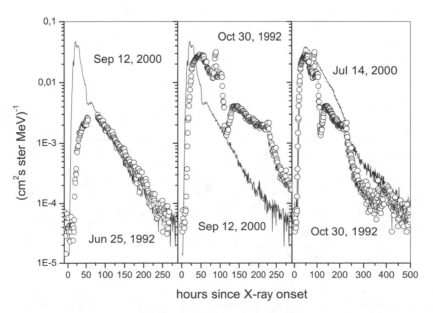

Figure 8. Comparison of the polar SEP events (14 July and 12 September 2000) with the events observed at 5 AU (25 June and 30 October 1992).

McKibben (1972) points towards a unique process for particle escape into the outer heliosphere, which effectively operates in different solar events and even solar cycles.

4. DISCUSSION

4.1. Source Distribution of SEPs

Adding the results from the analysis of the high latitude events, the in and out-of ecliptic observations suggest an effective propagation of SEPs in longitude as well as in latitude by more than 120° close to the Sun. This finding is supported by the conclusions of *Dalla et al.* (2003b) and *Lario et al.* (2004a). They exclude that the particles have been accelerated by coronal mass ejections driven shocks for most of the events presented in this paper.

As suggested by *Hamilton* (1981), observations at large distances might help separate interplanetary propagation effects from coronal storage and propagation effects. This separation would be possible, if the propagation time of SEP is greater than the coronal storage time. The longest time of pion production by protons is 8 hours from the gamma-ray observations on June 11, 1991 (*Kanbach et al.*, 1993). This time is consistent with duration of proton injection describing the proton time profile observed on June 11, 1991 (*Struminsky*, 2003). Therefore, the intensity-time profiles observed by Ulysses at ~5 AU are determined mainly by propagation effects and do not provide any reliable information on temporal properties of the solar source.

We may discuss the time of injection to high heliolatitudes, where the Ulysses heliocentric distances have been smaller and magnetic field lines are not twisted. Figure 9 from *Struminsky et al.* (2006) displays the measured time profiles of proton and X-ray intensities for 12 September 2000 and 22 November 2001. Both events occurred in the western hemisphere, so that the Earth was magnetically well connected to the particle source. The time profiles of the 22 November event are shifted by 13 hours to later times in order to get a coincidence of both particle event onsets at Ulysses. *Struminsky et al.* (2006) found a good temporal coincidence of the particle release time close to the Sun with the second and third peak in the X-ray intensities for the 22 November 2001 and the 12 September 2000 events using the 1 AU data. If they correlate the X-ray data with the Ulysses proton intensities this coincidence is not present.

This leads the authors to the 'speculation' that a delayed injection of solar protons to high heliographic latitudes is associated with some particular events during the late decay phase of large gradual X-ray events. Because the maximum particle intensities at high latitudes are the same within a factor of three for all events the majority of particles are injected first into the ecliptic plane and later to higher heliographic latitudes by a peculiar transport process close to the Sun. This transport leads to a universal distribution of SEPs close to the Sun, which are released into the interplanetary medium.

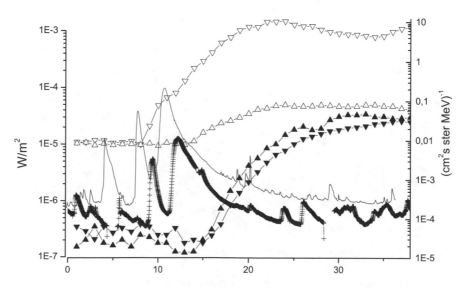

Figure 9. Hourly averaged intensities of GOES EPS (open symbols), Ulysses KET (solid simbols) and X-ray flux (lines) for the September, 12, 2000 (gray) and the November, 22, 2001 (black) events (adopted from *Struminsky et al.*, 2006). The data for the latter event are shifted in time by 13 hours to get the simultaneous proton onset at Ulysses.

4.2. SEP Propagation in the Inner Heliosphere

Since the correlation between Earth and Ulysses measurements is minor, we conclude that magnetic field lines have been connected to different parts of an extended particle source with different intensities. In this scenario SEP propagate along magnetic field lines filling up quickly a part of the intermediate heliosphere (5-6 AU) connected to the source. The similarity of proton intensity time profiles observed (Figures 1, 2) is a remarkable feature indicating a unique particle propagation process at different interplanetary positions during the rising phase of some events.

An outer boundary exists at 5-6 AU, where cross field diffusion starts to dominate D_{rr}, due to the fractal properties of the solar wind (*Milovanov and Zelenyi*, 1994). This boundary however does not exist at polar latitudes. Therefore the majority of particles are concentrated close to the ecliptic at ~5 AU along tangential magnetic field lines slowly escaping to high heliographic latitudes and the outer heliosphere. Thus the large spatial gradients observed in the event beginning in the heliosphere are reduced by cross-field diffusion leading to the reservoir effect at high latitudes and in the ecliptic (*McKibben et al.*, 2003). The magnetic field lines connected to the source would form the reservoir. For a case of multiple events it would be interesting to know how many reservoirs do we have and what should we expect to observe at high latitudes?

Regions, which are not connected to the source, are filled only by the across field diffusion. The reservoir effect is not observed there. Possibly in-ecliptic events, when minimal fluences and delayed arrival of protons have been observed, correspond to such a case. The ratio of fluences would give an estimate of the ratio of across and along field diffusion coefficients. For the 6 November 1997 event we have:

$$\frac{D_\perp}{D_\parallel} = \frac{F_{5AU} \cdot R_U^2}{F_{1AU}} = \frac{7.3 \cdot 10^3}{1.8 \cdot 10^5} = 0.041$$

The magnetic connection between Earth and the flare site on 24 August 1998 was worse, so the ratio would be larger.

4.3. SEP Release From the Inner Heliosphere

In what follows we discuss three different mechanisms for particle release into the outer heliosphere:

1. The cross-field diffusion supplies particles to high latitudes during the whole event, but it would be observable only later in the decay phase as reported by *Zhang et al.*, (2003a, b), because of the large background due to along field diffusion. Only a few percent of particles use this way. *Stuminsky et al.* (2006) have determined the

fluences observed during the South and North Pole Passes, respectively and their ratios to the in-ecliptic fluences, which are 0.003 (0.008) and 0.001 (0.006), respectively. The values in brackets have been normalized to 1 AU. It should be noted that this ratio varies from SEP event to SEP event from 0.0024 (0.01) to 0.275 (2) and reflects the strength of the source close to the ecliptic.

2. Particle convection from the inner heliosphere naturally explains the exponential time decay of particle intensity. For a volume V filled with particles of a density n we find $V \cdot dn/dt \sim -n \cdot U \cdot S$, where S is the effective area of the convection and U the solar wind speed. For the events presented in Figure 6 (the cases of slow solar wind (8 November U=350 km/s), normal solar wind (12 September 2000, u=460 km/s), fast solar wind (4 November 2001, U>700 km/s)) we find the expected reverse proportionality of T_R to the solar wind velocity U. The above equation does not consider the later events, which might be included as external source supplying particles into the reservoir, so T_R would increase.

3. In the early seventies *Syrovatsky* (1971) proposed a mechanism of effective particle transport inside a trap with decreasing volume and a continuous particle leakage. If the number of trapped particles would decrease faster than the square of the IMF strength, the average energy of these particles should increase. The time history of the particle events in October-November 1992 and 2003 (Figure 6) can be explained when the volume between a CIR and a traveling shock is decreasing fast enough as a result of large velocity dispersion. The moving trap, which opens twisted magnetic field lines, provides the third and most effective mechanism for a particle release from the reservoir. Such a trap should be created between the CIR and shock of the 28 October 2003 event in the Sun-Earth direction and might contain the majority of particles observed by GOES but not Ulysses. Therefore, in the outer heliosphere large enhancements of the solar proton intensity observed aboard Voyager 1 (72 AU, 33N) and Voyager 2 (56 AU, 18S) were associated with large-scale heliospheric disturbances, such as merged interaction regions (*Decker et al.*, 1999).

The two different time constants, found by *McKibben* (1972) during the decay phase of SEP events, can be understood when, the first constant, comparable with the diffusive propagation time along magnetic field lines to ~5 AU (10-20 hours), shows the dependence on how fast the reservoir is filled by SEP's but the second reflects the rate of particle release from the reservoir region. The rate of particle release from the reservoir should be rather small, so that the cross-field propagation could maintain there a uniform flux independent of longitude, latitude, or radius.

5. SUMMARY

We discussed observations of eight and six solar energetic particle (SEP) events at high heliographic latitudes and at distances of about 5 AU close to the ecliptic. Ulysses KET observations and observations by the GOES spacecraft close to Earth are put into context with the current paradigm of particle acceleration and propagation in the heliosphere. These events can be identified with episodes of solar activity in 1992, 1997, 2000-2001 and 2003.

The observational results suggest that the Sun is a complex source for energetic protons with energies >30 MeV extended by more than 120° both in longitude and latitude, and prolonged in time up to 9 hours. To our opinion the model of SEP acceleration by coronal mass ejection driven shocks can not account for the Ulysses observations at large heliocentric distances and at high heliographic latitudes.

The time profiles of most of the events observed by Ulysses show three distinct phases of particle propagation:

- During the rising phase of the Ulysses SEP events (first 30-50 hours), energetic particles propagate along magnetic field lines connected to different regions of the solar source up to heliocentric distances of 5-6 AU. The radial propagation of SEP becomes ineffective at such distances close to the ecliptic.

- Large spatial gradients observed in the inner heliosphere between different magnetic field tubes are reduced by cross-field diffusion leading to the reservoir effect at high latitudes and in the ecliptic. Therefore, the reservoir is a part of the heliosphere up to 5-6 AU connected to the solar source by magnetic field lines. Particles need about 100 hours to fill this region uniformly by cross-field diffusion.

- The last phase is characterized by particle escape from the intermediate heliosphere either by cross-field diffusion to open magnetic field lines in polar regions, convection close to the ecliptic, or moving and expanding traps where available.

Acknowledgments. The Ulysses/KET project is supported by the grant 50ON0105 BMBF. AS thanks the DAAD Foundation for the opportunity to work in the Osnabrück University in September to December 2003 and acknowledges the AGU travel grant enabling him to participate in the Chapman Conference on Solar Energetic Plasmas and Particles, Turku, Finland, August 2004. The authors thank two anonymous reviewers for their criticism and valuable comments improving the paper.

REFERENCES

Balogh A., T.J. Beek, and R.G. Forsyth, *et al.*, The magnetic field investigation on the Ulysses mission: instrumentation and preliminary scientific results, *Astron. Astrophys. Suppl. Ser.*, 92, 221-236, 1992.

Burger, R.A., M.S. Potgieter, and B. Heber, Rigidity dependence of cosmic ray proton latitudinal gradients measured by the Ulysses spacecraft: Implications for the diffusion tensor, *J. of Geophys. Res.*, 105, 27447-27456, 2000.

Conlon, T.F., The interplanetary modulation and transport of Jovian electrons, *J. Geophys. Res.*, 83, 541-552, 1978.

Dalla S., A. Balogh, and S. Krucker, *et al.*, Delay in solar energetic particle onsets at high heliographic latitudes, *Ann. Geophysicae*, 21, 1367-1375, 2003a.

Dalla S., A. Balogh, and S. Krucker, *et al.*, Properties of high heliolatitude solar energetic particle events and constraints on models of acceleration and propagation, *Geophys. Res. Lett.*, 309190, 8035, doi:10.1029/ 2003GL017139, 2003b.

Decker, R.B., E.C. Roelof, and S.M. Krimigis, Solar energetic particles from the April 1998 activity: observations from 1 to 72 AU, *26th ICRC*, 6, 328-331, 1999.

Ferrando, P., R. Ducros, and C. Rastoin, *et al.*, Propagation of Jovian electrons in and out of the ecliptic, *Adv. of Space Res.*, 13, 107-110, 1993.

Ferreira, S.E.S., M.S. Potgieter, and Burger, *et al.*, Modulation of Jovian and galactic electrons in the heliosphere: 1. Latitudinal transport of a few MeV electrons, *J. of Geophys. Res.*, 106, A11, 24979-24988, 2001.

Hamilton, D.C., The Radial transport of Energetic Solar Flare Particles from 1 to 6 AU, *J. of Geophys. Res.*, 82, 2157-2169, 1977.

Hamilton, D.C., Dynamics of Solar Cosmic Ray Events: Processes at Large Heliocentric Distances (>>1 AU), *Adv. Space Res.*, 1, 25-40, 1981.

Kanbach, G., D.L. Berssch, and C.E. Fichtel, *et al.*, Detection of a long-duration solar gamma-ray flare on June 11, 1991 with EGRET on COMPTON-GRO, *Astron Astrophys. Suppl. Ser.*, 97, 349-353, 1993.

Kallenrode, M.-B., Current views on impulsive and gradual solar energetic particle events, *Journal of Physics G: Nuclear and Particle Physics*, 29, 5, 965-981, 2003.

Lanzerotti L.J., S.M. Krimigis, and R.B. Decker, *et al.*, Low energy particles in the global heliosphere 2001-2004: 1-90 AU, *Space Science Reviews*, 97, 243-248, 2001.

Lario, D., R.G. Marsden, and T.R. Sanderson, *et al.*, Energetic proton observations at 1 and 5 AU 2. Rising phase of the solar cycle 23, *J. of Geophys. Res*, 105, 18251-, 2000.

Lario, D., R.B. Decker, and E.C. Roelof, *et al.*, Low-energy particle response to CMEs during the Ulysses solar maximum northern polar passage, *J. of Geophys. Res*, 109, A01107, doi:10.1029/2003JA010071, 2004a.

Lario, D., S. Livi, and E.C. Roelof, *et al.*, Heliospheric energetic particle observations by the Cassini spacecraft: Correlation with 1AU observations, *J. of Geophys. Res*, 109, A09S02, doi:10.1029/2003JA010107, 2004b.

Lim, T.L., J.J. Quenby, and M.K. Reuss, *et al.*, Ulysses observations of energetic particle acceleration and the superposed CME and CIR events of November 1992, *Ann. Geophysicae*, 14, 400-410, 1996.

McKibben, R.B., Azimuthal propagation of low-energy solar-flare protons as observed from spacecrafts very widely separated in solar azimuth, *J. of Geophys. Res*, 77, 3957-3983, 1972.

McKibben R.B., J.J. Connell, and C. Lopate, *et al.*, Ulysses Cospin observations of cosmic rays and solar energetic particles from the South Pole to the North Pole of the Sun during solar maximum, *Ann. Geophysicae*, 21, 1217-1228, 2003.

Milovanov A.V., and L.M. Zelenyi, Fractal clusters in the solar wind, *Adv. Space Res.*, 14, 7, (7)123-(7)133, 1994.

Ryan, J.M., J.A. Lockwood, and H. Debrunner, Solar energetic particles, *Space Sci. Rev.*, 93, 35-53, 2000.

Reames, D.V., L.M. Barbier, and C.K. Ng, The spatial distribution of particles accelerated by coronal mass ejection-driven shocks, *ApJ*, 466, 473-486, 1996.

Reames, D.V., Particle acceleration at the Sun and in the heliosphere, *Space Sci. Rev.*, 90, 413-491, 1999.

Sanderson, T.R., R.G. Marsden, and C. Tranquille, *et al.*, Propagation of energetic particles in the high-latitude high-speed solar wind, *Geophys. Res. Lett.*, 30, 8036, doi:10.1029/ 2003gl017306, 2003.

Simpson, J., J. Anglin, and A. Balogh, *et al.*, The Ulysses Cosmic-Ray and Solar Particle investigation, *Astron. and Astrophys. Suppl.*, 92, 365-399, 1992.

Struminsky, A., Evidence of prolonged existence of >100 MeV solar protons in coronal structures, *Astronomy Reports*, 47, 11, 916, 2003.

Struminsky, A., B. Heber, and M.-B. Kallenrode, *et al.*, Injection and Propagation of Solar Protons to High Heliospheric Latitudes: KET/Ulysses Observations, *Adv. Space Res.*, in press, 2006.

Syrovatsky, S.I., On shock waves associated with solar energetic particles, *Proc. of seminar "Generation of Cosmic Rays on the Sun"*, pp. 39-51, Moscow State University, Moscow, 1971, Russia (in Russian).

Zhang, M., R.B. McKibben, and C. Lopate, *et al.*, Ulysses observations of solar energetic particles from the 14 July 2000 event at high heliographic latitudes, *J. of Geophys. Res.*, 108(A4), 1154, doi: 10.1029/2002JA009531, 2003a.

Zhang, M., J.R. Jokipi, and R.B. McKibben, Perpendicular Transport of Solar Energetic Particles in Heliospheric Magnetic Fields, *ApJ*, 595, 493-499, 2003b.

Bernd Heber, Institute für Experimentelle und Angewandte Physik Christian–Albrechts–Universität Kiel, Leibnizstraße 19, Kiel, 24118, Germany

Alexei Struminsky, Space Research Institute, Profsoyuznaya st., 84/32, Moscow, 117997, Russia

Geoeffective Coronal Mass Ejections and Energetic Particles

Eino Valtonen

Space Research Laboratory, Department of Physics, University of Turku

Coronal mass ejections (CMEs) can be geoeffective and they are believed to cause major non-recurrent geomagnetic storms. Energetic particles are closely related to the shocks driven by coronal mass ejections in the interplanetary medium. In this paper, the geoeffectiveness of CMEs is reviewed and the means how energetic particle observations can be used to identify interplanetary CMEs approaching the Earth are discussed. A summary is given of the observed geoeffectiveness of halo CMEs and, in particular, of those ejecta, which are identified as magnetic clouds. The principal mechanism and the prerequisites for a geomagnetic storm to occur are very briefly outlined. Relations between CMEs and energetic particles are discussed, and the possibilities are examined to use solar energetic particles and energetic storm particles in the investigation of the properties of interplanetary CMEs and shocks. As a feasibility demonstration, two examples are given of methods using energetic particle observations for geomagnetic storm forecasting.

1. INTRODUCTION

Solar processes are said to be geoeffective, if they cause an observable effect in the geospace. Here geospace, the domain of Sun-Earth interactions, is considered to be the near-Earth space inside the magnetosphere. Effects can be seen in the space environment and ultimately even on ground [*Bothmer*, 1999; *Baker*, 2000; *Lanzerotti*, 2001]. The two main sources of geoeffects are coronal mass ejections (CMEs) and energetic particles. More precisely, the geoeffects are caused by the ejecta or interplanetary counterparts of CMEs (ICMEs) [*Gosling et al.*, 1991] and by the interactions of energetic particles [e.g., *Baker*, 1998]. Energetic particles observed in the interplanetary space and in the geospace are to a large extent related to the CMEs and ICMEs and their interactions with their surroundings. Most notably, the particles are related to the shocks, which the ICMEs are

driving [*Reames*, 1999] and to the geomagnetic storms that they cause when encountering the Earth's magnetosphere [*Reeves et al.*, 2003].

In this paper, the geoeffectiveness of coronal mass ejections is briefly reviewed. A particular emphasis is on the associated energetic particles and how the observations of these particles can be employed in deducing certain properties of the ICMEs and potentially predicting their geoeffects before they reach the Earth. Consequently, the discussion is limited to energetic particles produced at the Sun or in the interplanetary space. Only geomagnetic storms will be considered as geoeffects. Related topics, such as energetic particles in the magnetosphere, birth of new radiation belts, energetic particle hazards to man in space, and influence of solar energetic particles on atmospheric chemistry have been discussed elsewhere in this monograph. Furthermore, a broader view of space weather, addressing also magnetospheric storms, has been given by *Koskinen and Huttunen* [2005].

In Section 2, the relations between coronal mass ejections, energetic particles, and geomagnetic storms are briefly outlined. The geoeffectiveness of coronal mass

Solar Eruptions and Energetic Particles
Geophysical Monograph Series 165
10.1029/165GM31

ejections is discussed in more detail in Section 3, and the possibilities to use energetic particles as probes of ICMEs are explored in Section 4. Section 5 introduces two methods developed for forecasting geomagnetic storms by employing energetic particle observations. Section 6 summarizes the paper.

2. CORONAL MASS EJECTIONS, ENERGETIC PARTICLES, AND GEOMAGNETIC STORMS

Coronal mass ejections are believed to be the main source of major non-recurrent geomagnetic storms [*Gosling et al.*, 1991; *Tsurutani and Gonzalez*, 1997; *Webb et al.*, 2000; *Richardson et al.*, 2002; *Zhang et al.*, 2003]. Geomagnetic storms are produced, when mass and momentum is transferred from the solar wind into the magnetosphere. As a consequence, the magnetosphere falls in a strongly disturbed state, leading to intensification of the ring current [e.g., *Daglis et al.*, 1997] and frequently to injection of energetic ($\gtrsim 1$ MeV) electrons and ions into the inner magnetosphere [*Baker et al.*, 1998; *Li et al.*, 2003]. The strength of geomagnetic storms is often measured by the Dst or Kp index [e.g., *Gonzalez et al.*, 1994]. It should be noted, however, that due to the different response of these indices to the magnetospheric current systems, the behavior of Dst and Kp can be different depending on the type of the solar wind driver causing the storm [*Huttunen et al.*, 2002]. The effects of large magnetic storms can be seen both at ground and in near-Earth space. Variations in the ionospheric and magnetospheric current systems create electric fields driving geomagnetically induced currents causing various ground effects [e.g., *Lanzerotti et al.*, 2000], while disturbed magnetic fields and enhanced levels of energetic particles can cause interference in space systems [*Valtonen*, 2005 and references therein].

Energetic particles are accelerated in solar flare processes low in the corona and by shock waves driven by fast coronal mass ejections high in the corona and in the interplanetary space [e.g., *Reames*, 1999]. These particles can reach energies up to the GeV-range, but do not have a direct role in the initiation of geomagnetic storms. Energetic particles accelerated at the Sun or in the interplanetary space can be geoeffective only if they are able to penetrate into the magnetosphere. During quiet times, the Earth's geomagnetic field is an effective shield against charged particles arriving form outside of the magnetosphere. Magnetic cutoff rigidity is defined as the lowest rigidity a charged particle can have and still arrive at a specific point at the surface of the Earth. Normally, only very high-energy particles can reach low latitudes. During geomagnetic storms, however, sudden suppressions of the cutoff rigidities can occasionally occur [*Leske et al.*, 2001] due to a restructuring of the

magnetosphere, and particles from interplanetary space can access. In this sense, they can contribute to the effects of geomagnetic storms. Once inside the magnetosphere, energetic particles can pose a serious radiation hazard to man and technological systems in space, influence the conditions in the ionosphere and thermosphere, cause ground-level enhancements of secondary cosmic rays, and even contribute to the birth of new radiation belts [*Lorentzen et al.*, 2002].

Only those CMEs, which are directed towards the Earth, can cause geomagnetic storms. Full halo CMEs are the most probable candidates. However, not all full halo CMEs are geoeffective. All ICMEs related to halo CMEs observed by coronagraphs at the Sun do not necessarily encounter the Earth. Furthermore, even if they do, they may not posses the necessary properties to cause a geomagnetic storm, as discussed in Section 3. Energetic particles associated with solar flares and CME-driven interplanetary shocks can be helpful in identifying the solar sources of CMEs and may provide an early warning of an approaching, potentially geoeffective ICME. A prerequisite is that the occurrences of halo CMEs and energetic particle events correlate with geomagnetic storms. Figure 1 presents the 6-month averaged occurrence rates of full halo CMEs, large solar energetic particle (SEP) events, and intense geomagnetic storms between July 1996 and June 2004. The halo CME rates are based on the SOHO/LASCO [*Brueckner et al.*, 1995] and the particle events on the SOHO/ERNE [*Torsti et al.*, 1995] observations. For the particle events, a peak intensity of $\gtrsim 10$ cm^{-2}sr^{-1}s^{-1}MeV^{-1} was required in the energy range 20.8–40.5 MeV. A Dst minimum value of ≤ -100 nT was required for a geomagnetic storm to be counted as an intense storm [*Gonzalez et al.*, 1994]. Data gaps in the LASCO and ERNE observations were taken into account in calculating the average rates of full halo CMEs and energetic particle events, respectively. For clarity, the statistical errors are only shown for the ERNE observations. One can clearly see that the rates of full halo CMEs, large energetic particle events, and intense geomagnetic storms correlate with each other. The error bars are, however, relatively large, particularly in the last half of 1998 due to the long interruption of SOHO observations. Throughout the entire observation period, the rate of intense storms is relatively low, but indicates maximum geomagnetic activity at the time of the solar cycle maximum and at the beginning of the declining phase of the cycle. The occurrence rates of full halo CMEs and energetic particle events follow quite closely each other. During the first part of 2000 and the last part of 2002, the rate of energetic particle events exceeds the full halo CME rate. This is probably due to the production of energetic particles by other than full halo CMEs, which during those two time periods had exceptionally high rates.

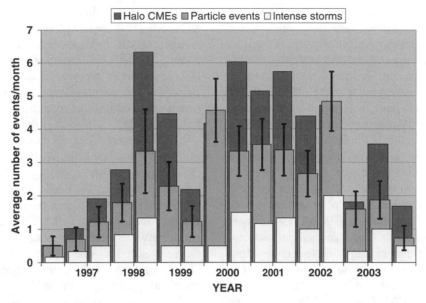

Figure 1. Monthly occurrence rates of full halo CMEs, solar energetic particle events, and intense geomagnetic storms averaged over 6-month periods from 1996 to 2004.

3. GEOEFFECTIVENESS OF CORONAL MASS EJECTIONS

Statistical properties of the CMEs observed by LASCO have been presented by *St. Cyr et al.* [2000] and *Yashiro et al.* [2004]. *Yashiro et al.* [2004] report nearly 7000 CME observations by LASCO between 1996 and 2002. Probably only a small fraction of the CMEs observed at the Sun can reach 1 AU before ceasing to exist as entities different from the ambient solar wind [*Gopalswamy*, 2004]. CMEs have widely varying initial speeds [*Yashiro et al.*, 2004], and it has been shown that CMEs with any projected speed above ~300 km/s are capable of producing severe geomagnetic storms [*Wang et al.*, 2002; *Zhang et al.*, 2003]. However, it has also been shown by *Gosling et al.* [1991] (see also [*Huttunen et al.*, 2002]) that fast CMEs capable of driving interplanetary shocks are most often geoeffective.

The average rate of coronal mass ejections is solar cycle dependent. The average rate over Carrington rotation periods increased from less than 1 per day in 1996 to slightly more than 6 during the maximum solar activity in 2000 [*Gopalswamy*, 2004]. Detection of a full halo CME indicates the possibility that the CME is traveling towards the Earth, and is expected to be potentially geoeffective, if it is able to reach 1 AU. The fraction of full halo CMEs of all coronal mass ejections is only ~3.5% with a broad peak in the occurrence rate during the solar activity maximum phase [*Gopalswamy*, 2004] (see also Figure. 1). *Yashiro et al.* [2004] found that the average speed of halo CMEs is twice

that of the normal (width between 20° and 120°) CMEs. Most of the full halo CMEs seem to belong to the fast and wide (speed >900 km/s, width >60°) population of CMEs capable of driving interplanetary shocks [*Gopalswamy*, 2004].

From a study of 36 ejecta associated with frontside halo or partial halo CMEs in 1996–1999, *Cane et al.* [2000] concluded that only about half of the frontside halo CMEs encountered the Earth. Observation of a halo CME at the Sun is not a sufficient indication of the launch of an Earthward-directed ejecta. The probability of detecting an ejecta at Earth depends on the source location. Ejecta from halo CMEs originating far from central meridian do not necessarily encounter the Earth. Even in some central meridian events the ejecta may pass the Earth from the north or south. *Cane et al.* [2000] found that ejecta were only observed following events within ~40° of central meridian. The majority of geoeffective halo CMEs between 1997 and 2000 occurred in the solar latitude range ±(10°–30°) and within ±30° of central meridian [*Wang et al.*, 2002]. The longitude distribution was found to be slightly asymmetric towards western longitudes with events up to west 70°, but no events beyond east 40°. Corresponding results for the longitude distribution of Earth-encountering ICMEs were obtained by *Cane and Richardson* [2003] and of source locations of halo or partial halo CMEs causing major geomagnetic storms by *Zhang et al.* [2003].

Near solar minimum very good correlation between halo CMEs and geomagnetic storms has been found [*Brueckner et al.*, 1998; *Webb et al.*, 2000; *St. Cyr et al.*, 2000].

St. Cyr et al. [2000] however also noted that a large number of "false alarms" is produced if one only relies on the detection of a halo CME as an indication of a forthcoming geomagnetic storm. *Zhao and Webb* [2003] studied the association of halo CMEs with geomagnetic storms in the ascending phase of the present solar cycle. The result was that in 1997 91% of the frontside full halo CMEs was associated with storms. The association decreased by 1999 to only 38%, but increased again being 70% in 2000, yielding the overall average of 64%. The average association was found better (71%) for events centered within ±45° of central meridian. *Zhang et al.* [2003] used a comprehensive set of solar, interplanetary, and geomagnetic observations to identify solar sources of 27 major (Dst ≤ –100 nT) geomagnetic storms between 1996 and 2000. They were able to associate 22 of the 27 major storms during the period to unique or multiple halo CMEs. In addition, in four cases they identified the source as a partial-halo gradual east limb CME. They noted that only 8.4% of the halo CMEs observed during the study period caused major geomagnetic storms. They also found that there was no preference for the geoeffective CMEs to be fast or to be associated with major flares and erupting filaments. In a statistical study of geoeffectiveness of Earth-directed CMEs between 1997 and 2000, *Wang et al.* [2002] found that 45% of the total 132 halo CMEs caused geomagnetic storms with Kp ≥5. *Yermolaev et al.* [2005] have pointed out that the reason for diverging estimates of CME geoeffectiveness may lie in the different methods of analysis and in the direction (storm–CME or CME–storm), in which the correlation is made. *Yermolaev et al.* [2005] came to the conclusion that it is paradoxical that the origin of almost all strong geomagnetic storms can be explained in retrospect, but the occurrence cannot be predicted with a sufficient degree of reliability.

In general terms, an interplanetary CME is observed as a transient fast interplanetary plasma stream. There are several in situ signatures of ICMEs [e.g., *Neugebauer and Goldstein*, 1997; *Burlaga et al.*, 2001; *Richardson and Cane*, 2004; and references therein]. Typical solar wind plasma, magnetic field, and energetic particle signatures used to identify ICMEs passing a spacecraft in interplanetary space include low kinetic temperatures of ions and electrons, bidirectional flows of suprathermal electrons, pronounced anisotropy of proton distribution, unusually high helium-to-proton ratio, anomalies in composition and exceptionally high charge states of minor ions, quiet, strong magnetic fields, and cosmic ray flux depressions. However, all the signatures are not consistently always observed and none of the single signatures is sufficient to identify an ICME.

Magnetic clouds are specific types of ICMEs. They are ejecta in which the magnetic field direction rotates slowly and smoothly through a large angle, the magnetic field strength is higher than average, and the proton temperature and plasma beta are low [*Burlaga*, 1991; *Burlaga et al.*, 2001; *Huttunen et al.*, 2005; and references therein]. The smoothly rotating magnetic field makes the distinction between magnetic clouds and other types of ejecta with disordered magnetic fields. In their simplest form, magnetic clouds can be interpreted as cylindrical flux rope structures with helical magnetic field lines [e.g., *Mulligan and Russell*, 2001].

In early observations, about one third of the ICMEs were estimated to be magnetic clouds [*Klein and Burlaga*, 1982]. Based on the observations by Helios 1 and 2 spacecraft between December 1974 and end of 1979, *Cane et al.* [1997] found that on the average 50% of the ICMEs were magnetic clouds. *Gopalswamy et al.* [2000] listed 28 interplanetary ejecta between December 1996 and June 1998, 20 of which were classified as magnetic clouds. *Burlaga et al.* [2001] have reported that 4 of the 9 fast ejecta observed by the ACE spacecraft between 1998 and 1999 were magnetic clouds and the rest were complex ejecta. *Cane and Richardson* [2003] compiled a detailed list of all types of interplanetary coronal mass ejections observed in the near-Earth solar wind during 1996–2002. They found that the average rate of ICMEs increased by about an order of magnitude from ~1 every three months at solar minimum to ~1 per week at solar maximum. The fraction of magnetic clouds decreased from 100% in 1996 to ~16% in 2000–2001 with a trend to recover in 2002 with declining solar activity. This indicates a solar cycle variation of the fraction of magnetic clouds from all ICMEs.

Magnetic clouds are considered to be particularly geoeffective. *Webb et al.* [2000] found all the six magnetic cloud events, which they identified during the period January 1997 and May 1997, to be associated with magnetic storms with the Dst minimum values between –41 nT and –115 nT. By using the Wind observations in 1995–1998, *Wu and Lepping* [2002a] identified 34 magnetic clouds and associated 30 of these with at least weak (Dst ≤ –30 nT) geomagnetic storms. With a higher statistics of 104 magnetic clouds between January 1998 and April 2002, *Zhang et al.* [2004] found that nearly 80% of magnetic clouds resulted in geomagnetic storms (Dst ≤ –30 nT). They also noted that magnetic clouds dominate intense storm occurrences, but only 30% of all geomagnetic storms during the study period resulted from magnetic clouds. Comparable results were obtained by *Li and Luhman* [2004] from a study covering the years 1978–2002. Magnetic clouds were the primary cause of 1/3 of the total number of storms analyzed and they were responsible for the most intense storms. *Huttunen et al.* [2005] examined 73 magnetic clouds observed by the Wind and ACE spacecraft during the seven-year period 1997–2003. In about 70% of the cases either the magnetic cloud or its preceding sheath region caused a moderate magnetic storm (Dst ≤ –50 nT). Magnetic clouds themselves were inclined to cause more intense storms than the sheath regions. For the special case of magnetic clouds associated with eruptions of sigmoidal structures, *Leamon et al.* [2002] found that at least moderate storms were produced in 85% of the studied cases between October 1991 and October 2000.

It has been fully established that the major single factor contributing to the geoeffectiveness of ICMEs is the existence of strong, long duration southward magnetic field component in some part of the ejecta or in the sheath region ahead of the ejecta [*Gonzalez and Tsurutani*, 1987; *Gonzalez et al.*, 1994; *Gonzalez et al.*, 1999 and references therein]. The physical mechanism for the solar wind energy transfer into the magnetosphere is magnetic reconnection between the strong southward interplanetary magnetic field and the northward dipole field of the Earth [*Dungey*, 1961]. Interconnection of interplanetary fields and magnetospheric dayside fields leads to enhanced reconnection of fields on the nightside with concomitant deep injection of plasma sheet plasma in the nightside leading to the enhanced storm-time ring current [*Gonzalez et al.*, 1999]. The speed and density of ICMEs also contribute to their geoeffectiveness [*Burton et al.*, 1975]. The effect of speed comes through the solar wind electric field, which controls the merging rate of the magnetic field lines at the boundary of the magnetosphere. The overall contribution to the storm strength is not, however, large because speed varies much less than the strength of the southward magnetic field [*Crooker*, 2000]. *Wu and Lepping* [2002b] found that the role of the magnetic cloud speed in storm prediction is not important, except for high-speed events. For a sample of 15 events with speeds between 600 and 750 km/s they obtained a correlation coefficient of 0.99 between Dst and the southward magnetic field component. The density of the ejecta has been found to increase the storm strength indirectly through the increase of the magnetospheric plasma density and consequently increase of the ring current [*Crooker*, 2000]. A factor contributing to the geoeffectiveness of ICMEs may be the occurrence of multiple CMEs in quick succession interacting with each other in the interplanetary medium and leading to strengthening of the magnetic field through compression. *Xue et al.* [2005] found that half (4) of the great storms (Dst $\leq$ –200 nT) during the years 2000–2001 were related to successive halo CMEs.

A southward magnetic field component can be embedded both in the sheath region between the shock and the ejecta and within the ejecta itself [*Tsurutani et al.*, 1990; *Gonzalez et al.*, 1999; *Gonzalez et al.*, 2001]. *Cane et al.* [2000] found that in the 86 ejecta identified in the ascending phase of the present solar cycle, the strongest southward magnetic fields were embedded in the ejecta in most of the cases. A strong southward magnetic field in the sheath region was significant in less than 25% of the events. Studying the geoeffectiveness

of magnetic clouds only, *Wu and Lepping* [2002a] found that in 6 cases out of 34 events the sheath region was solely responsible for inducing a storm, and that the leading field of the magnetic cloud was the most efficient in producing storms related to magnetic clouds. *Zhang et al.* [2004] obtained results in agreement with *Wu and Lepping* [2002a]. These statistics may, however, be dependent on the parameter used for defining the geomagnetic storm. *Huttunen et al.* [2002] have shown that the sheath regions generate more storms defined by using the Kp index, while the ejecta generate more Dst storms.

The primary requirement of the presence of a strong and long-duration southward magnetic field component for an interplanetary disturbance to produce a storm can explain why magnetic clouds should be particularly geoeffective. If the flux rope structure corresponding to the magnetic cloud is moving with its axis in the direction of the ecliptic plane, the slowly rotating helical out-of-ecliptic magnetic field provides favorable conditions for a storm to occur. *Bothmer and Rust* [1997] studied the magnetic field configurations of magnetic clouds over a long time period from 1965 to 1993. Magnetic clouds were classified according to their field rotation in the direction normal to the ecliptic (from north to south, or from south to north) and according to the azimuth direction of their magnetic field near the cloud center with respect to the ecliptic. They found a dominance of south-to-north clouds for the years 1974 to 1981 and a dominance of north-to-south clouds before ~1970 and after ~1981. The results of *Huttunen et al.* [2005] confirmed the solar cycle evolution of the leading polarity of magnetic clouds found by *Bothmer and Rust* [1997]. These results have also been confirmed by *Li and Luhman* [2004]. Their results suggested that the polarity change of magnetic clouds occurs during the later declining phase of the solar cycle. *Li and Luhman* [2004] found no substantial difference in geoeffectiveness of magnetic clouds with the two types of magnetic field configurations, and that regardless of the polarity, magnetic clouds are important intense storm producers.

Finally, it should be noted that the discussion in this paper has been limited to the geoeffectiveness of CMEs and their interplanetary counterparts. Although CMEs certainly are the most important sources of intense non-recurrent geomagnetic storms, they are not the primary source of geomagnetic activity even during solar maximum. It is well known that other factors, such as high speed solar wind streams, also contribute to geomagnetic storm occurrence, particularly at the time of solar minimum. The contributions of various solar wind components to geomagnetic activity at different phases of the solar cycle have been extensively studied [e.g., *Richardson et al.*, 2002 and references therein].

4. ENERGETIC PARTICLES AS PROBES OF ICMES

As briefly described in Section 2, the main manifestations of the geoeffectiveness of energetic particles are the radiation hazard to man and technological systems in space and interactions with the upper atmosphere [e.g., *Lanzerotti*, 2001]. Energetic particles (≳1 MeV) are not considered significant in commencing geomagnetic storms. However, for a long time cosmic rays have been known to provide signatures of geomagnetic storms [*Forbush*, 1938]. In the following, some possibilities are outlined to use energetic particles as probes of certain properties of CMEs and their interplanetary counterparts, the actual causes of the storms.

Association of energetic particle events with coronal mass ejections is well known. The current main paradigm is that all the large solar energetic particle events are associated with CMEs and that the particles are accelerated in the CME-driven shocks [e.g., *Reames*, 1999 and references therein]. Alternative views have also been presented [e.g., *Klein and Trottet*, 2001; *Torsti et al.*, 2001; and references therein] stressing the contribution of solar flares in high-energy particle production also in large SEP events. Shock acceleration of electrons, in particular, has been studied, e.g., by *Klassen et al.* [2002]. Various properties of solar energetic particles and their relations to flares and coronal mass ejections have been discussed in several papers in this monograph.

SEP events are often associated with fast and wide CMEs. Based on observations of decameter-hectometric type II radio bursts, *Gopalswamy et al.* [2001] concluded that fast but narrow CMEs are not efficient accelerators of particles. For a sample of 41 major SEP events *Gopalswamy et al.* [2003] gave the average CME initial speed of 1455 km/s. *Kocharov et al.* [2001] have, however, pointed out that in addition to the initial speed the near-Sun dynamics of a CME may also be important, SEP producing CMEs being those with fast acceleration close to the Sun. In agreement with *Kahler* [2001], *Gopalswamy et al.* [2003] found that the SEP intensity correlates with the CME speed, but for a given speed the intensity variation can be several orders of magnitude from event to event. In a more recent study, *Gopalswamy et al.* [2004] found that the correlation between the large SEP event intensities and the speed of associated CMEs is much better for events with preceding large-scale CMEs. Overall, however, the SEP intensity does not seem to be a reliable signature, which could be used, e.g., as an early warning of a fast potentially geoeffective ejecta. Additional difficulties may arise from the possible flare contributions to SEP intensities. As the SEP-associated CMEs, also the flare-associated CMEs tend to be in general faster than other CMEs [*Moon et al.*, 2002]. Therefore, statistically most of the CMEs producing SEPs are also associated with flares, and it is difficult to untangle the contributions from flare and shock sources [*Gopalswamy*, 2004].

Solar energetic particles could be one method to verify, whether an observed halo CME is frontside or backside. The intensity-time profiles of the SEP events can help to indicate the onset time and the location of the event on the solar disk, and to relate shocks and ICMEs with specific solar events. However, energetic particles do not give any information on the possible existence of the important southward magnetic field within the ejecta. Furthermore, while only fast and wide CMEs produce large SEP events, many slow halo CMEs are known to be geoeffective. The classical case is the January 1997 CME causing a major geomagnetic storm with no detectable particle fluxes in interplanetary space. On the other hand, *Cane et al.* [2000] have noted that even the absence of energetic particles can provide an upper limit for the CME transit speed from the Sun to the Earth, because CMEs, which create fast (>650 km/s) interplanetary shocks always have associated energetic particle enhancements.

It has been shown in several studies that solar energetic particles are useful in identifying ICMEs [e.g., *Richardson and Cane*, 1993; *Cane and Richardson*, 2003] and probing their magnetic topology [*Kahler and Reames*, 1991; *Reames*, 1997; *Richardson*, 1997 and references therein]. Identification of the magnetic topology of ICMEs would be useful in the sense that complex ejecta with disordered magnetic structure do not necessarily cause major magnetic storms, while magnetic clouds are often geoeffective. There is evidence of rapid access of solar energetic particles to the interiors of magnetic clouds [*Kahler and Reames*, 1991] indicating that the cloud field lines are connected to the Sun. *Richardson and Cane* [1996] investigated particle flows in 39 SEP events, which occurred while the observing spacecraft was inside an ejecta. They observed unusual particle flows from the east, in contrast to observations outside ejecta with consistent flows along the near-Parker spiral. Bidirectional flows were observed in ~70% of the events inside ejecta, while less than 10% of the events outside ejecta showed such signatures. *Richardson and Cane* [1996] interpreted the unusual flows as evidence of the presence of looped field lines rooted to the Sun in at least a subset of the ejecta. Based on ion composition measurements by the Wind spacecraft at energies up to ~MeV/n, *Mazur et al.* [1998] concluded that in 4 out of 13 magnetic clouds studied, ions from impulsive solar flares arrived at 1 AU while the spacecraft was inside the cloud. They concluded that in these cases the magnetic field lines of the cloud were connected to the flaring region at the Sun. Recently, *Torsti et al.* [2004] studied the anisotropy of 17–22 MeV proton fluxes from SOHO/ERNE in the May 2-3, 1998 SEP event. While the spacecraft was inside a magnetic cloud they found very high intensity of protons in the magnetic field direction compared to the perpendicular direction. The result indicated that inside the magnetic flux rope structure, the parallel mean free path of the protons was at least 10 AU providing a "highway" for proton transport inside the cloud.

Specific signatures of interplanetary CMEs and the associated shocks are the sudden decreases of cosmic ray intensities. These signatures are seen not only in space, but also on ground by neutron monitors [*Cane et al.*, 1996; *Kudela et al.*, 2000], and are considered as a reliable indication of the presence of an ejecta. Cosmic ray intensity starts to decrease after the passage of the shock (if it exists) and further decreases when the ejecta is encountered [*Cane*, 2000], leading to the classical two-step Forbush decrease. The time profile of the cosmic ray depression and the percentage of intensity decrease depend on which part of the ejecta is intercepted or if only the longitudinally more extensive shock is encountered. It has also been demonstrated by *Badruddin* [2002] that quasi-parallel shocks are more effective in producing Forbush decreases than quasi-perpendicular shocks, and that the turbulence in the sheath region is an important factor. *Cane et al.* [1997] identified 84 short-term decreases of ≥4% from Helios 1 and 2 observations of >60 MeV/n particles, and found that 88% were associated with an ejecta and 70% of these with a shock. *Cane et al.* [1997] found no difference in the particle response to ejecta with or without magnetic cloud signatures. They also concluded that magnetic clouds may be a substructure of ejecta and that whether or not the flux rope geometry is seen depends on where the ejecta is intercepted by the observing spacecraft.

As noted above, cosmic rays and SEPs can be useful for probing the ICMEs. In addition to cosmic rays and SEPs, there are energetic particles in the interplanetary space with still another origin. CMEs moving faster than the local Alfven speed can drive fast mode magnetohydrodynamic shocks in the interplanetary medium. These shocks are precursors of the arrival of the ICMEs. The shocks can accelerate particles locally, which are observed in association with the passage of the shock past a spacecraft. These particle enhancements are known as energetic storm particle (ESP) events [*Palmeira et al.*, 1971; *Datlowe*, 1972]. ESPs are observed as significant intensity enhancements of protons up to some tens of MeV, and occasionally even in the >100 MeV range [*Reames*, 1997]. ESPs can be distinguished from SEPs relatively easily, because they do not show the energy dispersion typical for SEPs. The proton intensity can rise several hours before the arrival of interplanetary shock waves, and are characterized by a steepening of the spectra at energies ≥1 MeV. Detection of ESPs can be a forewarning of approaching shocks and their drivers. The arrival of the drivers themselves can be identified as abrupt decreases of intensities of shock-accelerated energetic particles, when the observing spacecraft enters the ejecta. In events, where ESPs are observed, they can serve as an additional source of information in identfying the potentially geoeffective ICMEs.

5. STORM FORECASTING BY USING ENERGETIC PARTICLE OBSERVATIONS

Non-recurrent geomagnetic storms are produced 1–4 days after the launch of an Earth-directed CME from the Sun. After the launch of the CME, the goals of the storm prediction are to answer the following questions: (i) will the CME cause a geomagnetic storm, (ii) when exactly will the storm commence, and (iii) how strong the storm will be. The problems involved in the prediction are: will the ICME actually encounter the Earth, when will it arrive and at what speed, and how strong the associated southward magnetic field will be and what duration.

As discussed in Section 3, halo CMEs are the most probable candidates for causing major geomagnetic storms. Relying only on halo CME observations would, however, lead to a high rate of false alarms (cf. Figure 1). A number of sophisticated models based on simulations of magnetohydrodynamic shocks in the interplanetary medium or real time observations of interplanetary magnetic fields and solar wind properties have been developed to reach the goals (i)-(iii). Others rely on solar observations, such as evolution of large-scale solar magnetic fields or detailed observations of CME properties. In the following, however, only two simple methods based on observations of energetic particles by two different spacecraft at L1 are briefly described to demonstrate the feasibility of energetic particle observations for storm forecasting.

Smith et al. [2004] used ACE/EPAM observations from February 1998 to December 2000 to evaluate the potential of energetic ion enhancements for forecasting geomagnetic storms. Storms with both transient and high speed solar wind stream origins were studied. *Smith et al.* [2004] used proton data in the relatively low-energy 47–65 keV channel. Such protons are expected to be mainly accelerated by shocks in the interplanetary medium. An energetic ion enhancement event was defined as a gradual rise in the flux to a level that exceeded 10^4 particle flux units. They compared the occurrence of energetic ion enhancement events and geomagnetic storms defined by the Kp index. By defining a threshold level of $\geq 1 \times 10^5$ particle flux units, *Smith et al.* [2004] found that 95% of such energetic ion enhancement events were followed by activity with Kp > 4. At the same threshold, 80% hit rate (correct prediction) of large storms (Kp ≥ 7) was reached, but with the corresponding false alarm rate of 52%. The nature of the interplanetary driver, whether transient or high speed stream, was found to be useful in lowering the false alarm rate. All but one of the Kp ≥ 7 events were associated with transients.

Valtonen et al. [2005] studied associations of geomagnetic storms, coronal mass ejections, and energetic particle events with the goal of finding specific particle signatures, which

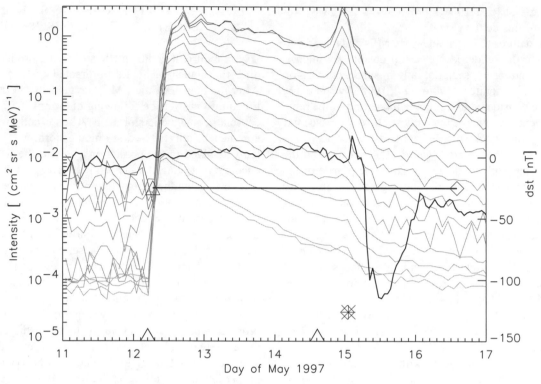

Figure 2. An example of a clear case of a solar energetic particle event and an energetic storm particle event associated with a full halo CME. The thin grey-scale curves are the proton intensities from 1 to 43 MeV. The triangles on the time axis give the SEP and ESP event onset times. The thick black curve shows the time behavior of the Dst index. The coincident diamond and asterisk indicate the time of arrival of the shock as derived from two sets of solar wind data. For details, see *Valtonen et al.* [2005].

could be used to predict the CME geoeffectiveness. They used SOHO/ERNE proton observations in the range 1–100 MeV from August 1996 to July 2000 excluding the period June–October 1998, when SOHO data were not available. SEP and ESP events were searched for from the SOHO/ERNE data and correlated with halo CME observations and magnetic storms defined by the Dst index. The onset of the ESP events was assumed to indicate the arrival of shocks at L1. As an example, the event of May 12, 1997 is shown in Figure 2 demonstrating a clear case of SEP and ESP events. In Figure 2, proton intensities in the range 1–43 MeV are shown by the thin curves and the Dst index by the thick curve. The onset of the SEP and ESP events are indicated by the triangles on the time axis. In accordance with *Gosling et al.* [1991], *Valtonen et al.* [2005] found that the most intense storms (average Dst = −115 nT in 24 storms) were caused by events, which were associated with both a halo CME and an interplanetary shock. By combining particle and CME observations it was possible to improve the identification of intense storms before the associated ICME reached the Earth. Moreover, *Valtonen et al.* [2005] showed that the time difference between the onsets of the SEP and ESP

events (Figure 2) provides a proxy for the shock transit time from the Sun to 1 AU, and that this transit time correlates with the storm strength. These results demonstrate the feasibilty of energetic particle observations as a source of information for geomagnetic storm forecasting.

6. SUMMARY

Coronal mass ejections are believed to be the principal source of major non-recurrent geomagnetic storms. Halo CMEs form only a small fraction (~3.5%) of all CMEs. The majority of geoeffective CMEs are launched from the solar latitude range ±(10°–30°) concentrating in the longitude range ±30° with respect to central meridian. Variable results have been obtained of the geoeffectiveness of halo CMEs. The fraction of geoeffective frontside halo CMEs range from less than 40% up to 100% with a possible solar cycle dependence.

Interplanetary CMEs can be identified by a number of in situ signatures, but no single signature exists, which is consistently always present and could be used for a definite identification of an ICME. Among the various types of

ICMEs, magnetic clouds are considered to be particularly geoeffective. Estimates of the fraction of magnetic clouds of all ICMEs range from 16% to 100% possibly depending on the phase of the solar cycle. Magnetic clouds dominate intense storm occurrences, but altogether they are the source of geomagnetic activity in only about 1/3 of the events.

The major single factor contributing to the geoeffectiveness of ICMEs is the existence of a strong, long duration southward magnetic field in some part of the ejecta or the sheath region ahead of the ejecta. The speed and density of ICMEs have only secondary contributions to their geoeffectiveness. Magnetic clouds change polarity during the solar cycle, but there seems to be no difference in the geoeffectiveness of the clouds with the two polarities.

Solar energetic particle events are usually associated with fast and wide CMEs. SEP intensity correlates with the CME speed, but with a given speed the intensity variation can be several orders of magnitude from event to event. The correlation improves, if the preconditioning of the ambient medium overlying the source region by preceding CMEs is taken into account. SEPs are useful in identifying CMEs and probing the large-scale magnetic topology of ICMEs. There are indications that the field lines of magnetic clouds are connected to the Sun giving a rapid access of SEPs to the interior of the clouds and along the field lines to 1 AU. Energetic particles and cosmic ray depressions can be used to reliably indicate the presence of interplanetary shocks and ejecta, and can provide information on the structure and longitudinal spread of the ejecta. In addition to SEPs, energetic storm particles can be used to probe interplanetary shocks. The arrival of shocks and ejecta can be identified by the behavior of the ESP intensities.

Although coronal mass ejections, geomagnetic storms and production of energetic particles are different physical phenomena, certain possibilities exist to use observations of energetic particles for alerts or even forecasting major geomagnetic storms. However, the full potential of energetic particle observations for storm forecasting has still to be demonstrated.

For space weather applications it would be relevant to be able to predict the occurrence of major geomagnetic storms well (at a minimum a few hours) before their commencement. The basic issues to be resolved in order to successfully predict geoeffectiveness of coronal mass ejections include: what is the initial magnetic configuration of the CME, what is the effect of the interplanetary environment into which the CME is launched and how will the properties of the ICME change on its way from the Sun to the Earth, will the ejecta (including the shock and the sheath region) encounter the Earth, and ultimately, what will be the magnetic configuration of the ICME at the time of encounter. To work through these problems towards reliable predictions requires detailed observations of the Sun and real-time monitoring of the conditions and response of the interplanteary space to solar disturbances.

Acknowledgments. The use of the CME catalog generated and maintained by the Center for Solar Physics and Space Weather, The Catholic University of America in cooperation with the Naval Research Laboratory and NASA is acknowledged. The use of Dst index from the Dst index service, WDC for Geomagnetism, Kyoto is acknowledged. SOHO is an international cooperation project between ESA and NASA. The reviewers are thanked for useful comments.

REFERENCES

Badruddin, Shock orientations, magnetic turbulence and Forbush decreases, *Sol. Phys.*, 209, 195-206, 2002.

Baker, D.N., What is space weather?, *Adv. Space Res.*, 22, 7-16, 1998.

Baker, D.N., Effects of the Sun on the Earth's environment, *J. Atmos. Solar Terr. Phys.*, 62, 1669-1681, 2000.

Baker, D.N., T.I. Pulkkinen, X. Li, *et al.*, Coronal mass ejections, magnetic clouds, and relativistic magnetospheric electron events: ISTP, *J. Geophys. Res.*, 103, 17,279-17,291, 1998.

Bothmer, V., Solar corona, solar wind structure and solar particle events, *Proc. of the ESA Workshop on Space Weather, ESA WPP*-155, pp. 117-126, 1999.

Bothmer, V., and D.M. Rust, The field configuration of magnetic clouds and the solar cycle, in *Coronal Mass Ejections, Geophys. Monogr. Ser.*, vol. 99, edited by N. Crooker, J.A. Joselyn, and J. Feynman, pp. 139-146, AGU, Washington, D.C., 1997.

Brueckner, G.E., J.-P. Delaboudiniere, R.A. Howard, *et al.*, Geomagnetic storms caused by coronal mass ejections (CMEs): March 1996 through June 1997, *Geophys. Res. Lett.*, 25, 3019-3022, 1998.

Brueckner, G.E., R.A. Howard, M.J. Koomen, *et al.*, The Large Angle Spectroscopic Coronagraph (LASCO), *Sol. Phys.*, 162, 357-402, 1995.

Burlaga, L.F.E., Magnetic clouds, in *Physics of the Inner Heliosphere II, Physics and Chemistry in Space*, vol. 21, edited by R. Schwenn and E. Marsch, pp. 1-22, Springer-Verlag, Berlin Heidelberg, 1991.

Burlaga, L.F., R.M. Skoug, C.W. Smith, D.F. Webb, T.H. Zurbuchen, and A. Reinard, Fast ejecta during the ascending phase of solar cycle 23: ACE observations, 1998-1999, *J. Geophys. Res.*, 106, 20,957-20,977, 2001.

Burton, R.K., R.L. McPherron, and C.T. Russell, An empirical relationship between interplanetary conditions and Dst, *J. Geophys. Res.*, 80, 4204-4214, 1975.

Cane, H.V., Coronal mass ejections and Forbush decreases, *Space Sci. Rev.*, 93, 55-77, 2000.

Cane, H.V., and I.G. Richardson, Interplanetary coronal mass ejections in the near-Earth solar wind during 1996-2002, *J. Geophys. Res.*, 108, 1156, doi: 10.1029/2002JA009817, 2003.

Cane, H.V., I.G. Richardson, and O.C.S. Cyr, Coronal mass ejections, interplanetary ejecta and geomagnetic storms, *Geophys. Res. Lett.*, 27, 3591-3594, 2000.

Cane, H.V., I.G. Richardson, and T.T. von Rosenvinge, Cosmic ray decreases: 1964-1994, *J. Geophys. Res.*, 101, 21,561-21,572, 1996.

Cane, H.V., I.G. Richardson, and G.Wibberenz, Helios 1 and 2 observations of particle decreases, ejecta, and magnetic clouds, *J. Geophys. Res.*, 102, 7075-7086, 1997.

Crooker, N.U., Solar and heliospheric geoeffective disturbances, *J. Atmos. Solar Terr. Phys.*, 62, 1071-1085, 2000.

Daglis, I.A., W.I. Axford, E.T. Sarris, S. Livi, and B. Wilken, Particle acceleration in geospace and its association with solar events, *Sol. Phys.*, 172, 287-296, 1997.

Datlowe, D., Association between interplanetary shock waves and delayed solar particle events, *J. Geophys. Res.*, 77, 5374-5384, 1972.

Dungey, J.W., Interplanetary magnetic field and the auroral zones, *Phys. Rev. Lett.*, 6, 47-48, 1961.

Forbush, S.E., On world-wide changes in cosmic-ray intensity, *Phys. Rev.*, 54, 975-988, 1938.

Gonzalez, W.D., A.L. Clúa de Gonzalez, J.H.A. Sobral, A. Dal Lago, and L.E. Vieira, Solar and interplanetary causes of very intense geomagnetic storms, *J. Atmos. Solar Terr. Phys.*, 63, 403-412, 2001.

Gonzalez, W.D., J.A. Joselyn, Y. Kamide, H.W. Kroehl, G. Rostoker, B.T. Tsurutani, and V.M. Vasyliunas, What is a geomagnetic storm?, *J. Geophys. Res.*, 99, 5771-5792, 1994.

Gonzalez, W.D., and B.T. Tsurutani, Criteria of interplanetary parameters causing intense magnetic storms (D_{st} −100 nT), *Planet. Space Sci.*, 35, 1101-1109, 1987.

Gonzalez, W.D., B.T. Tsurutani, and A.L. Clúa de Gonzalez, Interplanetary origin of geomagnetic storms, *Space Sci. Rev.*, 88, 529-562, 1999.

Gopalswamy, N., A global picture of CMEs in the inner heliosphere, in *The Sun and the Heliosphere as an Integrated System, Astrophysics and Space Science Library*, edited by G. Poletto and S. Suess, pp. 201-250, Kluwer, Boston, 2004.

Gopalswamy, N., A. Lara, R.P. Lepping, M.L. Kaiser, D. Berdichevsky, and O.C. St. Cyr, Interplanetary Acceleration of Coronal Mass Ejections, *Geophys. Res. Lett.*, 27, 145-148, 2000.

Gopalswamy, N., S. Yashiro, M.L. Kaiser, R.A. Howard, and J.-L. Bougeret, Characteristics of coronal mass ejections associated with long-wavelength type II radio bursts, *J. Geophys. Res.*, 106, 29,219-29,229, 2001.

Gopalswamy, N., S. Yashiro, S. Krucker, G. Stenborg, and R.A. Howard, Intensity variation of large solar energetic particle events associated with coronal mass ejections, *J. Geophys. Res.*, 109, A12105, doi: 10.1029/ 2004JA010602, 2004.

Gopalswamy, N., S. Yashiro, A. Lara, M.L. Kaiser, B.J. Thompson, P.T. Gallagher, and R.A. Howard, Large solar energetic particle events of cycle 23: A global view, *Geophys. Res. Lett.*, 30(12), 8015, doi: 10.1029/ 2002GL016435, 2003.

Gosling, J.T., D.J. McComas, J.L. Phillips, and S.J. Bame, Geomagnetic activity associated with earth passage of interplanetary shock disturbances and coronal mass ejections, *J. Geophys. Res.*, 96, 7831-7839, 1991.

Huttunen, K.E.J., H.E.J. Koskinen, and R. Schwenn, Variability of magnetospheric storms driven by different solar wind perturbations, *J. Geophys. Res.*, 107, 20-1, doi: 10.1029/2001JA900171, 2002.

Huttunen, K.E.J., R. Schwenn, V. Bothmer, and H.E.J. Koskinen, Properties and geoeffectiveness of magnetic clouds in the rising, maximum and early declining phases of solar cycle 23, *Ann. Geophys.*, 23, 625-641, 2005.

Kahler, S.W., The correlation between solar energetic particle peak intensities and speeds of coronal mass ejections: Effects of ambient particle intensities and energy spectra, *J. Geophys. Res.*, 106, 20,947-20,955, 2001.

Kahler, S.W., and D.V. Reames, Probing the magnetic topologies of magnetic clouds by means of solar energetic particles, *J. Geophys. Res.*, 96, 9419-9424, 1991.

Klassen, A., V. Bothmer, G. Mann, M.J. Reiner, S. Krucker, A. Vourlidas, and H. Kunow, Solar energetic electron events and coronal shocks, *Astron. Astrophys.*, 385, 1078-1088, 2002.

Klein, K., and G. Trottet, The origin of solar energetic particle events: Coronal acceleration versus shock wave acceleration, *Space Sci. Rev.*, 95, 215-225, 2001.

Klein, L.W., and Burlaga, L.F., Interplanetary magnetic clouds at 1 AU, *J. Geophys. Res.*, 87, 613-624, 1982.

Kocharov, L., J. Torsti, O.C. St. Cyr, and T. Huhtanen, A relation between dynamics of coronal mass ejections and production of solar energetic particles, *Astron. Astrophys.*, 370, 1064-1070, 2001.

Koskinen, H.E.J., and K.E.J. Huttunen, Space Weather: From solar eruptions to magnetospheric storms, 2005 (*this monograph*).

Kudela, K., M. Storini, M.Y. Hofer, and A. Belov, Cosmic rays in relation to space weather, *Space Sci. Rev.*, 93, 153-174, 2000.

Lanzerotti, L.J., Space weather effects on technologies, in *Space Weather, Geophys. Monogr. Ser.*, vol. 125, edited by P. Song, H.J. Singer, and G.L. Siscoe, pp. 11-22, AGU, Washington, D.C., 2001.

Lanzerotti, L.J., D.S. Sayers, L.V. Medford, C.G. Maclennan, R.P. Lepping, and A. Szabo, Response of large-scale geoelectric fields to identified interplanetary disturbances and the equatorial ring current, *Adv. Space Res.*, 26, 21-26, 2000.

Leamon, R.J., R.C. Canfield, and A.A. Pevtsov, Properties of magnetic clouds and geomagnetic storms associated with eruption of coronal sigmoids, *J. Geophys. Res.*, 107, 1234, doi: 10.1029/2001JA000313, 2002.

Leske, R.A., R.A. Mewaldt, E.C. Stone, and T.T. von Rosenvinge, Observations of geomagnetic cutoff variations during solar energetic particle events and implications for the radiation environment at the Space Station, *J. Geophys. Res.*, 106, 30,011-30,022, 2001.

Li, X., D.N. Baker, S. Elkington, *et al.*, Energetic particle injections in the inner magnetosphere as a response to an interplanetary shock, *J. Atmos. Solar Terr. Phys.*, 65, 233-244, 2003.

Li, Y., and J. Luhman, Solar cycle control of the magnetic cloud polarity and the geoeffectiveness, *J. Atmos. Solar Terr. Phys.*, 66, 323-331, 2004.

Lorentzen, K.R., J.E. Mazur, M.D. Looper, J.F. Fennell, and J.B. Blake, Multisatellite observations of MeV ion injections during storms, *J. Geophys. Res.*, 107, 1231, doi:10.1029/2001JA000276, 2002.

Mazur, J.E., G.M. Mason, J.R. Dwyer, and T.T. von Rosenvinge, Solar energetic particles inside magnetic clouds observed with the Wind spacecraft, *Geophys. Res. Lett.*, 25, 2521-2524, 1998.

Moon, Y.-J., G.S. Choe, H. Wang, Y.D. Park, N. Gopalswamy, G. Yang, and S. Yashiro, A statistical study of two classes of coronal mass ejections, *Astrophys. J.*, 581, 694-702, 2002.

Mulligan, T., and Russell, C.T., Multispacecraft modeling of the flux rope structure of interplanetary coronal mass ejections: Cylindrically symmetric versus nonsymmetric topologies, *J. Geophys. Res.*, 106, 10581-10596, 2001.

Neugebauer, M., and R. Goldstein, Particle and field signatures of coronal mass ejections in the solar wind, in *Coronal Mass Ejections, Geophys. Monogr. Ser.*, vol. 99, edited by N. Crooker, J.A. Joselyn, and J. Feynman, pp. 245-251, AGU, Washington, D.C., 1997.

Palmeira, R.A.R., F.R. Allum, and U.R. Rao, Low-energy proton increases associated with interplanetary shock waves, *Sol. Phys.*, 21, 204-224, 1971.

Reames, D.V., Energetic particles and the structure of coronal mass ejections, in *Coronal Mass Ejections, Geophys. Monogr. Ser.*, vol. 99, edited by N. Crooker, J. A. Joselyn, and J. Feynman, pp. 217-226, AGU, Washington, D.C., 1997.

Reames, D.V., Particle acceleration at the Sun and in the heliosphere, *Space Sci. Rev.*, 90, 413-491, 1999.

Reeves, G.D., K.L. McAdams, R.H.W. Friedel, and T.P. O'Brien, Acceleration and loss of relativistic electrons during magnetic storms, *Geophys. Res. Lett.*, 30(10), 36-1 doi: 10.1029/2002GL016513, 2003.

Richardson, I.G., Using energetic particles to probe the magnetic topology of ejecta, in *Coronal Mass Ejections, Geophys. Monogr. Ser.*, vol. 99, edited by N. Crooker, J.A. Joselyn, and J. Feynman, pp. 189-196, AGU, Washington, D.C., 1997.

Richardson, I.G., and H.V. Cane, Signatures of shock drivers in the solar wind and their dependence on the solar source location, *J. Geophys. Res.*, 98, 15,295-15,304, 1993.

Richardson, I.G., and H.V. Cane, Particle flows observed in ejecta during solar event onsets and their implication for the magnetic field topology, *J. Geophys. Res.*, 101, 27,521-27,532, 1996.

Richardson, I.G., and H.V. Cane, Identification of interplanetary coronal mass ejections at 1 AU using multiple solar wind plasma composition anomalies, *J. Geophys. Res.*, 109, A09104, doi: 10.1029/2004JA010598, 2004.

Richardson, I.G., H.V. Cane, and E.W. Cliver, Sources of geomagnetic activity during nearly three solar cycles (1972-2000), *J. Geophys. Res.*, 107, 8-1, doi:10.1029/ 2001JA000504, 2002.

Smith, Z., W. Murtagh, and C. Smithtro, Relationship between solar wind low-energy energetic ion enhancements and large geomagnetic storms, *J. Geophys. Res.*, 109, A01110, doi: 10.1029/ 2003JA010044, 2004.

St. Cyr, O.C., R.A. Howard, N.R. Sheeley, *et al.*, Properties of coronal mass ejections: SOHO LASCO observations from January 1996 to June 1998, *J. Geophys. Res.*, 105, 18,169-18,185, 2000.

Torsti, J., L. Kocharov, D.E. Innes, J. Laivola, and T. Sahla, Injection of energetic protons during solar eruption on 1999 May 9: Effect of flare and coronal mass ejection, *Astron. Astrophys.*, 365, 198-203, 2001.

Torsti, J., E. Riihonen, and L. Kocharov, The 1998 May 2-3 magnetic cloud: An interplanetary "highway" for solar energetic particles observed with SOHO/ERNE, *Astrophys. J.*, 600, L83-L86, 2004.

Torsti, J., E. Valtonen, M. Lumme, *et al.*, Energetic Particle Experiment ERNE, *Sol. Phys.*, 162, 505-531, 1995.

Tsurutani, B.T., B.E. Goldstein, E.J. Smith, W.D. Gonzalez, F. Tang, S.I. Akasofu, and R.R. Anderson, The interplanetary and solar causes of geomagnetic activity, *Planet. Space Sci.*, 38, 109-126, 1990.

Tsurutani, B.T., and W.D. Gonzalez, The interplanetary causes of magnetic storms: A review, in *Magnetic Storms, Geophys. Monogr. Ser.*, vol. 98, edited by B.T. Tsurutani, W.D. Gonzalez, Y. Kamide, and J.K. Arballo, pp. 77-89, AGU, Washington, D.C., 1997.

Valtonen, E., Space weather effects on technology, in *Space Weather: The Physics Behind the Slogan, Lecture Notes in Physics*, vol. 656, edited by K. Scherer, H. Fichtner, B. Heber, and U. Mall, pp. 241-274, Springer Verlag, Berlin Heidelberg, doi: 10.1007/b100037, 2005.

Valtonen, E., T. Laitinen, and K. Huttunen-Heikinmaa, Energetic particle signatures of geoeffective coronal mass ejections, *Adv. Space Res.*, 2005 (*in press*).

Wang, Y.M., P.Z. Ye, S. Wang, G.P. Zhou, and J.X. Wang, A statistical study on the geoeffectiveness of Earth-directed coronal mass ejections from March 1997 to December 2000, *J. Geophys. Res.*, 107, 1340, doi: 10.1029/2002JA009244, 2002.

Webb, D.F., E.W. Cliver, N.U. Crooker, O.C. St. Cyr, and B.J. Thompson, Relationship of halo coronal mass ejections, magnetic clouds, and magnetic storms, *J. Geophys. Res.*, 105, 7491-7508, 2000.

Wu, C.-C., and R.P. Lepping, Effects of magnetic clouds on the occurrence of geomagnetic storms: The first 4 years of Wind, *J. Geophys. Res.*, 107, 1314, doi:10.1029/ 2001JA000161, 2002a.

Wu, C.-C., and R.P. Lepping, Effect of solar wind velocity on magnetic cloud-associated magnetic storm intensity, *J. Geophys. Res.*, 107, 1346, doi: 10.1029/2002JA009396, 2002b.

Xue, X.H., Y. Wang, P.Z. Ye, S. Wang, and M. Xiong, Analysis on the interplanetary causes of the great magnetic storms in solar maximum (2000–2001), *Planet. Space Sci.*, 53, 443-457, 2005.

Yashiro, S., N. Gopalswamy, G. Michalek, O.C. St. Cyr, S.P. Plunkett, N.B. Rich, and R.A. Howard, A catalog of white light coronal mass ejections observed by the SOHO spacecraft, *J. Geophys. Res.*, 109, A07105, doi: 10.1029/2003JA010282, 2004.

Yermolaev, Y.I., M.Y. Yermolaev, G.N. Zastenker, L.M. Zelenyi, A.A. Petrukovich, and J.-A. Sauvaud, Statistical studies of geomagnetic storm dependencies on solar and interplanetary events: a review, *Planet. Space Sci.*, 53, 189-196, 2005.

Zhang, J., K.P. Dere, R.A. Howard, and V. Bothmer, Identification of solar sources of major geomagnetic storms between 1996 and 2000, *Astrophys. J.*, 582, 520-533, 2003.

Zhang, J., M.W. Liemohn, J.U. Kozyra, B.J. Lynch, and T.H. Zurbuchen, A statistical study of the geoeffectiveness of magnetic clouds during high solar activity years, *J. Geophys. Res.*, 109, A09101, doi: 10.1029/2004JA010410, 2004.

Zhao, X.P., and D.F. Webb, Source regions and storm effectiveness of frontside full halo coronal mass ejections, *J. Geophys. Res.*, 108, 1234, doi: 10.1029/2002JA009606, 2003.

E. Valtonen, Space Research Laboratory, Department of Physics, University of Turku, FI-20014 Turku University (eino.valtonen@utu.fi)

The Creation of New Ion Radiation Belts Associated With Solar Energetic Particle Events and Interplanetary Shocks

J.E. Mazur, J.B. Blake, and P.L. Slocum

The Aerospace Corporation, Los Angeles, California

M.K. Hudson

Dartmouth College, Hanover, New Hampshire

G.M. Mason

The University of Maryland, College Park, Maryland

We surveyed measurements of the inner magnetosphere during solar cycle 23 for the signature of a particular form of the Sun-Earth connection: the creation of new ion radiation belts in the inner magnetosphere that form during intense solar particle events and associated interplanetary shocks. We extended the database of such events that were previously reported and found that new ion belts appeared after geomagnetic storms in approximately one-quarter of the cases where solar energetic particles were present below an L shell of 4. Suppression of the nominal particle cutoffs was only one requirement for the creation of new belts. The presence of ions up to and including iron clearly indicated their source was solar energetic particles. Ions with larger mass (and therefore rigidity) appeared on lower L shells. Lifetimes varied from event to event, and in the 11/23/2001 event were consistent with a loss due to ionization energy loss in the residual atmosphere.

1. INTRODUCTION

The outer radiation belt has long-been known as a highly dynamic region. The large number of space particle and fields instruments and monitors in geostationary orbit over the years has yielded insights into the processes responsible for the dynamics. Even so, we still lack a unified understanding of the acceleration, loss, and transport processes for the MeV outer zone electrons.

In comparison, the inner magnetosphere (L shells less than about 3) has been probed or monitored much less

Solar Eruptions and Energetic Particles
Geophysical Monograph Series 165
Copyright 2006 by the American Geophysical Union
10.1029/165GM32

frequently. Accordingly, our understanding of its dynamics and knowledge of its extremes is even more limited. We can go back to the first Explorer missions to find observations of large changes to the trapped particle intensity near L~2. For example, *Pizella, McIlwain, and Van Allen* [1962] discussed Explorer-7 measurements using a GM-tube in 1959-1960. They were able to show, using the inclination of the spacecraft in low-Earth orbit, how increases in the trapped particle intensity as large as a factor of 10 occurred in association with solar particle events and geomagnetic storms. The response of the instrument prevented the unique identification of the particles responsible for the signatures (>18 MeV protons and >1.1 MeV electrons). The source of these particles was also not clear: were they newly-injected somehow from the solar particle event, or energized redistributions of previously trapped particles?

More recently, *Lorentzen et al.* [2002] reviewed the published cases of new radiation belts in the slot region and presented recent cases observed on highly-elliptical orbiting and low-Earth orbiting spacecraft. One result of the *Lorentzen et al.* [2002] summary was that there may be several processes responsible for the appearance of new trapped ion and electron populations below L~3. However, most cases were associated with sudden compressions of the magnetosphere during transient interplanetary shocks. This is in contrast to the outer zone, where high-speed solar wind streams are also an important initiator for enhanced trapped particles.

Hudson et al. [2004] modeled the radial transport and trapping of solar energetic particles in shock-related compressions of the magnetosphere in 2001. They also compared the solar wind drivers for a set of 11 solar particle events with associated interplanetary shocks; 8 events resulted in new ion belts below an L shell of 4.

In this paper we address the question of how frequently did new ion belts form in association with solar particle events and shocks in the last solar cycle. We took advantage of the continuous monitoring of the inner magnetosphere made available from low-Earth orbit with the NASA/ SAMPEX mission. With SAMPEX, we have a collection of charged particle instruments and a long-lifetime mission that has afforded a new look at the inner magnetosphere dynamics. We used heavy-ion identification to clearly indicate the source of the new radiation belts and thus removed any question that these particles originated outside the magnetosphere. Some of these events were included in the *Lorentzen et al. and Hudson et al.* studies, but we focused here on the heavy-ion measurements and extended the survey period to cover the entire cycle 23 maximum. We begin with an overview of the instrumentation and analysis, and discuss the results in the context of inner magnetosphere dynamics.

2. HEAVY-ION SURVEY

2.1. Methodology

Protons and electrons are always present throughout the Earth's magnetosphere. These particles thus lack a unique identity and call into question whether a newly observed population of trapped protons and/or electrons was due to a redistribution of previously trapped particles or some other process. *Li et al.* [1993] successfully modeled the 3/24/1991 event with radial transport of outer-zone electrons and solar protons while conserving their first adiabatic invariant, so in that case the resulting particles observed below L~3 could have originated from a pre-existing magnetospheric population and from the solar particle event.

In this survey, we used heavy-ions as a unique identifier for the existence of new radiation belts. We did this because ions above a few MeV/nucleon with Z > 2 are relatively rare in the magnetosphere [*e.g., Hovestadt et al.*, 1978] and because in fact we have found cases where new heavy-ion belts appeared in association with solar particle events and interplanetary shocks. There are heavy-ions that exist between L~2-3 whose source is anomalous cosmic rays (ACR) [*Cummings et al.*, 1993; *Looper et al.*, 1996; *Kleis et al.*, 1995; and *Mazur et al.*, 2000]. The trapped ACR was low enough in intensity near solar maximum that it was not a factor in this analysis of new ion belts.

2.2. Instrumentation

SAMPEX has sampled the low-Earth trapped particle environment since its launch in July 1992 [*Baker et al.*, 1993]. The combination of a long lifetime, spanning more than half a magnetic solar cycle, and high-sensitivity instruments yields a valuable dataset of inner magnetosphere dynamics. This is especially the case because its low-Earth orbit (LEO) cuts across roughly all L shells four times every ~96 minutes. While LEO data cannot be used to determine the particle intensity at the magnetic equator, the observations do allow us to observe changes in the trapped particles that are presumably communicated all along the field lines [*e.g., Kanekal et al.*, 2001].

The Low-energy Ion Composition Analyzer (LICA) provided the ion measurements for this study [*Mason et al.*, 1993]. The instrument uses time-of-flight and residual energy to measure ion composition and intensity from H to Fe over ~0.25 to 5 MeV/nucleon. Its data consist of both count rates as well as pulse-height analyzed events that satisfy the time-of-flight and silicon detector coincidence. The time-tag of each ion event along with spacecraft attitude yields the particle pitch angle, which we used below to verify the presence of locally-mirroring particles in new radiation belts.

2.3. Data Analysis

We constructed a database of LICA pulse-height analyzed events for the analysis of new ion radiation belts in the following manner. We began by selecting the data between L = 1.7 and 4.0 to cover the outer edge of the inner zone and the entire slot region. Even with its triple-coincidence requirement, the LICA sensor had accidentals and other backgrounds from penetrating protons while in the heart of the inner zone, thus our minimum L of 1.7. We placed constraints on minimum energy measured with the solid-state detectors (1.7 MeV) as well as a cut in time-of-flight versus measured energy, all to minimize the background from penetrating protons and at low L. The resulting energy ranges were: He, 0.7-3 MeV/nucleon; O, 1-5 MeV/nucleon; Ne-S, 0.8-4 MeV/nucleon; and Fe, 0.5-3 MeV/nucleon.

We applied these constraints to the data collected from July 1992 to July 2004, a full 12 years of continuous monitoring of the inner magnetosphere. Figure 1 summarizes part of the database that we analyzed below for the presence of new radiation belts. Each point in the figure is the L-shell and arrival time (UT) of an ion in the Ne-S mass range. The actual database contained all ions measured between H-Fe. We used the spacecraft ephemeris to derive the L-shell; this is based on an epoch-driven IGRF magnetic field model. The solid trace at the bottom is the smoothed sunspot number obtained from the NGDC web site (ftp.ngdc.noaa.gov/STP/ SOLAR_DATA/SUNSPOT_NUMBERS/MONTHLY).

There are several features of the data in Figure 1 that tell much about the response of the inner magnetosphere during the solar cycle. First, the incursions of heavy ions below L = 4 during intense solar particle events appeared as vertical swaths with this long time scale. The fact that particles of this energy (and rigidity ~<100 MV assuming typical charge states) had access to below L = 4 is a result of suppressed particle cutoffs during geomagnetic storms (e.g., *Leske et al.*, [2001]). These solar particle events were more frequent on the approach to and during solar maximum, as was to be expected. The other obvious population was seen between L~2 and 3, most apparently near solar minimum in 1996. These were trapped anomalous cosmic rays, and this mass selection included Mg-S ions that may also have been a singly-ionized component of interplanetary ions at solar minimum [*Mazur et al.*, 2000 and references therein]. The trapped ACR gradually decreased in 1998-1999 as a result of their energy loss in the residual atmosphere and the weakening source as solar activity increased.

While the LICA instrument operated for essentially the entire time period in Figure 1, the spacecraft attitude was not optimized to measure locally mirroring particles in the vicinity of the South Atlantic anomaly until mid-1994. We thus concentrated on the 110 months between May 1994 and July 2004 for the survey below. There were relatively few solar particle incursions in the months up to 1994 (the most obvious was 2/21/1994), so the survey includes the bulk of the events of interest.

We searched this heavy-ion database for the appearance of new populations of trapped ions in the L range of 1.7 to 4. This was done by visually scanning the data on time intervals a few days wide and visually noting whether newly trapped ions appeared. The trapped anomalous cosmic rays in Figure 1 amounted to a count or two per week and therefore were not a significant background for the search for solar-related injections. Because if its low-Earth orbit, SAMPEX passes through the vicinity of the South-Atlantic anomaly at these L shells only approximately twice per day. Thus, the orbit constrains our ability to time the appearance of newly trapped ions.

As an example of a new population, we show in Figure 2 the L shell and time of arrival of ions during and for the few

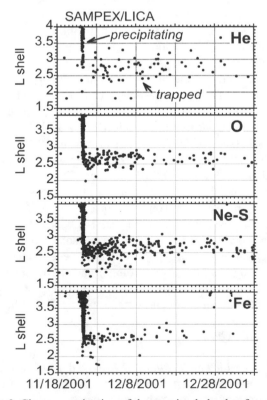

Figure 2. Closer examination of the new ion belts that formed during the 24 November 2001 solar particle event and shock. The LICA instrument is not equally efficient for He and heavies so relative intensities are difficult to estimate from the figure.

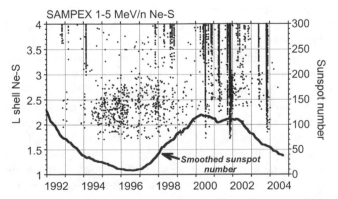

Figure 1. Time history of 0.8-4 MeV/nucleon Ne-S ions as measured on SAMPEX below L = 4. Each symbol is one ion event measured by the LICA sensor. The solid trace is the smoothed sunspot number.

weeks after the 11/23/2001 solar particle event. In this figure we have added the He, O, and Fe species. The trail of ions between L = 2.5 and 3 was clearly associated with the solar particle incursion. Their pitch angles (not shown) were peaked at 90 degrees, indicating locally mirroring particles. These were thus a subset of the entire trapped distribution that mirrored at the SAMPEX altitude.

Finally, we incorporated the hourly Dst index (*ftp.ngdc.noaa.gov/STP/GEOMAGNETIC_DATA/INDICES/ DST/*) into the database to check for the presence of a geomagnetic storm during the solar particle incursions and other events of interest.

2.4. Survey Results

There were 51 incursions of solar particles below L = 4 from May 1994 to June 2004. Every incursion occurred during a geomagnetic storm as indicated by the Dst index. Of these 13 (25%) resulted in detectable trapped ion populations. Table 1 lists the approximate time that ions were first seen below L = 4 in 51 incursions, the date and time of the geomagnetic storm as indicated by the minimun of Dst, as well as details of new ion belts (if applicable).

It was not the case that every solar particle event and storm resulted in a new population of trapped heavy ions. We motivate this point in Figure 3, where the solid trace shows the count rate of >0.7 MeV ions (mostly protons) measured with LICA above the polar caps and the dashed trace shows the Dst index. We used the rate of >0.7 MeV ions to simply indicate the intensity of the solar particle events. A new belt formed after the interplanetary shock of 9/25/2001. Precipitating solar particles appeared again below L = 4 on 10/2/2001 and 10/3/2001 (see Table 1). The storms on 10/1/2001 and 10/3/2001 had more intense ring currents, with correspondingly higher suppression of the solar particle cutoffs, yet there were no signatures of newly trapped populations. The solar particle event that peaked on 10/2/2001 was a factor of 50 less intense at its maximum than the peak on 9/25/2001.

To summarize the findings, for every solar particle incursion we plot in Figure 4 the solar particle event intensity (as indicated by the >0.7 MeV protons) and the minimum Dst index. The dashed line was drawn by eye to suggest a division between incursions with and without new radiation belts.

2.5. Specific Questions

2.5.1. New ion belt location.
We looked in detail at several of the most intense trapped ion events to determine where in L they appeared. Figure 5 shows the mean L shell versus ion atomic number Z for several events. The solid lines guide the eye between species in individual events. The mean location of the trapped ions was lower in L for increasing Z, by an

amount that varied from event to event. A typical full-width at half-maximum was 0.4 L. Even though the locations versus species and event changed by a few tenths of an L shell, they were nevertheless well defined. In this subset the mean L shells were lower for all species in events with larger Dst.

2.5.2. New ion belt lifetime.
Table 1 includes an approximate lifetime for each of the 13 new ion belts determined by visually inspecting the data for pulse-height analyzed events that were clearly associated with the original injection. To better quantify the lifetime in one case, we analyzed in detail the apparent decay of the new ion belts formed in the 11/23/2001 event. Figure 6 plots the number of ions per 4 days versus time along with exponential fits to the data. The figure also lists the e-folding times in days. The decay rates were higher for increasing Z.

3. DISCUSSION

The inner magnetosphere is dynamic. With the long-lifetime of the SAMPEX mission we were able to focus on the details of a particular form of the Sun-Earth connection, namely the creation of new ion belts associated with solar particle events and geomagnetic storms. We found 51 cases where the nominal geomagnetic cutoffs of a few MeV/nucleon solar energetic particles (approximately 100 MV) were suppressed to below L = 4. Solar particles as a consequence were seen to precipitate along these inner magnetosphere field lines to the SAMPEX altitude. These storms were all associated with transient interplanetary shocks. In 13 of these cases, we detected trapped ions in the slot region where there were none previous to the solar particle event and shock/ geomagnetic storm. We conclude that these heavy-ions were tracers of solar particles that were trapped during non-adiabatic processes associated with the geomagnetic storms.

That these were trapped solar particles is clear given (1) the presence of ions up to and including iron, species that are rare in the magnetosphere even at thermal energies, and (2) the coincidence in time between the detection of the new belts and incursions of solar particles below L = 4 during geomagnetic storms. A detailed composition survey of these events is the subject of a future study.

Our detection of a new ion belt appearing in the energy ranges of this study was not determined alone by the geometry factor of the LICA sensor. If that was the case, and if it was equally likely that every incursion below L = 4 led to a new belt, then the division between storms with and without belts would have been a horizontal line in Figure 4, perhaps only dependent on the solar particle intensity at the time of the storm. We suggest that a combination of the solar particle intensity and the cutoff suppression are factors in determining the creation of a new ion belt.

Table 1. Solar particle incursions below L = 4.

Date & time of first ions observed below L = 4 (UT, nearest hour)	Date & time of minimum Dst (UT)	Minimum Dst (nT)	Solar particle event intensity [a]	New ion belt [b]	Approximate new belt lifetime (days) [c]	Mean L shell for trapped iron [d]	New belt intensity [e]
11/7/1997 1:00:00	11/7/1997 5:00:00	−104	5.5e + 03	N			
4/24/1998 2:00:00	4/24/1998 8:00:00	−87	1.7e + 02	N			
5/4/1998 3:00:00	5/4/1998 6:00:00	−216	2.5e + 02	N			
8/26/1998 11:00:00	8/27/1998 10:00:00	−188	1.1e + 03	N			
9/24/1998 22:00:00	9/25/1998 10:00:00	−233	2.2e + 02	N			
9/30/1998 23:00:00	10/1/1998 4:00:00	−69	9.4e + 03	N			
11/6/1998 12:00:00	11/6/1998 16:00:00	−82	3.0e + 03	N			
11/8/1998 0:00:00	11/8/1998 7:00:00	−148	8.8e + 02	N			
11/8/1998 13:00:00	11/7/1998 17:00:00	−92	1.2e + 03	N			
11/9/1998 7:00:00	11/9/1998 19:00:00	−145	4.4e + 01	N			
4/6/2000 18:00:00	4/7/2000 1:00:00	−321	1.7e + 02	Y	92		0.63
5/17/2000 5:00:00	5/17/2000 6:00:00	−88	6.6e + 02	N			
5/24/2000 3:00:00	5/24/2000 9:00:00	−147	2.7e + 01	N			
6/8/2000 10:00:00	6/8/2000 20:00:00	−87	7.5e + 03	N			
7/14/2000 16:00:00	7/15/2000 22:00:00	−300	6.2e + 03	Y	24	2.0	0.12
8/12/2000 9:00:00	8/12/2000 10:00:00	−237	3.1e + 01	N			
9/12/2000 16:00:00	9/12/2000 19:00:00	−70	1.5e + 03	N			
9/15/2000 21:00:00	9/16/2000 3:00:00	−30	1.0e + 03	N			
9/16/2000 21:00:00	9/17/2000 0:00:00	−57	5.8e + 02	N			
9/17/2000 20:00:00	9/18/2000 0:00:00	−172	9.4e + 02	N			
11/9/2000 2:00:00	11/10/2000 13:00:00	−104	2.0e + 04	Y[f]	12	3.1	0.04
11/26/2000 16:00:00	11/27/2000 2:00:00	−72	4.4e + 03	Y	17		0.03
3/31/2001 3:00:00	3/31/2001 22:00:00	−285	8.1e + 02	Y	11		0.02
4/11/2001 16:00:00	4/12/2001 0:00:00	−256	6.2e + 03	Y	96	2.6	0.66
4/18/2001 4:00:00	4/18/2001 7:00:00	−101	3.1e + 03	N			
4/18/2001 4:00:00	4/18/2001 7:00:00	−101	3.0e + 03	N			
8/17/2001 12:00:00	8/17/2001 22:00:00	−103	4.4e + 02	N			
9/25/2001 12:00:00	9/26/2001 2:00:00	−101	7.8e + 04	Y	7	2.3	0.13
10/2/2001 0:00:00	10/2/2001 14:00:00	−101	4.1e + 03	N			
10/3/2001 8:00:00	10/3/2001 15:00:00	−182	7.8e + 02	N			
10/21/2001 17:00:00	10/22/2001 22:00:00	−166	1.2e + 03	N			
10/28/2001 7:00:00	10/28/2001 12:00:00	−160	4.7e + 02	N			
11/5/2001 11:00:00	11/6/2001 6:00:00	−277	1.9e + 05	Y	15	2.0	0.35
11/23/2001 9:00:00	11/24/2001 15:00:00	−212	2.7e + 04	Y	174	2.6	1.00
4/17/2002 12:00:00	4/17/2002 18:00:00	−106	1.3e + 03	N			
4/18/2002 2:00:00	4/18/2002 8:00:00	−126	1.3e + 03	N			
4/19/2002 11:00:00	4/19/2002 19:00:00	−122	1.9e + 03	N			
5/23/2002 12:00:00	5/23/2002 18:00:00	−108	1.1e + 04	Y	23		0.06
8/20/2002 20:00:00	8/21/2002 7:00:00	−96	9.4e + 01	N			
9/7/2002 22:00:00	9/8/2002 1:00:00	−170	2.0e + 03	N			
11/10/2002 5:00:00	11/10/2002 7:00:00	−22	3.9e + 03	N			
5/29/2003 17:00:00	5/30/2003 0:00:00	−130	4.7e + 03	Y	36	2.6	0.10
6/18/2003 9:00:00	6/18/2003 10:00:00	−145	4.7e + 01	N			
10/24/2003 13:00:00	10/24/2003 14:00:00	−65	1.0e + 04	N			
10/26/2003 22:00:00	10/27/2003 5:00:00	−72	6.9e + 02	N			
10/28/2003 17:00:00	10/29/2003 10:00:00	−180	1.6e + 05	N			
10/28/2003 17:00:00	10/30/2003 1:00:00	−363	3.4e + 04	N			
10/28/2003 17:00:00	10/30/2003 23:00:00	−401	1.4e + 04	Y	20	2.4	0.43
11/2/2003 23:00:00	11/3/2002 2:00:00	−44	1.0e + 04	N			
11/4/2003 7:00:00	11/2/2003 11:00:00	−89	1.1e + 04	N			
11/20/2003 10:00:00	11/20/2003 20:00:00	−472	5.0e + 02	Y	134	2.1	0.16

[a] Solar proton event intensity at the time of the minimum Dst, measured with SAMPEX/LICA above L~8, >0.7 MeV protons, #/cm²-sec-sr-MeV.
[b] H-Fe ons with energies ~ 0.5-3 MeV/nucleon.
[c] Lifetime approximated by eye from a plot of trapped H-Fe ion L shell versus time.
[d] Where iron was clearly visible.
[e] Sum of H-Fe observed during entire belt lifetime, normalized to the number in the 11/23/2001 event.
[f] New ion belt was missed in the study of *Hudson et al.* (2004).

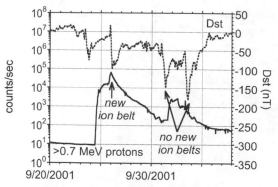

Figure 3. Interplanetary solar proton count rate from SAMPEX (left axis) and the Dst index (right scale) for 18 days in September/October 2001. The first geomagnetic storm resulted in a new ion belt, while the other two major storms in the figure did not.

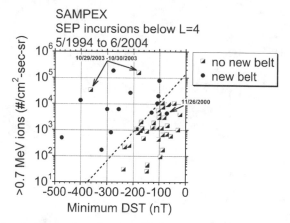

Figure 4. Scatter plot of solar particle event intensity as measured with >0.7 MeV proton rate above L~8 versus minimum Dst. The proton rate corresponds to the intensity of precipitating protons at the time of storm maximum. The dashed line was drawn by eye to separate the events with and without detectable ion belts.

There were three cases of ions below L = 4 in Figure 4 (10/29/2003, 10/30/2003, and 11/26/2000) that were exceptions to the apparent division between events with belts and those without belts. It is not obvious why these events were exceptions. The October 2003 events were complex with 3 consecutive major geomagnetic storms in a 2-day period. Also, in this analysis we used the Dst index to order the data. *Leske et al.* [2001] found a good but imperfect correlation between solar particle cutoffs and Dst. The particle cutoffs are probably dependent upon higher-latitude current systems that operate during the storm, that are not reflected in Dst, and that are certainly dependent on local time. Also, particles may have access to lower latitudes via the geomagnetic tail, in which case the more local ring current may not be important in determining their trajectories. In any case, we may be

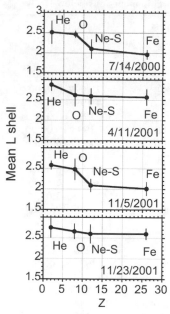

Figure 5. Mean L shell versus element for selected events.

more surprised that the events were ordered by Dst as well as they were.

It was also the case that a new ion belt could disappear at the onset of another large geomagnetic storm. The Halloween storms of 2003 are an example, shown in Figure 7. In the figure we show the trapped and precipitating iron at 0.5-3 MeV/nucleon in the bottom panel, and the interplanetary iron at about the same energy as measured on ACE (*Mason et al.* [1998]). A new ion belt formed after the shock and storm on 30 October 2003 and was observable for approximately 20 days. This was in contrast to the 8.8 day e-folding time of the 11/23/2001 event. Notice that the belt disappeared on 11/20/2003, but because of the sampling time of SAMPEX we cannot verify that the ions disappeared exactly at storm maximum.

There was a trend for higher-Z ions to reside at lower L shells, consistent with a rigidity-dependent access to lower L. The differences in location were often small, being less than half an L shell, but were nevertheless well defined. Once trapped, the particle lifetimes were a function of atomic number as well, indicating a loss due to ionization energy loss in the residual atmosphere.

In the 3-dimensional MHD simulations of *Hudson et al.* [2004] the increased dynamic pressure of the solar wind at the arrival of a shock suppressed the solar particle cutoffs below L = 3.5. We find that this suppression routinely occurred during the last solar maximum. *Hudson et al.* [2004] also discussed how subsequent trapping of the solar particles was inherently a non-adiabatic process, as applies

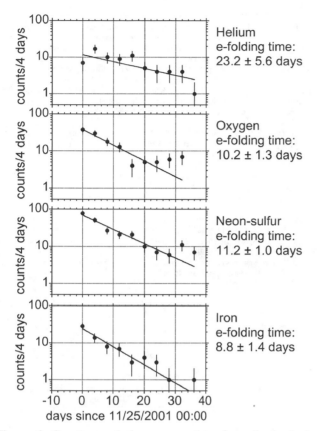

Figure 6. Counts per 4 days versus time from the beginning of 25 November 2001 for the 4 mass groups shown. The solid lines are exponential fits to the data with best-fit e-folding times indicated. The energy ranges were as follows: He, 0.7-3 MeV/nucleon; O, 1-5 MeV/nucleon; Ne-S, 0.8-4 MeV/nucleon; and Fe, 0.5-3 MeV/nucleon.

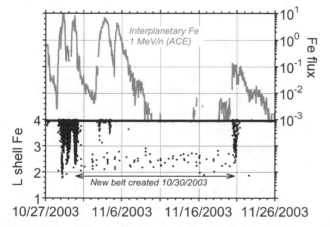

Figure 7. Example of where a new heavy-ion belt appearance and disappearance coincided with solar particle events and geomagnetic storms. The ACE spacecraft measured the iron flux (#/cm²-sec-sr-MeV/n) upstream of the Earth.

during a violation of the drift invariant with rapid radial transport. In the *Hudson et al.* picture, it was the time rate of change of the compression of the dayside magnetosphere that seemed to distinguish the events with new, long-lived ion belts (3/24/1991 and 11/23/2001). We have yet to order this dataset by the rate of change of the magnetic field as measured by ground-based magnetometers.

Another question is the relationship between the interplanetary and trapped particle intensities. While there are many variables to consider, the data in Table 1 show no clear correlation (consider 11/23/2001 versus 5/23/2002). It may be, for example, that the interplanetary flux at the time of minimum Dst is not the appropriate input to consider given the varying relationship between the time of maximum cutoff suppression and Dst [*Leske et al.*, 2001]. We can revisit this question in a later analysis, possibly including higher-altitude measurements from other vehicles to better establish the trapped particle intensity nearer to the magnetic equator.

With the data in hand, we do plan to explore the energy range over which the new belts exist. Preliminary LICA energy spectra suggest that the trapped ion spectra were much softer than the interplanetary ions at the times of the storms, also suggesting an energy-dependence to the access as well as the loss processes. Even with a relatively soft spectrum, the trapped ions may be intense enough to be observed with the MAST sensor also on SAMPEX above ~10 MeV/nucleon, but this remains to be verified. The higher energies are of interest for understanding the source and loss processes, as well as for any relevance for the single-event effect environment in microelectronics. Since we have no reliable way of extrapolating the intensity at SAMPEX to the magnetic equator, it remains to be seen if the trapped ion intensities are high enough to be important for single-event effects or other space weather impacts. It is clear that the belts can exist for weeks, thereby producing a fluence that might be comparable to a solar particle event outside the magnetosphere.

Acknowledgments. We thank the organizers and participants of the Chapman Conference on Solar Energetic Plasmas and Particles held in Turku, Finland in August 2004 for a successful meeting and inviting venue. We also thank N. Gopalswamy for inviting this paper. This work was supported in part under The Aerospace Corporation's Mission Oriented Investigation and Experimentation program, funded by the U.S. Air Force Space and Missile Systems Center under Contract No. FA8802-04-C-0001, and under contract Z667103 between the University of Maryland and The Aerospace Corporation.

REFERENCES

Baker, D.N., G.M. Mason, O. Figueora, G. Colon, J.G. Watzin, and R.M. Aleman, An overview of the Solar, Anomalous, and Magnetospheric Particle Explorer (SAMPEX) mission, *IEEE Trans. Geosci. Remote Sens.*, 31, 531-541, 1993.

Cummings, J.R., A.C. Cummings, R.A. Mewaldt, R.S. Selesnick, E.C. Stone, and T.T. von Rosenvinge, New evidence for geomagnetically trapped anomalous cosmic rays, *Geophys. Res. Lett.*, 20, 2003-2006, 1993.

Hudson, M.K., B.T. Kress, J.E. Mazur, K.L. Perry, and P.L. Slocum, 3D modeling of shock-induced trapping of solar energetic particles in the Earth's magnetosphere, *J. Atmos. and Solar-Terrestrial Phys.*, 66, 1,389-1,397, 2004.

Hovestadt, D., G. Gloeckler, C.Y. Fan, L.A. Fisk, F.M. Ipavich, B. Klecker, J.J. O'Gallagher, and M. Scholer, Evidence for solar wind origin of energetic heavy ions in the Earth's radiation belt, *Geophys. Res. Lett.*, 5, 1055-1057, 1978.

Kanekal, S.G., D.N. Baker, and J.B. Blake, Multisatellite measurements of relativistic electrons: Global coherence, *J. Geophys. Res.*, 106, 29,721-29,732, 2001.

Kleis, T., A.J. Tylka, J.H. Adams Jr., L.P. Beahm, P.R. Boberg, R. Beaujean, S. Barz, & W. Enge, Trapped low-energy heavy ions results from LDEF, *Proc. 24th Int. Cosmic Ray Conf.*, 4, 481-484, 1995.

Leske, R.A., R.A. Mewaldt, and E.C. Stone, Observations of geomagnetic cutoff variations during solar energetic particle events and implications for the radiation environment at the Space Station, *J. Geophys. Res.*, 107, 30,011-30,022, 2001.

Li, X., I. Roth, M. Temerin, J.R. Wygant, M.K. Hudson, and J.B. Blake, Simulations of the prompt energization and transport of radiation belt particles during the March 24, 1991 SSC, *Geophys. Res. Lett.*, 20, 2423-2426, 1993.

Looper, M.D., J.B. Blake, B. Klecker, and D. Hovestadt, Trapped anomalous cosmic rays near the geomagnetic cutoff, *J. Geophys. Res.*, 101, 24,747-24,753, 1996.

Lorentzen, K.R., J.E. Mazur, M.D. Looper, J.F. Fennell, and J.B. Blake, Multi-satellite observations of MeV ion injections during storms, *J. Geophys. Res.*, 107(A9), 1231, doi:10.1029/2001JA000276, 2002.

Mason, G.M., D.C. Hamilton, P.H. Walpole, K.F. Heuerman, T.L. James, M.H. Lennard, and J.E. Mazur, LEICA: A low energy ion composition analyzer for the study of solar and magnetospheric heavy ions *IEEE Trans. Geosci. Remote Sens.*, 31, 549-556, 1993.

Mason, G.M., R.E. Gold, S.M. Krimigis, J.E. Mazur, G.B. Andrews, K.A. Daley, J.R. Dwyer, K.F. Heuerman, T.L. James, M.J. Kennedy, T. Lefevere, H. Malcolm, B. Tossman, and P.H. Walpole, The ultra-low energy isotope spectrometer (ULEIS) for the ACE spacecraft, *Space Science Reviews*, 86, 409-448, 1998.

Mazur, J.E., G.M. Mason, J.B. Blake, B. Klecker, R.A. Leske, M.D. Looper, and R.A. Mewaldt, Anomalous cosmic ray argon and other rare elements at 1-4 MeV/nucleon trapped within the Earth's magnetosphere, *Journal of Geophysical Research*, 105, 21015-21023, 2000.

Pizzella, G., C.E. McIlwain, and J.A. Van Allen, Time variations of intensity in the Earth's inner radiation zone, October 1959 through December 1960, *J. Geophys. Res.*, 67, 1,235-1,253, 1962.

J.B. Blake, Space Sciences Department, The Aerospace Corporation, 2350 E. El Segundo Blvd., El Segundo, CA 90245

M.K. Hudson, Physics and Astronomy Department, Dartmouth College, Hanover, NH 03755

G.M. Mason, Johns Hopkins University, Applied Physics Laboratory, Johns Hopkins Road, Laurel, MD 20723

J.E. Mazur, Space Sciences Department, The Aerospace Corporation, 15049 Conference Center Drive, Chantilly, VA 20151

P.L. Slocum, Space Sciences Department, The Aerospace Corporation, 2350 E. El Segundo Blvd., El Segundo, CA 90245

Energetic Particles in the Magnetosphere and Their Relationship to Solar Wind Drivers

I. Roth

Space Sciences Laboratory, University of California, Berkeley, California

M.K. Hudson, B.T. Kress, and K.L. Perry[1]

Department of Physics and Astronomy, Dartmouth College, Hanover, New Hampshire

Enhancements in fluxes of energetic protons, heavy ions and relativistic electrons in planetary environments are initiated by solar and heliospheric processes: (1) directly, as a result of a propagating electromagnetic disturbance which impacts the magnetosphere, modifying the planetary magnetic configuration on the impulse propagation time scale; (2) indirectly, by exciting electromagnetic oscillations in the magnetosphere which diffuse particles across field lines or energize them on given field lines over hours to days. In a magnetized plasma energization of long-term trapped particles is due to a set of different physical processes which violate one or more of the adiabatic invariants. We survey geomagnetic modifications due to solar/heliospheric drivers, geomagnetic eigenoscillations and the mechanisms which break down invariants of trapped particle dynamics, and investigate the resulting effects on observed fluxes. The mechanisms include (a) radial diffusion due to ultra low-frequency (ULF) electromagnetic oscillations and fluctuations in the convection electric field, (b) transit-time damping due to fast mode waves, (c) diffusion due to electromagnetic ion-cyclotron or whistler waves and (d) sudden deformation of the magnetic field configuration. The latter can cause trapping of Solar Energetic Particles (SEPs) on a drift time scale to form transient proton and heavy ion belts deep in the magnetosphere ($L = 2$–3). Radial and energy diffusion time scales become comparable for MeV electrons around the plasmapause ($L = 4$–5).

INTRODUCTION

The formation of energetic populations of ions and electrons in terrestrial/planetary environments is caused by a combination of physical processes which take place in the flaring solar corona, the perturbed interplanetary medium and within magnetospheres. The long-term trapping of energetic ions and relativistic electrons in planetary magnetospheres is controlled by a core magnetic field which can be described by a distorted, shifted and tilted dipole. These magnetic fields are produced by dynamo processes, which modify structure over periods of 10s to 100s of thousands of years, to external impacts (due to solar/interplanetary processes) and internal electromagnetic effects involving magnetosphere-ionosphere coupling which modify electric and magnetic fields and particle characteristics over time scales of seconds to hours,

---

[1] Now at Boston College, Boston, Massachusetts.

Solar Eruptions and Energetic Particles
Geophysical Monograph Series 165
Copyright 2006 by the American Geophysical Union
10.1029/165GM33

setting parts of the magnetosphere in oscillation with eigenfrequencies of similar periods.

Enhancements of radiation belt fluxes depend on the time scale of electromagnetic perturbations, on the characteristic eigenfrequencies of electron or ion dynamics and on losses due to interaction with the atmosphere, with macroscopic bodies or due to other radiative processes. If the perturbation time scale is much longer than that of the quasiperiodic motion of a particle, the corresponding adiabatic invariant is conserved; the respective actions (adiabatic invariants) may be violated by perturbations on corresponding time scales. Such processes can violate one or more invariants while preserving the other(s). We survey the experimental evidence for the enhancement of energetic particles in the terrestrial magnetosphere and describe the relevant processes of violation of adiabatic invariants, resulting in formation of energetic populations.

ENERGIZATION MECHANISMS

An enhancement in the fluxes of magnetically trapped relativistic electrons generally occurs as a result of processes which are initiated by: (i) an external impulse (ii) external catalyst or (iii) internal source. Mechanism (i) consists of a direct, strong electromagnetic impulse which abruptly deforms the magnetic configuration and energizes the electrons (and ions) by breaking their third invariant associated with longitudinal drift, when a subset of particles is in phase with a single coherent wave [Li et al., 1993]. Such drift time-scale accleration occurs infrequently and requires a sudden impulse [Blake et al., 2005], typically triggering a geomagnetic storm and called a Storm Sudden Commencement (SSC), excited by a fast interplanetary shock resulting from a Coronal Mass Ejection (CME). Mechanism (ii) applies external perturbations which enhance the radial diffusion of a distribution function with a positive radial gradient, violating the third invariant by a random walk due to broad-band, small-amplitude, low-frequency electromagnetic or electrostatic perturbations [Schultz and Lanzerotti, 1974; Selesnick et al., 1997]. This mechanism tends to flatten the radial distribution of electrons f(L), where L denotes the equatorial distance in units of planetary radius, but cannot describe separately a peak in phase space density at lower L shells [Green and Kivelson, 2004], which is observed in some events during geomagnetically active periods. Mechanism (iii) applies resonant interaction with higher frequency waves on the order of gyration or bounce time scales which violates one or both of the first two invariants [Summers et al., 1998; Summers and Ma, 2000a; Roth et al., 1999; Horne et al., 2003, 2005]. It requires an increase in the power of waves which interact with a seed population of electrons [Meredith et al., 2003] diffusing them in energy and pitch-angle. Additionally, intense, fast Alfven waves may be able to enhance the flux of relativistic electrons by perturbing their parallel motion due to the mirror force, thus violating the second invariant [Summers and Ma, 2000b].

The loss of relativistic electrons in the radiation belts is important in the evaluation of steady state conditions (when the average losses are balanced by radial diffusion); it also forms an important diagnostic method based on observations of electromagnetic emission processes in which the electrons participate. The main loss mechanisms include: a) precipitation into the loss cone due to pitch-angle scattering when the equatorial pitch-angle is such that the mirror force does not prevent the particle from reaching low altitudes; b) Coulomb collisions when an electron, in the presence of a dense plasma, loses a small fraction of its energy to the ambient plasma; (c) X-ray Bremsstrahlung when a small fraction of electron energy is emitted as radiation; d) synchrotron emission in the presence of a strong magnetic field when a relativistic electron is accelerated (e.g., by gyration or motion along the field line) and radiates. In the solar system the main location of this process is the Jovian magnetosphere. e) Absorption by macroscopic bodies applies mainly to the outer planets which are surrounded by numerous moons.

The main scientific questions regarding the enhancements in the fluxes of relativistic electrons are: (1) what are the triggers for the acceleration mechanisms? (2) what are the time scales involved? and (c) what are the detectable loss processes? The balance between the energization processes and the loss of the trapped particles renders the values of fluxes which can be compared to observations [Summers et al., 2004]. The analysis of the physical mechanisms must include boundary conditions and may be helped by comparative magnetosphere studies, since different magnetospheres, due to their structure and size, correspond to a variety of physical conditions and parameters.

PHASE SPACE

Trapped energetic particles undergo quasi-periodic motion (gyration, bounce and drift) determined by the structure of the planetary magnetic field, generally at very disparate frequencies. Each one of these motions is associated with a specific adiabatic invariant which is approximately conserved along the particle trajectory. The gyration Ω, bounce ω_b and drift ω_d frequencies satisfy $\omega_d \ll \omega_b \ll \Omega$, and the motion of a magnetically trapped particle can be described with the help of the action-angle variables: μ, θ_g; J, ϕ_b; α, β, where $\mu = p_\perp^2 / 2m_0 B$ denotes the first adiabatic invariant and θ_g the gyrophase, J is the $\oint p_\parallel ds$ action (proportional to the second adiabatic invariant) and ϕ_b the bounce phase related to the bounce frequency; α, β are the Euler potentials, defining the magnetic field line on which the guiding center of the particle is instantaneously located. For a dipole configuration, in spherical coordinates (r, θ, ϕ),

$r = LR \sin^2\theta$ and $\alpha = - M_o / L \sim \Phi$ is the "flux" or third adiabatic invariant, where M_o is the strength of the planetary dipole and $\beta = \phi$ is the azimuth.

Electric field oscillations are effective in μ, J or α violation when they occur on time scales of gyrofrequency, bounce frequency or corotational and gradient-curvature drifts, respectively. Effectively, due to the different time scales of the eigenfrequencies, perturbations over longer time scales (lower frequencies) generally will not affect the adiabatic invariant conjugate to the higher eigenfrequency. However, those particles which undergo diffusion due to low-frequency modes can simultaneously interact with the magnetized plasma and undergo additional processes which can affect the higher-frequency adiabatic invariant. Without these additional processes, the modifications in the distribution function due to the violation of any one of the adiabatic invariants J_i is given approximately by

$$\frac{\partial F}{\partial t} = \Sigma \Sigma \frac{\partial}{\partial J_i}\left[D_{ij} \frac{\partial F}{\partial J_j}\right] \qquad (1)$$

where J_i denote the three adiabatic invariants. This Fokker-Planck equation determines the diffusion time for the invariant J_i:

$$\tau_i \sim J_i^2 / D_{ii} \qquad (2)$$

The violation of an adiabatic invariant occurs when the particle and the wave interact strongly by satisfying a particular resonance condition. For a gyrating particle, the local resonance condition equates the Doppler-shifted frequency with the harmonics of the relativistic gyrofrequency,

$$\omega - k_\| v_\| = n\Omega/\gamma \qquad (3)$$

Applying it to the bounce-drift motion, we convert the gyration Ω/γ into the bounce frequency ω_b, the parallel wavenumber into the wavenumber for a drift at radius r, m/r, while its parallel velocity converts to the drifting velocity $\omega_d r$, hence Equation 3 becomes

$$\omega - m\omega_d = n\omega_b \qquad (4)$$

Eq. 4 describes the resonance condition with either both of the drift and bounce motion, or with each one separately.

RADIAL DIFFUSION

Radial diffusion dominates the transfer of electrons across magnetic field lines. The physical mechanism is based on breaking the third or flux invariant. Since the distribution

function has generally a positive gradient in L, the diffusion tends to bring the electrons towards lower L-shells; preservation of the first two invariants increases the energy of the electrons. A peak in phase space density due to internal heating at lower L can, however, lead to diffusion both to lower and higher L [Selesnick and Blake, 2000]. Due to the different time scales of the three eigenfrequencies, this physical process is almost time independent over the gyro/bounce time scales, so the gyro/bounce phases become ignorable coordinates. Therefore, the remaining phase space variables include μ, J, α, β, and the modifications in the distribution function are given by [Birmingham et al., 1974],

$$\frac{\partial F}{\partial t} + \frac{\partial}{\partial \alpha}(\dot{\alpha}F) + \frac{\partial}{\partial \beta}(\dot{\beta}F) + \frac{\partial}{\partial \mu}(\dot{\mu}F) + \frac{\partial}{\partial J}(\dot{J}F) = S \qquad (5)$$

where $S(\alpha, \beta, \mu, J, t)$ denotes the source term, which is also related to the boundary conditions. The last two terms on the left hand side denote the changes in phase space density due to radiation processes. Performing an ensemble average over time scales longer than the azimuthal drift period, neglecting azimuthal variations $\partial_\beta = 0$, any average cross-shell drift $\langle\dot{\alpha}\rangle = 0$, and the small effect of radiation on J (i.e. parallel momentum), $(\dot{J}F) = 0$ results in

$$\frac{\partial F}{\partial t} + \frac{\partial}{\partial \mu}(\dot{\mu}F) = \frac{\partial}{\partial \alpha}\left[D_{\alpha\alpha} \frac{\partial F}{\partial \alpha}\right] + S \qquad (6)$$

Changing the variables from α to L and lumping the collision term into an effective time scale τ_{coll} results in

$$\frac{\partial F}{\partial t} + \frac{\partial}{\partial \mu}(\dot{\mu}F)_{rad} = L^2 \frac{\partial}{\partial L}\left[\frac{D_{LL}}{L^2} \frac{\partial F}{\partial L}\right] - \frac{F}{\tau_{coll}} \qquad (7)$$

Eq. 7 describes the most commonly used equation for radial diffusion. Without pitch angle scattering and radiation processes it is equivalent to a one-dimensional Equation 1. The crucial step for solving it requires the parametrization of the diffusion operator D_{LL} with respect to L and geomagnetic activity.

A direct approach to calculating the changes in the distribution function propagates a large number of electrons in prescribed electric and magnetic fields which are taken either from analytical models or from global MHD simulations [Elkington et al., 2002; 2004; Perry et al., 2005]. Generally, in the parametrization of Equation 7 one assumes $D_{LL} = D_0 L^n$. For a dipolar background magnetic field the electrostatic and electromagnetic contributions to D_{LL} give $n = 6$, 10, respectively [Falthammar, 1965], while later corrections due to the magnetic activity index Kp were included for electrostatic [Lyons and Thorne, 1973; Lyons and Schulz, 1989] and electromagnetic [Lanzerotti et al., 1978] perturbations. A numerical fit of Equation 7 to Polar data gave n = 11.7(±1.3) [Selesnick et al., 1997], while particle simulations with the inclusion of an Ultra Low Frequency (ULF) model wave field in the mHz range, superimposed on a compressed dipole correction to the background magnetic field, were shown to fit n = 11 [Elkington et al., 2003]. The large values of n signify that the diffusion slows down significantly at low L values. On the other hand, fits to the Jovian synchrotron radiation at L = 1.5 using Equation 7 show that at low L-values n = 1.8 − 3.0 [Birmingham et al., 1974; de Pater and Goertz, 1994]. A terrestrial study with a time-dependent D_{LL} due to changing Kp and measured geosynchronous fluxes as an outer boundary condition [Brautigam and Albert, 2000] gave a good fit to the observed electron fluxes at L = 4 at low first adiabatic invariant ($\mu = 100 − 300$ MeV/G), but a significant discrepancy at higher μ values, indicating that an additional process which may violate the first or second adiabatic invariant operates for higher energy electrons.

O'Brien et al. [2003] performed an extensive correlative study of solar wind data, magnetospheric indices and in situ geosynchronous particle measurements. They concluded that high-speed solar wind velocity and high recovery phase ULF wave power are closely associated with the production of relativistic electrons at geosynchronous, confirming earlier observations [Paulikas and Blake, 1979; Baker et al., 1998; Rostoker et al., 1998; Hudson et al., 1999; Mathie and Mann, 2001]. They found a typical two-day lag time between solar wind drivers and geosynchronous response which suggests a diffusive process. Vassiliades et al. [2004] applied a correlative technique which distinguished the two-day lag time MeV population at geosynchronous from a faster response population in the electron slot region (where pitch angle diffusion produces rapid loss), probably associated with impulsive injection following SSCs, and a third population at lower energy extending outside geosynchronous associated with substorm injections.

Since ULF wave periods of minutes are comparable to the electron drift period in the 100's of keV-MeV range at geosynchronous, they provide a means of violating third invariant conservation, enabling radial diffusion and changing energy due to first invariant conservation. ULF wave activity and its effects on particle motion are therefore linked via radial diffusion theory. Enhanced radial diffusion resulting from asymmetries in the magnetic field can also cause increases in the radiation belt fluxes over a period of hours [Elkington et al., 1999; 2002; 2003; 2004] instead of days as typically assumed. The time that it takes for diffusion to occur and how effective it is depends on the duration of increased ULF wave power [Mathie and Mann, 2001] and the level of enhanced power as discussed by O'Brien et al.

[2003]. Not all storms with enhanced ULF wave power show relativistic electron flux enhancements [Green and Kivelson, 2001], indicating that diffusion and loss processes compete.

Recently, the influence on radiation belt electrons in a compressed dipole magnetic field of ULF waves in the frequency range corresponding to a few mHz oscillations and drift periods of 100's of keV to MeV electrons at geosynchronous, has been examined for the first time in 3D [Perry et al., 2005]. This analysis utilizes model ULF wave electric and magnetic fields applied to the guiding center trajectories of relativistic electrons. A model was developed to describe magnetic and electric fields associated with poloidal-mode Pc5 ULF waves which have an azimuthal electric field component and compressional (parallel to the background field) magnetic component in the equatorial plane [Hughes, 1974], the latter previously neglected by Elkington et al. [2003]. The frequency and L dependence of the ULF wave power was included for the first time in this model by incorporating published ground-based magnetometer data. This ULF wave model is used as input to a three dimensional guiding center test particle code from which the L, energy, and pitch angle dependence of the diffusion rates can be calculated.

Plate 1 compares results from a dipole magnetic field model with radial diffusion coefficients used by Braughtigam and Albert [2000]. In a compressed dipole study, diffusion coefficients are slightly higher [Elkington et al., 2003; Perry, 2004] because additional resonances contribute besides the simple drift resonance $\omega = m\omega_d$, where ω is the ULF wave frequency, ω_d is the electron longitudinal drift frequency and m is the azimuthal mode number, Equation 4. The ULF wave power spectral density in the poloidal mode, which determines the amplitude of the radial diffusion coefficient D_{LL}, as determined from MHD simulations [Elkington et al., 2004] and ground based magnetometer measurements [Mathie and Mann, 2000], typically peaks around $m \sim 1–2$. Maximum radial diffusion rates were obtained, in the realistic case of an L and frequency dependent power spectral density, for equatorially mirroring electrons, so previous calculations restricted to the equatorial plane provide a reasonable upper limit to diffusion rates. Note the stronger L dependence evident in Plate 1 for an L and frequency dependent power spectral density than the L^{10} dependence plotted in black, based on earlier work going back to Falthammar [1965, 1968]. The magnitude of the diffusion coefficient is most sensitive to the L dependence assumed for the power spectral density, i.e. the blue curve assumes an L dependence consistent with groundbased measurements by Mathie and Mann [2000], while the earlier calculations in black assume $D_{LL} \sim L^{10}$, which does not explicitly incorporate a dependence of power spectral density on L. Satellite studies of ULF wave occurence vs. L [Anderson et al., 1994; Lessard et al., 1999; Hudson et al., 2004] show an increase with L, as do MHD simulations [*Elkington et al.*, 2004], consistent with excitation by the solar wind either through velocity shear instability at the magnetopause [*Miura*, 1992] or direct coupling of solar wind perturbations in the Pc5 (mHz) frequency range [*Kepko et al.*, 2002].

ENERGY AND PITCH ANGLE DIFFUSION

Radial diffusion, which preserves the first and the second adiabatic invariants, operates at low ULF wave frequencies (0.1 mHz - few 10's mHz) and energizes the particles while transporting them to lower L values. This *L*-diffusion explains the slow increase in energetic electron flux on time scales from a few days at Earth to many months at Jupiter. Since radial diffusion rates depend strongly on L through D_{LL}, as seen in Plate 1, diffusion into low L shells requires increasingly longer times. Generally, for times much shorter than typical radial diffusion time scales, or for local electron enhancements with a peak in phase space density over a narrow range around L = 4 [*Selesnick and Blake*, 1997], the energization requires violation of the first and/or second adiabatic invariant. Therefore, additional *in situ*, i.e. localized along given field lines, acceleration may be required [*Brautigam and Albert*, 2000; *Meredith et al.*, 2003; *Green and Kivelson*, 2004].

The acceleration processes are intimately connected with the loss mechanisms and the balance between both is critical [*Shprits and Thorne*, 2004]. *Horne and Thorne* [1998] identified the main modes capable of interacting with 100 keV-5 MeV electrons in different magnetospheric regions: left hand electromagnetic ion cyclotron (EMIC) and right hand whistler waves. *Summers et al.* [1998] presented diffusion curves in momentum space for fully relativistic electrons interacting with waves, concluding that the EMIC waves may induce pitch angle scattering of MeV electrons, while whistler waves can exchange energy resonantly with relativistic electrons in low plasma density regions. In almost all analyses, it was assumed that the waves propagate parallel to the background magnetic field, hence the resonant interaction between electron and whistler wave is given by Equation 3 with n = - 1. This assumption is well justified at low magnetic latitudes, however waves which propagate into high latitudes acquire a significant perpendicular wavenumber, therefore oblique whistler waves are an additional natural candidate for acceleration of electrons off equator. Since electron distribution functions measured at the equator are often isotropic, off-equatorial acceleration and loss process must be included. Whistler waves acquire significant oblique wavenumbers along their propagation paths due to the changing magnetic field and density profile [*Thorne and Horne*, 1995], and therefore are able to interact resonantly at several regular and anomalous gyroharmonic resonances. The

former correspond to propagation of electrons and whistlers in the same direction, and the latter to propagation in the opposite direction.

The interaction at higher and anomalous gyroharmonics is particularly efficient for relativistic electrons with gyroradii of the order of the whistler wavelengths. These modes of interaction violate the first and second adiabatic invariants, and the resulting diffusion in pitch angle and energy results in hardening of the spectrum (energization) of relativistic electrons. The mechanism involves resonant interaction with electrons bouncing and gyrating along inhomogeneous dipole magnetic field lines, satisfying the resonance condition given by Equation 3. The integer *n* denotes the harmonic of the cyclotron interaction. For high harmonic interaction the wave frequency is much smaller than the harmonics of the gyrofrequency and the resonance involves the gyrofrequency, parallel wavenumber and parallel velocity. Therefore, EMIC waves are also able to resonate with relativistic electrons [*Thorne and Kennel*, 1971; *Horne and Thorne*, 1998; *Albert*, 2003; *Summers and Thorne*, 2003].

Meredith et al. [2001] found experimental justification for scattering and acceleration of relativistic electrons by the EMIC and whistler waves. They showed that during active times ($A_e > 300$ nT) enhanced chorus emissions are observed outside of the plasmasphere, indicating a substorm correlation. The chorus emissions may propagate to high latitudes without substantial damping. During strong geomagnetic storms the magnetopause can be compressed inside L = 4, increasing the magnetic field and decreasing the electron plasma/gyrofrequency ratio at L = 4–5, allowing interaction with whistlers. *Meredith et al.* [2003] have shown that gradual acceleration of relativistic electrons is correlated with prolonged substorm activity which provides: 1) a source population with substorm injection of plasmasheet electrons, which can excite whistler mode chorus via electron loss-cone driven cyclotron resonance at tens of keV; 2) then energization of the tail of the electron distribution to MeV energies follows due to enhanced chorus emissions. Electromagnetic chorus whistler waves were observed over a broad range of L shells between the plasmapause and magnetopause. The high-latitude chorus generation mechanism may be due to loss-cone distributions around the minima of magnetic field formed due to compression of the dayside magnetosphere [*Tsurutani and Smith*, 1977]. At low latitudes numerous geostationary satellites have observed injections of electrons with moderate energy (10–100 keV) during magnetic substorms [*Reeves et al.*, 1998]. Whistler chorus, which may be considered an additional signature of the substorm growth and expansion phase, may interact with the relativistic electrons off equator. EMIC waves are produced by ring current ions, primarily protons in the tens of keV energy range, which are injected into the inner magnetosphere during increased convection periods. They are observed almost exclusively on the dusk side where the plasmasphere overlaps ring current drift trajectories, lowering the threshold for EMIC wave excitation.

The basic approach for electron energization involves an analysis of the availability in phase space to reach high energies through quasilinear interaction with the waves, i.e. to determine the curves of $(v_{\parallel}, v_{\perp})$ in the plasma frame, along which the particle diffuses in velocity space. The electrons can move on diffusion curves, conserving their energy in the wave frame, while satisfying the resonance condition (Eq 3) with the waves whose properties are determined by the relevant dispersion relation. The resulting traces in momentum space determine the energization path [*Summers et al.*, 1998]; the important missing parameter in this analysis is the time scale of the process, as well as loss processes. In order to obtain approximate time scales and calculate a time dependent distribution function at VLF perturbation frequencies of Hz to 100 kHz (EMIC or whistler), the main analytical approach is based on a modified Equation 7, with diffusion in L replaced by the other adiabatic invariants. The simplification of solving a separate equation for velocity space and radial diffusion confines the analysis to a localized (equatorial or high latitude) region and the adiabatic invariant variables are transformed into pitch angle and energy. Since pitch angle diffusion operates on times shorter than energy diffusion, one often uses the resulting Fokker-Planck equation [*Roberts and Miller*, 1998; *Summers and Ma*, 2000; *Summers et al.*, 2004]:

$$\frac{\partial F}{\partial t} = -\frac{\partial}{\partial E}[(A(E) - \dot{\varepsilon})F] + \frac{\partial^2}{\partial E^2}[D_E F] - \frac{F}{\tau_{esc}} + S(E) \quad (8)$$

where D_E is the energy diffusion rate, A(E) is the convection coefficient due to the whistler waves, $\dot{\varepsilon}$ is the energy loss due to Coulomb collisions, synchrotron radiation and other losses, while τ_{sec} denotes escape time due to pitch angle losses. In this scenario, the main coupling between the faster pitch angle diffusion and energization is due to the escape term (leaky box approximation). Hence momentum (energy) diffusion energizes the electrons while pitch angle diffusion (embedded in the escape time) removes them by scattering into the loss cone. S(E) denotes any possible sources of particle injection due to substorms. The changes in the flux of the energetic electrons results from a balance between the input (due mainly to D_E) and losses described by τ. *Summers et al.* [2004] obtained time-stationary fluxes by taking approximate values for these two parameters, yielding conditions for enhancement and decreases of electron fluxes in magnetic storms. However, most of the physics which determines the evolution of the energetic fluxes is embedded in

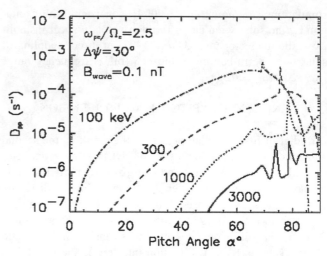

Figure 1. Local momentum diffusion coefficients for different energies for plasma/gyrofrequency ratio $\omega_{pe}/\Omega_e = 2.5$, where whistler mode wave angular width is assumed to be 30° about the background magnetic field direction [*Horne et al.*, 2003, Figure 5]. Whistler mode chorus amplitdes of 100 pT are assumed in the model.

the diffusion coefficient which is sensitive to the external plasma parameters as well as wave spectra. *Horne et al.* [2003] calculated the diffusion coefficient by assuming a specific form of wave power confined in frequency and wave normal angles. The resulting diffusion coefficients show the largest values for the lowest ω_{pe}/Ω_e, which occurs just outside of the plasmasphere, with strong sensitivity to the electron energies. Figure 1 shows the local momentum diffusion coefficient for a set of given energies and plasma/gyrofrequency ratio $\omega_{pe}/\Omega_e = 2.5$. More accurate bounce-averaged pitch angle and energy diffusion rates have recently been obtained by *Horne et al.* [2005, Figure 6], however energy/momentum diffusion rates are not substantially different from those shown in Figure 1. Comparing directly with Plate 1, corresponding to a 1 MeV electron at L = 4.2, one sees that the minimum radial diffusion time $1/D_{LL}$ increases above 1 day between L = 4–5 for the assumed ULF wave spectrum, while a momentum diffusion coefficient of $3 \times 10^{-5} s^{-1}$ in Figure 1 corresponds to a diffusion time of 0.38 days for a 1 MeV electron at 90°. The value of plasma/gyrofrequency ratio assumed in Figure 1 is appropriate for L = 4–5, during disturbed times when the plasmapause moves radially inward due to enhanced convection. As ω_{pe}/Ω_e increases inside the plasmasphere, the momentum diffusion coefficient decreases [*Horne et al.*, 2003, Figure 3]. Thus, just outside the storm time plasmapause, momentum diffusion rates are faster than radial diffusion rates which, however, increase rapidly with L.

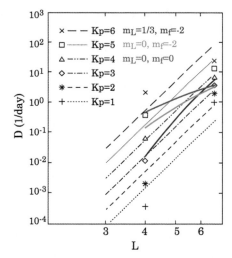

Plate 1. D_{LL}^B in day^{-1} as a function of L, for K_p = 1 to 6 after [*Brautigam, and Albert*, 2000]. Discrete values of D_{LL}^B at L = 4.0 are adapted from [*Lanzerotti and Morgan*, 1973]; those at L = 6.6, from [*Lanzerotti et al.*, 1978]. Continuous curves of D_{LL}^B are proportional to L^{10}. The legend associates a symbol (discrete D_{LL}^B) with its corresponding linestyle (continuous D_{LL}^B) for a given K_p. Plotted on top of this data are the results from guiding center test particle simulations in model 3D poloidal mode ULF wave fields [*Perry et al.*, 2005] at α_{eq} = 90°. The red curve is for no power spectral density (PSD) dependence on L or frequency, m_L = 0 and m_f = 0; the green curve is m_L = 0 and m_f = −2, an inverse square dependence of PSD on frequency only; the blue curve is for PSD ~$10^{m_L L}$ for m_L = 1/3 and m_f = −2 [After *Perry et al.*, 2005, Figure 4, with ULF wave amplitudes in the equatorial plane at geosynchronous normalized to the same magnetic power level at L ~ 6].

In another direct approach, one follows test particles in the presence of a magnetic field and oblique whistler waves [*Roth et al.*, 1999]. The background magnetic field model approximates the inhomogeneous dipole field and includes the effect of the mirror force, implying explicitly bounce motion along the magnetic field. As the electron traverses long distances along the field line its gyrofrequency changes substantially, and for relativistic energies, the electron can resonate efficiently multiple times in each bounce, increasing the effective energy diffusion. The wavenumbers and refractive index change along the electron trajectory due to the changing magnetic field and density. Between resonances the electron moves adiabatically, but when it crosses a resonance defined by Equation 3, an irreversible change in energy, in first adiabatic invariant and in equatorial pitch angle can occur. Successive resonant interactions are a source of stochastic diffusion in energy and pitch angle. For relativistic electrons the obliqueness of waves allows much stronger diffusion vs. parallel waves, since the resonance criteria (for higher harmonics) is satisfied over more of the bounce trajectory.

Finally, a wave mode which was borrowed from solar physics [*Miller*, 1997] and applied to the magnetosphere [*Summers and Ma*, 2000*b*] is the oblique magnetosonic mode; this mode has a finite compressional magnetic field component. The interaction between the electron magnetic moment and the parallel gradient of the magnetic field $(-\mu\nabla_{\parallel}B)$ may be described as a magnetic analogue of Landau damping, when the roles of the charge and the electric potential are replaced by μ (first adiabatic invariant) and the absolute value of the total magnetic field, respectively. The interaction applies mainly to electrons and is strongest when the period of the compression of the parallel magnetic component is equal to the transit time of the electron. In the frame of the wave the particles are reflected by the magnetic compression (μ-grad B force); in the plasma frame this is equivalent to head-on or trailing collisions, similar to Fermi acceleration. This interaction violates the second adiabatic invariant. For oblique waves the n = 0 resonance condition (Equation 3) is easily satisfied for the tail of the electron distribution, and together with the fast-mode dispersion relation, $\omega = kV_A$, results in a threshold for energy exchange between waves and particles, $E > E_{th} \sim \beta_A^2 / 2$, where β_A is the Alfven speed in units of c. This condition requires a seed population of several hundred keV electrons for bounce resonant acceleration by oblique magnetosonic modes, and may energize electrons at a fixed L shell over a time scale of hours.

ADDITIONAL LOSS PROCESSES

In addition to pitch angle scattering and loss to the atmosphere via resonant interaction with whistler mode chorus and EMIC waves, Coulomb collisions contribute to losses at lower energies [*Abel and Thorne*, 1998] and current sheet scattering does so at higher energies [*Anderson et al.*, 1987], as electrons fail to conserve their first invariant in a stretched tail geometry typical of substorms. Magnetopause shadowing results in radial loss when an electron drift trajectory encounters the magnetopause. This loss mechanism is enhanced by ring current buildup. The relativistic electron response is the so-called Dst effect, or dropout in relativistic electron fluxes associated with increased ring current plasma pressure and local reduction of magnetic field, causing electrons to move radially outward, decreasing their perpendicular energy while conserving all three invariants [*Kim and Chan*, 1997]. Synchrotron radiation is an important energy loss process at ultra-relativistic energies at Jupiter but not at Earth, where electron energies are lower in a spatially smaller and weaker magnetic field.

Hard X-ray microbursts provide a measurement of energetic electrons which are pitch-angle diffused into the loss cone and are able to reach low altitudes and interact with the denser ionospheric plasma where they are observed through X-ray emissions at balloon altitudes of 40 km. While the electrons lose a significant fraction of their energy via Coulomb collisions, the Bremsstrahlung is the most important observational method for their detection in the atmosphere. Balloons form a stationary platform for observations of a particular field line (in contrast to satellites which cross the field lines with typical velocity of ~10 km/s at a few hundred km altitude). Recent balloon measurements from Antarctica [*Lorentzen et al.*, 2000] confirm the existence of electron microbursts at MeV energies, providing losses from the radiation belts at a rate comparable to expected lifetimes [*Millan et al.*, 2002]. Satellite observations of electron microbursts from the low altitude polar orbiting SAMPEX satellite have been used as a proxy for VLF waves in the magnetosphere responsible for the measured precipitation [*O'Brien et al.*, 2003].

SHOCK INDUCED ACCELERATION

The unexpected appearance of new electron and proton radiation belts on a particle drift time scale during the March 24, 1991 geomagnetic storm [*Blake et al.*, 1992] proved that neither diffusive nor internal heating mechanisms are always adequate to explain rapid changes in the radiation belts. This storm was initiated by an interplanetary shock travelling at 1400 km/s, based on propagation time from the sun, with no *in situ* solar wind measurements available at that time. While this was a high shock speed event, there have been others with higher shock speed at the recent solar maximum, including the Bastille Day 2000 and Halloween 2003 storms, where *in situ* WIND and ACE measurements were available. Fortuitous measurement by the CRRES satellite in a near equatorial plane orbit detected the March 1991 injection and acceleration of electrons from the outer zone trapping region into L~2.5, producing a new trapped population with energy > 13 MeV occurring on an electron drift time scale. This injection was modelled using a simple constant-velocity solitary wave impulse in azimuthal electric field and compressional magnetic field, reflected from the ionosphere with reduced amplitude [*Li et al.*, 1993]. An MHD simulation of this event using a 3D global MHD code reproduced the 13 MeV peak measured by CRRES [*Elkington et al.*, 2002], as did the *Li et al.* model. This type of event has been shown to be extremely rare for electrons [*Looper et al.*, 2005]. The rarity for electrons results from the extreme SSC impulse required, as measured by ground magnetometers [*Blake et al.*, 2005], in order to produce an adequate *dBz/dt* and corresponding azimuthal electric field impulse which propagates inward as a mangetosonic wave inside the magnetosphere, requiring resonant azimuthal velocities of electrons comparable to the fast mode speed of ~1000 km/s inside the magnetosphere. Multi-MeV electrons at geosynchronous orbit are required for drift resonance, for an electron to see an approximately constant electric field in its drift frame. Thus, both a very strong solar wind perturbation is required, also an energetic source population. Buildup of the source population may require an extended quiet period, as was the case for the March 1991 event, allowing transport and energization by radial diffusion prior to the shock arrival. Since energization occurs while conserving the first invariant, both for shock drift acceleration and radial diffusion, electrons transported from higher L values gain more energy. Radial diffusion dominates the transfer of trapped magnetospheric electrons and ions across magnetic field lines on diffusive time scales.

Shock induced trapping of Solar Energetic Particles was first demonstrated for the March 24, 1991 CME-driven storm which produced a new proton as well as electron belt around L ~ 2.5 [*Blake et al.*, 1992]. This event was simulated for guiding center proton test particles using output from a 3D MHD code [*Hudson et al.*, 1997]. Subsequent analysis of this class of events has shown that they occur more frequently for protons and heavy ions such as Fe than for electrons [*Lorentzen et al.*, 2002; *Hudson et al.*, 2004; *Blake et al.*, 2005; *Looper et al.*, 2005]. 3D modelling of SEP access and trapping of 25 MeV protons was carried out using snapshot [*Kress et al.*, 2004] and time-dependent [*Kress et al.*, 2005] MHD simulation fields that utilize upstream solar wind measurements as input. Plate 2 shows results from these simulations for the November 24, 2001 event, which produced a new tens of MeV proton belt around L = 2.5, via deformation of the magnetic field topology, determining access on Stormer orbits [*Stormer*, 1950]. The time-dependent magnetic field perturbation allows SEP protons to become

trapped once the magnetic field relaxes from the SSC impulse. Thus protons (and heavier ions) reach lower L because of the magnetopause compression. They become trapped because the process is time dependent on a time scale comparable to the drift period, which is typically transitory prior to arrival of the interplanetary shock, i.e. most SEPs observed into L ~ 4 are not trapped prior to shock arrival. *Hudson et al.* [2004] examined eleven SEP events in 2000–2002 of which eight produced transient trapping. The most long-lived of the new belts was produced by the November 6 and 24, 2001 SSC events, with *dBz/dt* smaller but comparable to the March 1991 event, as observed by ground magnetometers [*Blake et al.*, 2005]. These 2001 injections lasted until the November 20, 2003 storm, with Dst = − 472 nT (Kyoto Provisional Dst Index), causing loss of adiabatic trapping. Such a strong ring current perturbation has been shown to reduce the ratio of ion gyroradius to magnetic gradient-curvature scale length such that the first adiabatic invarient is no longer conserved [*Young et al.*, 2002], similar to current sheet scattering which affects outer zone electron loss [*Anderson et al.*, 1997].

DISCUSSION AND CONCLUSIONS

Different types of solar wind drivers produce energetic magnetospheric particle populations. High speed CMEs, which transport magnetic cloud structures earthward and impose an extended interval of southward IMF B_z on the magnetosphere produce the most dramatic geomagnetic storm effects. The higher the velocity of the associated interplanetary shock, the greater the overall effect. There is a tradeoff between the effectiveness of a high speed shock in compressing and eroding the dayside magnetosphere, and the effectiveness of an extended period of southward IMF B_z in driving buildup of the ring current, increasing the frequency of substorm injections into the inner magnetosphere and increasing the time interval of enhanced ULF and VLF wave activity. With these tradeoffs, one can obtain both increases and decreases of relativistic electron flux levels in response to solar wind drivers [*Reeves et al.*, 2003].

CME-driven changes in relativistic electron fluxes are more common, as are CMEs, at solar maximum. Major reconfiguration of the coronal magnetic field launches CMEs which produce interplanetary shocks within 1 AU responsible both for SEP acceleration and reconfiguration of the planetary magnetic field upon arrival at Earth. High speed CMEs produce interplanetary shocks which, upon arrival at earth, can transport MeV electrons and protons earthward on a particle drift timescale. The compression of the magnetopause launches an azimuthal electric field which provides the transport. The most dramatic example was the March 24, 1991 event which produced a new > 13 MeV

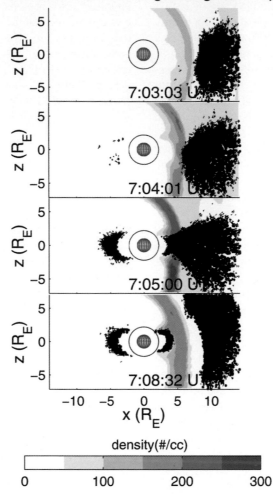

Plate 2. Sampled 25 MeV proton trajectories in time dependent MHD magnetospheric model fields in solar magnetic (SM) coordinates at several times from the 24 Nov 2001 storm simulation. The circle at $2.2R_E$ is the inner boundary of the magnetospheric model [*kress et al.*, 2005].

electron belt and tens of MeV proton belt at L = 2.5, which lasted for years [*Looper et al.*, 2005; *Blake et al.*, 2005]. This type of event is unusual for electrons, requiring an MeV outer zone electron source population to be drift resonant with the electric field impulse, while the more common several MeV ion source population characterizes SEP events, making the production of transient proton belts by this mechanism more common [*Hudson et al.*, 2004].

During the declining phase of the solar cycle recurring high speed streams, which map to coronal hole structures at the sun, cause relativistic electron flux enhancements with the 27 day period of solar rotation. With an increase in the area of coronal holes, the average speed of the solar wind emanating from these regions and crossing the ecliptic plane increases to 600–750 km/s for fast streams vs. 300–350 km/s for the slow solar wind; interaction between streams and the ambient background solar wind results in the formation of large compressional Alfven waves, the precursor of Corotating Interaction Regions which form at 2–5 AU. These CIR provide an additional source of energetic particles which are observed *in situ* and in trapped magnetospheric orbits [*Mazur et al.*, 2002].

The correlation of geosynchronous relativistic electron fluxes with solar wind velocity exceeding 500 km/s has been well established [*Paulikas and Blake*, 1979; *O'Brien et al.*, 2001], consistent with excitation by the solar wind either through velocity shear instability at the magnetopause [*Miura*, 1992] or direct coupling of solar wind perturbations in the Pc5 frequency range [*Kepko et al.*, 2002]. Increased ULF wave power, along with fluctuations in magnetospheric convection, proportional to solar wind velocity as well as strength of IMF B_z, leads to enhanced radial diffusion of electrons (and ions) on time scales measured in days. The intensity of this radial diffusion and the relevant time scale of energization depends on the diffusion coefficient, which is a strong function of L shell and the power spectral density of electromagnetic fields, depending in turn on the level of geomagnetic activity.

The response of a prolonged load on the magnetotail during periods of enhanced magnetospheric convection also leads to frequent (3–4 hours) release of energy through substorms. Substorms inject low energy (10's – 100 keV) electrons from the plasmasheet into geosynchronous orbit, providing both an enhanced source population for acceleration to relativistic energies, and driver for localized wave activity. Whistler mode chorus is excited by the electron loss cone gradient, along the electron drift trajectory around the dawn side outside the plasmasphere, while ring current ions excite EMIC waves along the dusk side, just inside the plasmasphere. Both the whistler mode chorus and EMIC waves cause pitch angle diffusion and loss, while the former is able to energize the electrons, forming a localized phase space density enhancement in L. The time scale for this process can

be much shorter, of order 1 day or less inside L = 4–5, than radial diffusion at low L. Details depend on the energy diffusion coefficient which increases with lower plasma/gyrofrequency ratio, maximizing outside the plasmapause, which moves to lower L value during periods of enhanced magnetospheric convection. Both radial and localized energy diffusion depend on the level of wave activity in the frequency range which violates respective adiabatic invariants, the third invariant for radial diffusion and gyration and bounce invariants for localized (in L) energization. Both ULF and VLF wave activity is elevated during magnetically active times, with ULF wave power and effectiveness greater at higher L values and VLF wave power and effectiveness peaking around the plasmapause. This tradeoff results in a likely dual acceleration of electrons [*O'Brien et al.*, 2003] with peak fluxes observed around L = 4 [*Li et al.*, 2002].

The systematic variation of radiation belt electrons and trapped ions of SEP origin has now been measured over a solar cycle by the SAMPEX spacecraft, with continuous *in situ* measurements of solar wind parameters beginning in 1994 with the launch of the WIND spacecraft, and continuing with the Advanced Composition Explorer (ACE). Our understanding of solar wind control of the magnetospheric energetic particle environment, greatly advanced by these and complimentary measurements, for example of electron pitch angle distribution from the Polar spacecraft, enabled calculation of electron phase space density using an assumed magnetic field model [*Selesnick and Blake*, 2000]. From these and related studies [*Green and Kivelson*, 2004], it has been shown that peaks in phase space density occur around the plasmapause during active times, confirming the contribution of internal heating to the overall picture of radiation belt formation. These studies necessarily assume a magnetic field model to map electron fluxes measured locally to their adiabatic invariant space, since only the first invariant is a local quantity. Future multipoint satellite measurements such as the Living With a Star Geospace Storm investigations will test magnetic field model assumptions and improve our understanding of the relative control of diffusion processes in respective invariants, and their corresponding solar wind drivers.

Acknowledgments. This work was supported in part by the STC Program of the National Science Foundation under Agreement Number ATM-0120950. Additional funding is from NASA grants NAG5-11733, NAG5-11944, FDNAG5-11733, and FDNAG5-11944 to UC, and NASA LWS NAG5-12202 and NSF ATM-0201624 grants to Dartmouth.

REFERENCES

Abel, B., and R.M. Thorne (1998), Electron scattering loss in Earth's inner magnetosphere: 1, Dominant physical processes, *J. Geophys. Res.*, 103, 2385.
Albert, J.M. (2003), Evaluation of quasi-linear diffusion coefficients for EMIC waves in a multispecies plasma, *J. Geophys. Res.*, 108, 1249, doi:10.1029/2002JA009792.

Anderson, B., J.R. Decker, and N.P. Paschalidas (1987), Onset of nonadiabatic particle motion in the near-earth magnetotail, *J. Geophys. Res.*, 102, 17,553.

Anderson, B.J. (1994), An overview of spacecraft observations of 10 s to 60 s period magnetic pulsations in the Earth's magnetosphere, in *Solar Wind sources of Ultra-Low- Frequency Waves*, M.J. Engebretson, K. Takahashi, and M. Scholer, ets., AGU, Washington, D.C.

Baker, D.N., T.I. Pulkinen, X. Li, S.G. Kanekal, J.B. Blake, R.S. Selesnick, M.G. Henderson, G.D. Reeves, H.E. Spence, and G. Rostoker (1998), Coronal mass ejections, magnetic coulds, and relativistic magnetospheric electron events: ISTP, *J. Geophys. Res.*, 103, 17,279.

Birmingham, T.J. (1969), Convection electric fields and the diffusion of trapped magnetospheric radiation, *J. Geophys. Res.*, 100, 14,853.

Birmingham, T., W. Hess, T. Northrop, R. Baxter, and M. Lojko (1974), The electron diffusion coefficient in Jupiter's magnetosphere, *J. Geophys. Res.*, 79, 87.

Blake, J.B., W.A. Kolasinski, R.W. Fillius, and E.G. Mullen (1992), Injection of electrons and protons with energies of tens of MeV into L < 3 on 24 March, 1991, *Geophys. Res. Lett.*, 19, 821.

Blake, J.B., P. Slocum, J.E. Mazur, R.S. Selesnick, and K. Shiokawa (2005), Geoeffectiveness of shocks in populating the radiation belts, in *Multiscale Coupling of Sun-Earth Processes*, A.T.Y. Lui, Y. Kamide, and G. Consolini, eds., Elsevier press, in press.

Brautigam, D.H., and J.M. Albert (2000), Radial diffusion analysis of outer radiation belt electrons during the October 9, 1990, magnetic storm, *J. Geophys. Res.*, 105, 291.

de Pater, I. and C.K. Goertz (1994), Radial diffusion models of energetic electrons and Jupiter's synchrotron radiation: Time variability, *J. Geophys. Res.*, 99, 2271.

Elkington, S.R., M.K. Hudson, and A.A. Chan (1999), Acceleration of relativistic electrons via resonant interaction with toroidal-mode Pc-5 ULF oscillations, *Geophys. Res. Lett.*, 26, 3273.

Elkington, S.R., M.K. Hudson, J.G. Lyon, and M.J. Wiltberger (2002), MHD/Particle simulations of radiation belt dynamics, *J. Atmos. Solar Terr. Phys.*, 64, 607.

Elkington, S.R., M.K. Hudson, and A.A. Chan (2003), Resonant acceleration and diffusion of outer zone electrons in an asymmetric geomagnetic field, *J. Geophys. Res.*, 108, 1116.

Elkington, S.R., M. Wiltberger, A.A. Chan, and D.N. Baker (2004), Physical models of the geospace radiation environment, and the Center for Integrated Space-Weather Modeling, *J. Atmos. Solar Terr. Phys.*, 66, 1371.

Falthammar, C.G. (1965), Effects of time-dependent electric fields on geomagnetically trapped radiation, *J. Geophys. Res.*, 70, 2503.

Falthammar, C.G. (1968), Radial diffusion by violation of the third adiabatic invariant, in *Earth's Particles and Fields*, edited by B.M. McCormac, NATO Adv. Stud. Inst., Reinhold, New York.

Green, J.C., and M.G. Kivelson (2001), A tail of two theories: How the adiabatic response and ULF waves affect relativistic electrons, *J. Geophys. Res.*, 106, 25,777.

Green, J.C., and M.G. Kivelson (2004), Relativistic electrons in the outer radiation belt: Differentiating between acceleration mechanisms, *J. Geophys. Res.*, 109, AO3213, doi:1029/2003JA010153.

Horne, R.B. and R.M. Thorne (1998), Potential waves for relativisrtic electron scattering and stochastic acceleration during magnetic storms, *Geophys. Res. Lett.*, 25, 3011.

Horne, R.B., S.A. Glauert, and R.M. Thorne (2003), Resonant Diffusion of radiation belt electrons by whistler-mode chorus, (2003), *Geophys. Res. Lett.*, 30, 1493.

Horne, R.B., R.M. Thorne, S.A. Glauert, J.M. Albert, N.P. Meredith, and R.R. Anderson (2005), Timescale for radiation belt electron acceleration by whistler mode chorus waves, *J. Geophys. Res.*, 110, 3225, doi: 10.1029/2004JA0

Hudson, M.K., S.R. Elkington, J.G. Lyon, V.A. Marchenko, I. Roth, M. Temerin, J.B. Blake, M.S. Gussenhoven and J.R. Wygant (1997), Simulations of radiation belt formation during storm sudden commencement, *J. Geophys. Res.*, 102, 14,087.

Hudson, M.K., S.R. Elkington, J.G. Lyon, C.C. Goodrich, and T.J. Rosenberg (1999), Simulation of radiation belt dynamics driven by solar wind variation, in *Sun-Earth Plasma Connections*, vol. 109, ed. by J.L. Burch, R.L. Carovillano, and S.K. Antiochos, p. 171, AGU, Washington D.C.

Hudson, M.K., B.K. Kress, J.E. Mazur, K.L. Perry, and P.L. Slocum (2004), 3D modeling of shock-induced trapping of solar energetic particles in the earth's magnetosphere, *J. of Atmospheric & Solar Terrestrial Phys.*, 66, 1389.

Hughes, W.J. (1994), Magnetospheric ULF waves: A tutorial with a historical perspective, in *Solar Wind sources of Ultra-Low- Frequency Waves*, M.J. Engebretson, K. Takahashi, and M. Scholer, ets., AGU, Washington, D.C.

Kepko, L., H.E. Spence, and H.J. Singer (2002), ULF waves in the solar wind as direct drivers of magnetospheric pulsations, *Geophys. Res. Lett.*, 29, doi:10.1029/2001GL014405.

Kim, H.J., and A.A. Chan (1997), Fully adiabatic changes in storm time relativistic electron fluxes, *J. Geophys. Res.*, 102, 22107.

Kress, B.K., M.K. Hudson, K.L. Perry, and P.L. Slocum (2004), Dynamic Modeling of Geomagnetic Cutoff for the 23-24 November 2001 Solar Energetic Particle Event, *Geophys. Res. Lett.*, 31, doi:10.1029/2003GL018599.

Kress, B.K., M.K. Hudson, and P.L. Slocum (2005), Impulsive solar energetic ion trapping in the magnetosphere during geomagnetic storms, *Geophys. Res. Lett.*, 32, No.6, L06108, doi:10.1029/2005GL022373.

Lanzerotti, L.J., and C.G. Morgan (1973), ULF geomagnetic power near L = 4-2, Temporal variation of the radial diffusion coefficient for relativistic electrons, *J. Geophys. Res.*, 78, 4600.

Lanzerotti, L.J., D.C. Webb, and C.W. Arthur (1978), Geomagnetic field fluctuations at synchronous orbit, 2, Radial diffusion, *J. Geophys. Res.*, 83, 3866.

Lessard, M., M.K. Hudson, and H. Luhr (1999), A statistical study of Pc3-Pc5 magnetic pulsations observed by the AMPTE/Ion Release Module satellite, *J. Geophys. Res.*, 104, 4523.

Li, X., I. Roth, M. Temerin, J.R. Wygant, M.K. Hudson, and J.B. Blake (1993), Simulation of the prompt energization and transport of radiation belt particles during the March 24, 1991 SSC, *Geophys. Res. Lett.*, 20, 2423.

Li, X., D.N. Baker, S.G. Kanekal, M. Looper, and M. Temerin (2001), Long term measurements of radiation belts by SAMPEX and their variations, *Geophys. Res. Lett.*, 28, 3827, 2001.

Looper, M.D., J.B. Blake, and R.A. Mewaldt (2005), Response of the inner radiation belt to the violent Sun-Earth Connection events of October–November 2003, *Geophys. Res. Lett.*, 32 L03S06, doi:10.1029/2004GL021502, 2005.

Lorentzen, K.R., M.P. McCarthy, G.K. Parks, J.E. Float, R.M. Millan, D.M. Smith, and R.P. Lin (2000), Precipitation of relativistic electrons by interaction with electromagnetic ion cyclotron waves, *J. Geophys. Res., 105*, 5381.

Lyons, L.R., and R.M. Thorne, Equilibrium structure of radiation belt electrons, *J. Geophys. Res.*, 78, 2142, 1973.

Lyons, L.R., and M. Schulz (1989), Access of energetic particles to storm time ring current through enhanced radial diffusion, *J. Geophys. Res.*, 94, 5491, 1989.

Mathie, R.A., and I.R. Mann (2000), Observations of Pc5 field line resonance azimuthal phase speeds: A diagnostic of their excitation mechanism, *J. Geophys. Res*, 27, 3261.

Mathie, R.A., and I.R. Mann (2001), On the solar wind control of Pc5 ULF pulsation power at mid-latitudes: Implications for MeV electron acceleration in the outer radiation belt, *J. Geophys. Res.*, 106, 29,783.

Mazur, J.E., G.M. Mason, and R.A. Mewaldt (2002), Charge states of energetic particles from corotating interaction regions as constraints on their sources, *Astrophys. Jour.*, 566, 551.

Meredith, N.P., R.B. Horne, R.H. A. Iles, R.M. Thorne, D. Heyndrickx, and R. Anderson (2002), Outer zone relativistic electron acceleration associated with substorm-related whistler mode chorus, *J. Geophs. Res.* 107, 1144, doi:10.1029/2001JA900146.

Meredith, N.P., M. Cain, R.B. Horne, R.M. Thorne, D. Summers, and R.A. Anderson (2003), Evidence for chorus-driven electron acceleration to relativistic energies from a survey of geomagnetically disturbed periods, *Jour. Geophys. Res.*, 108, 1248.

Millan, R.M., R.P. Lin, D.M. Smith, K.R. Lorentzen, and M.P. McCarthy (2002), X-ray observations of MeV electron precipitation with a balloon-borne Germanium spectrometer, *Geophys. Res. Lett.*, 29, 2194.

Miller, J. (1997), Electron acceleration in solar flares by fast mode waves: quasilinear theory and pitch-angle scattering, *Ap. J.*, 491, 939.

Miura, A. (1992), Kelvin-Helmholz Instability at the Magnetospheric Boundary: Dependence on the Magnetosheath Sonic Mach Number, *J. Geophys. Res.*, 97, 10,655.

O'Brien, T.P., R.L. McPherron, D. Sornette, G.D. Reeves, R. Friedel, and H.J. Singer (2001), Which magnetic storms produce relativistic electrons at geosynchronous orbit?, *J. Geophys. Res.*, 106, 15,553.

O'Brien, T.P., K.R. Lorentzen, I.R. Mann, N.P. Meredith, J.B. Blake, J.F. Fennell, M.D. Looper, D.K. Milling, and R.R. Anderson, Energization of relativistic electrons in the presence of ULF power and MeV microbursts: Evidence for dual ULF and VLF acceleration, (2003) *J. Geophys. Res.*, 108(A8), 1329, doi:10.1029/2002JA009784.

Paulikas, B., and J. Blake (1979), Effects of the solar wind on magnetospheric dynamics: energetic electrons at the synchronous orbit, in *Quantitative Modeling of Magnetospheric Processes*, ed. by W. Olson, AGU, Washington DC.

Perry, K.L., M.K. Hudson, and S.R. Elkington (2005), Incorporating spectral characteristics of Pc5 waves into three dimensional radiation belt modelling and the diffusion of relativistic electrons, *J. Geophys. Res.*, 110, No. A3, A03215, doi:10.1029/2004JA010760.

Perry, K.L. (2004), Radial diffusion of radiation belt electrons in three dimensions, Ph.D. thesis, Dartmouth College, Hanover, NH.

Reeves, G.D., R.H.W. Friedel, M.G. Henderson, R.D. Belian, M.M. Meier, D.N. Baker, T. Onsager, and H.J. Singer (1998), The relativistic electron response at geosynchronous orbit during the January 1997 magnetic storm, *J. Geophys. Res.*, 103, 17559.

Reeves, G.D., K.L, McAdams, R.H.W. Friedel, and T.P. O'Brien, (2003), Acceleration and loss of relativistic electrons during geomagnetic storms, *Geophys. Res. Lett.*, 30, 1529, doi:10.1029/2002GL106513.

Roberts, D.A., and J.A. Miller (1998), Generation of nonthermal electron distributions by turbulent waves near the Sun, *Geophys. Res. Lett.*, 25,607.

Rostoker, G., S. Skone, and D.N. Baker (1998), On the origin or relativistic electrons in the magnetosphere associated with some geomagnetic storms, *Geophys. Res. Lett.*, 25, 3701.

Roth I., M. Temerin, and M.K. Hudson (1999), Resonant enhancement of relativistic electron fluxes during geomagnetically active periods, *Annales. Geophys.*, 17, 631.

Shprits, Y.Y., and R.M. Thorne(2004), Time dependent radial diffusion modelling of relativistic electrons with realistic loss rates, *Geophys. Res. Lett.*, 31, doi:10.1029/2004GL019591.

Schulz, M., and L.J. Lanzerotti (1974), *Particle Diffusion in the Radiation Belts, Physics and Chemistry in Space*, vol. 7, Springer-Verlag.

Selesnick, R.S., J.B. Blake, W.A. Kolasinski, and T.A. Fritz (1997), A quiescent state of 3 to 8 MeV radiation belt electrons, *Geophys. Res. Lett.*, 24, 1343.

Selesnick, R.S., and J.B. Blake (2000), On the source location of radiation belt electrons, *J. Geophys. Res.*, 105, 2,607.

Stormer, C. (1950), "The Polar Aurora", Oxford University Press, pg. 236.

Summers, D., R.M. Thorne, and F. Xiao (1998), Relativistic theory of wave-particle resonant diffusion with application to electron acceleration in the magnetosphere, *J.Geophys. Res.*, 103, 20,487.

Summers, D., and C.Y. Ma (2000a), A model for generating relativistic electrons in the Earth's inner magnetosphere based on gyroresonant wave-particle interactions, *J. Geophys. Res.*, 105, 2,625.

Summers, D., and C.Y. Ma (2000b), Rapid acceleration of electrons in the magnetosphere by fast mode MHD waves, *J. Geophys. Res.*, 105, 15,887.

Summers, D., and R.M. Thorne (2003), Relativistic electron pitch-angle scattering by electromagnetic ion cyclotron waves during geomagnetic storms, *J. Geophys. Res.*, 108, 1143, doi:10.1029/2002JA009489.

Summers, D., C.Y. Ma, and T. Mukai(2004), Competition between acceleration and loss mechanisms of relativistic electrons during geomagnetic storms, *J. Geophys. Res.*, 109, doi:10.1029/2004JA010437.

Thorne, R.M., and C.F. Kennel (1971), Relativistic electron pitch-angle scattering by electromagnetic ion cyclotron waves during geomagnetic storms, *J. Geophys. Res.*, 76, 4446.

Tsurutani, B.T., and E.J. Smith (1977), Two types of magnetospheric chorus and their substorm dependences, *J. Geophys. Res.*, 82, 5112.

Vassiliades, D., A.J. Klimas, R.S.Weigel, D.N.Baker, J.igler, S.J. Kanekal, T. Nagai, S.F. Fung, R.W.H. Friedel, and T.E. Cayton (2003), Structure of Earth's outer radiation belt inferred from long-term electron flux dynamics, *Geophys. Res. Lett.*, 30, 2015, doi:10.1029/2003GL017328.

Young, S.L., R.E. Denton, M.K. Hudson, and B.J. Anderson, A new empirical μ-scattering model (2002), *J. Geophys. Res.*, 107

M. Hudon, B. Kress, and K. Perry, Dartmouth College, Hanover, NH 03755-3528. (maryk@gaia.dartmouth.edu, brian.kress@dartmouth.edu and kara.perry@hanscom.af.mil)

I. Roth, Space Sciences Laboratory, University of California, Berkeley, CA 94720. (ilan@peace.ssl.berkeley.edu)

Space Weather Challenges Intrinsic to the
Human Exploration of Space

Ronald E. Turner

ANSER, Arlington, Virginia

There is renewed emphasis on human exploration of the solar system, starting with a U.S. initiative to return to the Moon in the 2015-to-2020 timeframe and subsequent human missions to Mars and beyond. Exposure to radiation is one of the primary threats to the health of the astronauts who embark on exploration missions. While the steady exposure to galactic cosmic radiation provides the largest contribution to the cumulative mission dose for long-duration missions, the threat posed by solar particle events (SPEs), particularly for missions to the Moon, can be severe. This paper provides an overview of the impact of SPEs on human missions and emphasizes the multidisciplinary character of research needed to reduce the threat that SPEs pose. Predicting an SPE involves the solar physics community (predicting the eruption and character of a sufficiently large coronal mass ejection), the plasma/particle community (determining the time evolution of the event and its energy spectrum) and the heliospheric physics community (characterizing the propagation of emerging particles through the ambient solar wind background).

1. MOON, MARS, AND BEYOND

On January 14, 2004, in a speech at NASA headquarters, the U.S. President announced an ambitious change in U.S. civil space policy, declaring ("A Renewed Spirit of Discovery," Jan. 2004, and "The Vision for Space Exploration," Feb. 2004):

> In preparation for future human exploration, we must advance our ability to live and work safely in space and, at the same time, develop the technologies to extend humanity's reach to the Moon, Mars, and beyond.

In June 2004, the U.S. President's Commission on Implementation of the United States Space Exploration Policy affirmed this vision and noted:

Solar Eruptions and Energetic Particles
Geophysical Monograph Series 165
Copyright 2006 by the American Geophysical Union
10.1029/165GM34

> These goals lay out an astonishingly ambitious vision. Our objectives in space have been radically transformed … a successfully implemented vision will create a sustained capability in space for the nation.
>
> —President's Commission, 2004

NASA's response has been to transform the agency to meet these goals. On June 24, 2004, NASA revealed the first steps in its most ambitious reorganization in years (effective August 1, 2004). NASA programs are now managed through four Directorates: Exploration Systems, Science, Space Operations, and Aeronautics Research.

> This transformation will be an evolutionary process, exploring new ways to move forward and implement change.... Doing so will enable us to take the next bold steps into space and rekindle the innovation and entrepreneurial skills that is our legacy to humankind.
>
> —Sean O'Keefe, press release, June 24, 2005

These changes are not simply cosmetic or politically expedient words. The new vision must be implemented in a tightly constrained budget environment. This demands that tough decisions be made. Planned cuts and reductions included in the President's vision (Feb. 2004) include

- Retiring the Space Shuttle after completion of the International Space Station (ISS)
- Significant reductions in use of ISS
- Potential end to the Hubble Space Telescope
- And most important to the science community, programs that do not meaningfully contribute to the vision will be increasingly difficult to sustain.

The Presidential Commission (June 2004) noted, "The journey will need to be managed within available resources using a 'go as you can pay' approach."

For the space physics community, this should not be seen as all bad news. Radiation exposure will be a significant and serious hazard during any human expedition transiting deep space. This conclusion is affirmed by multiple studies of the science requirements that must be met prior to extended human space exploration (see, for example, the NASA Bioastronautics Critical Path Roadmap, 2004).

In a presentation to a broad audience of scientists engaged in a range of studies intended to address biological risks to spaceflight, *Guy Fogleman*, NASA's Associate Director of Human Health and Performance, said that two priorities would stand above all the challenges facing NASA and would receive priority in funding (*Fogleman*, 2005):

- Understanding the risk of space radiation (including a better ability to forecast solar events)
- Collecting health data from astronauts during the limited time remaining in the ISS program

In spite of these and similar comments from the exploration community, to ensure that appropriate resources are allocated to the scientific effort to extend our fundamental understanding of the physics of solar storms, such research must continuously be shown to be both relevant and necessary to reduce the operational risk to astronauts undertaking exploration missions.

Figure 1 shows several pending space weather missions in the context of key milestones for the space exploration vision. A key observation is that there is only one more solar cycle left (cycle 24) to learn what we need to know to protect astronauts embarking on lunar exploration missions that are planned for a period of high solar activity (cycle 25).

2. RADIATION THREAT TO ASTRONAUTS

There are two distinct natural sources of space radiation that affect astronauts: (1) the background, high-energy, low-flux galactic cosmic radiation (GCR) flux and (2) the periodic, lower-energy, extremely high-flux SPEs. The cumulative effect

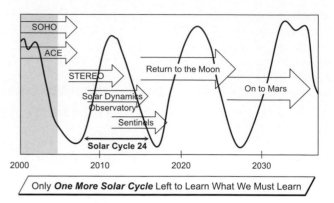

Figure 1. Significant space weather missions, mapped into the "Moon, Mars, and Beyond" vision.

of exposure to ionizing radiation is a function of the total dose, the location and distribution of the dose, the rate of accumulation of the dose, and the types of ionizing radiation that produce the dose. There are two broad categories of effects: prompt and delayed.

High dose over a short period can lead to acute effects such as headaches, dizziness, and nausea. In extreme cases, the effects of high dose rates can be severe, either directly through radiation sickness or indirectly, as from vomiting in a space suit.

Table 1 describes the impact of high doses of radiation over short periods, as defined by the United States Armed Forces Radiological Research Institute (AFRRI, 2003). The measure of dose used in the table is the Gray, a measure of the physical absorbed dose (defined as absorbed energy per unit mass: 1 Gray = 1 joule/kilogram). Note, however, that since high-energy solar energetic particles affect the body in

Table 1. High dose rate radiation impacts.

Dose	Impact
0–35 cGy	No impact on performance
35–75 cGy	Nausea; mild headache
75–125 cGy	5 to 30% experience nausea and vomiting within 3 to 5 hrs.
125–300 cGy	20 to 70% experience nausea and vomiting within 2 to 3 hrs. 5 to 10% probability of death with no treatment.
300–530 cGy	50 to 90% experience nausea and vomiting within 2 hrs. 10 to 50% probability of death with no treatment.
530–800 cGy	50 to 90% mild to severe nausea and vomiting within 2 hrs. 50 to 90% probability of death with no treatment.

Source: Medical Management of Radiological Casualties Handbook, 2nd Edition AFRRI, April 2003.

a qualitatively different way from gamma radiation, there is some concern that these limits, derived from studies of gamma exposure, may not translate easily to the biologically effective dose from an SPE, which considers not just the energy deposited, but the type of particles transiting the cell or tissue. See NCRP 132 (2000) for a detailed discussion.

A lower dose over a prolonged period can have long-term impact, including increased risk of cancer, effects on genetics or fertility, development of cataracts, and cumulative damage to tissue (particularly the central nervous system, digestive system, cardiovascular system, and immunological system). There is considerable uncertainty in quantifying the biological impact of long-term radiation (*Cucinotta et al.*, 2002). There is a major NASA program to reduce these uncertainties using ground-based facilities established at Brookhaven National Laboratory and Loma Linda University (see www.bnl.gov/medical/NASA/LTSF.asp).

For further discussion of the radiation hazards, see the National Academies Press (NRC 1996, NRC 1999, and NRC 2002), NCRP 98 (1989), NCRP 132 (2000), and Turner, 2001.

3. NASA GUIDELINES FOR RADIATION EXPOSURE LIMITS

NASA has a legal and moral responsibility to minimize astronauts' exposure to radiation. Its approach to radiation safety is a three-tier system. Lifetime limits are designed to reduce the long-term risks, especially the risk of developing fatal cancer. Short-term limits (annual and monthly) are established to reduce lifetime dose and to minimize risk of acute effects. Finally, NASA applies the philosophy "as low as reasonably achievable" (ALARA), reducing where practical any exposure to radiation, to ensure that astronauts do not come close to the monthly or annual limits.

NASA's primary science guidance in the area of radiation protection comes from the National Council on Radiation Protection and Measurements (NCRP), and recently from NCRP 132 (2000). NCRP 132 (2000) findings and recommendations supercede the findings and recommendations of NCRP 98 (1989).

The career, annual, and monthly limits recommended for low Earth orbit (LEO) from these two documents are compared in Figure 2. For a detailed discussion, the reader is referred to the referenced source documents. Briefly, the lifetime limits are based on a nominal three percent excess lifetime risk of fatal cancer, for males and females. Recently, NASA has begun to consider basing the LEO annual limits on the ninety-fifth percentile estimates to account for the inherent uncertainty in the biological risk assessments (*Cucinotta et al.*, 2005). This would decrease the allowed lifetime exposures and would affect the annual limits.

NCRP* 132 (2000)

Time Period	Blood Marrow (cGy-Eq)	Eye (cGy-Eq)	Skin (cGy-Eq)
Career	*see Table*	400	600
Annual	50	200	300
30 Days	25	100	150

Ten-Year Career Limits

Age	Female (cSv)	Male (cSv)
25	40	70
35	65	100
45	90	150
55	170	300

- NASA limits apply only to low Earth orbit. There are no approved limits for exploration missions.
- Age is age at first exposure.
- Based on three percent excess lifetime risk of fatal cancer.
- In addition, these limits incur 0.6 percent risk of heritable effects and 0.6 percent risk of nonfatal cancer.

* National Council on Radiation Protection and Measurements

Figure 2. NCRP recommended monthly, annual, and lifetime limits for low Earth orbit from NCRP 132 (2000) which superceded NCRP 98 (1989).

There are no approved limits for exploration-class missions, and it is not clear what philosophy will be recommended. An NCRP committee is addressing the question and is expected to make recommendations to NASA in 2005 or early 2006. There are competing drivers. On the one hand, the likely incorporation of higher confidence intervals would tend to reduce lifetime and annual exposure limits. Since a Mars mission will be years long, these limits, dominated by the GCR exposure, would certainly become a factor in mission planning, especially since non-cancer risks will become more of a factor. On the other hand, exploration-class missions will likely be considered to be more risky overall than LEO missions and may entail accepting higher limits to the acceptable excess lifetime risk. At any rate, the expected exposure and the principle of ALARA will make it even more important to limit the added dose by infrequent SPEs.

4. HOW BAD CAN AN SPE BE?

This leads naturally to the question "How bad can an SPE be?" The answer depends on the character of the SPE, total fluence, spectral character ("spectral hardness") and composition.

Figure 3 shows estimates of the event-integrated lunar-surface dose-equivalent for the blood-forming organs (BFO) under various levels of shielding for selected historical events, using the NASA-developed particle transport code, BRYNTRN (*Lande*, 2004; *Wilson*, 1989). Note that most of these events (representing about one to two events every ten years) exceeded the 30-day limits, and some exceeded the annual limits for at least some shield thicknesses.

However, many of these events occurred over several days of elevated flux, giving astronauts plenty of time to go to a shelter. Figure 4 gives the time evolution of the rate of lunar surface BFO dose equivalent of the most severe event on record, the August 1972 event that occurred between Apollos 16 and 17. The rate is shown for several shielding thicknesses.

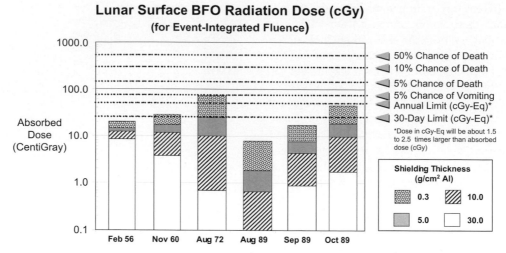

Figure 3. Event-integrated lunar surface absorbed dose for the BFO under various levels of shielding for selected historical SPEs.

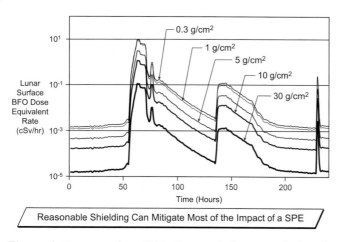

Figure 4. Lunar surface BFO dose-equivalent rate during the August 72 event under various levels of shielding. (T = 0 is approximately 0:00 UT 2 August 1972. The optical flare for this event began at 6:21 UT 4 August 1972.)

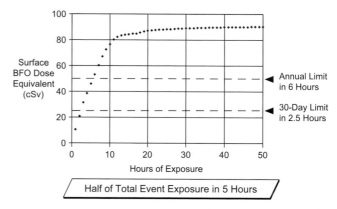

Figure 5. Peak BFO dose-equivalent vs. length of exposure on lunar surface excursions during the August 1972 event, assuming 0.3 g/cm^2 aluminum-equivalent shielding.

For reference, up to 1 g/cm^2 aluminum-equivalent is a nominal space suit thickness; 5 g/cm^2 is representative of a vehicle with limited shielding (a covered rover or the Apollo Lunar Lander); 10 g/cm^2 is a nominal value for a spacecraft or shielded rover; 30 g/cm^2 is representative of a reasonable shielding thickness for a designated safe haven within a spacecraft or a surface habitat. The key message of Figure 4 is that reasonable shielding can mitigate most of the impact of an SPE, given adequate time to reach such a shelter.

So how much time is reasonable for attaining shelter if caught out on a surface excursion? Figure 5 shows the worst-case BFO exposure on a lunar surface excursion versus time

exposed during the August 1972 event. During this extreme event, half of the total event exposure occurs within a five-hour period. The 30-day limit would have been exceeded in two and a half hours, the annual limit in six hours.

5. SPE-RELATED "FLIGHT RULES"

To mitigate this risk, mission operations are rule-driven. Astronaut activities are managed against a set of "Flight Rules" (*Turner*, 2000). These rules define the overall concept of operations, which should reflect the best science available to the mission planners. However, the implementation of research results to operational applications is not trivial and needs informed scientist dialog with the operations community.

Two examples of the type of flight rules that are directly affected by space weather illustrate the importance of the best available science input. During mission development, the architecture of rovers, habitats, and surface expeditions will have to be established. A key input will be the maximum distance astronauts should be permitted to travel away from a safe haven. Overly restrictive rules limit the science that can be accomplished. Too-lenient rules put the astronauts at risk. During a mission, under what conditions must astronauts abort an excursion, with how much urgency, based on what observations and forecasts? NASA recognizes the need for answers to these questions, but current efforts do not draw effectively upon all the resources that could be applied. In particular, the space physics community should seek to be more engaged in the development and implementation of operational rules.

6. SPACE WEATHER CONTRIBUTION

One of the greatest challenges to the space weather community is to improve our ability to forecast SPEs. While on average significantly lower in energy than the GCR flux, the proton flux of SPEs can be orders of magnitude greater for hours to days. Even under circumstances when much of the radiation can be shielded, the radiation management principle of ALARA requires that exposure to SPEs be minimized. The potential to be caught away from shelter on the lunar or Martian surface will impose operational rules that will limit flexibility and reduce efficiency.

For astronaut radiation safety, the important components of SPEs are protons with energy from 30 MeV up to 100-200 MeV (*Turner*, 2001). Generally, lower-energy protons are not sufficiently penetrating to contribute to the astronaut dose, and the flux of higher-energy protons is too low to affect the dose by more than a few percent. That said, understanding the relative contribution of the higher-energy to lower-energy protons within this range is very important.

In general it is commonly held, but not well documented, that solar energetic ions with atomic number greater than two do not contribute significantly to the astronaut dose because they are too sparse and are relatively easy to shield against as they are not very penetrating relative to the protons (at the same velocity, the range is reduced by a factor of approximately A/Z^2).

A 1996 NASA-sponsored SPE risk mitigation workshop with operational community input considered operational requirements for SPE forecasts, and found that optimal SPE forecasts should be (*Turner*, 1996):

- Ten- to twelve-hour forecast prior to a likely event
- Six- to eight-hour forecast of magnitude and spectral slope after event onset
- Three- to four-hour rolling forecast as SPE progresses

These are extremely challenging goals. Realistic near-term possibilities may be

- Eight-hour rolling forecast as an SPE progresses
- Predict, at event onset, the time of arrival and magnitude of shock-enhanced peak
- Reliably forecast one- to three-day "all clear"

Forecasting SPEs is a multidiscipline challenge, not just the purview of the plasma-shock particle physicists. Figure 6 is a "waterfall" chart of the steps involved. Each stage presents its own unique challenges and is best addressed by related but distinct members of the space physics community.

In addition, exploration missions to Mars have added challenges to provide forecasts that protect astronauts that are far from Earth and at risk from solar activity that is not visible from Earth-based observatories or near-Earth spacecraft. An example is shown in Figure 7, showing a comparison of SPE observation near Mars taken by the MARIE instrument on the Odyssey spacecraft with SPE observations taken by the GOES spacecraft in geostationary Earth orbit.

Routine solar monitoring is the necessary first step to forecast and characterize SPEs. For lunar missions, it would be helpful to image the Sun beyond the limbs, while to support Mars missions it may be critical to provide imaging of the "far side" of the Sun. For a more detailed discussion of how Mars-mission scenarios impact monitoring requirements, see *Turner and Levine*, 1998. Near-real-time observations of solar active regions and emerging coronal mass ejections

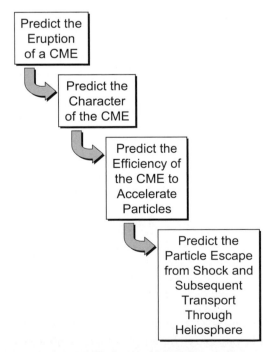

Figure 6. Forecasting SPEs is a multidiscipline challenge.

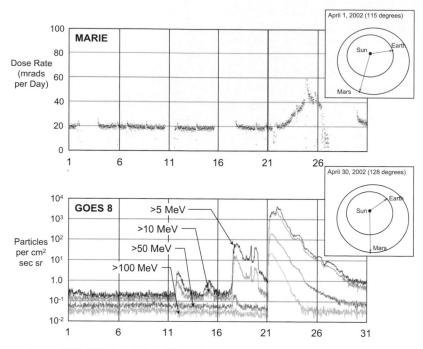

Figure 7. April 2002 comparison of MARIE data (primarily 20-200 MeV protons collected in Mars orbit (see *Zeitlin et al.*, 2004) with GOES data (collected in Earth orbit) illustrates the challenge of forecasting events at multiple heliospheric radii and longitude. The large event observed by GOES beginning April 21 was centered on the Sun at approximately W84 relative to the Earth. GOES data courtesy of NOAA National Geophysical Data Center.

(CMEs) may provide data useful to forecast the progress of an ongoing SPE over a period of hours to days. Additional progress in understanding the physics of CMEs may lead to a multi-day forecast of the probability and characteristics of an SPE.

Heliospheric observations provide information necessary to model or monitor the propagation of solar energetic particles from the source to the astronaut. The data that may be necessary for SPE propagation models include

- State of the ambient solar wind plasma
- Interplanetary magnetic field
- Local disturbances moving through the inner heliosphere
- Specification of significant heliospheric structure, including locations of corotating interaction regions.

Frequently the most severe episodes are created by interactions of a series of CMEs occurring in rapid succession (three or more events within a few days) when the ambient heliospheric magnetic field and associated solar wind plasma properties are not well represented by assuming a classic "Parker spiral." The potential complexity of specifying the state of the heliosphere is illustrated in Figure 8, after similar illustrations in *Dryer et al.*, 2004.

It may be possible within a few years to reliably predict periods of no SPE threat. If the challenge of predicting an event in advance proves too formidable, then the community should recognize that mission planners would also benefit from knowing when an SPE is not going to occur (*Turner*, 2001, and *Turner*, 2000). A success criterion that may be achievable within the next few years could be to, more than 95 percent of the time, identify conditions that would not lead to an SPE over a period of days; to ensure that fewer than five percent of the SPEs occur while "all clear" is forecast; and all the while to work to reduce the period of uncertainty about the prospects of an event (when one cannot issue a definitive forecast).

7. SUMMARY

SPEs pose a significant risk to exploration missions, primarily during excursions on the surface of the Moon and Mars. The space physics community can contribute substantially by providing science-based input to the development of exploration system architectures and operational rules.

A better understanding of solar dynamics will lead to improved forecasting of CMEs and to improved forecasting of intense SPEs.

A better understanding of heliospheric dynamics will lead to improved forecasting of the ambient solar wind, contributing to improved forecasting of the character of SPEs (time evolution and spectral character).

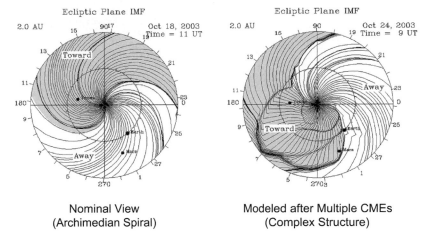

Nominal View
(Archimedian Spiral)

Modeled after Multiple CMEs
(Complex Structure)

Figure 8. Illustration of the challenge associated with heliospheric plasma forecasting during multiple large events. The complexity of the magnetic field structure continued to increase through the end of the month of October as intense solar activity continued. Details of this figure and the technique used for solar wind modeling and real-time forecasting of the complex event period of late October 2003 are discussed in *Dryer et al.*, 2004.

A better understanding of SPEs, including better climatology (historical rates of events by intensity and spectral character) will add confidence to projections of worst-case exposure and how this exposure varies with distance from the Sun. This in turn will provide designers of future missions with better guidelines, leading to improved spacecraft, habitats, and shelters, and will lead to higher confidence in mission planning.

Better forecasts of SPE evolution after event onset, including more reliable estimates of the time to and magnitude of peak flux, will lead to a higher confidence in exposure forecast and will enable the implementation of more flexible flight rules by removing much of the ambiguity that flight directors must contend with today.

High-reliability predictions of SPEs before event onset, if feasible, will lead to a higher confidence in event-specific exposure forecasts, substantially simplifying and extending day-to-day operations. This in turn will support greater mission schedule assurance.

In lieu of advance predictions of SPEs, prediction of "all clear" periods are potentially valuable to the operational community, as these periodic windows of safe operations would give mission operations more flexibility.

As our understanding increases and our forecasts become mature, the operational community will begin to move beyond a "cope and avoid" philosophy and progress to an "anticipate and react" philosophy of risk management. Taken together, the contributions of the space physics community will lead to improved astronaut safety and enhanced mission assurance.

Acknowledgments. The work described in this paper was supported by NASA research grant NAG9-1427. In addition, the author would like to thank Joshua Lande and John Starcher for their contributions to this paper.

REFERENCES

Armed Forces Radiological Research Institute, *Medical Management of Radiological Casualties Handbook*, 2nd edition, April 2003.

Cucinotta, F.A., G.D. Badhwar, P.B. Saganti, W. Schimmerling, J.W. Wilson, L.E. Peterson, and J.F. Dicello, "Space radiation cancer risk projections for exploration missions: Uncertainty reduction and mitigation," NASA TP 2002-210777, 2002.

Cucinotta, F.A., M.Y. Kim, and L. Ren, "Managing Lunar and Mars Mission Radiation Risks Part I: Cancer Risks, Uncertainties and Shielding Effectiveness," Draft NASA JSC technical document, NASA/TP-2005-210XXX, April 2005.

Dryer, M., Z. Smith, C.D. Fry, W. Sun, C.S. Deehr, and S.I. Akasofu, "Real-time shock arrival predictions during the Halloween 2003 Epoch," Space Weather, S09001, doi: 10.1029/2004 SW000,087,2004.

Executive Office of the President, "A Renewed Spirit of Discovery," U.S. Government Printing Office, January 2004.

Fogleman, G., presentation to NASA Bioastronautics Investigators' Workshop, Galveston, TX, January 11, 2005.

Lande, J., "Solar Particle Event Risk Analysis," George Washington University Science and Engineering Apprentice Program Final Report, George Washington University, August 2004.

NASA, "The Vision for Space Exploration," U.S. Government Printing Office, February 2004.

NASA, "Bioastronautics Critical Path Roadmap (BCPR): An Approach to Risk Reduction and Management for Human Space Flight," NASA JSC 62577, April 2004.

National Council on Radiation Protection and Measurements, NCRP Report Number 98, Guidance on Radiation Received in Space Activities, Washington, D.C., 1989.

National Council on Radiation Protection and Measurements, NCRP Report Number 132, Radiation Protection Guidance for Activities in Low Earth Orbit, Washington, D.C., 2000.

National Research Council, Space Studies Board, "Safe on Mars: Precursor Measurements Necessary to Support Human Operations on the Martian Surface," National Academy Press, Washington, D.C., 2002.

National Research Council, Space Studies Board "Radiation and the International Space Station: Recommendations to Reduce Radiation Risk During Solar Maximum," National Academy Press, Washington, D.C., 1999.

National Research Council, "Radiation Hazards to Crews of Interplanetary Missions—Biological Issues and Research Strategies," National Academy Press, Washington, D.C., 1996.

O'Keefe, S., NASA Administrator, press release dated June 24, 2004.

President's Commission on Implementation of United States Space Exploration Policy, "A Journey to Inspire, Innovate, and Discover," U.S. Government Printing Office, June 2004.

Turner, R.E., "What We Must Know About Solar Particle Events to Reduce the Risk to Astronauts," AGU Monograph 125, Space Weather, 2001.

Turner, R.E., "Ensuring Radiation Safety on International Space Station and Beyond," AIAA-2000-5206, AIAA Space 2000 Conference, Sept 2000.

Turner, R., editor, "Foundations of solar particle event risk management strategies: findings of the risk management workshop for solar particle events," ANSER Technical Report, Arlington, Virginia, July 1996.

Turner, R., and J. Levine, "Orbit selection and its impact on radiation warning architecture for a human mission to Mars," Acta Astronautica, Vol. 42, Number 1-8, 1998.

Wilson, J.W., L.W. Townsend, S.Y. Chun, S.L. Lamkin, B.D. Ganapol, B.A. Hong, W.W. Buck, F. Khan, F. Cucinotta, and J.E. Nealy, 1989, BRYNTRN: A Baryon Transport Model, NASA TP-2887.

Zeitlin, C., T. Cleghorn, F. Cucinotta, P. Saganti, V. Andersen, K. Lee, L. Pinsky, W. Atwell, R. Turner, and G. Badhwar. "Overview of the Martian radiation environment experiment," *Adv. Space Res.*, 33, 2204-2210, 2004.

Ronald E. Turner; ANSER; Suite 800; 2900 South Quincy Street; Arlington, Virginia 22206

Space Weather: From Solar Eruptions to Magnetospheric Storms

Hannu E.J. Koskinen[1] and K. Emilia J. Huttunen[2]

University of Helsinki, Department of Physical Sciences, Helsinki, Finland

Space weather can be defined as a topic dealing with the variable conditions in the Sun, solar wind, magnetosphere and ionosphere that may be hazardous to man-made systems in space and on ground and endanger human health and life. This definition has a strongly application-oriented flavor. Positive consequences of space weather activities during the last 10–15 years have been increasing dialogues between various fields of solar-terrestrial physics and between the science and technology communities. At the same time expectations of increased funding have contributed to opportunism and promises of useful space weather services that have not always been based on sound physical understanding. If we have sufficient in situ and remote observations, we can make useful specifications of the environmental conditions practically in real time. However, forecasting is difficult, in particular if we want to forecast the conditions at the Earth based on solar observations. As space weather involves a wide range of scientific topics from solar eruptions to current induction in conductors buried in ground, the space weather community comprises scientists with very different backgrounds. Some are specialists in interaction of solar wind shocks with the Earth's magnetopause but ignorant of how the Sun drives the shock. Others may know all details of twisted flux tubes on the Sun but think that the magnetospheric response may be characterized by one single parameter. This tutorial review aims at advancing the understanding of space weather as an integrated system, from its origins in the Sun, to the near-Earth environment.

1. WHAT IS SPACE WEATHER

Space weather became a popular concept during the 1990s. The past research in solar-terrestrial physics had progressed to a level where reliable forecasting of space environment conditions seemed to be within reach. New spacecraft focusing on the atmosphere of the Sun (Yohkoh, SOHO, TRACE) and the solar wind (WIND, ACE, SOHO), and the unprecedented fleet of satellites inside the magnetosphere had, for the first time, allowed for comprehensive characterization of severe space weather from its origins on the Sun to its consequences near the Earth. At the same time the growing dependence of the modern society on various space assets brought up the worries of wide reaching consequences of hazards caused by environmental conditions in space. Two specific events, the March 1989 power blackout in Canada and the loss of Telstar 401 in January 1997, were most efficient in spreading the knowledge of space weather risks to the general public and policy makers.

Based on an initiative from the American space science community in 1993 several U.S. governmental agencies established the National Space Weather Program whose Strategic

---

[1] Also at Finnish Meteorological Institute, Helsinki, Finland
[2] Presently at Space Science Laboratory, University of California, Berkeley, USA

Solar Eruptions and Energetic Particles
Geophysical Monograph Series 165
Copyright 2006 by the American Geophysical Union
10.1029/165GM35

Plan was published in 1995 [*NSWP*, 1995]. According to the document "Space Weather" refers to conditions on the Sun and in the solar wind, magnetosphere, ionosphere and thermosphere that can influence the performance and reliability of space-borne or ground-based technological systems and can endanger human health or life. This definition underlines the consequences of space weather. On the other hand, it also emphasizes the need for an integrated approach to the problem and this has already been positive to basic space sciences bringing the different science communities closer to each other.

1.1. Some Historical Milestones

The first reports on space weather related problems on technological systems are from events around the year 1850 when electric telegraph communications were disturbed and in some cases completely stopped during strong auroral activations [*Prescott*, 1860]. For a long time telegraph and later telephone communication lines were the most space weather sensitive technological systems. The first reported effect on power transmission systems took place on March 24, 1940 [*Davidson*, 1940]. A great geomagnetic storm caused voltage dips, large swings in reactive power, and tripping of transformer banks in the United States and Canada. The effects of the storm were also felt on telephone lines, e.g., 80% of long-distance telephone connections out of Minneapolis, Minnesota, were out of operation.

From the viewpoint of the present article a decisive scientific milestone was the first observation of a solar flare on September 1, 1859, by *Carrington* [1859] and *Hodgson* [1859]. According to Carrington a strong magnetic storm followed the flare. Later the time from the flare to the storm commencement has been estimated to have been 17 hours 40 minutes. In fact, this storm belongs to the strongest events in practically all records of solar-terrestrial activity collected thereafter [*Cliver and Svalgaard*, 2004, and references therein]. Gradually, this and subsequent coincidences of solar flares and magnetic storms led to the conclusion that the solar flares indeed were the sources of strong magnetic activity on Earth [*Hale*, 1931]. By the end of the 1930's this idea was widely known, as illustrated, e.g., on pages 332–337 of *Chapman and Bartels* [1940]. A suspected candidate to transmit the flare effect to the Earth was corpuscular radiation, but it was not yet understood how.

Lindemann [1919] was the first to suggest that quasi-neutral charged particle ejections related to solar activity were responsible for non-recurrent magnetic storms at the Earth. Much later *Biermann* [1951, 1957] recognized that the structured almost radially anti-sunward flowing parts of cometary tails could not be caused by the radiation pressure, contrary to the dust tails. As this was independent of solar eruptions, the corpuscular outflow from the Sun was not intermittent, but there was a continuous solar wind.

Consequently, the Sun was capable of driving the Earth's magnetic environment also independently of strong solar activity. As the solar wind is highly-conductive plasma, it was understood to carry the solar magnetic field to the interplanetary space. The critical role of the direction of this interplanetary magnetic field (IMF) in the solar wind-magnetosphere interaction was proposed by *Dungey* [1961] in his description of magnetospheric convection in terms of magnetic reconnection.

During the Apollo era it was vital for the astronauts to be aware of potentially dangerous eruptions on the Sun. Many of the still applied practices in space environment predictions were introduced and the solar flares were considered as the main cause of severe space weather near the Earth. However, a process leading to change of thinking began in 1971.

On December 14, 1971, a coronal mass ejection (CME) was for the first time observed using the white-light coronagraph of the OSO 7 satellite [*Tousey*, 1973]. (Figure 1 is an example of modern CME observations with the LASCO coronagraph onboard the SOHO spacecraft.) The pioneering OSO 7 observations of 20 CMEs were followed by observations on Skylab, P78-1 (Solwind), and SMM in the 1970s and 1980s. These data conclusively showed that, although flares and CMEs often were related to each other, there was no one-to-one correspondence. It also became evident that CMEs, expelling some $10^{12} - 10^{13}$ kg plasma and often exhibiting

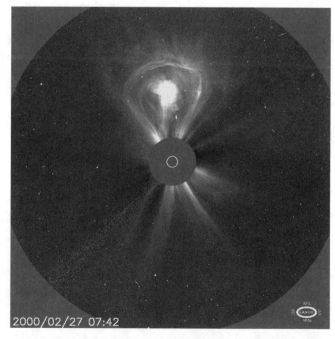

Figure 1. A CME observed by the LASCO instrument on SOHO on 27 February, 2000. The picture illustrates the typical light-bulb structure of a CME, consisting of a bright outer loop, followed by a void region and a bright inner kernel (Figure ESA).

flux-rope structures with very strong magnetic field intensities, were, by far, the most efficient drivers of magnetospheric perturbations. While an increasing fraction of solar-terrestrial scientists began to appreciate the fundamental role of the CMEs and the interplanetary shocks driven by them as the main drivers of major non-recurrent magnetospheric storms, a large fraction of the magnetospheric and ionospheric research community remained ignorant to this development and continued to refer to flares as the primary cause of magnetic storms. This lasted until the 1990s, as noted by *Gosling* [1993] in his landmark article "The Solar Flare Myth".

A breakthrough with high publicity took place in January 1997. The ESA-NASA Solar and Heliospheric Observatory (SOHO) had been launched in December 1995. A group of scientists were following real-time data from SOHO instruments when they observed a CME halo in the LASCO coronagraph images on January 6, 1997. They alerted the space weather service at NOAA of a possibly earthward directed CME, but as there were no other indications of an approaching storm, in particular no flare X-rays, the alert was ignored. The interplanetary shock at 1 AU was observed by WIND on January 10 at 00.10 UT. Thus the CME was rather slow and caused just a medium-strength storm in the magnetosphere ($Dst_{min} = -78$ nT and $Kp_{max} = 6$). What made this event so important was the following accumulation of a large flux of relativistic electrons in the outer radiation belt, which was suggested to have been the cause of the loss of the geostationary telecommunication satellite Telstar 401 on January 11. Next day, this was in the news world wide, and the importance of CMEs was finally recognized throughout the magnetospheric research community. The event has been discussed in a series of papers in a special issue of *Geophysical Research Letters* (vol 25(14), 1998).

The "paradigm shift" from flares to CMEs as the cause of major magnetospheric storms may well become to be remembered as the most important step in solar-terrestrial physics of the 1990s. However, it does not mean that CMEs would be the only efficient drivers of severe space weather. Also the flare-accelerated particles have space weather consequences. They contribute to large solar energetic particle events and cause sudden ionospheric disturbances leading to shortwave fades on HF communication systems. Furthermore, storms associated with high-speed solar wind from large coronal holes become particularly important near the solar minimum when the CME activity is smaller and the polar coronal holes reach to low solar latitudes.

1.2. Science or Technology

Spacecraft engineers and operators, people responsible for human space flights, and operators of various ground-based technological systems had been concerned with space environment effects much before the term space weather was coined.

One motivation for space weather initiatives has been to bring the scientific and engineering communities together to educate each other on the different aspects of space weather. One may wonder how many times the radiation belts and in particular the South Atlantic Anomaly have been "rediscovered" with surprise by scientists and engineers applying sensitive optical instruments or poorly-protected electronics.

However, it would be too narrow to consider space weather only as a tool for engineering and applications. Space weather is a system of interacting complex physical phenomena involving a large number of insufficiently understood scientific problems. For us scientists the technological requirements for exact specification, either after-the-fact or in real time, and reliable forecasting of future conditions are welcome, as they force us to advance the research from conceptual explanations to high accuracy in the details.

1.3. Space Weather vs. Atmospheric Weather

It can be instructive to compare space weather with the familiar atmospheric weather. The physics of atmospheric weather is based on hydrodynamics of neutral gas in local thermal equilibrium, whereas in space weather we have to consider non-equilibrium plasmas with long-range interactions. Thus while regional modeling of ordinary weather is meaningful, space weather is always global. This contraposition is further emphasized by the vastly different observational possibilities. The everyday weather reports and forecasts are based on continuous global satellite coverage and dense local observation grids, including regular vertical balloon soundings. In space weather the observational network is sparse and not all observations are continuous. At the same time the domain to be monitored is huge.

At practical level the benefits of reliable weather services are evident. The emerging space weather activities have brought the relation of space weather to economic benefit and safety to the attention of the public and policy makers, but the required measures to avoid the problems have remained unclear. We must also honestly admit that a tropical hurricane hitting a major city causes much more damage than any space weather event so far. While sobering, this does not mean that we should not be concerned with space weather. There is potential for more damaging consequences and, of course, we should do everything we can to protect people and technological assets even against less fatal disturbances.

2. FROM THE SUN TO THE EARTH

2.1. Sun-CME-Magnetospheric Storm

The term "coronal mass ejection" describes the observation: Large amount of mass is detected in the corona to leave the Sun. However, the material content of a CME is not coronal plasma but must come from a much denser region of the solar atmosphere.

After leaving the corona the CMEs are often called interplanetary CMEs (ICMEs). When a fast enough ICME propagates in the solar wind, a shock front is formed. ICMEs and shocks are the most important causes of severe space weather. Their effects are twofold. First, the shocks are efficient accelerators of protons and helium nuclei. The enhanced levels of energetic ions caused by both solar flares and CME-shocks provide hazardous environment not only near Earth but also elsewhere, including the road to Mars.

From the near-Earth viewpoint the most important feature of the ICMEs and shocks is their capability of shaking the entire magnetosphere through variable plasma pressure and by the dynamics related to the direction of the interplanetary magnetic field (IMF). The size and shape of the magnetosphere are determined by the balance between the dynamic plasma pressure of the solar wind, pv^2, and the magnetic pressure of the magnetosphere, $B^2/2\mu0$. This defines the location of the magnetopause, which is normally around 10 R_E (Earth radii) in the solar direction. A fast plasma cloud can increase the plasma pressure by a large factor pushing the dayside magnetopause inside the geostationary orbit at 6.6 R_E. This can already be serious for spacecraft that are designed to remain inside the magnetosphere and, e.g., use the geomagnetic field for attitude control. However, much more dramatic consequences are related to the direction of the IMF.

Dungey [1961] proposed that the magnetospheric convection was driven by magnetic reconnection on the dayside magnetopause. As the Earth's magnetic field in this region is directed northward, a southward pointing IMF would be favorable for reconnection. *Fairfield and Cahill* [1966] were the first to establish a statistical correlation between the direction of the IMF and the intermittent magnetospheric activations observed best in the auroral zone, the substorms.

In elementary reconnection theory the reconnection rate is expressed in terms of the reconnection electric field, which in this case is $E = vB_s$, where v is the solar wind speed and B_s the southward component of the IMF. In fact vB_s has turned out to correlate with magnetospheric activity almost as well as various more refined functions of the upstream solar wind parameters. Thus, in addition to the magnitude and direction of the IMF, the solar wind velocity is important.

A fast ICME can fulfill both the velocity and magnetic field requirements for strong magnetospheric activity. The velocities of the shock, the ICME and the post-shock streams can easily be more than twice the background slow solar wind speed. The IMF in the sheath region between the shock and the magnetic cloud is strongly compressed. If the IMF ahead of a fast ICME already has a southward component, the shock increases it typically by a factor of 3–4. This way

the sheath region can drive a storm even if the ICME itself does not hit the magnetosphere.

The southward IMF component may be further amplified by draping of the magnetic field around the ICME [*Gosling and McComas*, 1987], which can lead to a southward IMF component even in cases where the pre-existing IMF is slightly northward. This mechanism works also when the ICME is so slow that it does not drive a shock. Another agent to enhance the magnetic field impinging upon the magnetopause is the formation of a plasma depletion layer when the decreased plasma pressure is compensated by increased magnetic energy density [*Zwan and Wolf*, 1976].

Finally, the strongest southward magnetic fields are found if the magnetic cloud of the ICME has a well-formed fluxrope structure. While the jump across an MHD shock can increase the magnetic field at most by a factor of 4, the magnetic field of the magnetic cloud is determined by the eruptive magnetic structure on the Sun and can be much more intense. For example, on November 20, 2003, the southward magnetic field component of an ICME reached 53 nT, which is an order of magnitude more than typical IMF at 1 AU.

Figure 2 illustrates the ICME-storm relation. The arrival of the shock is seen as the compression of the magnetic flux and plasma densities and as a jump in the solar wind speed. In this particular case the IMF turned southward within the sheath region and the storm main phase began as illustrated by the rapid decrease of the *Dst*-index in the bottom panel. The arrival of the magnetic cloud was characterized by the decrease of the plasma density and temperature as well as by the smooth rotation of the magnetic field direction.

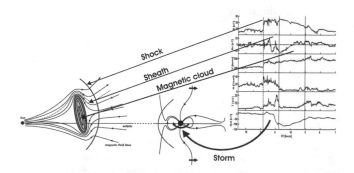

Figure 2. A sketch illustrating the ICME-magnetosphere interaction. The five first data panels (magnetic field magnitude, Z-component of the magnetic field, velocity, plasma density and temperature) show how the different parts of the ICME are seen in the WIND satellite data. The arrival of the shock is indicated by the dashed line and the magnetic cloud interval is between the solid lines. The lowermost panel is the *Dst*-index, which shows the storm main phase to begin when the southward IMF has propagated from L1 to the magnetopause and enhanced the magnetospheric convection. The data interval is from 14 May, 1997, 12 UT to 16 May, 1997, 16 UT.

2.2. Fast Solar Wind Streams and Recurrent Storms

Other important sources of large vB_s are the fast solar wind streams from the coronal holes. Although B_s may not be very large during such events, the fast stream can impinge upon the magnetopause for a long time, thus maintaining the energy input to the magnetosphere at a high level. In fact, there is clear relationship between the solar wind speed and accumulation of relativistic electrons in the magnetosphere [e.g., *Baker et al.*, 1986; *Blake et al.*, 1997]. As the large coronal holes are long-lived structures, the same fast stream reappears after about 27 days giving clear recurrence to storm and substorm activity. These recurrent storms become particularly important toward the end of the descending sunspot cycle when the coronal holes expand to low latitudes and the Earth is more frequently embedded in fast solar wind [see, e.g., *Baker and Li*, 2003, and references therein].

The high solar wind speed related relativistic electrons are of particular interest for geostationary satellite operators, as these electrons accumulate charge in the spacecraft subsystems and can cause strong discharge phenomena. In fact, a large number of serious operational anomalies, even total loss of spacecraft have been attributed to these "killer electrons" [*Wrenn*, 1995].

2.3. Characterizing the Strength of a Storm

As the storms in the magnetosphere cause magnetic variations on ground, several indices have been developed to characterize the strength of the magnetic perturbations [*Mayaud*, 1980]. The large number of useful indices illustrates the large variety of storm features; sometimes the effects are stronger at high latitudes, sometimes at low, sometimes the background current systems are strong already before the main perturbation, different current systems may decay at different rates, etc. To characterize the global storm level two indices are most widely used today: *Dst* and *Kp*.

The *Dst*-index is a weighted average of the deviation from the quiet level of the horizontal (*H*) magnetic field component measured at four low-latitude stations around the globe. The westward ring current flowing around the Earth at the distance of about 3-4 R_E is the main source of the *Dst*-index. During a magnetospheric storm the ring current is enhanced, which causes a negative deviation in *H*. Consequently, the more negative the *Dst*-index is, the stronger the storm is said to be. The threshold between weak and moderate storms is typically set to −40 or −50 nT, moderate storms range from −50 to −100 nT. Storms stronger than −100 nT can be called intense and stronger than −200 nT big. The *Dst*-index is calculated once an hour. A similar 1-minute index derived from a partly different set of six low-latitude stations (*SYM-H*) is also in use.

However, a magnetometer reacts to all current systems. High solar wind pressure pushes the magnetopause closer to the Earth increasing the magnetopause currents. The effect is strongest on the dayside where the magnetopause current flows in the direction opposite to the ring current. Thus a pressure pulse causes a positive deviation in the *H*-component. In fact, this is an excellent indicator of the time when a shock hits the magnetopause. If the solar wind parameters are known, the effect can be cleaned away from the *Dst*-index (so-called pressure corrected *Dst*-index). More difficult is the role of the dawn-to-dusk directed current separating the northern and southern lobes of the magnetotail. During strong activity this current enhances strongly and moves closer to the Earth, enhancing also the *Dst*-index. How to handle this effect is a controversial issue in magnetospheric physics.

Another widely used index is the planetary *K*-index *Kp*. Each magnetic observatory has its own *K*-index and the *Kp* is an average of *K*-indices from 13 mid-latitude stations. It is a range index expressed in a scale of one-thirds: 0,0 +, 1 −, 1, ..., 8 +, 9 −, 9. As *Kp* is based on mid-latitude observations, it is more sensitive to high-latitude auroral current systems and substorm activity than the *Dst*-index. On the other hand *Kp* is a 3-hour index and does not reflect very short term activity.

There is no one-to-one correlation between the storm strengths given by *Kp* and *Dst*. Moderate storms have typically $5 \leq Kp_{max} \leq 6$ and big storms have $Kp \geq 8$. *Huttunen et al.* [2002b] investigated the difference of the *Kp*- and *Dst*-responses to different solar wind perturbations during the years 1996-1999. They found that the fast post-shock streams and sheath regions had a relatively stronger effect on the *Kp*-index whereas the effects of ejecta favored the *Dst*-index. This is an interesting, though not fully-understood, result, which should be kept in mind when deciding which storm index one wants to correlate with solar or solar wind observations.

Figure 3 shows an example of a storm which was intense in the *Kp*-index but very weak in terms of the *Dst*-index. Note that in this case no magnetic cloud structure can be found but the magnetic field in the sheath region was highly variable and the dynamic pressure quite large.

2.4. Space Weather in the Ionosphere and on the Ground

There are various space weather phenomena in the Earth's ionosphere that affect the radiowave propagation and lead to induction effects on ground-based systems. Some of them, such as the signal scintillations due to the so-called equatorial spread-F, are only weakly correlated to solar activity. A directly solar driven phenomenon is the so called solar flare effect (SFE), which is caused by ionization due to soft X-ray

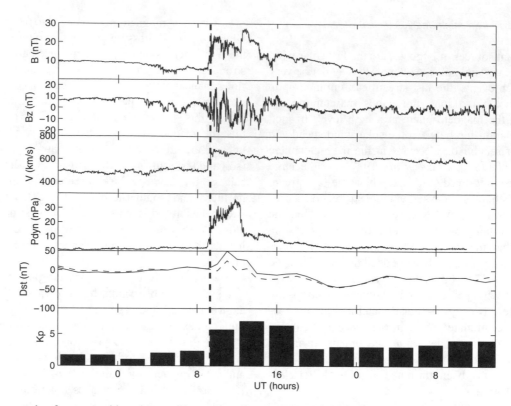

Figure 3. An example of a storm with an intense *Kp* response (lowermost panel) but only a weak signature in the *Dst*-index (second panel from the bottom). The solar wind data are from ACE at L1 showing the IMF magnitude, the IMF Z-component, velocity and dynamic pressure. The dashed line indicates the arrival of the shock. The data interval is from 12 July, 2000, 20 UT to 14 July, 12 UT.

and EUV radiation from a strong enough flare. It has a crochet-like appearance on the magnetograms following the shape of flare evolution. The SFE has not had a significant role in space weather context, but historically it is interesting to note that its amplitude in Greenwich following the 1859 Carrington flare was 110 nT, which was about the same as during the large flares in October and November 2003 discussed later in section 3.4.4 [*Cliver and Svalgaard*, 2004].

Several effects of changing solar activity are evident in the high-latitude ionosphere where changes in the ionizing solar EUV radiation as well as the storm and substorm related particle precipitation make the ionospheric plasma density to deviate from the quiet time conditions. These lead, e.g., to communication problems for airplanes on polar routes where they cannot use geostationary communication satellites and their routes have to be redirected to lower latitudes. Furthermore the increased energetic particle radiation may force them to lower altitudes. Effects on ground-to-ground and satellite-to-ground communications as well as on satellite positioning systems can also reach harmful levels.

Finally, the ionosphere is coupled to the magnetosphere through complicated field-aligned current systems that close through horizontal high-latitude ionospheric currents at altitudes just above 100 km. During magnetospheric storms and substorms these currents intensify and give rise to magnetic perturbations on ground, which sometimes reach to several thousand nanoteslas, i.e., to a considerable fraction of the background horizontal component of about 10,000 nT in the auroral zone. However, the mega ampere direct current at an altitude of 100 km does not itself cause much trouble, but sometimes the current changes very rapidly inducing a strong electric field in the ground. When such electric fields are imposed over long conductor systems, they drive so-called geomagnetically induced currents. They sometimes lead to disruptions in power grids and have also corrosion effects on long pipelines buried in earth.

3. IMPORTANT SCIENCE QUESTIONS

In this section we discuss some of the major problem areas in the physics of space weather. The limited space does not allow any in depth analysis to be included. We also leave many interesting problems associated with the ionosphere-magnetosphere coupling, including the ionospheric feedback effects, e.g., the role of oxygen ions in storm development, outside the present discussion.

3.1. When and Where on the Sun

The solar variability covers a wide range of time scales. For space weather the most relevant range is from 10 minutes (flare eruptions) to the magnetic cycle of 22 years (the Hale cycle). The climatic appearance of space weather is determined by the 11-year sunspot cycle (the Schwabe cycle). Although the sunspot cycle is repetitive, its details are hard to predict. While the sunspots have been regularly observed since the early 17th century, a clear cycle cannot be reconstructed before the 18th century. During the last 300 years the time between two consecutive minima has varied roughly between 7 and 17 years, the sunspot number at maximum has varied by more than a factor of 4, and the shapes of the maxima have been widely different. Time-series analyses indicate also longer cycles superposed on the basic cycle (e.g., the roughly 100-year Gleissberg cycle) suggesting that the next several maxima may exhibit smaller sunspot numbers than the maxima during the second half of the 20th century.

For many practical purposes the sunspot cycle predictions are useful. For example, the accumulated radiation exposure of a spacecraft strongly depends on the phase of solar cycle during the mission. This needs to be taken into account when calculating the estimated lifetimes of spacecraft subsystems and required radiation shielding. However, this is sufficient for integrated radiation doses only, as one single solar eruption with an ensuing space storm may lead to a dose corresponding to several years of low solar activity and be responsible for intense intermittent radiation environment from which a damaged system may no longer be able to recover.

The fundamental space weather processes on the Sun are the flares and CMEs. It is understandable that the flares have been popular in space storm forecasting. They are relatively easy to detect and with the space-borne X-ray detectors their intensity has become possible to classify in a uniform way. However, solar flares are unreliable predictors of space storms in the magnetosphere, as they lead to a large number of both false alarms and misses due to their incomplete correlation with the main drivers of magnetospheric storms. Most flares are not associated with observable CMEs and most ICMEs do not hit the Earth's magnetosphere. On the other hand, less than half of the CMEs have a clear flare association. Finally, not all earthward directed ICMEs cause significant storms, as the magnetic field structure of the ICME is the decisive factor for its magnetospheric consequences.

The SOHO mission has been an invaluable asset for space weather research, as it has provided almost uninterrupted CME observations throughout solar cycle 23 with its LASCO instrument. However, there is no guarantee that this state of matters will last forever and future space weather warning systems may have to rely on less comprehensive input data.

A topic to study further is certainly the flare-CME relationship, not only in a statistical sense but also the physics that determines whether a flare is associated with a CME, or not, and toward what kind of magnetic field structure the ICME will develop. This problem cannot be solved by looking at flares and CMEs alone but all associated phenomena on the Sun, in particular the development of magnetic structures leading to the eruption, must be carefully taken into account.

3.2. What Happens on the Way

Even a relatively complete knowledge of the CME properties in the corona is only the first step toward a reliable forecast. We know very little of the evolution of the ICMEs after they leave the LASCO field-of-view. An example of this is the disappearance of the bright kernel of the "light bulb" structure in Figure 1. The kernel is likely cool matter from the erupted prominence. However, this structure is difficult to identify from in situ measurements at 1 AU. Sometimes, e.g., in the January 1997 event, dense cool plasma is observed near the trailing edge of the CME, which has been interpreted as remnants of the filament [Webb et al., 2000]. It is possible that the filament matter often falls back to the Sun but we do not know it for sure.

The further evolution of an ICME is a formidable problem. With in-situ satellite observations we get data along a single line through a moving ICME. This leaves free hands to "space cartoonists" to produce sketches of the plasma and magnetic clouds. With more than one spacecraft the situation is improved, but owing to the very large size of the ICMEs the measurement points need to be far enough from each other.

Radio wave observations in long wavelengths (e.g., WAVES onboard WIND observing at frequencies below 14 MHz) provide global means to follow the propagation of CME-driven shocks into the solar wind [e.g., Reiner and Kaiser, 1999; Gopalswamy et al., 2001]. While this method gives valuable information of the early motion of the shocks, it has not been exploited very much in space weather context.

The upcoming STEREO mission of NASA will be very important in this respect. The CMEs will be followed visually for longer distances from the Sun than earlier and the stereoscopic view will give much better data of the three-dimensional structure of individual CMEs. It is also advantageous to be able to see the Earthward directed CMEs from an angle, as the halo observations are diffuse causing large uncertainties to the initial velocity determination. The in situ observations on both sides of the Earth will lead to much better information of the global structure of ICMEs hitting the Earth. Unfortunately, when the two STEREO spacecraft will have moved far enough from the Earth, they will no longer encounter the solar wind affecting the Earth's magnetosphere. Thus it will be of utmost importance to maintain the in situ upstream observations as well.

3.3. What Determines the Geoeffectivity of ICMEs

As discussed in section 2 the IMF B_z is the most critical parameter to determine the geoeffectiveness of an ICME. Of course, high speed and high pressure also perturb the magnetosphere, but they do not cause storm development if the IMF does not turn to the south. Depending both on the background solar wind conditions and on the magnetic structure of the ICME a large number of alternative storm evolutions can take place [for a systematic study, see *Huttunen et al.*, 2005].

The background IMF can have northward or southward orientation. In the first case the sheath region is not yet geoeffective and the storm main phase will not begin until a southward IMF arrives with the magnetic cloud. On the other hand, if the IMF already has even a weak southward component, it is enhanced in the sheath and, in case of strong enough southward IMF, the sheath can drive a strong storm alone.

In cases of well-defined flux-ropes the orientation of the flux-rope axis and the direction in which the magnetic field is wound varies [*Bothmer and Schwenn*, 1994]. If the inclination of the flux-rope from the ecliptic plane is small, the magnetic structure can arrive with northward or southward magnetic field ahead, which obviously lead to different storm evolutions. For example, if a southward sheath field is followed first by the northward orientation of the flux-rope field and thereafter by the southward orientation, the storm most likely develops a double-peaked signature in the *Dst*-index. The first main phase occurs during the sheath impinging the magnetopause, followed by a short recovery and then by the second main phase. Note that double- or multiple-peaked storms can also appear if several consecutive CMEs are released toward the Earth from the same active region on the Sun.

A flux-rope may have a large inclination with respect to the ecliptic. In such cases the IMF can be either northward or southward throughout the passage of the flux-rope. In the northward case the ICME will most likely pass the Earth with only minor perturbations, whereas the continuously southward case may lead to a really strong storm.

It would be of great value to be able to reliably predict the magnetic structure of an Earthward directed ICME from solar observations. However, the task is quite challenging as seen, e.g., from a careful event study by *McAllister et al.* [2001].

3.4. Forecasting Issues

Forecasting of space weather and its consequences is difficult [for a comprehensive review, see *Feynman and Gabriel*,

2000, and references therein]. If we just want to be alerted of possibly severe space weather, various means are available. The recurrent storm periods related to fast streams from large polar coronal holes repeat with a period of about 27 days. However, without more information it is almost impossible to tell how severe the situation will be after one solar rotation, as also the strength of the southward IMF component plays a role here.

Medium-term alerts of non-recurrent storms are based on monitoring the development of the active regions on the surface of the Sun. This can provide roughly one-week alerts based on structures becoming visible on the east limb. Recent progress on helioseismology [*Lindsey and Braun*, 2000] and on imaging of Lyman alpha radiation backscattered off the neutral hydrogen behind the Sun [*Bertaux et al.*, 2000] have introduced new means to recognize developing activity also on the invisible side of the Sun. While these alerts of a longer than one week lead-time may not have practical value for near-Earth applications, they may turn out to be quite useful for future interplanetary missions.

Below we briefly review some of the storms during solar cycle 23 from the forecasting point of view.

3.4.1. April 2000. The storm on 6–7 April, 2000, was analyzed in detail by *Huttunen et al.* [2002a]. The storm was preceded by a moderate (C9.8) flare near the western limb on April 4. It did not create significant energetic particle fluxes at 1 AU. The initial speed of the CME measured from LASCO images was 980 km s^{-1}. The USAF/NOAA forecast at 22 UT on April 5 for April 6 gave a 35–40-% probability for a minor storm and only 6-11-% probability for a major or severe storm.

The shock arrived to 1 AU after a 47.5-h travel time at 16 UT on April 6 with a speed of 819 km s^{-1}. A few hours later one of the strongest magnetospheric storms of solar cycle 23 was in progress. The maximum of Kp was 8+ and the minimum of *Dst* −288 nT. This storm was driven by the southward IMF in the sheath region. In fact, no clear evidence of the ejecta itself was found in the upstream solar wind data. *Huttunen et al.* [2002a] concluded that the main part of the ejecta did not hit the Earth's magnetopause.

3.4.2. July 2000. Sometimes forecasting may look easy. The famous Bastille-day event, beginning with a large (X5.7) flare on July 14, 2000, exhibited all classic signatures of a strong storm, including large fluxes of flare- and shock-accelerated energetic particles at 1 AU. The travel time of the shock was only 28 hours. The initial CME speed exceeded 1775 km s^{-1} and the shock speed at 1 AU was still 900 km s^{-1}. The storm indices reached high levels: $Kp_{max} = 9$ and $Dst_{min} = -301$ nT.

3.4.3. March–April 2001. Sometimes two CMEs are released after each other from the same region on the Sun.

In such cases the latter ICME often has a higher velocity than the former and catches it on the way. An example of such a sequence happened in March 2001. The CMEs were observed by LASCO on March 28 at 1300 UT and on March 29 at 1030 UT. A strong (X1.7) flare was associated with the latter (March 29 at 1015 UT). The initial speed difference was estimated to be 350 km/s and the interval between the arrival times at 1 AU was about 8.5 h (Figure 4).

These ICMEs led to the third strongest storm of cycle 23: $Kp_{max} = 9-$ and $Dst_{min} = -387$ nT. This was a classic example of a double-peaked storm. The minimum Dst was obtained during the passage of the first ejecta. When the Dst-index was already recovering, the southward IMF in the second ejecta intensified the storm again. In this case also the Kp-index exhibited a double-peaked structure.

3.4.4. October–November 2003.

Statistically the Earth's magnetic environment is most active somewhat before or after the sunspot maximum. In October 2003, when the overall solar activity was descending, a period of very strong activity took place (the so-called Halloween events). Several big flares, the largest one reaching X17 on October 28 with an associated halo CME, led to very strong storm activity.

The maximum Kp was 9 and the minimum of the (preliminary) Dst-index was -401 nT.

The energetic particle fluxes were so intense that several spacecraft, including SOHO, were switched to the safe mode in order to protect sensitive electronics. A large number of satellite anomalies were reported and the effects reached all the way to the ground. For example, in the Gothenburg region in Sweden the electric distribution system went down and anomalously strong geomagnetically induced current was observed in the Finnish natural gas pipeline system.

The source region on the Sun remained active and erupted again one week later on November 4, 2003, now on the western limb. This was the strongest recorded X-ray flare (X28) so far. The consequences of this event did not reach the Earth and we will never know how strong the storm would have been in the magnetosphere.

After these magnificent events the strongest Dst storm of cycle 23 came as a little surprise. A large filament disappeared from the solar disk between 0740 and 0800 UT on November 18, 2003. At the same time two medium size (M-class) flares were detected at 0752 and 0831 UT and two fast CMEs could be identified in LASCO images (0806 UT, 1223 km s^{-1}; 0850 UT, 1660 km s^{-1}).

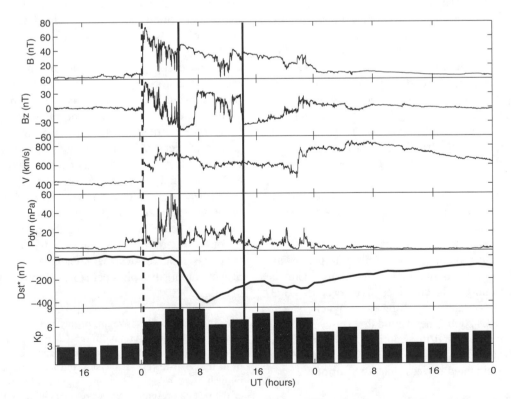

Figure 4. An example of a double-peaked storm driven by two consecutive ICMEs. The shock is indicated by the dashed line and the arrival of the magnetic clouds by the solid lines. The data are in the same format as in Figure 3, except that the storm is indicated by the pressure corrected Dst-index. The data cover the interval from 30 March, 2001, 12 UT to 2 April, 2001, 0 UT.

The shock arrived at 1 AU on November 20, 2003, at 0727 UT. It was followed by a particularly strong magnetic cloud. The maximum magnetic field intensity was about 60 nT and the southward component reached 53 nT. In addition the density of the cloud was large. The resulting storm intensity was 9- in the *Kp*-index (interval 15–21 UT) and $Dst_{min} = -472$ nT (at 20 UT; preliminary value).

Note that although the *Dst*-index reached its lowest value of cycle 23 during the November 20 storm, the Halloween storm was much more severe in the sense that the *Kp*-index was 9 during 9 hours, 9– during another 9 hours and remained above 5– for 60 hours (2.5 days). During the November 20 storm *Kp* was at its highest level for 6 hours only.

3.5. Rapid Particle Acceleration in the Magnetosphere

The storm-time magnetosphere is an efficient particle accelerator capable of energizing electrons up to relativistic (several MeV) energies. In the extreme cases the slot region between the inner and outer radiation belts can trap a large amount of relativistic electrons, which was the case during the strong storm on March 24, 1991, as observed by the CRRES satellite. Under normal conditions the slot region is an excellent magnetic bottle and the fluxes remained high beyond the end of the CRRES mission half a year later.

The relativistic electrons must have undergone a significant acceleration from a seed population of much lower energy. It is quite feasible that several mechanisms contribute to the final product, including both gyro- and drift-resonant wave-particle interactions. Also strong nonresonant electric field pulses may be involved. This was demonstrated for the March 24, 1991, event by *Li et al.* [1993]. In their model the solar wind shock compresses the magnetosphere rapidly causing a strong induced electric field. This field accelerates electrons to several MeV energies in the time scale of one minute. While being efficient the mechanism requires the existence of a sufficient seed population that must already have gone through significant acceleration before the arrival of the electric field pulse.

Large fluxes of relativistic electrons in the inner magnetosphere often appear around the *Dst* minimum, i.e., toward the end of the storm main phase. However, also here the storms show their many different faces. *Reeves et al.* [2003] investigated 276 moderate and strong geomagnetic storms during the years 1989-2000. They found that only 53% of the investigated storms increased the relativistic electron fluxes and 19% actually decreased the fluxes. In the remaining 28% of cases the fluxes changed by less than a factor of two.

3.6. Extreme Induction Effects

Important space weather effects on ground-based technologies are caused by electromagnetic induction. A frequently asked question is how an ionospheric current of a few MA can cause damage on the ground. The source is not the current itself but its large and rapid time variations. They cause strong variations in the magnetic field on ground, which results in an induced geoelectric field according to Faraday's law $\partial B/\partial t = -\nabla \times E$. It is this electric field that drives currents in long conductors either above or inside the ground.

A real ionospheric space weather challenge is to understand how so strong current fluctuations can take place. The most intense fluctuations are rather localized, and physics-based forecasting of the most hazardous induction events is not yet in sight.

4. CONCLUDING REMARKS

If we want to develop reliable physics-based forecasting methods for space storms originating from the Sun, much challenging basic research needs to be done in all domains involved. We need to understand much better the evolution of the plasma and magnetic field structures on the Sun that may lead to flares and CMEs. At present we have fairly good observational data and there is, at least, some theoretical understanding of the underlying physics, but more efforts must be put in the physical analysis and model development. For storm forecasting we need to understand the evolution of the sheath region and magnetic cloud structures during the motion of the ICME from the Sun to the Earth in order to be able to tell what part of the ICME will drive the storm. The error can easily be of the order of one day depending on whether the storm is driven already by the sheath or only by the southward IMF in the rear part of the magnetic cloud.

In the near-Earth space some of the toughest issues are how to predict the final intensity of a given storm, the acceleration of electrons to relativistic energies, and the creation of extremely strong induction effects between the ionosphere and the ground. Furthermore, we do not yet understand why the unsteady conditions in the sheath region cause relatively more high-latitude activity than the smooth magnetic clouds that are more efficient in driving the ring current activity.

For future progress it is also essential to have a continuous coverage of all critical parameters of the system. Without proper input the physical forecasting models are quite useless.

Acknowledgments. We are deeply indebted to *R. Schwenn* and *V. Bothmer* for collaboration in our CME studies based on LASCO data and on their previous work on magnetic clouds. The support from the solar-terrestrial physics and space weather groups at the Finnish Meteorological Institute is gratefully acknowledged. Much of the results presented here were obtained through the SWAP project supported by the Academy of Finland within the space research programme Antares. The data plots in this review are based on data available in the CDAWeb (WIND and ACE) and at the WDC C2 in Kyoto.

REFERENCE

Baker, D.N., J.B. Blake, R.W. Klebesadel, and P.R. Higbie, Highly relativistic electrons in the Earth's outer magnetosphere, I. Lifetimes and temporal history 1979-1984, *J. Geophys. Res.*, 91, 4265, 1986.

Baker, D.N., and X. Li, Relativistic electron flux enhancements During Strong Geomagnetic Activity, in *Disturbances in Geospace: The Storm-Substorm Relationship*, A.S. Sharma, Y. Kamide, and G.S. Lakhina (eds), Geophysical Monograph 142, p 217, American Geophysical Union, 2003.

Bertaux, J. L., E. Quemerais, R. Lallement, E. Lamassoure, W. Schmidt, and E. Kyrölä, Monitoring solar activity on the far side of the Sun from sky reflected Lyman α radiation, *Geophys. Res. Lett.*, 27, 1331, 2000.

Biermann, L., Kometenschweife und solare Korpuscularstrahlung, *Z. Astrophys.*, 29, 274, 1951.

Biermann, L., Solar corpuscular radiation and the interplanetary gas, *Observatory*, 77, 109, 1957.

Blake, J.B., D.N. Baker, N. Turner, K.W. Ogilvie, and R.P. Lepping, Correlation of changes in the outer-zone relativistic electron population with upstream solar wind and magnetic field measurements, *Geophys. Res. Lett.*, 24, 927, 1997.

Bothmer, V., and R. Schwenn, Eruptive prominences as sources of magnetic clouds in the solar wind, *Space Science Reviews*, 70, 215, 1994.

Carrington, R.C., Description of a singular appearance seen in the Sun on September 1, 1959, *Mon. Not. R. Astron. Soc.*, XX, 13, 1859.

Chapman, S., and J. Bartels, *Geomagnetism*, vol. I, Oxford Univ. Press., New York, 1940.

Cliver, E.W., and L. Svalgaard, The 1859 solar-terrestrial disturbance and the current limits of extreme space weather activity, *Solar Physics*, 224, 407, 2004.

Davidson, W.F., The magnetic storm of March 24, 1940 - Effects in the power system, *Edison Electric Institute Bulletin*, p. 365, July, 1940.

Dungey, J.W., Interplanetary magnetic field and the auroral zones, *Phys. Rev. Lett.*, 6, 47, 1961.

Fairfield, D.H., and L.J. Cahill, Jr., Transition region magnetic field and polar magnetic disturbances, *J. Geophys. Res.*, 71, 155, 1966.

Feynman, J., and S.B. Gabriel, On space weather consequences and predictions, *J. Geophys. Res.*, 105, 10,543, 2000.

Gopalswamy, N., S. Yashiro, M.L. Kaiser, R.A. Howard, and J. L. Bougeret, Characteristics of coronal mass ejections associated with long-wavelength type II radio bursts, *J. Geophys. Res.*, 106(A12), 29,219-29,229, 2001.

Gosling, J.T., The solar flare myth, *J. Geophys. Res.*, 98, 18,937, 1993.

Gosling, J.T., and D.J. McComas, Field line draping about fast coronal mass ejecta: a source of strong out-of-the-ecliptic interplanetary magnetic fields, *Geophys. Res. Lett.*, 14, 335, 1987.

Hale, G.E., The spectroheliscope and its work, Part III. Solar eruptions and their apparent terrestrial effects, *Astrophys. J.*, 73, 379, 1931.

Hodgson, R., On a curious appearance seen in the Sun, *Mon. Not. R. Astron. Soc.*, XX, 15, 1859.

Huttunen, K.E.J., H.E.J. Koskinen, T.I. Pulkkinen, A. Pulkkinen, M. Palmroth, E.G.D. Reeves, and H.J. Singer, April 2000 magnetic storm: Solar wind driver and magnetospheric response, *J. Geophys. Res.*, 107(A12), 1440, doi:10.1029/2001JA009154, 2002a.

Huttunen, K.E.J., H.E.J. Koskinen, R. Schwenn, Variability of magnetospheric storms driven by different solar wind perturbations, *J. Geophys. Res.*, 107(A7), doi:10.1029/2001JA900171, 2002b.

Huttunen, K.E.J., R. Schwenn, V. Bothmer, and H.E.J. Koskinen, Properties and geoeffectiveness of magnetic clouds in the rising, maximum and early declining phases of solar cycle 23, *Annales Geophysicae*, 23, 625, 2005.

Li, X., I. Roth, M. Temerin, J.R. Wygant, M.K. Hudson, and J.B. Blake, Simulation of the prompt energization and transport of radiation belt particles during the March 24, 1991 SSC, *Geophys. Res. Lett.*, 20, 2423, 1993.

Lindemann, F.A., Note on the theory of magnetic storms, *Philos. Mag.*, 38, 669, 1919.

Lindsey, C., and D.C. Braun, Seismic images of the far side of the Sun, *Science*, 287, 1799, 2000.

Mayaud, P.N., Derivation, Meaning, and Use of Geomagnetic Indices, *Geophys. Monogr*, 22, AGU, Washington, D.C., 1980.

McAllister, A.H., S.F. Martin, N.U. Crooker, R.P. Lepping, and R.J. Fitzenreiter, A test of real-time prediction of magnetic cloud topology and geomagnetic storm occurrence from solar signatures, *J. Geophys. Res.*, 106, 29,185, 2001.

NSWP, *National Space Weather Program, Strategic Plan*, FCM-P30-1995, Office of the Federal Coordinator for Meteorological Services and Supporting Research, Silver Spring, MD, 1995.

Prescott, G.B., *History, Theory, and Practice of the Electric Telegraph*, Ticknor and Fields, Boston, 1860.

Reeves, G.D., K.L. McAdams, and R.H.W. Friedel, Acceleration and loss of relativistic electrons during geomagnetic storms, *Geophys. Res. Lett*, 30(10), 1529, doi:10.1029/2002GL016513, 2003.

Reiner, M.J., and M.L. Kaiser, High-frequency type II radio emissions associated with shocks driven by coronal mass ejections, *J. Geophys. Res.*, 104(A8), 16,979-16,991, 1999.

Tousey, R., The solar corona, *Space Research*, 13, 713, 1973.

Webb, D.F., E.W. Cliver, N.U. Crooker, O.C. St. Cyr, and B.J. Thompson, Relationship of halo coronal mass ejections, magnetic clouds, and magnetic storms, *J. Geophys. Res.*, 105, 7491, 2000.

Wrenn, G.L., Conclusive evidence for internal dielectric charging anomalies in geosynchronous communication spacecraft, *J. Spacecraft*, 32, 3, 514, 1995.

Zwan, R.D., and R.A. Wolf, Depletion of solar wind plasma near a planetary boundary, *J. Geophys. Res.*, 81, 1636, 1976.

H.E.J. Koskinen, University of Helsinki, Department of Physical Sciences, P.O. Box 64, FIN-00014 Helsinki. (Hannu.E.Koskinen@helsinki.fi)

K.E.J. Huttunen, Space Science Laboratory, University of California, 7 Gauss Way, Berkeley, California 94720, USA. (huttunen@ssl. berkeley.edu)